Springer Series on Environmental Management

Series Editors
Lawrence R. Walker
University of Nevada, Las Vegas, Department of Biological Sciences,
Las Vegas, Nevada, USA

Robert W. Howarth
Cornell University, Department of Ecology and Evolutionary Biology,
Ithaca, New York, USA

Lawrence A. Kapustka
LK Consultancy, Turner Valley, Alberta, Canada

More information about this series at http://www.springer.com/series/412

The Springer Series on Environmental Management is concerned with humanity's stewardship of the environment, our use of natural resources, and the ways in which we can mitigate environmental hazards and reduce risks. The main focus of the series is on applied ecology in the widest sense of the term, in theory and in practice, and above all in the marriage of sound principles with pragmatic innovation. It focuses on the definition and monitoring of environmental problems and the search for solutions to them at scales that vary from global to local according to the scope of analysis. No particular academic discipline dominates the series, for environmental problems are interdisciplinary almost by definition. The volumes include a wide variety of specialties, from oceanography to economics, sociology to silviculture, toxicology to policy studies.

The series contributes to the immense effort by ecologists of all persuasions to nurture an environment that is both stable and productive. Increasing rate of resource use, population growth, and armed conflict have tended to magnify and complicate environmental problems that were already difficult to solve a century ago. Attempts to modify nature for the benefit of humankind have often had unintended consequences, especially in the disruption of natural equilibria. Yet, at the same time human ingenuity has developed a new range of sophisticated and powerful techniques for solving environmental problems, such as pollution monitoring, restoration ecology, landscape planning, risk management, and impact assessment. The Springer Series on Environmental Management sheds light on the problems of the modern environment and contributes to the further development of solutions.

David D. Briske

Editor

Rangeland Systems

Processes, Management and Challenges

 Springer Open

Editor
David D. Briske
Ecosystem Science and Management
Texas A&M University
College Station, TX, USA

ISSN 0172-6161
Springer Series on Environmental Management
ISBN 978-3-319-83568-6   ISBN 978-3-319-46709-2   (eBook)
DOI 10.1007/978-3-319-46709-2

Printed on acid-free paper

This Springer imprint is published by Springer Nature
The registered company is Springer International Publishing AG
The registered company address is: Gewerbestrasse 11, 6330 Cham, Switzerland

Preface

This book summarizes the current status of scientific and management knowledge regarding global rangelands and the major challenges that confront them. It originated from discussions to update the well-received book entitled *Grazing Management: An Ecological Perspective* that was first published in 1991. However, it became apparent that rangeland science had advanced so rapidly in the 25 years since this book had been published that its scope was no longer sufficient to accommodate the newly created knowledge. Consequently, it was decided that a new book with an expanded scope and greater relevance to contemporary rangeland challenges was required. This book also assesses why these major advances occurred so rapidly following a half century of limited conceptual change. Major advances have primarily been represented by scientific contributions such as nonequilibrium ecology and resilience theory, but socio-political events, including new policy and increasing societal demand for rangeland services, have also been important.

Collectively, this book represents an attempt to achieve these broader and more contemporary objectives by emphasizing three major themes. The first summarizes recent conceptual advances for rangeland science and management. The second addresses the implications of these conceptual advances with respect to management recommendations and policy decisions. The third evaluates some of the major challenges confronting global rangelands in the twenty-first century. This book is intended to complement applied range management textbooks by evaluating the conceptual foundation of the profession and recommending changes to promote future development and greater effectiveness.

The book is organized in three major sections addressing each of the primary themes.

- An ecological processes section includes six chapters highlighting major advances in ecological knowledge and theory regarding the function and dynamics of rangeland systems. Specific chapters emphasize woody plant encroachment, ecohydrology, soils and belowground processes, structural heterogeneity, nonequilibrium ecology and resilience theory, and potential consequences of climate change.

- A management section consists of five chapters describing major advances that have originated, at least in part, from the knowledge previously described in the ecological processes part. Management is used in a broad context that references landscape and regional scales, in addition to smaller "pasture" scales, to optimize land use decisions for both landowners and society at large. Specific chapters focus on social-ecological systems, state-and-transition models, livestock production systems, adaptive management, and wildlife–livestock interactions.
- A challenges section contains five chapters that focus on emerging, high profile issues that will substantially impact the ability of rangelands to continue to provide ecosystem services to human societies. These chapters emphasize invasive plant species, ecosystem services, climate change adaptation, ecological monitoring, and rangelands in developing nations.

Consolidation of these diverse and complex concepts within a single, readily accessible volume is intended to improve communication among rangeland professionals as well as to better inform stakeholders from other sectors of current rangeland concepts and challenges. This volume is designed for a broad audience, including ecosystem managers in the private sector, state and federal agencies, and nongovernmental organizations; and policy makers in local, state, and national government. This content also has value for educational instruction and research, especially in multidisciplinary academic programs.

Collectively, the content of this book confirms that a more comprehensive framework is necessary to address the complex challenges confronting global rangelands in the twenty-first century. Rapid human population growth, climate change, land tenure modification, landscape fragmentation, food security, biodiversity loss, and globalized markets represent some of the major challenges that have minimized the effectiveness of traditional range management. Many "complex" problems have gone unrecognized or have been inappropriately identified as "simple" problems that have been addressed with narrowly framed approaches. The complexity of contemporary challenges requires that rangelands be re-envisioned as integrated social-ecological systems, in which societal values, goals, and capacities are given equal consideration to those of ecological processes. A more comprehensive framework of rangeland systems may enable management agencies and educational, research, and policy-making institutions to more effectively assess complex problems and develop appropriate solutions.

This book represents the most comprehensive and diverse compilation of knowledge regarding rangeland systems to date. It is founded upon the collective experience, knowledge, and commitment of 80 authors who have worked in rangelands throughout the world. The constructive contributions of 35 reviewers, some of whom also served as authors of related chapters, improved the presentation and quality of content and enhanced content integration among chapters. The USDA-ARS Jornada Experimental Range generously provided funds to publish this book in an open access venue and to partially support the summer salary of the editor during book preparation. The guidance of the Springer Nature senior editor for environmental science, Melinda Paul, and book project manager, Silembarasan Panneerselvam, is gratefully acknowledged. I extend my sincere appreciation to all who were involved in the inception, writing, and publication of this volume.

College Station, TX   David D. Briske

Contents

Contributors

Martín R. Aguiar Faculty of Agronomy, Institute for Agricultural Plant Physiology and Ecology (IFEVA), University of Buenos Aires and CONICET, Buenos Aires, Argentina

Craig R. Allen U.S. Geological Survey, Nebraska Cooperative Fish and Wildlife Research Unit, School of Natural Resources, University of Nebraska, Lincoln, NE, USA

J. Marty Anderies School of Human Evolution and Social Change, and School of Sustainability, Arizona State University, Tempe, AZ, USA

Erik M. Andersen School of Natural Resources and the Environment, University of Arizona, Tucson, AZ, USA

David G. Angeler Department of Aquatic Sciences and Assessment, Swedish University of Agricultural Sciences, Uppsala, Sweden

Steven R. Archer School of Natural Resources and the Environment, University of Arizona, Tucson, AZ, USA

José Tulio Arredondo Instituto Potosino de Investigación Científica y Tecnológica, San Luis Potosi, Mexico

Andrew Ash CSIRO, St Lucia, QLD, Australia

Vanessa Bailey Biological Sciences Division, Pacific Northwest National Laboratory, Richland, WA, USA

Derek W. Bailey Animal and Range Sciences Department, New Mexico State University, Las Cruces, NM, USA

Brandon T. Bestelmeyer USDA-ARS, Jornada Experimental Range, New Mexico State University, Las Cruces, NM, USA

David D. Breshears School of Natural Resources and the Environment, University of Arizona, Tucson, AZ, USA

David D. Briske Department of Ecosystem Science and Management, Texas A&M University, College Station, TX, USA

Joel R. Brown USDA-NRCS, Jornada Experimental Range, New Mexico State University, Las Cruces, NM, USA

Mark Brunson Department of Environment & Society, Utah State University, Logan, UT, USA

Judith Capper Livestock Sustainability Consultancy, Oxford, UK

D. Layne Coppock Department of Environment & Society, Utah State University, Logan, UT, USA

Paul C. Cross U.S. Geological Survey, Northern Rocky Mountain Science Center, Bozeman, MT, USA

Bulgamaa Densambuu Green Gold Mongolia, Sky Plaza Business Centre, Ulaanbaatar, Mongolia

Justin D. Derner USDA-ARS, High Plains Grassland Research Station, Cheyenne, WY, USA

Joseph M. DiTomaso Department of Plant Sciences, University of California, Davis, CA, USA

R. Dave Evans School of Biological Sciences and WSU Stable Isotope Core Laboratory, Washington State University, Pullman, WA, USA

Valerie T. Eviner Department of Plant Sciences, University of California, Davis, CA, USA

María Fernández-Giménez Department of Forest and Rangeland Stewardship, Colorado State University, Fort Collins, CO, USA

Kepler Euclides Filho Department of Research and Development, Brazilian Agricultural Research Corporation (EMBRAPA), Brasilia, Brazil

Jennifer Firn School of Earth, Environmental, and Biological Sciences, Queensland University of Technology, Brisbane, QLD, Australia

Joseph J. Fontaine U.S. Geological Survey, Nebraska Cooperative Fish and Wildlife Research Unit, School of Natural Resources, University of Nebraska, Lincoln, NE, USA

Samuel D. Fuhlendorf Natural Resource Ecology & Management, Oklahoma State University, Stillwater, OK, USA

Richard W.S. Fynn Okavango Research Institute, University of Botswana, Maun, Botswana

Ahjond S. Garmestani National Risk Management Research Laboratory, U.S. Environmental Protection Agency, Cincinnati, OH, USA

Richard A. Gill Department of Biology and Evolutionary Ecology Laboratory, Brigham Young University, Provo, UT, USA

Guodong Han Department of Grassland Science, College of Ecology and Environmental Science, Inner Mongolia Agricultural University, Hohhot, China

Noelle M. Hart Nebraska Cooperative Fish and Wildlife Research Unit, School of Natural Resources, University of Nebraska, Lincoln, NE, USA

Kris Havstad USDA-ARS, Jornada Experimental Range, New Mexico State University, Las Cruces, NM, USA

Jeffrey E. Herrick USDA-ARS, Jornada Experimental Range, New Mexico State University, Las Cruces, NM, USA

Pierre Hiernaux Centre National de la Recherche Scientifique, Géosciences Environnement Toulouse, Toulouse, France

Tracy Hruska Department of Environmental Science, Policy, and Management, University of California, Berkeley, CA, USA

Elisabeth Huber-Sannwald Instituto Potosino de Investigación Científica y Tecnológica, San Luis Potosi, Mexico

Leigh Hunt CSIRO Ecosystem Sciences, Darwin, NT, Australia

Lynn Huntsinger Department of Environmental Science, Policy, and Management, University of California, Berkeley, CA, USA

Andrew W. Illius Department of Animal Ecology, University of Edinburgh, Edinburgh, UK

Michael Jacobs Oikos Services LLC, Fortine, MT, USA

Jeremy J. James Sierra Foothill Research and Extension Center, University of California, Browns Valley, CA, USA

Esteban Jobbagy Grupo de Estudios Ambientales, University of San Luis, San Luis, Argentina

Jamin Johanson USDA-NRCS, Dover-Foxcroft, ME, USA

Linda A. Joyce USDA-FS Rocky Mountain Research Station, Fort Collins, CO, USA

Jason W. Karl USDA-ARS, Jornada Experimental Range, New Mexico State University, Las Cruces, NM, USA

Matthew Levi USDA-ARS, Jornada Experimental Range, New Mexico State University, Las Cruces, NM, USA

Wenjun Li Department of Environmental Management, College of Environmental Sciences and Engineering, Peking University, Beijing, China

Dardo Lopez INTA Estación Forestal Villa Dolores, Unidad de Investigación en Bosque Nativo, Córdoba, Argentina

David Le Maitre Council for Scientific and Industrial Research (CSIR), Stellenbosch, South Africa

Nadine A. Marshall CSIRO, Land and Water, James Cook University, Townsville, QLD, Australia

Devan Allen McGranahan Range Science Program, North Dakota State University, Fargo, ND, USA

Thomas A. Monaco USDA-ARS Forage and Range Research Laboratory, Logan, UT, USA

Robert S. Nowak Department of Natural Resources and Environmental Science, University of Nevada Reno, Reno, NV, USA

José L. Oviedo Institute of Public Goods and Policies (IPP), Spanish National Research Council (CSIC), Madrid, Spain

Raul Peinetti Facultad de Agronomía, Universidad Nacional de La Pampa, Santa Rosa, La Pampa, Argentina

H. Wayne Polley USDA-Agricultural Research Service, Grassland, Soil & Water Research Laboratory, Temple, TX, USA

Kevin L. Pope U.S. Geological Survey, Nebraska Cooperative Fish and Wildlife Research Unit, School of Natural Resources, University of Nebraska, Lincoln, NE, USA

Katharine I. Predick School of Natural Resources and the Environment, University of Arizona, Tucson, AZ, USA

David A. Pyke U.S. Geological Survey, Forest and Rangeland Ecosystem Science Center, Corvallis, OR, USA

John Ritten Department of Agricultural and Applied Economics, University of Wyoming, Laramie, WY, USA

Libby Rumpff School of Biosciences, University of Melbourne, Melbourne, VIC, Australia

Osvaldo E. Sala School of Life Sciences and School of Sustainability, Arizona State University, Tempe, AZ, USA

Catherine Schloeder Oikos Services LLC, Fortine, MT, USA

Susanne Schwinning Biology Department, Texas State University, San Marcos, TX, USA

Mark Stafford-Smith CSIRO Land and Water, Canberra, ACT, Australia

Patrick Shaver Rangeland Management Services L.L.C., Woodburn, OR, USA

Robert J. Steidl School of Natural Resources and the Environment, University of Arizona, Tucson, AZ, USA

Johan T. du Toit Department of Wildland Resources, Utah State University, Logan, UT, USA

Cecilia Turin Division of Crop System Intensification and Climate Change, International Potato Center, Lima, Peru

Matthew Turner Department of Geography, University of Wisconsin, Madison, WI, USA

Dirac Twidwell Department of Agronomy and Horticulture, University of Nebraska, Lincoln, NE, USA

Corinne Valdivia Department of Agricultural and Applied Economics, University of Missouri, Columbia, MO, USA

Marion Valeix Laboratoire de Biométrie et Biologie Evolutive, CNRS, Lyon, France

Lixin Wang Department of Earth Sciences, Indiana University-Purdue University Indianapolis (IUPUI), Indianapolis, IN, USA

Hilary Whitcomb U.S. Fish & Wildlife Service, Utah Ecological Services Field Office, West Valley City, UT, USA

Bradford P. Wilcox Ecosystem Science and Management, Texas A&M University, College Station, TX, USA

Steven R. Woods School of Natural Resources and the Environment, University of Arizona, Tucson, AZ, USA

Laura Yahdjian Faculty of Agronomy, Institute for Agricultural Plant Physiology and Ecology (IFEVA), University of Buenos Aires and CONICET, Buenos Aires, Argentina

Chapter 1
Rangeland Systems: Foundation for a Conceptual Framework

David D. Briske

Abstract This book describes the conceptual advances in scientific and management knowledge regarding global rangelands in the past 25 years. This knowledge originated from a substantial shift in underlying ecological theory and a gradual progression of natural resource management models. The progression of management models reflects a shift from humans as resource users to humans as resource stewards and it represents the backdrop against which this book has been written. The most influential scientific and sociopolitical events contributing to transformation of the rangeland profession in the past quarter century were recognition of nonlinear vegetation dynamics that solidified dissatisfaction with the traditional rangeland assessment procedure, the introduction of resilience theory and state-and-transition models that provided a conceptual framework for development of an alternative assessment procedure, and the National Research Council's report on Rangeland Health that provided the political support to implement these changes in federal agencies. The knowledge created by this series of interrelated events challenged the traditional concepts developed decades earlier and provided the space and creativity necessary for development of alternative concepts. In retrospect, these conceptual advances originated from the ability of the rangeland profession to progress beyond the assumptions of equilibrium ecology and steady-state management that directly contributed to its inception 100 years ago. A more comprehensive framework of rangeland systems may enable management agencies and educational, research, and policy-making institutions to more effectively develop the capacity to address the challenges confronting global rangelands in the twenty-first century.

Keywords Drylands • Natural resource management • Rangelands • Range science • Resilience-based management • Steady-state management

D.D. Briske (✉)
Department of Ecosystem Science and Management, Texas A&M University,
College Station, TX, USA
e-mail: dbriske@tamu.edu

© The Author(s) 2017
D.D. Briske (ed.), *Rangeland Systems*, Springer Series on Environmental
Management, DOI 10.1007/978-3-319-46709-2_1

1

1.1 Introduction

This book summarizes the current state of scientific and management knowledge regarding global rangelands and the major challenges that confront them. Current knowledge is assessed relative to changes that have occurred within rangeland ecology, management applications, and, more broadly, global events that have influenced rangelands. A widely accepted philosophical interpretation of scientific advancement notes that progress is often gradual and incremental as prevailing theories are explored and refined (Kuhn 1996). These periods of incremental progress, however, are periodically interrupted by major changes in underpinning theories that are termed scientific revolutions. This proved to be the case for range ecology and the discipline of ecology in the 1970s and 1980s when the prevailing theory of ecological equilibrium was challenged by a more dynamic nonequilibrium interpretation (Briske et al. 2003). Whether or not this represented a scientific revolution remains in dispute, but there is no question that it introduced a period of rapid conceptual change for the rangeland profession.

Perhaps more pertinent to the goal of this book is that the development of this new knowledge broadly paralleled the progression of natural resource management models based on human–natural resource interactions. These models are envisioned to sequentially progress with time following human settlement and societal development from humans as natural resource users to humans as natural resource stewards (Chapin et al. 2009). Consequently, changes in the perception of how humans interact with nature contribute to different knowledge needs and management strategies to maintain the supply of desired natural resources.

Natural resource exploitation is an anticipated outcome following a long period of low-impact preindustrial human use (Fig. 1.1). Exploitation of US rangelands, prompted by the perception of limitless open-access resources, did occur in response to excessive livestock grazing in the late nineteenth and early twentieth centuries. This period of exploitation and subsequent natural resource degradation was termed the "range problem" in the southwest USA, and it directly contributed to development of the rangeland profession (Sayre et al. 2012; Sayre 2017). Exploitation was followed by development of steady-state management that attempts to maximize sustainable yield of specific goods that are most highly valued. This model is implemented through the control of ecosystem variation—fire suppression, predator control, and fencing—to optimize production of desired goods, on the basis of broad ecological principles that are administered through command and control management by various state or national agencies (Table 1.1).

Recognition that effective management needed to consider entire ecosystems, including their inherent variation, and a societal demand for more diverse ecosystem services promoted development of the ecosystem management model. The ecosystem management model—focused on planning for integrated ecosystems as well as solicitation of more diverse stakeholder feedback—originated in the 1970s and was

Fig. 1.1 Progression of natural resource management models following human settlement (redrawn from Chapin et al. 2009)

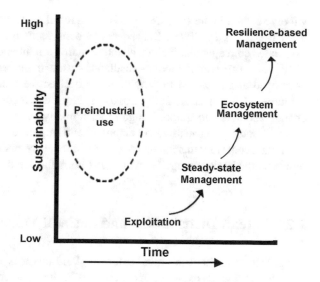

Table 1.1 Seven distinguishing attributes of steady-state, ecosystem, and resilience-based natural resource management models[a]

Attribute	Steady-state management	Ecosystem management	Resilience-based management
Ecological models	Succession-retrogression	State-and-transition, rangeland health	Multiple social–ecological systems/novel ecosystems
Reference condition	Historic climax plant community	Historic climax plant community, including historical range of variation	Landscapes with maximum options for ecosystem services
Role of humans	Use ecosystems	Part of ecosystems	Direct trajectories of ecosystem change
Ecosystem services	Meat and fiber products	Several ecosystem services	Options for diverse ecosystem services
Management goals	Sustain maximum yield of commodities	Sustain multiple uses	Sustain capacity of social–ecological systems to support human well-being
Science-management linkages	Top-down from management agencies	Top-down from management agencies	Multi-scaled social learning institutions
Knowledge systems	Management experience and agricultural experiments	Multidisciplinary science and ecological experiments	Collaborative groups, spatially referenced, updatable databases

[a]From Bestelmeyer and Briske (2012)

widely adopted in the 1990s, especially by natural resource management agencies in the USA (Quigley 2005). Subsequently, ecosystem management has introduced associated concepts that include adaptive management and ecosystem services (Nie 2013). A more recent model—resilience-based management—is currently being developed and explored to provide a more effective means for managing natural resources (Chapin et al. 2009, 2010). This model recognizes the inevitability of change and seeks to guide change to sustainably provide multiple ecosystem services for society. Successive development and implementation of steady-state management, ecosystem management, and, most recently, resilience-based management represent the backdrop against which this book has been written.

1.2   Extent, Distribution, and Societal Value

Rangelands represent the most extensive land cover type on Earth. Many definitions of rangelands exist, but most address both a land cover type, associated with vegetation or biome, and a land use that primarily emphasizes grazing or pastoralism (Lund 2007) (Text Box 1.1). Although varying definitions of rangelands are presented in the following chapters, they all contain one or both of these characteristics. Rangelands were placed within the drylands category of the Millennium Ecosystem Assessment that includes cultivated land, scrublands, shrublands, grasslands, savannas, semideserts, and true deserts (MA 2005). Drylands are defined as being limited by soil water, the result of low rainfall and high evaporation, and show a gradient of increasing primary productivity, ranging from hyperarid, arid, and semiarid to dry subhumid areas (Fig. 1.2). The ratio of annual precipitation to annual potential evapotranspiration is termed the aridity index and it is less than 0.65 for drylands. Although the majority of rangelands exist within the dryland category, a portion also occur in wetter regions, and high-latitude and high-elevation grasslands and tundra.

Drylands are estimated to occupy 41 % of the Earth's land area (6 billion hectares), 69 % of which are rangelands, and support 2 billion humans and 50 % of global livestock (MA 2005). This is an area 1.5 times larger than all forests combined and nearly three times greater than cropland (Reid et al. 2008). Given their expansiveness and heterogeneity, rangelands provide numerous ecosystem services including biodiversity, carbon sequestration, and cultural values, in addition to the provisioning services of food, fiber, and fuel. Limited and highly variable resource availability, both ecological and socioeconomic, makes these systems and their human inhabitants highly vulnerable to both ecological and social disruption. Approximately 73 % of drylands are affected by accelerated soil degradation and 10–20 % of drylands are currently degraded (MA 2005). Human populations inhabiting drylands lag far behind the rest of the world in terms of human well-being and development indicators; 90 % of the inhabitants reside in developing countries (MA 2005; Chapter 17, this volume).

Text Box 1.1: Chronology of Major Rangeland Definitions
"From the 100th meridian to the Pacific," "one of the most important economic uses is the grazing of livestock," and "climatic conditions do not, in most localities, favor the production of farm crops" are the phrases that A.W. Sampson used to refer to rangelands in his book entitled "Range and Pasture Management," 1923, p. 4.

Rangelands are those areas of the world, which by reason of physical limitations—low and erratic precipitation, rough topography, poor drainage, or cold temperatures—are unsuited to cultivation and which are a source of forage for free-ranging native and domestic animals, as well as a source of wood products, water, and wildlife. Range Management, Stoddard et al. (1975, p. 3).

All territories presently used as grazing lands, which are accounted for in yearly FAO statistics as well as other nonagricultural, largely unoccupied, drylands which are used only occasionally by nomadic pastoralists or are presently unused at all. United Nations Environment Program 1991 (cited in Lund 2007).

Rangeland is a type of land that supports different vegetation types including shrublands such as deserts and chaparral, grasslands, steppes, woodlands, temporarily treeless areas in forests, and wherever dry, sandy, rocky, saline, or wet soils; and steep topography precludes the growing of commercial farm and timber crops. Rangeland Ecology and Management, Heady and Child (1994, p. 1).

An area where wild and domestic animals graze or browse on uncultivated vegetation. FAO (2000).

Rangeland is defined as "uncultivated land that provides the necessities of life for grazing and browsing animals." Range Management: Principles and Practices. Holechek et al. (2011, p. 1).

Rangelands are a type of land (not just land grazed by livestock) on which natural vegetation is dominated by grasses and shrubs and the land is managed as a natural ecosystem. UNCCD (2011).

"Land supporting indigenous vegetation that either is grazed or that has the potential to be grazed, and is managed as a natural ecosystem. Range includes grassland, grazable forestland, shrubland and pastureland." SRM Glossary of Terms 1998, updated 2015.

1.3   Events Contributing to Rapid Conceptual Advancement

A large number of conceptual advances began to occur in the late 1980s, after a 50-year period of minimal conceptual change, to markedly transform the rangeland profession. These advances originated from scientific events both internal and external to the profession, as well as sociopolitical events motivated by dissatisfaction with the prevailing method of rangeland assessment. These major events and conceptual advances, along with their contribution to the

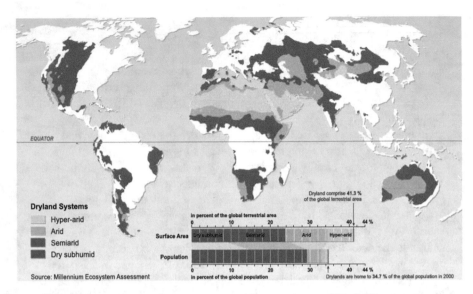

Fig. 1.2 Distribution of global drylands as classified by the UNCCD (Millennium Ecosystem Assessment (2005))

progression of natural resource management models previously described, are summarized in the following sections (Table 1.1, Fig. 1.3). This section examines the professional legacies that have influenced our current, and potentially our future, perceptions of rangelands.

1.3.1 Internal to the Profession

Scientific—A foundational concept of the US rangeland profession in the twentieth century was range condition and trend analysis (the range model). It was founded on plant successional theory developed by the influential early American ecologist Fredric E. Clements in 1916. The range model broadly assumed that livestock grazing counteracted plant succession to establish the species composition of plant communities in a linear response to the severity of livestock grazing (Briske et al. 2005). Arthur Sampson, a former student of Clements at the University of Nebraska, introduced succession as a conceptual framework for rangeland assessment in 1917. The adoption of successional theory—an equilibrium concept—was considered a major conceptual advance by rangeland professionals in the early twentieth century. It had a profound influence on the rangeland profession by directly linking it to equilibrium ecology and by indirectly contributing to the steady-state management model of natural resource management. Dyksterhuis (1949) further secured succession in the foundation of rangeland science by operationalizing the range model on a quantitative basis. The range model provided the standard assessment procedure for approximately 50 years (Fig. 1.3).

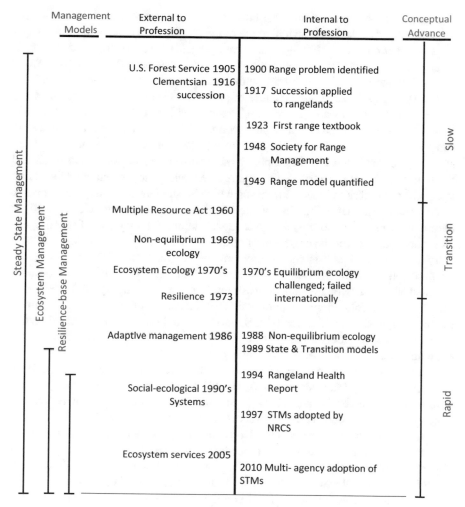

Fig. 1.3 Timeline of major events and conceptual advances that have contributed to development of the rangeland system framework described in this chapter and throughout the entire book. The relative rate of conceptual advances is shown on the *right* and the successive emergence of natural resource management models on the *left*

However, both the range model and successional theory encountered severe criticism by both Australian and US rangeland scientists in the 1970s and 1980s. This criticism was primarily founded on recognition that the rate and extent of woody plant expansion was not solely a consequence of grazing intensity and that removal of grazing did not necessarily prevent or reverse woody plant encroachment (Westoby et al. 1989; Laycock 1991). This ecological outcome was inconsistent with the assumptions of the range model and it provided a strong justification for development of an alternative model to more accurately interpret observed vegetation dynamics and to more effectively support rangeland management.

A second source of criticism occurred when the concepts of range science that had been developed in the USA, including the range model, were applied to pastoral systems on other continents. International development programs recognized that these Western concepts had limited application to pastoral systems in the late 1970s (Sayre 2017). The international scope of this knowledge proved to be extremely valuable by assessing range science through the lens of pastoral societies where private lands and market-oriented goals were of limited relevance (Reid et al. 2014). Research in arid pastoral systems indicated that plant production and livestock numbers were seldom in equilibrium because periodic multiyear drought prevented livestock numbers from attaining the maximum carrying capacity established by plant production (Ellis and Swift 1988). This also contributed to management and policy recommendations that rejected the equilibrium-based concepts on which Western range science was founded.

Sociopolitical—Growing dissatisfaction with the range model, as well as inconsistent rangeland assessment procedures among the major US federal agencies, contributed to the development of political pressure to devise a more ecologically relevant and consistent assessment procedure. The National Research Council (NRC) published the "Rangeland Health Report" in 1994 that broadly outlined an alternative rangeland assessment to replace the highly criticized range model. Shortly following the publication of this influential report, a group of rangeland specialists within the NRCS made the bold decision to adopt this alternative assessment procedure and began development of state-and-transition models within the framework of Ecological Site Descriptions.

A memorandum of agreement was signed by the NRCS, US Forest Service (USFS), and Bureau of Land Management (BLM) in 2010 to use Ecological Site Descriptions as the major procedure for rangeland assessment to successfully fulfill the recommendation of the NRC report. The NRC Report provided political motivation for change, especially within the federal agencies. However, the conceptual framework for development of an alternative assessment procedure was based on the influential paper of Westoby, Walker, and Noy-Meir entitled "Opportunistic Management for Rangelands not at Equilibrium" that was published in the Journal of Range Management in 1989. It is this series of events that propelled nonequilibrium ecology beyond equilibrium ecology as the dominant theory underpinning the rangeland profession.

Interest in range science educational programs in Western universities of the USA began to wane during this period and student enrollment declined drastically in some cases. Greater societal demands from rangelands and increasing complexity of natural resource management had exceeded the capacity of the traditional range science curriculum to effectively address them. This trend was also occurring in other natural resource disciplines that had originated with the assumptions of simplicity, predictability, and manageability that characterized steady-state management last century (Holling and Meffe 1996; Thurow et al. 2007).

The close association of range science with livestock grazing, both real and perceived, further minimized the value of range science as multiple resource use and ecosystem management began to develop. Declining student enrollment in many natural resource management disciplines led to consolidation of academic programs, including many range science departments that were merged with those of related disciplines (Abbott et al. 2012). However, these events contributed to the integration and broadening of academic curricula and research programs that were more inclusive of multiple disciplines. Although the rangeland profession was multidisciplinary from its inception, the contributions of the allied disciplines were often narrowly confined by prevailing perceptions within the rangeland profession.

Development of a more diverse multidisciplinary range science curriculum that included systems ecology, landscape ecology, spatial sciences, and biogeochemistry contributed to a knowledge base that has brought range science into closer alignment with the concepts and theories of its related disciplines. The Grassland Biome Project of the International Biological Program promoted ecosystem ecology as a research focus in the late 1960s and early 1970s and introduced subsequent generations of US and Canadian grassland and rangeland ecologists to this systems-oriented approach (Smith 1968). The past 25 years have witnessed an important generational turnover of researchers and managers that has introduced broader, multidisciplinary perspectives, and new scientific, technological, and communication skills that extend far beyond the traditional perspectives of the twentieth-century rangeland profession.

In some cases, professionals with degrees in disciplines other than range science were employed to further expand this knowledge base. It was also during this period that social scientists who had been studying pastoral societies and peoples began to interact with biophysical scientists to create a more comprehensive interdisciplinary framework for investigating rangeland systems (Reid et al. 2014). This provided necessary knowledge of the people that inhabit rangelands, including their culture, social structure, and livelihoods. These events collectively created both the scientific capacity and creative space for reassessment and exploration of alternative perspectives, interpretations, and concepts that contributed to this period of rapid change (Fig. 1.3).

1.3.2 External to Profession

Scientific—Resilience is undoubtedly the major scientific theory that contributed to the transformation of range science. Resilience was introduced in 1973 by C.S. Holling in an attempt to reconcile ecosystem dynamics with the prevailing concept of ecological stability. Resilience recognizes that ecosystems can exhibit dynamic behavior, and yet retain their general structure and function, and that alternative stable ecosystems may be formed in cases where resilience of the initial ecosystem has been exceeded. However, it took another 16 years before resilience was introduced to the rangeland profession (Westoby et al. 1989) and nearly another 10 years before it was incorporated into rangeland assessment. It continues to be developed as a central component of the rangeland profession (Bestelmeyer and Briske 2012; Herrick et al. 2012). Resilience theory is currently replacing nonequilibrium ecology as the dominant theoretical concept because what was previously considered nonequilibrium is now more appropriately interpreted as multiple equilibria in many cases (Petraitis 2013; Chapter 6, this volume). Resilience is also used to describe "a way of thinking," in addition to a property of ecological systems, especially in reference to social–ecological systems.

Recognition of the importance of spatial scales to ecological processes and ecosystem dynamics also had a profound influence by shifting emphasis from small plots to a broader landscape perspective (Turner 1989). The concept of social–ecological systems emerged in the 1990s to emphasize the strong linkage that existed among ecological and social components (Berkes and Folke 1992). This goes beyond simply stating that humans are dependent upon nature to emphasize that ecological and social systems are tightly integrated with many complex and poorly understood interactions that directly influence natural resource management.

Rangeland research priorities were further modified by major changes in funding sources beginning in the early 1990s. Research programs shifted from single-scientist projects funded by land-grant institutions and agricultural experimental stations to much larger, multidisciplinary programs that emphasized broader and more contemporary natural resource management issues identified by federal funding agencies (Thurow et al. 2007). A portion of these federal grant programs required that research include extension or education personnel, and some required direct stakeholder engagement to further ensure that research outcomes had practical application. This provided a strong incentive that moved range science toward more effectively integrated, multidisciplinary research programs that ultimately contributed to a portion of the conceptual advances summarized in this book.

Although this shift in research funding has enhanced the scientific capacity of the rangeland profession, concern has been expressed that it may have reduced management emphasis and expertise (Abbott et al. 2012). However, the rangeland CEAP assessment that was organized by the NRCS to evaluate the effectiveness of rangeland conservation practices indicated that the benefits of these practices that had been developed decades earlier were largely undocumented (Briske 2011). This was primarily a consequence of minimal monitoring of practice outcomes and the assessment further concluded that previous research had provided only modest support for management and policy recommendations. Consequently, this shift in research emphasis driven by research funding may potentially introduce greater, rather than less, management-relevant science, especially when federal grant programs require stakeholder involvement and demonstration of research application. This indicates that the management model in which knowledge is implemented is as important as the knowledge itself.

Sociopolitical—The steady-state model of natural resource management was developed early last century and it is still widely implemented today (Table 1.1, Fig. 1.3). This model attempts to maintain ecosystems in a single state through the implementation of management practices and policies that are applied in a command and control manner to efficiently optimize production of one or a few select ecosystem services

(Holling and Meffe 1996). The steady-state model began to be challenged in response to an increasing incidence of natural resource management failures and societal demand for more diverse ecosystem services. The publication of academic papers entitled "The pathology of natural resource management" (Holling and Meffe 1996) and "The era of management is over" (Ludwig 2001) was a direct challenge to steady-state management that has continued to the present. In many respects the major conceptual advances that range science has made in the past 25 years are a consequence of its ability to progress beyond the assumptions of equilibrium ecology and steady-state management that directly contributed to its inception 100 years ago.

Expanding societal awareness of the value of rangelands and greater demands for diverse services from them, especially those held in the public domain, required that federal land management agencies develop more comprehensive objectives following passage of the Multiple Use Act in 1960 (Holechek et al. 2011). The ecosystem management model emerged in the late 1970s in response to recognition that entire ecosystems, including their inherent variation, were appropriate units for natural resource management (Koontz and Bodine 2008; Nie 2013) (Table 1.1). By the mid-1990s all four of the major federal natural resource management agencies in the USA had adopted this model. This raised a new set of social, as well as ecological, questions and challenges regarding natural resource management that had not previously been considered. Consider that the initial definitions of range management provided by Stoddart and Smith in the first (1943) and second (1955) editions of the text book "Range Management" make reference to "obtaining maximum livestock production" which is consistent with steady-state management previously described (Text Box 1.2). It wasn't until the third edition in 1975 that this definition was modified to "optimize returns from rangelands in those combinations most desired and suitable to society." Heady (1975) introduced a similar definition in the text "Rangeland Management" that same year. These expanded definitions are indicative of a shift from the steady-state management to the ecosystem management model.

Use of the term ecosystem management rapidly declined in the early 2000s likely in response to its ambiguous definition, multiple interpretations, and numerous barriers encountered in its implementation (Nie 2013). However, several of its major components—stakeholder engagement, adaptive management, and restoration—continue to shape natural resource management and planning. Resilience-based management appears to have adopted some of the most effective components of the ecosystem management model (Table 1.1). This management model embraces the inevitability of ecological and social change and emphasizes that management should anticipate and guide change, rather than minimize it, to sustainably provide society with desired ecosystem services (Chapin et al. 2009, 2010; Bestelmeyer and Briske 2012). It seeks to address uncertainty and incomplete knowledge through the involvement of diverse stakeholders to develop adaptive capacity, rather than static management prescriptions and regulations, to maintain resilient systems.

Text Box 1.2: Chronology of Major Range Management Definitions
"The science and art of planning and directing range use so as to obtain the maximum livestock production consistent with conservation of the range resource"—Range Management. Stoddart and Smith (1943, p. 2).

"The science and art of obtaining maximum livestock production from range land consistent with the conservation of the land resources"—Range Management. Stoddart and Smith (1955, p. 1).

"The science and art of optimizing the returns from rangelands in those combinations most desired by and suitable to society through the manipulation of range ecosystems"—Range Management. Stoddard et al. (1975, p. 3).

"Land management discipline that skillfully applies an organized body of knowledge know as range science to renewable natural resource systems for two purposes: (1) protection, improvement, ad continued welfare of the basic range resource, which may include soils, vegetation, and animals; and (2) optimum production of goods and services in combinations needed by mankind"—Rangeland Management, Heady (1975, p. 4).

"The manipulation of rangeland components to obtain optimum combination of goods and services for society on a sustained basis"—Holechek et al. (1989, p. 5). This definition has been retained in all subsequent editions.

"A distinct discipline founded on ecological principles and dealing with the use of rangelands and range resources for a variety of purposes. These purposes include use as watersheds, wildlife habitat, grazing by livestock, recreation, and aesthetics, as well as other associated uses." SRM Glossary of Terms 1998, updated 2015.

1.4   Section Perspectives

All three sections of this book—processes, management, and challenges—summarize concepts that did not exist 25 years ago, and those few that did have been greatly modified and refined, which is indicative of the rate at which science and global events have advanced. This rapid change also parallels the progression of natural resource models previously described through greater understanding of human dependence and impact on natural resources (Table 1.1, Fig. 1.3).

1.4.1   Processes Section

The processes section outlines a comprehensive and in-depth understanding of ecological knowledge that has challenged and, in some cases, replaced traditional assumptions and concepts. The chapters in this section reflect the global significance of rangeland processes and indicate that accelerating global change will further amplify the inherent variability and uncertainty inherent to rangelands. Ecological processes are the source of multiple ecosystem services that society demands from rangelands, including the provisioning services of food, fiber, and fuel.

A greater understanding of belowground processes, including the structure and function of microbial communities, has increased insight into the contribution of soils in rangeland systems, and their significance in global biogeochemical cycles. Hydrological processes are central to the function of arid and semiarid rangelands and their close coupling with vegetation makes them very sensitive to natural disturbances and human activities. Woody plant encroachment has occurred in many rangelands throughout the globe to modify not only vegetation structure, but also ecological processes that create trade-offs among important ecosystem services. Heterogeneity, diversity, and variability are envisioned to possess inherent value and the occurrence of ecological processes over diverse spatial and temporal scales has been recognized to produce important ecological outcomes. Resilience theory has provided an interpretation of how ecosystems can be dynamic, but persist as self-organized systems, and climate change science provides valuable projections of how ecosystems may be impacted by these changes in the future.

1.4.2 Management Section

The management section emphasizes the transition of humans from users to stewards of natural resources within the context of social–ecological systems. Management is used in a broad context to reference landscape and regional scales, in addition to smaller "pasture" scales, to optimize land-use decisions for both landowners and society at large. The content of these chapters cautions that we must learn from past professional experiences, but stand ready to move beyond them to explore alternative approaches and to create innovative solutions. For example, rapidly increasing global demand for animal protein requires development of more efficient livestock production systems while minimizing their adverse ecological consequences. State-and-transition models are widely used to support rangeland management and they continue to undergo further refinement to increase their management utility.

Management decisions are often made under conditions of inherent uncertainty and risk that require systematic approaches to inform decision-making processes under these circumstances. Adaptive management has been developed to address these challenges, but its application has been limited by both insufficient management-relevant science and the inability of institutions to support its implementation. A consensus is emerging that collaborative learning and collective action are required among diverse stakeholders to produce useable knowledge, increase adaptive capacity, and maintain resilience of rangeland systems.

1.4.3 Challenges Section

The challenges section emphasizes that future events may surpass previous human experience regarding adaptation to changing climatic, ecological, and socioeconomic conditions. Surprises occur when unanticipated outcomes originate from threshold conditions, extreme events, and unrecognized drivers external to the system. A primary challenge will be to determine the types of changes that are desirable and beneficial and to implement them in a manner that does not create other problems or degrade rangeland resources. Invasive species continue to alter the structure and function of rangeland systems and ecosystem services supplied, and in some cases these transformations may be irreversible. Invasive plant management has adopted an ecosystem perspective to contend with this accelerating biotic challenge. Ecosystem services, including those that do not currently possess economic market value, are being explored as a means to recognize and evaluate trade-offs and create win-win outcomes regarding land-use decisions.

Changing socioeconomic conditions have required that pastoralists throughout the world become more sedentary which undermines the traditional risk aversion strategy of livestock mobility. Knowledge of rangelands and their human inhabitants in developing countries has rapidly increased, but numerous barriers exist to its implementation to improve rangeland resources and human well-being. The development of cost-effective, large-scale monitoring of rangeland resources will be a considerable challenge for implementing effective management and policy decisions.

1.5 Foundation for a Rangeland Systems Framework

Range management has focused on prescribed management practices to a much greater extent than management approaches or strategies to achieving desired outcomes (Text Box 1.2). The limited development of management approaches has been highlighted by the introduction of adaptive management—an approach to management that emphasizes structured learning through decision making for situations where knowledge is incomplete and managers must act despite uncertainty (Chapter 11, this volume). Limited management strategies may be a consequence of the heterogeneous environmental and managerial conditions encountered on rangelands that necessitates development of broad principles. However, the application of prescribed practices appears to be more consistent with administrative regulation than it does with an effective approach for addressing heterogeneity.

Limited development of well-defined management strategies may have partially resulted from the regulatory origins of the profession to minimize rangeland exploitation, initially by the US Forest Service (Sayre 2017). Emphasis on "prescribed" management practices—stocking rates, fencing of pastures, and grazing seasons—by land management agencies enabled them to retain authority over users of public land in a manner that is consistent with command and control management.

These practices were initially designed to minimize livestock impacts based on the assumption that they were the key variable influencing rangeland—controlling live-stock equated to controlling ecosystems. In addition, the economic benefits that fencing and predator control provided by reducing labor costs for herders may have also reinforced "management by practice" (Sayre 2015). Consequently, the need for management to control rangeland exploitation, support agency authority, and pro-duce economic value directly contributed to the development of range science—not the other way around (Sayre 2017).

The occurrence of these events early in the rangeland profession may partially explain why "management practices" have to some extent become synonymous with range management. This perspective is evident in the phrase "manipulation of rangeland components" that is used to define range management in a widely used textbook (Holechek et al. 2011). Even though it is obvious that practices do not equate to management—the process of *deciding* how to allocate finite resources—consider how prevalent practices are in a management context. For example, the USDA-NRCS Environmental Quality Assessment Program (EQIP)—a voluntary cost share program to enhance conservation on private rangelands—is primarily organized around the selection and implementation of practices in the context of broader conservation planning (Briske 2011). The implementation of prescribed practices and their management have made major contributions to rangeland con-servation in the twentieth century, but the Rangeland CEAP Assessment indicated that documentation of these outcomes was extremely limited (Briske 2011). Another important constraint of "practice-based" management is that it enforces a small-scale "pasture" focus that precludes assessment within landscapes and regions where many of the most pressing challenges exist.

Provenza (1991) cautioned that range *science* was dominated by managerial issues that limited progression of the science in an editorial written 40 years after the profes-sion had been formally founded. This is consistent with the interpretation that the range-land profession originated from the need for management action to resolve immediate practical problems, rather than from the establishment of sound scientific principles (Sayre 2017). However, the conceptual advances that have occurred in the past 25 years may have differentiated management and science to the greatest extent in the history of the profession. Ironically, these concepts have also provided the approaches and justifi-cation—social–ecological systems, adaptive management, and resilience-based man-agement—for integrating these two important knowledge sources. This represents a pressing challenge for which no solution or approach has yet emerged.

Collectively, these considerations make a compelling case for development of a more comprehensive framework to assess rangelands and to implement manage-ment. Definitions emphasizing land cover type and land use are narrowly focused on biophysical systems and do not recognize the social component of these systems (Reid et al. 2014; Chapter 17, this volume). The content of these chapters collectively indicates that neither ecological nor social knowledge alone is sufficient to effectively assess or manage rangeland systems because of the highly integrated nature of the social and ecological subsystems (Chapter 8, this volume). A new management framework is required to place greater emphasis on social components, including cultural values, land tenure, governance systems, and markets, that interact with ecological systems to influence the value, availability, and use of rangeland resources. Adaptive management and collaborative adaptive management—adaptive management among multiple stakeholder groups—represent essential approaches to a more comprehensive management framework.

The inability of the rangeland profession to resolve debates concerning intensive rotational grazing, shrub removal versus water yield, and wild horse and burro dilemma on public lands in the western USA is symptomatic of a narrow management framework that does not possess the capacity to effectively address the social components of these systems (Briske et al. 2011; Boyd et al. 2014). Similarly, the implementation of well-intended, but inappropriate, policy throughout the world has contributed to rangeland fragmentation, degradation, and conversion (MA 2003). These adverse consequences occur for numerous reasons, but they often result because the potential trade-offs and consequences were not recognized prior to policy implementation.

This questions whether current management approaches possess sufficient capacity to contend with future challenges confronting global rangelands. These seemingly intractable management dilemmas demonstrate that a framework is required that can keep pace with the increasing scope and complexity of natural resource management. In contrast to "practice-based" management, the comprehensive approach to collaborative adaptive management that was used to address the proposed listing of the greater sage grouse (*Centrocercus urophasianus*) as a threatened species in the western USA has been viewed as being highly successful (Boyd et al. 2014). A successful outcome to this complex, regional natural resource concern lends credibility to the emerging natural resource management model of resilience-based management.

Resilience-based management involves the development and implementation of strategies that support human well-being via adaptation and transformation of social–ecological systems to sustain the supply of ecosystem services in changing environments (Chapin et al. 2010; Chapter 6, this volume). This management model acknowledges both the dependence and impact that humans can have on natural resources and the ecosystem services they provided. It also cautions managers and scientists to exhibit greater humility regarding the management of natural resources, than that conveyed by the linear and predictable outcomes inherent to the steady-state management model. The emerging reality of natural resource management is one of increasing management complexity, disputed values, and incomplete knowledge (Benson and Craig 2014).

Incorporation of the concepts presented in this book—namely, ecosystem services, structural heterogeneity, and social–ecological systems—into a management context has been slow—although resilience is a clear exception. The primary challenge may reside in the fact that these concepts are more consistent with resilience-based management than they are with the steady-state management model. Therefore, management must learn to not only adopt new concepts, but also transition between natural resource management models to effectively incorporate new knowledge. However, concepts and experience with the ecosystem management model may serve to bridge this transition in some cases (Nie 2013). In addition, resilience-based management empowers diverse stakeholders to bring unique knowledge, goals, and values to decision-making processes. Integrating human dimensions into natural resource management will require expertise and methodology from the social sciences, which is currently underrepresented in the rangeland profession, and this has likely slowed concept adoption as well.

The Drylands Development Paradigm describes a set of management and policy recommendations that are consistent with resilience-based management (Reynolds et al. 2007). Five principles of this paradigm follow: (1) social–ecological systems are coupled, dynamic, and co-adapting with no single target equilibrium point; (2) critical system dynamics are determined by several slow or controlling variables; (3) controlling variables possess thresholds that, if crossed, cause the system to reorganize as a new state; (4) stakeholders are networked across multiple organizational levels in social–ecological systems to produce cross-scale interactions; and (5) "hybrid" knowledge that integrates management and policy experience with scientific knowledge must be developed and legitimized by relevant social institutions. Collectively, these principles are in direct contrast to those of the steady-state management model that is interwoven with the origins of the rangeland profession (Table 1.1). This provides further insight into the magnitude of the challenge associated with the transition from the steady-state management to the resilience-based management model.

This compilation of major conceptual advances provides an opportunity to envision a more comprehensive framework for rangeland systems that is capable of designing and implementing management strategies for landscape and regional applications. The following definitions of rangeland systems and management could provide the foundation for an alternative framework. *Rangeland systems* represent ecological systems supporting native or naturalized vegetation characterized as grasslands, shrub steppe, shrublands, savannas, and deserts that are managed as adaptive social–ecological systems to provision multiple ecosystem services to benefit human well-being. These systems function through complex interactions among social and ecological subsystems, at multiple scales, to influence supply, demand, and preferences for ecosystem services (Chapters 8 and 14, this volume). *Rangeland system management* is based on the iterative development of management strategies through collaborative adaptive management among diverse stakeholders, representing management and scientific knowledge, to provision multiple ecosystem services required by society. The outcomes of management strategies are collaboratively monitored and evaluated to provide information feedbacks to enhance subsequent management effectiveness and to promote adaptive capacity of multiple stakeholder groups to support resilient rangeland systems (Chapters 6 and 11, this volume). Development of a framework that can accommodate the concepts that have emerged in the past 25 years to support rangeland systems in the twenty-first century may be the primary challenge confronting the global rangeland community.

1.6   Summary

This book describes the advances that have occurred in scientific and management concepts regarding global rangelands in the past 25 years. This knowledge originated from two interwoven themes—a substantial shift in underlying ecological theory and a gradual progression of natural resource management models—the former appears to have been most influential, but the latter may prove most significant over the longer term. The conceptual advances that occurred in the 1980s and 1990s were a reaction to what can be considered the initial conceptual advance in the rangeland profession—the introduction of Clementsian successions as a conceptual framework for rangeland assessment. This had a profound influence on the rangeland profession by directly linking it to equilibrium ecology and by indirectly contributing to the steady-state management model of natural resource management.

However, both the initial procedure for rangeland assessment and successional theory, on which it was founded, encountered two broad categories of criticism in the 1970s and 1980s. The first was recognition of nonlinear vegetation dynamics which was inconsistent with both concepts. The second category of criticism occurred in response to the failure of range science concepts that had been applied to pastoral systems on other continents. These criticisms provided a strong justification for development of an alternative ecological theory to more accurately interpret the dynamics of rangeland vegetation and to more effectively support rangeland management. Two alternative models emerged simultaneously, but independently, in the late 1980s—the nonequilibrium and state-and-transition models.

The state-and-transition framework was adopted by the rangeland profession in the late 1990s and it has become an important management tool replacing the range model that had been introduced 80 years earlier. However, resilience theory is currently replacing nonequilibrium as the dominant theory because what was previously considered nonequilibrium is more appropriately interpreted as multiple equilibria. It is somewhat ironic that rangeland systems are now considered to have an equilibrial component after the severe criticism that the concept had previously received.

The conceptual advances described above were broadly paralleled by the progression of natural resource management models that reflected a shift from humans as resource users to humans as resource stewards. Although these models are not always obvious—they have a pronounced influence by shaping the perception of human interactions with nature. A major objective of range science in the twentieth century was to develop knowledge in support of the steady-state management model that emphasized the maximum sustainable production of forage and livestock. Recognition that management needed to consider entire ecosystems, including their inherent variation, promoted development of the ecosystem management model. The most recent management model—resilience-based management—is currently being developed and investigated as an extension of resilience theory. Currently, elements of all three management models are in operation to varying degrees.

The conceptual advances presented in this book make a compelling case for development of a more comprehensive framework to assess rangelands and to implement more effective management strategies. This framework could be organized around the following definitions of rangeland systems and rangeland management. *Rangeland systems* represent ecological systems supporting native or naturalized vegetation characterized as grasslands, shrub steppe, shrublands, savannas, and deserts that are managed as adaptive social–ecological systems to provision multiple ecosystem services to benefit human well-being. These systems function through complex interactions among social and ecological subsystems, at multiple scales, to influence supply, demand, and preferences for ecosystem services. *Rangeland system management* is based on the iterative development of management strategies through collaborative adaptive management among diverse stakeholders representing management and scientific knowledge to provision multiple ecosystem services required by society. Development of a framework that is capable of incorporating the concepts that have emerged in the past 25 years to support rangeland systems in the twenty-first century may represent the major challenge confronting the global rangeland community.

Acknowledgements Jeff Herrick, Mitch McClaran, Robin Reid, and Nathan Sayre provided insightful comments that improved clarity and enhanced chapter content. Layne Coppock provided a perspective that enriched the chapter in both depth and breadth. Nathan Sayre graciously provided prepublication content of his book that was in press at the time this chapter was written. Joeanna Brooks redrew Fig. 1.1 and created Fig. 1.3.

References

Abbott, L.B., K.L. Launchbaugh, and Susan Edinger-Marshall. 2012. Range education in the 21st century: Striking the balance to maintain a relevant profession. *Rangeland Ecology & Management* 65: 647–653.

Benson, M.H., and R.K. Craig. 2014. The end of sustainability. *Society and Natural Resources* 27: 777–782.

Berkes, F., and C. Folke. 1992. A systems perspective on the interrelations between natural, human-made and cultural capital. *Ecological Economics* 5: 1–8.

Bestelmeyer, B.T., and D.D. Briske. 2012. Grand challenges for resilience-based management of rangelands. *Rangeland Ecology & Management* 65: 654–663.

Boyd, C.S., D.D. Johnson, J.D. Kerby, T.J. Svejcar, and K.W. Davies. 2014. Of grouse and golden eggs: Can ecosystems be managed within a species-based regulatory framework? *Rangeland Ecology & Management* 67: 358–368.

Briske, D.D. 2011. *Conservation benefits of rangeland practices assessment, recommendations, and knowledge gaps.* Washington, DC: U.S. Department of Agriculture, Natural Resources Conservation Service.

Briske, D.D., S.D. Fuhlendorf, and F.E. Smeins. 2003. Vegetation dynamics on rangelands: A critique of the current paradigms. *Journal of Applied Ecology* 40: 601–614.

———. 2005. State-and-transition models, thresholds, and rangeland health: A synthesis of ecological concepts and perspectives. *Rangeland Ecology & Management* 58:1–10

Briske, D.D., N.F. Sayre, L. Huntsinger, M. Fernandez-Gimenez, B. Budd, and J.D. Derner. 2011. Origin, persistence, and resolution of the rotational grazing debate: Integrating human dimensions into rangeland research. *Rangeland Ecology & Management* 64: 325–334.

Bureau of Land Management (BLM). 2010. *Rangeland interagency ecological site manual.* Washington, DC: BLM Manual 1734-1.

Chapin, F.S. III, G.P. Kofinas, C. Folke, S.R. Carpenter, P. Olsson, N. Abel, R. Biggs, R.L. Naylor, E. Pinkerton, D.M. Stafford Smith, W. Steffen, B. Walker and O.R. Young. 2009. Resilience-based stewardship: Strategies for navigating sustainable pathways in a changing world. In *Principles of ecosystem stewardship: Resilience-based natural resource management in a changing world*, eds. F.S. Chapin III, G.P. Kofinas, and C. Folke, 319–337. New York, NY: Springer.

Chapin, F.S. III, S.R. Carpenter, G.P. Kofinas, C. Folke, N. Abel, W.C. Clark, P. Olsson, D.M. Stafford Smith, B. Walker, O.R. Young, F. Berkes, R. Biggs, J.M. Grove, R.L. Naylor, E. Pinkerton, W. Stephen and F. J. Swanson. 2010. Ecosystem stewardship: sustainability strategies for a rapidly changing planet. Trends in Ecology and Evolution. 25:241–249.

Clements, F.E. 1916. *Plant succession: An analysis of the development of vegetation.* Carnegie Institution of Washington Pub. 242. Washington, DC. 512 p.

Dyksterhuis, E.J. 1949. Condition and management of rangeland based on quantitative ecology. *Journal of Range Management* 2: 104–105.

Ellis, J.E., and D.M. Swift. 1988. Stability of African pastoral ecosystems: Alternate paradigms and implications for development. *Journal of Range Management* 41: 450–459.

Food and Agriculture Organization (FAO). 2000. *Pastoralism in the new millennium.* Animal production and health paper 150. Rome, Italy, 93 p.

Heady, H.F. 1975. *Rangeland management.* New York: McGraw-Hill Book Company.

Heady, H.F., and R.D. Child. 1994. *Rangeland ecology and management.* Boulder, CO: Westview Press.

Herrick, J.E., J.R. Brown, B.T. Bestelmeyer, S.S. Andrews, G. Baldi, et al. 2012. Revolutionary land use change in the 21st century: Is (rangeland) science relevant? *Rangeland Ecology & Management* 65: 590–598.

Holechek, J.L., R.D. Pieper, and C.H. Herbel. 1989. *Range management: Principles and practices.* Upper Saddle River, NJ: Prentice-Hall, Inc.

———. 2011. *Range management: Principles and practices.* 6th ed., New Jersey: Prentice-Hall, Inc. Upper Saddle River.

Holling, C.S. 1973. Resilience and stability of ecological systems. *Annual Review of Ecology and Systematics* 4: 1–23.

Holling, C.S., and G.K. Meffe. 1996. Command and control and the pathology of natural resource management. *Conservation Biology* 10: 328–337.

Koontz, T.M., and J. Bodine. 2008. Implementing ecosystem management in public agencies: Lessons from the U.S. Bureau of Land Management and the Forest Service. *Conservation Biology* 22: 60–69.

Kuhn, T.S. 1996. *The structure of scientific revolutions*, 3rd ed. Chicago, IL: University of Chicago Press.

Laycock, W.A. 1991. Stable states and thresholds of range condition on North American rangelands: A viewpoint. *Journal of Range Management* 44: 427–433.

Ludwig, D. 2001. The era of management is over. *Ecosystems* 4: 758–764.

Lund, H.G. 2007. Accounting for the world's rangelands. *Rangelands* 29: 3–10.

Millennium Ecosystem Assessment [MA]. 2005. *Current state and trends*, 917. Washington, DC: Island Press.

National Research Council. 1994. *Rangeland health: New methods to classify, inventory and monitor rangelands.* Washington, DC: National Academy Press. 180 p.

Nie, M. 2013. Whatever happened to ecosystem management and federal lands planning? In *The laws of nature: Reflections on the evolution of ecosystem management law and policy*, ed. Kalyani Robbins, 67–94. Akron, OH: University of Akron Press.

Petraitis, P. 2013. *Multiple stable states in natural ecosystems*, 188. Oxford, UK: Oxford University Press.

Provenza, F.D. 1991. Viewpoint: Range science and range management are complementary but distinct endeavors. *Journal of Range Management* 44: 181–183.

Quigley, T.M. 2005. Evolving views of public land values and management of natural resources. *Rangelands* 27(3): 37–44.

Reid, R.S., K.A. Galvin, and R.S. Kruska. 2008. Global significance of extensive grazing lands and pastoral societies: An Introduction. In *Fragmentation in semi-arid and arid landscapes: Consequences for human and natural systems*, eds. K.A. Galvin et al., 1–24, Springer.

Reid, R.S., M.E. Fernandez-Gimenez, and K.A. Galvin. 2014. Dynamics and resilience of range-lands and pastoral peoples around the globe. *Annual Review of Environment and Resources* 39: 217–242.

Reynolds, J.F., D.M. Stafford-Smith, E.F. Lambin, B.L. Turner, M. Mortimore, S.P.J. Batterbury, T.E. Downing, H. Dowlatabadi, R. Fernandez, J.E. Herrick, E. Huber-Sannwald, H. Jiang, R. Leemans, T. Lynam, F.T. Maestre, M. Ayarza, and B. Walker. 2007. Global desertification: Building a science for dryland development. *Science* 316: 847–851.

Sampson, A.W. 1923. *Range and pasture management*. New York: Wiley.

Sayre, N.F. 2015. The coyote-proof pasture experiment: How fences replaced predators and labor on US rangelands. *Progress in Physical Geography* 39: 576–593.

———. 2017. *The politics of scale: A history of rangeland science*. Chicago/London: University of Chicago Press.

Sayre, N.F., W. deBuys, B. Bestelmeyer, and K. Havstad. 2012. 'The Range Problem' after a century of rangeland science: New research themes for altered landscapes. *Rangeland Ecology & Management* 65: 545–552.

Smith, F.E. 1968. The international biological program and the science of ecology. *Proceedings of the National Academy of Science U S A* 60: 5–11.

SRM Glossary of Terms 1998, updated 2015. *Society for range management*. http://globalrange-lands.org/rangelandswest/glossary.

Stoddart, L.A., and A.D. Smith. 1943. *Range management*. New York: McGraw-Hill Book Company.

———. 1955. *Range management*. 2nd ed., New York: McGraw-Hill Book Company.

Stoddart, L.A., A.D. Smith, and T.W. Box. 1975. *Range management*, 3rd ed. New York: McGraw-Hill Book Company.

Thurow, T.L., M.M. Kothmann, J.A. Tanaka, and J.P. Dobrowolski. 2007. Which direction is forward: Perspectives on rangeland science curricula. *Rangelands* 29: 40–51.

Turner, M.G. 1989. Landscape ecology: The effect of pattern on process. *Annual Review of Ecology and Systematics* 20: 171–197.

United Nations Convention to Combat Desertification (UNCCD). 2011. *National report for the United States on efforts to mitigate desertification in the Western U.S.* p. 36.

Westoby, M., B.H. Walker, and I. Noy-Meir. 1989. Opportunistic management for rangelands not at equilibrium. *Journal of Range Management* 42: 266–274.

Section I
Processes

Chapter 2
Woody Plant Encroachment: Causes and Consequences

Steven R. Archer, Erik M. Andersen, Katharine I. Predick, Susanne Schwinning, Robert J. Steidl, and Steven R. Woods

Abstract Woody vegetation in grasslands and savannas has increased worldwide over the past 100–200 years. This phenomenon of "woody plant encroachment" (WPE) has been documented to occur at different times but at comparable rates in rangelands of the Americas, Australia, and southern Africa. The objectives of this chapter are to review (1) the process of WPE and its causes, (2) consequences for ecosystem function and the provision of services, and (3) the effectiveness of management interventions aimed at reducing woody cover. Explanations for WPE require consideration of multiple interacting drivers and constraints and their variation through time at a given site. Mean annual precipitation sets an upper limit to woody plant cover, but local patterns of disturbance (fire, browsing) and soil properties (texture, depth) prevent the realization of this potential. In the absence of these constraints, seasonality, interannual variation, and intensity of precipitation events determine the rate and extent of woody plant expansion. Although probably not a triggering factor, rising atmospheric CO_2 levels may have favored C_3 woody plant growth. WPE coincided with the global intensification of livestock grazing that by reducing fine fuels, hence fire frequency and intensity, facilitated WPE. From a conservation perspective, WPE threatens the maintenance of grassland and savanna

S.R. Archer (✉) • E.M. Andersen • K.I. Predick • R.J. Steidl • S.R. Woods
School of Natural Resources and the Environment, University of Arizona,
Tucson, AZ 85721-0137, USA
e-mail: sarcher@email.arizona.edu; erikandersen@email.arizona.edu; kipredick@email.arizona.edu; steidl@email.arizona.edu; srwoods@email.arizona.edu

S. Schwinning
Biology Department, Texas State University, San Marcos, TX 78666, USA
e-mail: schwinn@txstate.edu

© The Author(s) 2017
D.D. Briske (ed.), *Rangeland Systems*, Springer Series on Environmental
Management, DOI 10.1007/978-3-319-46709-2_2

ecosystems and its endemic biodiversity. Traditional management goals aimed at restoring forage and livestock production after WPE have broadened to support a more diverse portfolio of ecosystem services. Accordingly, we focus on how WPE and management actions aimed at reducing woody plant cover influence carbon sequestration, water yield, and biodiversity, and discuss the trade-offs involved when balancing competing management objectives.

Keywords Brush management • Mortality • Recruitment • Roots • Seedling establishment • Soil depth/texture

2.1   Introduction

The relative abundance or dominance of grasses and woody vegetation is highly dynamic at timescales ranging from decades to centuries to millennia (Fig. 2.1). Over the past 100 years or so, there has been a directional shift toward increased abundance of woody vegetation worldwide (Sala and Maestre 2014). The phenomenon of woody plant encroachment (WPE) in grasslands and savannas contrasts with deforestation and dieback occurring in many forested systems. The proliferating trees and shrubs can be non-native species that were introduced purposely or accidentally or native species that have either increased in abundance within their historic ranges or expanded their geographic range. Woody plants have been displacing grasses across bioclimatic zones. Trees proliferate in humid regions while unpalatable shrubs replace grasses in more arid regions, which is regarded as a type of desertification. In both cases, the proliferation of trees and shrubs threatens the maintenance of grassland and savanna ecosystems and the plants and animals that are endemic to these systems.

Proliferation of woody plants has long been of concern to range managers where grazing by cattle and sheep is the primary land use. Where funds and equipment were available, management was focused narrowly on reversing WPE with the goal of enhancing livestock production. Aggressively applied since the 1940s, "brush management" results have been mixed and their sustainability and cost-effectiveness questionable. As we gain a broader appreciation of how woody plants influence ecosystem processes and how changes in their abundance affect a broad portfolio ecosystem services, we are better positioned to evaluate trade-offs that must be considered as their abundance changes.

In this chapter we (1) review the rates, dynamics, causes, and consequences of woody plant proliferation over the past 100 years, (2) evaluate the extent to which interventions aimed at reducing woody vegetation have effectively restored lost or

a

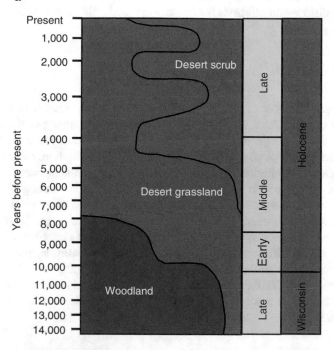

Fig. 2.1 (**a**) Holocene changes in woodland, desert scrub, and grassland in the southwestern USA (modified from Van Devender 1997) and (**b**) photographic record of increases in woody plant abundance at the Santa Rita Experimental Range, Arizona, USA, against the backdrop of Huerfano Butte (images are from public domain available from http://cals.arizona.edu/srer/photos.html; compiled by R. Wu)

altered key ecosystem services, and (3) assess trade-offs influencing ecological and socioeconomic decisions and priorities for managing woody plants in rangelands.

2.2 Rates of Change

Substantial increases in cover of woody plants can occur over decades. In North America, rates of encroachment vary by an order of magnitude among ecoregions (0.1–2.3 % cover year^{-1}, Barger et al. 2011) (Fig. 2.2). Their review indicated that rates of tree proliferation typically exceeded those of shrub proliferation, ostensibly reflecting the higher precipitation in areas where tree encroachment occurs. We might expect that differences in encroachment rates would differ among woody functional types, but the Barger et al. (2011) review found that rates were highest and comparable among scale-leaved evergreen (*Juniperus virginiana*) and N$_2$-fixing deciduous (*Prosopis glandulosa*) arborescents. Reported rates of change in woody cover across savannas and forest-savanna boundaries in Africa, Australia, and South America are comparable to those observed in North America (range = 0.1–1.1 % cover year^{-1}, Stevens et al. 2016), though maximum rates reported in their synthesis were much lower than those reported by Barger et al. (2011) for North America (1.1 vs. 2.3 % cover year^{-1}).

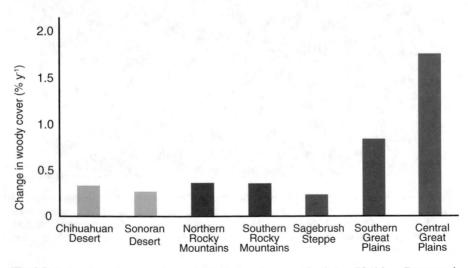

Fig. 2.2 Rates of woody encroachment in North American rangelands (modified from Barger et al. 2011)

Typically, research has targeted localities where encroachment is known to have occurred or is occurring. Estimates of encroachment rates, therefore, are likely biased. Furthermore, rates of encroachment tend to decline as woody proliferation progresses (Text Box 2.1). Variation in the rate and extent of encroachment is also mediated by local or regional differences in environmental factors, disturbance regimes, and land use as discussed in the following sections.

Text Box 2.1: Rates and Drivers of Woody Plant Encroachment

Absolute encroachment rates[1] range from nil to 3.3 % cover year^{-1} and average 0.85 % cover year^{-1}. Generally, rates of encroachment are highest in the early stages of encroachment, and then decline (e.g., Fensham et al. 2005) or fluctuate (Browning et al. 2008) as maximum cover thresholds are approached. Accordingly, studies based on long-term observations tend to report low rates of encroachment. Barger et al. (2011) found that rates of woody plant encroachment in North America were highest in Great Plains grasslands (1–2 % cover year^{-1}) and lowest in hot and cold deserts (<0.5 % cover year^{-1}). Trees and shrubs exhibited similar mean encroachment rates (0.62 and 0.52 % cover year^{-1}, respectively). Rates of increase for Great Plains species representing contrasting plant functional types (e.g., evergreen vs. deciduous; N$_2$ fixation potential) and dispersal mechanisms were comparable as well.

(continued)

---

[1] A database of peer-reviewed research papers was compiled by searching for the terms "bush encroach*," "brush encroach*," "desertification," "shrub grazing," "shrub encroach," "shrub invasion," "shrub expansion," "woody encroach*," and "woody plant invasion" on the ISI Web of Knowledge. This search produced 865 unique references that were then subdivided into papers that quantified encroachment rates ($n = 289$) or relationships between shrub encroachment and grazing ($n = 149$).

Text Box 2.1: (continued)

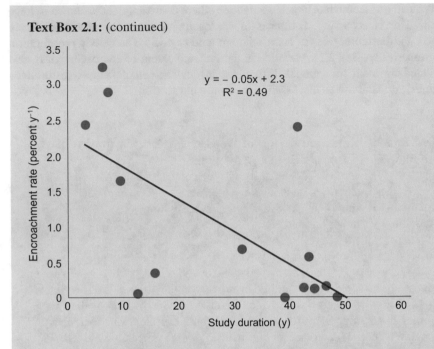

Among papers reporting relationships between shrub encroachment and grazing, mean (±SE) shrub cover was statistically comparable on grazed sites (21 % ± 0.9) and sites protected from grazing (24 % ± 0.9). Overall, the presence or absence of grazing did not predict changes in shrub cover over time. Variation within many of these studies was high, indicating that the role of grazing is complex, even at the ranch level. Weighted regression analysis further indicated that precipitation, continent (North and South America, Australia, Africa), and grain size (i.e., plot/pixel size) were not significant predictors of grazing importance. Interestingly, there was a significant relationship between the data source (field sampling vs. remote sensing) and grazing importance. Assessments based on broad-scale remote sensing (aerial photos, satellite imagery) were more likely to conclude that grazing promotes shrub encroachment, whereas field-based studies were more likely to conclude that grazing has no effect on shrub encroachment. This may reflect the fact that studies of shrub encroachment and grazing based on field data focus on, and are restricted to, the outcomes of short-term grass-woody plant interactions at plant and patch scales, whereas remote sensing assessments reveal the longer term, landscape-scale outcomes of patch-scale dynamics (e.g., Milne et al. 1996). Photo credit: E. Andersen

2.3 Factors Influencing Abundance of Woody Plants

Given the global scale of the WPE phenomenon, deriving robust generalizations about the causes of woody encroachment has been challenging, as species adaptations, land-use history, and climate trends differ markedly among bioclimatic zones. Numerous factors (including climate, fire, and grazing/browsing regimes, concentrations of atmospheric CO_2, and levels of N deposition) co-occur and interact to promote or constrain increases in woody dynamics at local scales, with their relative importance and interaction strength differing markedly among locations (Archer 1994; Bond and Midgley 2000; D'Odorico et al. 2012). In any location, it may be difficult to distinguish between "necessary" and "sufficient" conditions. For example, it may be necessary for a given biotic or abiotic environmental condition to change for woody plants to gain an advantage over grasses (e.g., higher atmospheric CO_2 concentrations), but a change in that condition may not by itself be sufficient to trigger woody plant proliferation unless accompanied by other changes (e.g., reductions in fire and browser populations). Accordingly, assigning primacy to the potential drivers of woody plant encroachment remains a topic of active debate and research.

Because woody plant encroachment has occurred across a wide range of climates from tropical to arctic and arid to humid, drivers likely vary among climate zones. Grazing effects on fire regimes and competitive interactions among plants may predominate in humid regions, whereas grazing effects on levels of plant stress and erosional processes (reducing ground cover and increasing wind/water erosion) may predominate in more arid regions. Disturbance is superimposed against a backdrop of climate and soils to further modify the local abundance of shrubs or trees. Where climate and soils are capable of supporting an abundance of woody vegetation, the occurrence of periodic fire or an abundance of browsers utilizing woody vegetation can prevent them from attaining dominance. Conversely, preferential utilization of herbaceous vegetation by grazers may create opportunities for woody plants to establish (via reductions in competition) and persist (via reductions in fine fuel mass and continuity needed to carry fires). Woody plant cover at a given locale within a bioclimatic region is the net outcome of these interrelated and potentially interacting factors (Fig. 2.3). In the following sections, we review briefly some of the key drivers and their mediation by geomorphology, soils, and topography. Ultimately, the challenge for land managers will be to apply these perspectives appropriately and creatively to their local settings and situations.

2.3.1 Herbivory: Grazers and Browsers

Livestock grazing is a primary use of grasslands worldwide (Asner et al. 2004) and is often associated with WPE. The arrival of livestock with Anglo-European settlers in the Americas, Australia, and Southern Africa, although occurring at different times, coincided with dramatic and swift changes in woody abundance in grasslands

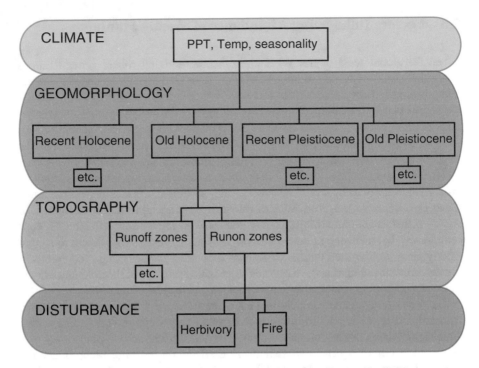

Fig. 2.3 The abundance of woody and herbaceous vegetation is determined by interactions across a hierarchy of drivers and constraints operating across a range of spatial and temporal scales. Changes in climate and atmospheric chemistry (e.g., increased CO_2 concentrations) determine grass-woody plant abundance at broad scales and over long time periods. Vegetation composition at local scales is mediated, and in some cases constrained, by geomorphology, soils, and topography via their effects on water and nutrient distribution. Soils and topography, in turn, mediate vegetation responses to disturbances associated with drought, fire, grazing or browsing pressure, and land use

and savannas (Archer 1994). Grazing by livestock removes fine fuels, which reduces fire frequency and intensity and also enhances woody plant recruitment (Madany and West 1983). The advantages for woody plants may be magnified where livestock are effective dispersers of their seeds. In addition, livestock introductions can be associated with displacement of native browsers and seed predators, releasing woody plants from top-down controls.

Reported effects of livestock on rates of woody plant encroachment have been variable due to differences in the inherent characteristics of study sites or the intensity, duration, or timing of grazing. Grazing has been associated with both substantial increases (Roques et al. 2001; Valone et al. 2002) and moderate or no increases in the cover of woody plants (Allen et al. 1995; Fensham et al. 2005). Further, grazing may even limit or retard shrub encroachment in some systems (Altesor et al. 2006; Batista et al. 2014). It is unclear to what extent these contrasting patterns might reflect differences in stocking rates and season(s) of use through time. Interpretation of grazing effects on shrub encroachment can vary with spatial and temporal scale. For example, conversion of grassland to shrub-dominated dune land in the Chihuahuan Desert occurred within large areas free of livestock, suggesting that factors other than livestock grazing were driving the change. However, closer inspection revealed that historically heavy livestock grazing had reduced ground cover and accelerated wind erosion in upwind areas. Aeolian deposition accelerated grass mortality via burial and promoted shrub recruitment to drive the conversion from grassland to shrubland in the downwind area excluded from livestock grazing (Peters et al. 2006). This example shows that grazing effects must be evaluated considering spatial context as well as land-use history.

Drivers of change must also be considered in the context of time. At a site in the Sonoran Desert, woody cover increased both within 74-year-old livestock exclosures *and* in the surrounding grazed landscapes, suggesting that factors other than grazing were responsible. However, heavy grazing in the late 1800s and early 1900s may have altered ecological processes in ways that predisposed the site to shrub encroachment prior to the time exclosures were established in 1932 (Browning and Archer 2011). In addition, cessation or relaxation of grazing subsequent to degradation may have promoted WPE by enabling a degree of grass recovery that then facilitated shrub recruitment (e.g., de Dios et al. 2014).

Preferential utilization of woody plants by wild browsers (e.g., Staver et al. 2009) or seed and seedling predators (Weltzin et al. 1997; Dulamsuren et al. 2008) may help maintain grassland and savanna communities. Activities of these herbivores can prevent shrubs and trees from establishing, prevent them from exerting dominance, and maintain them at a stature vulnerable to fire. Types and abundances of wildland herbivores can vary spatially and temporally and this can lead to highly variable effects on WPE. Understanding mechanisms that contribute to WPE can be especially difficult in areas where livestock grazing occurs in conjunction with native herbivores whose activities are also influencing plant composition and abundance (e.g., Heske et al. 1993). In some cases, native herbivores may be displaced by livestock or removed by managers if viewed as competing with livestock for forage (Weltzin et al. 1997). In those cases, the livestock grazing effects described earlier would be amplified by removal of native browsers. Maintaining populations of native herbivores in systems managed for livestock grazing may help maintain grass-woody populations in desired configurations while concurrently enhancing biodiversity and creating opportunities for lease hunting, game farming, and ecotourism revenue.

2.3.2 Climate

Grasslands of the world are situated between desert shrublands and woodlands/forests with respect to annual rainfall, annual temperature, and potential evapotranspiration. In the future, if climate becomes warmer and drier or if the frequency, magnitude, and duration of drought increase, present-day grasslands in some areas may become desert shrubland. In contrast, woodlands and forests could also shift to savanna or grassland (e.g., Allen et al. 2010; Anadón et al. 2014a) and increases in woody cover realized in recent decades may be reduced by a higher frequency of "hot droughts" (Bowers 2005; Breshears et al. 2005; Twidwell et al. 2014). Climate-change simulations under elevated atmospheric CO_2 predict pronounced shifts toward tree-dominated biomes (Scheiter and Higgins 2009). Changes in dry-season duration or precipitation seasonality will also influence the balance between grass and woody vegetation (Neilson et al. 1992; Bailey 2014).

Mean annual precipitation (MAP) determines the potential "carrying capacity" for woody plants and upper limit for woody plant cover (Sankaran et al. 2005). As MAP increases, the potential for landscapes to support woody cover increases linearly, becoming asymptotic at ca. 650 mm (Fig. 2.4). Shrub or tree savanna or open woodland communities may therefore characterize regions where MAP is below this threshold, whereas the tendency to develop woodland or forest communities to the general exclusion of grasses occurs above this threshold. Managers contemplating actions to regulate woody plant cover should first determine their sites' MAP in relation to this potential. Expensive interventions may not be warranted in areas where maximum cover potentials are relatively low.

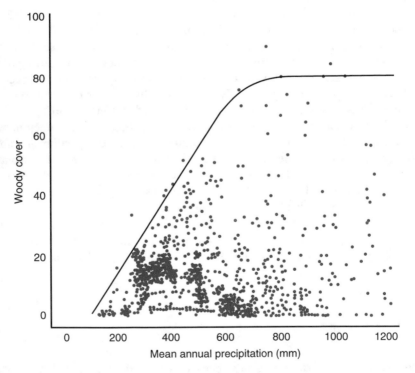

Fig. 2.4 Relationship between mean annual precipitation (MAP) and maximum woody plant cover in Africa. Maximum potential woody cover increases linearly with increases in MAP to ~650 mm, and then levels off at ~80 %. Note that many sites are well below their potential, ostensibly owing to constraints imposed by geomorphology, soils, topography, disturbance, and land use. Modified from Sankaran et al. (2005)

Depth of rainfall infiltration and seasonal timing of rainfall can interact with MAP to locally constrain the extent to which maximum potential woody cover might be realized at a given location. Rainfall that percolates deep into the soil is typically more assessable to deeper rooted woody plants than to shallow-rooted grasses. Accordingly, frequent low-intensity events (Good and Caylor 2011), large rainfall events (Kulmatiski and Beard 2013), and precipitation delivered during the period of grass dormancy (Walter 1979; Bond et al. 1994; Gao and Reynolds 2003) are more likely to recharge soil moisture at depths benefitting woody plants. Grasslands would therefore be favored in climates characterized by summer rainfall and small rainfall events that moisten only upper horizons (Neilson et al. 1992). At local scales, however, rainfall is redistributed by topography and the extent to which it infiltrates and percolates is influenced strongly by soil texture and depth (Sect. 2.3.3).

Precipitation variability influences grass-woody dynamics via its effects on plant recruitment, growth, and mortality. Interannual and spatial variability in rainfall is high in the semiarid zone. Given the potential longevity of woody plants, exceptionally good recruitment years can set the stage for seed production and opportunities for recruitment decades into the future, whereas only exceptionally strong drought years can significantly reduce mature tree cover (Fensham and Holman 1999; Twidwell et al. 2014). Models incorporating these dynamics predict that decadal or longer deviations from mean tree density may result (Fig. 2.5). Few empirical data have been available to verify this nonstationary concept of savanna dynamics, due to the paucity of long-term data. However, where data are available, they support the notion that history matters and that the current state of the system does not necessarily reflect recent events or current ecological processes (Staver et al. 2011). These long stochastic return times make it difficult in practice to distinguish natural fluctuation from a regime shift, or a temporary upturn in woody plant abundance from directional, persistent woody encroachment.

The globally widespread proliferation of woody plants in arid and semiarid grasslands suggests the importance of broad-scale factors, such as climate change and increases in atmospheric CO_2, as do recent increases in shrub abundance in high-latitude systems where climate change effects on ecosystem processes have been pronounced (Myers-Smith et al. 2011). The grasslands encountered by the Anglo-European settlers of southwestern North America in the mid-1800s may have established and flourished under the conditions of the Little Ice Age. These grasslands were only marginally supported under the climate of the 1800s–early 1900s and were in the process of transitioning to desert scrub with the advent of warmer, drier conditions, with changes in vegetation lagging well behind the changes in climate driving them (Neilson 1986) (Fig. 2.1a). Broad-scale factors such as climate, however, cannot account for "fence-line contrasts" and local variation in rates and patterns of woody plant increases. These local dynamics ostensibly reflect changes in land use and spatial variation of disturbance regimes, such as livestock grazing and the abundance of browsers. In these cases, climate may not be the driver *per se*, but it will influence the rates and dynamics of woody cover change and may increase the susceptibility of the herbaceous vegetation to other agents of change.

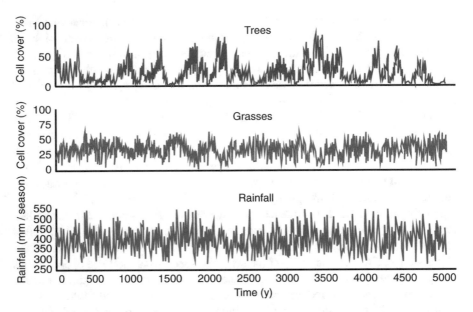

Fig. 2.5 Modeled fluctuations in grass and woody plant cover at decadal and longer time scales, assuming stationary stochastic rainfall distribution, in a savanna in Texas, USA (from van Wijk and Rodriguez-Iturbe 2002)

2.3.3 Topography and Soils

At the catena (hillslope) scale, edaphic properties—primarily soil depth and texture—mediate broad-scale climate and atmospheric chemistry effects. These effects, in turn, are mediated by topographic setting, which dictates radiant energy regimes (e.g., slope aspect effects), cold air drainage, and patterns of rainfall redistribution via run-off and runon (McAuliffe 2003). Grasses and woody plants possess different adaptations to exploit soil resources. Root mass decreases exponentially with depth in both life forms, but woody plants typically have a greater root mass at deeper depths and greater maximum rooting depths (Canadell et al. 1996; Jackson et al. 1996). Grasses, by contrast, have a dense, fibrous root system of limited depth, well suited to exploit soil resources in the upper 20–30 cm of the soil profile, where water and nutrients reach peak concentrations. Hence, grasses are generally favored by fine-textured surface soils and shallow soils that retain water and nutrients near the surface.

Woody plants are favored by deep, coarse soils that facilitate percolation and nutrient leaching. They are at a disadvantage on shallow soils where bedrock or claypan horizons restrict taproot extension. Many woody species have both a shallow, laterally extensive root system and deep taproots (Schenk and Jackson 2002). This reflects a generalist strategy for soil resource capture that allows them to use small rainfall events and the nutrients concentrated in the upper soil layers (Fravolini et al. 2005), but to also access water and nutrients (e.g., NO_3^-) percolated below the depths effectively exploited by grasses. Woody plants with this dimorphic root system can therefore exploit a wide range of growing season conditions (Scott et al. 2006; Priyadarshini et al. 2015).

The contrasting grass and woody plant rooting patterns are the basis for the "two-layer hypothesis," which characterizes the differential use of shallow and deeper soil resources by grasses and woody plants and grasses. The hypothesis appears to be widely applicable in a variety of dryland systems (Ward et al. 2013), but less so in mesic savannas with a shallow water table, where woody plants and grasses often have similar rooting depths and compete for moisture from the same soil horizons throughout the year (Rossatto et al. 2014). Interactions between topoedaphic properties and grass vs. woody plant rooting patterns help explain why some grassland sites are resistant to WPE and others are more susceptible (Knoop and Walker 1985). The two-layer hypothesis is a niche-based perspective, which helps explain how the amount of precipitation and its seasonality interact with soil properties (texture and depth) to influence the proportion of grasses and woody plants on a given site.

Grasses tend to dominate shallow soils, where lateritic or argillic horizons, bedrock, or limestone are near the surface; water and nutrient resources "perch" and concentrate above these impermeable layers (Molinar et al. 2002). However, if there are fissures or gaps in the impermeable layers that allow resources and woody plant roots to pass through, woody plants may thrive. Aboveground patterns in distribution, size, and mortality rates of woody plants that accompany drought may reflect variation in these edaphic heterogeneities (Bestelmeyer et al. 2011; Rossatto et al. 2014; Twidwell et al. 2014).

On playas and dry lake beds, where precipitation and runon accumulate in poorly drained, fine-textured topographic low points, conditions may become periodically anaerobic. These conditions tend to favor grasses to the exclusion of trees and shrubs regardless of grazing or fire regimes. Subtle, local variation in micro-topography within such sites may, however, provide refuges for woody plants and influence local patterns of woody plant composition and abundance (e.g., Sklar and Valk 2003).

Distribution, size, and density of woody vegetation are also influenced by topography. In the Northern Hemisphere, south-facing slopes are warmer and drier than north-facing slopes and typically support less woody plant cover (Bailey 2014). Runoff from slopes concentrates water and nutrients in downslope areas and augments incoming precipitation, potentially enabling arroyos, washes, and intermittent drainages to support higher densities of larger-sized woody plants than upslope portions of the landscape (Coughenour and Ellis 1993). Runoff and runon relationships and their substantive influences on woody plant abundance are also evident on gently sloping landscapes (Tongway et al. 2001). Landscape-scale variation in rates and patterns of WPE in recent decades are therefore related to and constrained by topoedaphic variation (Wu and Archer 2005; Naito and Cairns 2011; Browning et al. 2012; Rossatto et al. 2014) (Text Box 2.2).

Text Box 2.2: Soils and Topography Influence Susceptibility to Woody Plant Encroachment

Woody plant encroachment on the Santa Rita Experimental Range in the North American Sonoran Desert dates back to the early 1900s and has been well documented (McClaran 2003). However, most of the shrub encroachment (primarily *Prosopis velutina*) has occurred on Holocene-age sandy soils. Within the Holocene-age portions of the landscape, shrub cover appears to have peaked at about 30–35 %, consistent with predictions of the model in Fig. 2.4, but sites on the landscape with a subsurface clay content of 17 % at 33 cm depth reached this cover asymptote about 30 years sooner than sites where the subsurface clay content was 25 % at 23 cm depth (Browning et al. 2008).

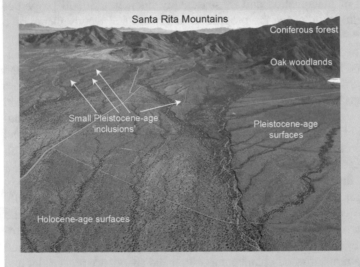

(continued)

> **Text Box 2.2   (continued)**
> Pleistocene-age surfaces, with their well-developed claypan horizons (39 % clay at 10 cm depth), have experienced similar climate and levels of atmospheric CO_2 enrichment and have experienced similar land-use (livestock grazing) and disturbance regimes (heavy grazing in the early to mid-1900s, and lack of fire) as the Holocene-age landscapes, and yet have persisted as C_4 grassland. Note that shrub abundance is also higher in runon areas (arroyos and intermittent drainages) in both geomorphic settings and that shrubs give way to trees as elevation increases.
>
> In this bioclimatic zone, it appears that a clay content threshold for the occurrence and persistence of an "edaphic grassland" occurs somewhere between 25 % at 23 cm depth and 39 % at 10 cm depth. It remains to be seen whether the edaphic grasslands on the Santa Rita Experimental Range will persist under the predicted changes in climate. See McAuliffe (1997) for details on the geomorphology of this site. Photo credits: W. Cable, aerial image; J. Fehmi, ground-level photo.

2.3.4 Increased Atmospheric CO_2

Atmospheric CO_2 concentrations have increased over the time period that WPE has occurred, from ~290 ppm at the beginning of the twentieth century to ~380 at the end. In this range, photosynthesis in C_3 plants is CO_2 limited, so it is possible that rising atmospheric CO_2 has benefited C_3 woody plants more than C_4 grasses. The response of plants to elevated CO_2 has been reviewed extensively elsewhere, but as a rule of thumb, a doubling of atmospheric CO_2 from 350 to 700 ppm typically results in a 30–50 % increase in the carbon assimilation rate of C_3 plants under optimal conditions. In contrast, C_4 plants are not affected directly by atmospheric CO_2 because they concentrate CO_2 at the carboxylation sites to substrate saturation. Still, C_4 plants often receive a growth advantage through partial stomatal closure, which increases their water-use efficiency in water-limited environments. At the whole-plant level, elevated atmospheric CO_2 can elicit a wide range of growth responses depending on other co-limitations including other resource limitations (light, nitrogen, water), stress conditions (heat, frost), crowding, and species differences in growth and reproductive strategies (Körner 2006). As a result, a community may contain many species that show no response to elevated CO_2 at all. Projections of CO_2 enrichment effects should therefore be made cautiously and in the context of other drivers and constraints.

Woody encroachers are composed overwhelmingly of C_3 plants. By contrast, grasslands and savannas in tropical, subtropical, and warm-temperate biomes often are dominated by C_4 grasses. This pattern led to the hypothesis that woody encroachment might be a consequence of a CO_2-mediated correction in the competitive relationships between C_3 and C_4 plants (Idso 1992; Polley 1997). However, this cannot entirely explain WPE at the global scale, as woody plants also encroach into grasslands dominated by C_3 grasses. Woody plants have other structural and functional advantages over herbaceous vegetation, which increase their ecological opportunities under accelerated growth conditions (Poorter and Navas 2003). Whereas herbaceous plants lose most annual biomass accumulation to herbivory, combustion, or decomposition, woody plants build up woody biomass and carbohydrate storage over decades, thereby strengthening their ability to persist in the face of stress and disturbance. Woody plants are most vulnerable to injury, physiological stress, and competition when they are small, and faster growth would expedite their transition to more resilient and competitive life stages.

Global vegetation models have solidified support for the connection between atmospheric CO_2 and "woody thickening," both within woodlands and forests and through the expansion of woodlands into grasslands. Importantly, these models have set WPE into the context of a millennial-scale global transition that started during the last glacial maximum when atmospheric CO_2 was at a low point (Prentice et al. 2011). Examination of sediment records in the Chihuahuan Desert concluded that woody encroachment during the past 200 years is unprecedented in the context of the preceding 5500 years, that it was not related to droughts or changes in ENSO event frequency, and that it was contemporaneous with the rise in atmospheric CO_2 and known grazing impacts (Brunelle et al. 2014). However, it has been argued that WPE clearly outpaced the gradual increase in atmospheric CO_2 and the modestly elevated concentrations present in the early- to mid-1900s, by which time substantial encroachment had occurred (Archer et al. 1995). This suggests that while changes in atmospheric CO_2 might have been contributed to WPE in the early to mid-1900s, it was not a driver *per se*. Continuing increases in atmospheric CO_2, however, may increasingly favor woody plants. For example, dynamic global vegetation models suggest that with fire multiple stable biome states are possible across broad areas of Africa, but that the potential for multiple stable states will decline with further increases in atmospheric CO_2 as biomes will become deterministically tree dominated (Moncrieff et al. 2014).

Growth advantages realized by woody plants under high CO_2 conditions may enable them to minimize the time during which they are vulnerable to disturbance. For example, frequent fires are a major limitation to tree recruitment in subtropical savannas. These fires may kill saplings outright, necessitating recruitment from seed, or they may force saplings to regenerate from basal sprouts. In either case, the woody plants are kept in a fire-susceptible size class. In this scenario shrub or tree recruits are able to mature into tall savanna trees only during rare periods of infrequent fires when saplings can grow large to escape the flame zone. All else equal, an acceleration of sapling growth by CO_2 fertilization would increase the probability of escaping the flame zone and increase tree density (Bond and Midgley 2000).

2.4   Population Interactions Between Grasses and Woody Plants

In previous sections we focused on environmental drivers of woody plant encroachment. We now turn to mechanisms that govern the ecological interactions between grasses and trees or shrubs. There is a large body of ecological and range-management literature on the effects of woody plants on grasses (Scholes and Archer 1997; Blaser et al. 2013; Dohn et al. 2013). Here, we focus on factors that influence recruitment and abundance of woody plants into grass-dominated communities. Grass-woody plant interactions affecting the proliferation of woody plants are quite complex, involving multiple plant functional types with numerous contrasting traits and important differences in life history. Generally, population interactions are governed by nonlinear density effects (both intra- and interspecific) on species' vital rates and environment effects on those rates. In the context of WPE, three questions are especially relevant. First, how do populations of grasses resist invasion by woody plants and how do drivers of WPE lower resistance? This is the key question for explaining where and when WPE occurs. Second, beyond establishment, how do populations of grasses affect growth of woody plants and development from seedling to sapling to seed-producing mature tree? This question is relevant to explaining the rates of woody plant invasion, after having established a presence in grasslands. Third, what are the interactions that limit woody plant cover and establish an upper limit, or carrying in encroached ecosystems? In Sect. 2.4.4 we examine what, if anything, sets woody encroachers apart from the large number of woody species in a flora that have not proliferated in grasslands.

2.4.1   Establishment of Woody Plant Seedlings

The seedling and early establishment stage of the woody plant life cycle is typically the most vulnerable. Once past this stage, woody plants capable of vegetative regeneration (resprouting) may be highly persistent in the face of climatic events (drought, frost) or disturbances (browsing, fire) that top-kill them. Environment and neighbor interactions control population growth through effects on establishing seedlings, modifying their survivorship odds. This stage is therefore often described as a recruitment "bottleneck" constraining the proliferation of woody plants in grasslands (Bond 2008). The implication is that if individuals survive this stage, their odds of surviving to maturity are greatly improved.

Woody plant encroachment begins with deposition of seed within grassland communities. In instances where seed must be transported from distant seed sources, woody species dispersed by wind and birds would likely be the first colonizers. Species distributed by water are more likely to encroach from upstream or upslope to downstream or downslope locations than from lowland to upland locations. Some woody plants are dispersed readily by native ungulates and livestock. Examples include leguminous species whose hard seeds are encased in nutrient-rich pods (e.g., some acacia and mesquite species). The pods are eaten but the hard seeds may escape mastication, become scarified during passage through the digestive tract, and deposited in a moist, nutrient-rich media away from parent plants harboring seed predators. Furthermore, foraging ungulates would deposit seeds in areas where defoliated grasses have diminished capacity to suppress seedlings by fueling fire or preempting water and nutrients. Secondary dispersal agents, such as dung beetles, may disperse seeds further and bury them at depths conducive to germination and establishment. In North America for example, mesquite may have been "dispersal limited" during the Holocene, owing to extinctions of Pleistocene megafauna, but introduction of livestock by Anglo-European settlers facilitated dispersal of mesquites into upland grasslands (Brown and Archer 1987). Seed produced by woody plants that are already established in grasslands can additionally be dispersed locally by a variety of vectors, including ants and rodents. These processes, however, may involve trade-offs with seed predation (Nicolai et al. 2010). Though seldom considered, dispersal has important implications for the rate of WPE, for as the dispersal of viable, germinable seed increases, so too do opportunities for establishment (Groom et al. 2000).

When woody plant seedlings germinate in grasslands, they face intense competition for light, water, and soil nutrients. In lightly grazed, high-productivity grasslands, grasses will initially be taller than woody plant seedlings, reducing light availability (de Dios et al. 2014). Typically, grasses and woody seedlings in water-limited environments share the same shallow soil horizon (Kambatuku et al. 2013), so that grasses may furthermore monopolize soil resources to near exclusion of woody plant recruits, especially under environmental conditions that favor grasses: fine-texture or shallow soil sites with a summer rainy season characterized by small rainfall events that wet only the near-surface soils (Fravolini et al. 2005). However, grazing reduces grass leaf area, root density, and depth and therefore competitive effects on seedlings above and below ground. The intensity of grazing required to induce this response is likely to vary among sites, and may vary with soil condition according to their favorability for grasses. Thus, critical grazing levels may be relatively low on sandy, deep sites and higher on clayey or shallower sites (Knoop and Walker 1985).

Ground cover of many grasslands is characterized by a matrix of grass patches and bare ground. Grazing does not typically reduce grass biomass homogeneously and can contribute to increases in bare ground cover. These gaps in grass cover, which occur even in lightly grazed grasslands dominated by late seral, productive grasses, provide opportunities for woody seedlings to establish (Jurena and Archer 2003; Wakeling et al. 2015). In woody species that develop taproots, seedlings may establish during periods when soil water content is high and belowground competition is minimal. Under such conditions, which can occur in years of average rainfall, taproots grow quickly beyond the zone exploited by grasses thereby reducing below-ground competition with grasses (Brown and Archer 1990; Weltzin and McPherson 1997). Drought-induced reductions in grass density or cover, perhaps amplified by grazing, may create additional opportunities for establishment of woody plant seedlings when rains return. Once established, these seedlings may then persist through subsequent dry periods residual soil moisture is available below the grass root zone. This is a possible explanation for the "stair-step" or "ratchet" pattern of woody plant encroachment that has been observed in some areas. Collectively, these mechanisms help explain how some woody species can establish (1) under light grazing when grass competition should be highest (Brown and Archer 1989; Brown and Archer 1999, and references therein), and (2) under typical (non-episodic) climatic conditions (Watson and Westoby 1997), and (3) persist through periods of drought.

The relationship between grasses and woody plant recruits is not necessarily antagonistic. Grasses can in turn compete with and facilitate woody seedlings. A grass patch may increase water infiltration and reduce evaporation from the soil surface and subsequently deplete soil moisture by transpiration. The net effect on woody seedling survival depends on multiple factors including species, soil texture, rainfall amount/intensity, and temperature. Net effects of grasses on woody plant seedlings are more likely to be facilitative in arid or semiarid regions and competitive in more mesic grasslands and savannas (Good et al. 2014). In semiarid and arid grasslands, small-scale heterogeneity may be such that there are patches where woody seedling establishment is high and patches where it is low (Maestre et al. 2003a), as well as settings where facilitation by grasses more than offsets even strong belowground competition (Maestre et al. 2001; 2003b). Grasses can enhance microenvironmental conditions for woody seedlings by increasing root turnover and litter deposition, which function to improve soil organic matter, soil structure, fertility, and moisture retention. Grass stems can also capture surface-water runoff and sediment, increasing inputs of moisture and nutrients to the soil. In addition, grass shoots provide shade, reducing daytime temperature stress levels and evapotranspiration. In dry years, the radiative protection afforded by grass litter can significantly reduce woody seedling mortality (de Dios et al. 2014). Even in tropical and subtropical savannas, woody seedling growth and survival rates can be markedly higher in grass patches than in areas of bare soil. Consequently, levels of establishment can be higher on protected sites than on grazed sites (e.g., O'Connor 1995), especially if the protected sites are recovering from past brush management (e.g., Browning and Archer 2011).

2.4.2   Transitioning from Saplings to Adults

Once woody plants progress into the sapling stage, they have become far less vulnerable to competition, drought, and herbivory; they have passed through their most vulnerable stage. Saplings have better developed root systems, are taller, and have higher leaf area and carbohydrate reserves than seedlings. Unfavorable climate conditions and competition will affect their growth rates, but not necessarily their survival (Cardoso et al. 2016). Belowground competition grasses can slow sapling growth particularly during periods of higher than average rainfall (February et al.

2013) or if mineral availability is increased (Vadigi and Ward 2012) and can also be amplified by browsing (Vadigi and Ward 2014). Accordingly, competition, nutrient limitations, and herbivory can combine to slow sapling development and prolong the time they require to achieve a size that allows them to competitively dominate grasses and begin to influence microclimate and soil properties that will alter future patterns of community development.

Both browsing and fire constrain the progression from sapling to mature shrub or tree (Norton-Griffiths 1979; Augustine and McNaughton 2004; Vadigi and Ward 2014). The frequency and intensity of fire are coupled strongly to grassland productivity (Krawchuk and Moritz 2011) and to grazing (Anderies et al. 2002; Fuhlendorf et al. 2008). Grasslands that develop a high density of standing biomass generate litter capable of fueling hot fires that top-kill or kill saplings. Further, reliability of dry-season fire in more productive systems reduces the occurrence of temporal refuges or fire-free periods that would permit some tree cohorts to pass into a fire-tolerant life stage. Similarly, high spatial connectivity of grass cover would reduce the occurrence of spatial refuges or patches that escape fire during a burn event. Fire is therefore considered the main factor limiting tree cover in warm, semiarid to subhumid savannas that would, without fire, transition to a community dominated by woody plants (Bond 2008).

Saplings of many grassland and savanna species can regenerate vegetatively (resprout) after fire. However, even if saplings survive, repeated fires would prevent them from reaching maturity. Grasslands and savannas may therefore have "seedling banks" or "sapling banks" where woody plants persist in a diminutive state caused by fire or browsing events that occur with sufficient frequency to prevent them from growing past the flame or browse zone. These plants would be "waiting in the wings" for an opportunity to "escape"—an opportunity that may come when populations of browsers decline or when fires are suppressed or when grazers reduce the fine fuel density.

Herbivory and fire are linked so inextricably that some consider them a single disturbance regime: *pyric herbivory* (Archibald et al. 2005; Fuhlendorf et al. 2009; Fuhlendorf et al. 2012). In this view, when fire occurs randomly and herbivores roam freely, the two disturbances become spatially and temporally interdependent and the landscape is composed of a shifting mosaic of woody and herbaceous vegetation (Fuhlendorf and Engle 2001). In contrast, the traditional, independent management of fire and herbivory where livestock movements are regulated and relatively inflexible gives rise to a "fuel vs. forage paradox" (i.e., at a given time and place, grass biomass can be one but not both). Coupling the two, as pyric herbivory, averts this paradox because herbivores are attracted to, and concentrate their foraging on, recently burned areas, which allows other areas to accumulate the fuel mass needed to enable future fires that would keep woody plants in check. Subsequent prescribed burns conducted on these areas would then attract grazing animals and alleviate grazing pressure on the previously burned area to allow fuel to accumulate for a follow-up prescribed fire. The net result is a shifting mosaic of vegetation states that provides habitat for a variety of species with contrasting habitat requirements.

The prevalence of fire will determine which woody species in a local flora are more likely to pass from sapling to maturity. Among tree species of tropical Africa, seedlings that allocated resources preferentially to growth and resource-capture traits (e.g., height, leaf area, root-shoot ratios) survived better in ecotones between forests and savannas where fire frequency was low; species that allocated preferentially to carbohydrate storage in leaves and roots had better survivorship in fire-prone savannas (Cardoso et al. 2016). However, larger saplings survived better than smaller saplings, irrespective of allocation traits in either plant community. These results have implications for WPE and highlight a question we have not yet addressed: If environmental conditions change to favor "woody plant" proliferation, why have so few of the species comprising the woody plant flora in an area become encroachers? We return to the question of species selection in the context of WPE in Sect. 2.4.4.

2.4.3   Woody Plant Carrying Capacity

A population is at carrying capacity when strong negative density feedbacks on recruitment or positive density feedbacks on mortality (i.e., self-thinning) prevent further population increases. In general, these feedbacks are mediated by resource competition or simply by patch occupancy, in the sense that a tree or shrub seedling cannot mature in a patch already occupied by a mature tree, or, if it could, would not actually increase woody plant cover.

Greater resource inputs into an ecosystem shift carrying capacities toward higher biomass densities. We have already noted that limits of woody plant cover increase with MAP up to a point when presumably other resources become more limiting (Fig. 2.4). However, in regions where annual precipitation is highly variable—a characteristic of many water-limited environments—it is challenging to pinpoint an absolute carrying capacity for woody plants, as mortality and recruitment in any given year are tied to that year's or the recent series of years' precipitation, not the long-term average. Precipitation deficits will decrease recruitment and increase adult mortality (Bowers 2005; Twidwell et al. 2014), but density-dependent mortality may also occur during more benign conditions (Meyer et al. 2008; Dwyer et al. 2010). Precipitation-induced fluctuations in recruitment and mortality rates (Fig. 2.5) may keep woody plants from reaching their MAP potential in some areas (Fig. 2.4).

Though the theory of density dependence or self-thinning is clear-cut, it has been difficult to find evidence of it in field studies. If density dependence is at play, it should leave an imprint on tree or shrub spatial distribution, such as a decrease in spatial aggregation with tree size or age (Meyer et al. 2008; Belay and Moe 2012). These patterns indicate that survivorship probabilities of woody plants decrease in the vicinity of woody plants. In savannas, the maintenance of long inter-canopy distance between mature trees is additionally mediated by grasses suppressing the seedling growth (Sea and Hanan 2012).

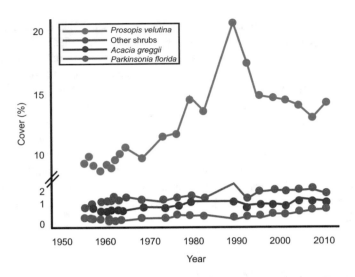

Fig. 2.6 Canopy cover of three shrub species and "all other shrubs" at a Sonoran Desert grassland (USA) where woody plant encroachment has been well documented. *Prosopis velutina* cover increased markedly from the 1950s to the 1990s, whereas that of *Acacia greggii* and *Parkinsonia florida* (both potentially N$_2$-fixing) and all other shrubs has remained low (note break in *y*-axis). Data are from the Santa Rita Experimental Range Digital Database, Pasture 8 (http://ag.arizona.edu/SRER/longterm/ltcover.xls)

Cover of velvet mesquite (*Prosopis velutina*) into grassland at a Sonoran Desert site is near the maximum level predicted by MAP in Fig. 2.4. The only woody encroacher at this site (Fig. 2.6), velvet mesquite has well-developed, shallow lateral roots extending well beyond their canopies. Intraspecific, shrub-shrub competition could therefore potentially explain why cover appears to have reached its maximum. However, an analysis of spatial patterns over a 74-year period failed to exhibit changes indicative of self-thinning (Browning et al. 2014). Are there explanations other than those related to plant spatial patterns that might set upper limits to shrub cover on a site? One hypothesis is related to hydraulic constraints on shrub size (e.g., Sperry and Hacke 2002; Hacke et al. 2006). As shrubs approach their upper size limit for a site with a given soil texture, depth, topographic setting, etc., their ability to maintain continuity in transport of xylem water may become increasingly jeopardized and lead to higher probabilities of branch or whole-canopy mortality. This loss of plant branch systems or canopies would reduce canopy cover that subsequently would be compensated by recruitment of new plants or growth of other, smaller plants if stand-level canopy cover were to be maintained. This more subtle form of density-dependent interaction manifests itself via canopy reductions rather than whole-plant mortality. Support for this proposition comes from observations of shrub height asymptotes and shifts in leaf-stem biomass allocation (Martinez and Lopez-Portillo 2003) and shrub size-abundance relationships (Allen et al. 2008).

Some woody encroachers can generate positive density dependence by facilitating the encroachment of other woody species. In these instances, the initial encroaching species, perhaps arriving via dispersal from wind, water, ungulates, rodents, or ants, adds vertical structure to the grassland community and modifies soils and microclimate subsequent to its establishment. Seeds of other woody species concentrated in other parts of the landscape may then arrive via birds attracted to this new vertical structure, and their germination, growth, and establishment would be enhanced through modifications of microclimate and soils from pioneer plants (Archer 1995; Stokes and Archer 2010). Facilitation, therefore, may have a combination of passive and active components: adding structure and altering local ecosystem processes. Active processes would include hydraulic lifting of soil moisture from deep to shallow layers (Zou et al. 2005) and modification of soil-nutrient pools and radiant-energy regimes (Barnes and Archer 1996; Barnes and Archer 1999). Where woody encroachment reduces grass biomass and cover, fire frequency and intensity can be reduced, enabling increased establishment of woody seedlings and clonal reproduction (Ratajczak et al. 2011a; Brandt et al. 2013). Accordingly, in fire-prone grasslands and savannas, encroachment by a relatively fire-tolerant or fast-growing woody species may facilitate the spread of fire-intolerant or slow-growing woody species. These processes have an important temporal component, as changes initiated by the initial encroaching species may occur gradually over decades (Throop and Archer 2008; Liu et al. 2013).

2.4.4 Why Do So Few Woody Species Proliferate in Grasslands?

The diverse mechanisms proposed to explain woody plant encroachment in Sect. 2.3 are united by being general enough to pertain to many woody plant species occurring in any biogeographic province. The treatment of "woody plants" as a *de facto* functional group befits investigation of woody plant encroachment as a global phenomenon, but ignores another important aspect of WPE: that very few woody species in a regional flora have actually become aggressive encroachers or have spearheaded the encroachment process (Stokes and Archer 2010; Barger et al. 2011). However, there are dozens of woody species with growth forms that should have benefited from the changes in drivers, yet have not proliferated (Fig. 2.6).

The apparent selectivity of woody encroachment suggests that it may be useful to examine the phenomenon under a different light: (1) which, if any, traits unify woody encroachers around the world and (2) what might this potentially tell us about the relative importance of various potential drivers? In addressing these questions we borrow from community theory the perspective of viewing the landscape distribution and abundance of species as the result of a regional species pool passing through a sequence of abiotic environmental and biotic community "filters", such that species with mismatched trait combinations are excluded from a community (Keddy 1992). Applied to encroachment of tree and shrub species, we propose that there are a sequence of barriers for entering and proliferating in a grassland or savanna community (Fig. 2.7). One or more of these barriers may be made progres-

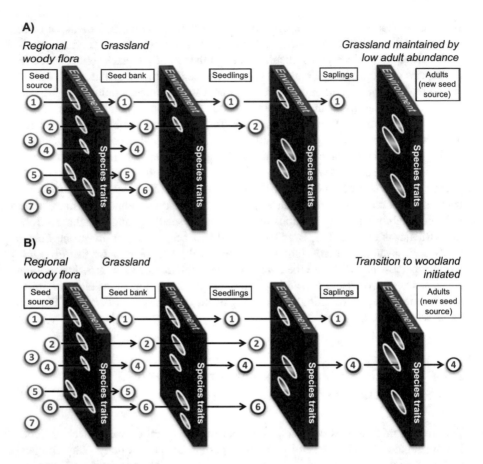

Fig. 2.7 Conceptual model of the species selection process in woody plant encroachment into grasslands. *Numbered circles* represent woody species and *blocks* represent barriers or trait filters constraining advancement to the next life stage. Each barrier may be comprised of several independent and interactive challenges to growth and survivorship (e.g., dispersal, predation, nutrient scarcity, disturbance). In this hypothetical example, (**a**) one or more barriers prevent any of the woody species in the flora from recruiting in the grassland. (**b**) Land-use change makes one barrier (in red) more surmountable, now permitting two additional species to get to the seedling stage (#4 and #6). However, one of these (#6) is constrained by the next life stage barrier, whereas the other (#4) is not. This framework explains how grazing, fire suppression, elevated CO_2, climate change, etc. could have nonselective positive effects on many woody plant species in a flora, and yet only a very narrow subset of those would be capable of developing a viable population in grassland. Research should seek to identify the combination of woody plant traits required for passage through all barriers. Doing so would help us explain past encroachment and predict encroachment under future environmental conditions

sively "leaky" or "porous" by changes in drivers of WPE. We furthermore integrate the concept with population biology to highlight the fact that the exclusion of species is most likely to occur during the more vulnerable and uncertain stages of population growth, specifically seed survivorship and dispersal, germination and seedling establishment. Each of these is necessary for grassland invasion and each is a potential bottleneck for WPE.

In this framework, encroaching woody species must have heightened responsiveness to at least one encroachment driver, but must also overcome all other barriers to surviving a precarious life stage. Non-encroachers either may not be responsive to drivers or remain limited by other barriers. Prior to woody plant encroachment, every woody species in the regional flora must have been limited by at least one environmental or community barrier in at least one life stage (Fig. 2.7a). For example, the high productivity and flammability of grasslands may have universally blocked recruitment of woody plants in grasslands, but individual species could have been excluded by any number of additional barriers, such as low-seed production or survivorship, shade intolerance, or slow growth.

The key effect of the historic drivers of WPE was to modify one or more of the filters in such a way that at least one species could pass through a former barrier. A *necessary* condition for a species in the regional pool to encroach would be the release from at least one recruitment bottleneck by a shift in environmental conditions. Nevertheless, many species meeting this requirement would have been prevented from encroaching through unyielding restrictions in other life-stage transitions. Environmental regime change would have been a *sufficient* condition to trigger woody encroachment only for species not constrained by additional recruitment bottlenecks (Fig. 2.7b). The relative paucity of species in the worldwide set of recognized "woody encroachers" suggests that most woody species remain excluded from grasslands through demographic barriers affecting recruitment, growth, or reproduction that have been essentially unchanged by regime shift.

What then are the traits that distinguish woody encroachers from non-encroachers? Table 2.1 lists a variety of functional attributes of woody encroachers on different continents. The list is not meant to be exhaustive, but only to provide examples. The list shows that woody encroachers are not consistently represented by, or restricted to, one or a few functional traits or groups. For example, it might be reasonable to expect that plants that fix N_2, are deciduous, and are livestock-dispersed would be aggressive encroachers—and they certainly can be. But so too can species that are evergreen, bird-dispersed, and lacking the capability of symbiotic N_2 fixation. Similarly, encroaching species can be subshrubs, shrubs, or treelike in stature, and may or may not be capable of vegetative regeneration following disturbance.

The conceptual framework in Fig. 2.7 paints the broader picture for organizing questions of woody encroachment based on species traits and provides a basis for developing and testing hypotheses regarding woody plant encroachment systematically. The main point is that not necessarily all traits, but certainly several

Table 2.1 Woody plants proliferating in grasslands and savannas encompass a wide variety of functional traits and taxonomic families

Functional traits	North America	South America	Africa	Australia
Stature				
Fruticose (shrubby)	x^1	x^2	x^3	x^4
Arboreal (treelike)	$x^{5,6}$	x^9	x^7	x^8
Leaf Habit				
Evergreen	$x^{1,5,6}$	x^9	x^{10}	x^8
Deciduous	x^{11}	x^2	x^{12}	x^{23}
Potential N_2 fixation				
Yes	x^{11}	x^2	$x^{7,10}$	x^8
No	$x^{1,5,6}$	x^9	x^{12}	x^{13}
Dispersal				
Livestock	x^{11}	x^2	$x^{7,20}$	x^8
Wind/water	x^1	x^9	x^{14}	x^{15}
Bird	x^5	x^{22}	x^{12}	x^{16}
Recruitment				
Readily generates from seed	x^{11}	x^2	$x^{7,10}$	x^8
Vegetative regeneration	x^{11}	x^2	$x^{7,14}$	x^8
Deep or dimorphic root system	x^{11}	x^2	x^{17}	x^{18}
Nativity				
Native species				
Exotic (non-native) species	x^{19}	x^{21}	x^{20}	$x^{8,15,18}$

An 'X' denotes that a functional trait is represented by a species on a given continent. Superscripts link a given trait to the species exhibiting that trait (bottom of table). Species list is not intended to be comprehensive

[1]Creosote bush, *Larrea tridentata*, Zygophyllaceae (Grover and Musick 1990)
[2]Mesquite, *Prosopis* spp., Fabaceae (Cabral et al. 2003)
[3]Blackthorn *Acacia mellifera,* Fabaceae (Kraaij and Ward 2006)
[4]Coastal wattle, *Acacia sophorae*, Fabaceae (Costello et al. 2000)
[5]Eastern red cedar, *Juniperus virginiana*, Cupressaceae (Barger et al. 2011)
[6]Ponderosa pine, *Pinus ponderosa*, Pinaceae (Barger et al. 2011)
[7]Karroo thorn, *Acacia karroo*, Fabaceae (O'Connor 1995)
[8]Prickly acacia, *Acacia nilotica*, Fabaceae (Kriticos et al. 2003)
[9]Quebracho blanco, *Aspidosperma quebracho-blanco*, Apocynaceae (Morello and Saravia-Toledo 1959)
[10]Paperbark thorn, *Acacia sieberiana*, Fabaceae (Mitchard and Flintrop 2013)
[11]Velvet mesquite, *Prosopis glandulosa*, Fabaceae (Bahre and Shelton 1993)
[12]African myrrh, *Commiphora Africana*, Burseraceae (Oba et al. 2000)
[13]Rubber vine, *Cryptostegia grandiflora*, Asclepiadaceae (Grice 1996)
[14]Sickle bush, *Dichrostachys cinerea*, Fabaceae
[15]Catclaw mimosa, *Mimosa pigra*, Fabaceae (Lonsdale 1993)
[16]Chinee apple, *Ziziphus mauritiana*, (Rhamnaceae) (Grice 1996)
[17]Blackthorn *Acacia mellifera*, Fabaceae (Kambatuku et al. 2013)
[18]Mesquite, *Prosopis* spp., Fabaceae (Robinson et al. 2008)
[19]Chinese tallow, *Sapium sebiferum* (Euphorbiaceae) (Bruce et al. 1995)
[20]Mesquite, *Prosopis* spp., Fabaceae (Shackleton et al. 2015)
[21]Paraiso, *Melia azedarach*, Meliaceae (Ruiz Selmo et al. 2007; Batista et al. 2014)
[22]Glossy privet, *Ligustrum lucidum*, Oleaceae (Tecco et al. 2006)
[23]Mulga, *Acacia aneura*, Fabaceae (Noble 1997)

key traits, could distinguish encroachers from non-encroachers in a given bioregion. Furthermore, common trait trade-offs could be influencing the selection of woody encroachers in interesting ways. Most species in a regional pool could be prevented from encroaching by a trade-off between seed dispersal and seedling survivorship such that some small-seeded species may readily disperse into grassland but not survive as seedlings, whereas large-seeded species could potentially establish but lack adequate dispersal. Seen in this light, it is clearer why there does not seem to be a universally applicable set of encroacher characteristics, but also why taxonomic groups that may be less constrained by dispersal-survivorship trade-offs do seem to contribute more species to the global set of woody encroachers. Knowledge of the trade-offs in trait combinations could help to explain changes and patterns of WPE observed to date and also to predict future changes in woody species or functional group composition.

2.5 Ecosystem Services

Maintenance of a desirable mixture of herbaceous and woody vegetation is a key component of sustainable ecosystem management in grazed rangelands. Over the past century, this balance has been disrupted and shifted in favor of unpalatable shrubs in many areas of the world. Widespread conversion of grasslands and savannas to shrublands or woodlands has long been of concern to those whose livelihoods depend on livestock production; but the recent realization that this land cover change has significant implications for a myriad of other ecosystem services is now challenging us to adopt a broader perspective on this global phenomenon. Here, we review the effects of WPE on a subset of ecosystem services related to carbon sequestration, hydrology, and biodiversity. Management actions aimed at reducing woody cover also influence ecosystem service portfolios and these are reviewed in Sect. 2.6. The effects discussed here should be further considered in the context of the supply and demand perspectives presented in Chap. 14.

2.5.1 Carbon Sequestration: Plant and Soil Pools

The global phenomenon of WPE has resulted in a significant redistribution of carbon (C) among major terrestrial pools. Trees and shrub proliferation across a range of bioclimatic regions (Fig. 2.2) constitute a potentially significant, but highly uncertain component of the North American C budget (Barger et al. 2011). Presently we cannot confidently predict the magnitude, let alone the direction, of change (Eldridge et al. 2011). Robust generalizations about WPE impacts on ecosystem C balance are elusive because of insufficient quantification of woody plant productivity in encroached ecosystems. Definitive conclusions have been further constrained by confounding methodologies used to estimate soil organic carbon pools, and how those pools change with disturbance (e.g., drought, wildfire) and land management practices (e.g., prescribed burning, brush management). These knowledge gaps are amplified at regional scales where quantifying the net effects of WPE on regional carbon balance would require an accounting of the area undergoing WPE, the stages of encroachment, and the area recovering from past disturbances (Asner et al. 2003).

Studies quantifying herbaceous production in drylands in relation to climate, land use, and disturbance are numerous, but relatively few have simultaneously quantified woody plant production and even fewer have quantified plant *and* soil pools. Accordingly, we know very little about how ecosystem (plant + soil) carbon pools change with changes in grass-woody plant abundance. Scenarios where aboveground net primary production (ANPP) increases, decreases, or remains unchanged can be logically theorized following woody plant encroachment (House et al. 2003). At broad scales, if encroaching woody plants are *less* productive than the grass communities they replace net ANPP will *decrease*. Conversely, if encroaching woody plants are *more* productive than the replaced grass communities net ANPP will *increase*. Lastly, if grassland and woody plant communities are equally productive then no change in ANPP would be expected. So, which of these three scenarios is most likely to occur? As it turns out, the answer depends on rainfall.

Recent syntheses suggest that ANPP scales linearly with MAP in landscapes where woody plants have displaced grasses. At an MAP of ~340 mm the ANPP contribution to the C pool in woody plant-encroached landscapes switches from being a net C source to a net C sink (Fig. 2.8a). Whereas grassland ANPP stabilizes at MAP > 500 mm, woody plant ANPP continues to increase linearly with increases in MAP. This presumably reflects the ability of woody plants, with their more complex canopy architecture, to utilize greater leaf area than grasses (Knapp et al. 2008a). However, the belowground soil organic carbon (SOC) pool typically dwarfs the aboveground pool in drylands. Given its large size, even small changes in the SOC pool could have big impacts on ecosystem C balance, especially given the expansiveness of grasslands and savannas. So how and to what extent do these aboveground changes in plant production affect belowground C pools?

The SOC pool reflects long-term inputs from plant leaves, stems, and roots. This suggests that changes in the amount of SOC would vary with changes in the plant production. However, a survey of studies quantifying changes in SOC with WPE revealed no consistent patterns—it increased markedly in some cases, and remained unchanged, or decreased in others and had no correlation with MAP (Fig. 2.8b). This indicates that when grass communities are replaced by woody plant communities, there is a major difference between ANPP and belowground carbon pools: ANPP scales with MAP while SOC has no apparent relation to it. Reasons for this disconnect are unclear, but may (1) be an artifact of different soil sampling methodologies (see discussion in Barger et al. 2011; Throop et al. 2012), (2) reflect the nonequilibrium status of many landscapes experiencing WPE and the fact that changes in soils lag well behind the changes in the vegetation that drive them, and (3) plant species or functional group differences in allocation of carbon for aboveground vs. belowground growth.

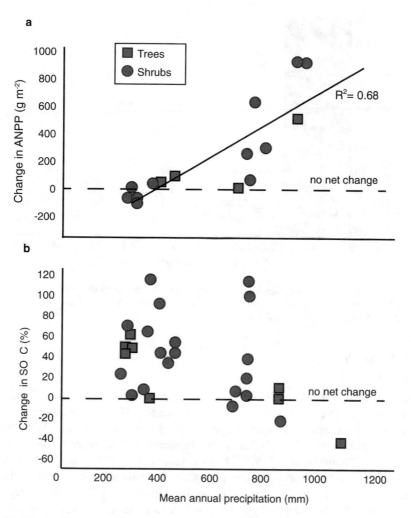

Fig. 2.8 Changes in (**a**) aboveground net primary productivity (g biomass m^{-2} year^{-1}) and (**b**) soil organic carbon with woody encroachment as a function of mean annual precipitation. Data span a range of species and ecoregions and encompass a range of sample collection and processing methodologies (from Barger et al. 2011)

Where landscape effects of both ANPP and SOC responses to have been taken into account in North America it appears that arid zones are likely to become net sources of carbon when WPE occurs, whereas higher rainfall areas will become net sinks (Fig. 2.9). Given that WPE has been occurring since the late 1800s in many of these regions, the sites depicted in Fig. 2.9 may have been at relatively advanced stages of woody plant stand development. Accordingly, the reported values may represent potential envelopes between the lower and upper limits of an ecological site. However, natural disturbances (e.g., drought, wildfire, pathogen outbreaks) and land management (Sect. 2.6) will alter the extent to which these potentials may be realized or maintained.

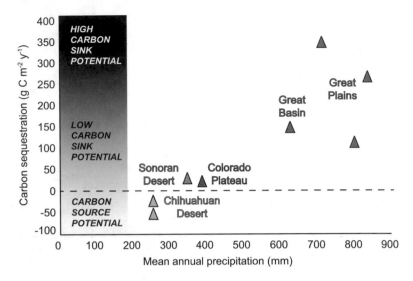

Fig. 2.9 Carbon source–sink potential with woody plant encroachment in North American ecoregions. Values are the mean of changes in aboveground net primary production plus soil organic carbon. Modified from Barger et al. (2011)

2.5.2 Hydrology

The hydrological impact of WPE has been of intense interest, as climate change and human demand for freshwater have increased, inciting global concerns about water security for communities (Vorosmarty et al. 2010). The question that generally concerns the public most is whether WPE decreases groundwater recharge and/or streamflow (Tennesen 2008). WPE has the potential to interfere with all components of the water budget equation: precipitation, evapotranspiration (ET), runoff (R), and deep drainage (D, recharge below the rhizosphere). Structural differences between woodlands and grasslands suggest that, in general, woodlands should have higher ET and lower R than grasslands (Bonan 2008). Four major mechanisms follow. First, woody plants can take up water stored in deeper soil layers (Sect. 2.3.3). Second, woodlands have lower albedo and greater air turbulence in the canopy boundary layer, which increases their potential ET (PET). Third, protracted periods of dormancy limit the number of days over which transpiration occurs in grasslands, whereas shrubs and trees, particularly if they are evergreen, have longer periods of transpiration (Donohue et al. 2007). Fourth, canopy interception of rainwater, a component of ET, is lower in grasslands especially when compared with needle- or scale-leaf conifers (e.g., *Pinus*, *Juniperus*) (Owens et al. 2006).

WPE also can influence runoff by changing soil infiltration rates. In water-limited systems, runoff comes during intense rainfall events, when the precipitation input rates exceed the infiltration rate. Water begins to pond and run off, eventually flowing into streams (Dunne 1978). Woody cover may change the infiltration characteristics of soil through effects on soil quality and spatial heterogeneity of plant

cover. A recent meta-analysis showed that these effects are highly context dependent (Eldridge et al. 2011). Shrub encroachment into grasslands is often classified
as a "trigger" for soil degradation and "desertification" (Schlesinger et al. 1990).
But a study conducted in a semiarid Mediterranean grassland in Spain showed that
the effect can also be opposite (Maestre et al. 2009). In this example, shrubs establishing in degraded pastures created "islands of fertility" that enhanced vascular
plant richness, microbial biomass, soil fertility, and nitrogen mineralization. In this
sense, shrubs may be seen as reversing, rather than causing, desertification.

Regardless of changes in vegetation and soil structure, there are physical limits
to the magnitude with which WPE can modify the hydrological budget. Potential
effects are greatest where precipitation approximately equals PET (Zhang et al.
2001); above or below this threshold, ET is constrained either by precipitation or
PET. Grassland and savanna biomes occur under both climate conditions. The Great
Plains of North America, for example, straddle regions with precipitation surplus to
the east and precipitation deficit to the west (Fig. 2.10). Therefore, WPE should
have maximal hydrological consequences in central regions of the USA.

There are several caveats to these generalizations and we mention two: first,
there are hydrological systems with large bypass-flow components. Bypass flow is

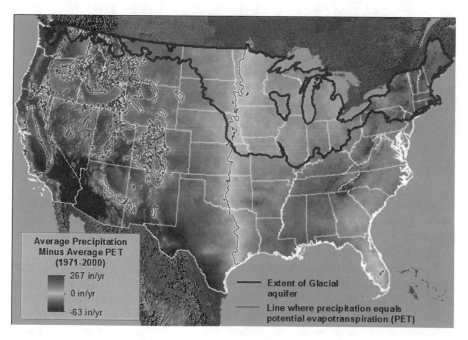

Fig. 2.10 A map of average precipitation (P) minus average potential evapotranspiration (PET)
for the contiguous USA. P-PET decreases prominently from east to west and less so from north to
south. Impacts of woody plant encroachment on ET are expected to be maximized near the 97th
degree west longitude, where P approximately equals PET. Where P exceeds PET, ET is energy
limited approaching PET; where PET > P, ET is water limited, approaching P, irrespective of
woody cover

the rapid transport of water through the root zone by way of macropore conduits (e.g., channels left by large dead roots, cracks, and fissures in bedrock). Bypass flow expedites recharge of aquifers or spring-fed streams and the brief residence time for water in the rhizosphere means that vegetation has practically no influence on the volume of bypass flow. This minimizes the effect that WPE can have on the water budget. This was demonstrated in a series of rainfall simulation experiments in the karst region of Central Texas, a semiarid area where P is not far below PET. Shallow caves at the field site made it possible to capture drainage out of the root zone as cave drip. Juniper removal had no significant effect on the amount of water captured as cave drip (Bazan et al. 2013). Decades of controlled experiments in this region have generally returned the same result, that the effect of removing encroaching woody plants on ET and/or spring flow is small and short-lived (Wilcox et al. 2005) (Chapter 3, this volume).

A second important exception to the general pattern occurs in systems with shallow water tables, in which the incursion of deeply rooted trees can fundamentally alter the hydrological cycle, including precipitation. For example, a regional increase in the woody cover of the African Sahel zone has recently been linked to a precipitation feedback: as woody plant cover increased, more moisture (from groundwater) was cycled into the atmosphere, which increased cloud formation and rainfall. The positive-feedback loop closes when higher rainfall in turn increases woody cover (Scheffer et al. 2005). It has been suggested that this regional vegetation-precipitation feedback may be locally enhanced by a vegetation-infiltration feedback, in which infiltration is improved as a consequence of higher vegetation cover, enabled by WPE. Together, these two feedbacks are a powerful force of self-organization of the hydrological system, which can either be locked into an arid, low-productivity state or a mesic/high-productivity state (Dekker et al. 2007).

2.5.3   Biodiversity

Biodiversity, whether quantified as richness of species, plant functional groups, or animal guilds, is influenced strongly by WPE. From the perspective of vegetation structure, WPE is transformative: grasslands become shrub or tree savannas and shrub and tree savannas become shrublands or woodlands.

Grassland ecosystems are among the most endangered in North America, with most having been reduced to small remnants of their original distribution (Noss et al. 1995; Hoekstra et al. 2004). Initially, colonization of grasslands by woody plants involves new species that increase the biodiversity pool directly. Subsequently, modification of soil properties, vegetation structure, and microclimate may facilitate establishment by other novel plant and animal species. Maximum diversity in savanna-like configurations occurs often where woody and herbaceous plants are both well represented or where gains in new woody and herbaceous species outweigh losses of the initial grassland-obligate species (Fig. 2.11). As abundance of woody plants increases, grassland components eventually decrease and are

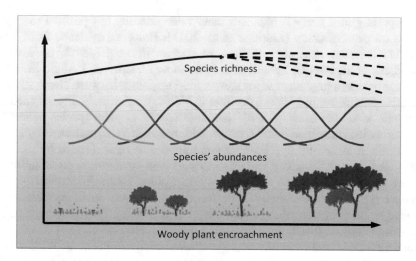

Fig. 2.11 Conceptual model of community changes in species abundances and richness with woody plant encroachment. Species richness is likely to be highest where both shrub-associated and grassland-associated species co-occur, with the endpoints varied, depending on the encroaching species

replaced by plants and animals adapted to shrublands or woodlands. In tropical and subtropical areas with large and diverse regional species pools, there may be a net increase in diversity along with concomitant changes in community structure. In other settings, there may be little or no net change in numerical diversity, but profound changes in community structure. In settings where the number of encroaching woody species is low, their proliferation may create virtual monocultures with little or no understory that will result in profound decreases in the diversity of plants and animals. Examples of the latter in North America include both native (juniper, ponderosa pine) and non-native (salt cedar, Chinese tallow) species. No matter the numerical changes in biodiversity, persistence of plants and animals endemic to grassland and open savanna ecosystems is jeopardized. Some grassland obligates are immediately lost at the initial stages of encroachment (e.g., Fuhlendorf et al. 2002; Lautenbach et al. 2016), whereas others may persist until woody plant cover reaches about 15 % (reviewed in Archer 2010).

2.5.3.1   Herbaceous Vegetation

Encroaching woody plants may have an immediate, adverse effect on herbaceous vegetation in some cases or a positive, facilitative effect in others. In the latter, woody plants may eventually suppress herbaceous plants as their density increases. These overstorey-understory relationships are influenced strongly by soil type, such that herbaceous plants may be suppressed on lowlands and facilitated on uplands (Hughes et al. 2006). Local- and landscape-scale diversity perspectives should therefore be kept in mind when generalizations are made.

A recent global analysis indicates that WPE generally has positive to neutral effects on plant diversity (Eldridge et al. 2011). However, evidence from North America indicates consistent declines in species richness (45 %, on average) (Ratajczak et al. 2011b). Variation in evolutionary history and Anglo-European land-use practices may account for these varied responses between North America and other continents. In addition, declines in North America plant diversity seem to vary with MAP. For example, long-term assessments of plant species richness in desert grasslands revealed linear declines with time since encroachment by an arid land shrub. Additionally, species-poor communities in areas invaded by the same shrub were less stable (more variable in time) than species-rich communities in nearby grassland-dominated areas (Baez and Collins 2008). In contrast, species richness declined exponentially with woody plant cover in humid grasslands invaded by an evergreen arborescent (Knapp et al. 2008b). These contrasts in arid and humid regimes (linear vs. exponential declines, respectively) suggest that the future magnitude and dynamics of vegetation diversity response to WPE will be mediated by climate change. Changes in species composition should not be lost in discussions of diversity. As we mentioned earlier, substantial changes in species, functional groups, or guilds, as well as changes in relative species abundances (evenness), can occur with small, or even no, net changes in species richness. Furthermore, these changes in species composition impact ecosystem processes related to primary production, nutrient cycling, and structure of tropic pyramids. Accordingly, measures of species richness alone provide a limited metric of changes in diversity.

As summarized in the next section, plant diversity changes have a multiplier effect on animal diversity by adding keystone structures and increasing vegetation heterogeneity (Tews et al. 2004).

2.5.3.2 Animals

Changes in the plant community associated with WPE have affected many grassland animals principally by reducing the quantity or quality of habitat and by altering a suite of fundamental ecological processes. Consequently, the abundance and distribution of many organisms that inhabit grassland ecosystems have decreased markedly (Samson 1994; Sauer and Link 2011). During the last 30 years, for example, grassland birds have declined more rapidly than any other group of birds in North America (Knopf 1994; Peterjohn and Sauer 1999; Vickery et al. 1999; Brennan and Kuvlesky 2005; Sauer and Link 2011).

Although long-term declines in the abundance and distribution of many grassland species have been relatively well documented, linkages between changes in grassland plant communities and their effects on animals are less clear. Vegetation structure is a key determinant of animal diversity, and because a principal consequence of WPE is a marked increase in vertical and horizontal structure, populations and communities of many resident animals shift markedly in response to woody encroachment (Skowno and Bond 2003; Coppedge 2004; Sirami and

Monadjem 2012). Although some species respond to changes in vegetation at broader scales, animals that function at smaller scales, such as small mammals and arthropods, are more likely to respond to changes in vegetation that alter local environmental characteristics (Wiens and Milne 1989). Consequently, some taxa, including birds, mammals, and reptiles, are more likely to respond to the structural changes in the plant community that accompany WPE, whereas other taxa, especially arthropods, are also likely to respond to changes in species composition that interfere with coevolved relationships with specific plant species (Litt and Steidl 2010). Relative to vertebrates, many arthropods are less mobile, depend on a narrower range of plants for food, cover, and sites for reproduction, and can have specialized relationships with specific plant species (Kremen et al. 1993). This makes them especially vulnerable to compositional changes in the plant community (Steidl et al. 2013). Changes in the arthropod community may feed back to influence multiple ecological processes, including pollination, decomposition, and nutrient cycling, as well as food resources for insectivores, including breeding grassland birds, small mammals, and reptiles.

Responses of animals to WPE vary broadly by taxa, plant community, and geographic region, but ultimately responses can vary by species (Ayers et al. 2001; Meik et al. 2002; Blaum et al. 2007a; Blaum et al. 2007b; Blaum et al. 2009). Species-specific responses are expressed frequently as sharp transitions in the probability of occupancy (i.e., changes in distribution) or as changes in demographic rates such as density, survival, or reproductive success at specific levels of woody plant cover (Grant et al. 2004; Sirami et al. 2009). For example, verdins (*Auriparus flaviceps*) and eastern meadowlarks (*Sturnella magna*), species common throughout grassland and shrublands of southern Arizona, respond strongly and oppositely to changes in the abundance of woody vegetation (Fig. 2.12). For verdins, as the amount of woody vegetation increases, the probability of them selecting an area for breeding increases; in contrast, the probability of eastern meadowlarks selecting an area for breeding decreases sharply as the amount of woody vegetation increases. Species-specific responses such as these explain why the effects of WPE on animal populations and communities vary with stage of encroachment (Fig. 2.11); composition of these communities shifts as density of woody plants changes. In early stages of encroachment when cover of woody plants is relatively low, vertical structure in the plant community increases. These structural changes increase the diversity of niche spaces available for exploitation by animals. Therefore, species capable of exploiting these niches are added to the initial animal community, increasing species richness and diversity. Overall richness and diversity of these areas increase as shrub-associated species join the existing community of grassland-associated species (Tews et al. 2004). In the southwestern USA, for example, increases in species richness of several taxa were associated with increased cover of woody plants (Arnold and Higgins 1986; Lloyd et al. 1998; Bestelmeyer 2005; Block and Morrison 2010).

As encroachment advances and shrub cover continues to increase, habitat for grassland-associated species declines, so their abundances decline. This pattern has been well documented for grassland birds (Coppedge et al. 2001; Cunningham and

Johnson 2006; Winter et al. 2006; Block and Morrison 2010), but has also been observed for mammals (Krogh et al. 2002; Blaum et al. 2007a) and reptiles (Mendelson and Jennings 1992; Pike et al. 2011). When woody cover exceeds species-specific thresholds, which as yet have been poorly established, populations of grassland-associated species are displaced (Grant et al. 2004; Sirami et al. 2009) and animal communities shift from being dominated by grassland-associated species to shrubland-associated species (Igl and Ballard 1999; Rosenstock and Van Riper 2001; Skowno and Bond 2003; Sirami and Monadjem 2012). Overall, richness of animal communities is likely maximized where cover of woody plants is below the threshold levels that displace grassland specialists but above levels where habitat becomes more exclusively suitable for shrub-associated species (Fig. 2.11); that is, where gains of new species outweigh losses of existing species (Archer 2010). This pattern of peak species richness at intermediate levels of woody cover has been documented for mammalian carnivores (Blaum et al. 2007a), arthropods (Blaum et al. 2009), and birds (Grant et al. 2004; Sirami and Monadjem 2012). Regardless of how encroachment affects animal diversity at local scales, animal diversity is ultimately reduced at broader scales if grassland-associated species are displaced.

Although systematic patterns in responses of animals to WPE are becoming clearer, the mechanisms governing them are not. Specifically, we do not understand clearly how WPE and other vegetation transitions influence demographic processes at the population scale or the behavior of individuals, particularly those related to habitat selection. In general, WPE influences populations and communities of animals directly by reducing both the quantity and the quality of habitat. Many animals rely on vegetation-based cues to indicate the presence of habitat—that is, to identify areas that provide the suite of resources necessary for survival and reproduction (Mannan and Steidl 2013). Therefore, as vegetation composition and structure change in response to WPE, areas that once provided habitat for a species may no longer provide that function. Specifically, as WPE proceeds, species will continue to persist in patches that provide habitat; as the vegetation transition continues, the same species could be displaced entirely.

WPE can lower habitat quality for animals that continue to inhabit encroached areas and reduce their survival or reproductive success. Changes in habitat quality may reflect changes in rates of predation or brood parasitism or changes in the types, abundance, or availability of food resources. WPE can alter predation risk by influencing the types, densities, and behaviors of predators in a community. For example, predation is often the primary cause of nest failure in grassland birds (Martin 1992) and is thought to be responsible for decreases in reproductive success of birds in areas encroached by woody plants (With 1994; Mason et al. 2005; Graves et al. 2010). Further, for songbirds nesting in grassland patches, the risk of nest predation increases with proximity to woody plants (Johnson and Temple 1990; Mason et al. 2005). WPE could affect food resources available to herbivores through changes in the composition or biomass of vegetation and subsequently to carnivores through changes in herbivore populations and communities (Maurer 1985). Among birds, declines in food availability can delay nest initiation or lead to nest failure (Ortega et al. 2006), and increase rates of nestling starvation (Maron and Lill 2005; Granbom et al. 2006) and predation (Dewey and Kennedy 2001; Zanette et al. 2003).

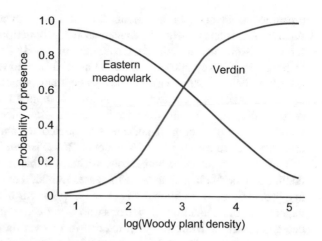

Fig. 2.12 Predicted probabilities of eastern meadowlark and verdin presence as a function of woody plant density in a Sonoran Desert grassland (R.J. Steidl, unpublished)

Additionally, food availability for nestlings could be affected by rates of brood parasitism by brown-headed cowbirds (*Molothrus ater*), which are correlated positively with woody plant cover (e.g., Johnson and Temple 1990; Shaffer et al. 2003).

Despite the global scale of the encroachment phenomenon and the tremendous number of grassland-associated animals that might be affected, only a modest amount of research has explored and quantified responses of animals to WPE. A variety of perspectives exist, but unifying, robust generalizations are still elusive. Some of the variation in results among studies might be attributable to artifacts of study design. For example, many studies simply contrast areas as "encroached" vs. "not encroached." These coarse classifications likely occlude biologically meaningful variation in animal responses along complex gradients of vegetation structure, composition, and dominance, as well as their continuous changes throughout the encroachment process (e.g., Thompson et al. 2009). Many of the studies that have explored broader encroachment gradients use space-for-time substitutions as a way to compare areas with different amounts of woody cover. Although these approaches can be useful in understanding how animals respond to structural changes in habitat resources—especially when gradients span large areas—they are predicated on the assumption that animals respond to vegetation changes in space in the same way they respond to changes in time (Sirami and Monadjem 2012). In addition, areas that have been encroached by shrubs may differ inherently from areas that have not been encroached in ways that are not apparent to researchers, but that may be important to animals.

2.6   Management Perspectives

Proliferation of woody plants has long been of concern in areas where the primary land use is cattle and sheep grazing. WPE on these lands typically reduces production of valued forages, complicates animal handling, and improves habitat for ectoparasites. Furthermore, and despite limited supporting evidence, WPE is often presumed to adversely affect stream flow and groundwater recharge (Sect. 2.5.2). As a result, management of rangelands for production of cattle and sheep has focused historically on reducing the amount of woody vegetation using a variety of technologies (Bovey 2001; Hamilton et al. 2004). Known as "brush management" (North and South America), "woody weed management" (Australia), and "bush clearing" (Africa), these technologies may be applied singly, in combination, or sequentially. As a result, rangelands are complex mosaics of areas undergoing woody plant encroachment and areas subjected to, and transitioning from, past efforts to reduce woody cover (Asner et al. 2003; Browning and Archer 2011).

Cover and biomass of herbaceous vegetation that is valued as forage typically decline as woody plant abundance increases (Anadón et al. 2014b). This loss of forage production has traditionally been the impetus for brush management, with the expectation that reductions in tree or shrub cover would promote recovery of herbaceous production. More recently, interest in recovering grassland biodiversity has become a priority (Sect. 2.5.3.1). A synthesis of research on this topic indicates that responses of herbaceous vegetation to brush management are highly variable (Fig. 2.13). Although 64 % of investigations reported increases in forage production following brush management, those gains were, on average, short-lived, typically less than 5–7 years. Furthermore, herbaceous production and diversity remained unchanged, or even decreased—sometimes substantially—in 36 % of the studies. This range of herbaceous responses to brush control begs several questions. First, why is the response of herbaceous vegetation short-lived on some sites and longer lived on others? Second, why is herbaceous vegetation unresponsive to reductions in cover of woody plants at many sites? Third, what caused herbaceous vegetation at some sites to respond so negatively? Answers to such questions are needed if we are to identify where, when, how, and under what circumstances to intervene with a given brush management practice (Archer et al. 2011).

Integrated brush management systems (IBMS) (e.g. Noble and Walker 2006) are the hallmark of progressive, modern brush management. The IBMS approach advocates consideration of the type and timing of a given brush management technology and makes explicit allowances for the type and timing of follow-up treatments. This approach benefits from knowledge of how woody and herbaceous plants are likely to respond and how climate, soils, topography, and livestock and wildlife management might mediate plant responses. These considerations are crucial for long-term cost-benefit analysis of these treatments (e.g., Torell et al. 2005a). The conceptual model in Fig. 2.14 represents the kinds of ecological data that will be needed to evaluate the feasibility and sustainability of brush management practices from a forage production standpoint. Rangeland ecologists should develop families of curves for ecological sites in a given bioclimatic zone (e.g., McDaniel et al. 2005).

Historically, brush management treatments were often applied across entire landscapes and watersheds. However, it would be more effective to treat portions of a landscape and distribute treatments across landscapes in both time and space to create mosaics of vegetation structure, patch sizes, shapes, and age states (Scifres et al. 1988; Fulbright 1996) that would increase diverse habitats to potentially increase biodiversity (Jones et al. 2000) (Sect. 2.5.3.2). This would enable a

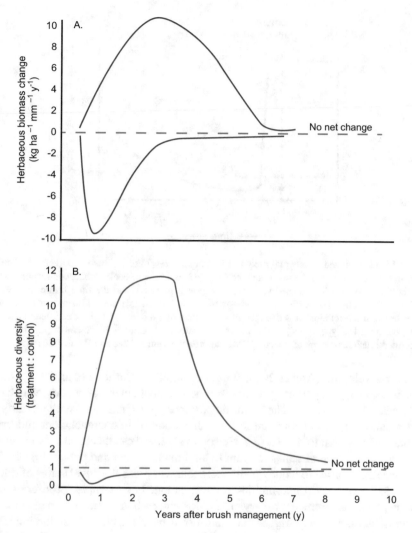

Fig. 2.13 Response envelope depicting the upper and lower limits of herbaceous (**a**) production and (**b**) diversity responses to brush management (based on data in Archer and Predick 2014)

low-diversity shrubland or woodland developing on a grassland site to be transformed into a patchwork of grassland-savanna-shrubland or woodland communities that promotes diversity at multiple scales (Chapter 5, this volume).

Economic analyses of brush management suggest that assessments based solely on increased forage and livestock performance may not be economically justified, especially when external subsidies are not available (Torell et al. 2005b; Tanaka et al. 2011). Full and explicit consideration of other ecosystem services may, however, change the cost-benefit assessment. Knowledge gaps remain, but a large and growing body of work on woody plant encroachment impacts on ecosystem ser-

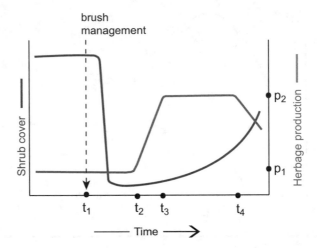

Fig. 2.14 Generalized conceptual model of herbaceous response to brush management. The lag time in response (t_1 to t_2), the magnitude of (p_1 to p_2) and time to peak herbaceous response (t_1 to t_3), the duration of peak elevated production response (t_3 to t_4), and the time frame over which herbaceous productions decline as shrubs reestablish (t_4 onward) vary with numerous factors. Knowledge of the relationships depicted in this conceptual model for a given ecological site will help determine the type, timing, and appropriate sequencing of brush management practices in an integrated brush management system (IBMS) approach (from Archer and Predick 2014)

vices is developing (Archer 2010; Barger et al. 2011; Eldridge et al. 2011). Much less is known about how post-encroachment management of woody vegetation influences those services. The scientific community is challenged with quantifying and monitoring the concomitant impacts of woody plant encroachment and brush management so that trade-offs (e.g. Nelson et al. 2009) can be objectively evaluated at spatial and temporal scales relevant to land management and policy (Fig. 2.15).

Rangelands prone to woody plant encroachment present a novel series of dilemmas, challenges, and opportunities for mitigation. For example, proliferation of woody plants can promote primary production and carbon sequestration under some circumstances, and may trigger new land-use drivers for biofuel production (Park et al. 2012) or as industries seek opportunities to offset CO_2 emissions. Woody plant proliferation in grasslands and savannas managed traditionally for grazing may therefore shift from being an economic liability to a source of income and economic diversification. However, under this scenario, grasslands and savannas and the plants and animals endemic to them would be at risk and their influences on hydrology, tropospheric chemistry (such as non-methane hydrocarbons, Guenther et al. 1999), and mesoscale meteorology altered. At present, our ability to evaluate and weigh these trade-offs, and their potentially synergistic interactions, is limited owing to variable, and often conflicting, results, and by limited scientific information (Archer and Predick 2014). These ecosystem-science challenges are magnified when placed in the human dimension context of cultural traditions, stakeholder preferences and priorities, market externalities, and climate change (Chapter 14, this volume). Given the cost and short longevity of brush management treatments, the adage "an ounce of prevention is worth a pound of cure" is applicable. In areas where WPE is at advanced stages, grassland restoration may not be economically feasible or sustainable and alternative land uses should be considered.

Grasslands and savannas are integral to the global carbon, water, and nitrogen cycles, and to human well-being (Campbell and Stafford Smith 2000; Reynolds et al. 2007; Peters et al. 2015). Their extensive airsheds and watersheds provide habitat for wildlife and a variety of ecosystem goods and services important to both local and distant settlements and cities. As such, they have considerable multipurpose value. A key component of dryland ecosystem management is maintaining the proportions of herbaceous and woody plants within a range that satisfies a given set of objectives and values, some of which may be conflicting (e.g., wildlife vs. livestock, Du Toit et al. 2010; Augustine et al. 2011). Perspectives on woody plants in rangelands vary widely depending on cultural traditions and land-use goals and objectives. In many regions of the world, woody plants are a valued source of food (e.g., honey, fruits, seeds), fuel, charcoal, and construction materials and an important source of fodder for browsing livestock (e.g., goats, camels), and wildlife. Additionally, there is growing recognition that woody plants on rangelands can provide products with potential commercial (e.g., gums, resins) or medicinal value.

Policy and management issues related to rangeland conservation have evolved to extend well beyond the traditional concerns of livestock production and game management (wildlife valued for sport hunting) to include potential effects on hydrology, carbon sequestration, biological diversity, atmospheric chemistry, and climate system (Archer 2010; Eldridge et al. 2011). The research community is challenged with quantifying and monitoring these varied impacts so that trade-offs (Fig. 2.15) can be assessed objectively and used as the foundation for science-based decision making. The management community is challenged with devising approaches for creating or maintaining woody-herbaceous mixtures in spatial arrangements that negotiate and balance competing land use and conservation objectives.

2.7 Future Perspectives

The woody plant encroachment phenomenon highlights the challenges of integrating stochastic and deterministic drivers of environmental change and plant trait representations to predict vegetation change. Vegetation models that account for the complexity of these interactions will be better suited to predict how changes in climate and atmospheric conditions will influence the future structure, function, and distribution of grasslands, savannas, woodlands, and forests (e.g., Scheiter and Higgins 2009). Among the philosophical differences that remain are the long-standing controversies regarding the influence of equilibrial dynamics, based upon the persistent properties of mature plants, and the influence of random environmental events and externalities on recruitment, mortality, and mutable competitive hierarchies of species during establishment.

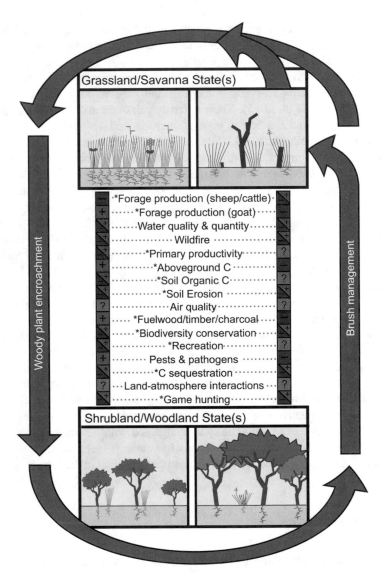

Fig. 2.15 Potential outcomes of woody plant encroachment and associated "brush management" activities. Symbols in boxes denote potential decreases (−), increases or improvements (+), mixed, context-dependent results (−\+), or insufficient information (?). From Archer and Predick (2014)

Research on WPE should draw from and contribute to the area of trait-based ecology. Discussions framed in terms of "woody plants" and their proliferation in "grasslands" do not help explain why only a few of the many woody species in a flora have become encroachers. What specific traits allowed these species to proliferate in grasslands after livestock introduction, while most other woody species could not? How do those traits determine the varied ecosystem effects of WPE on carbon stocks, soil fertility, and water budget? What traits may explain the idiosyncratic responses of herbaceous vegetation to brush management? Such questions are germane to those being asked in modern evolutionary ecology. Their answers have real-world implications for human welfare, rural economies, and climate change readiness.

There is broad consensus in the Earth sciences that the regulation of global water and carbon cycles by terrestrial vegetation is a critical aspect upon which the climate future of our planet depends. There remain significant knowledge gaps not the least of which center on vegetation change in the world's herbaceous communities. The influence of WPE on local water budgets, we are now learning, can influence the hydrological cycle at regional scales.

A better understanding of the controls over woody plant "carrying capacity" is needed to position us to predict how community dynamics and ecological processes will respond to changing environmental conditions. The upper limits of woody cover in rangelands seem to be dictated by mean annual precipitation, but mechanisms contributing to the MAP constraint are not clear. Density-dependent mechanisms would be a logical expectation, but studies quantifying interactions among woody plants in rangelands are uncommon and should receive more emphasis. The limited evidence available for density-dependent control over woody plant density or cover is equivocal. Our understanding of the extent to which the upper limits of woody plant cover are governed by the traits of seedlings influencing recruitment patterns and the traits of adult plants that influence ecosystem processes is limited. Alternative conceptual models highlight stochastic spatial processes, in which the equivalent of a carrying capacity is an emergent property of recruitment and disturbance probabilities.

Woody plant proliferation in grasslands and savannas has been ongoing for decades and is approaching or exceeding 100 years in some areas. Our focus has been on understanding the encroachment process, its rates, causes, and consequences. But "encroachment" is not the end of the story. We know relatively little of the dynamics of the shrubland or woodland communities that have developed on former grassland and how they might change through time. Understanding post-encroachment dynamics is important if we are to predict how ecosystem structure and function might continue to unfold over time. With accelerating rates of climate change and other anthropogenic disturbances, the potentially novel and dynamic communities of plants and animals created in the wake of Anglo-European settlement may be a natural laboratory for studying vegetation dynamics in the Anthropocene.

Responses of herbaceous vegetation to brush management are highly variable. Herbaceous production and diversity increase on some sites, but decrease on others, and positive responses, when they occur, vary greatly in their longevity. Improvements in our ability to explain these varied responses will enable us to identify (1) where and when brush management intervention might be most likely to achieve the outcomes desired for a given set of management or policy goals, and (2) the combination and time series of intervention methods that are most likely to effect desired changes within socioeconomic constraints.

Uncertainties and knowledge gaps regarding the impact of WPE and subsequent brush management activities on carbon sequestration are substantial. Studies quantifying the herbaceous production responses to WPE and brush management are abundant, but robust predictions are elusive, particularly with brush management. Data quantifying woody plant productivity is a major data gap, as are estimates of belowground production. Flux-tower networks targeting WPE-brush management areas will enable us to better determine source-sink relationships. Recent advances in tools for gathering remote-sensing data (e.g., LIDAR; unmanned aerial vehicles; multispectral, hyperspectral, and thermal satellite-based sensor arrays) have given us new capabilities for quantifying aboveground vegetation structure and biomass over expansive and remote areas. Furthermore, these technologies have the potential to quantify cacti, an important and sometimes very abundant, plant functional type on rangelands. The contributions of cacti have been virtually ignored in biomass and ANPP estimates of the aboveground carbon pool. These synoptic perspectives will position us to inventory carbon stocks more accurately at regional scales, where landscapes are mosaics of areas in various stages of WPE and recovery from extreme events or management interventions. Studies quantifying changes in the soil organic carbon pool with WPE have been accumulating over the past 10–15 years, but there is an urgent need to balance these with data documenting brush management impacts. In both cases, there is a need for standardization of methodologies if we hope to develop robust, meaningful generalizations.

From a conservation perspective, WPE represents a major threat to grassland and savanna ecosystems and their endemic plants and animals. This perspective needs to be considered explicitly when evaluating ecosystem service portfolios that have focused traditionally on forage and livestock production, water quality/quantity, etc. Biodiversity perspectives should be broadened similarly to include organisms valued for their functional and charismatic roles, as well as animals valued for hunting and plants valued for forage.

Concerns over WPE will be complicated by the invasion and proliferation of non-native grasses. Acting as "transformer species," these exotic grasses can change the character, condition, function, and form of native ecosystems. Once established, non-native annual and perennial grasses can generate massive, high-continuity fine fuel loads that predispose grasslands to fires that can be more frequent and intense than those with which they evolved. The result is the potential for shrublands and woodlands developing on former grasslands to be quickly and radically transformed into exotic grass monocultures over large areas. This is well under way in the North American cold desert region (e.g., cheatgrass, *Bromus tectorum*) and is in its early stages in hot deserts. These transformations have profound effects on ecosystem processes (Betancourt 2015) and biodiversity (Steidl et al. 2013) and present unique management challenges. More research is needed to develop an understanding of how WPE and non-native grasses could be comanaged to conserve biodiversity and ensure the sustained provision of core ecosystem services.

Assessments of woody plan encroachment and actions taken to halt or reverse it must be broadly considered and evaluated in the context of plant and animal community dynamics, biodiversity, and ecosystem function. The near-term context will be largely determined by land-use priorities and socioeconomic externalities. Over the longer term, climate change will determine the context within which land-use and socioeconomic decisions are made. Management therefore needs to address ongoing and near-term challenges associated with WPE while positioning us to anticipate and adapt to changes on the horizon.

2.8   Summary

Woody plant encroachment (WPE) is an umbrella phrase describing increases in abundance and distribution of woody plants in grassland and savanna plant communities worldwide. WPE has been documented in arid, semiarid, and subhumid climate zones and in tropical, subtropical, temperate, and arctic regions. WPE has been traditionally associated with ecosystems degraded due to intensive grazing by cattle and sheep. We now appreciate, however, that woody plants play important roles in maintaining ecosystem processes on these degraded landscapes. Consequently, their proliferation is now viewed more appropriately as a symptom, rather than cause, of degradation.

2.8.1   Causes

Although numerous efforts have sought to elucidate the proximate causes of woody encroachment, robust generalizations remain elusive. The WPE process is highly context dependent and influenced by numerous, interacting location-specific factors related to climate, fire frequency and intensity, grazing/browsing regimes, soil properties, and functional traits of the encroaching species and native browsers.

MAP sets an upper limit to woody plant cover, which tends to plateau to a maximum above 650 mm. However, local patterns of disturbance (fire, browsing) and soil properties (texture, depth) may prevent this potential from being realized. In the absence of these constraints, interactions among the seasonality, interannual variation, and intensity of precipitation events will determine the rate and extent of woody plant recruitment. Precipitation in arid grasslands varies markedly in both space and time. This can cause cover of woody plants to wax and wane at decadal or longer time scales, which helps to explain the high variation observed in rates of WPE. Climate zones with higher precipitation have the capacity for rapid conversion from grassland to woodland, but decadal-scale variation in precipitation can make it difficult to distinguish natural fluctuations from directional changes in vegetation communities.

Paleoecological studies indicate that the balance between grass- and woody-plant-dominated communities has fluctuated over the last 10,000 years, suggesting climate as a long-term determinant. However, WPE in the "Anthropocene" is more complicated. Concentrations of atmospheric CO_2 have been increasing exponentially since the advent of the industrial revolution. Although probably not a triggering factor *per se*, rising CO_2 levels may well have been a supporting factor in that woody plants, which are characterized by the C_3 photosynthetic pathway, would have benefited more from CO_2 "fertilization" than the C_4 grasses that dominate tropical, subtropical, and warm-temperate regions. Further, woody plants can use higher assimilation rates to expedite the accumulation of woody biomass and carbohydrate storage. This would lower their mortality risks during the critical establishment phase while also enabling more rapid growth to sizes where they could escape constraints imposed by fire and browsers.

Woody plant encroachment has also coincided with the global intensification of livestock grazing. Prior to the introduction of domestic grazers, an abundance of fine fuels produced by grasses stimulated periodic fires that regularly suppressed woody plant recruitment and controlled the density of mature shrubs and trees. The introduction of large numbers and high concentrations of livestock reduced both the density and continuity of fine fuels, which reduced fire frequency and intensity, and facilitated development of woody plant communities. Locally, woody plants benefited from secondary factors, such as livestock dispersing seeds or by displacing native browsers and seed predators.

When woody plant seedlings germinate in the immediate proximity of mature grasses, they face potentially intense competition for light, water, and soil nutrients. Grazing reduces grass biomass both above- and belowground and therefore the ability of grasses to competitively suppress shrub seedlings. However, this does not explain why woody plants are also encroaching into areas protected from grazing livestock. In many grasslands, ground cover consists of bare and vegetated patches, and thus spatially variable levels of competition. In addition, where annual rainfall is monsoonal or bimodal, woody seedlings may germinate and establish during those periods when competition for soil moisture is low. Having survived the most vulnerable period immediately after germination, woody plants rapidly develop deep taproots below the primary root zone of grasses. This increases their access to water that has infiltrated more deeply and alleviates competition for water with grasses. As woody seedlings grow taller, they incrementally gain competitive dominance over their grass neighbors and may begin to displace grasses through resource competition. At some stage, grasses can substantially influence the dominance of woody saplings only through their influence on the fire cycle. However, woody encroachers capable of regenerating vegetatively (resprouting) often survive fire. Then if grazing reduces fire frequency, plants in these "seedling" or "sapling banks" are poised to grow quickly and escape the flame zone of future fires. Eventually, these plants will produce seed and intensify propagule pressure in grasslands. Long-term maintenance of grassland and savanna ecosystems is therefore contingent on maintaining a balance woody plants and grasses based on climate, disturbance, and species traits.

2.8.2   Consequences for Ecosystem Services

Traditional concerns related to the loss of forage production accompanying WPE have been broadened to include consequences for provision of services related to primary production and carbon sequestration. Because of the global extent and magnitude of the impact of WPE, these changes can potentially significantly affect the global carbon budget and energy balance. If encroaching woody plants are less productive than the grasses they replace, then ecosystem ANPP would decline; if they are more productive than the grasses they displace, then ANPP would increase; and if ANPP of encroaching woody plants is comparable to that of the grasses they are replacing, then there would be no net change. Evidence indicates that all three scenarios are at play, with changes in ANPP scaling linearly with MAP. Below an MAP of ~340 mm, ANPP will decline with WPE and above this level ANPP will increase. Our understanding of WPE effects on the soil organic C pool, which typically dwarfs the aboveground pool in grasslands, is poor. Some studies show large increases in soil organic C with WPE, whereas others show no change or large decreases. Reasons for this range of responses have yet to be explained. This is a major knowledge gap that needs to be filled if we are to understand fully the effects of WPE on the carbon cycle.

WPE has the potential to reduce streamflow and/or groundwater recharge by reducing deep recharge and runoff through increases in evapotranspiration (ET). However, evidence for the relationship between WPE and water yield has been equivocal and may depend on climate, edaphic factors, and traits of the encroaching woody species. WPE may impact the water budget only where MAP approximately equals PET. Where deep-rooted trees have encroached in grasslands on sites with shallow water tables, ET has increased, but where recharge and runoff are controlled strongly by physical properties of the soil, WPE has had little additional effect on the hydrological budget.

WPE markedly affects biodiversity and threatens the very existence of grassland and savanna ecosystems and their endemic plants and animals. In North America, diversity of herbaceous plants declines ~45 % when woody plants encroach. Changes in vegetation structure and species composition accompanying WPE contribute to the loss of grassland-adapted animals by reducing both the quantity and quality of their required habitat. Because a principal result of WPE is a marked increase in vertical and horizontal vegetation structure, composition of animal communities shifts to favor species that prefer woody vegetation. WPE can affect habitat quality for grassland-associated species that persist within encroached areas through changes in rates of predation or changes in the types, abundance, or availability of food resources. Ultimately, when woody cover exceeds species-specific thresholds, populations of grassland-associated species are displaced by shrubland- or woodland-associated species.

2.8.3   Management

Proliferation of woody plants has long been of concern in areas managed primarily for grazing cattle and sheep, where WPE typically reduces production of forage, complicates animal handling, and improves habitat for ectoparasites. As a result, multiple strategies have been developed to reduce cover of woody vegetation. Collectively known as "brush management," these approaches include prescribed burning, mechanical clearing, and herbicide application. Responses of herbaceous vegetation to brush management practices have been highly variable and typically short-lived. Early goals for brush management centered on eradicating shrubs to improve production of livestock, which gave way to efforts aimed at shrub "control," which gave way to integrated brush management systems (IBMS). IBMS is ecologically based and predicated on using location-specific knowledge of vegetation characteristics, climate, soils, and topography to determine the type, sequencing, and timing of initial and follow-up treatments. In the IBMS model, landscapes are comanaged for livestock and wildlife, and with consideration for the diverse portfolio of ecosystem services that rangelands provide.

Unless subsidized, brush management is rarely economically feasible based solely on increases in forage production and livestock performance. However, consideration of "intangibles" related to enhancements of other ecosystem services will influence the conclusions taken from traditional, narrowly focused cost-benefit calculations. For example, brush management contributions to the conservation of grassland ecosystems and the plants and animals unique to them constitute an important benefit that is largely unaccounted for. Conversely, increased potential for carbon sequestration may be a positive outcome of WPE that would have to be weighed against potential reductions in biodiversity, water yield, or changes in vegetation structure that affect key wildlife species adversely. The scientific community is challenged to quantify and monitor the concomitant impacts of WPE and brush management on the diverse components comprising an ecosystem service portfolio so that trade-offs can be evaluated objectively in the context of a clear set of goals and priorities.

Acknowledgments We thank David Briske for his vision and efforts in organizing this volume and his input on earlier drafts of this chapter. This effort was supported, in part, by NSF-DEB-1413900, -DEB-0531691, -DEB-1235828, and -DEB- 1557262; USDA NIFA-2015-67019-23314; the Bureau of Land Management; the Arizona Game and Fish Department; the Audubon Appleton-Whittell Research Ranch; and the Arizona Agricultural Experimentation Project ARZT-1360540-H12-199.

References

Allen, R.B., J.B. Wilson, and C.R. Mason. 1995. Vegetation change following exclusion of grazing animals in depleted grassland, Central Otago, New Zealand. *Journal of Vegetation Science* 6: 615–626.

Allen, A.P., W.T. Pockman, C. Restrepo, and B.T. Milne. 2008. Allometry, growth and population regulation of the desert shrub *Larrea tridentata. Functional Ecology* 22: 197–204.

Allen, C.D., A.K. Macalady, H. Chenchouni, D. Bachelet, N. McDowell, M. Vennetier, T. Kitzberger, A. Rigling, D.D. Breshears, E.T. Hogg, and P. Gonzalez. 2010. A global overview of drought and heat-induced tree mortality reveals emerging climate change risks for forests. *Forest Ecology and Management* 259: 660–684.

Altesor, A., G. Pineiro, F. Lezama, R.B. Jackson, M. Sarasola, and J.M. Paruelo. 2006. Ecosystem changes associated with grazing in subhumid South American grasslands. *Journal of Vegetation Science* 17: 323–332.

Anadón, J.D., O.E. Sala, and F.T. Maestre. 2014a. Climate change will increase savannas at the expense of forests and treeless vegetation in tropical and subtropical Americas. *Journal of Ecology* 102: 1363–1373.

Anadón, J.D., O.E. Sala, B.L. Turner, and E.M. Bennett. 2014b. Effect of woody-plant encroachment on livestock production in North and South America. *Proceedings of the National Academy of Sciences* 111: 12948–12953.

Anderies, M.J., A.M. Janssen, and H.B. Walker. 2002. Grazing management, resilience, and the dynamics of a fire-driven rangeland system. *Ecosystems* 5: 23–44.

Archer, S. 1994. Woody plant encroachment into southwestern grasslands and savannas: Rates, patterns and proximate causes. In *Ecological implications of livestock herbivory in the west*, ed. M. Vavra, W. Laycock, and R. Pieper. Denver, CO: Society for Range Management.

Archer, S. 1995. Tree-grass dynamics in a *Prosopis*-thornscrub savanna parkland: Reconstructing the past and predicting the future. *Ecoscience* 2: 83–99.

———. 2010. Rangeland conservation and shrub encroachment: New perspectives on an old problem. In *Wild rangelands: Conserving wildlife while maintaining livestock in semi-arid ecosystems*, eds. J. du Toit, R. Kock, and J. Deutsch. Oxford, England: Wiley-Blackwell Publishing.

Archer, S., and K. Predick. 2014. An ecosystem services perspective on brush management: Research priorities for competing land use objectives. *Journal of Ecology* 102: 1394–1407.

Archer, S., D.S. Schimel, and E.A. Holland. 1995. Mechanisms of shrubland expansion: Land use, climate or CO_2? *Climatic Change* 29: 91–99.

Archer, S., K. Davies, T. Fulbright, K. McDaniel, B. Wilcox, K.I. Predick, and D.D. Briske. 2011. Brush management as a rangeland conservation strategy: A critical evaluation. In *Conservation benefits of rangeland practices: Assessment, recommendations, and knowledge gaps*, ed. D. Briske. Washington, DC: United States Department of Agriculture, Natural Resources Conservation Service.

Archibald, S., W.J. Bond, W.D. Stock, and D.H.K. Fairbanks. 2005. Shaping the landscape: Fire–grazer interactions in an African savanna. *Ecological Applications* 15: 96–109.

Arnold, T.W., and K.E. Higgins. 1986. Effects of shrub coverages on birds of North Dakota mixed-grass prairies. *Canadian Field-Naturalist* 100: 10–14.

Asner, G.P., S. Archer, R.F. Hughes, R.J. Ansley, and C.A. Wessman. 2003. Net changes in regional woody vegetation cover and carbon storage in Texas Drylands, 1937–1999. *Global Change Biology* 9: 316–335.

Asner, G.P., A.J. Elmore, L.P. Olander, R.E. Martin, and A.T. Harris. 2004. Grazing systems, ecosystem responses, and global change. *Annual Review of Environment and Resources* 29: 261–299.

Augustine, J.D., and S.J. McNaughton. 2004. Regulation of shrub dynamics by native browsing ungulates on East African rangeland. *Journal of Applied Ecology* 41: 45–58.

Augustine, D.J., K.E. Veblen, J.R. Goheen, C. Riginos, and T.P. Young. 2011. Pathways for positive cattle-wildlife interactions in semiarid rangelands. *Smithsonian Contributions to Zoology* 632: 55–71.

Ayers D, G. Melville, J. Bean, and D. Beckers. 2001. *Woody weeds, biodiversity and landscape function in Western New South Wales.*

Baez, S., and S. Collins. 2008. Shrub invasion decreases diversity and alters community stability in northern Chihuahuan desert plant communities. *PLoS ONE* 3, e2332. doi:10.1371/journal.pone.0002332.

Bahre, C.J., and M.L. Shelton. 1993. Historic vegetation change, mesquite increases, and climate in southeastern Arizona. *Journal of Biogeography* 20: 489–504.

Bailey, R.G. 2014. *Ecoregions: The ecosystem geography of the oceans and continents.* New York: Springer.

Barger NN, S. Archer, J. Campbell, C. Huang, J. Morton, and A.K. Knapp. 2011. Woody plant proliferation in North American drylands: A synthesis of impacts on ecosystem carbon balance. *Journal of Geophysical Research: Biogeosciences* 116:G00K07. doi: 10.1029/2010JG001506.

Barnes, P.W., and S.R. Archer. 1996. Influence of an overstorey tree (*Prosopis glandulosa*) on associated shrubs in a savanna parkland: Implications for patch dynamics. *Oecologia* 105: 493–500.

Barnes, P., and S. Archer. 1999. Tree-shrub interactions in a subtropical savanna parkland: Competition or facilitation? *Journal of Vegetation Science* 10: 525–536.

Batista, W.B., A.G. Rolhauser, F. Biganzoli, S.E. Burkart, L. Goveto, A. Maranta, A.G. Pignataro, N.S. Morandeira, and M. Rabadán. 2014. Savanna plant community types at El Palmar National Park (Argentina). *Darwiniana, Nueva Serie* 2: 5–38.

Bazan, R.A., B.P. Wilcox, C. Munster, and M. Gary. 2013. Removing woody vegetation has little effect on conduit flow recharge. *Ecohydrology* 6: 435–443.

Belay, T.A., and S.R. Moe. 2012. Woody dominance in a semi-arid savanna rangeland—Evidence for competitive self-thinning. *Acta Oecologica* 45: 98–105.

Bestelmeyer, B.T. 2005. Does desertification diminish biodiversity? Enhancement of ant diversity by shrub invasion in southwestern USA. *Diversity and Distributions* 11: 45–55.

Bestelmeyer, B.T., D. Goolsby, and S.R. Archer. 2011. Spatial patterns in state-and-transition models: A missing link to land management? *Journal of Applied Ecology* 48: 746–757.

Betancourt, J.L. 2015. Energy flow and the "grassification" of desert shrublands. *Proceedings of the National Academy Sciences* 112: 9504–9505.

Blaser, W.J., J. Sitters, S.P. Hart, P.J. Edwards, and H. Olde Venterink. 2013. Facilitative or competitive effects of woody plants on understorey vegetation depend on N-fixation, canopy shape and rainfall. *Journal of Ecology* 101: 1598–1603.

Blaum, N., E. Rossmanith, A. Popp, and F. Jeltsch. 2007a. Shrub encroachment affects mammalian carnivore abundance and species richness in semiarid rangelands. *Acta Oecologica* 31: 86–92.

Blaum, N., E. Rossmanith, and F. Jeltsch. 2007b. Land use affects rodent communities in Kalahari savannah rangelands. *African Journal of Ecology* 45: 189–195.

Blaum, N., C. Seymour, E. Rossmanith, M. Schwager, and F. Jeltsch. 2009. Changes in arthropod diversity along a land use driven gradient of shrub cover in savanna rangelands: Identification of suitable indicators. *Biodiversity and Conservation* 18: 1187–1199.

Block, G., and M.L. Morrison. 2010. Large-scale effects on bird assemblages in desert grasslands. *Western North American Naturalist* 70: 19–25.

Bonan, G.B. 2008. Forests and climate change: Forcings, feedbacks, and the climate benefits of forests. *Science* 320: 1444–1449.

Bond, W.J. 2008. What limits trees in C_4 grasslands and savannas? *Annual Review of Ecology, Evolution, and Systematics* 39: 641–659.

Bond, W.J., and G.F. Midgley. 2000. A proposed CO_2-controlled mechanism of woody plant invasion in grasslands and savannas. *Global Change Biology* 6: 865–869.

Bond, W.J., W.D. Stock, and M.T. Hoffman. 1994. Has the Karoo spread? A test for desertification using carbon isotopes from soils. *South African Journal of Science* 90: 391–397.

Bovey, R.W. 2001. *Woody plants and woody plant management: Ecology, safety, and environmental impact*. New York: Marcel Dekker, Inc.

Bowers, J.E. 2005. Effects of drought on shrub survival and longevity in the northern Sonoran Desert. *The Journal of the Torrey Botanical Society* 132: 421–431.

Brandt, J.S., M.A. Haynes, T. Kuemmerle, D.M. Waller, and V.C. Radeloff. 2013. Regime shift on the roof of the world: Alpine meadows converting to shrublands in the southern Himalayas. *Biological Conservation* 158: 116–127.

Brennan, L.A., and W.P. Kuvlesky. 2005. Grassland birds—An unfolding conservation crisis. *Journal of Wildlife Management* 69: 1–13.

Breshears, D.D., N.S. Cobb, P.M. Rich, K.P. Price, C.D. Allen, R.G. Balice, W.H. Romme, J.H. Kastens, M.L. Floyd, J. Belnap, and J.J. Anderson. 2005. Regional vegetation die-off in response to global-change-type drought. *Proceedings of the National Academy of Sciences of the United States of America* 102: 15144–15148.

Brown, J.R., and S. Archer. 1987. Woody plant seed dispersal and gap formation in a North American subtropical savanna woodland: The role of domestic herbivores. *Vegetatio* 73: 73–80.

———. 1989. Woody plant invasion of grasslands: Establishment of honey mesquite (*Prosopis glandulosa* var. *glandulosa*) on sites differing in herbaceous biomass and grazing history. *Oecologia* 80: 19–26.

———. 1990. Water relations of a perennial grass and seedlings vs adult woody plants in a subtropical savanna, Texas. *Oikos* 57: 366–374.

———. 1999. Shrub invasion of grassland: Recruitment is continuous and not regulated by herbaceous biomass or density. *Ecology* 80: 2385–2396.

Browning, D., and S. Archer. 2011. Protection from livestock fails to deter shrub proliferation in a desert landscape with a history of heavy grazing. *Ecological Applications* 21: 1629–1642.

Browning, D., S. Archer, G. Asner, M. McClaran, and C. Wessman. 2008. Woody plants in grasslands: Post-encroachment stand dynamics. *Ecological Applications* 18: 928–944.

Browning, D.M., M.C. Duniway, A.S. Laliberte, and A. Rango. 2012. Hierarchical analysis of vegetation dynamics over 71 years: Soil-rainfall interactions in a Chihuahuan desert ecosystem. *Ecological Applications* 22: 909–926.

Browning, D.M., J. Franklin, S.R. Archer, J.K. Gillan, and D.P. Guertin. 2014. Spatial patterns of grassland-shrubland state transitions: A 74 year record on grazed and protected areas. *Ecological Applications* 24: 1421–1433.

Bruce, K., G. Cameron, and P. Harcombe. 1995. Initiation of a new woodland type on the Texas coastal prairie by the Chinese tallow tree (*Sapium sebiferum* (l) Roxb). *Bulletin of the Torrey Botanical Club* 122: 215–225.

Brunelle, A., T.A. Minckley, J. Delgadillo, and S. Blissett. 2014. A long-term perspective on woody plant encroachment in the desert southwest, New Mexico, USA. *Journal of Vegetation Science* 25: 829–838.

Cabral, A.C., J.M. de Miguel, A.J. Rescia, M.F. Schmitz, and F.D. Pineda. 2003. Shrub encroachment in Argentinean savannas. *Journal of Vegetation Science* 14: 145–152.

Campbell, B.D., and D.M. Stafford Smith. 2000. A synthesis of recent global change research on pasture and range production: Reduced uncertainties and their management implications. *Agriculture, Ecosystems and Environment* 82: 39–55.

Canadell, J., R.B. Jackson, J.R. Ehleringer, H.A. Mooney, O.E. Sala, and E.D. Schulze. 1996. Maximum rooting depth of vegetation types at the global scale. *Oecologia* 108: 583–595.

Cardoso, A.W., J.A. Medina-Vega, Y. Malhi, S. Adu-Bredu, G.K.D. Ametsitsi, G. Djagbletey, F. Langevelde, E. Veenendaal, and I. Oliveras. 2016. Winners and losers: Tropical forest tree seedling survival across a West African forest-savanna transition. *Ecology and Evolution* 6: 3417–3429.

Coppedge, B. 2004. Predicting juniper encroachment and CRP effects on avian community dynamics in southern mixed-grass prairie, USA. *Biological Conservation* 115: 431–441.

Coppedge, B.R., D.M. Engle, R.E. Masters, and M.S. Gregory. 2001. Avian response to landscape change in fragmented southern Great Plains grasslands. *Ecological Applications* 11: 47–59.

Costello, D.A., I.D. Lunt, and J.E. Williams. 2000. Effects of invasion by the indigenous shrub Acacia sophorae on plant composition of coastal grasslands in south-eastern Australia. *Biological Conservation* 96: 113–121.

Coughenour, M.B., and J.E. Ellis. 1993. Landscape and climatic control of woody vegetation in a dry tropical ecosystem: Turkana District, Kenya. *Journal of Biogeography* 20: 383–398.

Cunningham, M.A., and D.H. Johnson. 2006. Proximate and landscape factors influence grassland bird distributions. *Ecological Applications* 16: 1062–1075.

de Dios, V.R., J.F. Weltzin, W. Sun, T.E. Huxman, and D.G. Williams. 2014. Transitions from grassland to savanna under drought through passive facilitation by grasses. *Journal of Vegetation Science* 25: 937–946.

Dekker, S.C., M. Rietkerk, and M.F.P. Bierkens. 2007. Coupling microscale vegetation-soil water and macroscale vegetation-precipitation feedbacks in semiarid ecosystems. *Global Change Biology* 13: 671–678.

Dewey, S.R., and P.L. Kennedy. 2001. Effects of supplemental food on parental-care strategies and juvenile survival of Northern Goshawks. *The Auk* 118: 352–365.

D'Odorico, P., G.S. Okin, and B.T. Bestelmeyer. 2012. A synthetic review of feedbacks and drivers of shrub encroachment in arid grasslands. *Ecohydrology* 5: 520–530.

Dohn, J., F. Dembélé, M. Karembé, A. Moustakas, K.A. Amévor, et al. 2013. Tree effects on grass growth in savannas: Competition, facilitation and the stress-gradient hypothesis. *Journal of Ecology* 101: 202–209.

Donohue, R.J., M.L. Roderick, and T.R. McVicar. 2007. On the importance of including vegetation dynamics in Budyko's hydrological model. *Hydrology and Earth System Sciences* 11: 983–995.

Du Toit, J., R. Kock, and J. Deutsch. 2010. *Wild rangelands: Conserving wildlife while maintaining livestock in semi-arid ecosystems*. Oxford, England: Blackwell Publishing Ltd.

Dulamsuren, C., M. Hauck, and M. Muhlenberg. 2008. Insect and small mammal herbivores limit tree establishment in northern Mongolian steppe. *Plant Ecology* 195: 143–156.

Dunne, T. 1978. Field studies of hillslope flow processes. In *Hillslope hydrology*, ed. M.J. Kirby. New York, NY: John Wiley and Sons.

Dwyer, J.M., R.J. Fensham, R.J. Fairfax, and Y.M. Buckley. 2010. Neighbourhood effects influence drought-induced mortality of savanna trees in Australia. *Journal of Vegetation Science* 21: 573–585.

Eldridge, D.J., M.A. Bowker, F.T. Maestre, E. Roger, J.F. Reynolds, and W.G. Whitford. 2011. Impacts of shrub encroachment on ecosystem structure and functioning: Towards a global synthesis. *Ecology Letters* 14: 709–722.

February, E.C., S.I. Higgins, W.J. Bond, and L. Swemmer. 2013. Influence of competition and rainfall manipulation on the growth responses of savanna trees and grasses. *Ecology* 94: 1155–1164.

Fensham, R.J., and J.E. Holman. 1999. Temporal and spatial patterns in drought-related tree dieback in Australian savanna. *Journal of Applied Ecology* 36: 1035–1050.

Fensham, R.J., R.J. Fairfax, and S. Archer. 2005. Rainfall, land use and woody vegetation cover change in semi-arid Australian savanna. *Journal of Ecology* 93: 596–606.

Fravolini, A., K. Hultine, E. Brugnoli, R. Gazal, N. English, et al. 2005. Precipitation pulse use by an invasive woody legume: The role of soil texture and pulse size. *Oecologia* 144: 618–627.

Fuhlendorf, S.D., and D.M. Engle. 2001. Restoring heterogeneity on rangelands: Ecosystem management based on evolutionary grazing patterns. *Bioscience* 51: 625–632.

Fuhlendorf, S.D., A.J.W. Woodward, D.M. Leslie Jr., and J.S. Shackford. 2002. Multi-scale effects of habitat loss and fragmentation on lesser prairie-chicken populations of the US southern Great Plains. *Landscape Ecology* 17: 617–628.

Fuhlendorf, S., S. Archer, F. Smeins, D. Engle, and C. Taylor. 2008. The combined influence of grazing, fire and herbaceous productivity on tree-grass interactions. In *Western North American Juniperus communities*, ed. O.V. Auken. New York: Springer.

Fuhlendorf, S.D., D.M. Engle, J.A.Y. Kerby, and R. Hamilton. 2009. Pyric Herbivory: Rewilding Landscapes through the Recoupling of Fire and Grazing Herbivoría Pírica: Restablecimiento de Paisajes Silvestres Mediante la Combinación de Fuego y Pastoreo. *Conservation Biology* 23: 588–598.

Fuhlendorf, S.D., D.M. Engle, R.D. Elmore, R.F. Limb, and T.G. Bidwell. 2012. Conservation of pattern and process: Developing an alternative paradigm of rangeland management. *Rangeland Ecology and Management* 65: 579–589.

Fulbright, T.E. 1996. Viewpoint—A theoretical basis for planning woody plant control to maintain species diversity. *Journal of Range Management* 49: 554–559.

Gao, Q., and J.F. Reynolds. 2003. Historical shrub-grass transitions in the northern Chihuahuan Desert: Modeling the effects of shifting rainfall seasonality and event size over a landscape gradient. *Global Change Biology* 9: 1475–1493.

Good, S.P., and K.K. Caylor. 2011. Climatological determinants of woody cover in Africa. *Proceedings of the National Academy of Sciences of the United States of America* 108: 4902–4907.

Good, M.K., P.J. Clarke, J.N. Price, and N. Reid. 2014. Seasonality and facilitation drive tree establishment in a semi-arid floodplain savanna. *Oecologia* 175: 261–271.

Granbom, M., H.G. Smith, and M.T. Murphy. 2006. Food limitation during breeding in a heterogeneous landscape. *The Auk* 123: 97–107.

Grant, T., E. Madden, and G. Berkey. 2004. Tree and shrub invasion in northern mixed-grass prairie: Implications for breeding grassland birds. *Wildlife Society Bulletin* 32: 807–818.

Graves, B.M., A.D. Rodewald, and S.D. Hull. 2010. Influence of woody vegetation on grassland birds within reclaimed surface mines. *The Wilson Journal of Ornithology* 122: 646–654.

Grice, A.C. 1996. Seed production, dispersal and germination in *Cryptostegia grandiflora* and *Ziziphus mauritiana*, two invasive shrubs in tropical woodlands of northern Australia. *Australian Journal of Ecology* 21: 324–331.

Groom, P.K., B.B. Lamont, and I.W. Wright. 2000. Lottery (stochastic) and non-lottery (biological) processes explain recruitment patterns among eight congeneric shrub species in southwestern Australia. *Journal of Mediterranean Ecology* 2: 1–14.

Grover, H.D., and H.B. Musick. 1990. Shrubland encroachment in southern New Mexico, USA: An analysis of desertification processes in the American Southwest. *Climatic Change* 17: 305–330.

Guenther, A., S. Archer, J. Greenberg, P. Harley, D. Helmig, L. Klinger, L. Vierling, M. Wildermuth, P. Zimmerman, and S. Zitzer. 1999. Biogenic hydrocarbon emissions and land cover/climate change in a subtropical savanna. *Physics and Chemistry of the Earth, Part B: Hydrology, Oceans and Atmosphere* 24: 659–667.

Hacke, U.G., J.S. Sperry, J.K. Wheeler, and L. Castro. 2006. Scaling of angiosperm xylem structure with safety and efficiency. *Tree Physiology* 26: 689–701.

Hamilton, W.T., A. McGinty, D.N. Ueckert, C.W. Hanselka, and M.R. Lee. 2004. *Brush management: Past, present, future*. College Station, TX: Texas A&M University Press.

Heske, E.J., J.H. Brown, and Q.F. Guo. 1993. Effects of Kangaroo rat exclusion on vegetation structure and plant species diversity in the Chihuahuan Desert. *Oecologia* 95: 520–524.

Hoekstra, J.M., T.M. Boucher, T.H. Ricketts, and C. Roberts. 2004. Confronting a biome crisis: Global disparities of habitat loss and protection. *Ecology Letters* 8: 23–29.

House, J.I., S. Archer, D.D. Breshears, and R.J. Scholes. 2003. Conundrums in mixed woody-herbaceous plant systems. *Journal of Biogeography* 30: 1763–1777.

Hughes, R.F., S.R. Archer, G.P. Asner, C.A. Wessman, C. McMurtry, J.I.M. Nelson, and R.J. Ansley. 2006. Changes in aboveground primary production and carbon and nitrogen pools accompanying woody plant encroachment in a temperate savanna. *Global Change Biology* 12: 1733–1747.

Idso, S.B. 1992. Shrubland expansion in the American Southwest. *Climate Change* 22: 85–86.

Igl, L., and B. Ballard. 1999. Habitat associations of migrating and overwintering grassland birds in southern Texas. *Condor* 101: 771–782.

Jackson, R.B., J. Canadell, J.R. Ehleringer, H.A. Mooney, O.E. Sala, and E.D. Schulze. 1996. A global analysis of root distributions for terrestrial biomes. *Oecologia* 108: 389–411.

Johnson, R., and S. Temple. 1990. Nest predation and brood parasitism of tallgrass prairie birds. *Journal of Wildlife Management* 54: 106–111.

Jones, B., S.F. Fox, D.M. Leslie, D.M. Engle, and R.L. Lochmiller. 2000. Herpetofaunal responses to brush management with herbicide and fire. *Journal of Range Management* 53: 154–158.

Jurena, P.N., and S.R. Archer. 2003. Woody plant establishment and spatial heterogeneity in grasslands. *Ecology* 84: 907–919.

Kambatuku, J.R., M.D. Cramer, and D. Ward. 2013. Overlap in soil water sources of savanna woody seedlings and grasses. *Ecohydrology* 6: 464–473.

Keddy, P.A. 1992. Assembly and response rules: Two goals for predictive community ecology. *Journal of Vegetation Science* 3: 157.

Knapp, A., J. Briggs, S. Collins, S. Archer, M. Bret-Harte, B.E. Ewers, D.P. Peters, D.R. Young, G.R. Shaver, E. Pendall, and M.B. Cleary. 2008a. Shrub encroachment in North American grasslands: Shifts in growth form dominance rapidly alters control of ecosystem carbon inputs. *Global Change Biology* 14: 615–623.

Knapp, A.K., J.K. McCarron, A.M. Silletti, G.A. Hoch, J.C. Heisler, M.S. Lett, J.M. Blair, J.M. Briggs, and M.D. Smith. 2008b. Ecological consequences of the replacement of native grassland by *Juniperus virginiana* and other woody plants. In *Western North American Juniperus communities: A dynamic vegetation type*, ed. O.W. Van Auken. New York: Springer.

Knoop, W.T., and B.H. Walker. 1985. Interactions of woody and herbaceous vegetation in southern African savanna. *Journal of Ecology* 73: 235–253.

Knopf, F. 1994. Avian assemblages on altered grasslands. *Studies in Avian Biology* 15: 247–257.

Körner, C. 2006. Plant CO_2 responses: An issue of definition, time and resource supply. *New Phytologist* 172: 393–411.

Kraaij, T., and D. Ward. 2006. Effects of rain, nitrogen, fire and grazing on tree recruitment and early survival in bush-encroached savanna, South Africa. *Plant Ecology* 186: 235–246.

Krawchuk, M.A., and M.A. Moritz. 2011. Constraints on global fire activity vary across a resource gradient. *Ecology* 92: 121–132.

Kremen, C., R. K. Colwell, T. L. Erwin, D. D. Murphy, R. F. Noss, and M. A. Sanjayan. 1993. Terrestrial arthropod assemblages: their use in conservation planning. Conservation Biology 7:796–808.

Kriticos, D.J., R.W. Sutherst, J.R. Brown, S.W. Adkins, and G.F. Maywald. 2003. Climate change and the potential distribution of an invasive alien plant: Acacia nilotica ssp. indica in Australia. *Journal of Applied Ecology* 40: 111–124.

Krogh, S.N., M.S. Zeisset, E. Jackson, and W.G. Whitford. 2002. Presence/absence of a keystone species as an indicator of rangeland health. *Journal of Arid Environments* 50: 513–519.

Kulmatiski, A., and K.H. Beard. 2013. Woody plant encroachment facilitated by increased precipitation intensity. *Nature Climate Change* 3: 833–837.

Lautenbach, J.M., R.T. Plumb, S.G. Robinson, D.A. Haukos, and J.C. Pitman. 2016. Lesser prairie-chicken avoidance of trees in a grassland landscape. *Rangeland Ecology and Management*, in press http://dx.doi.org/10.1016/j.rama.2016.07.008.

Litt, A.R., and R.J. Steidl. 2010. Insect assemblages change along a gradient of invasion by a non-native grass. *Biological Invasions* 12: 3449–3463.

Liu, F., S. Archer, F. Gelwick, E. Bai, T. Boutton, and X.B. Wu. 2013. Woody plant encroachment into grasslands: Spatial patterns of functional group distribution and community development. *PLoS ONE* 8, e84364. doi:10.1371/journal.pone.0084364.

Lloyd, J., R.W. Mannan, S. Destefano, and C. Kirkpatrick. 1998. The effects of mesquite invasion on a southeastern Arizona grassland bird community. *The Wilson Bulletin* 110: 403–408.

Lonsdale, W.M. 1993. Rates of spread of an invading species—*Mimosa pigra* in northern Australia. *Journal of Ecology* 81: 513–521.

Madany, M.H., and N.E. West. 1983. Livestock grazing-fire regime interactions within montane forests of Zion National Park, Utah. *Ecology* 64: 661–667.

Maestre, F.T., S. Bautista, J. Cortina, and J. Bellot. 2001. Potential for using facilitation by grasses to establish shrubs on a semiarid degraded steppe. *Ecological Applications* 11: 1641–1655.

Maestre, T.F., J. Cortina, S. Bautista, J. Bellot, and R. Vallejo. 2003a. Small-scale environmental heterogeneity and spatiotemporal dynamics of seedling establishment in a semiarid degraded ecosystem. *Ecosystems* 6: 630–643.

Maestre, F., S. Bautista, and J. Cortina. 2003b. Positive, negative and net effects in grass–shrub interactions in Mediterranean semiarid grasslands. *Ecology* 84: 3186–3197.

Maestre, F.T., M.A. Bowker, M.D. Puche, M.B. Hinojosa, I. Martinez, P. García-Palacios, A.P. Castillo, S. Soliveres, A.L. Luzuriaga, A.M. Sánchez, and J.A. Carreira. 2009. Shrub encroachment can reverse desertification in semi-arid Mediterranean grasslands. *Ecology Letters* 12: 930–941.

Mannan, R.W., and R.J. Steidl. 2013. Habitat. In *Wildlife management and conservation: Contemporary principles and practices*, ed. P. Krausman and J. Cain. Baltimore, MD: John Hopkins University Press.

Maron, M., and A. Lill. 2005. The influence of livestock grazing and weed invasion on habitat use by birds in grassy woodland remnants. *Biological Conservation* 124: 439–450.

Martin TE. 1992. Breeding productivity considerations: What are the appropriate habitat features for management? In *Ecology and conservation of neotropical migrant landbirds*.

Martinez, A., and J. Lopez-Portillo. 2003. Allometry of *Prosopis glandulosa* var. *torreyana* along a topographic gradient in the Chihuahuan desert. *Journal of Vegetation Science* 14: 111–120.

Mason, L., M. Desmond, and M. Agudelo. 2005. Influence of grassland type, nest type, and shrub encroachment on predation of artificial nests in Chihuahuan Desert grasslands. *Western North American Naturalist* 65: 196–201.

Maurer, B.A. 1985. Avian community dynamics in desert grasslands: Observational scale and hierarchical structure. *Ecological Monographs* 55: 295–312.

McAuliffe, J.R. 1997. Landscape evolution, soil formation and Arizona's desert grasslands. In *The desert grassland*, ed. M.P. McClaran and T.R. Van Devender. Tucson: University of Arizona.

McAuliffe, J.R. 2003. The interface between precipitation and vegetation: The importance of soils in arid and semiarid environments. In *Changing precipitation regimes and terrestrial ecosystems*, ed. J.F. Weltzin and G.R. McPherson. Tucson, AZ: University of Arizona Press.

McClaran MP. 2003. A century of vegetation change on the Santa Rita experimental range. In *The Santa Rita experimental range: 100 years (1903 to 2003) of accomplishments and contributions*, eds. M.P. McClaran, P.F. Ffolliott, and C.B. Edminster. Tucson, AZ: Proc. RMRS-P-30, U.S. Department of Agriculture, Forest Service, Rocky Mountain Research Station, Ogden, UT.

McDaniel, K.C., L.A. Torell, and C.G. Ochoa. 2005. Wyoming big sagebrush recovery and understory response with tebuthiuron control. *Rangeland Ecology and Management* 58: 65–76.

Meik, J.M., R.M. Jeo, J.R. Mendelson Iii, and K.E. Jenks. 2002. Effects of bush encroachment on an assemblage of diurnal lizard species in central Namibia. *Biological Conservation* 106: 29–36.

Mendelson, J., and W. Jennings. 1992. Shifts in the relative abundance of snakes in a desert grassland. *Journal of Herpetology* 26: 38–45.

Meyer, K.M., D. Ward, K. Wiegand, and A. Moustakas. 2008. Multi-proxy evidence for competition between savanna woody species. *Perspectives in Plant Ecology, Evolution and Systematics* 10: 63–72.

Milne, B.T., A.R. Johnson, T.H. Keitt, C.A. Hatfield, J. David, and P.T. Hraber. 1996. Detection of critical densities associated with pinon-juniper woodland ecotones. *Ecology* 77: 805–821.

Mitchard, E.T.A., and C.M. Flintrop. 2013. Woody encroachment and forest degradation in sub-Saharan Africa's woodlands and savannas 1982–2006. *Philosophical Transactions of the Royal Society of London, Series B: Biological Sciences* 368: 20120406.

Molinar, F., J. Holechek, D. Galt, and M. Thomas. 2002. Soil depth effects on Chihuahuan Desert vegetation. *Western North American Naturalist* 62: 300–306.

Moncrieff, G.R., S. Scheiter, W.J. Bond, and S.I. Higgins. 2014. Increasing atmospheric CO_2 overrides the historical legacy of multiple stable biome states in Africa. *New Phytologist* 201: 908–915.

Morello, J., and C. Saravia-Toledo. 1959. Bosque Chaqueño I. Paisaje primitivo, paisaje natural y paisaje cultural en el oriente de Salta. *Revista Agronómica del Noroeste Argentina* 3: 5–81.

Myers-Smith, I.H., B.C. Forbes, M. Wilmking, M. Hallinger, T. Lantz, D. Blok, K.D. Tape, M. Macias-Fauria, U. Sass-Klaassen, L. Esther, and P. Ropars. 2011. Shrub expansion in tundra ecosystems: dynamics, impacts and research priorities. *Environmental Research Letters* 6: 045509.

Naito, A.T., and D.M. Cairns. 2011. Relationships between Arctic shrub dynamics and topographically derived hydrologic characteristics. *Environmental Research Letters* 6: 045506.

Neilson, R.P. 1986. High resolution climatic analysis and southwest biogeography. *Science* 232: 27–34.

Neilson, R.P., G.A. King, and G. Koerper. 1992. Toward a rule-based biome model. *Landscape Ecology* 7: 27–43.

Nelson, E., G. Mendoza, J. Regetz, S. Polasky, H. Tallis, D. Cameron, K. Chan, G.C. Daily, J. Goldstein, P.M. Kareiva, and E. Lonsdorf. 2009. Modeling multiple ecosystem services, biodiversity conservation, commodity production, and tradeoffs at landscape scales. *Frontiers in Ecology and the Environment* 7: 4–11.

Nicolai, N., R.A. Feagin, and F.E. Smeins. 2010. Spatial patterns of grass seedling recruitment imply predation and facilitation by harvester ants. *Environmental Entomology* 39: 127–133.

Noble, J.C. 1997. *The delicate and noxious scrub: Studies on native tree and shrub proliferation in semi-arid woodlands of Australia.* Canberra: CSIRO Division of Wildlife and Ecology.

Noble, J.C., and P. Walker. 2006. Integrated shrub management in semi-arid woodlands of eastern Australia: A systems-based decision support system. *Agricultural Systems* 88: 332–359.

Norton-Griffiths, M. 1979. The influence of grazing, browsing, and fire on the vegetation dynamics of the Serengeti, Tanzania, Kenya. In *Serengeti: Dynamics of an ecosystem*, ed. A.R.E. Sinclair and M. Norton-Griffiths. Chicago: University of Chicago Press.

Noss, R.F., E.T. LaRoe III, and J.M. Scott. 1995. *Endangered ecosystems of the United States: A preliminary assessment of loss and degradation.* National Biological Survey, Biological Report No. 28.

O'Connor, T.G. 1995. Acacia karroo invasion of grassland: Environmental and biotic effects influencing seedling emergence and establishment. *Oecologia* 103: 214–223.

Oba, G., E. Post, P.O. Syvertsen, and N.C. Stenseth. 2000. Bush cover and range condition assessments in relation to landscape and grazing in southern Ethiopia. *Landscape Ecology* 15: 535–546.

Ortega, Y. K., K. S. McKelvey, and D. L. Six. 2006. Invasion of an exotic forb impacts reproductive success and site fidelity of a migratory songbird. Oecologia 149: 340–351.

Owens, M.K., R.K. Lyons, and C.L. Alejandro. 2006. Rainfall partitioning within semiarid juniper communities: Effects of events size and canopy cover. *Hydrological Processes* 20: 3179–3189.

Park, S.C., R.J. Ansley, M. Mirik, and M.A. Maindrault. 2012. Delivered biomass costs of honey mesquite (prosopis glandulosa) for bioenergy uses in the south central USA. *BioEnergy Research* 5: 989–1001.

Peterjohn, B.G., and J.R. Sauer. 1999. Population status of North American species of grassland birds from the North American breeding bird survey, 1966–1996. *Studies in Avian Biology* 19: 27–44.

Peters, D.P.C., B.T. Bestelmeyer, J.E. Herrick, E.L. Fredrickson, H.C. Monger, and K.M. Havstad. 2006. Disentangling complex landscapes: New insights into arid and semiarid system dynamics. *Bioscience* 56: 491–501.

Peters, D., K. Havstad, S. Archer, and O. Sala. 2015. Beyond desertification: New paradigms for dryland landscapes. *Frontiers in Ecology and the Environment* 13: 4–12.

Pike, D., J. Webb, and R. Shine. 2011. Removing forest canopy cover restores a reptile assemblage. *Ecological Applications* 21: 274–280.

Polley, H.W. 1997. Implications of rising atmospheric carbon dioxide concentration for rangelands. *Journal of Range Management* 50: 562–577.

Poorter, H., and M.L. Navas. 2003. Plant growth and competition at elevated CO_2: On winners, losers and functional groups. *New Phytologist* 157: 175–198.

Prentice, I.C., S.P. Harrison, and P.J. Bartlein. 2011. Global vegetation and terrestrial carbon cycle changes after the last ice age. *New Phytologist* 189: 988–998.

Priyadarshini, K., H.H. Prins, S. de Bie, I.M. Heitkönig, G. Woodborne, G. Gort, K. Kirkman, F. Ludwig, T.E. Dawson, and H. de Kroon. 2015. Seasonality of hydraulic redistribution by trees to grasses and changes in their water-source use that change tree–grass interactions. *Ecohydrology*. doi:10.1002/eco.1624.

Ratajczak, Z., J. Nippert, J. Hartman, and T. Ocheltree. 2011a. Positive feedbacks amplify rates of woody encroachment in mesic tallgrass prairie. *Ecosphere* 2: 121.

Ratajczak, Z., J.B. Nippert, and S.L. Collins. 2011b. Woody encroachment decreases diversity across North American grasslands and savannas. *Ecology* 93: 697–703.

Reynolds, J., D. Stafford Smith, E. Lambin, B. Turner, M. Mortimore, S.P. Batterbury, T.E. Downing, H. Dowlatabadi, R.J. Fernández, J.E. Herrick, and E. Huber-Sannwald. 2007. Global desertification: Building a science for dryland development. *Science* 316: 847–851.

Robinson, T.P., R.D. van Klinken, and G. Metternicht. 2008. Spatial and temporal rates and patterns of mesquite (*Prosopis* species) invasion in Western Australia. *Journal of Arid Environments* 72: 175.

Roques, K.G., T.G. O'Connor, and A.R. Watkinson. 2001. Dynamics of shrub encroachment in an African savanna: Relative influences of fire, herbivory, rainfall and density dependence. *Journal of Applied Ecology* 38: 268–280.

Rosenstock, S.S., and C. Van Riper III. 2001. Breeding bird responses to juniper woodland expansion. *Journal of Range Management* 54: 226–232.

Rossatto, D.R., L.C.R. Silva, L.S.L. Sternberg, and A.C. Franco. 2014. Do woody and herbaceous species compete for soil water across topographic gradients? Evidence for niche partitioning in a neotropical savanna. *South African Journal of Botany* 91: 14–18.

Ruiz Selmo, F.E., P.G. Minotti, A. Scopel, and M.A. Parimbelli. 2007. Análisis de la heterogeneidad fisonómico-funcional de la vegetación del Parque Nacional El Palmar y su relación con la invasión por leñosas exóticas. In *Teledetección—Hacia un mejor entendimiento de la dinámica global y regional*. Buenos Aires, Argentina.

Sala, O.E., and F.T. Maestre. 2014. Grass–woodland transitions: Determinants and consequences for ecosystem functioning and provisioning of services. *Journal of Ecology* 102: 1357–1362.

Samson, F., and F. Knopf. 1994. Prairie conservation in North America. *Bioscience* 44: 418–421.

Sankaran, M., N.P. Hanan, R.J. Scholes, J. Ratnam, D.J. Augustine, B.S. Cade, J. Gignoux, S.I. Higgins, X. Le Roux, F. Ludwig, and J. Ardo. 2005. Determinants of woody cover in African savannas. *Nature* 438: 846–849.

Sauer, J.R., and W.A. Link. 2011. Analysis of the North American breeding bird survey using hierarchical models. *The Auk* 128: 87–98.

Scheffer, M., M. Holmgren, V. Brovkin, and M. Claussen. 2005. Synergy between small- and large-scale feedbacks of vegetation on the water cycle. *Global Change Biology* 11: 1003–1012.

Scheiter, S., and S.I. Higgins. 2009. Impacts of climate change on the vegetation of Africa: An adaptive dynamic vegetation modelling approach. *Global Change Biology* 15: 2224–2246.

Schenk, H.J., and R.B. Jackson. 2002. Rooting depths, lateral root spreads and below-ground/above-ground allometries of plants in water-limited ecosystems. *Journal of Ecology* 90: 480–494.

Schlesinger, W.H., J.F. Reynolds, G.L. Cunningham, L.F. Huenneke, W.M. Jarrell, R.A. Virginia, and W.G. Whitford. 1990. Biological feedbacks in global desertification. *Science* 247: 1043–1048.

Scholes, R., and S. Archer. 1997. Tree–grass interactions in savannas. *Annual Review of Ecological Systems* 28: 517–544.

Scifres, C.J., W.T. Hamilton, B.H. Koerth, R.C. Flinn, and R.A. Crane. 1988. Bionomics of patterned herbicide application for wildlife habitat enhancement. *Journal of Range Management* 41: 317–321.

Scott, R.L., T.E. Huxman, D.G. Williams, and D.C. Goodrich. 2006. Ecohydrological impacts of woody-plant encroachment: Seasonal patterns of water and carbon dioxide exchange within a semi-arid riparian environment. *Global Change Biology* 12: 311–324.

Sea, W.B., and N.P. Hanan. 2012. Self-thinning and tree competition in savannas. *Biotropica* 44: 189–196.

Shackleton, R.T., D.C. Le Maitre, and D.M. Richardson. 2015. Stakeholder perceptions and practices regarding Prosopis (mesquite) invasions and management in South Africa. *Ambio* 44: 569–581.

Shaffer, J.A., C.M. Goldade, M.F. Dinkins, D.H. Johnson, L.D. Igl, and B.R. Euliss. 2003. *Brown-headed Cowbirds in grasslands: Their habitats, hosts, and response to management*. USGS Northern Prairie Wildlife Research Center. p. 158.

Sirami, C., and A. Monadjem. 2012. Changes in bird communities in Swaziland savannas between 1998 and 2008 owing to shrub encroachment. *Diversity and Distributions* 18: 390–400.

Sirami, C., C. Seymour, G. Midgley, and P. Barnard. 2009. The impact of shrub encroachment on savanna bird diversity from local to regional scale. *Diversity and Distributions* 15: 948–957.

Sklar, F.H., and A.V.D. Valk. 2003. *Tree Islands of the Everglades*. Dordrecht: Kluwer Academic Publishers.

Skowno, A., and W. Bond. 2003. Bird community composition in an actively managed savanna reserve, importance of vegetation structure and vegetation composition. *Biodiversity and Conservation* 12: 2279–2294.

Sperry, J.S., and U.G. Hacke. 2002. Desert shrub water relations with respect to soil characteristics and plant functional type. *Functional Ecology* 16: 367–378.

Staver, C.A., W.J. Bond, W.D. Stock, S.J. vanRensburg, and M.S. Waldram. 2009. Browsing and fire interact to suppress tree density in an African savanna. *Ecological Applications* 19: 1909–1919.

Staver, A.C., W.J. Bond, and E.C. February. 2011. History matters: Tree establishment variability and species turnover in an African savanna. *Ecosphere* 2: 12.

Steidl, R.J., A.R. Litt, and W.J. Matter. 2013. Effects of plant invasions on wildlife in desert grasslands. *Wildlife Society Bulletin* 37: 527–536.

Stevens, N., C.E.R. Lehmann, B.P. Murphy, and G. Durigan. 2016. Savanna woody encroachment is widespread across three continents. *Global Change Biology*. doi:10.1111/gcb.13409.

Stokes, C., and S. Archer. 2010. Niche differentiation and neutral theory: An integrated perspective on shrub assemblages in a parkland savanna. *Ecology* 91: 1152–1162.

Tanaka, J., M. Brunson, and L. Torell. 2011. A social and economic assessment of rangeland conservation practices. In *Conservation benefits of rangeland practices: Assessment, recommendations, and knowledge gaps*, ed. D. Briske. Washington, DC: United States Department of Agriculture, Natural Resources Conservation Service.

Tecco, P.A., D.E. Gurvich, S. Diaz, N. Perez-Harguindeguy, and M. Cabido. 2006. Positive interaction between invasive plants: The influence of Pyracantha angustifolia on the recruitment of native and exotic woody species. *Austral Ecology* 31: 293–300.

Tennesen, M. 2008. When juniper and woody plants invade, water may retreat. *Science* 322: 1630–1631.

Tews, J., U. Brose, V. Grimm, K. Tielborger, M.C. Wichmann, M. Schwager, and F. Jeltsch. 2004. Animal species diversity driven by habitat heterogeneity/diversity: The importance of keystone structures. *Journal of Biogeography* 31(1): 79–92.

Thompson, T., C. Boal, and D. Lucia. 2009. Grassland bird associations with introduced and native grass Conservation Reserve Program fields in the Southern High Plains. *Western North American Naturalist* 69: 481–490.

Throop, H.L., and S.R. Archer. 2008. Shrub (*Prosopis velutina*) encroachment in a semidesert grassland: Spatial-temporal changes in soil organic carbon and nitrogen pools. *Global Change Biology* 14: 2420–2431.

Throop, H.L., S.R. Archer, H.C. Monger, and S. Waltman. 2012. When bulk density methods matter: Implications for estimating soil organic carbon pools in coarse soils. *Journal of Arid Environments* 77: 66–71.

Tongway, D.J., C. Valentin, and J. Seghier. 2001. *Banded vegetation patterning in arid and semiarid environments: Ecological processes and consequences for management*. New York: Springer.

Torell, L.A., K.C. McDaniel, and C.G. Ochoa. 2005a. Economics and optimal frequency of Wyoming big sagebrush control with tebuthiuron. *Rangeland Ecology and Management* 1: 77–84.

Torell, L.A., N.R. Rimbey, O.A. Ramirez, and D.W. McCollum. 2005b. Income earning potential versus consumptive amenities in determining ranchland values. *Journal of Agricultural and Resource Economics* 30: 537–560.

Twidwell, D., C.L. Wonkka, C.A. Taylor, C.B. Zou, J.J. Twidwell, and W.E. Rogers. 2014. Drought-induced woody plant mortality in an encroached semi-arid savanna depends on topoedaphic factors and land management. *Applied Vegetation Science* 17: 42–52.

Vadigi, S., and D. Ward. 2012. Fire and nutrient gradient effects on the sapling ecology of four Acacia species in the presence of grass competition. *Plant Ecology* 213: 1793–1802.

Vadigi, S., and D. Ward. 2014. Herbivory effects on saplings are influenced by nutrients and grass competition in a humid South African savanna. *Perspectives in Plant Ecology, Evolution and Systematics* 16: 11–20.

Valone, T.J., S.E. Nordell, and S.K.M. Ernest. 2002. Effects of fire and grazing on an arid grassland ecosystem. *Southwestern Naturalist* 47: 557–565.

Van Devender, T. 1997. Desert grassland history: Changing climates, evolution, biogeography, and community dynamics. In *The desert grassland*, eds. M.P. McClaran, and T.R.V. Devender. Tucson: The University of Arizona.

van Wijk, M.T., and I. Rodriguez-Iturbe. 2002. Tree-grass competition in space and time: Insights from a simple cellular automata model based on ecohydrological dynamics. *Water Resources Research* 38:18-1–18-15.

Vickery, P.D., P.L. Tubaro, J.M.C.D. Silva, B.G. Peterjohn, J.R. Herkert, et al. 1999. *Ecology and conservation of grassland birds of the Western Hemisphere*. Camarillo, CA: Cooper Ornithological Society.

Vorosmarty, C.J., P.B. McIntyre, M.O. Gessner, D. Dudgeon, A. Prusevich, P. Green, S. Glidden, S.E. Bunn, C.A. Sullivan, C.R. Liermann, and P.M. Davies. 2010. Global threats to human water security and river biodiversity. *Nature* 467: 555–561.

Wakeling, J.L., W.J. Bond, M. Ghaui, and E.C. February. 2015. Grass competition and the savanna-grassland 'treeline': A question of root gaps? *South African Journal of Botany* 101: 91–97.

Walter, H. 1979. *Vegetation of the earth and ecological systems of the geobiosphere*. New York: Springer.

Ward, D., K. Wiegand, and S. Getzin. 2013. Walter's two-layer hypothesis revisited: Back to the roots! *Oecologia* 172: 617–630.

Watson, I.W., and M. Westoby. 1997. Continuous and episodic components of demographic change in arid zone shrubs: Models of two Eremophila species from Western Australia compared with published data on other species. *Journal of Ecology* 85: 833.

Weltzin, J.F., and G.R. McPherson. 1997. Spatial and temporal soil moisture resource partitioning by trees and grasses in a temperate savanna, Arizona, USA. *Oecologia* 112: 156–164.

Weltzin, J.F., S. Archer, and R.K. Heitschmidt. 1997. Small-mammal regulation of vegetation structure in a temperate savanna. *Ecology* 78: 751–763.

Wiens, J.A., and B.T. Milne. 1989. Scaling of 'landscapes' in landscape ecology, or, landscape ecology from a beetle's perspective. *Landscape Ecology* 3: 87–96.

Wilcox, B.P., M.K. Owens, R.W. Knight, and R.K. Lyons. 2005. Do woody plants affect stream-flow on semiarid karst rangelands? *Ecological Applications* 151: 127–136.

Winter, M., D.H. Johnson, Shaffer Ja, T.M. Donovan, and W.D. Svedarsky. 2006. Patch size and landscape effects on density and nesting success of grassland birds. *Journal of Wildlife Management* 70: 158–172.

With, K.A. 1994. The hazards of nesting near shrubs for a grassland bird, the McCown's longspur. *The Condor* 96: 1009–1019.

Wu, X.B., and S. Archer. 2005. Scale-dependent influence of topography-based hydrologic features on vegetation patterns in savanna landscapes. *Landscape Ecology* 20: 733–742.

Zanette, L., J.N. Smith, H. van Oort, and M. Clinchy. 2003. Synergistic effects of food and predators on annual reproductive success in song sparrows. *Proceedings of the Royal Society of London, Series B: Biological Sciences* 270: 799–803.

Zhang, L., W.R. Dawes, and G.R. Walker. 2001. Response of mean annual evapotranspiration to vegetation changes at catchment scale. *Water Resources Research* 37: 701–708.

Zou, C.B., P.W. Barnes, S.R. Archer, and C.R. McMurtry. 2005. Soil moisture redistribution as a mechanism of facilitation in savanna tree-shrub clusters. *Oecologia* 145: 32–40.

Chapter 3
Ecohydrology: Processes and Implications for Rangelands

Bradford P. Wilcox, David Le Maitre, Esteban Jobbagy, Lixin Wang, and David D. Breshears

Abstract This chapter is organized around the concept of ecohydrological processes that are explicitly tied to *ecosystem services*. Ecosystem services are benefits that people receive from ecosystems. We focus on (1) the regulating services of water distribution, water purification, and climate regulation; (2) the supporting services of water and nutrient cycling and soil protection and restoration; and (3) the provisioning services of water supply and biomass production. Regulating services are determined at the first critical juncture of the water cycle—on the soil surface, where water either infiltrates or becomes overland flow. Soil infiltrability is influenced by vegetation, grazing intensity, brush management, fire patterns, condition of biological soil crusts, and activity by fauna. At larger scales, water-regulating services are influenced by other factors, such as the nature and structure of riparian zones and the presence of shallow groundwater aquifers. Provisioning services are those goods or products that are directly produced from ecosystems, such as water, food, and fiber. Work over the last several decades has largely overturned the notion

B.P. Wilcox (✉)
Ecosystem Science and Management, Texas A&M University,
College Station, TX 77803, USA
e-mail: bwilcox@tamu.edu

D. Le Maitre
Council for Scientific and Industrial Research (CSIR),
PO Box 320, Stellenbosch 7600, South Africa
e-mail: dlmaitre@csir.co.za

E. Jobbagy
Grupo de Estudios Ambientales, University of San Luis, San Luis, Argentina
e-mail: Jobbagy@gmail.com

L. Wang
Department of Earth Sciences, Indiana University-Purdue University Indianapolis (IUPUI),
723 W Michigan St., SL 118P, Indianapolis, IN 46202, USA
e-mail: Wang.iupui@gmail.com

D.D. Breshears
School of Natural Resources and the Environment, University of Arizona, Tucson, AZ, USA
e-mail: daveb@email.arizona.edu

© The Author(s) 2017
D.D. Briske (ed.), *Rangeland Systems*, Springer Series on Environmental
Management, DOI 10.1007/978-3-319-46709-2_3

that water supply can be substantially increased by removal of shrubs. In riparian areas, surprisingly, removal of invasive, non-native woody plants appears to hold little potential for increasing water supply. Here, the primary factor appears to be that non-native plants use no more water than the native vegetation they displace. Clearly there is a close coupling between biota (both fauna and flora) and water on rangelands—which is why water-related ecosystem services are so strongly dependent on land management strategies.

Keywords Ecosystem Services • Infiltration • Rangeland Hydrology • Riparian • Groundwater • Overland Flow • Soil Water • Climate • Water Supply • Climate Regulation • Erosion • Spatial Variability • Scale • Thresholds • Connectivity

3.1 Introduction

The distribution, quality, and provisioning of water are intimately related to how rangeland landscapes function and are managed, particularly with respect to land-use change. Understanding the linkages between vegetation and the water cycle is a major focus of ecohydrology, an emerging discipline that melds the sciences of hydrology and ecology as a means of addressing complex environmental issues. Its scientific heritage also embraces many other disciplines, including watershed management, plant physiology, soil science, geomorphology (Newman et al. 2006), and of course rangeland hydrology (Branson et al. 1981). In addition, the importance of interactions between fauna and the water cycle is increasingly being recognized.

Ecohydrology is very much an applied science with a focus on problem solving (Nuttle 2002; Jackson et al. 2009b; Wilcox et al. 2011), but at the same time it has a firm theoretical foundation (D'Odorico et al. 2012, 2013a; Turnbull et al. 2012; Saco and Moreno de las Heras 2013). Because of its strong intellectual roots in research conducted on drylands—including semiarid and subhumid rangelands (Rodriguez-Iturbe and Porporato 2004; D'Odorico and Porporato 2006; Newman et al. 2006)—and its "transdisciplinary" nature, ecohydrology has advanced our knowledge of rangelands (Wilcox and Newman 2005; Wilcox et al. 2012a). But much more needs to be done to take full advantage of the scientific strengths of ecology and hydrology (King and Caylor 2011).

In this chapter, we present some of the major ecohydrological advances that have occurred in rangelands in the last quarter century and discuss their importance for management. There has been extraordinary scientific progress on so many fronts that it will be impossible to adequately address all of them; but we aim to provide a comprehensive overview of those most relevant to rangeland systems. We rely extensively on the recent publication of several review papers and books dealing with the ecohydrology of rangelands (D'Odorico and Porporato 2006; Newman et al. 2006; D'Odorico et al. 2010; Asbjornsen et al. 2011; Wang et al. 2012a).

We have organized our chapter around the concept of *ecosystem services*—as elaborated in the Millennium Ecosystem Assessment (Millennium Ecosystem Assessment

2005). Ecosystem services are benefits that people receive from ecosystems. They can be categorized as *regulating services*, *supporting services*, *provisioning services*, and *cultural services*. We focus on (1) the regulating services of water distribution, water purification, and climate regulation; (2) the supporting services of water and nutrient cycling and soil protection and restoration; and (3) the provisioning services of water supply and biomass production. In addition, we review current conceptual, theoretical, and technical developments that will provide a foundation for future advances in rangeland ecohydrology—advances critical to informed management decisions and actions needed to meet the growing environmental challenges of rangeland systems.

3.2 Ecosystem Services

The provisioning of water to ensure that humans obtain the quantity and quality of water needed is the most fundamental service provided by ecosystems (Falkenmark and Rockstrom 2004; Brauman et al. 2007). Paradoxically, this is especially true of rangelands, even though most are considered "drylands," which by definition convert a relatively small percentage of precipitation into streamflow or groundwater (Wilcox et al. 2003b). Water produced on rangelands, whether drawn from aquifers or from surface sources, is vitally important to support the people, livestock, and wildlife that inhabit these regions (Le Maitre et al. 2007; Reynolds et al. 2007). Many dryland population centers are growing at alarming rates, and this growth brings with it numerous environmental stresses (D'Odorico et al. 2013a). The degradation of rangelands diminishes their ability to regulate and provide water (MEA).

Figure 3.1 illustrates an important conceptual advance in understanding water dynamics in rangelands: the explicit partitioning of water resources into "blue water" (liquid water) and "green water" (vapor- or water-produced evapotranspiration [ET]) (Falkenmark and Rockstrom 2004, 2006; Gordon et al. 2005; Falkenmark et al. 2009; Rockstrom et al. 2009; Hoff et al. 2010). To date, the water management community has focused almost exclusively on blue water resources and has failed to recognize the opportunity to effectively allocate green water. Maximizing the amount of green water used for plant production or transpiration and minimizing the amount lost as soil evaporation is an imperative. How rangelands are managed— especially their surface cover—has a tremendous effect on both the relative proportion of blue water to green water and the partitioning of green water between E and T. The ability of rangelands to regulate and provide water is strongly dependent on conditions at three critical junctures in the terrestrial water cycle (Falkenmark and Rockstrom 2004). The first and most critical is whether water infiltrates into the soil or becomes *overland flow*—which is mainly a function of rainfall intensity, slope, and soil infiltrability. The importance of soil infiltrability has long been recognized (Smith and Leopold 1941) and has been the focus of considerable research in the last half century or more. It is influenced by many factors, including management practices; for example, overgrazing that results in a loss of vegetation cover and an increased exposure of bare soil can dramatically reduce soil infiltrability (Blackburn et al. 1982; Snyman and du Preez 2005).

Water that does not infiltrate becomes overland flow on slopes, but the final outcome in terms of net water losses is highly scale dependent. At the hillslope scale, runoff–runon dynamics become important and are strongly influenced by the spatial variability of infiltration. For example, overland flow may be generated from some areas on the hillslope only to infiltrate the soil somewhere downslope (Bergkamp 1998a; Wilcox et al. 2003a), and can contribute to surface and groundwater recharge.

The second critical juncture is at the root zone: soil water may drain out of the root zone, and eventually be stored as groundwater or discharged into a stream as baseflow, or may stay in the root zone and eventually be transpired or evaporated from the soil surface. Although largely a function of climate, soil, and geological characteristics, this process can also be affected by management strategies, especially if the functional type of vegetation—and particularly its rooting depth—is changed. The linkage between vegetation and groundwater is very much influenced by the depth to groundwater. Recent work has highlighted the importance of rangelands where groundwater tables are shallow and strongly influenced by vegetation that are termed *groundwater-coupled rangelands* (Jobbágy and Jackson 2004a).

The third critical juncture is the fate of soil water: whether it is absorbed by plants and transpired or lost through evaporation from the soil surface, which is often described as the partitioning of E and T (Fig. 3.1). This juncture is critical because it dictates the amount of biologically available water on rangelands (Falkenmark and Rockstrom 2004; Newman et al. 2006). The portioning of E and T is central to water cycling and is discussed in more detail in the section *Supporting Services*, below.

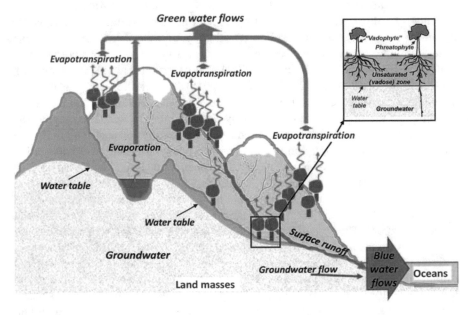

Fig. 3.1 Conceptual diagram of the water cycle, highlighting blue (liquid) and green (vapor) flows. Source: Figure 1 in D'Odorico et al. (2010)

3.2.1 Regulating Services: Water Distribution and Purification

Ecosystem services regulating water on rangelands include those that affect the amount, timing, and quality of blue water flows. These are to a large extent determined at the first critical juncture of the water cycle—on the soil surface, where water either infiltrates or becomes overland flow. For this reason, a great deal of research, most of it conducted at the point or plot scale, has focused on understanding the infiltration process and how it is affected by different management strategies (Pyke et al. 2002; Stavi et al. 2009).

3.2.1.1 Infiltration: Water Regulation at the Soil Surface

Infiltration of water into the soil is enhanced and maintained by the presence of vegetation, both by direct influences (soil protection, root action, etc.) and by modification of the soil through the addition of organic matter. This tight coupling between vegetation and soil infiltrability on rangelands was recognized many years ago (Smith and Leopold 1941; Woodward 1943; Dyksterhuis and Schmutz 1947; Dortignac and Love 1961); but recent research is adding greatly to our understanding by providing specifics concerning how management practices and disturbances (grazing, shrub management, fire) and vegetation cover types (shrubs vs. grasses, biological soil crusts) affect soil infiltrability, but also the contributions of spatial variability and scale. In addition, we now recognize that fauna—large and small—can significantly affect soil infiltrability.

Influence of Grazing. There is an extensive body of work examining the ecohydrological influence of grazing, and specifically its influence on soil infiltration. Much of this work was conducted in the USA in the 1970s and 1980s and has been summarized in several review papers (Gifford 1978; Wood et al. 1978; Wood and Blackburn 1981; Blackburn et al. 1982; Trimble and Mendel 1995). The findings consistently show that, irrespective of grazing systems, light-to-moderate grazing generally has little adverse effect on the ecohydrology of rangelands and may even have a positive effect, whereas heavy grazing generally significantly decreases soil infiltrability. These conclusions have been verified by more recent investigations conducted on rangeland throughout the globe (Hiernaux et al. 1999; Ludwig et al. 1999; Savadogo et al. 2007).

Influence of Shrubs. Over the past several decades, grasslands and savannas worldwide have been undergoing a process of woodland conversion, often described as woody plant encroachment (Archer 1994; Archer et al. 2011). For many rangelands, attempts to reverse this process or even to control it have met with minimal success (Archer et al. 2011). During the past quarter century, considerable research has been focused on understanding the ecohydrological implications of this conversion (Huxman et al. 2005; Wilcox et al. 2006). It has generally been found (though not always—see Moran et al. (2010)) that infiltration rates are higher beneath shrub canopies than in intercanopy areas (Lyford and Qashu 1969; Seyfried 1991;

Bergkamp 1998b; Schlesinger et al. 1999; Wilcox 2002; D'Odorico et al. 2007; Wilcox et al. 2008; Pierson et al. 2010; Daryanto et al. 2013; Eldridge et al. 2013), primarily owing to the accumulation of organic matter under shrubs, root activity (Joffre and Rambal 1993; Martinez-Meza and Whitford 1996; Jackson et al. 2000), and soil disturbance by fauna (see "Influence of Fauna" section). In some situations the chemical composition of the litter may cause water repellency (hydrophobicity), which reduces the infiltration capacity of soils beneath the canopy, at least in the short term (Doerr et al. 2000). In addition, burning can cause or aggravate hydrophobicity (Hester et al. 1997; Cammeraat and Imeson 1999).

Influence of Biological Soil Crusts. Biological soil crusts are the community of living organisms, including fungi, lichens, cyanobacteria, and algae, at the soil surface. The integrity of biological soil crusts, which are common in many drylands, is extremely sensitive to disturbance such as heavy grazing or off-road vehicle traffic (Belnap and Lange 2001). The relationship between biological soil crusts and processes of soil infiltrability is complex: their presence can increase, decrease, or have no effect on this process (Eldridge 2003; Warren 2003; Belnap 2006b). One factor that appears to determine local hydrological response is the successional stage, or status of crust development. As crusts mature, the biomass of cyanobacteria, mosses, and lichens increases—which in turn increases aggregate stability, shear strength, and roughness of the soil surface (Belnap 2003, 2006a). A six-level classification of level of crust development (LOD) was recently developed for biological soil crusts, based on (1) color (light to dark, visual assessment); (2) presence of mosses/lichens; and (3) soil surface roughness (Belnap et al. 2008). Soil crust classification was found to be strongly related to infiltration rates, with infiltration being highest where crusts were the most developed (Belnap et al. 2013).

Influence of Fauna. A recent review of ecohydrological studies revealed a strong emphasis on plant–hydrology interactions, with few studies of fauna–hydrology interactions (Westbrook et al. 2013). Only 17 % of the 339 papers reviewed considered fauna–hydrology interactions, and more than half of those focused on how hydrology affects fauna rather than how fauna function to influence ecohydrology. Fauna are usually seen as passive beneficiaries of ecohydrological changes rather than as playing a key role in the formation of vegetation patterns.

Fauna have both direct and indirect effects on ecohydrology, ranging from microperturbations to the macro-perturbation commonly described as ecosystem engineering (Whitford and Kay 1999; Jones et al. 2006; Butler 2007; Hastings et al. 2007; Jones 2012; Raynaud et al. 2013). These processes are critical for producing the organic matter that binds with mineral soil particles to form aggregates (peds), which facilitate the movement of water through soils and thereby increase infiltration and percolation rates and capacities (Weaver 1926; Coleman et al. 1992; Lavelle 1997; Angers and Caron 1998; Roth 2004; Jones et al. 2006). Soil fauna, particularly the mammals and macro-invertebrates (such as earthworms, termites, or cicadas), engineer ecosystems by creating openings at the soil surface and tunnels, also known as macropores, beneath the soil surface (Beven and Germann 1982; Lavelle 1997; Leonard et al. 2004; Roth 2004). These openings increase infiltration and percolation of water through the soil profile (Dean 1992; Angers and Caron 1998;

Whitford and Kay 1999; O'Farrell et al. 2010), in the same way as do the channels left by decayed plant roots (Beven and Germann 1982). Clearly, one cannot separate the roles played by animals from those played by plants; but, in combination, they significantly affect how water moves through the soil (Shafer et al. 2007)—including processes such as groundwater recharge, which in turn affect plant productivity and other ecosystem services.

Influence of Fire. The frequency and intensity of wildfires are increasing on rangelands as a result of several factors, including rising temperatures and the invasion of non-native grasses (Running 2006; Wilcox et al. 2012b). In addition, prescribed fire is now more commonly applied as a management tool for many rangelands (Twidwell et al. 2013). A number of recent reviews summarize the extensive literature on the hydrological consequences of fire on rangelands; in general, study results indicate that the infiltration capacity of soils is significantly reduced immediately following fires, but the extent of this reduction depends on fire severity, degree of hydrophobicity, antecedent soil moisture, and topographic position (Baker and Shinneman 2004; Shakesby and Doerr 2006; Pierson et al. 2011).

3.2.2 Overland Flow: Regulation at the Hillslope Scale

Water that does not infiltrate, of course, becomes overland flow or surface runoff. It is at the hillslope scale that important interactions take place between vegetation patches and runoff. Surface runoff may be captured and stored by vegetation patches or other surface obstructions, a process known as runoff–runon (Ludwig et al. 2005).

An important conceptual advance in describing and clarifying the linkages between surface runoff and vegetation patches is the trigger-transfer-reserve-pulse (TTRP) framework (Fig. 3.2) (Ludwig et al. 1997, 2005). This framework was originally proposed as a way of describing runoff–runon processes observed in areas of banded vegetation (Anderson and Hodgkinson 1997; Dunkerley and Brown 1999; Valentin and d'Herbes 1999; Tongway and Ludwig 2001); it was subsequently verified for other vegetation patch types in semiarid settings (Reid et al. 1999; Wilcox et al. 2003a; Ludwig et al. 2005). The framework assumes that the redistribution of resources from source areas (bare patches) to sink areas is a fundamental process within drylands, and that this process may be disrupted if vegetation patch structure is altered by disturbances such as overgrazing or multiyear drought. These dynamics govern how runoff and runon vary with scale in semiarid settings. In regions where runoff is efficiently captured down slope by vegetation patches, unit-area runoff and erosion diminish rapidly with increasing scale. But where vegetation patch structure has been disturbed and runoff is not efficiently captured, declines in runoff with increasing scale are much smaller (Fig. 3.3). Erosion may even increase as runoff increases with increasing scale, leading to rilling and gully formation (Wilcox et al. 2003a; Moreno de las Heras et al. 2010).

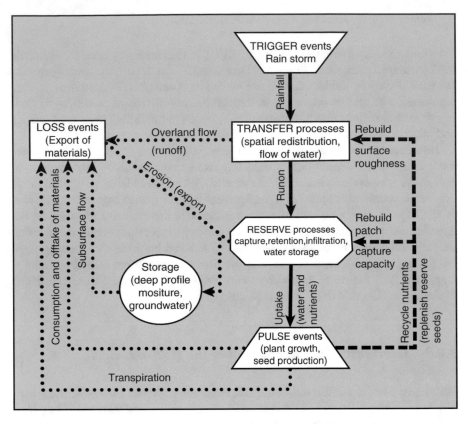

Fig. 3.2 The Trigger-Transfer-Reserve-Pulse framework illustrates how temporal events (e.g., water input from rainfall) initiate a number of other events. Solid arrows indicate direct action of flows of water, dashed arrows indicate feedbacks, and dotted arrows indicate losses (including transpiration, lateral subsurface flow, and groundwater recharge). Source: Figure 4 in Le Maitre et al. (2007)

Fig. 3.3 Hypothetical relationships demonstrating the relative changes in runoff and erosion with changes in scale and how these relationships are altered by disturbance. Source: Figure 9 in Wilcox et al. (2003)

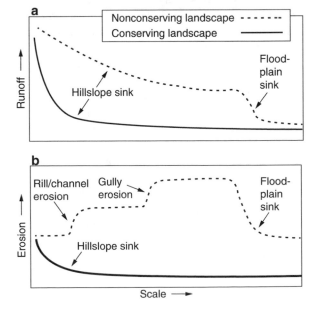

3.2.3   Drainage: Water Regulation Within the Soil

Water that enters the soil may either evaporate, be transpired by plants, or drain out of the root zone and ultimately contribute to groundwater and streamflow. On rangelands, drainage is generally (but not always) a small percentage (<5 %) of the water budget (Wilcox et al. 2003b). Vegetation management that alters the amount of woody plants may affect drainage because woody plants are deeper rooted than grasses or forbs and tend to transpire more water. Therefore, woodlands and forests generally use more water than grasslands (Zhang et al. 2001). The relationship is complex, however, especially for drylands, as it is modified by other factors— including climate, soils, and topographic position (Huxman et al. 2005).

The shrub–streamflow framework (Fig. 3.4) was developed to aid in determining which landscapes are most "hydrologically sensitive" to changes in woody plant cover. A hydrologically sensitive landscape is one in which a shift in functional vegetation type (woody to nonwoody or vice versa) causes an important shift in the water balance. Hydrological sensitivity is dictated or strongly influenced by how vegetation change affects drainage; and it is also influenced by factors such as depth to water table, soil and geological characteristics, and topographic position. The shrub–streamflow framework uses these concepts to predict where hydrologically sensitive shrublands might exist (Wilcox et al. 2006).

The first criterion for hydrologic sensitivity—the presence of shallow groundwater—is likely to be stronger where the groundwater table is within a few meters of the surface, as in riparian zones or groundwater-coupled rangelands. Obviously, this condition affords more opportunity for interaction between deep-rooted vegetation and groundwater.

Seasonality of precipitation is a second criterion in determining hydrologic sensitivity. Those rangelands having the greatest potential for water to move deeply into the soil—beneath the rooting zone of herbaceous plants—will be the most hydrologically sensitive. Such deep drainage occurs in regions where winter precipitation is high. It is no coincidence that the strongest linkage between woody plants and streamflow has been observed in Mediterranean climates where precipitation is often "out of phase" with potential ET. For example, in South Africa (van Wilgen et al. 1998), Spain (Puigdefabregas and Mendizabal 1998), Australia (Walker et al. 1993), and California (Hibbert 1983), dramatic changes in drainage have been observed following vegetation changes in native shrublands. Similarly, shrublands in which soil recharge comes mainly from snowmelt may be hydrologically sensitive; a large pulse of melting snow often produces enough water to saturate or exceed the water storage capacity of the upper soil (Baker 1984; Seyfried and Wilcox 2006).

Finally, soil or geological conditions also determine hydrologic sensitivity, by affecting the potential for deep drainage. We would expect higher hydrologic sensitivity where soils are sandy (Moore et al. 2012; Dzikiti et al. 2013), are deeply cracked (Richardson et al. 1979), or are shallow and overlie fractured bedrock (Huang et al. 2006).

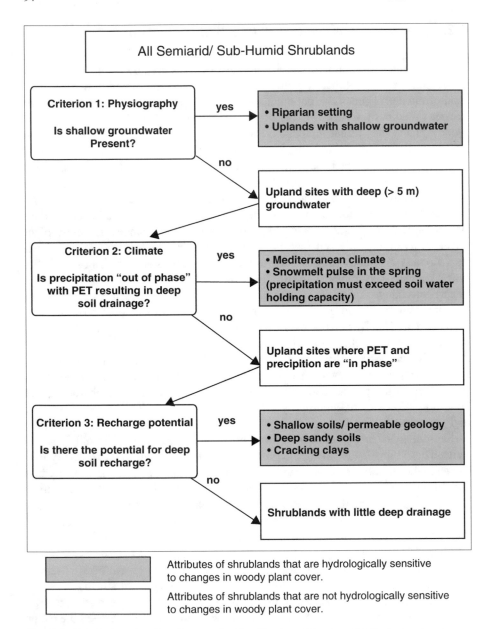

Fig. 3.4 The shrub–streamflow framework: Classification of the potential for increasing stream-flow in various shrublands. Source: Figure 1 in Wilcox et al. (2006)

For many rangelands, the opportunities for deep drainage are quite limited, because of either climate or soils. In these landscapes, shifting from a grassland to a shrubland will have little effect on the overall water balance, but may nevertheless affect drainage in important ways. For example, in areas where even small amounts of drainage can be significant (such as sites where hazardous wastes are buried), the presence of deep-rooted shrubs may ensure that drainage below the root zone seldom occurs (Scanlon et al. 2005b, c; Seyfried et al. 2005).

3.2.4   Riparian Systems: Regulation at the Watershed Scale

The common perception is that rangelands are exclusively dryland environments. Even when the presence of riparian environments is recognized, these are generally seen as a minor component of the entire landscape system. However, there is a growing body of research showing that riparian environments are not only key habitats for rangeland fauna and flora, but also critical providers of ecosystem services to rangeland inhabitants (Milton 1990; Dean et al. 1999, 2002; Naiman et al. 2002; Sabo et al. 2005; Le Maitre et al. 2007; Soykan and Sabo 2009; Jones et al. 2010; Merritt and Bateman 2012).

Many of the features and key processes in groundwater-coupled systems are likewise found in riparian environments. The principal differences are that (1) riparian zones receive inputs of surface water, often from areas far upstream, that play a major role in their ecology (Boulton and Hancock 2006; Bunn et al. 2006; Nagler et al. 2008) and (2) the dynamics of riparian zones are strongly influenced by flow regimes and fluvial processes (Ward 1998), particularly sediment movement (Naiman et al. 1999; Tabacchi et al. 2000).

Riparian environments are typically located in the lowest parts of a landscape, where surface water (overland flow) and groundwater (subsurface flow) collect; thus they integrate outputs from all watershed-scale processes (Naiman et al. 2002). Their structure is long and narrow with a very large perimeter-to-area ratio—which makes them highly connected to, and thereby highly influenced by, events originating in the adjacent drylands. The headwaters section of a river typically has steep gradients, and the river bed contains rocks or boulders with little accumulation of fine sediments, whereas the middle and lower reaches are characterized by extensive alluvial deposits and wide floodplains (e.g., Nile, Platte, Euphrates, Ganges) (Vannote et al. 1980; Wiens 2002). These deposits are typically heterogeneous, with coarse sediments that can capture, store, and transmit large volumes of water interspersed vertically and horizontally with fine sediments that have a lower storage capacity and low transmissivity (Blasch et al. 2006; Morin et al. 2009). In humid-to-semiarid landscapes, rainfall is sufficient to generate runoff and groundwater that sustain river systems, and the rivers are gaining water, albeit seasonally. But in arid landscapes the rivers are often hydrologically disconnected from the adjacent dryland areas—except for losing water to the floodplain, and gaining water during rainy seasons or after very high rainfall events.

Water use by riparian communities has been intensively studied in the southwestern USA, mainly to estimate transmission losses, but also to quantify the effects of woody species such as the invasive or saltcedar (*Tamarix* spp.) on these losses. Evapotranspiration rates can exceed annual rainfall in these arid environments (Doody et al. 2011). Early research on saltcedar suggested that its water use was very high: up to 200 m^3 ha^{-1} day^{-1} (Sala et al. 1996); but subsequent research has found that it is much lower and close to that of the native poplars and willows that grow in riparian systems of the southwest. Therefore, removal of these plants would have little effect on water loss if native riparian shrubs remain (Nagler et al. 2009; Shafroth et al. 2010b; Doody et al. 2011). Similar work in Australia found that although invasive *Salix* species in the river channel can use large quantities of water (±2000 mm year^{-1} vs. 1500 mm year^{-1} for open water), overall ET for this invasive species is very similar to that found for native riparian *Eucalyptus* forest (Doody and Benyon 2011; Doody et al. 2011).

In the floodplain of the perennial San Pedro River in Arizona, *Prosopis* woodlands have replaced native grasslands, increasing ET from 407 to 639 mm year^{-1} (Scott et al. 2006). Evapotranspiration from *Prosopis* woodlands in floodplains linked to perennial rivers ranges from about 350 to 750 mm year^{-1} (Scott et al. 2004, 2008), which suggests that other perennial river systems (such as those in South Africa where native tree species such as *Acacia karroo* are sparse or absent) could be similarly affected by invasion of non-native species.

The Working for Water program in South Africa, a national initiative for removal of invasive plants, emphasizes clearing to increase river flows (van Wilgen et al. 1998). Extensive invasions by *Acacia mearnsii*, Eucalyptus species, willows, and poplars have taken place along perennial rivers in the arid grasslands and savannas of the interior, where the native riparian species are mainly shrubs or small trees. If the difference between the annual ET from stands of these species and that from native species is as much as indicated by some studies (Dye and Jarmain 2004), or by data for willows and eucalypts from Doody and Benyon (2011) and Doody et al. (2011), removal could lead to an increase in river flows that would be substantial and very important for downstream water users and ecosystems. However, there may be other cases in which streamflows could be significantly reduced, such as invasions of species that are high water users along ephemeral streams (Doody et al. 2011; Hultine and Bush 2011).

The distinct species composition, structure, and dynamics of riparian environments generate a suite of ecosystem services very different from that of dryland environments. This makes them a key resource area, particularly in developing countries where they are less likely to have undergone extensive transformation by agriculture and other activities (Tockner and Stanford 2002; Kgathi et al. 2005; Brauman et al. 2007).

Recent work has also documented the extent to which large fauna can alter riparian processes. For example, in riparian ecosystems, large-scale earthworks are created mainly by the activities of fauna, particularly large mammals, which shape floodplains at a range of scales, from the microtopographical to that of river channels (Naiman and Rogers 1997; Moore 2006). Ecosystem modifications brought about by beavers, through the construction of dams, have been well studied; but much less is known about the ecological roles played by large mammals. In wetlands like the Okavango, large mammals (elephant, buffalo, hippopotamus) open up flow paths for water through reeds, changing water circulation patterns. Similarly, warthogs carve out feeding patches that form temporary pools during the wet season, creating habitats for many other species to complete their life cycles.

Riparian vegetation provides important feedbacks to the river system: it captures and stabilizes sediments, shapes river channels, and determines and regulates biotic processes (and, thus, water quality) (Tabacchi et al. 2000; Naiman et al. 2002). By these means, the vegetation creates its own habitat as well as that for animal species, and ensures its replacement through succession. Further, riparian vegetation serves as a buffer, shielding the aquatic ecosystems from the effects of land-use practices in adjacent environments—by filtering sediments, nutrients, and other pollutants (Naiman et al. 1999; Tabacchi et al. 2000; Brauman et al. 2007; Corenblit et al. 2009).

3.2.5 Regulation in Groundwater-Coupled Rangelands

In all rangelands the recharge, transport, and quality of groundwater depend on the nature of deep drainage and solute leaching. Where groundwater tables are shallow, reciprocal interactions between vegetation and groundwater are often observed (Le Maitre et al. 1999). This two-way exchange of water and solutes increases primary and secondary production, particularly under dry climatic conditions; at the same time; however, it renders water, soil, and vegetation resources more vulnerable to land management.

Groundwater-coupled rangelands—those in which shallow water tables are found, and the potential for a strong coupling between vegetation and groundwater exists—are increasingly recognized as important, and yet are poorly understood. These ecosystems have been categorized as "groundwater dependent"; yet the degree to which they are dependent varies greatly in time and space (Boulton and Hancock 2006; Eamus and Froend 2006). For this reason, we prefer the term "groundwater coupled" to describe the broad array of rangelands characterized by shallow water tables. We do know that vegetation has a major role in regulating groundwater resources in these systems, and significant strides have been made recently in understanding these ecohydrological interactions.

Many rangeland landscapes host, at their lowest topographic points, shallow groundwater zones that are sustained by local or distant recharge sources (Tóth 1999). In dry rangelands, where evapotranspiration recycles essentially all precipitation inputs back to the atmosphere, local recharge is negligible (Scanlon et al. 2006) and such shallow aquifers are rare. Regions where they do occur are characterized by sandy or rocky soils (such as sand dunes, fractured rock outcrops), highly seasonal and intense precipitation regimes, and zones of extensive lateral flow and intense runon. In such regions, at least some deep drainage into the saturated zone will eventually take place (Scanlon and Goldsmith 1997; Athavale et al. 1998; Seyfried et al. 2005; Small 2005; Gates et al. 2008). Recharge from more distant sources is particularly significant in arid regions located downstream of water-yielding mountains. For example, shallow water tables, wetlands, and lakes fed by mountain snowmelt are found at topographic lows within sand-dune rangelands such as the Great Sand Dunes of Colorado (Wurster et al. 2003), the Bahrain Jaram and Taklamakan deserts in China (Thomas et al. 2000; Chen et al. 2004; Gates et al. 2008), and the Monte desert in Argentina (Jobbágy et al. 2011).

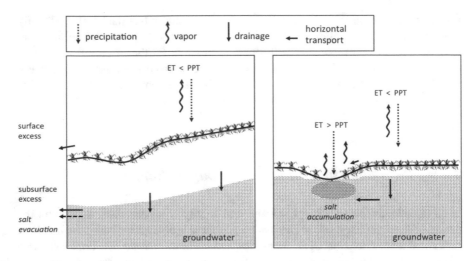

Fig. 3.5 Schematic of water balance for two landscape types in a subhumid climate. In sloped landscapes, vegetation regulates the rate of groundwater recharge. Groundwater gains that are eventually balanced by liquid discharge to streams (taking with it salts and dissolved nutrients). In flat landscapes, groundwater gains can be balanced only through higher evaporative discharge. When water leaves the landscape as vapor, its solute load is left behind. Transpiration results in salt build-up in the root zone, and direct evaporation results in salt build-up on the surface

In more humid rangelands, local groundwater recharge is more widespread and frequent, and shallow water tables are commonly found at topographic lows and along riparian corridors (Jobbágy and Jackson 2007). Finally, shallow water tables are a widespread feature in very flat, sedimentary rangelands (Fan et al. 2013) (Fig. 3.5); some conspicuous examples are the steppes of Western Siberia, the Northern Caspian plains in Asia, the grasslands of the Llanos and Pampas in South America, and the Miombo systems of sub-Saharan Africa (von der Heyden 2004).

3.2.5.1 Vegetation Dynamics Affect Groundwater Consumption

Groundwater consumption by rangeland plants is dictated by the interplay of water demand and accessibility. Most plants use groundwater in a dynamic and facultative manner, according to rainfall variability, preferring surface/shallow soil moisture when available (Engel et al. 2005). When soil moisture is low, the ability of plants to access groundwater depends, first, on the depth to the water table. Most rangeland plants access groundwater from the capillary fringe, where upflowing water and air coexist in the pore spaces of soils. Special adaptations include root aerenchyma tissue that enables species to survive and grow in saturated soils by providing air spaces to supply oxygen and facilitate groundwater consumption where water tables are very close to the surface (Visser et al. 2000). The optimum condition for plants is one in which the water table is deep enough to prevent waterlogging, but still high enough for easy access to groundwater (Jackson et al. 2009a). Groundwater use declines as water table levels drop, both in space—along topographic gradients

(Zencich et al. 2002; Gries et al. 2003; Nosetto et al. 2009)—and through time, e.g., as depth shifts seasonally (Stromberg et al. 1992; Naumburg et al. 2005; Cooper et al. 2006). Certain shrub and tree species can have maximum rooting depths of many meters (Schenk and Jackson 2002), and some observations suggest ground-water uptake from as deep as 20 m below the surface (Haase et al. 1996; Gries et al. 2003); but in dry ecosystems where water tables exceed depths of 10 m, it is rare to find significant groundwater use by plants (Zencich et al. 2002).

Other variables governing groundwater consumption by plants include water salinity and the fluid transport properties of soils and sediments. As the salinity of groundwater increases, the number of plant species able to tolerate the salinity declines, as does the rate at which the water is used. In groundwater-coupled range-lands, this relationship is commonly reflected by a series of drops in the diversity and productivity of vegetation along gradients of increasing salinity (Perelman et al. 2001). With regard to fluid transport properties, coarse-textured materials with high hydraulic conductivity favor groundwater recharge, whereas clay-dominated mate-rials limit it to negligible rates (Jobbágy and Jackson 2004b).

3.2.5.2 Land Use/Management Affects Groundwater Consumption

The way in which the vegetation and soils of groundwater-coupled rangelands are managed can have a strong influence on the exchange of water and solutes, and thereby the availability and quality of groundwater for human and livestock con-sumption, among other uses. Reciprocally, intense extraction of groundwater can significantly alter the structure and functioning of groundwater-coupled rangelands. Groundwater consumption often rises when deep-rooted woody species become abundant (Huxman et al. 2005) or when salt-tolerant species proliferate in areas where high salinity levels previously limited consumption of groundwater (Pataki et al. 2005). Conversely, where rangeland use contributes to a reduction in the den-sity of deep-rooted species the overall reduction in leaf area and transpiration decrease groundwater consumption (Meglioli et al. 2013).

This trade-off can lead to actions having different potential outcomes in different situa-tions. In the very dry, sand-dune landscapes of central Argentina, groundwater-coupled woodlands occupy less than 15 % of the land area, but represent the major source of forage for local herders. At the same time, these woodlands may consume up to 17% of the mountain-source recharge that sustains the aquifer—which is the only local source of water for humans and livestock (Jobbágy et al. 2011). If groundwater consumption by these woodlands were to be reduced, for example through clearing of the vegetation, the actual effect on groundwater availability would be very localized and minor, but the nega-tive effect on forage availability and the herding economy would be huge.

A contrasting example comes from groundwater-coupled rangelands along river banks in the southwestern USA; here, mesquite encroachment has resulted in a dou-bling of groundwater consumption—producing more biomass, but with little benefit to livestock production (Scott et al. 2006). Finally, there are situations in which groundwater consumption can be a desirable factor in hydrological regulation. In many rangelands in Australia, the removal of native vegetation for cultivation led to massive waterlogging and salinization of the soils (Turner and Ward 2002). The only means of reversing this process has been reforestation of large areas of the watershed (Barrett-Lennard 2002; Asseng et al. 2010)—the biomass gains and consequent water losses to lower the water table in this case both bringing benefits.

Rangeland use can affect not only the amount of available groundwater, but also its quality. When plants consume groundwater, they typically filter out salts at the root surface, which then accumulate in the absorption zone (Heuperman 1999), raising groundwater salinity (Jobbágy and Jackson 2007). Salinity levels tend to stabilize once the maximum tolerance of the consuming species is reached (Nosetto et al. 2008). It should be noted that where water tables are shallow enough to connect the capillary fringe with the surface, substantial amounts of groundwater can be lost through direct evaporation. If salinity is high, evaporation can seriously damage surface soils (Lavado and Taboada 1987). To reduce direct evaporation and restore transpiration, management methods such as halting grazing and creating means for retaining surface runoff appear to be effective (Alconada et al. 1993; Chaneton and Lavado 1996).

Groundwater availability and quality can also be compromised by rangeland uses involving animals, such as livestock. For example, continual livestock trampling has worn channels in groundwater-fed meadows. If the overall slope of the ground is somewhat steep, such channeling can rapidly lower the water table, leading to shifts in rangeland composition and productivity (Loheide and Booth 2011). The quality of groundwater is often affected as well, as has been documented in corrals and homestead areas in the groundwater-coupled woodlands of central Argentina. The combined effects of denudation from overgrazing and nutrient concentration from feces and urine have switched the net groundwater flux from discharge (losing water) to recharge (gaining water), at the same time placing soluble nitrogen contaminants into the groundwater (Meglioli et al. 2013).

Groundwater-coupled rangelands in many regions have been severely affected by direct human interventions—such as intensive pumping of groundwater—greatly drawing down the water table. Some of the most dramatic examples have been documented in the Owens Lake basin in California (Elmore et al. 2006; Pritchett and Manning 2012).

3.3   Regulating Services: Climate Regulation

The water cycle in rangelands is strongly influenced by vegetation dynamics, owing in part to the tight coupling between the water, energy, and biogeochemical cycles in these systems (Noy-Meir 1973; Austin et al. 2004; Wang et al. 2009b). In rangelands where water availability is typically low, the dominant factor controlling vegetation cover and interannual variability in vegetation productivity is mean annual precipitation. The effects of rainfall on vegetation productivity have been investigated in many parts of the world, such as the western USA (Nippert et al. 2006) and northern Africa (Le Houérou and Hoste 1977). For example, shrub encroachment has been shown to change the spatial patterns of water infiltration into soils (Daryanto et al. 2013), thus affecting local water balance. In the Mojave desert in the southwestern USA, paired lysimeter data showed that when vegetation productivity increased significantly following elevated winter precipitation, soil water storage was reduced by half, precluding drainage below the root zone (Scanlon et al. 2005a). Such vegetation-controlled soil water flow has been occurring for 10,000–15,000 years in this region (Scanlon et al. 2005a), as it most likely has in many other rangeland ecosystems across the globe. A contrasting example comes from southwestern Australia, where replacement of perennial vegetation with annual crops led to much higher groundwater recharge, which resulted in soil salinity problems (Turner and Ward 2002).

Vegetation dynamics not only influence local hydrological conditions, but they also affect local and regional climate. Recent studies have shown that invasive shrubs in rangelands modify surface energy fluxes, causing greater nighttime air temperatures near the soil surface—particularly during the winter—thus producing a positive feedback for further shrub encroachment (D'Odorico et al. 2013b). At the regional scale, the effect of vegetation changes on climate has been observed in the Sahel (West Africa); although rainfall variability in this region is mainly influenced by variations in the surface temperature of the oceans, it is also accompanied by variations in vegetation, as seen during the multi-decadal drying trend from the 1950s to the 1980s (Zeng et al. 1999; Hein and de Ridder 2006; Prince et al. 2007). Another modeling exercise showed, in addition, that vegetation dynamics in the late 1960s in the Sahel played a critical role in maintaining the drought through the following decades. The course of the drought has been marked by a forced shift from a self-sustaining wet climate equilibrium to a similarly self-sustaining, but dry climate equilibrium (Wang and Eltahir 2000). Other research has indicated the role vegetation plays in the dynamics of the West African monsoon (Zheng and Eltahir 1998; McAlpine et al. 2009).

3.4 Supporting Services: Water Cycling and Protection Against Erosion

Supporting services are those required for the production of other ecosystem services. Their effects on people are either indirect or manifest over a very long time. Examples of supporting services include soil formation, nutrient cycling, water cycling, and protection against erosion. Of these, water cycling and protection against erosion are most germane to ecohydrology.

3.4.1 Water Cycling: With a Focus on E vs. T

The cycling of water on rangelands is obviously driven by many factors, some of which have been discussed in the previous section. A fundamental factor is the process of evapotranspiration (ET), which on most rangelands accounts for more than

95 % of the water budget (Wilcox et al. 2003b). Evapotranspiration is the sum total of interception—water captured by vegetation or litter and subsequently evaporated, transpiration, and evaporation from the soil or surface of water bodies. Recently, ecohydrologists have recognized the importance of better understanding the dynamics of ET, and in particular have placed more emphasis on accurately partitioning ET into soil evaporation and transpiration (Newman et al. 2006). Soil evaporation, from an ecohydrological perspective, is not a productive use of water because it does not contribute to plant productivity and carbon sequestration, food, fiber, or fuel production (D'Odorico et al. 2013a). This insight indicates that the main focus of ecohydrology should be to develop methods for better partitioning of the green water resources (i.e., decrease soil evaporation and increase transpiration) in semiarid and subhumid landscapes (Falkenmark and Rockstrom 2004). The same insight is motivating ecohydrologists to better understand and quantify ET.

Evapotranspiration can be partitioned into three components: (1) water that is intercepted by foliage and then evaporates back to the air; (2) water that is intercepted by litter on the soil surface, infiltrates into that litter and into the soil, and then evaporates; and (3) water that infiltrates into soil, is absorbed by plants, and later transpired back to the atmosphere. An additional process, previously not taken into account, is the potential for plants to absorb foliar-intercepted rainfall (Breshears et al. 2008); this process can be important during protracted periods of water stress, allowing plants to take advantage of rainfall events that are just large enough to be intercepted, but not large enough to infiltrate soil (Loik et al. 2004; Owens et al. 2006). This process has not been fully investigated, and the degree to which it may affect multiple species of plants is not yet known.

The rate at which soil evaporation takes place depends on several variables, including soil texture, soil temperature, and near-surface wind; these in turn are affected by basic properties of rangeland structure, such as the amount and type of woody-plant canopy cover. Recently, considerable work has focused on ways to identify the linkages between vegetation characteristics, soil evaporation, and microclimates for a diverse set of rangeland vegetation types—including mesquite, piñon-juniper, ponderosa pine, eucalypt, and saguaro cactus (Breshears and Ludwig 2010; Royer et al. 2010; Villegas et al. 2010a, b; Zou et al. 2010; Royer et al. 2012). Other recent work has focused on understanding how changes in woody plant cover may affect the ratio of transpiration to ET (Wang et al. 2010b, 2012a).

3.4.2   Protection of Soils Against Erosion and Degradation

3.4.2.1   Understanding the Importance of Vegetation Patch Structure

Another important supporting service of healthy rangelands is that of soil protection from erosion—in other words, on healthy rangelands, soils are not eroding. The obvious reason for this is that vegetation cover is adequate. But what is adequate cover? Many rangelands, particularly in drier climates, have significant areas of bare ground and yet are not eroding. According to Ludwig et al. (1997), Vegetation patch structure is the key: vegetation patches must be numerous enough and large enough to be able to recapture soil eroded from bare areas. In fact, the transfer of water, soil, and nutrients from bare areas (sources) to vegetated areas (sinks) is a fundamental process within drylands that may be disrupted if the vegetation patch structure is disturbed. "Resource-conserving" drylands are organized such that run-off is quickly captured by, and concentrated in, vegetation patches—minimizing the loss of resources from the landscape. Resource concentration of resources increases the efficiency of their use, which translates to higher net primary productivity and the maintenance of rangeland functionality (Stavi et al. 2009).

If a disturbance, such as overgrazing, reduces the density and size of vegetation patches, the system will become "leaky" or "nonconserving"—less efficient at trapping runoff, leading to a loss of valuable water and nutrient resources (Ludwig and Tongway 2000). A positive-feedback loop may then reinforce the degradation process: the higher runoff rates will mean less water available to plants and higher erosion rates (Davenport et al. 1998; D'Odorico et al. 2013a). This degradation cycle may proceed to the point that overland-flow runoff increases in both amount and energy, erosion increases, and plant density and production declines, and the microclimate becomes more extreme (Fig. 3.6). Recognition of these processes is important not only for understanding how rangelands retain function, but also for how to devise more effective remediation strategies (Tongway and Ludwig 1997).

3.4.2.2 Wind and Water Erosion

Erosion research on rangelands has traditionally focused on water erosion and associated fluvial processes. One key advance in recent decades is recognition of the importance of wind-driven transport (aeolian) and its linkage with water erosion (Breshears et al. 2003; Belnap et al. 2011). Aeolian processes are much better understood now, thanks to improvements in measurement methods (Zobeck et al. 2003)—including relative humidity near the soil surface (Ravi et al. 2007a), the effects of vegetation patterns, and predictions of how vegetation structure influences horizontal sediment transport (Okin and Gillette 2001). Like water erosion, aeolian sediment transport is strongly influenced by the structure and arrangement of vegetation patches (Field et al. 2012). But when a grass patch is denuded (as can be caused by overgrazing) and the soil is exposed to wind action, there is a "double-whammy" effect: not only is the potential for recapturing the sediment lost, but also the wind causes the bare patch to generate additional sediment (Field et al. 2012). In the absence of disturbance, shrublands may inherently generate more wind-derived sediment than grasslands, as they have greater surface roughness as well as less intercanopy ground cover (Breshears et al. 2009). Aeolian erosional processes may also be interrelated with fire dynamics (Ravi et al. 2007b, 2009; Field et al. 2011a).

Under future climatic conditions, in regions where precipitation may become more intense while simultaneously drought frequency and intensity increase,

Fig. 3.6 Feedback loops
in the degradation process.
Positive feedbacks are
depicted between loss of
vegetation cover and (top
loop) decreased
precipitation and changes
in atmospheric conditions;
and (bottom loop) soil
erosion and loss of fertility.
Source: Figure 4 in
D'Odorico et al. (2013)

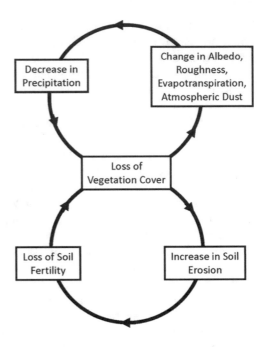

fluvial and aeolian erosion processes will be affected in different ways. A simple
but important point is that fluvial erosion occurs in rangelands only during the
infrequent precipitation events that have sufficient magnitude and intensity to
generate runoff, whereas small wind gusts occurring frequently can result in
regular, ongoing aeolian sediment transport even during less windy periods of
the year. Consequently, aeolian transport is a relatively constant and ongoing
process (Field et al. 2011b) and may even be highly interactive with fluvial processes (Belnap et al. 2011).

3.5 Provisioning Services: Water Supply

Provisioning services are considered as those goods or products—food, fiber, and
water—that are directly produced from ecosystems. Water supply, including its
magnitude, timing, and quality, is a fundamental service provided by rangelands,
even those having relatively dry climates. Given that most rangelands are in semiarid
settings, the amount of "blue water" is generally quite low—often less than 5 %
(Wilcox et al. 2003b). Nevertheless, given the extent of rangelands, even a relatively
small fraction of blue water can translate to a considerable amount of freshwater,
which is of particularly high value in regions where the quantity is very limited. In
rangelands having more humid climates, cold and snowy climates, or rocky or very
sandy soils, "blue water" outputs can be much higher (Wilcox et al. 2006).

Water supply as a provisioning service is essentially a product of the array of regulating services discussed above. The amount of "blue water" supplied by a given landscape, i.e., water flow to streams and aquifers, is fundamentally a function of climate, vegetation, soils, and geology. Of these factors, vegetation and—to a lesser extent—soils are the most affected by rangeland management. The concept of managing vegetation for the purpose of augmenting water supply has a long, complicated, and rich history. In fact, one could argue that it is a fundamental tenet of the science and art of watershed management (Wilcox 2010). The last decade in particular has seen a considerable refinement of our understanding of the linkage between vegetation and water supply on rangelands—especially concerning the effects of removing woody plants and invasive riparian species (Huxman et al. 2005; Edwards and Roberts 2006; Shafroth et al. 2010a; Doody et al. 2011; Hultine and Bush 2011; van Wilgen et al. 2012). With respect to the USA, the issue has been reviewed in detail in Archer et al. (2011). In general, large-scale woody plant removal has not resulted in measurable increases in streamflows or groundwater recharge, although increases would have been anticipated given (1) the long experience of similar manipulations (various levels of clear-cutting) carried out in forest watersheds (Bosch and Hewlett 1982) and (2) experience with the reductions in water yield brought about by the reverse type of intervention, i.e., the establishment of tree plantations in areas that were originally treeless (Farley et al. 2005; Jobbagy et al. 2013). The only areas in which there may be a true potential for enhancing water supply through woody plant removal appears to be those having annual precipitation above 500 mm (Zhang et al. 2001) along with at least one of the following conditions: (1) predominantly winter precipitation or significant snow accumulation; (2) permeable (sandy) and deep soils; and (3) karst geology (Huxman et al. 2005).

Surprisingly, the conventional wisdom has even been overturned in the case of riparian areas invaded by alien shrubs. Until recently it was widely accepted that removal or control of invasive riparian shrubs such as Russian olive (*Elaeagnus angustifolia*) and saltcedar (*Tamarix ramosissima*) would result in substantial water savings. A recent comprehensive review on the subject, however, concluded that there is in fact little evidence for large-scale water supply augmentation through these interventions (Shafroth et al. 2010a). The primary finding was that the invasive species do not appear to use more water than the native vegetation they displace (Doody et al. 2011; Hultine and Bush 2011).

Finally, studies of groundwater recharge in the sandy deserts of Central Argentina have yielded some paradoxical results. Certain regions that are highly degraded from constant wind erosion and dune formation, with severe loss of both forage and sediment, have nevertheless seen an improvement in freshwater supply as recharge gives rise to high-quality groundwater lenses (Jobbágy et al. 2011). Except for this peculiar example of vegetation denudation proceeding in concert with gains in groundwater, the region is characterized by low-quality groundwater. In undisturbed areas it exhibits high total salt and/or arsenic content, while in disturbed areas with high animal concentrations it is less salty but polluted with nitrogen (Aranibar et al. 2011; Meglioli et al. 2014).

3.6   Observational and Conceptual Advances

As noted in previous sections, the availability and distribution of water in the land-scape are of paramount importance for rangelands. Over the last few decades, a number of exciting developments have taken shape, both observational and conceptual. The former category includes *in situ* and remote-sensing monitoring tools—such as field-deployable, laser-based spectroscopy instruments that determine the ratios of hydrogen and oxygen isotopes (Lee et al. 2005; Wang et al. 2009a); portable 3D LIDAR systems for plant canopy analysis; electromagnetic imaging (EMI) devices for *in situ* soil water moisture monitoring; and distributed temperature-sensing (DTS) and remote-sensing tools, including drones and radio-controlled helicopters with lightweight digital cameras, that gather data for estimating key hydrological variables (Alsdorf et al. 2000). These and other recent developments are revolutionizing data gathering, in terms of both the scale and the precision of information used to inform ecohydrological measurement and investigation. It would be impractical to try to cover all the advances here; we have therefore selected a few that are closely related to topics already discussed in this chapter: the observational technologies of remote sensing and stable isotopes, and the conceptual advances in understanding nonlinear ecosystem behavior, scale and spatial variability, and hydrological connectivity. Discussions of other geophysical advances (e.g., EMI) can be found in the following sources (e.g., Robinson et al. 2008; Zreda et al. 2012).

3.6.1   Observational Advances

3.6.1.1   Remote Sensing for Investigating Components of the Water
Budget

Remote-sensing technology has a long history in rangeland management (e.g., Prince and Tucker 1986). One of its key advantages is that it enables extrapolation not just in space, but also temporally, offering insight into change of vegetation pattern and development through time. Rapid developments in remote-sensing-based hydrological monitoring are providing unprecedented temporal and spatial coverage in estimates of hydrological variables such as rainfall, soil moisture, ET (Kustas et al. 1994; Garcia et al. 2008), surface water level (Alsdorf et al. 2000), and ground-water storage (Yeh et al. 2006).

In rangelands, the irregular spatial and temporal distribution of rainfall imposes key constraints on ecosystem function and development. Remote measurement of precipitation has an extensive history, with numerous hydrological investigations being informed by the two-decade-long Tropical Rainfall Measuring Mission (TRMM) satellite system (Kummerow et al. 2000) and related sensors. Over the coming years, the next generation of satellite rainfall-measuring systems, referred to as the Global Precipitation Measurement (GPM) mission, will provide a much-needed update to the space-based rainfall monitoring capacity. The GPM Core Observatory is in the final stages of testing at the NASA Goddard Space Flight

Center; launch is scheduled for early 2014. The resolution of spatial and temporal rainfall data derived with CPM will exceed that possible with previous designs and it will enable a much greater range of ecohydrological investigations in rangelands.

Like other water-limited systems, rangelands are characterized by a strong coupling between the dynamics of soil moisture and vegetation productivity. Soil moisture can be estimated remotely, through either active or passive microwave-based systems—each of which involves a compromise between spatial and temporal resolution. Although passive microwave sensing can be used for routine, daily global-scale estimates of soil moisture (Njoku and Entekhabi 1996), which makes it an ideal technique for large-scale studies, it also has a clear limitation: the spatial resolution of retrievals is quite coarse (approximately 25 km) (McCabe et al. 2005). Active microwave sensing provides a higher spatial resolution (up to a few kilometers), but the repeat time is generally on the order of a few days. It is possible that improved data sets for large-scale ecohydrological investigations can be obtained by merging the best features from multiple systems and sensors (e.g., Liu et al. 2011).

3.6.1.2 In Situ Methods for Measuring Components of the Water Budget

Partitioning of Evapotranspiration

Evapotranspiration is a major component of the water budget and accounts for up to 95 % of the total water input (e.g., precipitation) in rangelands (Huxman et al. 2005). It has two distinct constituents (E and T), which are controlled by different mechanisms. Partitioning of ET is important not only for better understanding the water budget but also for predicting the biogeochemical fluxes driven by hydrological variations (Wang et al. 2010a). Efficient use of the limited water resources in rangelands requires maximizing the productive water loss (T) and minimizing the unproductive water loss (E) (Wang and D'Odorico 2008). Separating E from T, however, has always been a difficult task—especially from the observational point of view at larger scales.

A useful tool for separating E from T is stable isotopes of water, because E and T carry distinct isotopic signatures. Traditionally, the stable isotopic compositions of water samples are measured by isotope ratio mass spectrometry (IRMS), while the vapor-phase measurements are based on cryogenic water vapor collection coupled with IRMS. Such methods are labor intensive and time consuming. Over the past decade, a revolutionary change has taken place in water isotope measurement: the appearance of spectroscopy-based instruments capable of continuously measuring water vapor isotopic compositions (Fig. 3.7) (Lee et al. 2005; Wen et al. 2008; Wang et al. 2009a; Griffis et al. 2010).

Monitoring of Soil Moisture

Perhaps the most important recent innovation for measuring soil moisture *in situ* is the COSMOS monitoring system (Zreda et al. 2012). Based on both the release of fast and slow neutrons from interactions between water in the soil column and a

Fig. 3.7 Examples of recent advances in hydrological monitoring technology. (A) Eddy covariance system and scintillometer for ecosystem-scale measurements of sensible heat flux (evapotranspiration); (B) COSMOS system for monitoring ecosystem-scale soil moisture; (C) spectroscopy-based instrument for measuring the isotopic composition of water vapor in situ, which can be used in applications such as partitioning of evapotranspiration

regular flux of cosmic rays from space, the COSMOS system provides, for the first time, a reliable estimate of the soil wetness in a system. In addition, because the hydrogen in the top layer will have more sensitivity to the neutron counts, COSMOS, combined with modeling to separate the various hydrogen pools in the average measurement, has the potential to discriminate between moisture in the topsoil and that in the subsoil. The COSMOS installations are revolutionary in terms of bridging the spatial divide that often exists between remote-sensing and *in situ* measurement approaches. If a network of these systems can be distributed globally, our ability to monitor ecosystem change and development will be markedly improved.

3.7 Conceptual Advances

The last quarter century has seen considerable advances in our conceptual understanding of ecohydrological processes and interactions, particularly in regard to (1) spatial variability and scale, (2) ecosystem thresholds and feedbacks, and (3) hydrological connectivity of landscapes.

3.7.1 Spatial Variability and Scale

Understanding spatial dynamics and scale relationships has been a formidable challenge in both ecology and hydrology and, by extension, ecohydrology (Wood et al. 1990; Sivapalan and Kalma 1995; Sposito 1998; Grayson and Bloschl 2000; Western et al. 2001). Nevertheless, important advances have been made (Newman et al. 2006; Asbjornsen et al. 2011). For example, comparative studies across spatial scales have revealed the nonlinear nature of runoff and erosion with changing scales and how disturbance alters these relationships (Fig. 3.3) (Bergkamp 1998a; Puigdefabregas et al. 1999; Wilcox et al. 2003a; Favreau et al. 2009; Moreno de las Heras et al. 2010). Nonlinear responses in runoff and erosion are the result of redistribution across the landscape as well as alterations in runoff generation mechanisms with changing scale (Seyfried and Wilcox 1995).

Similarly significant strides have been made in quantifying the spatial variability of infiltration at the hillslope scale (Berndtsson and Larson 1987; Seyfried 1991; Pierson et al. 1994, 2001; Bhark and Small 2003; Daryanto et al. 2013). Infiltration capacities are generally higher under shrub canopies than in intercanopy areas, and these differences markedly influence patterns of soil moisture (Breshears and Barnes 1994, 1999). In addition, runon from intercanopy patches often contributes additional water to the shrub patches (Ludwig et al. 2005). Correspondingly, a number of conceptual advances have enhanced our understanding of the spatial variability of vegetation patterns on rangelands and how these are regulated by rainfall and runoff (HilleRisLambers et al. 2001; Rietkerk et al. 2002, 2004; Thompson et al. 2011).

Faunal activities also play an important role in structuring dryland landscapes as well. One feature of many arid landscapes is the formation of mounds, generally regularly dispersed, that range in diameter from a few meters to tens of meters. Known as *mima mounds* in the western USA, they are called *heuweltjies* in South Africa, where they cover from 14 to 25 % of the landscape (Lovegrove and Siegfried 1986, 1989; Whitford and Kay 1999). Their regular distribution is probably the result of competition among fauna for resources (Lovegrove and Siegfried 1986; Laurie 2002). Most authors agree that these enigmatic features are initiated by animals, whether mammals or invertebrates. One theory regarding the heuweltjies is that they developed over buried termite nests (Milton and Dean 1990; Moore and Picker 1991), but a recent paper argues that they are relicts of shrub-clump-controlled erosion processes (Cramer et al. 2012). Whatever their origin, the accumulation of transported organic matter, softer soil, and food remains they contain increases their fertility (Midgley and Musil 1990) and infiltration rates (Dean 1992), supports a distinctive suite of plant species (Knight et al. 1989), and attracts faunal activity—digging by termite-eating mammals, burrowing by rodents and/or nesting ostrich (Lovegrove and Siegfried 1986, 1989; Milton and Dean 1990), and foraging by game and domestic livestock (Armstrong and Siegfried 1990; Kunz et al. 2012). The movement of water across and between the vegetation mosaic and the heuweltjies has not been studied to determine whether these mounds contribute to groundwater recharge; but their higher infiltration rates (Dean 1992) suggest that their ecohydrological function may be analogous to that of the vegetation patches; that is, they may capture and filter runoff and act as foci for deep infiltration and recharge of groundwater.

The origins of the mima mounds in North America are no less controversial, but in this case small mammals (gophers) appear to be the primary drivers for the accumulation of materials (Whitford and Kay 1999; Jackson et al. 2003; Horwath and Johnson 2006; Johnson and Horwath-Burnham 2012). Whether or not that proves to be the sole explanation, these features also accumulate materials and alter the ecohydrology of the landscape. These important soil modifications justify the need for further research into the ecohydrological consequences of soil (Westbrook et al. 2013).

As noted by Vivoni (2012), our understanding of the role of scale and spatial variability in ecohydrological processes on rangelands will certainly increase in the future as remote-sensing and computational capabilities continue to progress.

3.7.2   Ecological Threshold and Feedback Mechanisms

Ecological thresholds and feedback loops are intimately related (Runyan et al. 2012; D'Odorico et al. 2013a). Threshold behavior occurs when a relatively small change in external drivers causes a disproportionally large response. A classic example of an ecological threshold is the transition between two stable states—such as the transition from a grassland or savanna to woodland or highly eroded state (D'Odorico et al. 2013a). The shift or change in state is induced and maintained by positive feedbacks that destabilize the system (Chapter 6, this volume). Examples of positive feedbacks are those between vegetation cover and (1) erosion, (2) soil moisture, and (3) climate (Runyan et al. 2012; D'Odorico et al. 2013a). The desertification feedback loop presented in D'Odorico et al. (2013a) (Fig. 3.6) illustrates these: A decrease in vegetation cover triggers the loss of water, nutrients, and soil that may as changes in albedo and evapotranspiration. All of these changes in turn create an environment that is less conducive to vegetation growth. In the last decade in particular, a considerable amount of work has been done that helps us better understand feedback loops and their important role in ecohydrological interactions (D'Odorico et al. 2007, 2012, 2013a, b; Stavi et al. 2009; Runyan et al. 2012; Turnbull et al. 2012).

3.7.3   Hydrological Connectivity

Hydrological connectivity refers to the water-mediated transfer of matter, energy, and organisms within or between elements of the hydrologic cycle (Pringle 2001). We now recognize that hydrological connectivity is essential for ecological integrity—and, more important, that activities by humans that disrupt this connectivity (dams, interbasin water transfers, etc.) can have dramatic negative consequences (Pringle 2003). "Connectivity" can be more broadly understood as the transfer of energy, matter, and organisms by not only water but also other vectors—such as wind and animals (Peters et al. 2006; Okin et al. 2009). One of the major benefits of studying connectivity in physical processes is that it identifies cross-scale interactions. For example, how do various different stomata in individual grass leaves, when under stress (e.g., from grazing or drought), function to modify water fluxes at the landscape scale? Answering such questions, on the basis of information from smaller scales, will significantly improve our ability to make predictions at larger scales (Peters et al. 2004). Hydrological connectivity has proved useful in explaining ecohydrological patterns on at the landscape scale as previously indicated (Wainwright et al. 2011). However, quantifying connectivity among different scales is still a major challenge, owing largely to a lack of a conceptual framework and modeling approaches applicable at multiple scales (Miller et al. 2012). Analogical models, which simulate the behaviors of complex physical systems using laws and theorems known to control components of those systems, may be able to fill some of these gaps. Recently, Wang et al. (2012b) developed a conceptual framework that uses electrical circuit analogies and Thévenin's theorem to upscale ecohydrological and biogeochemical processes from point scales to watershed scales. This conceptual work, by providing a means of representing concomitant processes at both small and large spatial scales, may prove useful for multi-scale rangeland management efforts.

A number of important conceptual advances have improved our understanding of hydrological connectivity and flows—longitudinal, lateral, and vertical—within river systems as well as between river systems and landscapes, and the importance of this connectivity for river ecosystem structure, functioning, and maintenance of ecosystem services (Naiman et al. 1999; Ward et al. 2001; Wiens 2002; Caylor et al. 2004; Boulton and Hancock 2006). Combined with hydrogeomorphology, connectivity processes play a vital role in the structuring of river systems and the ecosystem services they provide (Thorp et al. 2006, 2010) (Fig. 3.8). The implication, for those involved in land management and in water resource management—two traditionally separate policy and legislative domains—is important: the two are actually inseparable (Postel and Thompson 2005). In fact, rivers are complex social–ecological systems, and if we are to ensure continued delivery of the numerous essential ecosystem services they provide, including their traditional use as water conduits, we must advance our knowledge of not only the scientific but also the social and economic aspects of managing them (Chapter 8, this volume).

3.8 Future Perspectives

The past quarter century has seen impressive advances in our understanding of ecohydrological processes on rangelands, and new research is providing a much clearer picture of water dynamics (amounts and timing of both green and blue water and how these fluxes are affected by biota). These advances are attributable not only to the sheer number of new studies but also to the development of new observational methodologies, such as remote sensing and the use of stable isotopes. We anticipate that these advances will continue.

In addition, new conceptual and theoretical approaches, coupled with increases in computational power, have significantly improved our ability to predict and model ecohydrological processes. These approaches have and will continue to prove particularly useful for elucidating (1) spatial variability and scale, (2) ecosystem thresholds and feedbacks, and (3) hydrological connectivity of landscapes. We expect that the near future will bring further developments in all these areas, paving the way for more new and exciting insights into the ecohydrology of rangelands.

3.9 Summary

Our discussion of recent advances in the ecohydrology of rangelands has been organized around the concept of ecosystem services, especially those related to water. The fate of water in rangeland environments and, by extension, that of the flora and fauna that depend on this water are determined by conditions at three critical junctures: (1) The soil surface—will water infiltrate or run off? (2) The vadose zone—will water remain in the root zone or move beyond it? (3) The root zone—will water be transpired or evaporate?

Rangeland ecosystem services are categorized as regulating, supporting, and provisioning. Water-regulating services include those that affect the amount, timing, and quality of blue water flows. These are to a large extent determined at the first critical juncture of the water cycle—on the soil surface, where water either infiltrates or becomes overland flow, depending on the infiltrability of the soil. Soil infiltrability in turn depends on myriad factors, including vegetation, grazing intensity, brush management, fire patterns, condition of biological soil crusts, and activity by fauna. At larger scales, water-regulating services are influenced by other factors, such as the nature and structure of riparian zones and the presence of shallow groundwater aquifers. Finally, an important ecohydrological interaction that occurs at large scales is that between the land surface and the atmosphere. Climate regulation may result from feedbacks between rangeland vegetation and rainfall patterns.

Supporting services are those required for the production of other ecosystem services. Examples include the process of ET, which supports water cycling, and the

Fig. 3.8 (Continued) Thoms and Parsons 2003); and (2) the ecological measures of food chain length (FCL), nutrient spiraling (NS), and species diversity (SpD), the first two scaled from long to short and the third from low to high. The light bar within each box is the expected median, with the shading estimating the range of conditions. The size of each arrow reflects the magnitude of lateral, longitudinal, and vertical connectivity. Source: Figure 1.1 and color plate 1 (revised) in Thorp et al. (2008)

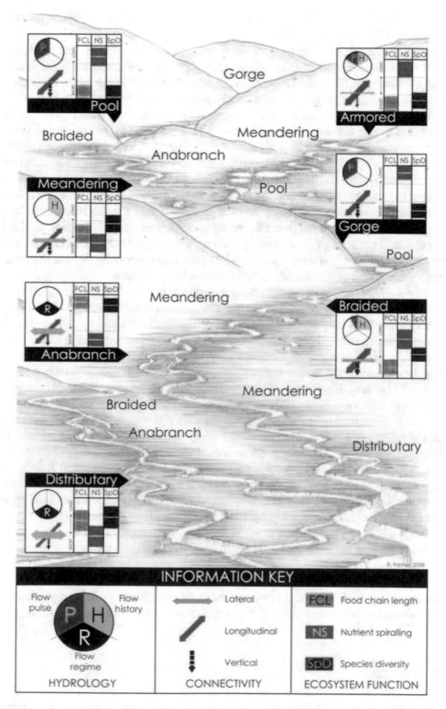

Fig. 3.8 A conceptual riverine landscape, depicting various functional process zones (FPZs) and their possible arrangement in the longitudinal dimension. Information contained in the boxes show the hydrological and ecological conditions predicted for that FPZ, including (1) the hydrological scale of greatest importance (scales being flow pulse, flow history, and flow regime, as defined by

processes by which soils are protected against erosion or degradation. The process of ET has become a subject of active inquiry in ecohydrological research—in particular, the partitioning of ET into soil evaporation and transpiration. From an ecohydrological perspective, soil evaporation is not a productive use of water because it does not contribute to plant productivity. An improved understanding of ET partitioning may lead to new management insights concerning methods for shifting more green water to productive uses. The processes that act to protect soils from erosion and/or degradation are also important ecohydrological support services. We now recognize that vegetation patch structure has a very large influence on soil erosion. Fundamentally, vegetation patches must be numerous enough and large enough to effectively capture water and sediment coming off of the bare patches. If such a patch structure is lost, rangelands begin eroding at rates that render the ecohydrological balance of the land unsustainable. Another factor affecting soil erosion, and which has been the focus of much recent research, is wind—and how it is related to water erosion. New measurement methods are yielding fresh insights into aeolian processes.

Provisioning services are those goods or products that are directly produced from ecosystems, such as water, food, and fiber. With respect to ecohydrology, the production of water from rangelands and how that production is affected by different management strategies are issues of paramount importance—but concerning which there has also been considerable misunderstanding. Work over the last several decades has largely overturned the notion that water supply can be substantially increased by removal of shrubs. Evidence of a true potential for enhancing water supply through woody plant removal has so far been found only in upland regions, and appears to be limited to those having annual precipitation above 500 mm, along with at least one of the following conditions: (1) predominantly winter precipitation or significant snow accumulation and (2) deep and permeable (sandy) soils (Huxman et al. 2005). But even where these conditions are met, in many cases the additional amount of water gained through manipulation of vegetation may be marginal. In riparian areas, surprisingly, removal of invasive, non-native woody plants appears to hold little potential for increasing water supply. Here, the primary factor appears to be that non-native plants use no more water than the native vegetation they displace.

We hope that by making an explicit linkage between ecohydrological processes and the ecosystem services concept, we have made it easier to grasp the multifaceted and complex nature of these processes on rangelands. Clearly there is a close coupling between biota (both fauna and flora) and water on rangelands—which is why water-related ecosystem services are so strongly dependent on land management strategies.

References

Alconada, M., O.E. Ansin, R.S. Lavado, V.A. Deregibus, G. Rubio, and F.H.G. Boem. 1993. Effect of retention of run-off water and grazing on soil and on vegetation of a temperate humid grassland. *Agricultural Water Management* 23: 233–246.

Alsdorf, D.E., J.M. Melack, T. Dunne, L.A.K. Mertes, L.L. Hess, and L.C. Smith. 2000. Interferometric radar measurements of water level changes on the Amazon flood plain. *Nature* 404: 174–177.

Anderson, V.J., and K.C. Hodgkinson. 1997. Grass-mediated capture of resource flows and the maintenance of banded mulga in a semi-arid woodland. *Australian Journal of Botany* 45: 331–342.

Angers, D.A., and J. Caron. 1998. Plant-induced changes in soil structure: Processes and feedbacks. *Biogeochemistry* 42: 55–72.

Aranibar, J.N., P.E. Villagra, M.L. Gomez, E. Jobbágy, M. Quiroga, R.G. Wuilloud, R.P. Monasterio, and A. Guevara. 2011. Nitrate dynamics in the soil and unconfined aquifer in arid groundwater coupled ecosystems of the Monte desert, Argentina. *Journal of Geophysical Research, Biogeosciences* 116, G04015. doi:10.1029/2010JG001618.

Archer, S. 1994. Woody plant encroachment into southwestern grasslands and savannas: Rates, patterns and proximate causes. In *Ecological implications of livestock herbivory in the west*, ed. M. Vavra, W.A. Laycock, and R.D. Pieper, 13–68. Denver, CO: Society for Range Management.

Archer, S., K.W. Davies, T.E. Fulbright, K.C. McDaniel, B.P. Wilcox, and K.I. Predick. 2011. Brush management as a rangeland conservation tool: A critical evaluation. In *Conservation effects assessment project*, ed. D.D. Briske, 105–170. Washington, DC: USDA-National Soil Conservation Service.

Armstrong, A.J., and W.R. Siegfried. 1990. Selective use of heuweltjies earth mounds by sheep in the Karoo. *South African Journal of Ecology* 1: 77–80.

Asbjornsen, H., G.R. Goldsmith, M.S. Alvarado-Barrientos, K. Rebel, F.P. Van Osch, M. Rietkerk, J.Q. Chen, S. Gotsch, C. Tobon, D.R. Geissert, A. Gomez-Tagle, K. Vache, and T.E. Dawson. 2011. Ecohydrological advances and applications in plant-water relations research: A review. *Journal of Plant Ecology* 4: 3–22.

Asseng, S., A. Dray, P. Perez, and X. Su. 2010. Rainfall–human–spatial interactions in a salinity-prone agricultural region of the Western Australian wheat-belt. *Ecological Modelling* 221: 812–824.

Athavale, R.N., R. Rangarajan, and D. Muralidharan. 1998. Influx and efflux of moisture in a desert soil during a 1 year period. *Water Resources Research* 34: 2871–2877.

Austin, A.T., L. Yahdjian, J.M. Stark, J. Belnap, A. Porporato, U. Norton, D.A. Ravetta, and S.M. Schaeffer. 2004. Water pulses and biogeochemical cycles in arid and semiarid ecosystems. *Oecologia* 141: 221–235.

Baker, M.B. 1984. Changes in streamflow in an herbicide-treated Pinyon-juniper watershed in Arizona. *Water Resources Research* 20: 1639–1642.

Baker, W.L., and D.J. Shinneman. 2004. Fire and restoration of pinon-juniper woodlands in the western United States: A review. *Forest Ecology and Management* 189: 1–21.

Barrett-Lennard, E.G. 2002. Restoration of saline land through revegetation. *Agricultural Water Management* 53: 213–226.

Belnap, J. 2003. Comparative structure of physical and biological soil crusts. In *Biological soil crusts: Structure, function, and management*, 177–192. Berlin: Springer.

———. 2006a. The potential role of biological soil crusts in dryland hydrologic cycles. *Hydrological Processes* 20(15): 3159–3178.

———. 2006b. The potential roles of biological soil crusts in dryland hydrologic cycles. *Hydrological Processes* 20: 3159–3178.

Belnap, J., and O.L. Lange (eds.). 2001. *Biological soil crusts: Structure, function, and management*. Berlin: Springer.

Belnap, J., S.L. Phillips, D.L. Witwicki, and M.E. Miller. 2008. Visually assessing the level of development and soil surface stability of cyanobacterially dominated biological soil crusts. *Journal of Arid Environments* 72: 1257–1264.

Belnap, J., S.M. Munson, and J.P. Field. 2011. Aeolian and fluvial processes in dryland regions: The need for integrated studies. *Ecohydrology* 4: 615–622.

Belnap, J., B.P. Wilcox, M.W. Van Scoyoc, and S.L. Phillips. 2013. Successional stage of biological soil crusts: An accurate indicator of ecohydrological condition. *Ecohydrology* 6: 474–482.

Bergkamp, G. 1998. A hierarchical view of the interactions of runoff and infiltration with vegetation and microtopography in semiarid shrublands. *Catena* 33: 201–220.

———. 1998b. Hydrological influences on the resilience of Quercus Spp. dominated geoecosystems in Central Spain. *Geomorphology* 23: 101–126.

Berndtsson, R., and M. Larson. 1987. Spatial variability of infiltration in a semiarid environment. *Journal of Hydrology* 90: 117–133.

Beven, K., and P. Germann. 1982. Macropores and water flow in soils. *Water Resources Research* 18: 1311–1315.

Bhark, E.W., and E.E. Small. 2003. Association between plant canopies and the spatial patterns of infiltration in shrubland and grassland of the Chihuahuan Desert, New Mexico. *Ecosystems* 6: 185–196.

Blackburn, W.H., R.W. Knight, and M.K. Wood. 1982. *Impact of grazing on watersheds: A state of knowledge*. MP 1496, Texas Agricultural Experiment Station, College Station, TX.

Blasch, K.W., T.P.A. Ferre, J.P. Hoffmann, and J.B. Fleming. 2006. Relative contributions of transient and steady state infiltration during ephemeral streamflow. *Water Resources Research* 42(8), W08405. doi:10.1029/2005WR004049.

Bosch, J.H., and J.D. Hewlett. 1982. A review of catchment experiments to determine the effect of vegetation changes and water yield and evapotranspiration. *Journal of Hydrology* 55: 3–23.

Boulton, A.J., and P.J. Hancock. 2006. Rivers as groundwater-dependent ecosystems: A review of degrees of dependency, riverine processes and management implications. *Australian Journal of Botany* 54: 133–144.

Branson, F.A., G.F. Gifford, K.G. Rennard, and R.F. Hadley. 1981. *Rangeland hydrology*. Dubuque, IA: Kendall/Hunt.

Brauman, K.A., G.C. Daily, T.K. Duarte, and H.A. Mooney. 2007. The nature and value of ecosystem services: An overview highlighting hydrologic services. *Annual Review of Environment and Resources* 32: 67–98.

Breshears, D.D., and F.J. Barnes. 1999. Interrelationships between plant functional types and soil moisture heterogeneity for semiarid landscapes within the grassland/forest continuum: A unified conceptual model. *Landscape Ecology* 14: 465–478.

———. 1994. *Spatial partitioning of water use by herbaceous and woody plants in semiarid woodlands*. Report submitted to Ecology for publication LA-UR-94-1806, Los Alamos National Laboratory, Los Alamos.

Breshears, D.D., and J.A. Ludwig. 2010. Near-ground solar radiation along the grassland-forest continuum: Tall-tree canopy architecture imposes only muted trends and heterogeneity. *Austral Ecology* 35: 31–40.

Breshears, D.D., J.J. Whicker, M.P. Johansen, and J.E. Pinder. 2003. Wind and water erosion and transport in semi-arid shrubland, grassland and forest ecosystems: Quantifying dominance of horizontal wind-driven transport. *Earth Surface Processes and Landforms* 28: 1189–1209.

Breshears, D.D., N.G. McDowell, K.L. Goddard, K.E. Dayem, S.N. Martens, C.W. Meyer, and K.M. Brown. 2008. Foliar absorption of intercepted rainfall improves woody plant water status most during drought. *Ecology* 89: 41–47.

Breshears, D.D., J.J. Whicker, C.B. Zou, J.P. Field, and C.D. Allen. 2009. A conceptual framework for dryland aeolian sediment transport along the grassland-forest continuum: Effects of woody plant canopy cover and disturbance. *Geomorphology* 105: 28–38.

Bunn, S.E., M.C. Thoms, S.K. Hamilton, and S.J. Capon. 2006. Flow variability in dryland rivers: Boom, bust and the bits in between. *River Research and Applications* 22: 179–186.

Butler, D.R. 2007. *Zoogeomorphology: Animals as geomorphic agents*. Cambridge, UK: Cambridge University Press.

Cammeraat, L.H., and A.C. Imeson. 1999. The evolution and significance of soil-vegetation patterns following land abandonment and fire in Spain. *Catena* 37: 107–127.

Caylor, K.K., T.M. Scanlon, and I. Rodriguez-Iturbe. 2004. Feasible optimality of vegetation patterns in river basins. *Geophysical Research Letters* 31(13), L13502. doi:10.1029/2004GL020260.

Chaneton, E.J., and R.S. Lavado. 1996. Soil nutrients and salinity after long-term grazing exclusion in a Flooding Pampa grassland. *Journal of Range Management* 49: 182–187.

Chen, J.S., L. Li, J.Y. Wang, D.A. Barry, X.F. Sheng, W.Z. Gu, X. Zhao, and L. Chen. 2004. Groundwater maintains dune landscape. *Nature* 432: 459–460.

Coleman, D.C., E.P. Odum, and D.A. Crossley. 1992. Soil biology, soil ecology, and global change. *Biology and Fertility of Soils* 14: 104–111.

Cooper, D.J., J.S. Sanderson, D.I. Stannard, and D.P. Groeneveld. 2006. Effects of long-term water table drawdown on evapotranspiration and vegetation in an arid region phreatophyte community. *Journal of Hydrology* 325: 21–34.

Cramer, M.D., S.N. Lnnes, and J.J. Midgley. 2012. Hard evidence that heuweltjies earth mounds are relictual features produced by differential erosion. *Palaeogeography, Palaeoclimatology, Palaeoecology* 350: 189–197.

D'Odorico, P., and A. Porporato. 2006. *Dryland ecohydrology*. Netherlands: Springer.

D'Odorico, P., K. Caylor, G.S. Okin, and T.M. Scanlon. 2007. On soil moisture-vegetation feedbacks and their possible effects on the dynamics of dryland ecosystems. *Journal of Geophysical Research, Biogeosciences* 112: G04010.

D'Odorico, P., F. Laio, A. Porporato, L. Ridolfi, A. Rinaldo, and I.R. Iturbe. 2010. Ecohydrology of terrestrial ecosystems. *BioScience* 60: 898–907.

D'Odorico, P., G.S. Okin, and B.T. Bestelmeyer. 2012. A synthetic review of feedbacks and drivers of shrub encroachment in arid grasslands. *Ecohydrology* 5: 520–530.

D'Odorico, P., A. Bhattachan, K.F. Davis, S. Ravi, and C.W. Runyan. 2013a. Global desertification: Drivers and feedbacks. *Advances in Water Resources* 51: 326–344.

D'Odorico, P., Y.F. He, S. Collins, S.F.J. De Wekker, V. Engel, and J.D. Fuentes. 2013b. Vegetation-microclimate feedbacks in woodland-grassland ecotones. *Global Ecology and Biogeography* 22: 364–379.

Daryanto, S., D.J. Eldridge, and L. Wang. 2013. Spatial patterns of infiltration vary with disturbance in a shrub-encroached woodland. *Geomorphology* 194: 57–64.

Davenport, D.W., D.D. Breshears, B.P. Wilcox, and C.D. Allen. 1998. Viewpoint—sustainability of piñon-juniper ecosystems—a unifying perspective of soil erosion thresholds. *Journal of Range Management* 51: 231–240.

Dean, W.R.J. 1992. Effects of animal activity on the absorption rate of soils in the southern Karoo, South Africa. *Journal of the Grasslands Society of South Africa* 9(4): 178–180.

Dean, W.R.J., S.J. Milton, and F. Jeltsch. 1999. Large trees, fertile islands, and birds in arid savanna. *Journal of Arid Environments* 41: 61–78.

Dean, W.R.J., M.D. Anderson, S.J. Milton, and T.A. Anderson. 2002. Avian assemblages in native Acacia and alien Prosopis drainage line woodland in the Kalahari, South Africa. *Journal of Arid Environments* 51: 1–19.

Doerr, S.H., R.A. Shakesby, and R.P.D. Walsh. 2000. Soil water repellency: Its causes, characteristics and hydro-geomorphological significance. *Earth Science Reviews* 51: 33–65.

Doody, T., and R. Benyon. 2011. Quantifying water savings from willow removal in Australian streams. *Journal of Environmental Management* 92: 926–935.

Doody, T.M., P.L. Nagler, E.P. Glenn, G.W. Moore, K. Morino, K.R. Hultine, and R.G. Benyon. 2011. Potential for water salvage by removal of non-native woody vegetation from dryland river systems. *Hydrological Processes* 25: 4117–4131.

Dortignac, E.J., and L.D. Love. 1961. *Infiltration studies on Ponderosa pine ranges of Colorado*. Station paper 59, USDA Forest Service, Fort Collins.

Dunkerley, D.L., and K.J. Brown. 1999. Banded vegetation near Broken Hill, Australia: Significance of surface roughness and soil physical properties. *Catena* 37: 75–88.

Dye, P., and C. Jarmain. 2004. Water use by black wattle (Acacia mearnsii): Implications for the link between removal of invading trees and catchment streamflow response. *South African Journal of Science* 100: 40–44.

Dyksterhuis, E.J., and E.M. Schmutz. 1947. Natural mulches or litter of grasslands: With kinds and amounts on a southern prairie. *Ecology* 28(2): 163–179.

Dzikiti, S., K. Schachtschneider, V. Naiken, M. Gush, G. Moses, and D.C. Le Maitre. 2013. Water relations and the effects of clearing invasive Prosopis trees on groundwater in an arid environment in the Northern Cape, South Africa. *Journal of Arid Environments* 90: 103–113.

Eamus, D., and R. Froend. 2006. Groundwater-dependent ecosystems: The where, what and why of GDEs. *Australian Journal of Botany* 54: 91–96.

Edwards, M.B.P., and P.J.T. Roberts. 2006. Managing forests for water: The South African experience. *International Forestry Review* 8: 65–71.

Eldridge, D.J. 2003. Biological soil crusts and water relations in Australian deserts. In *Biological soil crusts: Structure, function, and management*, ed. J. Belnap and O.L. Lange, 315–326. Berlin: Springer.

Eldridge, D.J., S. Soliveres, M.A. Bowker, and J. Val. 2013. Grazing dampens the positive effects of shrub encroachment on ecosystem functions in a semi-arid woodland. *Journal of Applied Ecology* 50: 1028–1038.

Elmore, A.J., S.J. Manning, J.F. Mustard, and J.M. Craine. 2006. Decline in alkali meadow vegetation cover in California: The effects of groundwater extraction and drought. *Journal of Applied Ecology* 43: 770–779.

Engel, V., E.G. Jobbágy, M. Stieglitz, M. Williams, and R.B. Jackson. 2005. Hydrological consequences of eucalyptus afforestation in the Argentine Pampas. *Water Resources Research* 41: W10409.

Falkenmark, M., and J. Rockstrom. 2004. *Balancing water for humans and nature: The new approach in ecohydrology*. London: Earthscan Publications.

———. 2006. The new blue and green water paradigm: Breaking new ground for water resources planning and management. *Journal of Water Resources Planning and Management: ASCE* 132: 129–132.

Falkenmark, M., J. Rockstrom, and L. Karlberg. 2009. Present and future water requirements for feeding humanity. *Food Security* 1: 59–69.

Fan, Y., H. Li, and G. Miguez-Macho. 2013. Global patterns of groundwater table depth. *Science* 339: 940–943.

Farley, K.A., E.G. Jobbágy, and R.B. Jackson. 2005. Effects of afforestation on water yield: A global synthesis with implications for policy. *Global Change Biology* 11: 1565–1576.

Favreau, G., B. Cappelaere, S. Massuel, M. Leblanc, M. Boucher, N. Boulain, and C. Leduc. 2009. Land clearing, climate variability, and water resources increase in semiarid southwest Niger: A review. *Water Resources Research* 45: W00A16.

Field, J.P., D.D. Breshears, J.J. Whicker, and C.B. Zou. 2011. Interactive effects of grazing and burning on wind- and water-driven sediment fluxes: Rangeland management implications. *Ecological Applications* 21: 22–32.

———. 2011b. On the ratio of wind- to water-driven sediment transport: Conserving soil under global-change-type extreme events. *Journal of Soil and Water Conservation* 66: 51A–56A.

———. 2012. Sediment capture by vegetation patches: Implications for desertification and increased resource redistribution. *Journal of Geophysical Research: Biogeosciences* 117: G01033.

Garcia, M., C. Oyonarte, L. Villagarcia, S. Contreras, F. Domingo, and J. Puigdefabregas. 2008. Monitoring land degradation risk using ASTER data: The non-evaporative fraction as an indicator of ecosystem function. *Remote Sensing of Environment* 112: 3720–3736.

Gates, J.B., W.M. Edmunds, J. Ma, and B.R. Scanlon. 2008. Estimating groundwater recharge in a cold desert environment in northern China using chloride. *Hydrogeology Journal* 16: 893–910.

Gifford, G.F. 1978. Hydrologic impact of grazing on infiltration: A critical review. *Water Resources Research* 14: 305–313.

Gordon, L.J., W. Steffen, B.F. Jonsson, C. Folke, M. Falkenmark, and A. Johannessen. 2005. Human modification of global water vapor flows from the land surface. *Proceedings of the National Academy of Sciences of the United States of America* 102: 7612–7617.

Grayson, R., and G. Bloschl. 2000. *Spatial patterns in catchment hydrology*. New York: Cambridge University Press.

Gries, D., F. Zeng, A. Foetzki, S.K. Arndt, H. Bruelheide, F.M. Thomas, X. Zhang, and M. Runge. 2003. Growth and water relations of Tamarix ramosissima and Populus euphratica on Taklamakan desert dunes in relation to depth to a permanent water table. *Plant, Cell and Environment* 26: 725–736.

Griffis, T.J., S.D. Sargent, X. Lee, J.M. Baker, J. Greene, M. Erickson, X. Zhang, K. Billmark, N. Schultz, W. Xiao, and N. Hu. 2010. Determining the oxygen isotope composition of evapotranspiration using eddy covariance. *Boundary-Layer Meteorology* 137: 307–326.

Haase, P., F.I. Pugnaire, E.M. Fernández, J. Puigdefábregas, S.C. Clark, and L.D. Incoll. 1996. An investigation of rooting depth of the semiarid shrub Retama sphaerocarpa (L.) Boiss. by labelling of ground water with a chemical tracer. *Journal of Hydrology* 177: 23–31.

Hastings, A., J.E. Byers, J.A. Crooks, K. Cuddington, C.G. Jones, J.G. Lambrinos, T.S. Talley, and W.G. Wilson. 2007. Ecosystem engineering in space and time. *Ecology Letters* 10: 153–164.

Hein, L., and N. de Ridder. 2006. Desertification in the Sahel: A reinterpretation. *Global Change Biology* 12: 751–758.

Hester, J.W., T.L. Thurow, and C.A. Taylor. 1997. Hydrologic characteristics of vegetation types as affected by prescribed burning. *Journal of Range Management* 50: 199–204.

Heuperman, A. 1999. Hydraulic gradient reversal by trees in shallow water table areas and repercussions for the sustainability of tree-growing systems. *Agricultural Water Management* 39: 153–167.

Hibbert, A.R. 1983. Water yield improvement potential by vegetation management on western rangelands. *Water Resources Bulletin* 19: 375–381.

Hiernaux, P., C.L. Bielders, C. Valentin, A. Bationo, and S. Fernandez-Rivera. 1999. Effects of livestock grazing on physical and chemical properties of sandy soils in Sahelian rangelands. *Journal of Arid Environments* 41: 231–245.

HilleRisLambers, R., M. Rietkerk, F. van den Bosch, H.H.T. Prins, and H. de Kroon. 2001. Vegetation pattern formation in semi-arid grazing systems. *Ecology* 82: 50–61.

Hoff, H., M. Falkenmark, D. Gerten, L. Gordon, L. Karlberg, and J. Rockstrom. 2010. Greening the global water system. *Journal of Hydrology* 384: 177–186.

Horwath, J.L., and D.L. Johnson. 2006. Mima-type mounds in southwest Missouri: Expressions of point-centered and locally thickened biomantles. *Geomorphology* 77: 308–319.

Huang, Y., B.P. Wilcox, L. Stern, and H. Perotto-Baldivieso. 2006. Springs on rangelands: Runoff dynamics and influence of woody plant cover. *Hydrological Processes* 20: 3277–3288.

Hultine, K.R., and S.E. Bush. 2011. Ecohydrological consequences of non-native riparian vegetation in the southwestern United States: A review from an ecophysiological perspective. *Water Resources Research* 47: WO7542.

Huxman, T., B. Wilcox, D.D. Breshears, R.L. Scott, K.A. Snyder, E.E. Small, K. Hultine, J. Pittermann, and R. Jackson. 2005. Ecohydrological implications of woody plant encroachment. *Ecology* 86: 308–319.

Jackson, R.B., H.J. Schenk, E.G. Jobbágy, J. Canadell, G.D. Colello, R.E. Dickinson, C.B. Field, P. Friedlingstein, M. Heimann, K. Hibbard, D.W. Kicklighter, A. Kleidon, R.P. Neilson, W.J. Parton, O.E. Sala, and M.T. Sykes. 2000. Belowground consequences of vegetation change and their treatment in models. *Ecological Applications* 10: 470–483.

Jackson, E.C., S.N. Krogh, and W.G. Whitford. 2003. Desertification and biopedturbation in the northern Chihuahuan Desert. *Journal of Arid Environments* 53: 1–14.

———. 2009a. Ecohydrology bearings-invited commentary. Ecohydrology in a human-dominated landscape. *Ecohydrology* 2: 383–389.

———. 2009b. Ecohydrology in a human-dominated landscape. *Ecohydrology* 2: 383–389.

Jobbágy, E.G., and R.B. Jackson. 2004. Groundwater use and salinization with grassland afforestation. *Global Change Biology* 10: 1299–1312.

———. 2004b. Groundwater use and salinization with grassland afforestation. *Global Change Biology* 10: 1299–1312.

———. 2007. Groundwater and soil chemical changes under phreatophytic tree plantations. *Journal of Geophysical Research G: Biogeosciences* 112: G02013.

Jobbágy, E.G., M.D. Nosetto, P.E. Villagra, and R.B. Jackson. 2011. Water subsidies from mountains to deserts: Their role in sustaining groundwater-fed oases in a sandy landscape. *Ecological Applications* 21: 678–694.

Jobbagy, E.G., A.M. Acosta, and M.D. Nosetto. 2013. Rendimiento hídrico en cuencas primarias bajo pastizales y plantaciones de pino de las sierras de Córdoba (Argentina). *Ecología Austral* 23: 87–96.

Joffre, R., and S. Rambal. 1993. How tree cover influences the water balance of Mediterranean rangelands. *Ecology* 74: 570–582.

Johnson, D.L., and J.L. Horwath-Burnham. 2012. Introduction: Overview of concepts, definitions and principles of soil mound studies. *Geological Society of America Special Papers* 490: 1–19.

Jones, C.G. 2012. Ecosystem engineers and geomorphological signatures in landscapes. *Geomorphology* 157: 75–87.

Jones, C.G., J.L. Gutierrez, P.M. Groffman, and M. Shachak. 2006. Linking ecosystem engineers to soil processes: A framework using the Jenny State Factor Equation. *European Journal of Soil Biology* 42: S39–S53.

Jones, K.B., E.T. Slonecker, M.S. Nash, A.C. Neale, T.G. Wade, and S. Hamann. 2010. Riparian habitat changes across the continental United States (1972–2003) and potential implications for sustaining ecosystem services. *Landscape Ecology* 25: 1261–1275.

Kgathi, D.L., G. Mmopelwa, and K. Mosepele. 2005. Natural resources assessment in the Okavango Delta, Botswana: Case studies of some key resources. *Natural Resources Forum* 29: 70–81.

King, E.G., and K.K. Caylor. 2011. Ecohydrology in practice: Strengths, conveniences, and opportunities. *Ecohydrology* 4: 608–612.

Knight, R.S., A.G. Rebelo, and W.R. Siegfried. 1989. Plant assemblages on mima-like earth mounds in the Clanwilliam District. *South Africa South African Journal of Botany* 55: 465–472.

Kummerow, C., J. Simpson, O. Thiele, J. Barnes, A.T.C. Chang, E. Stocker, R.F. Adler, and A. Hou. 2000. The status of the Tropical Rainfall Measuring Mission (TRMM) after two years in orbit. *Journal of Applied Meteorology* 39: 1965–1982.

Kunz, N.S., M.T. Hoffman, and B. Weber. 2012. Effects of heuweltjies and utilization on vegetation patterns in the Succulent Karoo, South Africa. *Journal of Arid Environments* 87: 198–205.

Kustas, W., E. Perry, P. Doraiswamy, and M. Moran. 1994. Using satellite remote sensing to extrapolate evapotranspiration estimates in time and space over a semiarid rangeland basin. *Remote Sensing of Environment* 49: 275–286.

Laurie, H. 2002. Optimal transport in central place foraging, with an application to the overdispersion of heuweltjies. *South African Journal of Science* 98: 141–146.

Lavado, R.S., and M.A. Taboada. 1987. Soil salinization as an effect of grazing in a native grassland soil in the Flooding Pampa of Argentina. *Soil Use & Management* 3: 143–148.

Lavelle, P. 1997. Faunal activities and soil processes: Adaptive strategies that determine ecosystem function. In *Advances in ecological research*, eds. M. Begon, and A.H. Fitter, vol. 27, 93–132.

Le Houérou, H., and C. Hoste. 1977. Rangeland production and annual rainfall relations in the Mediterranean Basin and in the African Sahelo-Sudanian zone. *Journal of Range Management* 30: 181–189.

Le Maitre, D.C., D.F. Scott, and C. Colvin. 1999. A review of information on interactions between vegetation and groundwater. *Water SA* 25: 137–152.

Le Maitre, D.C., S.J. Milton, C. Jarmain, C.A. Colvin, I. Saayman, and J.H.J. Vlok. 2007. Linking ecosystem services and water resources: Landscape-scale hydrology of the Little Karoo. *Frontiers in Ecology and the Environment* 5: 261–270.

Lee, X., S. Sargent, R. Smith, and B. Tanner. 2005. In situ measurement of the water vapor $^{18}O/^{16}O$ isotope ratio for atmospheric and ecological applications. *Journal of Atmospheric and Oceanic Technology* 22: 555–565.

Leonard, J., E. Perrier, and J.L. Rajot. 2004. Biological macropores effect on runoff and infiltration: a combined experimental and modelling approach. *Agriculture, Ecosystems & Environment* 104: 277–285.

Liu, Y.Y., R.M. Parinussa, W.A. Dorigo, R.A.M. De Jeu, W. Wagner, A.I.J.M. Van Dijk, M.F. McCabe, and J.P. Evans. 2011. Developing an improved soil moisture dataset by blending passive and active microwave satellite-based retrievals. *Hydrology and Earth System Sciences* 15: 425–436.

Loheide, S.P., and E.G. Booth. 2011. Effects of changing channel morphology on vegetation, groundwater, and soil moisture regimes in groundwater-dependent ecosystems. *Geomorphology* 126: 364–376.

Loik, M.E., D.D. Breshears, W.K. Lauenroth, and J. Belnap. 2004. A multi-scale perspective of water pulses in dryland ecosystems: Climatology and ecohydrology of the western USA. *Oecologia* 141: 269–281.

Lovegrove, B.G., and W.R. Siegfried. 1986. Distribution and formation of Mimalike earth mounds in the western Cape province of South Africa. *South African Journal of Science* 82: 432–436.

———. 1989. Spacing and origin of mima-like mounds in the Cape province of South Africa. *South African Journal of Science* 85: 108–112.

Ludwig, J.A., and D.J. Tongway. 2000. Viewing rangelands as landscape systems. In *Rangeland desertification*, ed. O. Arnalds and S. Archer, 39–52. Dordrecht, The Netherlands: Kluwer Academic Publishers.

Ludwig, J.A., D.J. Tongway, D. Freudenberger, J. Noble, and K. Hodgkinson. 1997. *Landscape ecology function and management: Principles from Australia's rangelands.* Collingwood, Australia: CSIRO Publications.

Ludwig, J.A., R.W. Eager, R.J. Williams, and L.M. Lowe. 1999. Declines in vegetation patches, plant diversity, and grasshopper diversity near cattle watering-points in Victoria River District, northern Australia. *Rangeland Journal* 21: 135–149.

Ludwig, J.A., B.P. Wilcox, D.D. Breshears, D.J. Tongway, and A.C. Imeson. 2005. Vegetation patches and runoff-erosion as interacting ecohydrological processes in semiarid landscapes. *Ecology* 86: 288–297.

Lyford, F., and H.K. Qashu. 1969. Infiltration rates as affected by desert vegetation. *Water Resources Research* 5: 1373–1377.

Martinez-Meza, E., and W.G. Whitford. 1996. Stemflow, throughfall and channelization of stemflow by roots in three Chihuahuan Desert shrubs. *Journal of Arid Environments* 32: 271–287.

McAlpine, C.A., J. Syktus, J.G. Ryan, R.C. Deo, G.M. McKeon, H.A. McGowan, and S.R. Phinn. 2009. A continent under stress: interactions, feedbacks and risks associated with impact of modified land cover on Australia's climate. *Global Change Biology* 15: 2206–2223.

McCabe, M.F., H. Gao, and E.F. Wood. 2005. Evaluation of AMSR-E-derived soil moisture retrievals using ground-based and PSR airborne data during SMEX02. *Journal of Hydrometeorology* 6: 864–877.

Meglioli, P.A., J.N. Aranibar, P.E. Villagra, J.A. Alvarez, and E.G. Jobbágy. 2013. Livestock stations as foci of groundwater recharge and nitrate leaching in a sandy desert of the Central Monte, Argentina. *Ecohydrology.* doi: 10.1002/eco.1381.

Meglioli, P. A., J. N. Aranibar, P. E. Villagra, J. A. Alvarez, and E. G. Jobbagy. 2014. Livestock stations as foci of groundwater recharge and nitrate leaching in a sandy desert of the Central Monte, Argentina. Ecohydrology 7:600–611

Merritt, D.M., and H.L. Bateman. 2012. Linking stream flow and groundwater to avian habitat in a desert riparian system. *Ecological Applications* 22: 1973–1988.

Midgley, G.F., and C.F. Musil. 1990. Substrate effects of zoogenic soil mounds on vegetation composition in the Worchester-Robertson Valley, Cape Province. *South African Journal of Botany* 56: 158–166.

Millennium Ecosystem Assessment. 2005. *Ecosystems and human well-being: Syntheses.* Washington, DC: Island Press.

Miller, G., J. Cable, A. McDonald, B. Bond, T. Franz, L. Wang, S. Gou, A. Tyler, C. Zou, and R. Scott. 2012. Understanding ecohydrological connectivity in savannas: A system dynamics modeling approach. *Ecohydrology* 5: 200–220.

Milton, S.J. 1990. Aboveground biomass and plant cover in a succulent shrubland in the southern Karoo. *South Africa South African Journal of Botany* 56: 587–589.

Milton, S.J., and W.R.J. Dean. 1990. Mima-like mounds in the southern and western cape-are origins so mysterious. *South African Journal of Science* 86: 207–208.

Moore, J.W. 2006. Animal ecosystem engineers in streams. *BioScience* 56: 237–246.

Moore, J.M., and M.D. Picker. 1991. Heuweltjies (earth mounds) in the Clanwilliam District, Cape province, South Africa--4000 year old termite nests. *Oecologia* 86: 424–432.

Moore, G.W., D.A. Barre, and M.K. Owens. 2012. Does shrub removal increase groundwater recharge in Southwestern Texas semiarid rangelands? *Rangeland Ecology & Management* 65: 1–10.

Moran, M.S., E.P. Hamerlynck, R.L. Scott, J.J. Stone, C.D.H. Collins, T.O. Keefer, R. Bryant, L. DeYoung, G.S. Nearing, Z. Sugg, and D.C. Hymer. 2010. Hydrologic response to precipitation pulses under and between shrubs in the Chihuahuan Desert, Arizona. *Water Resources Research* 46: W10509.

Moreno de las Heras, M., J.M. Nicolau, L. Merino-Martin, and B.P. Wilcox. 2010. Plot-scale effects on runoff and erosion along a slope degradation gradient. *Water Resources Research* 46, W04503. doi:10.1029/2009WR00787.

Morin, E., T. Grodek, O. Dahan, G. Benito, C. Kulls, Y. Jacoby, G. Van Langenhove, M. Seely, and Y. Enzel. 2009. Flood routing and alluvial aquifer recharge along the ephemeral arid Kuiseb River, Namibia. *Journal of Hydrology* 368: 262–275.

Nagler, P.L., E.P. Glenn, O. Hinojosa-Huerta, F. Zamora, and K. Howard. 2008. Riparian vegetation dynamics and evapotranspiration in the riparian corridor in the delta of the Colorado River, Mexico. *Journal of Environmental Management* 88: 864–874.

Nagler, P.L., K. Morino, K. Didan, J. Erker, J. Osterberg, K.R. Hultine, and E.P. Glenn. 2009. Wide-area estimates of saltcedar (Tamarix spp.) evapotranspiration on the lower Colorado River measured by heat balance and remote sensing methods. *Ecohydrology* 2: 18–33.

Naiman, R.J., and K.H. Rogers. 1997. Large animals and system level characteristics in river corridors. *BioScience* 47: 521–529.

Naiman, R.J., S.R. Elliott, J.M. Helfield, and T.C. O'Keefe. 1999. Biophysical interactions and the structure and dynamics of riverine ecosystems: The importance of biotic feedbacks. *Hydrobiologia* 410: 79–86.

Naiman, R.J., S.E. Bunn, C. Nilsson, G.E. Petts, G. Pinay, and L.C. Thompson. 2002. Legitimizing fluvial ecosystems as users of water: An overview. *Environmental Management* 30: 455–467.

Naumburg, E., R. Mata-Gonzalez, R.G. Hunter, T. McLendon, and D.W. Martin. 2005. Phreatophytic vegetation and groundwater fluctuations: A review of current research and application of ecosystem response modeling with an emphasis on great basin vegetation. *Environmental Management* 35: 726–740.

Newman, B.D., B.P. Wilcox, S.R. Archer, D.D. Breshears, C.N. Dahm, C.J. Duffy, N.G. McDowell, F.M. Phillips, B.R. Scanlon, and E.R. Vivoni. 2006. Ecohydrology of water-limited environments: A scientific vision. *Water Resources Research* 42, W06302. doi:10.1029/2005WR004141.

Nippert, J.B., A.K. Knapp, and J.M. Briggs. 2006. Intra-annual rainfall variability and grassland productivity: Can the past predict the future? *Plant Ecology* 184: 65–74.

Njoku, E.G., and D. Entekhabi. 1996. Passive microwave remote sensing of soil moisture. *Journal of Hydrology* 184: 101–129.

Nosetto, M.D., E.G. Jobbágy, T. Toth, and R.B. Jackson. 2008. Regional patterns and controls of ecosystem salinization with grassland afforestation along a rainfall gradient. *Global Biogeochemical Cycles* 22(2), GB2015. doi:10.1029/2007GB003000.

Nosetto, M.D., E.G. Jobbágy, R.B. Jackson, and G.A. Sznaider. 2009. Reciprocal influence of crops and shallow ground water in sandy landscapes of the Inland Pampas. *Field Crops Research* 113: 138–148.

Noy-Meir, I. 1973. Desert ecosystems: Environment and producers. *Annual Review of Ecology and Systematics* 4: 25–51.

Nuttle, W.K. 2002. *Ecohydrology's past and future focus.* EOS, pp. 205–216.

O'Farrell, P.J., J.S. Donaldson, and M.T. Hoffman. 2010. Vegetation transformation, functional compensation, and soil health in a semi-arid environment. *Arid Land Research and Management* 24: 12–30.

Okin, G.S., and D.A. Gillette. 2001. Distribution of vegetation in wind-dominated landscapes: Implications for wind erosion modeling and landscape processes. *Journal of Geophysical Research-Atmospheres* 106: 9673–9683.

Okin, G.S., A.J. Parsons, J. Wainwright, J.E. Herrick, B.T. Bestelmeyer, D.C. Peters, and E.L. Fredrickson. 2009. Do changes in connectivity explain desertification? *BioScience* 59: 237–244.

Owens, M.K., R.K. Lyons, and C.L. Alejandro. 2006. Rainfall partitioning within semiarid juniper communities: Effects of event size and canopy cover. *Hydrological Processes* 20: 3179–3189.

Pataki, D.E., S.E. Bush, P. Gardner, D.K. Solomon, and J.R. Ehleringer. 2005. Ecohydrology in a Colorado River riparian forest: Implications for the decline of Populus fremontii. *Ecological Applications* 15: 1009–1018.

Perelman, S.B., R.J.C. León, and M. Oesterheld. 2001. Cross-scale vegetation patterns of Flooding Pampa grasslands. *Journal of Ecology* 89: 562–577.

Peters, D., S.R. Pielke, B. Bestelmeyer, C. Allen, S. Munson-McGee, and K. Havstad. 2004. Cross-scale interactions, nonlinearitites, and forecasting catastrophic events. *Proceedings National Academy of Sciences, U S A* 101: 15130–15135.

Peters, D.P.C., B.T. Bestelmeyer, J.E. Herrick, E.L. Fredrickson, H.C. Monger, and K.M. Havstad. 2006. Disentangling complex landscapes: New insights into arid and semiarid system dynamics. *BioScience* 56: 491–502.

Pierson, F.B., W.H. Blackburn, S.S. Vanvactor, and J.C. Wood. 1994. Partitioning small scale spatial variability of runoff and erosion on sagebrush rangeland. *Water Resources Bulletin* 30: 1081–1089.

Pierson, F.B., P.R. Robichaud, and K.E. Spaeth. 2001. Spatial and temporal effects of wildfire on the hydrology of a steep rangeland watershed. *Hydrological Processes* 15: 2905–2916.

Pierson, F.B., C.J. Williams, P.R. Kormos, S.P. Hardegree, P.E. Clark, and B.M. Rau. 2010. Hydrologic vulnerability of sagebrush steppe following pinyon and juniper encroachment. *Rangeland Ecology & Management* 63: 614–629.

Pierson, F.B., C.J. Williams, S.P. Hardegree, M.A. Weltz, J.J. Stone, and P.E. Clark. 2011. Fire, plant invasions, and erosion events on western rangelands. *Rangeland Ecology & Management* 64: 439–449.

Postel, S.L., and B.H. Thompson. 2005. Watershed protection: Capturing the benefits of nature's water supply services. *Natural Resources Forum* 29: 98–108.

Prince, S., and C. Tucker. 1986. Satellite remote sensing of rangelands in Botswana II. NOAA AVHRR and herbaceous vegetation. *International Journal of Remote Sensing* 7: 1555–1570.

Prince, S.D., K.J. Wessels, C.J. Tucker, and S.E. Nicholson. 2007. Desertification in the Sahel: A reinterpretation of a reinterpretation. *Global Change Biology* 13: 1308–1313.

Pringle, C.M. 2001. Hydrologic connectivity and the management of biological reserves: A global perspective. *Ecological Applications* 11: 981–998.

Pringle, C. 2003. What is hydrologic connectivity and why is it ecologically important? *Hydrological Processes* 17: 2685–2689.

Pritchett, D., and S.J. Manning. 2012. Response of an intermountain groundwater-dependent ecosystem to water table drawdown. *Western North American Naturalist* 72: 48–59.

Puigdefabregas, J., and T. Mendizabal. 1998. Perspectives on desertification—Western Mediterranean. *Journal of Arid Environments* 39: 209–224.

Puigdefabregas, J., A. Sole, L. Gutierrez, G. del Barrio, and M. Boer. 1999. Scales and processes of water and sediment redistribution in drylands: Results from the Rambla Honda field site in Southeast Spain. *Earth-Science Reviews* 48: 39–70.

Pyke, D.A., J.E. Herrick, P. Shaver, and M. Pellant. 2002. Rangeland health attributes and indicators for qualitative assessment. *Journal of Range Management* 55: 584–597.

Ravi, S., P. D'Odorico, and G.S. Okin. 2007a. Hydrologic and aeolian controls on vegetation patterns in arid landscapes. *Geophysical Research Letters* 34: 5. doi:10.1029/2007GL031023.

Ravi, S., P. D'Odorico, T.M. Zobeck, T.M. Over, and S.L. Collins. 2007b. Feedbacks between fires and wind erosion in heterogeneous arid lands—Art. no. G04007. *Journal of Geophysical Research: Biogeosciences* 112: 4007.

Ravi, S., P. D'Odorico, T.M. Zobeck, and T.M. Over. 2009. The effect of fire-induced soil hydrophobicity on wind erosion in a semiarid grassland: Experimental observations and theoretical framework. *Geomorphology* 105: 80–86.

Raynaud, X., C.G. Jones, and S. Barot. 2013. Ecosystem engineering, environmental decay and environmental states of landscapes. *Oikos* 122: 591–600.

Reid, K.D., B.P. Wilcox, D.D. Breshears, and L. MacDonald. 1999. Runoff and erosion in a pinon-juniper woodland: Influence of vegetation patches. *Soil Science Society of America Journal* 63: 1869–1879.

Reynolds, J.F., D.M. Stafford Smith, E.F. Lambin, B.L. Turner, M. Mortimore, S.P.J. Batterbury, T.E. Downing, H. Dowlatabadi, R.J. Fernandez, J.E. Herrick, E. Huber-Sannwald, H. Jiang, R. Leemans, T. Lynam, F.T. Maestre, M. Ayarza, and B. Walker. 2007. Global desertification: Building a science for dryland development. *Science* 316: 847–851.

Richardson, C.W., E. Burnett, and R.W. Bovey. 1979. Hydrologic effects of brush control on Texas rangelands. *Transactions of the ASAE* 22(2): 315–319.

Rietkerk, M., M.C. Boerlijst, F. van Langevelde, R. HilleRisLambers, J. van de Koppel, L. Kumar, H.H.T. Prins, and A.M. de Roos. 2002. Self-organization of vegetation in arid ecosystems. *American Naturalist* 160: 524–530.

Rietkerk, M., S.C. Dekker, P.C. de Ruiter, and J. van de Koppel. 2004. Self-organized patchiness and catastrophic shifts in ecosystems. *Science* 305: 1926–1929.

Robinson, D.A., C.S. Campbell, J.W. Hopmans, B.K. Hornbuckle, S.B. Jones, R. Knight, F. Ogden, J. Selker, and O. Wendroth. 2008. Soil moisture measurement for ecological and hydrological watershed-scale observatories: A review. *Vadose Zone Journal* 7: 358–389.

Rockstrom, J., M. Falkenmark, L. Karlberg, H. Hoff, S. Rost, and D. Gerten. 2009. Future water availability for global food production: The potential of green water for increasing resilience to global change. *Water Resources Research* 45(7), W00A12. doi:10.1029/2007WR006767.

Rodriguez-Iturbe, I., and A. Porporato. 2004. *Ecohydrology of water-controlled ecosystems: Soil moisture and plant dynamics*. Cambridge: Cambridge University Press.

Roth, C.H. 2004. A framework relating soil surface condition to infiltration and sediment and nutrient mobilization in grazed rangelands of northeastern Queensland, Australia. *Earth Surface Processes and Landforms* 29: 1093–1104.

Royer, P.D., D.D. Breshears, C.B. Zou, N.S. Cobb, and S.A. Kurc. 2010. Ecohydrological energy inputs in semiarid coniferous gradients: Responses to management- and drought-induced tree reductions. *Forest Ecology and Management* 260: 1646–1655.

Royer, P.D., D.D. Breshears, C.B. Zou, J.C. Villegas, N.S. Cobb, and S.A. Kurc. 2012. Density-dependent ecohydrological effects of pinon–juniper woody canopy cover on soil microclimate and potential soil evaporation. *Rangeland Ecology & Management* 65: 11–20.

Running, S.W. 2006. Is global warming causing more, larger wildfires? *Science* 313: 927–928.

Runyan, C.W., P. D'Odorico, and D. Lawrence. 2012. Physical and biological feedbacks of deforestation. *Reviews of Geophysics* 50, RG4006.

Sabo, J.L., R. Sponseller, M. Dixon, K. Gade, T. Harms, J. Heffernan, A. Jani, G. Katz, C. Soykan, J. Watts, and A. Welter. 2005. Riparian zones increase regional species richness by harboring different, not more, species. *Ecology* 86: 56–62.

Saco, P.M., and M. Moreno de las Heras. 2013. Ecogeomorphic coevolution of semiarid hillslopes: Emergence of banded and striped vegetation patterns through interaction of biotic and abiotic processes. *Water Resources Research* 49: 115–126.

Sala, A., S.D. Smith, and D.A. Devitt. 1996. Water use by Tamarix ramosissima and associated phreatophytes in a Mojave Desert floodplain. *Ecological Applications* 6: 888–898.

Savadogo, P., L. Sawadogo, and D. Tiveau. 2007. Effects of grazing intensity and prescribed fire on soil physical and hydrological properties and pasture yield in the savanna woodlands of Burkina Faso. *Agriculture, Ecosystems & Environment* 118: 80–92.

Scanlon, B.R., and R.S. Goldsmith. 1997. Field study of spatial variability in unsaturated flow beneath and adjacent to playas. *Water Resources Research* 33: 2239–2252.

Scanlon, B.R., D.G. Levitt, R.C. Reedy, K.E. Keese, and M.J. Sully. 2005a. Ecological controls on water-cycle response to climate variability in deserts. *Proceedings of the National Academy of Sciences* 102: 6033–6038.

Scanlon, B.R., D.G. Levitt, R.C. Reedy, K.E. Keese, and M.J. Sully. 2005b. Ecological controls on water-cycle response to climate variability in deserts. *Proceedings of the National Academy of Sciences of the United States of America* 102: 6033–6038.

Scanlon, B.R., R.C. Reedy, K.E. Keese, and S.F. Dwyer. 2005c. Evaluation of evapotranspirative covers for waste containment in arid and semiarid regions in the southwestern USA. *Vadose Zone Journal* 4: 55–71.

Scanlon, B.R., K.E. Keese, A.L. Flint, L.E. Flint, C.B. Gaye, W.M. Edmunds, and I. Simmers. 2006. Global synthesis of groundwater recharge in semiarid and arid regions. *Hydrological Processes* 20: 3335–3370.

Schenk, H.J., and R.B. Jackson. 2002. Rooting depths, lateral root spreads and below-ground/above-ground allometries of plants in water-limited ecosystems. *Journal of Ecology* 90: 480–494.

Schlesinger, W.H., A.D. Abrahams, A.J. Parsons, and J. Wainwright. 1999. Nutrient losses in run-off from grassland and shrubland habitats in Southern New Mexico: I. Rainfall simulation experiments. *Biogeochemistry* 45: 21–34.

Scott, R.L., E.A. Edwards, W.J. Shuttleworth, T.E. Huxman, C. Watts, and D.C. Goodrich. 2004. Interannual and seasonal variation in fluxes of water and carbon dioxide from a riparian wood-land ecosystem. *Agricultural and Forest Meteorology* 122: 65–84.

Scott, R.L., T.E. Huxman, D.G. Williams, and D.C. Goodrich. 2006. Ecohydrological impacts of woody-plant encroachment: seasonal patterns of water and carbon dioxide exchange within a semiarid riparian environment. *Global Change Biology* 12: 311–324.

Scott, R.L., W.L. Cable, T.E. Huxman, P.L. Nagler, M. Hernandez, and D.C. Goodrich. 2008. Multiyear riparian evapotranspiration and groundwater use for a semiarid watershed. *Journal of Arid Environments* 72: 1232–1246.

Seyfried, M.S. 1991. Infiltration patterns from simulated rainfall on a semiarid rangeland soil. *Soil Science Society of America Journal* 55: 1726–1734.

Seyfried, M.S., and B.P. Wilcox. 1995. Scale and the nature of spatial variability: Field examples having implications for hydrologic modeling. *Water Resources Research* 1: 173–184.

———. 2006. Soil water storage and rooting depth: key factors controlling recharge on rangelands. *Hydrological Processes* 20: 3261–3275.

Seyfried, M.S., S. Schwinning, M.A. Walvoord, W.T. Pockman, B.D. Newman, R.B. Jackson, and E.M. Phillips. 2005. Ecohydrological control of deep drainage in arid and semiarid regions. *Ecology* 86: 277–287.

Shafer, D.S., M.H. Young, S.F. Zitzer, T.G. Caldwell, and E.V. McDonald. 2007. Impacts of inter-related biotic and abiotic processes during the past 125,000 years of landscape evolution in the Northern Mojave Desert, Nevada, USA. *Journal of Arid Environments* 69: 633–657.

Shafroth, P.B., C.A. Brown, and D.M. Merrit, eds. 2010a. *Saltcedar and Russian olive control demonstration act science assessment*. U.G. Geological Survey.

Shafroth, P.B., A.C. Wilcox, D.A. Lytle, J.T. Hickey, D.C. Andersen, V.B. Beauchamp, A. Hautzinger, L.E. McMullen, and A. Warner. 2010a. Ecosystem effects of environmental flows: Modelling and experimental floods in a dryland river. *Freshwater Biology* 55: 68–85.

Shakesby, R.A., and S.H. Doerr. 2006. Wildfire as a hydrological and geomorphological agent. *Earth-Science Reviews* 74: 269–307.

Sivapalan, M., and J.D. Kalma. 1995. Scale problems in hydrology—Contributions of the Robertson workshop. *Hydrological Processes* 9: 243–250.

Small, E.E. 2005. Climatic controls on diffuse groundwater recharge in semiarid environments of the southwestern United States. *Water Resources Research* 41: 1–17.

Smith, H.L., and L.B. Leopold. 1941. Infiltration studies in the Pecos River Watershed, New Mexico and Texas. *Soil Science* 53: 195–204.

Snyman, H.A., and C.C. du Preez. 2005. Rangeland degradation in a semi-arid South Africa— II: Influence on soil quality. *Journal of Arid Environments* 60: 483–507.

Soykan, C.U., and J.L. Sabo. 2009. Spatiotemporal food web dynamics along a desert riparian-upland transition. *Ecography* 32: 354–368.

Sposito, G. 1998. *Scale dependence and scale invariance in hydrology*. New York: Cambridge University Press.

Stavi, I., H. Lavee, E.D. Ungar, and P. Sarah. 2009. Ecogeomorphic feedbacks in semiarid range-lands: A review. *Pedosphere* 19: 217–229.

Stromberg, J.C., J.A. Tress, S.D. Wilkins, and S.D. Clark. 1992. Response of velvet mesquite to groundwater decline. *Journal of Arid Environments* 23: 45–58.

Tabacchi, E., L. Lambs, H. Guilloy, A.M. Planty-Tabacchi, E. Muller, and H. Decamps. 2000. Impacts of riparian vegetation on hydrological processes. *Hydrological Processes* 14: 2959–2976.

Thomas, F.M., S.K. Arndt, H. Bruelheide, A. Foetzki, D. Gries, J. Huang, M. Popp, G. Wang, X. Zhang, and M. Runge. 2000. Ecological basis for a sustainable management of the indige-nous vegetation in a Central-Asian desert: Presentation and first results. *Journal of Applied Botany* 74: 212–219.

Thompson, S.E., C.J. Harman, P.A. Troch, P.D. Brooks, and M. Sivapalan. 2011. Spatial scale dependence of ecohydrologically mediated water balance partitioning: A synthesis framework for catchment ecohydrology. *Water Resources Research* 47, W00J03. doi:10.1029/2010WR009998.

Thorp, J.H., M.C. Thoms, and M.D. Delong. 2006. The riverine ecosystem synthesis: Biocomplexity in river networks across space and time. *River Research and Applications* 22: 123–147.

Thorp, J.H., J.E. Flotemersch, M.D. Delong, A.F. Casper, M.C. Thoms, F. Ballantyne, B.S. Williams, B.J. O'Neill, and C.S. Haase. 2010. Linking ecosystem services, rehabilitation, and river hydrogeomorphology. *BioScience* 60: 67–74.

Tockner, K., and J.A. Stanford. 2002. Riverine flood plains: Present state and future trends. *Environmental Conservation* 29: 308–330.

Tongway, D.J., and J.A. Ludwig. 1997. Chapter 5. The nature of landscape dysfunction in range-lands. In *Landscape ecology, function and management: Principles from Australia's range-lands*, eds. J. Ludwig, D. Tongway, D. Freudenberger, J. Noble, and K. Hodgkinson, 49–61. Collingwood, VIC: CSIRO Publishing.

———. 2001. Theories on the origins, maintenance, dynamics and functioning of banded land-scapes. In *Banded vegetation patterning in arid and semiarid environments: Ecological pro-cesses and consequences of management*, eds. D.J. Tongway, C. Valentin, and J. Segheri, 20–31. New York, NY: Springer.

Tóth, J. 1999. Groundwater as a geologic agent: An overview of the causes, processes, and manifestations. *Hydrogeology Journal* 7: 1–14.

Trimble, S.W., and A.C. Mendel. 1995. The cow as a geomorphic agent—A critical-review. *Geomorphology* 13: 233–253.

Turnbull, L., B.P. Wilcox, J. Belnap, S. Ravi, P. D'Odorico, D. Childers, W. Gwenzi, G. Okin, J. Wainwright, K.K. Caylor, and T. Sankey. 2012. Understanding the role of ecohydrological feedbacks in ecosystem state change in drylands. *Ecohydrology* 5: 174–183.

Turner, N.C., and P.R. Ward. 2002. The role of agroforestry and perennial pasture in mitigating water logging and secondary salinity: Summary. *Agricultural Water Management* 53: 271–275.

Twidwell, D., W.E. Rogers, S.D. Fuhlendorf, C.L. Wonkka, D.M. Engle, J.R. Weir, U.P. Kreuter, and C.A. Taylor. 2013. The rising Great Plains fire campaign: Citizens response to woody plant encroachment. *Frontiers in Ecology and the Environment* 11: 64–71.

Valentin, C., and J.M. d'Herbes. 1999. Niger tiger bush as a natural water harvesting system. *Catena* 37: 231–256.

van Wilgen, B.W., D.C. Le Maitre, and R.M. Cowling. 1998. Ecosystem services, efficiency, sus-tainability and equity: South Africa's Working for Water programme. *Trends in Ecology & Evolution* 13: 378–378.

van Wilgen, B.W., G.G. Forsyth, D.C. Le Maitre, A. Wannenburgh, J.D.F. Kotze, E. van den Berg, and L. Henderson. 2012. An assessment of the effectiveness of a large, national-scale invasive alien plant control strategy in South Africa. *Biological Conservation* 148: 28–38.

Vannote, R.L., G.W. Minshall, K.W. Cummins, J.R. Sedell, and C.E. Cushing. 1980. River continuum concept. *Canadian Journal of Fisheries and Aquatic Sciences* 37: 130–137.

Villegas, J.C., D.D. Breshears, C.B. Zou, and D.J. Law. 2010a. Ecohydrological controls of soil evaporation in deciduous drylands: How the hierarchical effects of litter, patch and vegetation mosaic cover interact with phenology and season. *Journal of Arid Environments* 74: 595–602.

Villegas, J.C., D.D. Breshears, C.B. Zou, and P.D. Royer. 2010b. Seasonally pulsed heterogeneity in microclimate: Phenology and cover effects along deciduous grassland-forest continuum. *Vadose Zone Journal* 9: 537–547.

Visser, E.J.W., G.M. Bogemann, H.M. Van de Steeg, R. Pierik, and C.W.P.M. Blom. 2000. Flooding tolerance of Carex species in relation to field distribution and aerenchyma formation. *New Phytologist* 148: 93–103.

Vivoni, E.R. 2012. Spatial patterns, processes and predictions in ecohydrology: Integrating technologies to meet the challenge. *Ecohydrology* 5: 235–241.

von der Heyden, C.J. 2004. The hydrology and hydrogeology of dambos: A review. *Progress in Physical Geography* 28: 544–564.

Wainwright, J., L. Turnbull, T.G. Ibrahim, I. Lexartza-Artza, S.F. Thornton, and R.E. Brazier. 2011. Linking environmental regimes, space and time: Interpretations of structural and functional connectivity. *Geomorphology* 126: 387–404.

Walker, J., F. Bullen, and B.G. Williams. 1993. Ecohydrological changes in the Murray-Darling Basin. I. The number of trees cleared over two centuries. *Journal of Applied Ecology* 30: 265–273.

Wang, L., and P. D'Odorico. 2008. The limits of water pumps. *Science* 321: 36–37.

Wang, G., and E.A. Eltahir. 2000. Ecosystem dynamics and the Sahel drought. *Geophysical Research Letters* 27: 795–798.

Wang, L., K. Caylor, and D. Dragoni. 2009a. On the calibration of continuous, high-precision $\delta18O$ and $\delta2H$ measurements using an off-axis integrated cavity output spectrometer. *Rapid Communications in Mass Spectrometry* 23: 530–536.

Wang, L., P. D'Odorico, S. Manzoni, A. Porporato, and S. Macko. 2009b. Carbon and nitrogen dynamics in southern African savannas: The effect of vegetation-induced patch-scale heterogeneities and large scale rainfall gradients. *Climatic Change* 94: 63–76.

Wang, L., K.K. Caylor, J.C. Villegas, G.A. Barron-Gafford, D.D. Breshears, and T.E. Huxman. 2010a. Evapotranspiration partitioning with woody plant cover: Assessment of a stable isotope technique. *Geophysical Research Letters* 37, L09401.

Wang, L.X., K.K. Caylor, J.C. Villegas, G.A. Barron-Gafford, D.D. Breshears, and T.E. Huxman. 2010b. Partitioning evapotranspiration across gradients of woody plant cover: Assessment of a stable isotope technique. *Geophysical Research Letters* 37, L09401. doi:10.1029/2010GL043228.

Wang, L., P. D'Odorico, J.P. Evans, D.J. Eldridge, M.F. McCabe, K.K. Caylor, and E.G. King. 2012a. Dryland ecohydrology and climate change: Critical issues and technical advances. *Hydrology and Earth System Sciences* 16: 2585–2603.

Wang, L., C. Zou, F. O'Donnell, S. Good, T. Franz, G. Milller, K. Caylor, J. Cable, and B. Bond. 2012b. Characterizing ecohydrological and biogeochemical connectivity across multiple scales: A new conceptual framework. *Ecohydrology* 5: 221–233. doi:10.1002/eco.1187.

Ward, J.V. 1998. Riverine landscapes: Biodiversity patterns, disturbance regimes, and aquatic conservation. *Biological Conservation* 83: 269–278.

Ward, J.V., F. Malard, and K. Tockner. 2001. Landscape ecology: A framework for integrating pattern and process in river corridors. *Landscape Ecology* 17: 35–45.

Warren, S.D. 2003. Synopsis: Influence of biological soil crusts on arid land hydrology and soil stability. In *Biological soil crusts: Structure, function, and management*, ed. J. Belnap and O.L. Lange, 349–362. Berlin: Springer.

Weaver, J. 1926. *Root development of field crops*. New York: McGraw-Hill Book Company.

Wen, X., X. Sun, S. Zhang, G. Yu, S. Sargent, and X. Lee. 2008. Continuous measurement of water vapor D/H and $^{18}O/^{16}O$ isotope ratios in the atmosphere. *Journal of Hydrology* 349: 489–500.

Westbrook, C.J., W. Veatch, and A. Morrison. 2013. Ecohydrology bearings—Invited commentary is ecohydrology missing much of the zoo? *Ecohydrology* 6: 1–7.

Western, A.W., G. Bloschl, and R.B. Grayson. 2001. Toward capturing hydrologically significant connectivity in spatial patterns. *Water Resources Research* 37: 83–97.

Whitford, W.G., and F.R. Kay. 1999. Biopedturbation by mammals in deserts: A review. *Journal of Arid Environments* 41: 203–230.

Wiens, J.A. 2002. Riverine landscapes: Taking landscape ecology into the water. *Freshwater Biology* 47: 501–515.

Wilcox, B.P. 2002. Shrub control and streamflow on rangelands: A process based viewpoint. *Journal of Range Management* 55: 318–326.

———. 2010. Transformative ecosystem change and ecohydrology: Ushering in a new era for watershed management. *Ecohydrology* 3: 126–130.

Wilcox, B.P., and B.D. Newman. 2005. Ecohydrology of semiarid landscapes. *Ecology* 86: 275–276.

Wilcox, B.P., D.D. Breshears, and C.D. Allen. 2003a. Ecohydrology of a resource-conserving semiarid woodland: Effects of scale and disturbance. *Ecological Monographs* 73: 223–239.

Wilcox, B.P., M.S. Seyfried, and D.D. Breshears. 2003b. The water balance on rangelands. *Encyclopedia of Water Science* 791–795.

Wilcox, B.P., M.K. Owens, W.A. Dugas, D.N. Ueckert, and C.R. Hart. 2006. Shrubs, streamflow, and the paradox of scale. *Hydrological Processes* 20: 3245–3259.

Wilcox, B.P., P.I. Taucer, C.L. Munster, M.K. Owens, B.P. Mohanty, J.R. Sorenson, and R. Bazan. 2008. Subsurface stormflow is important in semiarid karst shrublands. *Geophysical Research Letters* 35, L10403.

Wilcox, B.P., M.G. Sorice, and M.H. Young. 2011. Drylands in the anthropocene: Taking stock of human-ecological interactions. *Geography Compass* 5(3): 112–127.

Wilcox, B.P., M.S. Seyfried, D.D. Breshears, and J.J. McDonnell. 2012a. Ecohydrologic connections and complexities in drylands: New perspectives for understanding transformative landscape change preface. *Ecohydrology* 5: 143–144.

Wilcox, B.P., L. Turnbull, M.H. Young, C.J. Williams, S. Ravi, M.S. Seyfried, D.R. Bowling, R.L. Scott, M.J. Germino, T.G. Caldwell, and J. Wainwright. 2012b. Invasion of shrublands by exotic grasses: Ecohydrological consequences in cold versus warm deserts. *Ecohydrology* 5: 160–173.

Wood, M.K., and W.H. Blackburn. 1981. Grazing systems: Their influence on infiltration rates in the Rolling Plains of Texas. *Journal of Range Management* 34: 331–335.

Wood, M.K., W.H. Blackburn, F.E. Smeins, and W.A. McGinty. 1978. Hydrologic impacts of grazing systems. In *First international rangeland congress*, 288–291.

Wood, E.F., M. Sivapalan, and K. Beven. 1990. Similarity and scale in catchment storm response. *Reviews of Geophysics* 28: 1–18.

Woodward, L. 1943. Infiltration capacities of some plant-soil complexes on Utah range watershedlands. *Transactions of American Geophysical Union* 24: 468–473.

Wurster, F.C., D.J. Cooper, and W.E. Sanford. 2003. Stream/aquifer interactions at Great Sand Dunes National Monument, Colorado: Influences on interdunal wetland disappearance. *Journal of Hydrology* 271: 77–100.

Yeh, P.J.Ä., S. Swenson, J. Famiglietti, and M. Rodell. 2006. Remote sensing of groundwater storage changes in Illinois using the Gravity Recovery and Climate Experiment (GRACE). *Water Resources Research* 42, W12203. doi:10.1029/2006WR005374.

Zencich, S.J., R.H. Froend, J.V. Turner, and V. Gailitis. 2002. Influence of groundwater depth on the seasonal sources of water accessed by Banksia tree species on a shallow, sandy coastal aquifer. *Oecologia* 131: 8–19.

Zeng, N., J.D. Neelin, K.-M. Lau, and C.J. Tucker. 1999. Enhancement of interdecadal climate variability in the Sahel by vegetation interaction. *Science* 286: 1537–1540.

Zhang, L., W.R. Dawes, and G.R. Walker. 2001. Response of mean annual evapotranspiration to vegetation changes at catchment scale. *Water Resources Research* 37: 701–708.

Zheng, X., and E.A. Eltahir. 1998. The role of vegetation in the dynamics of West African monsoons. *Journal of Climate* 11: 2078–2096.

Zobeck, T.M., G. Sterk, R. Funk, J.L. Rajot, J.E. Stout, and R.S. Van Pelt. 2003. Measurement and data analysis methods for field-scale wind erosion studies and model validation. *Earth Surface Processes and Landforms* 28: 1163–1188.

Zou, C.B., P.D. Royer, and D.D. Breshears. 2010. Density-dependent shading patterns by Sonoran saguaros. *Journal of Arid Environments* 74: 156–158.

Zreda, M., W.J. Shuttleworth, X. Zeng, C. Zweck, D. Desilets, T. Franz, R. Rosolem, and T.P.A. Ferre. 2012. COSMOS: The cosmic-ray soil moisture observing system. *Hydrology and Earth System Sciences Discussions* 9: 4505–4551.

Chapter 4
Soil and Belowground Processes

R. Dave Evans, Richard A. Gill, Valerie T. Eviner, and Vanessa Bailey

Abstract Soil characteristics and functions are critical determinants of rangeland systems and the ecosystem services that they provide. Rangeland soils are extremely diverse, but an emerging understanding is that paradigms developed in more mesic forest ecosystems may not be applicable. Vascular plants, biological soil crusts, and the soil microbial community are the three major functional groups of organisms that influence rangeland soils through their control over soil structure and soil carbon, water, and nutrient availability. Rangelands occur across a broad range of precipitation regimes, but local water status can be modified by management and land use. Important processes in carbon and nutrient cycling can be unique to arid rangelands. Physical drivers such as UV radiation and soil–litter mixing can be important factors for decomposition. Precipitation, vascular species composition and spatial pattern, presence of biological soil crusts, and surface disturbance interact to determine rates of carbon and nutrient cycling. The low resource availability in rangeland soils makes them very vulnerable to drivers of global change, and also excellent indicators of small changes in resource availability. Recent large-scale experiments demonstrate that rangelands are very susceptible to changes in precipitation regimes, warming, and atmospheric carbon dioxide. Growth of molecular tools in combination with other techniques has allowed scientists to increasingly link microbial community composition and function, thereby shedding light on what was formerly

R.D. Evans (✉)
School of Biological Sciences and WSU Stable Isotope Core Laboratory, Washington State University, Pullman, WA, USA
e-mail: rdevans@wsu.edu

R.A. Gill
Department of Biology and Evolutionary Ecology Laboratory, Brigham Young University, Provo, UT, USA
e-mail: rgill@byu.edu

V.T. Eviner
Department of Plant Sciences, University of California, Davis, CA, USA
e-mail: veviner@ucdavis.edu

V. Bailey
Biological Sciences Division, Pacific Northwest National Laboratory, Richland, WA, USA
e-mail: Vanessa.Bailey@pnnl.gov

© The Author(s) 2017
D.D. Briske (ed.), *Rangeland Systems*, Springer Series on Environmental Management, DOI 10.1007/978-3-319-46709-2_4

viewed as the black box of microbial dynamics in soils. Concurrent technological advances in environmental sensors and sensor arrays allow more mechanistic understanding of soil processes while also offering new opportunities to develop questions at the landscape scale.

Keywords Atmospheric deposition • Biological soil crusts • Grazing • Soil carbon • Soil community composition and function • Soil nitrogen

4.1   Introduction

The maintenance of healthy, productive soils in rangelands is a critical management issue, determining vegetation production, cover, and composition, and the livelihood of over 2.5 billion people, most of whom live in poverty and directly rely upon the ecosystem services of rangelands for their well-being and survival (MEA 2005; Reynolds et al. 2007). Their inherently low soil carbon (C) and nutrient levels cause rangelands to be among the most responsive terrestrial land cover type to even small changes in resource availability caused by drivers of global change. This sensitivity to change, coupled with their extensive global coverage, is the primary reason that rangelands are now recognized as critical components of global biogeochemical cycles (Donohue et al. 2013; Evans et al. 2014; Poulter et al. 2014).

Their importance for a large fraction of the human population despite their low resource availability is a reason for an increased emphasis on the function and global significance of rangeland soils over the past 25 years. Prior to this scientists had developed detailed understanding of soil classification and physical characteristics, especially those properties contributing to water-holding capacity and plant water availability. However, our functional understanding of soil microbial community composition and biologically mediated processes in the carbon and nitrogen cycles had not progressed as rapidly as our knowledge of aboveground dynamics, or even soil function, in more mesic ecosystems. Recent emphases on community and ecosystem-level questions as well as development of new technologies have greatly facilitated our understanding and led to the following conceptual understandings: (1) increased awareness that paradigms developed in more mesic ecosystems may be less applicable in rangeland ecosystems, and in some cases major drivers of ecosystem function may be unique to rangelands; (2) identification of unifying principles in soil ecology is difficult because rangelands encompass many diverse ecosystems with wide variation in temperature, precipitation, and disturbance regimes; (3) rangeland ecosystems respond rapidly to perturbations in resource availability caused by anthropogenic drivers of change. The conceptual advances reaffirm that proper management and restoration are necessary to ensure ecosystem health into the future.

The broad goals are to identify the primary advances in our understanding of soil and belowground processes and function that have occurred during the past 25 years, and discuss the potential consequences of future drivers on rangeland soils.

4.2 Major Conceptual Advances

4.2.1 Community Composition and Function

4.2.1.1 Soil–Plant Interactions

Soils have a large influence on shifts in vegetation composition that are common in rangelands (Bestelmeyer et al. 2003), particularly since rangeland vegetation is extremely sensitive to soil degradation (van de Koppel et al. 2002). Vegetation composition not only responds to soils, but can also cause large changes in soil properties and functions (Diaz et al. 2002; Eviner and Chapin 2003; Bestelmeyer and Briske 2012). The magnitude and direction of soil changes, and which attributes are modified, depend on the magnitude of species composition change, as well as ecological site conditions. The most common vegetation shifts that occur in rangelands include shifts from palatable to unpalatable vegetation; shifts from herbaceous to woody dominance; changes in grass species composition; shifts from grass to forb dominance (Vetter 2005). Each of these categories are summarized below.

Shifts from palatable to unpalatable vegetation. Preferential grazing can decrease the prevalence of palatable plants and increase unpalatable plants (Bestelmeyer et al. 2003; Seymour et al. 2010). This shift tends to occur under continuous grazing regimes, particularly under lower soil nutrient availability (Wardle 2002; Harrison and Bardgett 2008), and more arid conditions (Diaz et al. 2007). Unpalatable plants represent a variety of growth forms, including other grasses, forbs, or woody species. These plants usually have lower nutrient concentrations and higher chemical defenses than palatable species, so their litter inputs decrease rates of nutrient recycling, further decreasing plant nutrient availability (Bardgett and Wardle 2003; Harrison and Bardgett 2008). Such changes in soils can lead to delayed or inhibited recovery of palatable vegetation, even after grazing has ceased (Bestelmeyer et al. 2003; Seymour et al. 2010).

Shifts from herbaceous to woody dominance. Many rangelands across the globe have experienced increasing woody encroachment over the past century (Fig. 4.1), which has been attributed to a variety of factors, including: overgrazing, decreased human disturbance, lower fire frequency, nitrogen (N) deposition, climate change, and predator suppression (Eldridge et al. 2011; Chapter 2, this volume). The encroachment of shrubs and trees into herbaceous-dominated rangeland has extremely variable impacts on multiple soil properties depending upon the plant species, local climate, and soil type (Schuman et al. 2002; Barger et al. 2011). Generally, woody plant encroachment leads to more patchy vegetation cover causing higher heterogeneity in soil conditions (Van Auken 2009; Barger et al. 2011) and lower water infiltration and increased vulnerability to erosion, particularly after fires and during droughts, in plant interspaces (Ravi et al. 2009; Barger et al. 2011). This is why in some areas, shrub encroachment has been linked with significant soil degradation and even desertification (Bestelmeyer et al. 2003; Ravi et al. 2009), but this is not true everywhere (Eldridge et al. 2011). In general, woody encroachment increases surface soil C, total N, mineralizable N, soil aggregate stability, and available

Low honey mesquite canopy cover

Moderate honey mesquite canopy cover

High honey mesquite canopy cover

Fig. 4.1 Changes in community structure that occur during a shift in dominance from herbaceous to woody species (from Mohamed et al. 2011)

calcium (Archer et al. 2001; Schuman et al. 2002). Carbon accumulation tends to be higher in arid sites than in wetter sites (Jackson et al. 2002; Barger et al. 2011), but this strongly depends on species of shrub, soil bulk density (higher bulk density associated with more C loss), and % clay content (higher C accumulation associated with more clay) (Barger et al. 2011). In general, woody encroachment has stronger impacts on soil C pools than soil N (Van Auken 2009).

Woody encroachment into arid systems can enhance surface soil water through a combination of reduced evapotranspiration and hydraulic lift (Gill and Burke 1999). Shrub encroachment in arid ecosystems is generally patchy, leading to many areas of little to no plant cover, with low infiltration (Eviner and Chapin 2003). The effects of woody encroachment on streamflow vary with site hydrology. For example, a decrease in shrub cover can significantly enhance streamflow on sites with springs and groundwater flow, or that receive greater than 500 mm of rain per year, due to decreased transpiration rates. However, shrub removal may have no effect or even decrease streamflow in drylands or in systems where most streamflow is delivered through overland flow (Wilcox et al. 2006; Chapter 3, this volume). Woody species may also decrease infiltration by depositing hydrophobic substances (e.g., waxes), which are very common in species such as oaks, but tend to be low in soils associated with grasses and mosses (Doerr et al. 2000; Schnabel et al. 2013).

Shifts in grass species dominance. Grazing can induce compositional shifts in some regions from perennial to annual dominated grasslands (Fernandez-Gimenez and Allen-Diaz 1999; Diaz et al. 2007). An increased prevalence of annual grasses can also be due to increased N deposition and introduction of non-native annuals (Bai et al. 2009; Ravi et al. 2009). Transition to annual plant dominance can lead to high temporal variability in plant production and cover paralleling climate variability, and little to no vegetation cover during droughts (Bai et al. 2004; Ravi et al. 2009). During these drought periods, low vegetation cover can lead to massive erosion and irreversible desertification (Ravi et al. 2009). Domination by annual plants can also homogenize spatial distribution of soil C, N, and water, eliminating the islands of fertility created by continuous litter inputs from perennial species that are important in facilitating reestablishment of other vegetation. Soil C has been shown to be higher under perennial vs. annual grasses across a range of sites in the western USA (Gill and Burke 1999). Shifts in grass composition within the annual or perennial groups can also have significant soil impacts. For example, in the shortgrass steppe, microbial biomass C and C mineralization rates were higher under perennial bunchgrasses than under rhizomatous grasses (Vinton and Burke 1995). In a tallgrass prairie, soil N mineralization rates differed tenfold across five different perennial grasses (Wedin and Tilman 1990). In California's annual grasslands, annual grass species differed in their impacts on net N mineralization and nitrification (Eviner et al. 2006).

Shifts between grass and forb dominance. Rangeland herbaceous vegetation commonly contains both grasses and broad-leafed species (forbs and legumes), with their relative dominance shifting over space and time (Fig. 4.2). Forbs tend to

Fig. 4.2 The relative dominance of grasses and broad-leaved species in many rangelands often shifts both temporally and spatially. For example, in this grassland in California the community is grass dominated (*left*) in some years, while in others forbs and legumes are dominant (*right*) (photographs by V. Eviner)

increase after disturbances that remove grass biomass (e.g., fire, grazing, gopher mounds, and herbicides) (Fernandez-Gimenez and Allen-Diaz 1999; Stahlheber and D'Antonio 2013). The effects of increasing forbs on soil processes will vary strongly by species. In general, forbs have lower C:N ratios than grasses, which could lead to higher rates of decomposition and nutrient recycling. However, many forbs contain secondary compounds that make plant litter more resistant to break-down, and these species can slow rates of nutrient recycling (Eviner and Chapin 2003). Legumes generally increase total soil N and rates of N cycling through their fixation of atmospheric N (Eviner and Chapin 2003), and can also enhance soil organic C (Derner and Schuman 2007).

4.2.1.2 Biological Soil Crusts

Biological soil crusts are composed of cyanobacteria, lichens, mosses, fungi, and algae that intermix and form a consolidated matrix with surface soils (Fig. 4.3). The dominant organisms are autotrophic; thus they reach their greatest development in ecosystems where low water availability prevents continuous plant cover; under these conditions they can be a dominant feature of the landscape and coverage can approach 100 % in plant interspaces. The occurrence of biological soil crusts on the soil surface directly influences hydrology and soil surface stability, and autotrophs and organisms capable of biological N_2 fixation within crusts play a central role in ecosystem biogeo-chemistry. An extensive review of the composition and ecological roles of biological soil crusts in diverse biomes is provided by Belnap and Lange (2001).

The impacts of biological soil crusts on surface hydrology and soil stability are complex and variable (Warren et al. 2001; Rodriguez-Caballero et al. 2013). Biological soil crusts decrease erosion by protecting the soil surface from the direct impact of rain drops in a similar manner as vegetation. Greater microtopography of well-developed soil crusts can also increase pooling of water on the crust surface,

Fig. 4.3 Biological soil crusts are a consolidated matrix of organisms that stabilize soil surfaces (*left*) and can be important components of ecosystem carbon and nitrogen cycles. Coverage can be 100 % in plant interspaces. The soil crusts are the dark covering on the soil surface in the photo-graph on the *right*; the light shade is an artificial disturbance treatment illustrating the color of the soil underlying the soil crusts (photographs by R.D. Evans)

thus decreasing surface runoff (Rodriguez-Caballero et al. 2013). Wetting can also cause swelling of crust organisms effectively blocking infiltration into the soil (Belnap 2006; Bowker et al. 2013; Rodriguez-Caballero et al. 2013). Organisms within the crust also exude a gelatinous sheath that binds fine soil particles into larger aggregates (Belnap and Gardner 1993). This can have a negative impact on water infiltration in sandy soils because polysaccharide secretions, filaments, and accumulated fine soil particles may block pores at the soil surface reducing porosity (Warren et al. 2001). Infiltration may also be lower because the organic component of soil crusts is slightly hydrophobic. In contrast, in fine textured soils binding of soil fine particles into larger aggregates may increase effective porosity and infiltration. In general, crusts in cool and cold environments increase infiltration and reduce runoff, while crusts in very dry environments generally reduce infiltration and increase runoff, and results are mixed in intermediate locations (Belnap 2006).

Biological soil crusts have long been recognized as critical components of biogeochemical cycles in most arid rangelands. Autotrophic organisms within the crusts contribute to ecosystem C cycles (Evans et al. 2001a), and N_2 fixation by crust organisms can be the dominant source of N into many arid ecosystems (Evans and Ehleringer 1993; Evans and Belnap 1999; Evans et al. 2001a). Recent synthesis efforts estimate that biological soil crusts provide 7 % of the net primary production of terrestrial vegetation and nearly half of terrestrial biological N_2 fixation worldwide (Elbert et al. 2012). Lichens and mosses can also directly impact phosphorous availability and can promote chemical weathering of substrates (Cornelissen et al. 2007; Porada et al. 2014).

Understanding the fate of C and N assimilated by soil crusts is critical because of their central role in ecosystem C and N cycles. Rates of N_2 fixation by crust organisms are much greater than necessary for their own growth; thus it appears that a majority of the N assimilated during N_2 fixation is either emitted back to the atmosphere as nitric oxide (NO), nitrous oxide (N_2O), or ammonia (NH_3) during nitrification, denitrification, and volatilization or leached into underlying soils where they could be assimilated by microbes and vascular plants (Evans and Johansen 1999; Strauss et al. 2012). Recent studies combining molecular and microsensor approaches (Johnson et al. 2005, 2007; Strauss et al. 2012) indicate that despite significant variation in community structure across arid lands of North America, the relative magnitude of N_2 fixation, N gaseous loss, and N inputs into underlying soils was similar for all sites, with low flux rates of N to the atmosphere, indicating that biological soil crusts are net exporters of NH_4^+, NO_3^-, and organic N into underlying soils at rates commensurate with rates of N_2 fixation (Johnson et al. 2007; Strauss et al. 2012).

4.2.1.3 Soil Microbial Diversity and Function

Bacteria. Microbial diversity in rangeland soils remains poorly characterized (An et al. 2013) but this is rapidly changing with recent advancements in molecular and bioinformatics techniques. Recent studies have revealed unexpectedly large bacterial diversities despite low microbial biomass (An et al. 2013). These diversities are described by culture-independent, DNA-based approaches, such as sequencing the 16S gene to profile bacterial communities. Generally, the composition of soil bacterial communities are site specific, but at the phylum level, the most abundant members include Proteobacteria, Bacteroidetes, and Actinobacteria (An et al. 2013; Kim et al. 2013; Steven et al. 2014); these phyla are typical of soils around the world. In a global analysis of the biogeography of soil bacteria, arid soils were observed to have the highest proportions of Actinobacteria (Fierer et al. 2009). This abundance is particularly notable as the degree of ecosystem aridity increases, the relative abundances of the Proteobacteria decrease (Neilson et al. 2012). Abundances of beta-Proteobacteria and Bacteroidetes are often correlated with carbon availability as measured by carbon mineralization (Fierer et al. 2007). Total bacterial richness has also been observed to increase with increasing C:N (An et al. 2013), suggesting that increased diversity may result from nitrogen limitations. Overall, soil texture is a good predictor of community structure (Kim et al. 2013; Pasternak et al. 2013), suggesting that the local environmental conditions that underlie habitat quality—pore structure, mineral surfaces, and water-holding capacity—influence microbial composition.

DNA-based approaches identify organisms that are present in the soil including those that are dormant when environmental conditions are not favorable for their activity. In contrast, RNA-based approaches provide a more functional approach by describing only the active members of the microbial community. This is an important distinction in rangelands where short-term stresses such as water limitations or high temperatures can cause the active microbial community to be very different from the dormant, and thus total, community. For example, in a series of sites in the Mediterranean, the community structure of the total Bacteria and Archaea revealed by DNA was driven by large-scale landscape patterns and less strongly correlated with local transient variables such as soil water (Angel et al. 2013). The RNA-based profiles were significantly more correlated to soil water, though this active subset of the microbial communities was less diverse indicating that a large portion of the community was dormant. The effect of water on these communities was stronger in Bacteria than in Archaea (Angel et al. 2013), which is consistent with other observations of the greater tolerance of Archaea to extreme conditions (Sher et al. 2013). Ammonia-oxidizing bacteria and archaea both had relatively stable community structures and were able to withstand the extreme conditions in arid systems; however the archaeal communities were more associated with the dry and hot periods, whereas the bacterial community was more responsive during the wetter periods (Sher et al. 2013).

Fungi. The fungal community is less well studied than the bacterial community in rangeland systems; yet fungi are potentially more able to withstand stressful conditions (Jin et al. 2011). Fungi are more drought tolerant than bacteria in arid soils, as fungal:bacterial ratios reported by phospholipid fatty acid profiling are negatively correlated with soil water content (Jin et al. 2011). Arbuscular mycorrhizal (AM) fungi can stimulate plant growth and drought tolerance (Querejeta et al. 2007); therefore, these plant symbionts are likely very important in arid ecosystems characterized by patchy plant distribution. AM fungal diversity is largely controlled by local soil nutrients with lower levels of phosphorus causing more diverse AM fungal communities, and yet less diverse microbial communities, because competition for phosphorus excluded several types of microbes (Martinez-Garcia et al. 2011). Comparison of rhizosphere with bulk soil showed greater AM fungal diversity in the bulk soil, while the plant host selected for the most fit symbiont, reducing diversity in the rhizosphere (Martinez-Garcia et al. 2011). Free-living fungal communities in the bulk soils of very arid systems have relatively low diversity dominated by Basidiomycota and Ascomycota (Steven et al. 2014). Within the Ascomycota, the Dothideomycetes were highly abundant in the bulk soil (Steven et al. 2014).

4.2.2 Ecosystem Processes

4.2.2.1 Water

Water availability in rangeland systems is a critical determinant of soil community composition and function (Campbell et al. 1997). In most rangelands, precipitation inputs are low and episodic often occurring as short pulses (Austin et al. 2004). Soil physical parameters and vegetation condition are the major determinants of the fate of this water. Precipitation can evaporate from the soil or vegetation surface, run off from the surface through overland flow, or infiltrate into the soil, where the water can be stored, evapotranspired, or leached into groundwater (Wilcox 2002; Chapter 3, this volume).

Soil water-holding capacity is influenced by particle sizes, soil organic matter, and soil depth. Soil texture is critical in determining available soil water, and soil water-holding capacity can increase up to 3.7 % for every 1 % increase in soil organic matter (Bot and Benites 2005). In areas receiving less than 360 mm of precipitation per year, coarse soils decrease evaporative loss and enhance plant water availability. In areas receiving greater than 370 mm of precipitation per year, fine-textured soils increase plant water availability by enhancing water-holding capacity (Austin et al. 2004).

Soil water infiltration is critical for plant water supply and to prevent erosion through overland flow. High soil pore volume can greatly increase infiltration, and is enhanced by soil aggregation, and channels created by roots and through the activity of soil organisms. Vegetation can greatly enhance infiltration by slowing down overland flow, allowing more time for water to infiltrate into the soil. Vegetation cover is also critical in enhancing infiltration by minimizing soil compaction, decreasing the impacts of raindrops directly on the soil surface, increasing porosity through root growth, and increasing soil organic matter to promote aggregation (Li et al. 2011). For example, when 60–75 % of the ground is covered with live plants and litter, surface runoff can be as low as 2 % of rainfall. However, with 37 % plant cover, surface runoff increases to 14 % of rainfall, and when vegetation cover is less than 10 %, up to 73 % of rainfall can run off the surface (Bailey and Copleand 1961). Biological processes can also induce soil water repellency which can decrease infiltration rates by 6–25-fold (Doerr et al. 2000) and is caused by factors that include waxy plant materials, hydrophobic compounds produced by fungi, and hot surface temperatures and fires which melt or volatilize these compounds, causing them to coat the soil surface (Doerr et al. 2000; Schnabel et al. 2013). Water repellency of soil is particularly common in semiarid and seasonally dry conditions, and in fire-prone areas, which are typical in rangeland systems. This repellency can last from seconds to weeks (Doerr et al. 2000), and is particularly high in the dry season, and in the transitions between dry and wet periods (Schnabel et al. 2013). This water repellency can lead to heterogeneity in infiltration, and soil water content, causing preferential flow paths which can accelerate leaching losses and erosion (Doerr et al. 2000, Schnabel et al. 2013).

While most rangelands are constrained by water inputs, water limitation can be caused by meteorological drought resulting from low rainfall, or agricultural drought resulting from management practices that degrade the soil's ability to infiltrate and store water, thus limiting production of crops or forage (Rockstrom 2003; Mishra and Singh 2010). Intensive grazing can decrease soil water by compacting soil, which decreases infiltration, soil water (Trimble and Mendel 1995), and soil organic matter (Hutchens 2011). In general, low-to-moderate grazing has variable effects on soil water availability. While soil water availability generally mirrors grazing effects on soil organic matter, soil water decreases when grazing leads to compaction (Derner and Schuman 2007) that can cause increased runoff and overland flow (Wilcox 2002). High stocking rates can also decrease water infiltration because livestock trampling can destroy soil aggregates critical for maintaining soil porosity, and decrease plant cover, which decreases soil organic matter and increases the erosive impact of raindrops (Briske et al. 2011). These effects seem to be more associated with stocking rate than duration/seasonality of grazing, so that a high-density, short-duration grazing regime can have larger impacts than moderate, continuous grazing. However, rotational grazing that provides rest periods greater than 1 year can improve hydrologic function. The intensity of grazing necessary to disrupt water infiltration depends on vegetation cover (Briske et al. 2011).

4.2.2.2 Decomposition

Decomposition is the physical and chemical breakdown of dead biological materials. Decomposition is responsible for the formation of soil organic matter, and transformations result in organism-available energy and nutrient sources. The chemical composition of litter is a traditional metric used in studies of decomposition because it relates directly to the stoichiometric and chemical requirements of microbial decomposers. Predictors of quality of litter, C:N and lignin:N ratios, describe energy and nutrient source as well as the relative recalcitrance of available substrates. It is for these reasons measures of climate and litter quality are correlated with rates of net N mineralization at a global level (Manzoni et al. 2008) and are included in models of ecosystem biogeochemistry (Parton et al. 1987). However, in arid and semiarid rangelands the value of these predictors is more problematic (Throop and Archer 2009; Austin and Ballare 2010). Recent research in arid range-lands has focused on other determinants of litter to better predict rates and patterns of decomposition. The role of UV radiation has received considerable attention because litter on the soil surface is often exposed to direct solar radiation for long periods in ecosystems with low plant cover. Another key determinant is the rate of soil–litter mixing, which is relatively slow compared to more mesic ecosystems.

The role of UV radiation in aboveground decomposition has become increasingly apparent, especially in more xeric regions with lower plant cover. Photomineralization of litter by UV radiation can short-circuit the C cycle because it results in direct loss of C to the atmosphere (Austin and Vivanco 2006), and can alter the chemical composition of remaining litter (Austin and Ballare 2010). In contrast, in the more mesic shortgrass steppe, precipitation, litter chemistry, and UV radiation were all important factors, but precipitation was the best determinant of litter decomposition, followed by litter chemistry (Brandt et al. 2010). Intercepted radiation was the major factor observed to have a significant effect on decomposition rates in the Patagonian Steppe, where filtering of UV-B and total solar radiation decrease decomposition rates by 33 % and 60 %, respectively (Austin and Vivanco 2006). Significant mass loss was associated with solar radiation in a mesocosm experiment in North America, and rates of loss were not significantly different between UV radiation alone, microbial activity alone, or UV and microbial activity in combination. The authors conclude that UV radiation can be as effective at decomposing litter as microbial activity (Gallo et al. 2006). Lignin is often very resistant to microbial attack, so decomposition rates often decrease with increasing lignin content. However, rates of photodegradation have been observed to increase with lignin concentration because it is an effective light-absorbing compound (Austin and Ballare 2010), and exposure to sunlight leads to a decrease in litter lignin concentrations. Thus, while increased lignin will decrease microbial medi-ated decomposition, it actually enhances photodegradation. An added consequence of greater photodegradation of lignin is that it allows microbial decomposers access to labile compounds protected by lignified materials.

The importance of photodegradation decreases with increasing plant cover and water, and as litter is mixed with soil (Barnes et al. 2012). The development of the soil–litter matrix is critical in our understanding of rangeland decomposition as it mediates the transition from physical breakdown from abrasion and photodegrada-tion to biologically mediated decomposition by soil microbes. In a UV–soil mixing factorial experiment in a savanna ecosystem early decomposition was enhanced by UV radiation, but the effects decreased with increasing soil–litter mixing (Barnes et al. 2012). Nearly half of the surface area of litter was covered by a film of soil particles bound with fungal hyphae after 180 days. Others have also observed a positive relationship between soil litter mixing and mass loss of litter (Hewins et al. 2013). Three mechanisms lead to the increase in decomposition with soil–litter mixing (Throop and Archer 2009). First, soil is the vector for colonization of litter by microbial decomposers. Second, mixing with soil can ameliorate temperature and water extremes associated with the soil surface provide a more favorable micro-environment for microbial colonization. Finally, soil transport along the soil and litter interface may increase the physical abrasion of litter making it more accessible to microbial decomposers.

4.2.2.3 Rhizosphere Dynamics

The influence of plant communities and the formation of resource "islands" exert strong controls on microbial community composition in arid and semiarid soils (Angel et al. 2013; Wang et al. 2013; Steven et al. 2014), with effects being more pronounced in arid soils compared to semiarid (Ben-David et al. 2011). Plant roots create local environments that are enriched in organic C and N (Orlando et al. 2012), have improved water-holding capacity (Wang et al. 2013), and support increased microbial biomass (Marasco et al. 2012; Wang et al. 2013), including the obligate plant host for symbiotic fungi (Martinez-Garcia et al. 2011). These islands of fertility have significantly greater function as indicated by greater rates of gaseous N losses and respiration (McCalley et al. 2011, Wang et al. 2013).

Rhizosphere dynamics link multiple complex plant processes—photosynthate translocation, rhizodeposition, root demography, and mycorrhizal associations—with equally complex soil biotic factors—root herbivory, biofilm development and maintenance, and micro- and mesofaunal interactions. Further complicating our understanding of these interacting processes is the sensitivity of each of these processes to abiotic factors, including temperature, soil water, and atmospheric CO_2. Several exciting developments in rhizosphere research have shown that the microbial communities that develop around root-soil interfaces inhabit biofilms, the extracellular polymeric substance matrix typical of benthic microbial habitat everywhere, and that these biofilms mediate mineral-microbe-root chemical exchanges (Balogh-Brunstad et al. 2008) and create resilient microenvironments suitable for maintaining microbial life and sustaining terrestrial ecosystems (Schimel et al. 2007). Production by microbes within biofilms and the associated roots of higher plants can determine soil C accumulation, local pH, and rates of C and N transfer between roots and soil organisms. The development of mycorrhizospheric biofilms, by microbial and plant production of extracellular polymeric substances (EPS) and manipulation of biofilm chemistry, creates a microenvironment where microbial metabolism is buffered from extremes in soil conditions during drying and wetting.

4.2.2.4 Carbon Dynamics

The physical properties of soils exert strong controls on the soil C cycle. Soil structure created by aggregation of physical particles and organic matter increases C storage at fine scales. Greater concentrations of soil organic C are observed in soil aggregates (Wiesmeier et al. 2012; Fernandez-Ugalde et al. 2014) because aggregation decreases the surface area per unit mass of soil increasing physical protection of soil organic matter from decomposition. This stabilization of organic C is enhanced in calcareous soils typical of many arid rangelands because the deposition of calcite crystals can decrease the overall porosity of the aggregates (Fernandez-Ugalde et al. 2014). A comparison of the C mineralization rates between soil cores with intact aggregates and those with aggregates disrupted by sieving revealed that removal of structure increased decomposition and subsequent respiration of soil C from 11- to 16-fold (Norton et al. 2012).

At broader scales, community structure in ecosystems with bunchgrasses or shrubs can create resource islands under the canopies of dominant plant species. Microbial biomass and resulting biological activity are enhanced in these areas compared to plant interspaces because of greater organic matter inputs from above- and belowground plant litter (Bolton et al. 1993; Smith et al. 1994). Resource islands can have twice the total soil C and over five times the detectable microorganisms as soils located between plant canopies (Bolton et al. 1993) resulting in much greater respiration rates (Bolton et al. 1993; Smith et al. 1994; Wang et al. 2013).

Small changes in the timing and quantity of precipitation can significantly alter the C cycle in arid systems. Additions of 2–5 mm of precipitation each week can increase soil respiration as much as 50 % (Lai et al. 2013). Changes in the seasonal timing of precipitation can also be critical. Water additions during the dry season can greatly increase microbial activity and respiration demonstrating that soil organisms are able to respond rapidly to pulses to soil water, but similar treatments during the moist season had little effect (Song et al. 2012). However, more frequent but smaller events can have detrimental effects on soil organisms. Small events can immediately increase microbial activity and respiration, but short-term increases in soil water do not persist long enough to stimulate photosynthesis in autotrophic organisms such as cyanobacteria. This leads to carbon deficits and long-term declines in their population size (Johnson et al. 2012).

Of particular interest in rangeland ecosystems is the influence of drying and wetting soils on CO_2 fluxes. Soil scientists have documented that drying soils reduces microbial activity and rewetting soils rapidly increases respiration rates (Birch 1958; Miller et al. 2005; Borken and Matzner 2009). When a dry soil was rewetted, CO_2 production could be elevated by up to 500 % over soils continuously moist (Fierer and Schimel 2002). This "Birch effect" can result in much higher respiration-related soil C loss than in soils that have not experienced soil water pulses. There is abundant discussion about the mechanism responsible for the Birch effect (Placella et al. 2012), and hypotheses include accumulation of solutes in microbial cells during drying that are rapidly mineralized during rewetting (Harris 1981), hydration, lysis, and mineralization of dead microbial cells (Schimel et al. 2007), and an increase in the availability of nonmicrobial substrates (Unger et al. 2012), exoenzyme activity (Moorhead and Sinsabaugh 2000), or physical release of CO_2 from carbonates (Billings et al. 2004). A recent study (Blazewicz et al. 2014) provides insight by combining DNA stable isotope probing and measures of soil CO_2 fluxes. CO_2 production increased significantly within 3 h of a water pulse, but this was not accompanied by significant microbial growth. Total microbial abundance did not change significantly in the 7 days post-pulse, but substantial turnover of ~50% was observed for both bacteria and fungi. The authors suggest that cell death that occurs during dry-down periods or during rapid changes in water potential associated with pulses generates a pool of labile C that likely contributes to the large CO_2 fluxes observed following precipitation pulses.

4.2.2.5 Nitrogen Dynamics

The relative importance of processes within the N cycle is highly variable as a function of precipitation, plant species and phenology, and disturbance regime. The soil N cycle in more xeric sites is very "open" characterized by relatively rapid fluxes into and out of the soil (Peterjohn and Schlesinger 1990). The primary ecosystem input of N is N_2 fixation by biological soil crusts (Evans and Ehleringer 1993) and much of this is leached into soils in organic and inorganic forms (Johnson et al. 2005, 2007). Rates of gaseous loss from volatilization, nitrification, and denitrification are of comparable magnitude, leading to very slow accumulation of organic N over time. Retention and internal cycling of N increase in importance with increasing precipitation. For example, nitrogen-use efficiency of plants increased along a water gradient in North America resulting in greater plant biomass in spite of similar rates of N mineralization (McCulley et al. 2009). Decomposition rates and release of N from litter decreased with increasing precipitation, resulting in larger N pools, but also greater N limitations to plants, on the mesic end of the gradient.

While average precipitation patterns can have strong impacts on nutrient dynamics, rangelands are also characterized by strong seasonal variations in precipitation, often with distinct wet and dry seasons, and sometimes frequent dry-wet pulses (Austin et al. 2004; Borken and Matzner 2009). During dry periods, soil microbial activity may be low, but inorganic N tends to accumulate through slow mineralization rates, and low diffusion and plant uptake (Evans and Burke 2013). Nitrogen uptake may also be facilitated during these dry periods by hydraulic lift that promotes mineralization even when surface soils are dry (Cardon et al. 2013). When soils rewet during a rainfall event, there tends to be a spike in N availability. The wetting up of dry soils stimulates a short-term spike in C and N mineralization through the disruption of soil aggregates (releasing physically protected soil organic matter), through lysis of microbial cells, and through microbial release of solutes to prevent them from bursting due to wet up. The size and duration of this flush of mineralization vary depending on soils and rainfall patterns, temperatures, and whether soils are hydrophobic or not (Borken and Matzner 2009; Morillas et al. 2013). Pulses tend to be higher in fine-textured soils, due to their higher C and N contents (Austin et al. 2004). Frequent wet-dry pulses often lead to lower pulses of mineralization, due to the depletion of labile soil organic matter over time. While the rates of N and C mineralization spike during these rewetting periods, cumulatively, mineralization rates are usually higher on continuously wet soils than on dry soils during these wet-dry cycles, due to low rates during the dry periods (Borken

Fig. 4.4 Native (*left*) and invaded (*right*) grasslands on the Colorado Plateau. Cheatgrass (*Bromus tectorum*) invasion shades biological soil crusts located in plant interspaces and greatly increases aboveground litter biomass (photographs by R.D. Evans)

and Matzner 2009). These wet-dry cycles can also lead to high losses of N through leaching and trace gases, particularly since the wet-dry cycles stimulate N mineralization to a greater extent than C.

Changes in species composition as modified by local environment also impact soil N dynamics. This is exemplified by the contrasting impacts of invasion by the exotic annual cheatgrass (*Bromus tectorum* L.). In the Colorado Plateau of North America, cheatgrass invasion has established a series of positive feedbacks among biological soil crust disturbance, fire frequency, and cheatgrass abundance (Fig. 4.4), all of which result in declines in total and available soil N (Evans et al. 2001b). In this scenario cheatgrass shades autotrophic organisms within the biological soil crust decreasing N_2 fixation, which also favors cheatgrass establishment. Ecosystem N also accumulates in aboveground litter where it is not available for mineralization within the soil (Sperry et al. 2006), decreasing inorganic N availability. Subsequent range fires volatilized N within the litter, further decreasing soil fertility (Evans et al. 2001b; Sperry et al. 2006). In contrast, other studies from the Great Basin of North America (Booth et al. 2003) have observed greater total N, gross mineralization, and gross nitrification under cheatgrass than in neighboring native communities where differences in microenvironment promote greater mass loss of litter and N mineralization under cheatgrass. Nitrifier populations were also greater in invaded communities, and ultimately these differences in microenvironment and soil properties promoted greater N availability under cheatgrass compared to native communities.

Disturbance history can also influence soil N dynamics, especially in ecosystems that have evolved in the absence of large herbivores such as the Intermountain West of North America (Mack and Thompson 1982). In such cases, surface disturbance can disrupt the biological soil crust, greatly decreasing N_2 fixation. Gaseous loss continues, causing overall N loss from the ecosystem and decreases in soil fertility (Evans and Belnap 1999). For example, disturbed sites that had been allowed to recover 30 years still had total soil N content and potential N mineralization rates that were 30% and 70% lower, respectively, than adjacent undisturbed locations. This presents an interesting contrast on anthropogenic effects on the N cycle, because surface disturbance can decrease soil fertility across many rangelands, while a majority of terrestrial ecosystems are impacted by N deposition leading to N excess (Pardo et al. 2011). Grazing can have: no effect, enhance, or decrease N recycling rates in more mesic rangelands, depending on the conditions. In general, on high-fertility soils, grazing enhances recycling rates. However, under low-fertility conditions, livestock preferentially graze high-quality plants, selecting for unpalatable plants which decrease nutrient recycling rates (Bardgett and Wardle 2003). Similarly, grazing on low-fertility soils can decrease aboveground plant production, especially during drought (Augustine and McNaughton 2006). However, the impacts of grazing can vary due to the interactions of soil fertility and precipitation. In low-rainfall conditions, grazers reduce aboveground net primary production, regardless of soil fertility. However in wetter conditions, grazing increases aboveground net primary production on high-fertility soils, but decreases it in low-fertility soils (Augustine and McNaughton 2006).

4.3 Anthropogenic Impacts and Societal Implications

4.3.1 Responses to Land-Use Change

Grazing effects on soil characteristics have been a dominant area of research in rangeland ecology. The key responses most often measured include changes in soil organic matter (McSherry and Ritchie 2013), N transformations (Anderson et al. 2006; Ingram et al. 2008), and soil texture, with its impacts on infiltration and water-holding capacity (Zhao et al. 2007, 2010). A challenge in synthesizing the impacts of grazing on soils is the extreme variability in environmental conditions and grazing practices in rangelands. As a consequence, many reviews have concluded that grazing has mixed results (Milchunas and Lauenroth 1993; Derner et al. 2006) on key soil attributes. However, with the accumulation of additional studies and novel meta-analytical tools, there is a clearer understanding of how environmental variability may interact with grazing practices to produce predictable outcomes in soil processes. For example, there is an interaction between soil texture and precipitation in determining the effect of grazing on soil C (McSherry and Ritchie 2013). On fine-textured soils, as precipitation increases the effect of grazing decreases while over the same range in precipitation, coarse-textured soils had an increase in the grazer effects on soil C. Increasing grazing intensity in rangelands dominated by plant communities utilizing C4 or mixed C4–C3 photosynthetic pathways increased soil C, while increasing grazing intensity in C3-dominant communities decreased soil C. While it is challenging to recognize context-specific grazing effects on soil processes, we are moving into a period where we can begin to predict specific effects due to grazing.

For many rangeland soils, soil loss is driven by surface disturbance of physical and biological soil crusts (Belnap et al. 2014). Dust production is best understood as the interacting processes that control the entrainment, transport, and deposition of wind-borne sediments (Chepil 1951, 1953; Ravi et al. 2011). Initially, wind erosion research was motivated by a desire to understand the geomorphic and erosion processes associated with agriculture (Chepil 1951, 1953). In the past two decades, this research has been reinvigorated as it has become clear that airborne sediments strongly influence soil fertility, planetary energy balance (Goudie and Middleton 2001; Goudie 2008; Ravi et al. 2011), snow surface albedo, and thus melt rates on downwind mountain snowpack (Painter et al. 2010, 2012a, b). Drylands around the world are considered to naturally generate large amounts of sediment; many studies show that this is generally not true unless soil surfaces are disturbed (Belnap et al. 2014). Instead, most undisturbed soil surfaces, unless they are barren sand dunes or soils consisting of mostly fine sands, have some form of protective surface, whether physical crusts, biocrusts, rocks, or plant cover. However, almost all these protective covers are highly vulnerable to the forces associated with vehicle and animal traffic. Once disturbed, protective crusts are unable to stabilize soil surfaces. Surface disturbance due to animal traffic, off-road vehicle use, or other processes causes sediment generation to increase dramatically at most sites (Warren et al. 2001). As high winds commonly occur in rangeland regions, this destabilization often results in large dust storms originating from once stable areas. While there are some common predictors of sediment emission globally (e.g., silt content), other biophysical factors such as biocrust biomass and rock content are more important at the site scale. Soil surface disturbances are expected to increase in dryland regions, given the increasing demand for energy, minerals, recreational opportunities, and food production. With this disturbance, aeolian erosion is also expected to increase from the disturbed soils.

4.3.2 Responses to Invasive Species

Invasive, non-native plant species have become one of the most pressing rangeland management issues (Fig. 4.5). In the western USA, 51 million hectares of rangeland are now dominated by invasive plants considered to be noxious weeds (Duncan and Clark 2005), including diverse life forms such as annual and perennial grasses, annual and perennial forbs, shrubs, and trees. In over 2/3 of western rangelands, non-native annual grasses account for 50–85 % of vascular plant cover (DiTomaso 2000; Chapter 13, this volume). Invasions by exotic species are altering the structure and function of soils, including water availability and flow, C storage, nutrient availability, erosion rates, soil microbial communities, and disturbance regimes (Eviner et al. 2012; Eviner and Hawkes 2012).

While invasive plants have the potential to strongly impact soil, the specific impacts of any invasion strongly depend on environmental conditions and the unique functional traits of the invasive compared to native species. For example, the invasion of medusa head wildrye (*Taeniatherum caput-medusae* (L.) Neviski) into western US rangelands is associated with loss of soil C (Eviner and Hawkes 2012). In contrast, in New Zealand, the exotic mouse-eared hawkweed (*Hieracium pilosella* L.) increases soil organic matter and productivity in overgrazed pastures, but

Fig. 4.5 Senesced (background) grasses are naturalized annual species in California's rangelands. Medusa head and goatgrass are present in the foreground, and both are invasive noxious weeds (photograph by V. Eviner)

this effect depends on grazing intensity and aspect (Scott et al. 2001). Invasive species may enhance N availability through increases in decomposition and N mineralization rates (Ehrenfeld 2003; Corbin and D'Antonio 2004; Liao et al. 2008), although some invasive species decrease N availability, such as goatgrass (*Aegilops triuncialis*) into grasslands of California, USA (Drenovsky and Batten 2007), and crested wheatgrass (*Agropyron cristatum* L.) into the northern Great Plains of the USA (Christian and Wilson 1999). The effects of cheatgrass and spotted knapweed (*Centaurea maculosa*) on soil N vary with local soil and climate conditions, with different studies showing an increase, decrease, or no change (Eviner and Hawkes 2012). Even when invaders don't alter the total amount of soil resources, they can change the timing, location, and type of resource availability, restricting which plant species have access to these resources. For example, replacement of native perennial species by the annual cheatgrass alters the timing of soil N availability, with high N availability occurring after its senescence because perennials are not present to assimilate inorganic N released during decomposition of cheatgrass litter (Adair and Burke 2010). Subsequent leaching in the absence of active plant cover redistributes soil nitrate deep in the soil profile. Cheatgrass is a winter annual, while native grasses resume activity in spring; so root growth by cheatgrass during winter and early spring allows it to assimilate this leached nitrogen pool before native grasses are active later in spring. This enhances cheatgrass growth at the expense of the native grasses (Sperry et al. 2006). Invasive plants also can alter the form of N available. For example, in California grasslands, USA, invasive grasses increase the soil bacterial nitrifier populations responsible for nitrification (Hawkes et al. 2005). Conversely, the invasion of gamba grass (*Andropogon gayanus*) into Australian grasslands inhibits nitrification (Rossiter-Rachor et al. 2009). Changes in the form of available N can alter competition between species (Marschner 1986; Crabtree and Bazzaz 1993). The largest impacts of invasive plants on soils can be mediated through changes in disturbance regimes. For example, invasion of cheatgrass and medusa head can greatly increase fire frequency, which greatly increases potential loss of soil C and N. Soil nutrient availability may briefly spike after disturbances, stimulating reestablishment of invaders (Eviner et al. 2012; Eviner and Hawkes 2012). Increased fire frequency in response to annual grass invasion can also lead to higher erosion rates after fire, leading to potential desertification of rangelands (Ravi et al. 2009).

Many invasive species alter soil in ways that shift competitive balance from native to non-native species resulting in "invasional meltdown" (Simberloff and Von Holle 1999). For example, changes in the soil environment caused by the invasion of cheatgrass favor medusa head, and further modification of soils favors exotic forbs (Eviner et al. 2012). Similarly, changes in the soil microbial community caused by smooth brome (*Bromus inermis*) can favor the establishment of the invader leafy spurge (*Euphorbia esula*) (Jordan et al. 2008). Controlling the positive soil feedbacks of invaders requires a specific understanding of the mechanism of feedback, which can vary greatly by species, and even within a species across sites (Eviner and Hawkes 2012). Early eradication is critical, as the strength of soil feedbacks tends to increase with time since invasion. For example, extensive erosion as a result of invasion of spotted knapweed (*Centaurea maculosa*) (Lacey et al. 1989) can take decades to centuries to reverse via soil formation processes and the gradual buildup of organic matter by the restored plant community.

4.3.3   Responses to Global Climate Change

4.3.3.1   Precipitation Change

Climate is a major controlling factor over ecosystem structure and function and changes in the frequency and intensity of drought caused by anthropogenic activity are the most likely to negatively impact rangeland ecosystem processes. Drought restricts biological activity, primary production, timing of plant growth, patterns of mortality, organic matter dynamics, and nutrient availability (Borken and Matzner 2009; Evans and Burke 2013; DeMalach et al. 2014). For example, extended drought produced region-wide tree mortality in the southwestern USA (Breshears et al. 2005). There have been a number of studies that evaluate two critical elements of precipitation change: (1) the timing of precipitation, independent of total amount, and (2) changes in the amount of precipitation.

While soil water is a principal driver of soil processes in rangelands, it is very likely that precipitation change will interact strongly with rising temperatures—particularly extreme heat waves. One of the most promising advances in understanding rangeland soil responses to climate change is the addition of factorial experiments that look at the individual and combined effects of various climate change factors. Relatively modest changes in soil water content can produce much larger changes in ecosystem processes, including soil CO_2 efflux (Fay et al. 2011). A 15% decrease in soil water that accompanied precipitation variability treatments caused measured responses to decrease by up to 40%. In addition to modifying precipitation timing, year-round warming can cause a change in plant phenology and shift the seasonality of microbial activity with higher winter soil CO_2 flux and lower summer soil CO_2 flux. While the effects of changing precipitation variability and warming were modest compared to large between year variability, they are likely ecologically significant. While this experiment altered mean patterns, additional studies have begun looking at extreme events including the interactions between extreme drought and heat waves and found that even extreme warming had only modest effects on plant production and community composition and that the majority of observed changes were driven by soil water effects. In addition, drought and precipitation variability could impact ecosystem function. Drought has a greater impact when it occurs in the wettest portion of the year when organisms are most active, in contrast to typical seasonally dry periods. In addition, heat waves and droughts late in the growing season had no significant effects on ecosystem dynamics.

4.3.3.2   Elevated CO_2

Arid and semiarid soils are critical components of the Earth's C budget (Evans et al. 2014; Poulter et al. 2014). Recent increases in atmospheric $[CO_2]$ are unprecedented as emissions have accelerated from 1% year^{-1} from 1990 to 1999 to 2.5% year^{-1} from 2000 to 2009 (Friedlingstein et al. 2010). The increase in atmospheric $[CO_2]$ has the potential to dramatically alter the biogeochemistry of rangeland soils. The direct effects of increasing $[CO_2]$ on soils are negligible because soil $[CO_2]$ greatly exceeds that of the atmosphere (Drigo et al. 2008); however, increases in atmospheric $[CO_2]$ can indirectly impact soil elemental cycles by altering the amount and chemical composition of organic matter inputs in the form of above- and below-ground plant litter, rhizodeposition, and mycorrhizal associations. Carbon dioxide is also the dominant greenhouse gas influencing climate, so increases in atmospheric $[CO_2]$ impact ecosystem function through changes in ambient temperatures.

Early studies demonstrated that elevated $[CO_2]$ could increase rates of C assimilation and decrease stomatal conductance, causing increased water-use efficiency (Nowak et al. 2004; Chapter 7, this volume). This led to predictions that water-limited ecosystems, including many rangelands, would be most responsive to future atmospheric conditions. A conceptual model was proposed (Strain and Bazzaz 1983) that predicted that the magnitude of ecosystem responses to elevated $[CO_2]$ was controlled by water and N availability. Ecosystems with sufficient water, or where N was limiting, would see limited response to elevated $[CO_2]$. In contrast, water-limited ecosystems with sufficient N would exhibit the greatest responses. It has since been recognized that N limitations to net ecosystem productivity can increase over time with increased exposure to elevated $[CO_2]$ as organic matter is sequestered into recalcitrant materials such as woody biomass and soil pools with slow turnover (Gill et al. 2002), a process termed progressive nitrogen limitation (Luo et al. 2004).

Fig. 4.6 The Prairie Heating and CO_2 Enrichment (PHACE) study located manipulated temperature and CO_2 concentration in C3–C4 shortgrass steppe. Warming was controlled by infrared heaters at the top of each plot, while CO_2 was controlled by emitting gas from the ring at the bottom of the plot (photograph by Sam Cox, (former) USDA-ARS Biological Science Technician, (now) Natural Resource Specialist with Wyoming BLM)

Results from long-term elevated [CO_2] experiments in rangelands demonstrate complex responses to elevated [CO_2] especially in those studies that consider multiple environmental drivers. Three experiments in North American rangelands occur along a precipitation gradient from 134 mm year^{-1} for the Nevada Desert FACE Facility (NDFF) and 384 mm year^{-1} for the Prairie Heating and CO_2 Enrichment (PHACE) in the C3–C4 shortgrass steppe (Fig. 4.6) to 914 year^{-1} mm at the site of two experiments in the C3–C4 tallgrass prairie (Fig. 4.7). Biogeochemical cycling at the xeric end of the precipitation gradient is characterized by C limitations to microbial activity and very open C and N cycles, so significant increases in ecosystem C and N were observed over the life of the experiment (Evans et al. 2014). Greater soil C under elevated CO_2 resulted from death of enhanced plant growth that occurred during periods of adequate soil water that could not be supported during intervening drought, and significantly greater rhizodeposition (Jin and Evans 2010). Progressive N limitation has yet to be observed as increased C has accelerated rates of N transformations increasing gross N mineralization and organism N availability (Billings et al. 2004; Jin et al. 2011). Nutrient availability was further enhanced by increases in microbial diversity and abundance (Jin and Evans 2010), especially for fungi that are able to utilize the recalcitrant C and N that dominate these soils (Billings et al. 2004). Elevated [CO_2] also increased retention of N within the soil by promoting microbial immobilization of N and decreasing rates of gaseous N loss (Schaeffer et al. 2003; McCalley and Sparks 2008, 2009).

Fig. 4.7 The CO_2 concentration-gradient study in the tall-grass steppe. Air with an elevated CO_2 concentration simulating future conditions is introduced at one end and the concentration decreases through photosynthesis as it moves through the tunnel. This allows scientists to examine plant and soil responses ranging from preindustrial to expected levels in 2050 in the same experiment. The experimental infrastructure also allows manipulation of other variables such as soil type (photograph by F.A. Fay)

Similar results were observed in open top chamber (OTC) experiments in the shortgrass steppe. Elevated $[CO_2]$ stimulated above- and belowground production and increased rhizodeposition almost 100 % (Pendall et al. 2004). Results also suggest that available soil N remained high during the 5-year experiment presumably due to increased soil water promoting more rapid N mineralization (Dijkstra et al. 2008). No differences were observed in gaseous N loss (Mosier et al. 2002). Research on elevated $[CO_2]$ in the shortgrass steppe was subsequently expanded to include effects of warming through the PHACE experiment, and $[CO_2]$ and warming produced contrasting effects on soil processes. In contrast to earlier studies, elevated $[CO_2]$ resulted in a more closed N cycle by promoting microbial immobilization of N, decreasing soil inorganic N availability (Dijkstra et al. 2010; Carrillo et al. 2012). In contrast, warming enhanced inorganic N presumably because of greater rates of decomposition. Similar results have been observed for grasslands in Australia, where elevated $[CO_2]$ decreased available soil N, but warming of 2 °C plus elevated $[CO_2]$ prevented this effect (Hovenden et al. 2008). Results from PHACE also suggest that the interacting effects of warming and elevated CO_2 may decrease ecosystem C storage (Pendall et al. 2013). Microbes from the combined warming–elevated $[CO_2]$ treatments exhibited greater decomposition activity than other treatments suggesting greater rates of decomposition and C loss, a conclusion supported by direct measurements of soil C efflux (Pendall et al. 2013).

Two experiments at the most mesic end of the precipitation gradient strongly support progressive N limitation under elevated CO_2. In the first, intact C3–C4 grasslands were exposed to a gradient in $[CO_2]$ from 200 to 560 µmol CO_2 mol^{-1} (Gill et al. 2002; Gill et al. 2006). A threefold decrease in N availability was observed with increasing $[CO_2]$ (Gill et al. 2002) and N transferred from soil organic matter to plant pools (Gill et al. 2006). Similarly, transfer of soil C from recalcitrant to labile pools was observed resulting in greater C loss through respiration with increasing $[CO_2]$. Greater respiration was offset by enhanced NPP at higher $[CO_2]$ resulting in no change in the overall C balance, but future C sequestration may be limited by progressive N limitations. This experiment was followed by the establishment of the lysimeter CO_2 gradient study using the same gradient in $[CO_2]$ but incorporating three soil treatments. Results from this study strongly emphasize the importance of soil type in controlling responses to drivers of global change. For example, elevated $[CO_2]$ favored forbs over grasses on only one of the three soil types (Fay et al. 2012; Polley et al. 2012). Similar responses were observed for soils where elevated $[CO_2]$ enhanced soil water availability on only two of the three soil types (Fay et al. 2012). In contrast to earlier research, no changes were observed in soil N availability with increasing $[CO_2]$ across all soil types.

4.3.3.3 Atmospheric Deposition

Ecosystems affected by atmospheric deposition are often located downwind of large urban centers, near point emission sources, or in regions with a mix of urban and agricultural emission sources (Fenn et al. 2003a, b). Background rates of atmospheric deposition are generally less than 3 kg N ha^{-1} year^{-1}, but can approach and may even exceed 30 kg N ha^{-1} year^{-1} in the most heavily impacted areas (Fenn et al. 2003b). Research addressing the effects of atmospheric deposition on rangeland soils is rare compared to forest and aquatic ecosystems (Pardo et al. 2011). Assessing the regional importance is difficult because accurate sampling networks for wet and dry deposition are either widely separated or absent in most rangelands around the globe, and when studies do occur they often focus on plant or aquatic species rather than soils.

Attempting to extrapolate the effects of atmospheric deposition from many fertilization studies can be misleading. The major effects of elevated atmospheric deposition result from long-term, chronic addition of low amounts of N. Most fertilization studies apply N in one or a few applications that may overwhelm the assimilation capacity of plants and microbial populations, leading to artificially high rates of leaching and gaseous loss from the soil. Application rates can also greatly exceed rates of deposition observed even in the most impacted areas. Atmospheric deposition in most rangelands is still less than 10 kg N ha^{-1} year^{-1}, and rates exceeding 20 kg N ha^{-1} year^{-1} are found only in ecosystems adjacent to heavily industrialized areas. Critical loads for most ecosystem components are less than 10 kg N ha^{-1} year^{-1} (Pardo et al. 2011); thus application rates in excess of 30–40 kg N ha^{-1} year^{-1} have minimal value when attempting to understand or predict ecosystem responses to atmospheric deposition. Finally, experiments must not only apply realistic amounts spread throughout the year, but must also match the form of deposition most common in that region because oxidized and reduced forms can have very different effects on soil processes.

It is for these reasons that the best studies of the effects of atmospheric deposition are most often those that occur in currently impacted areas. Even here the responses are complex due to interactions between the form and amount of N deposition and their interaction with local soils making generalizations difficult (Greaver et al. 2012). A common observation is that increased atmospheric deposition relieves N limitations to growth and activity by plants and soil organisms (Aber et al. 1989). Initial increases in net primary production are often observed in N-limited soils. Further deposition can exceed rates of plant assimilation leading to increases in nitrification, denitrification, and emissions of greenhouse gases such as NO and N_2O (Fenn et al. 1996, 2003b; Templer et al. 2012). Chronic, high rates of atmospheric deposition may ultimately lead to significant leaching of nitrate from soils into surrounding aquatic ecosystems and groundwater, decreases in base saturation, acidification of soils, and cation limitations to plant and microbial activity (Aber et al. 1989). N deposition can greatly alter plant species composition, particularly promoting invasive species.

4.3.4   Restoration

Rangelands are extremely variable and multiple stable states may exist depending on precipitation patterns, grazing, fire, and land use (Campbell et al. 1997; Reynolds et al. 2007) (Chapter 6, this volume), but functioning soils are the key to resilience under these varying conditions (Briske et al. 2006). Soil functions that are critical for rangeland resilience include infiltration, water storage, erosion control, soil fertility, and net primary production (MEA 2005; Briske et al. 2006), and their degradation can limit restoration of productive communities and the ecosystem services they provide (Bestelmeyer et al. 2003; MEA 2005; Seymour et al. 2010; Li et al. 2013). The key factor to maintaining soil resilience is vegetation cover. This can be disrupted through overgrazing, fire suppression, and drought that may contribute to state shifts from grassland to shrublands, which often increases bare ground and erosion.

Degradation due to soil erosion can be difficult, if not impossible, to reverse (Li et al. 2013), so the best management options are those that prevent it from occurring. Grazing practices to maintain plant cover are the key to preventing erosion (Pulido-Fernandez et al. 2013), with low-to-moderate grazing intensities, and seasonal grazing tending to have minor impacts on soils, while continuous intensive grazing can deplete C and N and degrade soil structure (Kotze et al. 2013; Xu et al. 2014). Grassland productivity can often be maintained by leaving critical levels of plant biomass as residual dry matter. The amounts of residual biomass vary depending on slope and precipitation (Bartolome et al. 2007; Xu et al. 2014), but managing for this target can be difficult due to fluctuations in precipitation, and thus livestock-carrying capacity (Kiage 2013). Restoration of degraded soils often requires cessation of grazing during the recovery period.

4.4   Future Perspectives

The development and application of molecular tools are revolutionizing disciplines within the life sciences. This is especially apparent in soils, where microorganisms are responsible for a majority of the biological activity, but available technologies caused them to be treated as a black box termed "microbial biomass." Recent advances now allow for descriptions of microbial diversity and community structure, and continued development of high-throughput sequencing and metagenomics approaches permits discovery of new organisms and large-scale comparisons across diverse ecosystems (Zak et al. 2006; Weber et al. 2011; Dunbar et al. 2012). Stable isotope probing, where substrates with distinct isotope composition are assimilated by organisms that are later identified using DNA (Zak et al. 2006) or cell wall profiles (Jin and Evans 2010), now has the potential to directly link functional groups of organisms with their biogeochemical function. Quantifying functional genes that encode specific enzymes also allows grouping of organisms into functional groups focused on specific processes (Zak et al. 2006).

Efforts are also under way to describe the functional state of rangeland soil microbial communities using proteomics, the study of protein structure and function (Bastida et al. 2014). Proteomic profiling is potentially very powerful for soil ecology, as the proteome encompasses the overall metabolism of the system, and the cellular machinery that is poised to respond to new conditions. Key functions identified in this study included dehydrogenases, catalases, and superoxide dismutases. The presence of a large number of carboxysome proteins and other C fixation proteins (phycocyanins) in highly degraded, low-C soils suggests that there is large capacity for C fixation that is not favored for environmental reasons such as substrate availability or microbial vigor (Bastida et al. 2014). As proteomic analyses of soils continue to develop, new knowledge of how local conditions regulate soil metabolism is likely to emerge.

Technological advances are not confined to soil microbiology; development of new microelectronics, digital computation, and networking has created a data deluge in rangeland ecology. These new sensors are facilitating our understanding of water fluxes (Reeves and Smith 1992; Paige and Keefer 2008; Flerchinger and Seyfried 2014), soil electrical connectivity, plant and soil microenvironments, incoming radiation amount and quality, and soil CO_2 fluxes (Kao et al. 2012; Taylor and Loescher 2013; Loescher et al. 2014). Rangeland ecologists are now able to monitor microclimate with high spatial and temporal resolution with better precision at costs well below standard prices in past decades. Capacitance and time domain reflectometry measurements of soil water have become commonplace in many rangeland studies, allowing for direct connections between soil water availability and ecological processes. One of the key attributes of the new National Ecological Observatory Network in the USA is the high level of sensor integration into a continental-scale research network (Kao et al. 2012; Taylor and Loescher 2013; Loescher et al. 2014) that offers the potential to ask complex, landscape-scale ecological questions using wireless sensor networks (Simoni et al. 2011; Kerkez et al. 2012; Rosenbaum et al. 2012; Chaiwatpongsakorn et al. 2014). With the promise of a future of inexpensive, spatially and temporally expansive data comes the challenges of managing, integrating, analyzing, and interpreting these data.

Fortunately for soil biologists, many of the challenges that come in a data-rich science have been addressed by bioinformaticians trained to apply computational techniques to analyze information tied to biomolecules, including genomics, gene expression, structural biology, and other molecular applications. Environmental research is quickly requiring the same informatics tools to deal with sensor data (Michener and Jones 2012; Porter et al. 2012) leading to the new field of ecological informatics (Suri et al. 2006; Michener et al. 2007; Michener and Jones 2012).

4.5 Summary

The past 25 years have seen increased emphasis on understanding soil function and sustainability in rangelands, and their significance in global biogeochemical cycles. Research demonstrates that rangelands are extremely diverse and future studies should strive to further develop common unifying principles. It is also apparent that paradigms developed in more mesic ecosystems may not be applicable in rangelands because of their extreme environments and inherently low and often stochastic resource availability. Plant community composition is a major driver of soil function. Grazing can cause shifts to species with lower nutrient content, higher concentrations of defense compounds, or more woody vegetation, lowering nutrient availability and impacting soil water dynamics. Increased emphasis is also being placed on nonvascular components of the rangeland ecosystems. Biological soil crusts are common on surface soils of ecosystems with less than 100 % plant coverage. Soils crusts enhance soil stability and can be the dominant source of N through nitrogen fixation. Recent results demonstrate that N assimilated by crusts is subsequently leached into soils making it available for uptake by plants and soil microbes. Growth of molecular tools with high throughput has greatly expanded our understanding of the diversity and community composition of the archaea, fungi, and bacteria responsible for most transformations within soils. Combining molecular approaches with other techniques has allowed scientists to increasingly link composition and function in soils, thereby shedding light on what was formerly viewed as the black box of microbial dynamics in soils.

Decomposition is a major driver of nutrient availability in all ecosystems. Unlike more mesic ecosystems, photodegradation caused by UV radiation can be a major determinant of rates of decomposition, especially in more arid rangelands. The rate of mixing of litter into soils is also important as this enhances inoculation by microbial decomposers. Carbon and nutrients can also enter soils directly through rhizodeposition through plant roots and mycorrhizae. At small scales, the greatest soil C is associated with soil aggregates and significant amounts of soil C can be protected by soil structure. The unique patchiness of vegetation in many rangelands controls C dynamics at larger scales, and larger microbial populations, greater plant and microbial activity, and greater C fluxes are associated with the canopies of higher plants, rather than in plant interspaces. Soil nutrient cycles are controlled by precipitation, plant species, and disturbance regimes. Nutrient cycles in more xeric ecosystems are characterized as being very "open" with rates of input balanced by relatively equal rates of loss. In contrast, nutrient contents are greater with increasing precipitation, and cycling rates are more "closed" and characterized by internal cycling. Seasonal patterns of precipitation are also important and can lead to pulses of nutrient availability following rain events. Restoration efforts in rangelands must adapt to local conditions, but keys to maintaining soil condition and function are infiltration, water storage, erosion control, and maintaining soil C and nutrients.

Rangeland responses to land-use change are diverse matching their highly variable composition and associated grazing practices. Surface disturbance can increase soil loss in more xeric rangelands by disrupting the biological soil crusts. The impact of invasive plant species on rangeland soils can also be highly variable. For example, cheatgrass can increase, decrease, or not alter available nitrogen. Invasive species can also alter the timing, location, and type of resources available. One of the most profound responses of rangeland soils to changing atmospheric conditions is likely to be responses to the increase in the frequency and intensity of drought. Water is a primary driver of soil processes in rangelands and decreases in water availability can negatively impact microbial activity, nutrient availability, and subsequent plant growth. Of special concern is the interaction between drought and predicted increases in extreme temperatures, which will greatly exacerbate responses observed under drought alone. Precipitation regime will also determine ecosystem responses to increases in atmospheric CO_2. Experiments in more xeric ecosystems found that soil C and N cycling was energy limited and increasing $[CO_2]$ enhanced nutrient availability and C storage. In contrast, additional C in more mesic ecosystems led to microbial sequestration of N and progressive N limitation over time. Although soils have received less recognition than vegetation or climate the knowledge summarized in this chapter demonstrates that they are essential to the resilience of rangeland ecosystems.

References

Aber, J.D., K.J. Nadelhoffer, P. Steudler, and J.M. Melillo. 1989. Nitrogen saturation in northern forest ecosystems. *Bioscience* 39: 378–386.

Adair, E.C., and I.C. Burke. 2010. Plant phenology and life span influence soil pool dynamics: *Bromus tectorum* invasion of perennial C3–C4 grass communities. *Plant and Soil* 355: 255–269.

An, S., C. Couteau, F. Luo, J. Neveu, and M.S. DuBow. 2013. Bacterial diversity of surface sand samples from the Gobi and Taklamaken Deserts. *Microbial Ecology* 66: 850–860.

Anderson, R.H., S.D. Fuhlendorf, and D.M. Engle. 2006. Soil nitrogen availability in tallgrass prairie under the fire-grazing interaction. *Rangeland Ecology & Management* 59: 625–631.

Angel, R., Z. Pasternak, M.I.M. Soares, R. Conrad, and O. Gillor. 2013. Active and total prokaryotic communities in dryland soils. *FEMS Microbiology Ecology* 86: 130–138.

Archer, S.R., T.W. Boutton, and K.A. Hibbard. 2001. Trees in grasslands: Biogeochemical consequences of woody plant expansion. In *Global biogeochemical cycles in the climate system*, ed. E.D. Schulze, S.P. Harrison, M. Heimann, E.A. Holland, J. Lloyd, I.C. Prentice, and D. Schimel. San Diego, CA: Academic.

Augustine, D.J., and S.J. McNaughton. 2006. Interactive effects of ungulate herbivores, soil fertility, and variable rainfall on ecosystem processes in a semi-arid savanna. *Ecosystems* 9: 1242–1256.

Austin, A.T., and C.L. Ballare. 2010. Dual role of lignin in plant litter decomposition in terrestrial ecosystems. *Proceedings of the National Academy of Sciences of the United States of America* 107: 4618–4622.

Austin, A.T., and L. Vivanco. 2006. Plant litter decomposition in a semi-arid ecosystem controlled by photodegradation. *Nature* 442: 555–558.

Austin, A.T., L. Yahdijan, J.M. Stark, J. Belnap, A. Poporato, U. Norton, D.A. Ravetta, and S.M. Schaeffer. 2004. Water pulses and biogeochemical cycles in arid and semiarid ecosystems. *Oecologia* 141: 221–235.

Bai, Y.F., X.G. Han, J.G. Wu, Z.Z. Chen, and L.H. Li. 2004. Ecosystem stability and compensatory effects in the Inner Mongolia grassland. *Nature* 431: 181–184.

Bai, Y.F., J. Wu, C.M. Clark, S. Naeem, Q. Pan, J. Huang, L. Zhang, and X. Han. 2009. Tradeoffs and thresholds in the effects of nitrogen addition on biodiversity and ecosystem functioning: Evidence from inner Mongolia Grasslands. *Global Change Biology* 16: 358–372.

Bailey, R.W., and O.L. Copeland. 1961. Low flow discharges and plant cover relations on two mountain watersheds in Utah. *International Association of Science Hydrology Publication* 51: 267–278.

Balogh-Brunstad, Z., C.K. Keller, R.A. Gill, B.T. Bormann, and C.Y. Li. 2008. The effect of bacteria and fungi on chemical weathering and chemical denudation fluxes in pine growth experiments. *Biogeochemistry* 88: 153–167.

Bardgett, R.D., and D.A. Wardle. 2003. Herbivore-mediated linkages between above-ground and belowground communities. *Ecology* 84: 2258–2268.

Barger, N.N., S.R. Archer, J.L. Campbell, C. Huant, J.A. Morton, and A.K. Knapp. 2011. Woody plant proliferation in North American drylands: A synthesis of impacts on ecosystem carbon balance. *Journal of Geophysical Research, Biogeosciences* 116, G00K07.

Barnes, P.W., H.L. Throop, D.B. Hewins, M.L. Abbene, and S.R. Archer. 2012. Soil coverage reduces photodegradation and promotes the development of soil-microbial films on dryland leaf litter. *Ecosystems* 15: 311–321.

Bartolome, J.W., R.D. Jackson, A.D.K. Betts, J.M. Connor, G.A. Nader, and K.W. Tate. 2007. Effects of residual dry matter on net primary production and plant functional groups in California annual grasslands. *Grass and Forage Science* 62: 445–452.

Bastida, F., T. Hernandez, and C. Garcia. 2014. Metaproteomics of soils from semiarid environment: Functional and phylogenetic information obtained with different protein extraction methods. *Journal of Proteomics* 101: 31–42.

Belnap, J. 2006. The potential roles of biological soil crusts in dryland hydrologic cycles. *Hydrological Processes* 20: 3159–3178.

Belnap, J., and J.S. Gardner. 1993. Soil microstructure in soils of the Colorado Plateau—The role of the cyanobacterium *Microcoleus vaginatus*. *Great Basin Naturalist* 53: 40–47.

Belnap, J., and O.L. Lange. 2001. *Biological soil crusts: Structure, function, and management*.

Belnap, J., B.J. Walker, S.M. Munson, and R.A. Gill. 2014. Controls on sediment production in two U.S. deserts. *Aeolian Research* 14: 15–24.

Ben-David, E.A., E. Zaady, Y. Sher, and A. Nejidat. 2011. Assessment of the spatial distribution of soil microbial communities in patchy arid and semi-arid landscapes of the Negev Desert using combined PLFA and DGGE analyses. *FEMS Microbiology Ecology* 76: 492–503.

Bestelmeyer, B.T., and D.D. Briske. 2012. Grand challenges for resilience-based management of rangelands. *Rangeland Ecology & Management* 65: 654–663.

Bestelmeyer, B.T., J.R. Brown, K.M. Havstad, R. Alexander, G. Chavez, and J.E. Herrick. 2003. Development and use of state-and-transition models for rangelands. *Journal of Range Management* 56: 114–126.

Billings, S.A., S.M. Schaeffer, and R.D. Evans. 2004. Soil microbial activity and N availability with elevated CO_2 in Mojave desert soils. *Global Biogeochemical Cycles* 18, GB1011.

Birch, H. 1958. The effect of soil drying on humus decomposition and nitrogen availability. *Plant and Soil* 10: 9–31.

Blazewicz, S.J., E. Schwartz, and M.K. Firestone. 2014. Growth and death of bacteria and fungi underlie rainfall-induced carbon dioxide pulses from seasonally dried soil. *Ecology* 95: 1162–1172.

Bolton Jr., H., J.L. Smith, and S.O. Link. 1993. Soil microbial biomass and activity of a disturbed and undisturbed shrub-steppe ecosystem. *Soil Biology and Biochemistry* 25: 545–552.

Booth, M.S., J.M. Stark, and M.M. Caldwell. 2003. Inorganic N turnover and availability in annual- and perennial-dominated soils in a northern Utah shrub-steppe ecosystem. *Biogeochemistry* 66: 311–330.

Borken, W., and E. Matzner. 2009. Reappraisal of drying and wetting effects on C and N mineralization and fluxes in soils. *Global Change Biology* 15: 808–824.

Bot, A., and J. Benites. 2005. *Importance of soil organic matter: Key to drought-resistance food production. Iaea Tecdoc Series*. Rome: Food and Agriculture Organization.

Bowker, M.A., D.J. Eldridge, J. Val, and S. Soliveres. 2013. Hydrology in a patterned landscape is co-engineered by soil-disturbing animals and biological crusts. *Soil Biology & Biochemistry* 61: 14–22.

Brandt, L.A., J.Y. King, S.E. Hobbie, D.G. Milchunas, and R.L. Sinsabaugh. 2010. The role of photodegradation in surface litter decomposition across a grassland ecosystem precipitation gradient. *Ecosystems* 13: 765–781.

Breshears, D.D., N.S. Cobb, P.M. Rich, K.P. Price, C.D. Allen, R.G. Balice, W.H. Romme, J.H. Kastens, M.L. Floyd, J. Belnap, J.J. Anderson, O.B. Myers, and C.W. Meyer. 2005. Regional vegetation die-off in response to global-change-type drought. *Proceedings of the National Academy of Sciences of the United States of America* 102: 15144–15148.

Briske, D.D., S.D. Fuhlendorf, and F.E. Smeins. 2006. A unified framework for assessment and application of ecological thresholds. *Rangeland Ecology & Management* 59: 225–236.

Briske, D.D., J.D. Derner, D.G. Milchunas, and K.W. Tate. 2011. An evidence-based assessment of prescribed grazing practices. In *Conservation benefits of rangeland practices: Assessment, recommendations, and knowledge gaps*, ed. D.D. Briske, 22–74. Washington, DC: United States Department of Agriculture, Natural Resource Conservation Service.

Campbell, B.D., D.M. Stafford Smith, and G.M. McKeon. 1997. Elevated CO_2 and water supply interactions in grasslands: A pastures and rangelands management perspective. *Global Change Biology* 3: 177–187.

Cardon, Z.G., J.M. Stark, P.M. Herron, and J.A. Rasmussen. 2013. Sagebrush carrying out hydraulic lift enhances surface soil nitrogen cycling and nitrogen uptake into inflorescences. *Proceedings of the National Academy of Sciences of the United States of America* 110: 18988–18993.

Carrillo, Y., F.A. Dijkstra, E. Pendall, J.A. Morgan, and D.M. Blumenthal. 2012. Controls over soil nitrogen pools in a semiarid grassland under elevated CO_2 and warming. *Ecosystems* 15: 761–774.

Chaiwatpongsakorn, C., M.M. Lu, T.C. Keener, and S.J. Khang. 2014. The deployment of carbon monoxide wireless sensor network (CO-WSN) for ambient air monitoring. *International Journal of Environmental Research and Public Health* 11: 6246–6264.

Chepil, W.S. 1951. Properties of soil which influence wind erosion: IV. State of dry aggregate structure. *Soil Science* 72: 387–402.

———. 1953. Field structure of cultivated soils with special reference to erodibility by wind. *Soil Science Society of America Proceedings* 17: 185–190.

Christian, J.M., and S.D. Wilson. 1999. Long-term ecosystem impacts of an introduced grass in the northern Great Plains. *Ecology* 80: 2397–2407.

Corbin, J.D., and C.M. D'Antonio. 2004. Effects of exotic species on soil nitrogen cycling: Implications for restoration. *Weed Technology* 18: 1464–1467.

Cornelissen, J.H.C., S.I. Lang, N.A. Soudzilovskaia, and H.J. During. 2007. Comparative crypto-gam ecology: A review of bryophyte and lichen traits that drive biogeochemistry. *Annals of Botany* 99: 987–1001.

Crabtree, R.C., and F.A. Bazzaz. 1993. Seedling response of four birch species to simulated nitro-gen deposition—Ammonium vs. nitrate. *Ecological Applications* 3: 315–321.

DeMalach, N., J. Kigel, H. Voet, and E.D. Ungar. 2014. Are semiarid shrubs resilient to drought and grazing? Differences and similarities among species and habitats in a long-term study. *Journal of Arid Environments* 102: 1–8.

Derner, J.D., and G.E. Schuman. 2007. Carbon sequestration and rangelands: A synthesis of land management and precipitation effects. *Journal of Soil and Water Conservation* 62: 77–85.

Derner, J.D., T.W. Boutton, and D.D. Briske. 2006. Grazing and ecosystem carbon storage in the North American Great Plains. *Plant and Soil* 280: 77–90.

Diaz, S., D.D. Briske, and S. McIntyre. 2002. Range management and plant functional types. In *Global rangelands: Progress and prospects*, eds. A.C. Grice, and K.C. Hodgkinson, 81–100.

Diaz, S., S. Lavorel, S. McIntyre, V. Falczuk, F. Casanoves, D.G. Milchunas, C. Skarpe, G. Rusch, M. Sternberg, I. Noy-Meir, J. Landsberg, W. Zhang, H. Clark, and B.D. Campbell. 2007. Plant trait responses to grazing—A global synthesis. *Global Change Biology* 13: 313–341.

Dijkstra, F.A., E. Pendall, A.R. Mosier, J.Y. King, D.G. Milchunas, and J.A. Morgan. 2008. Long-term enhancement of N availability and plant growth under elevated CO_2 in a semi-arid grass-land. *Functional Ecology* 22: 975–982.

Dijkstra, F.A., D. Blumenthal, J.A. Morgan, E. Pendall, Y. Carrillo, and R.F. Follett. 2010. Contrasting effects of elevated CO_2 and warming on nitrogen cycling in a semiarid grassland. *New Phytologist* 187: 426–437.

DiTomaso, J.M. 2000. Invasive weeds in rangelands: Species, impacts, and management. *Weed Science* 48: 255–265.

Doerr, S.H., R.A. Shakesby, and R.P.D. Walsh. 2000. Soil water repellency: Its causes, character-istics and hydro-geomorphological significance. *Earth-Science Reviews* 51: 33–65.

Donohue, R.J., M.L. Roderick, T.R. McVicar, and G.D. Farquhar. 2013. CO_2 fertilisation has increased maximum foliage cover across the globe's warm, arid environments. *Geophysical Research Letters* 40(12): 3031–3035.

Drenovsky, R.E., and K.M. Batten. 2007. Invasion by *Aegilops triuncialis* (barb goatgrass) slows carbon and nutrient cycling in a serpentine grassland. *Biological Invasions* 9: 107–116.

Drigo, B., G.A. Kowalchuk, and J.A. van Veen. 2008. Climate change goes underground: effects of elevated atmospheric CO_2 on microbial community structure and activities in the rhizo-sphere. *Biology and Fertility of Soils* 44: 667–679.

Dunbar, J., S.A. Eichorst, L.V. Gallegos-Graves, S. Silva, G. Xie, N.W. Hengartner, R.D. Evans, B.A. Hungate, R.B. Jackson, J.P. Megonigal, C.W. Schadt, R. Vilgalys, D.R. Zak, and C.R. Kuske. 2012. Common bacterial responses in six ecosystems exposed to 10 years of ele-vated atmospheric carbon dioxide. *Environmental Microbiology* 14: 1145–1158.

Duncan, L., and J.K. Clark. 2005. *Invasive plants of range and wild lands and their environmental, economic and societal impacts*. Weed Science Society of America.

Ehrenfeld, J.G. 2003. Effects of exotic plant invasions on soil nutrient cycling processes. *Ecosystems* 6: 503–523.

Elbert, W., B. Weber, S. Burrows, J. Steinkamp, B. Budel, M.O. Andreae, and U. Poschl. 2012. Contribution of cryptogamic covers to the global cycles of carbon and nitrogen. *Nature Geoscience* 5: 459–462.

Eldridge, D.J., M.A. Bowker, F.T. Maestre, E. Roger, J.F. Reynolds, and W.G. Whitford. 2011. Impacts of shrub encroachment on ecosystem structure and functioning: Towards a global synthesis. *Ecology Letters* 14: 709–722.

Evans, R.D., and J. Belnap. 1999. Long-term consequences of disturbance on nitrogen dynamics in an arid ecosystem. *Ecology* 80: 150–160.

Evans, S.E., and I.C. Burke. 2013. Carbon and nitrogen decoupling under an 11-Year drought in the shortgrass steppe. *Ecosystems* 16: 20–33.

Evans, R.D., and J.R. Ehleringer. 1993. A break in the nitrogen cycle in arid lands—Evidence from $\delta^{15}N$ of soils. *Oecologia* 94: 314–317.

Evans, R.D., and J.R. Johansen. 1999. Microbiotic crusts and ecosystem processes. *Critical Reviews in Plant Sciences* 18: 183–225.

Evans, R.D., R. Rimer, L. Sperry, and J. Belnap. 2001a. Exotic plant invasion alters nitrogen dynamics in an arid grassland. *Ecological Applications* 11: 1301–1310.

Evans, R.D., J. Belnap, F. Garcia-Pichel, S.L. Phillips, J. Belnap, and O.L. Lange. 2001a. Global change and the future of biological soil crusts. In *Ecological studies. Biological soil crusts: Structure, function, and management*, 417–429.

Evans, R.D., A. Koyama, D.L. Sonderegger, T.N. Charlet, B.A. Newingham, L.F. Fenstermaker, B. Harlow, V.L. Jin, K. Ogle, S.D. Smith, and R.S. Nowak. 2014. Greater ecosystem carbon in the Mojave Desert after ten years exposure to elevated CO_2. *Nature Climate Change* 4: 394–397.

Eviner, V.T., and C.V. Hawkes. 2012. The effects of plant-soil feedbacks on invasive plants: Mechanisms and potential management options. In *Invasive plant ecology and management: Linking processes to practice*, 122–141.

Eviner, V.T., and F.S. Chapin. 2003. Functional matrix: A conceptual framework for predicting multiple plant effects on ecosystem processes. *Annual Review of Ecology, Evolution, and Systematics* 34: 455–485.

Eviner, V.T., F.S. Chapin, and C.E. Vaughn. 2006. Seasonal variations in plant species effects on soil N and P dynamics. *Ecology* 87: 974–986.

Eviner, V.T., K. Garbach, J.H. Baty, and S.A. Hoskinson. 2012. Measuring the effects of invasive plants on ecosystem services: Challenges and prospects. *Invasive Plant Science and Management* 5: 125–136.

Fay, P.A., J.M. Blair, M.D. Smith, J.B. Nippert, J.D. Carlisle, and A.K. Knapp. 2011. Relative effects of precipitation variability and warming on tallgrass prairie ecosystem function. *Biogeosciences* 8: 3053–3068.

Fay, P.A., V.L. Jin, D.A. Way, K.N. Potter, R.A. Gill, R.B. Jackson, and H.W. Polley. 2012. Soil-mediated effects of subambient to increased carbon dioxide on grassland productivity. *Nature Climate Change* 2: 742–746.

Fenn, M.E., M.A. Poth, and D.W. Johnson. 1996. Evidence for nitrogen saturation in the San Bernardino Mountains in southern California. *Forest Ecology and Management* 82: 211–230.

Fenn, M.E., J.S. Baron, E.B. Allen, H.M. Rueth, K.R. Nydick, L. Geiser, W.D. Bowman, J.O. Sickman, T. Meixner, D.W. Johnson, and P. Neitlich. 2003a. Ecological effects of nitrogen deposition in the western United States. *Bioscience* 53: 404–420.

Fenn, M.E., R. Haeuber, G.S. Tonnesen, J.S. Baron, S. Grossman-Clarke, D. Hope, D.A. Jaffe, S. Copeland, L. Geiser, H.M. Rueth, and J.O. Sickman. 2003b. Nitrogen emissions, deposition, and monitoring in the western United States. *Bioscience* 53: 391–403.

Fernandez-Gimenez, M.E., and B. Allen-Diaz. 1999. Testing a non-equilibrium model of rangeland vegetation dynamics in Mongolia. *Journal of Applied Ecology* 36: 871–885.

Fernandez-Ugalde, O., I. Virto, P. Barre, M. Apesteguia, A. Enrique, M.J. Imaz, and P. Bescansa. 2014. Mechanisms of macroaggregate stabilisation by carbonates: Implications for organic matter protection in semi-arid calcareous soils. *Soil Research* 52: 180–192.

Fierer, N., and J.P. Schimel. 2002. Effects of drying-rewetting frequency on soil carbon and nitrogen transformations. *Soil Biology & Biochemistry* 34: 777–787.

Fierer, N., M.A. Bradford, and R.B. Jackson. 2007. Toward an ecological classification of soil bacteria. *Ecology* 88: 1354–1364.

Fierer, N., M.S. Strickland, D. Liptzin, M.A. Bradford, and C.C. Cleveland. 2009. Global patterns in belowground communities. *Ecology Letters* 12: 1238–1249.

Flerchinger, G.N., and M.S. Seyfried. 2014. Comparison of methods for estimating evapotranspiration in a small rangeland catchment. *Vadose Zone Journal* 13. doi:10.2136/vzj2013.08.0152.

Friedlingstein, P., R.A. Houghton, G. Marland, J. Hackler, T.A. Boden, T.J. Conway, J.G. Canadell, M.R. Raupach, P. Ciais, and C. Le Quere. 2010. Update on CO_2 emissions. *Nature Geoscience* 3: 811–812.

Gallo, M.E., R.L. Sinsabaugh, and S.E. Cabaniss. 2006. The role of ultraviolet radiation in litter decomposition in and ecosystems. *Applied Soil Ecology* 34: 82–91.

Gill, R.A., and I.C. Burke. 1999. Ecosystem consequences of plant life form changes at three sites in the semiarid United States. *Oecologia* 121: 551–563.

Gill, R.A., H.W. Polley, H.B. Johnson, L.J. Anderson, H. Maherali, and R.B. Jackson. 2002. Nonlinear grassland responses to past and future atmospheric CO_2. *Nature* 417: 279–282.

Gill, R.A., L.J. Anderson, H.W. Polley, H.B. Johnson, and R.B. Jackson. 2006. Potential nitrogen constraints on soil carbon sequestration under low and elevated atmospheric CO_2. *Ecology* 87: 41–52.

Goudie, A.S. 2008. The history and nature of wind erosion in deserts. *Annual Review of Earth and Planetary Sciences* 36: 97–119.

Goudie, A.S., and N.J. Middleton. 2001. Saharan dust storms: Nature and consequences. *Earth-Science Reviews* 56: 179–204.

Greaver, T.L., T.J. Sullivan, J.D. Herrick, M.C. Barber, J.S. Baron, B.J. Cosby, M.E. Deerhake, R.L. Dennis, J.-J.B. Dubois, C.L. Goodale, A.T. Herlihy, G.B. Lawrence, L. Liu, J.A. Lynch, and K.J. Novak. 2012. Ecological effects of nitrogen and sulfur air pollution in the US: What do we know? *Frontiers in Ecology and the Environment* 10: 365–372.

Harris, R.F. 1981. The effect of water potential on microbial growth and activity. In *Water potential relations in soil microbiology*, ed. J.F. Parr, W.R. Gardner, and L.F. Elliott. Wisconsin: SSSA Special Publication.

Harrison, K.A., and R.D. Bardgett. 2008. Impacts of grazing and browsing by large herbivores on soils and soil biological properties. In *The ecology of browsing and grazing*, ed. I.J. Gordon and H.H.T. Prins, 201–216. New York: Springer.

Hawkes, C.V., I.F. Wren, D.J. Herman, and M.K. Firestone. 2005. Plant invasion alters nitrogen cycling by modifying the soil nitrifying community. *Ecology Letters* 8: 976–985.

Hewins, D.B., S.R. Archer, G.S. Okin, R.L. McCulley, and H.L. Throop. 2013. Soil-litter mixing accelerates decomposition in a Chihuahuan Desert grassland. *Ecosystems* 16: 183–195.

Hovenden, M.J., P.C.D. Newton, R.A. Carran, P. Theobald, K.E. Wills, J.K.V. Schoor, A.L. Williams, and Y. Osanai. 2008. Warming prevents the elevated CO_2-induced reduction in available soil nitrogen in a temperate, perennial grassland. *Global Change Biology* 14: 1018–1024.

Hutchens, C. 2011. *The role of soil organic matter in rangeland sustainability*. Chico: California State University.

Ingram, L.J., P.D. Stahl, G.E. Schuman, J.S. Buyer, G.F. Vance, G.K. Ganjegunte, J.M. Welker, and J.D. Derner. 2008. Grazing impacts on soil carbon and microbial communities in a mixed-grass ecosystem. *Soil Science Society of America Journal* 72: 939–948.

Jackson, R.B., J.L. Banner, E.G. Jobbagy, W.T. Pockman, and D.H. Wall. 2002. Ecosystem carbon loss with woody plant invasion of grasslands. *Nature* 418: 623–626.

Jin, V.L., and R.D. Evans. 2010. Microbial ^{13}C utilization patterns via stable isotope probing of phospholipid biomarkers in Mojave Desert soils exposed to ambient and elevated atmospheric CO_2. *Global Change Biology* 16: 2334–2344.

Jin, V.L., S.M. Schaeffer, S.E. Ziegler, and R.D. Evans. 2011. Soil water availability and microsite mediate fungal and bacterial phospholipid fatty acid biomarker abundances in Mojave Desert soils exposed to elevated atmospheric CO_2. *Journal of Geophysical Research: Biogeosciences* 116. doi:10.1029/2010JG001564.

Johnson, S.L., C.R. Budinoff, J. Belnap, and F. Garcia-Pichel. 2005. Relevance of ammonium oxidation within biological soil crust communities. *Environmental Microbiology* 7: 1–12.

Johnson, S.L., S. Neuer, and F. Garcia-Pichel. 2007. Export of nitrogenous compounds due to incomplete cycling within biological soil crusts of arid lands. *Environmental Microbiology* 9: 680–689.

Johnson, S.L., C.R. Kuske, T.D. Carney, D.C. Housman, L. Gallegos-Graves, and J. Belnap. 2012. Increased temperature and altered summer precipitation have differential effects on biological soil crusts in a dryland ecosystem. *Global Change Biology* 18: 2583–2593.

Jordan, N.R., D.L. Larson, and S.C. Huerd. 2008. Soil modification by invasive plants: Effects on native and invasive species of mixed-grass prairies. *Biological Invasions* 10: 177–190.

Kao, R.H., C.M. Gibson, R.E. Gallery, C.L. Meier, D.T. Barnett, K.M. Docherty, K.K. Blevins, P.D. Travers, E. Azuaje, Y.P. Springer, K.M. Thibault, V.J. McKenzie, M. Keller, L.F. Alves, E.L.S. Hinckley, J. Parnell, and D. Schimel. 2012. NEON terrestrial field observations: Designing continental-scale, standardized sampling. *Ecosphere* 3: 1–17.

Kerkez, B., S.D. Glaser, R.C. Bales, and M.W. Meadows. 2012. Design and performance of a wireless sensor network for catchment-scale snow and soil moisture measurements. *Water Resources Research* 48. doi:10.1029/2011WR011214.

Kiage, L.M. 2013. Perspectives on the assumed causes of land degradation in the rangelands of Sub-Saharan Africa. *Progress in Physical Geography* 37: 664–684.

Kim, M., B. Boldgiv, D. Singh, J. Chun, A. Lkhagva, and J.M. Adams. 2013. Structure of soil bacterial communities in relation to environmental variables in a semi-arid region of Mongolia. *Journal of Arid Environments* 89: 38–44.

Kotze, E., A. Sandhage-Hofmann, J.A. Meinel, C.C. du Preez, and W. Amelung. 2013. Rangeland management impacts on the properties of clayey soils along grazing gradients in the semi-arid grassland biome of South Africa. *Journal of Arid Environments* 97: 220–229.

Lacey, J.R., C.B. Marlow, and J.R. Lane. 1989. Influence of spotted knapweed (*Centaurea maculosa*) on surface runoff and sediment yield. *Weed Technology* 3: 627–631.

Lai, L.M., J.J. Wang, Y. Tian, X.C. Zhao, L.H. Jiang, X. Chen, Y. Gao, S.M. Wang, and Y.R. Zheng. 2013. Organic matter and water addition enhance soil respiration in an arid region. *PLoS ONE* 8, e77659.

Li, X.-Y., S. Contreras, A. Sole-Benet, Y. Canton, F. Domingo, R. Lazaro, H. Lin, B. Van Wesemael, and J. Puigdefabregas. 2011. Controls of infiltration-runoff processes in Mediterranean karst rangelands in SE Spain. *Catena* 86: 98–109.

Li, X.L., J. Gao, J. Brierley, Y.M. Qiao, J. Zhang, and Y.W. Yang. 2013. Rangeland degradation on the Qinghai-Tibet Plateau: Implications for rehabilitation. *Land Degradation & Development* 24: 72–80.

Liao, C., R. Peng, Y. Luo, X. Zhou, X. Wu, C. Fang, J. Chen, and B. Li. 2008. Altered ecosystem carbon and nitrogen cycles by plant invasion: A meta-analysis. *New Phytologist* 177: 706–714.

Loescher, H., E. Ayres, P. Duffy, H.Y. Luo, and M. Brunke. 2014. Spatial variation in soil properties among North American ecosystems and guidelines for sampling designs. *PLoS ONE* 9: e83216.

Luo, Y., B. Su, W.S. Currie, J.S. Dukes, A. Finzi, U. Hartwig, B. Hungate, R. E. McMurtrie, R. Oren, W.J. Parton, D.E. Pataki, M.R. Shaw, D.R. Zak and C.B. Field. 2004. Progressive nitrogen limitation of ecosystem responses to rising atmospheric carbon dioxide. *BioScience* 54: 731–739.

Mack, R.N., and J.N. Thompson. 1982. Evolution in steppe with few large, hoofed mammals. *American Naturalist* 119: 757–773.

Manzoni, S., R.B. Jackson, J.A. Trofymow, and A. Porporato. 2008. The global stoichiometry of litter nitrogen mineralization. *Science* 321: 684–686.

Marasco, R., E. Rolli, B. Ettoumi, G. Vigani, F. Mapelli, S. Borin, A.F. Abou-Hadid, U.A. El-Behairy, C. Sorlini, A. Cherif, G. Zocchi, and D. Daffonchio. 2012. A drought resistance-promoting microbiome is selected by root system under desert farming. *PLoS ONE* 7: e48479.

Marschner, H. 1986. *Mineral nutrition of higher plants*, 2nd ed. London: Academic.

Martinez-Garcia, L.B., C. Armas, J.D. Miranda, F.M. Padilla, and F.I. Pugnaire. 2011. Shrubs influence arbuscular mycorrhizal fungi communities in a semi-arid environment. *Soil Biology & Biochemistry* 43: 682–689.

McCalley, C.K., and J.P. Sparks. 2008. Controls over nitric oxide and ammonia emissions from Mojave Desert soils. *Oecologia* 156: 871–881.

———. 2009. Abiotic gas gormation drives nitrogen loss from a desert ecosystem. *Science* 326: 837–840.

McCalley, C.K., B.D. Strahm, K.L. Sparks, A.S.D. Eller, and J.P. Sparks. 2011. The effect of long-term exposure to elevated CO_2 on nitrogen gas emissions from Mojave Desert soils. *Journal of Geophysical Research: Biogeosciences* 116. doi:10.1029/2011JG001667.

McCulley, R.L., I.C. Burke, and W.K. Lauenroth. 2009. Conservation of nitrogen increases with precipitation across a major grassland gradient in the Central Great Plains of North America. *Oecologia* 159: 571–581.

McSherry, M.E., and M.E. Ritchie. 2013. Effects of grazing on grassland soil carbon: A global review. *Global Change Biology* 19: 1347–1357.

MEA. 2005. *Ecosystems and human well-being: Desertification synthesis*. Washington, DC: World Resources Institute.

Michener, W.K., and M.B. Jones. 2012. Ecoinformatics: Supporting ecology as a data-intensive science. *Trends in Ecology & Evolution* 27: 85–93.

Michener, W.K., J.H. Beach, M.B. Jones, B. Ludascher, D.D. Pennington, R.S. Pereira, A. Rajasekar, and M. Schildhauer. 2007. A knowledge environment for the biodiversity and ecological sciences. *Journal of Intelligent Information Systems* 29: 111–126.

Milchunas, D.G., and W.K. Lauenroth. 1993. Quantitative effects of grazing on vegetation and soils over a global range of environments. *Ecological Monographs* 63: 327–366.

Miller, A.E., J.P. Schimel, T. Meixner, J.O. Sickman, and J.M. Melack. 2005. Episodic rewetting enhances carbon and nitrogen release from chaparral soils. *Soil Biology & Biochemistry* 37: 2195–2204.

Mishra, A.K., and V.P. Singh. 2010. A review of drought concepts. *Journal of Hydrology* 391: 204–216.

Mohamed, A.H., J. L. Holechek, D. W. Bailey, C. L. Campbell, and M. N. DeMers, 2011. Mesquite encroachment impact on southern New Mexico rangelands: remote sensing and geographic information systems approach. Journal of. Applied. Remote Sensing. 5, 053514.

Moorhead, D.L., and R.L. Sinsabaugh. 2000. Simulated patterns of litter decay predict patterns of extracellular enzyme activities. *Applied Soil Ecology* 14: 71–79.

Morillas, L., M. Portillo-Estrada, and A. Gallardo. 2013. Wetting and drying events determine soil N pools in two Mediterranean ecosystems. *Applied Soil Ecology* 72: 161–170.

Mosier, A.R., J.A. Morgan, J.Y. King, D. LeCain, and D.G. Milchunas. 2002. Soil-atmosphere exchange of CH_4, CO_2, NO_x, and N_2O in the Colorado shortgrass steppe under elevated CO_2. *Plant and Soil* 240: 201–211.

Neilson, J.W., J. Quade, M. Ortiz, W.M. Nelson, A. Legatzki, F. Tian, M. LaComb, J.L. Betancourt, R.A. Wing, C.A. Soderlund, and R.M. Maier. 2012. Life at the hyperarid margin: Novel bacterial diversity in arid soils of the Atacama Desert, Chile. *Extremophiles* 16: 553–566.

Norton, U., P. Saetre, T.D. Hooker, and J.M. Stark. 2012. Vegetation and moisture controls on soil carbon mineralization in semiarid environments. *Soil Science Society of America Journal* 76: 1038–1047.

Nowak, R.S., D.S. Ellsworth, and S.D. Smith. 2004. Functional responses of plants to elevated atmospheric CO_2—Do photosynthetic and productivity data from FACE experiments support early predictions? *New Phytologist* 162: 253–280.

Orlando, J., M. Caru, B. Pommerenke, and G. Braker. 2012. Diversity and activity of denitrifiers of Chilean arid soil ecosystems. *Frontiers in Microbiology* 3: 101.

Paige, G.B., and T.O. Keefer. 2008. Comparison of field performance of multiple soil moisture sensors in a semi-arid rangeland. *Journal of the American Water Resources Association* 44: 121–135.

Painter, T.H., J.S. Deems, J. Belnap, A.F. Hamlet, C.C. Landry, and B. Udall. 2010. Response of Colorado River runoff to dust radiative forcing in snow. *Proceedings of the National Academy of Sciences of the United States of America* 107: 17125–17130.

Painter, T.H., S.M. Skiles, J.S. Deems, A.C. Bryant, and C.C. Landry. 2012a. Dust radiative forcing in snow of the Upper Colorado River Basin: 1. A 6 year record of energy balance, radiation, and dust concentrations. *Water Resources Research* 48, W07521.

Painter, T.H., A.C. Bryant, and S.M. Skiles. 2012a. Radiative forcing by light absorbing impurities in snow from MODIS surface reflectance data. *Geophysical Research Letters* 39. doi:10.1029/2012GL052457.

Pardo, L.H., M.E. Fenn, C.L. Goodale, L.H. Geiser, C.T. Driscoll, E.B. Allen, J.S. Baron, R. Bobbink, W.D. Bowman, C.M. Clark, B. Emmett, F.S. Gilliam, T.L. Greaver, S.J. Hall, E.A. Lilleskov, L. Liu, J.A. Lynch, K.J. Nadelhoffer, S.S. Perakis, M.J. Robin-Abbott, J.L. Stoddard, K.C. Weathers, and R.L. Dennis. 2011. Effects of nitrogen deposition and empirical nitrogen critical loads for ecoregions of the United States. *Ecological Applications* 21: 3049–3082.

Parton, W.J., D.S. Schimel, C.V. Cole, and D.S. Ojima. 1987. Analysis of factors controlling soil organic matter levels in Great Plains grasslands. *Soil Science Society of America Journal* 51: 1173–1179.

Pasternak, Z., A. Al-Ashhab, J. Gatica, R. Gafny, S. Avraham, D. Minz, O. Gillor, and E. Jurkevitch. 2013. Spatial and temporal biogeography of soil microbial communities in arid and semiarid regions. *PLoS ONE* 8, e69705.

Pendall, E., A.R. Mosier, and J.A. Morgan. 2004. Rhizodeposition stimulated by elevated CO_2 in a semiarid grassland. *New Phytologist* 162: 447–458.

Pendall, E., J.L. Heisler-White, D.G. Williams, F.A. Dijkstra, Y. Carrillo, J.A. Morgan, and D.R. LeCain. 2013. Warming reduces carbon losses from grassland exposed to elevated atmospheric carbon dioxide. *PLoS ONE* 8, e71921.

Peterjohn, W.T., and W.H. Schlesinger. 1990. Nitrogen loss from deserts in the southwestern United States. *Biogeochemistry* 10: 67–79.

Placella, S.A., E.L. Brodie, and M.K. Firestone. 2012. Rainfall-induced carbon dioxide pulses result from sequential resuscitation of phylogenetically clustered microbial groups. *Proceedings of the National Academy of Sciences of the United States of America* 109: 10931–10936.

Polley, H.W., V.L. Jin, and P.A. Fay. 2012. CO_2-caused change in plant species composition rivals the shift in vegetation between mid-grass and tallgrass prairies. *Global Change Biology* 18: 700–710.

Porada, P., B. Weber, W. Elbert, U. Poschl, and A. Kleidon. 2014. Estimating impacts of lichens and bryophytes on global biogeochemical cycles. *Global Biogeochemical Cycles* 28: 71–85.

Porter, J.H., P.C. Hanson, and C.C. Lin. 2012. Staying afloat in the sensor data deluge. *Trends in Ecology & Evolution* 27: 121–129.

Poulter, B., D. Frank, P. Ciais, R.B. Myneni, N. Andela, J. Bi, G. Broquet, J.G. Canadell, F. Chevallier, Y.Y. Liu, S.W. Running, S. Sitch, and G.R. van der Werf. 2014. Contribution of semi-arid ecosystems to interannual variability of the global carbon cycle. *Nature* 509: 600–603.

Pulido-Fernandez, M., S. Schnabel, J. Francisco Lavado-Contador, I. Miralles Mellado, and R.O. Perez. 2013. Soil organic matter of Iberian open woodland rangelands as influenced by vegetation cover and land management. *Catena* 109: 13–24.

Querejeta, J.I., M.F. Allen, M.M. Alguacil, and A. Roldan. 2007. Plant isotopic composition provides insight into mechanisms underlying growth stimulation by AM fungi in a semiarid environment. *Functional Plant Biology* 34: 683–691.

Ravi, S., P. D'Odorico, S.L. Collins, and T.E. Huxman. 2009. Can biological invasions induce desertification? *New Phytologist* 181: 512–515.

Ravi, S., P. D'Odorico, D.D. Breshears, J.P. Field, A.S. Goudie, T.E. Huxman, J.R. Li, G.S. Okin, R.J. Swap, A.D. Thomas, S. Van Pelt, J.J. Whicker, and T.M. Zobeck. 2011. Aeolian processes and the biosphere. *Reviews of Geophysics* 49. doi:10.1029/2010RG000328.

Reeves, T.L., and M.A. Smith. 1992. Time domain reflectometry for measuring soil water content in range surveys. *Journal of Range Management* 45: 412–414.

Reynolds, J.F., D.M. Stafford Smith, E.F. Lambin, B.L. Turner, M. Mortimore, S.P.J. Batterbury, T.E. Downing, H. Dowlatabadi, R.J. Fernandez, J.E. Herrick, E. Huber-Sannwald, H. Jiang, R. Leemans, T. Lynam, F.T. Maestre, M. Ayarza, and B. Walker. 2007. Global desertification: Building a science for dryland development. *Science* 316: 847–851.

Rockstrom, J. 2003. Resilience building and water demand management for drought mitigation. *Physics and Chemistry of the Earth* 28: 869–877.

Rodriguez-Caballero, E., Y. Canton, S. Chamizo, R. Lazaro, and A. Escudero. 2013. Soil loss and runoff in semiarid ecosystems: A complex interaction between biological soil crusts, microtopography, and hydrological drivers. *Ecosystems* 16: 529–546.

Rosenbaum, U., H.R. Bogena, M. Herbst, J.A. Huisman, T.J. Peterson, A. Weuthen, A.W. Western, and H. Vereecken. 2012. Seasonal and event dynamics of spatial soil moisture patterns at the small catchment scale. *Water Resources Research* 48. doi:10.1029/2011WR011518.

Rossiter-Rachor, N.A., S.A. Setterfield, M.M. Douglas, L.B. Hutley, G.D. Cook, and S. Schmidt. 2009. Invasive *Andropogon gayanus* (gamba grass) is an ecosystem transformer of nitrogen relations in Australian savanna. *Ecological Applications* 19: 1546–1560.

Schaeffer, S.M., S.A. Billings, and R.D. Evans. 2003. Responses of soil nitrogen dynamics in a Mojave Desert ecosystem to manipulations in soil carbon and nitrogen availability. *Oecologia* 134: 547–553.

Schimel, J., T.C. Balser, and M. Wallenstein. 2007. Microbial stress-response physiology and its implications for ecosystem function. *Ecology* 88: 1386–1394.

Schnabel, S., M. Pulido-Fernandez, and J.F. Lavado-Contador. 2013. Soil water repellency in rangelands of Extremadura (Spain) and its relationship with land management. *Catena* 103: 53–61.

Schuman, G.E., H.H. Janzen, and J.E. Herrick. 2002. Soil carbon dynamics and potential carbon sequestration by rangelands. *Environmental Pollution* 116: 391–396.

Scott, N.A., S. Saggar, and P.D. McIntosh. 2001. Biogeochemical impact of *Hieracium* invasion in New Zealand's grazed tussock grasslands: Sustainability implications. *Ecological Applications* 11: 1311–1322.

Seymour, C.L., S.J. Milton, G.S. Joseph, W.R.J. Dean, T. Ditlhobolo, and G.S. Cumming. 2010. Twenty years of rest returns grazing potential, but not palatable plant diversity, to Karoo rangeland, South Africa. *Journal of Applied Ecology* 47: 859–867.

Sher, Y., E. Zaady, and A. Nejidat. 2013. Spatial and temporal diversity and abundance of ammonia oxidizers in semi-arid and arid soils: Indications for a differential seasonal effect on archaeal and bacterial ammonia oxidizers. *FEMS Microbiology Ecology* 86: 544–556.

Simberloff, D., and B. Von Holle. 1999. Positive interaction of nonindigenous species: Invasional meltdown? *Biological Invasions* 1: 21–32.

Simoni, S., S. Padoan, D.F. Nadeau, M. Diebold, A. Porporato, G. Barrenetxea, F. Ingelrest, M. Vetterli, and M.B. Parlange. 2011. Hydrologic response of an alpine watershed: Application of a meteorological wireless sensor network to understand streamflow generation. *Water Resources Research* 47:doi:10.1029/2011WR010730.

Smith, J.L., J.J. Halvorson, and H. Bolton. 1994. Spatial relationships of soil microbial biomass and C and N mineralization in a semi-arid shrub-steppe ecosystem. *Soil Biology & Biochemistry* 26: 1151–1159.

Song, W.M., S.P. Chen, B. Wu, Y.J. Zhu, Y.D. Zhou, Y.H. Li, Y.L. Cao, Q. Lu, and G.H. Lin. 2012. Vegetation cover and rain timing co-regulate the responses of soil CO_2 efflux to rain increase in an arid desert ecosystem. *Soil Biology & Biochemistry* 49: 114–123.

Sperry, L.J., J. Belnap, and R.D. Evans. 2006. *Bromus tectorum* invasion alters nitrogen dynamics in an undisturbed arid grassland ecosystem. *Ecology* 87: 603–615.

Stahlheber, K.A., and C.M. D'Antonio. 2013. Using livestock to manage plant composition: A meta-analysis of grazing in California Mediterranean grasslands. *Biological Conservation* 157: 300–308.

Steven, B., L.V. Gallegos-Graves, C. Yeager, J. Belnap, and C.R. Kuske. 2014. Common and distinguishing features of the bacterial and fungal communities in biological soil crusts and shrub root zone soils. *Soil Biology & Biochemistry* 69: 302–312.

Strain, B.R., and F.A. Bazzaz. 1983. Terrestrial plant communities. In *CO_2 and plants: The response of plants to rising levels of atmospheric carbon dioxide*, 177–222. AAAS Selected Symposium 84. Washington, DC: AAAS.

Strauss, S.L., T.A. Day, and F. Garcia-Pichel. 2012. Nitrogen cycling in desert biological soil crusts across biogeographic regions in the Southwestern United States. *Biogeochemistry* 108: 171–182.

Suri, A., S.S. Iyengar, and E.C. Cho. 2006. Ecoinformatics using wireless sensor networks: An overview. *Ecological Informatics* 1: 287–293.

Taylor, J.R., and H.L. Loescher. 2013. Automated quality control methods for sensor data: A novel observatory approach. *Biogeosciences* 10: 4957–4971.

Templer, P.H., R.W. Pinder, and C.L. Goodale. 2012. Effects of nitrogen deposition on greenhouse-gas fluxes for forests and grasslands of North America. *Frontiers in Ecology and the Environment* 10: 547–553.

Throop, H.L., and S.R. Archer. 2009. Resolving the dryland decomposition conundrum: Some new perspectives on potential drivers. In *Progress in botany*, eds. U. Luttge, W. Beyschlag, B. Budel, and D. Francis, 171–194.

Trimble, S.W., and A.C. Mendel. 1995. The cow as a geomorphic agent—A critical review. *Geomorphology* 13: 233–253.

Unger, S., C. Maguas, J.S. Pereira, T.S. David, and C. Werner. 2012. Interpreting post-drought rewetting effects on soil and ecosystem carbon dynamics in a Mediterranean oak savannah. *Agricultural and Forest Meteorology* 154: 9–18.

Van Auken, O.W. 2009. Causes and consequences of woody plant encroachment into western North American grasslands. *Journal of Environmental Management* 90: 2931–2942.

van de Koppel, J., M. Rietkerk, F. van Langevelde, L. Kumar, C.A. Klausmeier, J.M. Fryxell, J.W. Hearne, J. van Andel, N. de Ridder, A. Skidmore, L. Stroosnijder, and H.H.T. Prins. 2002. Spatial heterogeneity and irreversible vegetation change in semiarid grazing systems. *American Naturalist* 159: 209–218.

Vetter, S. 2005. Rangelands at equilibrium and non-equilibrium: Recent developments in the debate. *Journal of Arid Environments* 62: 321–341.

Vinton, M.A., and I.C. Burke. 1995. Interactions between individual plant species and soil nutrient status in shortgrass steppe. *Ecology* 76: 1116–1133.

Wang, Y.G., H. Zhu, and Y. Li. 2013. Spatial heterogeneity of soil moisture, microbial biomass carbon and soil respiration at stand scale of an arid scrubland. *Environmental Earth Sciences* 70: 3217–3224.

Wardle, D. 2002. *Communities and ecosystems: Linking the aboveground and belowground components*. NJ: Princeton University Press.

Warren, S.D., J. Belnap, and O.L. Lange. 2001. Synopsis: Influence of biological soil crusts on arid land hydrology and soil stability. Ecological Studies. *Biological Soil Crusts: Structure, Function, and Management* 150: 349–360.

Weber, C.F., D.R. Zak, B.A. Hungate, R.B. Jackson, R. Vilgalys, R.D. Evans, C.W. Schadt, J.P. Megonigal, and C.R. Kuske. 2011. Responses of soil cellulolytic fungal communities to elevated atmospheric CO_2 are complex and variable across five ecosystems. *Environmental Microbiology* 13: 2778–2793.

Wedin, D.A., and D. Tilman. 1990. Species effect on nitrogen cycling—A test with perennial grasses. *Oecologia* 84: 433–441.

Wiesmeier, M., M. Steffens, C.W. Mueller, A. Kolbl, A. Reszkowska, S. Peth, R. Horn, and I. Kogel-Knabner. 2012. Aggregate stability and physical protection of soil organic carbon in semi-arid steppe soils. *European Journal of Soil Science* 63: 22–31.

Wilcox, B.P. 2002. Shrub control and streamflow on rangelands: A process based viewpoint. *Journal of Range Management* 55: 318–326.

Wilcox, B.P., M.K. Owens, W.A. Dugas, D.N. Ueckert, and C.R. Hart. 2006. Shrubs, streamflow, and the paradox of scale. *Hydrological Processes* 20: 3245–3259.

Xu, M.-Y., F. Xie, and K. Wang. 2014. Response of vegetation and soil carbon and nitrogen storage to grazing intensity in semi-arid grasslands in the agro-pastoral zone of northern China. *PLoS ONE* 9, e96604.

Zak, D.R., C.B. Blackwood, and M.P. Waldrop. 2006. A molecular dawn for biogeochemistry. *Trends in Ecology & Evolution* 21: 288–295.

Zhao, Y., S. Peth, J. Krummelbein, R. Horn, Z.Y. Wang, M. Steffens, C. Hoffmann, and X.H. Peng. 2007. Spatial variability of soil properties affected by grazing intensity in Inner Mongolia grassland. *Ecological Modelling* 205: 241–254.

Zhao, Y., S. Peth, X.Y. Wang, H. Lin, and R. Horn. 2010. Controls of surface soil moisture spatial patterns and their temporal stability in a semi-arid steppe. *Hydrological Processes* 24: 2507–2519.

Chapter 5
Heterogeneity as the Basis for Rangeland Management

Samuel D. Fuhlendorf, Richard W.S. Fynn, Devan Allen McGranahan, and Dirac Twidwell

*Most people view averages as basic reality and variation as a device for calculating a meaningful measure of central tendency .... **Central tendency is a harmful abstraction and variation stands out as the only meaningful reality***

—Stephen Jay Gould- Full House: The Spread of Excellence from Plato to Darwin

Abstract Rangeland management, like most disciplines of natural resource management, has been characterized by human efforts to reduce variability and increase predictability in natural systems (steady-state management often applied through a command-and-control paradigm). Examples of applications of traditional command and control in natural resource management include wildfire suppression, fences to control large ungulate movements, predator elimination programs, and watershed engineering for flood control and irrigation. Recently, a robust theoretical foundation has been developed that focuses on our understanding of the importance of variability in nature. This understanding is built upon the concept of heterogeneity, which originated from influential calls to consider spatial and temporal scaling in ecological research. Understanding rangeland ecosystems from a resilience perspective where we recognize that these systems are highly variable in space and time cannot be achieved without a focus on heterogeneity across multiple scales.

S.D. Fuhlendorf (✉)
Natural Resource Ecology & Management, Oklahoma State University, Stillwater, OK 74078, USA
e-mail: sam.fuhlendorf@okstate.edu

R.W.S. Fynn
Okavango Research Institute, University of Botswana, PVT Bag 285, Maun, Botswana

D.A. McGranahan
Range Science Program, North Dakota State University, Fargo, ND, USA

D. Twidwell
Agronomy & Horticulture, University of Nebraska-Lincoln, Lincoln, NE 68583, USA

© The Author(s) 2017
D.D. Briske (ed.), *Rangeland Systems*, Springer Series on Environmental Management, DOI 10.1007/978-3-319-46709-2_5

We highlight the broad importance of heterogeneity to rangelands and focus more specifically on (1) animal populations and production, (2) fire behavior and management, and (3) biodiversity and ecosystem function. Rangelands are complex, dynamic, and depend on the variability that humans often attempt to control to ensure long-term productivity and ecosystem health. We present an ecological perspective that targets variation in rangeland properties—including multiple ecosystem services—as an alternative to the myopic focus on maximizing agricultural output, which may expose managers to greater risk. Globally, rangeland science indicates that heterogeneity and diversity increase stability in ecosystem properties from fine to broad spatial scales and through time.

Keywords Scale • Landscape ecology • Hierarchy • Pattern • Disturbance • Resilience

5.1 Introduction

The modern era of natural resource management has been characterized by human efforts to reduce variability and increase predictability in natural systems. This command-and-control paradigm is an extension of societal attempts to identify problems and design and apply solutions to control or mitigate those problems (Holling and Meffe 1996). Examples of command and control in natural resource management include wildfire suppression, fences to control large ungulate movements, weed control for herbaceous native forbs, predator elimination programs, and watershed engineering for flood control and irrigation, including dams, terraces, and subsurface tile drainage. Each of these practices employs human technology to attempt to modulate and regulate the spatial and temporal distribution of resources and complexity of ecological processes. As a result of these attempts (and others), modern natural resource management has created more simple and homogeneous landscapes, which have been considered to be more economically productive due to their perceived increase in predictability. Alongside resource homogenization, land subdivision, fences, transport networks, growing human populations, and other forms of development are increasingly fragmenting ecosystems into smaller management units (Hobbs et al. 2008).

Paralleling command-and-control management is a scientific paradigm that likewise seeks to control or even eliminate variation. Generations of scientists have been trained to design experiments that control all variations except for that which is expected to drive the hypothesized differences. Extreme examples are greenhouse studies and small plot studies that attempt to control for weather and spatial variation, even though we are studying systems that are often described as non-equilibrial and interconnected with other systems. Data from such experiments have most often been subjected to Fisherian statistics, which describe differences between groups in terms of variation around mean values. These models have traditionally considered variation only as a nuisance parameter that is only useful to calculate an accurate mean, rather than a critical parameter in itself. These factors have contributed to a scientific discipline that has tendencies to reduce variation and study small plots, rather than embracing variation and studying complex landscapes at multiple scales.

Yet natural systems are subject to a host of biotic and abiotic processes that shape these landscapes ranging from broad-scale, long-term changes in climate to more localized, short-term events such as droughts, floods, and fires (Fig. 5.1). Over time these factors have created complex systems in nature with a high degree of spatio-temporal variability associated with topography, soils, climate, weather, and disturbance regimes overlain with a diversity of plant and animal communities. Consequently, management seeking to maintain homogeneity presents a major quandary, and often comes at substantial cost, because it is attempting to override the inherent heterogeneity of rangelands and the behavior of disturbances. Considerable money and energy are spent attempting to minimize heterogeneity. Rangeland managers have long sought to override variability in nature with management infrastructure (e.g., fencing, water provisioning) and by controlling disturbances (e.g., channeling or damming water courses and suppressing fires). Despite these efforts, disturbances such as fire, flooding, and drought continue to create variation in systems managed for equilibrium, although the variability is structured differently. This has led to a prevailing view that such disturbances are destructive threats to production systems.

Fig. 5.1 The complexity of rangeland landscapes is a consequence of varying topo-edaphic characteristics and disturbance patterns, including land use. Photo by Sam Fuhlendorf

A robust theoretical foundation now underlies our understanding of the importance of variability in nature. This understanding is built upon the concept of heterogeneity, which originated from influential calls to consider spatial and temporal scaling in ecological research (Wiens 1989; Levin 1992), that resulted in the field of landscape ecology (Urban et al. 1987; Turner 1989; 2005). As a consequence, heterogeneity has become a familiar concept in the study and management of landscapes. But heterogeneity has only recently become appreciated as a component of ecological systems, and adopting it as a guiding principle for ecosystem management has been slow. Obstacles to heterogeneity-based management and policy stem from problems associated with understanding the concept, inconsistent definitions and measurement, as well as a general affinity for homogeneity of landscapes associated with command-and-control management to optimize efficiency.

The intentional simplification and fragmentation of landscapes have contributed to a limited understanding of variability and complexity. In this chapter, we synthesize the current status of rangeland science and management demonstrating the importance of heterogeneity in rangeland ecosystems and the limitations of homogeneity-based management approaches. This chapter is organized into the following sections. First, we discuss how heterogeneity is defined and measured. Second, we use case examples from rangeland research in North American and sub-Saharan Africa to address the following questions: (1) How does heterogeneity support faunal diversity and abundance? (2) How is heterogeneity critical to ecosystem function? (3) How is heterogeneity featured in policy and management? We end by offering suggestions as to how heterogeneity may represent the cornerstone to rangeland management, which should support a large degree of spatial and temporal variability.

5.2 Heterogeneity and Scale: Concepts Linking Pattern and Process

Because heterogeneity is largely associated with spatial and temporal variation of pattern–process relationships, heterogeneity depends on the scale of measurement or observation. Thus, heterogeneity cannot be operationalized without explicit consideration of scale—both in time and space. A widely accepted approach is to measure and evaluate heterogeneity across several scales (Senft et al. 1985; Fuhlendorf and Smeins 1999). Still, studies that have actually evaluated hierarchical relationships between the scale of heterogeneity and the structure and function of rangeland ecosystems are very limited. In this section, we discuss (1) the different types of heterogeneity and (2) the sources of heterogeneity contributing to variation in rangeland ecosystems.

5.2.1 Types of Heterogeneity

5.2.1.1 Measured vs. Functional

Measured Heterogeneity. This is a measure of the variability of an ecological property or process without explicit relations to variability in animal behavior or ecological function (Li and Reynolds 1995). Measured heterogeneity is the product of the perspective of the observer and dependent on sampling protocols and arbitrary decisions of experimental design. For example, a study conducted with meter square sampling plots uniformly or randomly distributed that calculates the variation among this arbitrary plot size and arrangement should be described as measured heterogeneity. Studies that considered multiple levels of arbitrary or measured heterogeneity have demonstrated considerable differences in the measured response of an ecological property or process across multiple scales of studies (Wiens 1989; Fuhlendorf and Smeins 1999). These measures can be useful for understanding how important patterns and processes change with scale. However, measured heterogeneity can only be used to infer ecological function since relationships are arbitrarily established and may, in fact, have little relevance to the ecological questions of interest. Also, if the range of scales measured does not include the appropriate scale to describe the relationship, we can erroneously conclude an inappropriate value of heterogeneity in describing the process.

   Functional Heterogeneity. This is variability at a scale that influences the function of a specific ecological property or process (Li and Reynolds 1995). Because the ecosystem properties that are important to a beetle are not the same as those that are important to a fox or an elk, the scale of heterogeneity relevant to their behavior differs among species. Also, patterns driven by climate fluctuations occur at differing scales than topo-edaphic features or local pathogen outbreaks. Functional heterogeneity assumes that scale of variability is determined by the ecological entity of interest, and is based on the perspective of the participating ecological entities, not the perspective of the ecologist.

   The functional heterogeneity concept suggests that rather than asking if a species or process responds to heterogeneity, the more relevant question is what types, patterns, and scales of heterogeneity are important to a species or process of interest (Kolasa and Rollo 1991). Experiments demonstrate that functional heterogeneity has greater potential to explain variability in the relationship between pattern and process than measured heterogeneity (Gómez et al. 2004; Twidwell et al. 2009). But functional heterogeneity requires greater knowledge of pattern–process relationships and often demands more sampling effort. In the face of such limitations, measured heterogeneity and establishment of arbitrary sampling points across multiple spatial or temporal scales have the potential to identify likely scales of interaction between pattern and process. Measured heterogeneity can therefore be a useful step toward understanding the spatiotemporal scales at which functional heterogeneity emerges (Twidwell et al. 2009).

Linking pattern and process through a lens of functional heterogeneity is extraordinarily rare in rangeland research and monitoring. Random sampling points are used to satisfy assumptions of independence for commonly used statistical analyses (e.g., analysis of variance, ANOVA). But heterogeneity occurring within the study area can lead to misinterpretations from arbitrary sampling approaches and produce erroneous results. This occurs because a fundamental assumption of many sampling procedures and statistical analyses is that heterogeneity within experimental units is not present or unimportant and the ecological property or process of interest operates uniformly across experimental units. While small homogeneous plots can have some value, it is important to recognize that ecological processes rarely operate in this way.

5.2.1.2 Spatial vs. Temporal

Landscapes consist of variable patterns and processes that are dynamic in space and time and lead to complexity that is an essential characteristic of rangelands. *Spatial heterogeneity* refers to how an ecosystem property—nutrients, vegetation type, or amount of cover—varies among points within the landscape. *Temporal heterogeneity* is similar but refers to variability at one point in space over time. When we consider heterogeneity we often consider spatial and temporal heterogeneity separately for statistical and logistical reasons, but in nature they are largely inseparable. For example, if temporal heterogeneity differs between two locations, the locations are also spatially heterogeneous (Kolasa and Rollo 1991). Furthermore, when patch types change positions within the landscape—which often occurs at some spatial and temporal scale in nature because ecosystems are not static—then heterogeneity is changing over both space and time.

A third scenario is the shifting mosaic, in which a specific set of patch types shift across space over time, such that the same *type* of patch occurs in each time step but never in the same *space* in consecutive time steps. In such cases, spatial heterogeneity within the landscape is conserved over time. Although the pattern of bison following burned areas of the pre-European North American Great Plains has become a model for the shifting mosaic, the phenomenon has been repeatedly shown to drive the conservation of pattern–process relationships and the functioning of rangeland ecosystems (Fuhlendorf et al. 2012). Experimental and statistical norms limit our ability to understand landscapes that are highly dynamic in space and time and overcoming these norms is an important challenge to producing usable science on rangelands.

5.2.2   Sources of Heterogeneity

Heterogeneity in rangeland landscapes arises from two main sources. *Inherent heterogeneity* is variability driven by abiotic factors such as geology and topo-edaphic variation influenced by factors such as soil depth, soil fertility, and soil water availability that ultimately contribute to patterns of vegetation composition, productivity, and nutrient content (McNaughton and Banyikwa 1995; Fynn et al. 2014). *Disturbance-driven heterogeneity* is variability influenced by processes such as fire and grazing. These effects can be temporary or persistent and are strongly interactive. On rangelands, a heterogeneous patchwork of vegetation conditions can result from differential timing of disturbances and corresponding out-of-phase succession among patches (Fuhlendorf and Engle 2004); spatial heterogeneity of resources associated with rainfall (Sala et al. 1988; Hopcraft et al. 2010); topo-edaphic patterns (Acres et al. 1985; Scoones 1995); or competitive interactions among plant species (Fuhlendorf and Smeins 1998).

5.2.2.1 Inherent Heterogeneity

Rangelands are inherently heterogeneous in that community composition, productivity, and diversity can vary across scales ranging from centimeters to continents (Fuhlendorf and Smeins 1999; Fuhlendorf et al. 2009). Several environmental factors drive spatial heterogeneity in plant community composition (through competition or tolerance), which can in turn create functional heterogeneity. For example, soil fertility, as influenced by geology and landscape position, plays an important role in determining nutrient concentrations in grasses. In some cases higher clay fertility soils derived from weathered, mineral-rich rock (e.g., dolerite or basalt) promote higher concentrations of protein and minerals in grass tissue than sandy, leached soils derived from sandstone and granite (Hopcraft et al. 2010). High soil salinity can inhibit growth while providing an excess of minerals for uptake by grasses (e.g., McNaughton and Banyikwa 1995; Murray 1995; Grant and Scholes 2006). Another, illustrative example is that patterns in soil depth that can occur at fine or broad spatial scales can lead to differences in species composition that may be as important as and interactive with disturbance processes (Fuhlendorf and Smeins 1998).

In terms of local plant productivity, the effect of geology depends on landscape position. In some cases, deep moist and fertile soils in bottomland positions promote the growth of taller, more productive grasses (Briggs and Knapp 1995; McNaughton and Banyikwa 1995). On shallow but fertile soils in some uplands there may be strong moisture limitation of growth (McNaughton and Banyikwa 1995). From a herbivore perspective, short, leafy grasses often provide higher forage quality (digestibility and nutrient concentrations) than taller grasses (O'Reagain and Owen-Smith 1996; Coetsee et al. 2011), but taller grasses in wetter sites can provide an important source of biomass during the resource-limited dry season. Consequently, functional heterogeneity for herbivores is distributed along forage productivity gradients with high-quality forage needed to satisfy the high-resource demands of herbivores during calving, lactation, and growth occurring in less productive sites but forage to sustain livestock maintenance during the dry season occurring in more productive sites (Maddock 1979; Hopcraft et al. 2010; Fynn et al. 2014).

Inherent heterogeneity at the scale of variable ecological sites dominates discussions of heterogeneity within rangeland landscapes. These patterns of sites are characterized by differences in plant communities and different responses to disturbances such as fire and grazing. Ecologists and soil scientists working with these sites recognize that patterns exist within sites and at broader scales, but this resolution was based on the ability to map soils and other features. The spatial scale of mapping sites is arbitrary indicating that this is measured heterogeneity, rather than functional heterogeneity, which makes its relevance to management and society dubious.

5.2.2.2 Disturbance-Driven Heterogeneity

Ecologists understand that rangeland ecosystems evolved with disturbances, including fire and grazing, but the spatial patterns and heterogeneity of these disturbances were not recognized until recently (Fig. 5.2). Research on spatial and temporal heterogeneity in rangelands has been motivated by loss of habitat for species of conservation concern, as well as the recognition that animals need to be able to respond to extreme climate events through behavior (Allred et al. 2011). Many of the wildlife species that are declining on rangelands today likely evolved with conditions that are best described as heterogeneous across many spatiotemporal scales and are largely driven by disturbance.

Disturbances like fire or prairie dog colonization create feedbacks in which heterogeneity influences subsequent disturbance—the effect or condition of a patch in

Fig. 5.2 Fire interacts with topography and other disturbances to produce a shifting mosaic that is variable in fire severity and time since fire. This landscape in British Columbia was modified by prescribed fire. Photo by Sam Fuhlendorf

one spatial context at a given time not only depends upon the nature of other patches in previous times, but also influences patches at future times. Patterns created by feedbacks are either shifting continuously or relatively stable, and can even vary between the two at different scales. For example, the shifting mosaic created by fire-grazing interaction in tallgrass prairie induces great contrast between patches within a single season, but little permanent change in plant community composition at broad spatial scales through time (Fuhlendorf and Engle 2004; Fuhlendorf et al. 2009). Alternatively, if grazing is sustained in sufficient intensity on certain portions of the landscape or uniformly across the entire landscapes, shifts in plant communities—such as compositional changes from tall grass species to short grass species—are either permanent or at least persist through the duration of the disturbance and may require decades to change to their former composition (Knapp et al. 1999; Archibald et al. 2005). Examples include the white rhinoceros-moderated grazing lawns in South Africa's Hluhluwe-iMfolozi Park (Coetsee et al. 2011) as well as within spatially discrete bison patch grazing in tallgrass prairie (Knapp et al. 1999). With domestic herbivores, long-term changes in composition and structure are often the result of constant grazing distribution from water distribution patterns or promotion of more uniform grazing through cross-fencing.

5.3   Heterogeneity and Rangeland Function: Three Major Cases

In this section, three major cases are presented to demonstrate the importance of heterogeneity. These are just a few examples but understanding heterogeneity is essential for most major rangeland functions. These cases are (1) herbivore population productivity and stability, (2) fire and rangeland ecosystems, and (3) biodiversity and ecosystem function.

5.3.1   Heterogeneity and Herbivore Populations

Herbivores must be able to move across a landscape to deal with stressors associated with availability of resources (water and forage) as well as thermal stress or predation (Allred et al. 2011). Simplification or fragmentation can result in smaller units that will limit an animal's ability to use its behavior to deal with stress that can be cyclic and predictable or stochastic. Smaller pastures or land fragments result in less inherent variation within each pasture and potentially more inherent variability among pastures (Wiens 1989). This shift in variability results in a fundamental change in management required to sustain and match primary and secondary productivity. Small management units suggest a need for greater knowledge and management control of animal requirements and availability of forage quality and quantity, as well as refugia from thermal stress or predation. Consequently, smaller management units will often require greater economic and management inputs—supplementary feeding, licks, and controlled movement—to compensate for limited adaptive foraging options available to herbivores on large landscapes. Larger management units enhance an animal's ability and freedom to respond to variable requirements and the changing environment without management interference (Hobbs et al. 2008). Wild and domestic herbivores, whether grazers or browsers, must cope with elevated requirements for protein, energy, and minerals during certain life stages (Murray 1995; Parker et al. 2009) that may not match resource patterns due to weather and plant phenology, especially when landscapes are small and compartmentalized (Ellis and Swift 1988; Owen-Smith 2004). It is critical that we understand that periods of limited forage biomass and quality may be most important to herbivore populations, rather than average conditions across space and time (Hempson et al. 2015).

Landscape or regional-level variability in plant community composition and productivity—inherent and/or disturbance driven—is important on many rangelands. For example, on African rangelands high-quality short grass sites provide excellent wet-season grazing, but they generally provide little growth and forage during dry periods (McNaughton and Banyikwa 1995; Fynn et al. 2014). By contrast, greater soil moisture availability for dry-season forage production is found in low-lying, poorly drained positions in the landscape such as various wetland types and floodplains (Vesey-FitzGerald 1960; Pamo 1998), as well as in high-rainfall regions receiving significant rainfall during the dry season (Breman and de Wit 1983; McNaughton and Banyikwa 1995). Shallow water tables of wetland sites enable perennial grasses to regrow after fire in the dry season, thereby providing quality regrowth for herbivores (Vesey-FitzGerald 1960; Fynn et al. 2014). Access to green regrowth after fire in the dry season may greatly increase dry-season protein intake for herbivores (Parrini and Owen-Smith 2010). In the absence of fire, taller coarser grasses may be left uneaten, thereby forming a drought-refuge resource for herbivores if rains fail. Such uneaten resources of productive perennial grasses can buffer herbivore populations against the effects of drought, despite their low quality, and have been referred to as buffer resources (Owen-Smith 2002) or key resources (Illius and O'Connor 2000). Loss of access to these key resource areas can result in herbivore population crashes during droughts (Fynn and Bonyongo 2011). Soil texture also plays an important role in facilitating moisture available for growth during the dry season with sandy soils generally supporting growth later into the dry season than clay soils (Sala et al. 1988; McNaughton and Banyikwa 1995). Heterogeneity in clay and sandy soils across landscapes contributes to variation in soil water availability and habitat productivity on strongly developed catenas. When these inherent patterns interact with disturbances such as fire, functional heterogeneity and adaptive foraging options for herbivores are maximized allowing animals to deal with environmental stress. It is important to note that pastoralists of West Africa and wild herbivores that share landscapes follow similar seasonal patterns across large regions indicating that both may have converged on key ecological indicators reflecting seasonal functionality of habitats along various ecological gradients and importance of broad-scale spatial patterns to herbivore-dominated landscapes (Vesey-FitzGerald 1960; Jarman 1972; Pamo 1998; Bartlam-Brooks et al. 2011).

Similar patterns in grassland productivity-driven heterogeneity are seen in North American prairies on landscape-scale soil depth gradients (Briggs and Knapp 1995) and regional-scale rainfall gradients. Elevation and the associated temperature gradients provide another key source of functional heterogeneity by prolonging the length of time during which herbivores have access to forage at peak nutritional quality (Hobbs and Gordon 2010). Warmer conditions and less snow accumulation at lower elevations result in forage growing earlier in spring than at higher altitudes but also maturing and losing quality earlier such that the highest quality forage will move in a "green wave" up the altitudinal gradient over summer as the snowline recedes (Frank et al. 1998; Hobbs and Gordon 2010). Livestock in transhumance systems as well as wildlife such as bison, elk, and bighorn sheep track this high-quality green wave upslope into higher altitude regions during the growing season (Festa-Bianchet 1988; Albon and Langvatn 1992; Frank et al. 1998; Omer et al. 2006; Hebblewhite and Merrill 2007). If forage across the available landscape all matured at the same time then herbivores would have a much shorter period of access to optimal quality forage over the growing season (Hobbs and Gordon 2010). The ability of herbivores to migrate and track the early phenology peak-quality forage in relation to increasing altitude and variation of aspect has been demonstrated to result in increased body size of red deer compared to nonmigratory individuals (Albon and Langvatn 1992; Mysterud et al. 2001).

In the absence of disturbance-driven patches of high-quality forage patches, large regional- and landscape-level movements may be required for a foraging animal to be able to access alternate forage resources and sustain year-around diet quality and quantity. The spatial scale at which heterogeneity is distributed determines the distance that herbivores need to move to forage adaptively over the annual cycle. Disturbance can further enhance heterogeneity at community, landscape, and regional scales by modifying grassland structure and forage quality (Fuhlendorf and Engle 2004). As landscapes are made smaller from fragmentation and compartmentalization, promoting highly variable disturbance patterns to provide greater heterogeneity becomes even more important.

In conclusion, heterogeneity associated with large and complex landscapes enables herbivores to optimize energy and nutrient intake rates during key growth periods of pregnancy, lactation, and body growth while minimizing losses of gains in body mass or population size during resource-limited periods such as the dry season, hot summers, or extreme winters. In addition to diet, herbivores must simultaneously moderate thermal stress and maintain access to water, which is a function of the interaction between the type of animal and its grazing environment (Fig. 5.3; Allred et al. 2011). Similarly, empirical studies have demonstrated much lower mortality of wildlife and livestock during drought years if they have greater access

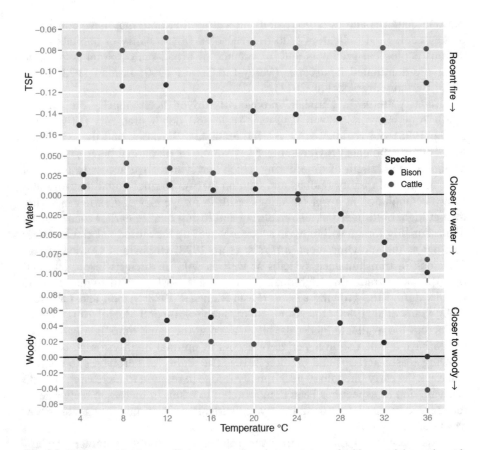

Fig. 5.3 Resource selection coefficients at varying air temperatures for bison and domestic cattle at the Tallgrass Prairie Preserve, USA. Environmental factors include time since fire (TSF), distance to water (Water), and distance to woody vegetation (Woody). Bison and cattle most strongly select for areas that minimize time since fire, but begin selecting sites nearer woody vegetation and water as the temperature increases. Domestic cattle-grazing behavior changes sooner and more dramatically than does bison behavior. Modified from Allred et al. (2013)

to functional heterogeneity (Walker et al. 1987; Scoones 1993). In addition, access to greater functional heterogeneity in rangelands with increasing spatial scale results in a decrease in the strength of density dependence in the relationship between stocking rate and animal growth rate (Hobbs et al. 2008) and also increases body size (Albon and Langvatn 1992; Mysterud et al. 2001). These conclusions are in direct contrast to management prescriptions that reduce functional heterogeneity by reducing the management scale and simplifying the landscape suggesting that greater managerial certainty is expected by a command-and-control perspective on management. Rangeland management should be designed to specifically acknowledge and address uncertainty and variability and we advocate that it should consider the importance of maintaining or enhancing heterogeneity at multiple scales and allow animals to effectively interact with their environment.

5.3.2   Fire and Rangeland Ecosystems

5.3.2.1   Heterogeneity and the Shifting Mosaic

Grasslands, shrublands, and savannas are often described by their dependence on herbivores and fire. Most studies have focused on these factors independent of each other and based on studies of relatively small experimental units that could be well replicated (Fuhlendorf et al. 2009; Fuhlendorf et al. 2012). Recently studies have investigated the landscape-level interaction of fire and herbivores across many continents and various herbivores Fuhlendorf and Engle 2004; Archibald et al. 2005; Allred et al. 2011). These large-scale patterns are best described as a shifting mosaic where fire and grazing interact through a series of feedback mechanisms. As herbivores select recently burned areas for foraging sites, unburned areas are subject to less grazing activity and accumulate fuel. When additional areas burn, grazing animals switch to the more recently burned areas and previously burned areas recover through a transitional stage eventually reaching a state that has accumulated its maximum fuel load. This fire-grazing interaction, or pyric herbivory (grazing driven by fire), results in a shifting mosaic across the landscape allowing herbivores to select from high-quality, recently burned sites and sites that have high biomass accumulation. Herbivores in rangelands of North America spend as much as 70 % of their time on recently burned areas and for domestic herbivores it can increase livestock gains (Limb et al. 2011) and stabilize productivity through drought years when compared to areas managed homogenously (Allred et al. 2014) (Fig. 5.4). For bison of North America, access to burned and unburned areas leads to increased selection of burned areas and higher reproductive rates compared to herds that do not have variable fire patterns (Fuhlendorf and Engle 2001; Fuhlendorf et al. 2009).

The effect of patch fires on forage available for herbivores is best described as a shifting mosaic with patches that vary in forage quantity and quality (Figs. 5.3 and 5.5). Averaged across pastures or experimental units, biomass may be similar between the shifting mosaic and a pasture more traditionally managed. But, the variability is much greater in terms of forage quality and quantity when the shifting mosaic is maintained. Recently burned patches produce forage of high quality and digestibility. Alternatively, biomass accumulation is higher on areas that have greater time since fires, resulting in an overall increase in heterogeneity of forage resources. Animals may respond to this variability differently depending on age, sex, and conditions pre- and postfire. In dry years unburned areas can serve as forage, albeit low quality, through the dry season. Following rain, rapid growth occurs in burned patches and herbivores, particularly females, can select high-quality diets. Heterogeneous landscapes that have been created by patch fires have greater functional heterogeneity as indicated by the high degree of deviation around the mean (Fig. 5.6, Panel B), than landscapes without fire or that are homogeneously managed (Fig. 5.6, Panel C). Smaller landscapes where animal movements are

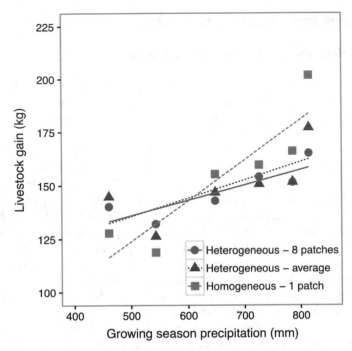

Fig. 5.4 Livestock gain (kg/head) in relation to growing season precipitation within pastures that varied in heterogeneity in tallgrass prairie of North America. Heterogeneous pastures had 2–8 patches that had been burned in a fire-grazing interaction while the homogeneous pasture was uniformly burned annually. The heterogeneous—eight patches—treatment had two patches burned annually (one in spring and one in summer) over 4 years and was the most heterogeneous treatment. Pastures that were managed to promote heterogeneity had more consistent livestock production and were less influenced by low-rainfall years. Modified from Allred et al. (2014)

Fig. 5.5 Bison are able to forage in different patches to meet their differing forage requirement. This photo is from the Tallgrass Prairie Preserve in Oklahoma and illustrates the increased heterogeneity in forage quality and quantity. Photo by Steve Winter

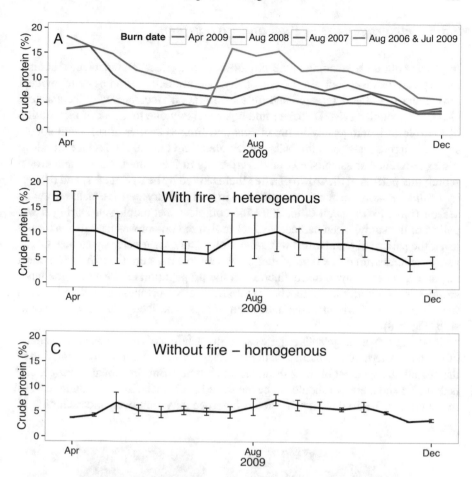

Fig. 5.6 Forage crude protein (%) as influenced by fire-grazing interactions in tallgrass prairie. (**a**) Forage quality of patches that vary in time since fire. Recently burned areas have much higher forage quality than unburned areas and remain high throughout the season as animals graze thereon. (**b**) Error bars reflect the heterogeneity in forage quality available to herbivores. Due to the high variation in forage resources when heterogeneous fires are applied to the landscape, animal choice is unrestricted so animals can seek out desired forage based on dietary needs and preferences. (**c**) Elimination of fire (mean of unburned patches) results in low forage quality through time and low patch variability. This low variation homogenizes available forage resources to restrict livestock choice

strongly limited by fences may require more application of alternative disturbances than large landscapes with free-roaming herds. It has been demonstrated that grazing operations on small pastures characteristic of many compartmentalized rangelands also benefit from the creation of fire-driven heterogeneity promoted by small burned patches that are rotated over several years (Limb et al. 2011; Allred et al. 2014).

5.3.3   Heterogeneity of Fuel and Fire Effects

Improving our understanding of the variability of fire effects on rangelands requires that researchers account for heterogeneity. Contradictory findings are evident throughout fire effects research and little information elucidates why inconsistencies among studies exist. Disparate findings are likely due to a lack of recognition of the role of heterogeneity in fire effects and limited studies at sufficient spatial scale to capture real-world fire behavior (Fuhlendorf et al. 2011). As an example, a fire experiment was established to link variability in fuels, fire behavior, and crown scorch (the portion of the crown of a tree that is killed by heat) (Fig. 5.7) and mortality of *Juniperus ashei* for fires conducted in wet and dry periods of the growing season (refer to Twidwell et al. 2009 for complete methodology). During a wet period of the study, when herbaceous fuel moisture content was near its maximum level, the pattern of area burned was a function of fine-scale patch dynamics. Of the parameters measured, the type of fuel patch and its size were the two factors most important in determining discontinuities in the propagation of fire across the landscape. Discontinuities in the fuel bed create fuel gaps—patches without herbaceous fuels that occurred within a continuous bed of grassland fuels—drove this relationship (Fig. 5.8).

Studying fire on rangelands requires an understanding of functional heterogeneity of fire and pattern of area burned on the landscape, which is ultimately a function of the spatial arrangement of these different patch types. But functional heterogeneity is dynamic and therefore should not be measured or characterized at a single scale of measurement. Understanding functional heterogeneity improves understanding of

Fig. 5.7 Variability in height of tree scorch on two adjacent Ashe juniper (*Juniperus ashei*) trees indicates that heterogeneity of fuel load and tree size may be critical in understanding fire effects on rangelands. Photo by Dirac Twidwell

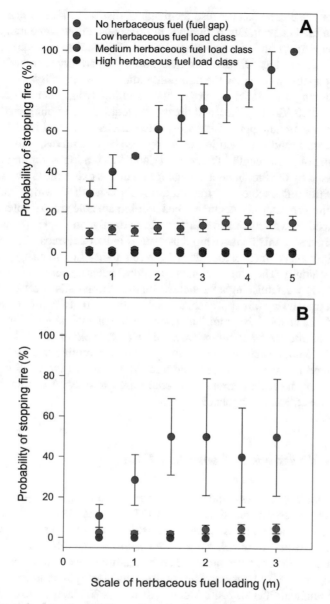

Fig. 5.8 Fine-scale determinants of area burned, shown here as the relationships between the amount of herbaceous fuel loading occurring at multiple spatial scales and the probability of stopping fire spread for fires conducted in two different fuel moisture conditions (**A**, high fine fuel moisture; **B**, low fine fuel moisture). Increasing gaps in herbaceous fuels increase the probability of stopping fire, but the relationship is less predictable in fuel conditions promoting more erratic fire behavior (e.g., in low fuel moisture conditions)

second-order fire effects that are also the result of discontinuities in fire spread and variability in fire behavior. Discontinuities in fire spread allow some juniper trees to escape damage from fire, whereas others directly adjacent to them are completely scorched and killed (Fig. 5.7). In other cases, trees are partially scorched, indicating that the heat applied was below the threshold required for mortality.

Understanding functional heterogeneity resulting from interactions between variability in fuels, fire behavior, and fire effects requires a sampling procedure that differs from traditional approaches. Most often, herbaceous fuel load is randomly sampled across a landscape and in the interspaces between trees, followed by random sampling of flame length (if done at all), and then random sampling of physical damage caused by the fire. In contrast, sampling in this experiment was stratified to account for the influence of the tree on the fuels beneath it, which subsequently influenced the fire intensity occurring beneath the specific tree of interest, and in turn influenced crown scorch and mortality. Importantly, no single spatial scale was used to characterize this relationship, but instead individual trees were of different sizes and influenced interrelationships between fuels, fire intensity, and crown scorch differentially. The contribution of functional heterogeneity to fire effects was best described by establishing a variable scale that accounted for differences in the size of each tree. Using this approach, a clear relationship of functional heterogeneity emerged that enabled the empirical detection and quantification of the fire intensity threshold required for juniper mortality (Twidwell et al. 2013a, b). This threshold could not be detected using simple random sampling that assumes homogeneity around the average in fuel load and fire behavior (Twidwell et al. 2009), which has been the most common procedure used to attempt to understand how fuels drive fire effects in rangelands.

5.3.4 Biodiversity and Ecosystem Function

Meeting variable forage demands and analyzing fire effects are just a couple of examples of enhanced functionality created by heterogeneity in rangelands. Heterogeneity can also increase habitat availability for different plant, insect, bird, and mammal species. Research worldwide describes how different species within major taxonomic groups have variable habitat requirements, and managing for spatially heterogeneous landscapes creates multiple habitat types simultaneously (Tews et al. 2004; Fuhlendorf et al. 2006; McGranahan et al. 2013a, b). Furthermore, plant biomass production varies less across seasons in spatially heterogeneous landscapes (McGranahan et al. 2016).

The fire–grazing interaction is especially important in the North American Great Plains, where more distinct habitat types are created in rangeland managed with pyric herbivory than in rangeland managed with fire or grazing alone (Fuhlendorf et al. 2009). This breadth of habitat types is essential for the conservation of grassland-obligate fauna, such as the Henslow's Sparrow (*Ammodramus henslowii*), which requires dense, moribund grass material for nesting, and the regal fritillary

Fig. 5.9 Greater prairie chickens require short vegetation for their breeding displays, but females build their nests in nearby dense vegetation that has not been burned or grazed for 1 year or more. Brood rearing requires open vegetation with high plant and insect diversity and structural heterogeneity required for protection from temperature extremes and ease of movement. Photo by Torre Hovick

(*Speyeria idalia*), a fire-dependent but grazing-sensitive butterfly, which might depend on spatial heterogeneity to persist in working rangeland landscapes that are managed specifically for grazing (Fuhlendorf et al. 2006; Moranz et al. 2014).

Perhaps the most illustrative example of a species that requires the full breadth of habitat types available under heterogeneity-based fire and grazing management is the greater prairie chicken (*Tympanuchus cupido*) (Fig. 5.9, Hovick et al. 2014). These grassland-obligate birds begin the breeding season on leks, areas of extremely short vegetation where males gather and display to attract females. Upon mating, females seek dense vegetation to hide their nests during incubation, and between hatching and fledging, prairie chicken broods benefit from transitional patches following focal disturbance from pyric herbivory, as these areas provide aerial plant cover with limited obstruction from litter and other vegetative debris. Importantly, prairie chickens require each of these habitat types to be accessible within relatively short distances. Greater prairie chickens select nesting sites at coarse scales to be near leks (frequently on burned sites) and far from trees and at very fine scales for specific sites best suited to moderate temperature extremes (Hovick et al. 2014).

Likewise, several African antelope species require patches of short grassland for grazing and adjacent taller grass patches for resting and concealing young (Everett et al. 1991). Short-grass grazers such as wildebeest and Thomson's gazelle occur in higher densities in the heavily grazed livestock areas outside the Masai-Mara Game Reserve in East Africa, whereas tall-grass grazers such as Cape buffalo (*Syncerus caffer*) are restricted to within the less heavily grazed taller grass areas within the game reserve (Bhola et al. 2012). Similarly, diverse communities of African herbivores require heterogeneity due to topo-edaphic patterns as well as disturbance.

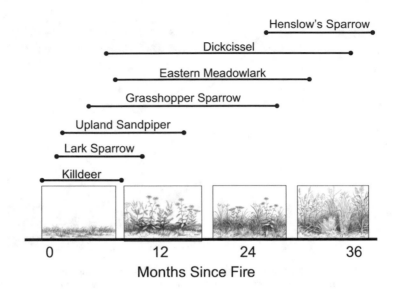

Fig. 5.10 Response of grassland birds to time since disturbance by fire and grazing at the Tallgrass Prairie Preserve, Oklahoma, from 2001 to 2003. Some birds native to the area require recently burned patches that are heavily grazed while others require habitats that are undisturbed for several years (Fuhlendorf et al. 2006)

Similar patterns of plant, insect, and bird compositional responses to burning and grazing frequency effects on grassland structural heterogeneity have been observed in both North American and African rangelands. On both continents, the interactive effects of fire and grazing on grassland structure and plant composition translated into differences in invertebrate communities (Chambers and Samways 1998; Engle et al. 2008; Dosso et al. 2010; Doxon et al. 2011). The community composition of several bird taxa—especially passerines—also varies with grassland structure in both continents (Fig. 5.10) (Fuhlendorf et al. 2006; Bouwman and Hoffman 2007; Krook et al. 2007; Gregory et al. 2010; Chalmandrier et al. 2013). Exceptions, however, do exist, and the community composition of some taxa is not associated with spatially heterogeneous disturbance patterns especially when those patterns are limited in spatial extent and largely based on measured rather than functional heterogeneity estimates (Pillsbury et al. 2011; Davies et al. 2012; Moranz et al. 2012; McGranahan et al. 2013a, b).

Beyond maximizing habitat heterogeneity across a landscape within a given season, maintaining a spatial mosaic of patches across several seasons increases stability of important ecosystem functions like aboveground biomass production and habitat. Ecologists have long recognized that biological diversity—often measured as plant species richness or functional types—can stabilize community composition and ecosystem function (McNaughton 1977; Tilman and Downing 1994; Tilman et al. 2006; Zimmerman et al. 2010; Isbell et al. 2011). Experiments now demonstrate that ecological diversity—measured as differences among patches within heterogeneous landscapes—can stabilize avian community composition, as well as both plant biomass and livestock production, over time (McGranahan et al. 2016; Allred et al. 2014; Hovick et al. 2014). This suggests that spatially heterogeneous disturbance regimes that reduce temporal variability in primary production might represent a land-use paradigm that enhances landscape-level diversity to promote rangeland conservation and resilience in a changing world (Fuhlendorf et al. 2012). So, we would argue that at broad scales, functioning rangelands should have variable disturbance patterns that interact with inherent topo-edaphic variability and are central to many aspects of landscape and population stability. This relationship among disturbance, pattern, diversity, heterogeneity, and stability is more realistic at broad scale than the more simplistic focus on diversity as a driver of stability as predicted by small-plot agronomic experiments.

Increasing spatial heterogeneity might buffer ecosystem function against climate change, which is consistent with the predictions of diversity-stability theory (Mori et al. 2013). In many rangelands, variation in primary production can destabilize ecosystem structure and function. Such variability is expected to increase under many climate change scenarios, which might make ecosystems more vulnerable to degradation of ecosystem function (Walker et al. 2004; Mori et al. 2013). Because rangeland management often depends upon a degree of dynamic equilibrium (Briske et al. 2003; Mori 2011), enhancing a spatial pattern of heterogeneity that buffers against temporal variability can enhance response to global change much in the way of portfolio effects of diversity-stability studies (Turner 2010).

5.4 Future Perspectives

Rangeland management has been slow to adopt a dynamic basis for ecosystem management that sustains multiple ecosystem services. Instead, livestock production systems continue to trump management for other rangeland services. For example, emergency haying programs in the USA permit the harvesting of grass biomass in grasslands prioritized for wildlife conservation when production is lower than optimal due to environmental stressors such as drought. The result is a strong reliance on command-and-control management approaches to rangeland management (e.g., attempting to minimize variation). To overcome natural rangeland variability, conventional grazing management relies upon a myopic focus on cross-fencing and controlled access to forage and water, seeking to minimize variability in disturbance intensity by promoting uniform, moderate grazing across the entire landscape. This attempt to override heterogeneity has been aptly described as management toward a uniform middle and has become the central theme to the discipline or rangeland management (Fuhlendorf and Engle 2001, 2004; Bailey and Brown 2011).

Mounting evidence suggests that heterogeneity enhances biodiversity in agricultural landscapes (Ricketts et al. 2001; Benton et al. 2003; Hobbs et al. 2008; Franklin and Lindenmayer 2009) where native biodiversity is threatened by the intensification and compartmentalization of land use (Reidsma et al. 2006; Flynn et al. 2009). But agricultural policy has been slow to respond. In the USA, no federal farm bill program specifically targets farmland or rangeland heterogeneity, although limited heterogeneity-based management practices have been recently allowed for use in long-standing programs like the Conservation Reserve Program (NRCS 2004; Hart 2006), and such heterogeneity-based management has been shown to increase the quality of CRP projects for wildlife (Matthews et al. 2012). In Europe, where agricultural conservation policies tend to place greater emphasis on landscape-level objectives than in the USA (Baylis et al. 2008), agri-environmental schemes that can increase wildlife habitat heterogeneity remain an unstated objective and incidental outcome (Vickery et al. 2004).

Understanding heterogeneity has been an important limitation to the application and principles of science and management on rangelands. Globally, rangeland science indicates that heterogeneity and diversity increase stability in ecosystem properties from a broad spatial and temporal perspective. Management should no longer consider fine-scale spatial and temporal variability as a threat to ecosystem structure and function. It is logistically critical to the science of rangelands because of the importance of scale in experimental design and the point that traditional experimental design was largely based on Fisherian statistics where small experimental units were used to minimize variation within treatments (Fuhlendorf et al. 2009). Embracing heterogeneity requires academics, practitioners, and policy makers to realize the fallacy in building a profession that relies on statistical replications of small-scale plots to represent complex rangeland landscapes that are dynamic in space and time. This is a fundamental fallacy of our profession that is mostly a social construct of the profession rather than a reflection of a true need to simplify landscapes. Understanding that this simplified focus on homogeneity is cultural to the profession suggests that we can work to overcome these biases through academia and natural resource agencies. The greatest challenge and opportunity in contemporary rangeland science and management are overcoming our traditional focus on uniformity and developing policies and an understanding that promotes rangelands as heterogeneous natural resources that are complex and capable of achieving many objectives by operating at the nexus of working and wild landscapes.

5.5 Summary

Understanding rangelands as complex, dynamic ecosystems that are highly variable in space and time cannot be achieved without a focus on heterogeneity as a critical and multiscale characteristic. Comparisons between the state of our current scientific knowledge and the application of management have often identified scale and heterogeneity as limitations to making our science applicable to land management and policy (Bestelmeyer et al. 2011, Fuhlendorf et al. 2011). While we have theoretically understood rangelands as dynamic and variable in space there has been minimal effort focusing on the variability as a critical and inherent characteristic of rangelands. One very important exception to this has been the efforts to connect variation in soil, landform, and climate to ecological sites through USDA-NRCS

(Chapter 9, this volume). This is a critical first step, but still limits variability to mapping units that are primarily viewed as static in space and time. This is often still focused on a single state or phase existing within each site rather than a dynamic and shifting condition that is variable in space and time (Twidwell et al. 2013a, b). The limited use of non-equilibrial concepts and landscape ecological principles is surprising because rangelands are disturbance-driven ecosystems that are clearly dynamic in space and time.

Ultimately, the science and management of rangelands need to advance beyond a focus on average conditions and the current paradigm of uniform and moderate disturbance. This simplistic focus leads to debates, such as wildlife vs. livestock, fuel vs. forage, and forests vs. grassland. Understanding heterogeneity in space and time should be central to the framework for advancing our discipline and progressing to solve problems that arise with changes in societal desires on rangelands. Perhaps the greatest challenge for applying heterogeneity-based science in rangeland management is overcoming a century-old vision of rangelands as simple ecosystems that sustainably provide forage for domestic livestock. Below are general principles for our profession to begin thinking of rangelands as highly dynamic in space and time that provide many goods and services to society that include livestock production, wildlife habitat, biodiversity, and water quality and quantity (Fuhlendorf et al. 2012):

1. Large continuous tracts of rangelands are critical for conservation so that disturbance processes can interact with inherent heterogeneity to form multiscaled mosaics that are capable of providing multiple goods and services. Large landscapes will include more heterogeneity than small landscapes and this will buffer ecosystems and populations from unexpected and stochastic perturbations (Ash and Stafford-Smith 1996). Rangeland fragmentation that results in many small land units precludes sufficient patch size or number for long-term conservation and land management objectives. Conservancies and landowner associations can help coordinate heterogeneity-based management at broad spatial scales (Toombs et al. 2010; McGranahan 2011).

2. Professionals and the general public have largely learned to promote uniformity in disturbance processes and minimize the occurrence of both undistributed and severely disturbed areas. The first step in managing for heterogeneity and multiple objectives is to place value on these disturbance-driven attributes and to minimize efforts to manage for homogeneity or uniformity. This will require us to develop approaches that promote variability in disturbance frequency and intensity across complex and large landscapes, preferably by recognizing, maintaining, and restoring broad-scale processes.

3. Shifting mosaics of landscape patches are necessary for maintaining ecosystem structure and function and achieving multiple objectives such as improved productivity and stability of livestock production (Limb et al. 2011) and conservation objectives (Fuhlendorf et al. 2009). Managing for a single condition, state, phase, or successional stage is incapable of sustaining livestock production and is not capable of promoting biodiversity or multiple uses.

4. Inherent heterogeneity, associated with soils, topography, and temporal variability from climate, is a defining characteristic of rangelands. Additionally, disturbance regimes, such as fire and grazing, are as vital to ecosystem structure and function as climate and soils and are capable of driving landscape-level heterogeneity. These disturbances must be viewed as interactive processes that are critical to heterogeneity of rangelands rather than mere optional management tools.

5. As policy developers and implementers recognize the importance of multiple land uses and the full suite of ecosystem services, a focus must be placed on maintaining large landscapes, in spite of fragmented ownerships, and conserving the processes that drive heterogeneity at multiple scales. Developing policies that move beyond the traditional command-and-control paradigm/steady-state management model will be our greatest challenge in the next century.

References

Acres, B., A.B. Rains, R. King, et al. 1985. African dambos: Their distribution, characteristics and use. *Zeitschrift für Geomorphologie* 52: 63–86.

Albon, S., and R. Langvatn. 1992. Plant phenology and the benefits of migration in a temperate ungulate. *Oikos* 65: 502–513.

Allred, B.W., S.D. Fuhlendorf, D.M. Engle, et al. 2011. Ungulate preference for burned patches reveals strength of fire-grazing interaction. *Ecology and Evolution* 1: 132–144.

Allred, B.W., S.D. Fuhlendorf, T.J. Hovick, R. Dwayne Elmore, D.M. Engle, and A. Joern. 2013. Conservation implications of native and introduced ungulates in a changing climate. Global *Change Biology* 19: 1875–1883.

Allred, B.W., J.D. Scasta, T.J. Hovick, et al. 2014. Spatial heterogeneity stabilizes livestock productivity in a changing climate. *Agriculture, Ecosystems & Environment* 193: 37–41.

Archibald, S., W.J. Bond, W.D. Stock, et al. 2005. Shaping the landscape: Fire–grazer interactions on an African savanna. *Ecological Applications* 15: 96–109.

Ash, A.J., and D.M. Stafford-Smith. 1996. Evaluating stocking rate impacts in rangelands: Animals don't practice what we preach. *The Rangeland Journal* 18: 216–243.

Bailey, D.W., and J.R. Brown. 2011. Rotational grazing systems and livestock grazing behavior in shrub-dominated semi-arid and arid rangelands. *Rangeland Ecology & Management* 64: 1–9.

Bartlam-Brooks, H., M. Bonyongo, and S. Harris. 2011. Will reconnecting ecosystems allow long-distance mammal migrations to resume? A case study of a zebra Equus burchelli migration in Botswana. *Oryx* 45: 210–216.

Baylis, K., S. Peplow, G. Rausser, et al. 2008. Agri-environmental policies in the EU and United States: A comparison. *Ecological Economics* 65: 753–764.

Benton, T.G., J.A. Vickery, and J.D. Wilson. 2003. Farmland biodiversity: Is habitat heterogeneity the key? *Trends in Ecology & Evolution* 18: 182–188.

Bestelmeyer, B.T., J.R. Brown, S.D. Fuhlendorf, G.A. Fults, and X.B. Wu. 2011. A landscape approach to rangeland conservation practices. In *Conservation benefits of rangeland practices: Assessment, recommendations, and knowledge*. Washington, DC: United States Department of Agriculture, Natural Resource Conservation Service.

Bhola, N., J.O. Ogutu, M.Y. Said, et al. 2012. The distribution of large herbivore hotspots in relation to environmental and anthropogenic correlates in the Mara region of Kenya. *Journal of Animal Ecology* 81: 1268–1287.

Bouwman, H., and R. Hoffman. 2007. The effects of fire on grassland bird communities of Barberspan, North West Province, South Africa. *Ostrich Journal of African Ornithology* 78: 591–608.

Breman, H., and C. De Wit. 1983. Rangeland productivity and exploitation in the Sahel. *Science* 221: 1341–1347.

Briggs, J.M., and A.K. Knapp. 1995. Interannual variability in primary production in tallgrass prairie: Climate, soil moisture, topographic position, and fire as determinants of aboveground biomass. *American Journal of Botany* 82: 1024–1030.

Briske, D.D., S.D. Fuhlendorf, and F.E. Smeins. 2003. Vegetation dynamics on rangelands: A critique of the current paradigms. *Journal of Applied Ecology* 40: 601–614.

Chalmandrier, L., G.F. Midgley, P. Barnard, et al. 2013. Effects of time since fire on birds in a plant diversity hotspot. *Acta Oecologica* 49: 99–106.

Chambers, B.Q., and M.J. Samways. 1998. Grasshopper response to a 40-year experimental burning and mowing regime, with recommendations for invertebrate conservation management. *Biodiversity and Conservation* 7: 985–1012.

Coetsee, C., W.D. Stock, and J.M. Craine. 2011. Do grazers alter nitrogen dynamics on grazing lawns in a South African savannah? *African Journal of Ecology* 49: 62–69.

Davies, A.B., P. Eggleton, B.J. van Rensburg, et al. 2012. The pyrodiversity-biodiversity hypothesis: A test with savanna termite assemblages. *Journal of Applied Ecology* 49: 422–430.

Dosso, K., S. Konaté, D. Aidara, et al. 2010. Termite diversity and abundance across fire-induced habitat variability in a tropical moist savanna (Lamto, Central Côte d'Ivoire). *Journal of Tropical Ecology* 26: 323–334.

Doxon, E.D., C.A. Davis, S.D. Fuhlendorf, et al. 2011. Aboveground macroinvertebrate diversity and abundance in sand sagebrush prairie managed with the use of pyric herbivory. *Rangeland Ecology & Management* 64: 394–403.

Ellis, J.E., and D.M. Swift. 1988. Stability of African pastoral ecosystems: Alternate paradigms and implications for development. *Journal of Range Management* 41: 450–459.

Engle, D.M., S.D. Fuhlendorf, A. Roper, et al. 2008. Invertebrate community response to a shifting mosaic of habitat. *Rangeland Ecology & Management* 61: 55–62.

Everett, P., M. Perrin, and D. Rowe-Rowe. 1991. Responses by oribi to different range management practices in Natal. *South African Journal of Wildlife Research* 21: 114–118.

Festa-Bianchet, M. 1988. Seasonal range selection in bighorn sheep: Conflicts between forage quality, forage quantity, and predator avoidance. *Oecologia* 75: 580–586.

Flynn, D.F.B., M. Gogol-Prokurat, T. Nogeire, et al. 2009. Loss of functional diversity under land use intensification across multiple taxa. *Ecology Letters* 12: 22–33.

Frank, D.A., S.J. McNaughton, and B.F. Tracy. 1998. The ecology of the earth's grazing ecosystems. *BioScience* 48: 513–521.

Franklin, J.F., and D.B. Lindenmayer. 2009. Importance of matrix habitats in maintaining biological diversity. *Proceedings of the National Academy of Science* 106: 349–350.

Fuhlendorf, S.D., and D.M. Engle. 2001. Restoring heterogeneity on rangelands: Ecosystem management based on evolutionary grazing patterns. *BioScience* 51: 625–632.

———. 2004. Application of the fire–grazing interaction to restore a shifting mosaic on tallgrass prairie. *Journal of Applied Ecology* 41: 604–614.

Fuhlendorf, S.D., and F.E. Smeins. 1998. The influence of soil depth on plant species response to grazing within a semi-arid savanna. *Plant Ecology* 138: 89–96.

———. 1999. Scaling effects of grazing in a semi-arid grassland. *Journal of Vegetation Science* 10: 731–738.

Fuhlendorf, S.D., W.C. Harrell, D.M. Engle, et al. 2006. Should heterogeneity be the basis for conservation? Grassland bird response to fire and grazing. *Ecological Applications* 16: 1706–1716.

Fuhlendorf, S.D., D.M. Engle, J. Kerby, et al. 2009. Pyric herbivory: Rewilding landscapes through the recoupling of fire and grazing. *Conservation Biology* 23: 588–598.

Fuhlendorf, S.D., R.F. Limb, D.M. Engle, R.F. Miller. 2011. Assessment of prescribed fire as a conservation practice. In *Conservation benefits of rangeland practices: Assessment, recommendations, and knowledge*. Washington, DC: United States Department of Agriculture, Natural Resource Conservation Service.

Fuhlendorf, S.D., D.M. Engle, R.D. Elmore, et al. 2012. Conservation of pattern and process: Developing an alternative paradigm of rangeland management. *Rangeland Ecology & Management* 65: 579–589.

Fynn, R.W., and M.C. Bonyongo. 2011. Functional conservation areas and the future of Africa's wildlife. *African Journal of Ecology* 49: 175–188.

Fynn, R.W., M. Chase, and A. Roder. 2014. Functional habitat heterogeneity and large herbivore seasonal habitat selection in northern Botswana. *South African Journal of Wildlife Research* 44: 1–15.

Gómez, J., F. Valladares, and C. Puerta-Piñero. 2004. Differences between structural and functional environmental heterogeneity caused by seed dispersal. *Functional Ecology* 18: 787–792.

Grant, C., and M. Scholes. 2006. The importance of nutrient hot-spots in the conservation and management of large wild mammalian herbivores in semi-arid savannas. *Biological Conservation* 130: 426–437.

Gregory, N.C., R.L. Sensenig, and D.S. Wilcove. 2010. Effects of controlled fire and livestock grazing on bird communities in East African savannas. *Conservation Biology* 24: 1606–1616.

Hart, J.L. 2006. *Patch burn grazing*. Kansas: USDA Natural Resource Conservation Service.

Hebblewhite, M., and E.H. Merrill. 2007. Multiscale wolf predation risk for elk: Does migration reduce risk? *Oecologia* 152: 377–387.

Hempson, G.P., S. Archibald, W.J. Bond, R.P. Ellis, C.C. Grant, F.J. Kruger, L.M. Kruger, C. Moxley, N. Owen-Smith, M.J.S. Peel, I.P.J. Smit, and K.J. Vickers. 2015. Ecology of grazing lawns in Africa. *Biological Reviews* 90: 979–994.

Hobbs, N.T., and I.J. Gordon. 2010. How does landscape heterogeneity shape dynamics of large herbivore populations. In *Dynamics of large herbivore populations in changing environments: Towards appropriate models*, ed. N. Owen-Smith, 141–164. Chicester: Wiley-Blackwell.

Hobbs, N.T., K.A. Galvin, C.J. Stokes, et al. 2008. Fragmentation of rangelands: Implications for humans, animals, and landscapes. *Global Environmental Change* 18: 776–785.

Holling, C.S., and G.K. Meffe. 1996. Command and control and the pathology of natural resource management. *Conservation Biology* 10: 328–337.

Hopcraft, J.G.C., H. Olff, and A. Sinclair. 2010. Herbivores, resources and risks: Alternating regulation along primary environmental gradients in savannas. *Trends in Ecology & Evolution* 25: 119–128.

Hovick, T.J., R.D. Elmore, and S.D. Fuhlendorf. 2014. Structural heterogeneity increases diversity of non-breeding grassland birds. *Ecosphere* 5: art62.

Illius, A.W., and T.G. O'Connor. 2000. Resource heterogeneity and ungulate population dynamics. *Oikos* 89: 283–294.

Isbell, F., V. Calcagno, A. Hector, et al. 2011. High plant diversity is needed to maintain ecosystem services. *Nature* 477: 199–202.

Jarman, P. 1972. Seasonal distribution of large mammal populations in the unflooded Middle Zambezi Valley. *Journal of Applied Ecology* 9: 283–299.

Knapp, A.K., J.M. Blair, J.M. Briggs, et al. 1999. The keystone role of bison in North American tallgrass prairie. *BioScience* 49: 39–50.

Kolasa, J., and C.D. Rollo. 1991. Introduction: The heterogeneity of heterogeneity: A glossary. In *Ecological heterogeneity*, 1–23. New York, NY: Springer.

Krook, K., W.J. Bond, and P.A. Hockey. 2007. The effect of grassland shifts on the avifauna of a South African savanna. *Ostrich* 78: 271–279.

Levin, S.A. 1992. The problem of pattern and scale in ecology. *Ecology* 73: 1943–1967.

Li, H., and J.F. Reynolds. 1995. On definition and quantification of heterogeneity. *Oikos* 73: 280–284.

Limb, R.F., S.D. Fuhlendorf, D.M. Engle, et al. 2011. Pyric-herbivory and cattle performance in grassland ecosystems. *Rangeland Ecology & Management* 64: 659–663.

Maddock, L. 1979. The "migration" and grazing succession. In *Serengeti: Dynamics of an ecosystem*, ed. A.R.E. Sinclair and M. Norton-Griffiths, 104–129. Chicago: University of Chicago Press.

Matthews, T.W., J.S. Taylor, and L.A. Powell. 2012. Mid-contract management of Conservation Reserve Program grasslands provides benefits for ring-necked pheasant nest and brood survival. *Journal of Wildlife Management* 76: 1643–1652.

McGranahan, D.A. 2011. Identifying ecological sustainability assessment factors for ecotourism and trophy hunting operations on private rangeland in Namibia. *Journal of Sustainable Tourism* 19: 115–131.

McGranahan, D.A., D.M. Engle, S.D. Fuhlendorf, et al. 2013a. Inconsistent outcomes of heterogeneity-based management underscore importance of matching evaluation to conservation objectives. *Environmental Science & Policy* 31: 53–60.

McGranahan, D.A., D.M. Engle, J.R. Miller, et al. 2013b. An invasive grass increases live fuel proportion and reduces fire spread in a simulated grassland. *Ecosystems* 16: 158–169.

McGranahan, D.A., T. Hovick, R.D. Elmore, D.M. Engle, S.D. Fuhlendorf, S. Winter, J.R. Miller, D.M. Debinski. 2016. Temporal variability in aboveground plant biomass decreases as spatial variability increases. *Ecology* 97: 555–560.

McNaughton, S.J. 1977. Diversity and stability of ecological communities: A comment on the role of empiricism in ecology. *American Naturalist* 111: 515–525.

McNaughton, S.J., and F.F. Banyikwa. 1995. Plant communities and herbivory. In *Serengeti II: Dynamics, management, and conservation of an ecosystem*, ed. A.R.E. Sinclair and P. Arcese, 49–70. Chicago: University of Chicago Press.

Moranz, R.A., D.M. Debinski, D.A. McGranahan, et al. 2012. Untangling the effects of fire, grazing, and land-use legacies on grassland butterfly communities. *Biodiversity and Conservation* 21: 2719–2746.

Moranz, R.A., S.D. Fuhlendorf, and D.M. Engle. 2014. Making sense of a prairie butterfly paradox: The effects of grazing, time since fire, and sampling period on regal fritillary abundance. *Biological Conservation* 173: 32–41.

Mori, A.S. 2011. Ecosystem management based on natural disturbances: Hierarchical context and non-equilibrium paradigm. *Journal of Applied Ecology* 48: 280–292.

Mori, A.S., T. Furukawa, and T. Sasaki. 2013. Response diversity determines the resilience of ecosystems to environmental change. *Biological Reviews* 88: 349–364.

Murray, M.G. 1995. Specific nutrient requirements and migration of wildebeest. In *Serengeti II: dynamics, management, and conservation of an ecosystem*, ed. A.R.E. Sinclair and P. Arcese, 231–256. Chicago: University of Chicago Press.

Mysterud, A., R. Langvatn, N.G. Yoccoz, et al. 2001. Plant phenology, migration and geographical variation in body weight of a large herbivore: the effect of a variable topography. *Journal of Animal Ecology* 70: 915–923.

NRCS. 2004. *Designing a patch burn grazing system*. Missouri: United States Department of Agriculture.

O'Reagain, P., and R. Owen-Smith. 1996. Effect of species composition and sward structure on dietary quality in cattle and sheep grazing South African sourveld. *Journal of Agricultural Science* 127: 261–270.

Omer, R., A. Hester, I. Gordon, et al. 2006. Seasonal changes in pasture biomass, production and offtake under the transhumance system in northern Pakistan. *Journal of Arid Environments* 67: 641–660.

Owen-Smith, N. 2002. A metaphysiological modelling approach to stability in herbivore–vegetation systems. *Ecological Modelling* 149: 153–178.

———. 2004. Functional heterogeneity in resources within landscapes and herbivore population dynamics. *Landscaoe Ecology* 19: 761–771.

Pamo, E.T. 1998. Herders and wildgame behaviour as a strategy against desertification in northern Cameroon. *Journal of Arid Environments* 39: 179–190.

Parker, K.L., P.S. Barboza, and M.P. Gillingham. 2009. Nutrition integrates environmental responses of ungulates. *Functional Ecology* 23: 57–69.

Parrini, F., and N. Owen-Smith. 2010. The importance of post-fire regrowth for sable antelope in a Southern African savanna. *African Journal of Ecology* 48: 526–534.

Pillsbury, F.C., J.R. Miller, D.M. Debinski, et al. 2011. Another tool in the toolbox? Using fire and grazing to promote bird diversity in highly fragmented landscapes. *Ecosphere* 2: art28.

Reidsma, P., T. Tekelenburg, M. Vandenberg, et al. 2006. Impacts of land-use change on biodiversity: An assessment of agricultural biodiversity in the European Union. *Agriculture, Ecosystems and Environment* 114: 86–102.

Ricketts, T.H., G.C. Daily, P.R. Ehrlich, et al. 2001. Countryside biogeography of moths in a fragmented landscape: Biodiversity in native and agricultural habitat. *Conservation Biology* 15: 378–388.

Sala, O.E., W.J. Parton, L.A. Joyce, et al. 1988. Primary production of the central grassland region of the United States. *Ecology* 69: 40–45.

Scoones, I. 1993. Why are there so many animals? Cattle population dynamics in the communal areas of Zimbabwe. In *Range ecology at disequilibrium*, ed. R.H. Behnke, I. Scoones, and C. Kerven, 62–76. London: Overseas Development Institute.

———. 1995. Exploiting heterogeneity: Habitat use by cattle in dryland Zimbabwe. *Journal of Arid Environments* 29: 221–237.

Senft, R.L., L.R. Rittenhouse, and R.G. Woodmansee. 1985. Factors influencing patterns of cattle grazing behavior on shortgrass steppe. *Journal of Range Management* 38: 82–87.

Tews, J., U. Brose, V. Grimm, et al. 2004. Animal species diversity driven by habitat heterogeneity/diversity: The importance of keystone structures. *Journal of Biogeography* 31: 79–92.

Tilman, D., and J.A. Downing. 1994. Biodiversity and stability in grasslands. *Nature* 367: 363–365.

Tilman, D., P.B. Reich, and J.M. Knops. 2006. Biodiversity and ecosystem stability in a decade-long grassland experiment. *Nature* 441: 629–632.

Toombs, T.P., J.D. Derner, D.J. Augustine, et al. 2010. Managing for biodiversity and livestock. *Rangelands* 32: 10–15.

Turner, M.G. 1989. Landscape ecology: The effect of pattern on process. *Annual Review of Ecological Systems* 20: 171–197.

———. 2005. Landscape ecology: What is the state of the science? *Annual Review of Ecological Systems* 36: 319–344.

———. 2010. Disturbance and landscape dynamics in a changing world. *Ecology* 91: 2833–2849.

Twidwell, D., S.D. Fuhlendorf, D.M. Engle, et al. 2009. Surface fuel sampling strategies: Linking fuel measurements and fire effects. *Rangeland Ecology & Management* 62: 223–229.

Twidwell, D., B.W. Allred, and S.D. Fuhlendorf. 2013. National-scale assessment of ecological content in the world's largest land management framework. *Ecosphere* 4: art94.

Twidwell, D., S.D. Fuhlendorf, C.A. Taylor, et al. 2013b. Refining thresholds in coupled fire-vegetation models to improve management of encroaching woody plants in grasslands. *Journal of Applied Ecology* 50: 603–613.

Urban, D.L., R.V. O'Neill, and H.H. Shugart Jr. 1987. Landscape ecology. *BioScience* 37: 119–127.

Vesey-FitzGerald, D.F. 1960. Grazing succession among East African game animals. *Journal of Mammalogy* 41: 161–172.

Vickery, J.A., R.B. Bradbury, I.G. Henderson, et al. 2004. The role of agri-environment schemes and farm management practices in reversing the decline of farmland birds in England. *Biological Conservation* 119: 19–39.

Walker, B.H., R. Emslie, R. Owen-Smith, et al. 1987. To cull or not to cull: Lessons from a southern African drought. *Journal of Applied Ecology* 24: 381–401.

Walker, B., C.S. Holling, S.R. Carpenter, et al. 2004. Resilience, adaptability and transformability in social–ecological systems. *Ecology and Society* 9: art5.

Wiens, J.A. 1989. Spatial scaling in ecology. *Functional Ecology* 3: 385–397.

Zimmerman, J.K., L.S. Comita, J. Thompson, et al. 2010. Patch dynamics and community meta-stability of a subtropical forest: Compound effects of natural disturbance and human land use. *Landscape Ecology* 25: 1099–1111.

Chapter 6
Nonequilibrium Ecology and Resilience Theory

David D. Briske, Andrew W. Illius, and J. Marty Anderies

Abstract Nonequilibrium ecology and resilience theory have transformed rangeland ecology and management by challenging the traditional assumptions of ecological stability and linear successional dynamics. These alternative interpretations indicate that ecosystem dynamics are strongly influenced by disturbance, heterogeneity, and existence of multiple stable states. The nonequilibrium persistent model indicates that plant production and livestock numbers are seldom in equilibrium in pastoral systems because reoccurring drought maintains livestock number below the ecological carrying capacity. However, it has recently been demonstrated that livestock are often in equilibrium with key dry-season resources, even though they may only be loosely coupled to abundant wet-season resources. Similarly, state-and-transition models were initially influenced by nonequilibrium ecology, but they have subsequently been organized around resilience theory to represent both equilibrial dynamics within states and existence of multiple states. Resilience theory was introduced to describe how ecosystems can be dynamic, but still persist as self-organized systems. It envisions that community structure is maintained by ecological processes representing feedback mechanisms and controlling variables to moderate community fluctuation in response to disturbance. Appropriate qualification of equilibrium ecology within resilience theory, rather than its complete replacement by nonequilibrium models, provides more realistic interpretations for both plant–herbivore interactions and vegetation dynamics than does complete reliance on disturbance-driven events. Resilience thinking represents a "humans-in-nature" perspective that emphasizes human values and goals and it seeks to guide change in social-ecological systems by creating opportunities for multiple stakeholders to adaptively design management strategies and policies.

D.D. Briske (✉)
Department of Ecosystem Science and Management, Texas A&M University,
College Station, TX, USA
e-mail: dbriske@tamu.edu

A.W. Illius
Department of Animal Ecology, University of Edinburgh, Edinburgh, UK
e-mail: A.Illius@ed.ac.uk

J.M. Anderies
School of Human Evolution and Social Change, and School of Sustainability, Arizona State University, Tempe, AZ, USA
e-mail: m.anderies@asu.edu

© The Author(s) 2017
D.D. Briske (ed.), *Rangeland Systems*, Springer Series on Environmental Management, DOI 10.1007/978-3-319-46709-2_6

Keywords Alternative stable states • Multiple equilibria • Rangeland ecology • Resilience-based management • Resilience thinking • Social resilience • Thresholds

6.1 Introduction

Humans interact with nature through the use of simplified and incomplete perceptions of its structure, interrelationships, and dynamics. These perceptions are based on experience, specific to place, and subject to change (Jones et al. 2011). They influence which problems are considered, how they are envisioned, and the potential solutions to address them (Lynam and Stafford Smith 2004). Consider the following questions regarding ecosystem dynamics. How stable are ecosystems? Do limits exist to ecosystem recovery following disturbance? What management actions are most likely to sustain desired ecosystems? A major shift in our perception of nature would greatly alter our responses to these questions, and the manner in which we interact with nature to promote sustainable ecosystem management and human well-being.

Nonequilibrium ecology and resilience theory represent such a change in the human perception of nature. Nonequilibrium ecology challenged the prevailing perception of ecosystem stability and rapid, linear recovery following natural or human disturbances. Equilibrium ecology is reflected in the "balance-of-nature" metaphor and is exemplified by the controversial Gaia hypothesis which suggests that the Earth system is in part self-regulated to maintain conditions for life. Equilibrial ecology was initially challenged by theoretical evidence of nonlinear system dynamics in the mid-twentieth century and, thereafter, by inconsistencies in natural resource management outcomes (Holling 1973; Folke 2006).

Nonequilibrium ecology represents a more dynamic and less predictable perception of ecosystem dynamics that recognizes the contributions of disturbance, spatial heterogeneity, and multiple stable states, in addition to internal biotic regulation (Wu and Loucks 1995). It further challenges the prevailing model of natural resource management—the steady-state management model—that was founded upon equilibrium ecology. This management model emphasizes the maximum sustainable yield of one or a few resources through implementation of management actions to minimize variability and redundancy that may interfere with maximum sustainable production (Holling and Meffe 1996). Practices that optimize harvest efficiency and reduce diversity and heterogeneity—fire prevention, plant control measures, and planting of monocultures—are representative of this management approach that often relies on technological solutions to increase production and reduce uncertainty. It is now recognized that this management model along with the command-and-control management strategy—top-down regulation by a centralized authority—can destabilize the very ecosystems that they were intended to sustain (Holling and Meffe 1996). The adverse outcomes originating from these management approaches have been termed the "pathology of natural resource management."

Resilience theory emerged in response to recognition that the prevailing concept of ecological stability was not a realistic interpretation of observed ecosystem dynamics. For example, ecosystems can exhibit wide fluctuations in species composition, but still be very resilient (Curtain and Parker 2014). This inconsistency was resolved by dividing stability into two components—resistance and resilience. Resistance describes the capacity of systems to remain unchanged by disturbance, while resilience is the capacity to return to a former configuration following disturbance (Holling 1973). Resilience also recognizes the existence of threshold conditions that contribute to the formation of alternative stable states. Grassland conversion to woodland and perennial shrub steppe conversion to annual grasslands are widely recognized examples of nonreversible dynamics that result in the formation of alternative ecosystems on the same site. Resilience-based management further provides an alternative to steady-state management that encourages managers to anticipate and guide the direction of change, rather to prevent change, so that ecosystems can sustainably provide ecosystem services to society (Holling 1973; Chapin et al. 2010).

Resilience theory has recently been extended to social systems to provide a "humans-in-nature" perspective to ecosystem management and policy. Adaptive management—learning by doing—and social learning—the capacity of groups of people to achieve goals—have emerged as essential components of resilient human-dominated systems. These resilience-based approaches are collectively termed "resilience thinking" and they are intended to provide a path toward greater sustainability by embracing uncertainty, variability, and recognition of incomplete knowledge (Walker and Salt 2012; Curtain and Parker 2014).

The goals of this chapter are to provide a synopsis of the origins and development of nonequilibrium ecology and resilience theory and to describe how these concepts have influenced the ecology, management, and governance of rangeland systems. Specific objectives are to:

(1) Summarize equilibrium and nonequilibrium ecology and resilience theory
(2) Assess the consequences of these concepts to rangeland ecology and management
(3) Explore the application and utility of resilience in social-ecological systems
(4) Describe future perspectives regarding further integration of resilience in rangeland systems

6.2 Conceptual Advances

6.2.1 Equilibrium and Nonequilibrium Ecology

Equilibrium ecology and its associated metaphor, "the balance of nature," is an ancient human concept, but the modern foundation was derived from systems theory in the 1960s. It is founded on the assumption that ecosystems are highly self-regulated by internal biotic processes, including intra- and interspecific competition and plant–animal interactions that restrict their dynamics to a single stable state

Table 6.1 Proposed characteristics of equilibrium and nonequilibrium systems (from Wiens 1984)

	Equilibrium systems	Nonequilibrium systems
Abiotic patterns	Relatively constant	Stochastic/variable
Plant–herbivore interactions	Tight coupling	Weak coupling
	Biotic regulation	Abiotic drivers
Population patterns	Density dependence	Density independence
	Populations track carrying capacity	Dynamic carrying capacity limits population tracking
Community/ecosystem characteristics	Competitive structuring of communities	Competition not expressed
	Internal regulation	External drivers

(Wu and Loucks 1995). It is further assumed that this state will return to its original pre-disturbance condition after a disturbance has ceased. The predictable and directional response of plant succession that passes through anticipated, sequential stages toward a single equilibrium point or stable state provides a well-known example (Pickett and Ostfeld 1995). Equilibrium ecology experienced growing criticism in the mid-twentieth century for several reasons, including (1) limited supporting evidence of equilibrium conditions in ecosystems, (2) an inability to account for the occurrence of alternative stable states in some ecosystems, and (3) slow or nonexistent recovery of alternative states when they had formed (Wu and Loucks 1995; Briske et al. 2003).

Nonequilibrium theory emerged from investigation of theoretical competition models in the mid-1950s (Petraitis 2013) and the potential existence of multiple ecological states was first described some 15 years later (Lewontin 1969). However, this theory did not enter the ecological mainstream until the following decade when several non-equilibrial systems, include rangelands, were described (May 1977). Nonequilibrium ecology and its associated metaphor, "the flux of nature" (Pickett and Ostfeld 1995), are founded on the assumption that ecosystems possess a finite capacity for internal regulation such that they may be strongly influenced by disturbances (Wiens 1984; Wu and Loucks 1995). This implies that nonequilibrium systems possess greater potential for change than do equilibrium systems, including the potential to exhibit multiple stable states (Table 6.1).

6.2.2   Engineering Versus Ecological Resilience

Holling (1973) initially envisioned resilience theory by recognizing the potential occurrence of multiple stable states associated with the nonlinear dynamics in theoretical predator–prey models. Resilience was initially defined as the "persistence of relationships within a system and is a measure of the ability of these systems to absorb changes of state variables, driving variables and parameters, and still persist" (Holling 1973, p. 17). Two expressions of resilience later emerged to describe unique categories of ecosystem dynamics (Gunderson 2000). Engineering and ecological resilience broadly correspond to, but do not originate from, equilibrium and non-equilibrium ecological models, respectively. Engineering resilience describes system behavior near an individual equilibrium point and, therefore, system dynamics are assumed to be more consistent and predictable. Engineering resilience represents the time required for a system to return to its original equilibrium state after it has been modified by a disturbance (Holling 1973; Folke 2006). In contrast, ecological resilience describes system dynamics far from an equilibrium point and it recognizes the possibility that ecosystems may not return to their original equilibrium point and that they may reorganize around alternative equilibrium points (Gunderson 2000). Ecological resilience is currently defined as the capacity of systems to absorb disturbances and reorganize *while undergoing change* so as to still retain essentially the same function, structure, identity, and feedbacks (Walker et al. 2004).

Resilience theory is often presented graphically in an attempt to clarify this abstract concept. The "basin-of-attraction" or "ball-and-cup" graphic is among the most commonly used. In this highly simplified presentation the ball represents the current state of the system (state variables; structural system characteristics) with respect to the slow or controlling variables (parameters; ecological processes) of the system, the limits of which are represented by the size and shape of the basin (Beisner et al. 2003; Walker et al. 2012). Engineering resilience reflects the shape of the basin—its depth and degree of inclination—that determines the rate of recovery following disturbance (i.e., rate at which the ball returns to the bottom of the same basin). Ecological resilience is signified by the width of the basin of attraction, rather than its depth and inclination, as in the case for engineering resilience. If a disturbance forces the ball (structural system or community) beyond the rim of the basin (threshold) or if the width of the basin is narrowed by the modification of a controlling variable, resilience is exceeded and an alternative state may be formed as the ball moves into an adjacent basin (Gunderson 2000; Beisner et al. 2003). Multiple basins of attraction are representative of ecological resilience indicating that an ecosystem may possess more than one equilibrium state. The total number and shape of the basins in which an ecosystem may reside are collectively termed the resilience landscape (Walker et al. 2004) (Fig. 6.1).

6.2.3 Drivers, Controlling Variables, and Feedback Mechanisms

As indicated in the previous section, resilience is influenced by interactions among several variables and these interactions can be modified by events both internal and external to the system (Walker et al. 2004, 2012). Drivers, controlling variables, and

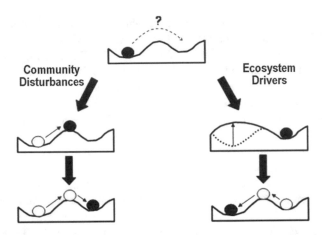

Fig. 6.1 The basin of attraction model illustrates ecological resilience as a ball (the community) that can reside in one or more basins of attraction (alternative states). Drivers may sufficiently modify controlling variables to force a community out of the original basin, beyond the ridge representing the threshold, and into an alternative basin forming a new stable state (see Table 6.1 for concept definitions) (from Beisner et al. 2003)

feedback mechanisms, in addition to the state variables previously introduced, are among the most important components of resilience (Text Box 6.1). Drivers are considered to be external to the system and they are not coupled to the system by feedback mechanisms (e.g., climatic regimes and extreme weather events). Controlling variables have a major influence on resilience and most systems are assumed to be regulated by a rather small number (3–5) of these variables (Chapin et al. 2010; Walker et al. 2012). They are often relatively stable through time, because they are buffered by feedback mechanisms (see below). Important controlling variables are major ecological processes—primary production and nutrient cycling; biodiversity—plant functional groups and woody plant encroachment; and historical disturbances—fire and grazing regimes. The modification of controlling variables directly by drivers or indirectly by feedback mechanisms represents the major way that resilience is altered (Beisner et al. 2003; Walker et al. 2012). These modifications are depicted as changes in the resilience landscape of the basin-of-attraction model (Fig. 6.1). Fast variables—annual plant and animal production—are more obvious than controlling (slow) variables because they fluctuate widely throughout an annual cycle. Critical interactions among these components occur when a driver—drought—modifies important controlling variables—grazing and fire regimes to influence numerous fast variables—grass growth and livestock gains.

Text Box 6.1: Concept Definition and Application to Resilience Theory

System or stable state—collection of multiple state variables and the feedback mechanisms that exist among them. State variables are broadly categorized as fast and slow (controlling).

Examples: Grassland, savanna, or shrubland communities.

Fast variables—variables characterized by dynamic and rapid responses to controlling variables and external drivers.

Examples: seasonal plant and animal production, compositional shifts in annual and transient species, soil water availability, pathogen, and insect dynamics.

Controlling (slow) variables—variables that often operate at slow rates and have a controlling influence on fast variables, feedbacks, and collectively system resilience; they are the central focus of resilience management.

Examples: dominant plant species, including plant functional groups and woody plant encroachment; grazing and fire regimes, soil characteristics, invasive species.

Drivers—events that are external to the system and do not possess feedbacks within the system; drivers may be of natural or human origin. They influence both fast and slow variables and their interactions within systems.

Examples: climatic regimes, extreme weather events, globalized markets, and human population growth.

Interpretation—drivers directly impact both fast and slow variables and the feedback mechanisms that exist between them. When a driver of sufficient magnitude modifies one or more slow (controlling) variables, threshold conditions may be established and existing stable states may transition to alternative stable states.

Application of resilience concepts to woody plant encroachment

System—grassland or savanna characterized by contiguous grass production that provides fine fuel to support regular fire regimes.

Fast variables—soil water availability, grass production, and fine fuel accumulation.

Controlling (slow) variables—dominant grass species, sustained intensive grazing regimes that reduce fine fuel accumulation, negative human perceptions and regulations limiting use of prescribed burning.

Drivers—severe drought that contributes to mortality of dominant grasses, human-induced land cover change, and increasing atmospheric carbon dioxide that increases woody plant growth.

Interpretation—interaction of natural and human drivers suppresses the controlling variable of fire frequency to enable threshold conditions to develop and grasslands or savannas to transition to alternative woodland states.

Feedback mechanisms are ecological processes that influence the rate of change among system variables. More specifically, they are secondary effects of one variable interacting with another to either enhance or dampen the rate of change of the initial variable. Stabilizing (negative) feedbacks reduce the rate of change of the initial variable (Gunderson 2000; Walker et al. 2012). For example, a high density of dominant grass species provides abundant, continuous fine fuel loads capable of supporting frequent fires to prevent woody plant encroachment and maintain grasslands. Amplifying (positive) feedbacks have the opposite effect and accelerate change of the initial variable. For example, increasing abundance of the invasive annual grass cheatgrass (*Bromus tectorum*) in the western USA increases fire frequency that contributes to the mortality of native vegetation to further increase cheatgrass dominance. In both examples, feedbacks interacted with a controlling variable—dominance of native grasses and invasion of an exotic plant species, respectively.

In the basin-of-attraction graphic, stabilizing feedbacks are greatest when communities reside near the bottom of the basin, while the relative strengths of stabilizing and amplifying feedbacks are assumed to be equivalent near the rim of the basin, which represents threshold conditions (Scheffer and Carpenter 2003; Walker et al. 2012). When the relative strength of amplifying feedbacks exceeds that of stabilizing feedbacks, one or more controlling variables may be sufficiently modified to create threshold conditions and initiate formation of an alternative state within a different basin of attraction (Fig. 6.2). Once an alternative stable state has been formed, resilience is established through a unique set of controlling variables and feedback mechanisms. The strengthening of stabilizing, relative to amplifying,

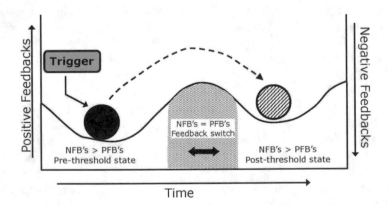

Fig. 6.2 The feedback switch model depicts thresholds as the point where feedbacks switch from a dominance of negative (stabilizing) feedbacks (NFB) that maintain resilience (solid ball) to a dominance of positive (amplifying) feedbacks (PFB) that decrease resilience. The dominance of positive feedbacks contributes to formation of an alternative state (cross-hatched ball) in a different basin of attraction. Resilience of the alternative state requires that NFBs exceed PFBs (from Briske et al. 2006)

feedbacks will support controlling variables and increase resilience of the alternative state. The potential for multiple interactions among external drivers, controlling variables, and feedback mechanisms over various time periods contributes to the difficulty of anticipating and describing thresholds and identifying ecological indicators of their occurrence (Briske et al. 2006; Bestelmeyer et al. 2011) (Text Box 6.1).

6.2.4 Threshold Indicators

The difficultly associated with threshold identification has focused attention on the search for early warning indicators. Indicators are assumed to signify modifications to state variables (structural characteristics), controlling variables, and to a lesser extent feedback mechanisms that determine the ecological resilience of a state. From a management perspective, indicators can be used to identify the trajectory of systems toward pending thresholds so that management strategies can be implemented or modified to prevent thresholds from being crossed (Briske et al. 2008; Standish et al. 2014). Alternatively, restoration ecologists may use threshold indicators to promote restoration of previous states that were considered more desirable (Suding and Hobbs 2009; Limb et al. 2014). Indicator effectiveness is a function of (1) the rate at which a system will respond to management actions to modify its resilience, (2) the amount and type of variability (noise) within systems that may mask indicator detection, and (3) the number of feedback mechanisms and controlling variables that contribute to system resilience (Contamin and Ellison 2009). In addition, threshold indicators are most relevant to systems where resilience is associated with gradual modification of controlling variables, rather than abrupt events that are difficult or even impossible to detect in advance (Hastings and Wysham 2010).

Two categories of theoretical early warning signals have been developed for threshold identification. The first emphasizes an increase in time required for recovery of system variables following disturbances that is termed "critical slowing down" (Scheffer et al. 2012; Dakos et al. 2012). The second category focuses on increasing variance and autocorrelation among system variables as thresholds are approached. It is assumed that both categories of indicators reflect a decrease in stabilizing feedback mechanisms as thresholds are approached (Walker et al. 2012). However, the limited scope of these early warning signals suggests that specific knowledge of systems dynamics, especially the major controlling variables, is still of greatest value (Dakos et al. 2012). Consequently, threshold identification on rangelands currently relies on ecological indicators that have been previously developed for evaluation of rangeland health and implementation of the range model last century and they are primarily implemented within the STM framework (Bestelmeyer et al. 2013) (Chapter 9, this volume).

6.2.5 Rethinking Rangeland Ecology

The concepts of nonequilibrium and resilience profoundly altered rangeland ecology by supporting development of the nonequilibrium persistent (NEP) model and state-and-transition model (STM), respectively, in the late 1980s. However, these two models functioned independently because they focused on unique aspects of rangeland systems. The NEP model is based on the occurrence of nonequilibrium dynamics among vegetation and livestock, but it does not reference the existence of multiple stable states as does the STM. The following three subsections contrast the traditional equilibrium range model with the nonequilibrium persistent and multi-equilibrium state-and-transition model.

6.2.5.1 Range Model

The assessment of rangeland vegetation in response to grazing was initially linked to successional theory by Arthur Sampson, a former student of Fredric Clements, shortly following the publication of Clements' influential work on succession in 1916 (Briske et al. 2005). However, a quantitative assessment of this procedure was not developed for another 30 years when Dyksterhuis (1949) published his classic paper outlining rangeland condition and trend analysis (here termed the range model). This procedure was adopted and applied to rangelands throughout the world during the last half of the twentieth century even though it encountered considerable criticism (Joyce 1993). The range model envisioned vegetation dynamics to occur along a single axis in which grazing intensity linearly counteracted secondary succession. The species composition of plant communities along a succession–grazing axis was compared to that of a single historic plant community to define a range condition rating. The more closely the species composition of a plant community approached that of the reference community, the higher the condition rating. These ratings were used to draw inferences for both production goals and ecological assessments (Joyce 1993). Range trend described the relative change in range condition ratings on specific sites through time. The adoption of Clementsian succession as the basis for vegetation assessment deeply embedded equilibrium ecology within the rangeland profession from its very beginnings (Fig. 6.3).

The expansion of woody plants and the persistence of these plant communities following the reduction or removal of grazing resulted in strong criticism of the range model in the 1970s and 1980s (Laycock 1991; Briske et al. 2005). However, in retrospect, the decision to use the grassland-savanna fire climax community (e.g., pre-European, Native American), as opposed to the climatic climax community (e.g., shrubland or woodland in wetter regions), as the primary reference state in the range model was a major contributor to these inconsistent outcomes. The selection of this reference state likely resulted from the recognized value of grasslands and savannas to livestock production, but *climatic* climax communities began to be expressed as historical fire regimes were minimized by grazing induced fuel reductions and direct fire suppression (Smith 1988; Westoby et al. 1989). However,

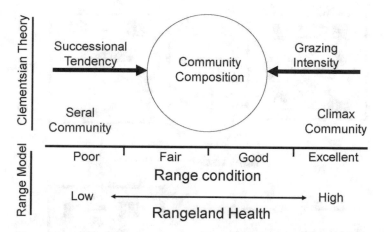

Fig. 6.3 The range model assumes that the species composition of plant communities is a result of the opposing forces of plant succession and grazing intensity. Grazing can slow, stop, or reverse secondary succession to produce communities that differ in species composition from the historical climax plant community that represents the single reference (equilibrium) point (from Briske et al. 2005)

the range model is still considered an appropriate interpretation of vegetation dynamics in more productive grasslands similar to those in which Dyksterhuis devised the range model (Fort Worth Prairie in north central Texas) (Dyksterhuis 1949). The occurrence of relatively linear vegetation dynamics in these grasslands is a result of intense plant competition and stronger plant–livestock interactions that are characteristic of equilibrium ecology (Díaz et al. 2001) (Table 6.1).

6.2.5.2 Nonequilibrium Persistent Model

The nonequilibrium persistent model was introduced by Ellis and Swift (1988) while conducting research in pastoral systems of the Turkana region of East Africa. This region is characterized by low annual and high interannual rainfall variability. This variability, especially when expressed as multiyear drought, frequently contributes to substantial livestock mortality in spite of attempts by nomadic pastoralists to track this variation. Livestock numbers recover less rapidly than plant production in the intervening favorable rainfall years such that they lag behind the availability of forage resources. This weak relationship between plant production and animal numbers contributed to the interpretation that these were nonequilibrium ecosystems. This interpretation gave rise to an alternative set of management and policy recommendations for pastoral systems in the early 1990s that was termed the "New Range Ecology" (Behnke et al. 1993). These recommendations rejected the equilibrial concepts of carrying capacity, stocking rate, and the potential for livestock to degrade rangeland resources that were inherent to traditional grazing management (Cowling 2000) (Fig. 6.4).

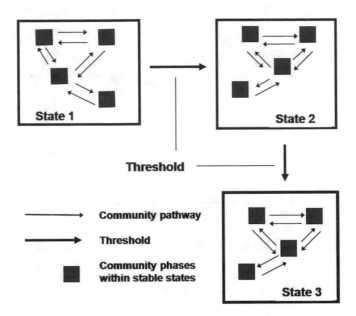

Fig. 6.4 The persistent nonequilibrium model indicates that multiyear droughts occurring on approximately decadal intervals minimize equilibrium between plant production and livestock numbers because vegetation recovers much more rapidly than livestock numbers following drought (redrawn from Ellis and Swift 1988)

Interannual variability in rainfall is negatively correlated with mean annual rainfall, and so the limitation on animal numbers imposed by low primary production in arid and semiarid rangelands is compounded by intra-annual rainfall variation. This makes it difficult to separate the effects of low annual rainfall from those of seasonal variability. A critical level of rainfall variability—an interannual coefficient of variation (CV, annual variability relative to the long-term mean) of ≥33 %—was established as the value at which animal numbers are no longer in equilibrium with plant production (Ellis and Swift 1988) (Fig. 6.4).

Reassessment of NEP. The NEP model was critically evaluated by Illius and O'Connor (1999, 2000) from the perspective of spatial and temporal forage availability to livestock. They concluded that the NEP model did not sufficiently consider livestock use of distinct vegetation resources between wet and dry seasons within an annual cycle. They reasoned that livestock numbers in environments characterized by wet and dry seasons are closely couple to a subset of "key" resources that are accessible in the dry season, while being largely uncoupled from more abundant forage in the wet season. The critical parameter establishing livestock herd size is often animal survival over the dry season, which is a function of forage availability during this period. Therefore, the ultimate determinant of animal numbers and their potential impacts on vegetation is the relative proportion of the grazed ecosystem that provides key resources during the dry season. It is this partitioning of forage resources, and not rainfall variability *per se*, that is the true determinant of livestock persistence and the potential for grazing to impact vegetation (llius and O'Connor 2000).

A reduction in access to key resource areas would cause livestock numbers to decline over the course of several years, especially during drought periods. In contrast, an abundance of non-key resources is likely to occur during the wet season, because animal numbers have been reduced by a scarcity of, and competition for, high-quality forage during the dry season. This interpretation establishes that livestock are closely coupled to forage resources, for at least part of the year, to meet their energy and nutrient requirements for survival, growth, and reproduction. This interpretation has been experimentally corroborated in a pastoral system located in an arid and highly stochastic environment (Hempson et al. 2015). In this investigation, livestock body condition followed density-dependent depletion of the limited dry-season riverine vegetation (key resources), and annual demographic parameters of animal populations tracked dry-season conditions. Dry-season length and previous population size were the main determinants of the animal population trajectory, with no clear evidence for an effect of growing season conditions over the vast area accessible to them. Therefore, wet-season rangeland can be categorized as nonequilibrium, because animal populations are only loosely coupled with it, but livestock do exist in equilibrium with dry-season resources.

Implications to Grazing Management. Reinterpretation of the NEP model has several important implications for management of livestock grazing. It indicates that the potential for grazing to modify vegetation and potentially degrade rangeland resources during the wet season increases as the proportion of key dry-season resource areas increases because it is these resources that establish the maximum number of livestock supported over the long term. Consequently, a high ratio of key dry:wet-season resources could support livestock numbers which are sufficient to produce high grazing intensities on wet-season resources, even though they may not be in equilibrium with them. Key resource areas themselves are obviously of considerable importance, and since they represent an equilibrial part of the grazed ecosystem they generally respond to increasing grazing intensity through reduced productivity and altered species composition (von Wehrden et al. 2012; Muthoni et al. 2014). These negative vegetation impacts will likely have direct, negative feedbacks on animal populations as described by traditional grazing management.

Commercial ranching represents a situation where livestock are often provided with supplemental feed during the dry or winter season to minimize animal mortality and weight loss. In these cases, livestock numbers would become uncoupled from both wet- and dry-season resources because grazing intensity is determined by animal numbers maintained by supplemental feeding. The maintenance of high livestock numbers during these periods increases the potential to adversely impact vegetation during the growing season and it reduces the opportunity for vegetation recovery following drought compared with less intensively managed systems. In principle, this interpretation would also apply to wild herbivores that migrate from wet-season (summer) resources to dry-season resources during the winter and then return to wet-season resources. Vegetation on summer range would be impacted to the extent that resource availability in winter range can support total animal numbers.

This reassessment strongly qualifies the NEP model by indicating that livestock will always maintain an equilibrial relationship with forage in key resource areas, even though this is not necessarily the case for abundant forage during the wet season. It also minimizes legitimacy of the "new" rangeland ecology by reaffirming that stocking rate and carrying capacity are valid concepts for grazing management, albeit in the context of larger landscapes and longer time periods (Cowling 2000).

6.2.5.3   The State-and-Transition Model

State-and-transition models were introduced as a "management language," rather than an ecological theory, to organize and interpret rangeland vegetation dynamics (Westoby et al. 1989). They provided an alternative to the range model that had been severely criticized in the 1970s and 1980s for being overly reliant on linear, directional vegetation dynamics that were unable to account for nonreversible vegetation change, especially woody plant encroachment and invasion by exotic invasive species (Briske et al. 2005). An influential report by the US National Research Council (NRC 1994) endorsed development and adoption of an alternative management model based on the STM framework and the Society for Range Management quickly supported this endorsement (SRM Task Group 1995). The US Department of Agriculture—Natural Resource Conservation Service (USDA-NRCS) formally adopted STMs for rangeland assessment in the late 1990s and established programs to develop and organize these models for all 50 states in the USA. A National Ecological Site Manual was developed and adopted in 2010 to standardize the use of ecological site descriptions and STMs among the NRCS, Bureau of Land Management, and US Forest Service (BLM 2010).

State-and-transition models are organized as a collection of all recognized or anticipated stable states that individual ecological sites may support (Bestelmeyer et al. 2003; Stringham et al. 2003). Individual stable states (e.g., grassland or shrubland) include transient and reversible shifts in species composition that occur in response to disturbances or self-regulating processes. These internal state dynamics are referred to as community phases and represent variation in species composition associated with wet and dry years, periodic intensive grazing, and fire frequency. In contrast, individual states are assumed to be separated by thresholds that are considered to be irreversible without management intervention. Ecological indicators of state variables, controlling variables, and to lesser extent feedback mechanisms that underpin state resilience are used to determine if a state is trending toward or away from pending thresholds (Briske et al. 2008). This information can inform managers of the need to implement actions to modify state resilience to achieve desired outcomes (Watson et al. 1996; Bagchi et al. 2013) (Fig. 6.5).

State-and-transition models have subsequently been organized around ecological resilience to link them to an accepted ecological theory and to accommodate scientific, in addition to management knowledge (Briske et al. 2008). In relation to resilience, individual states exist within a single basin of attraction that is consistent with engineering resilience and thresholds represent boundaries between multiple

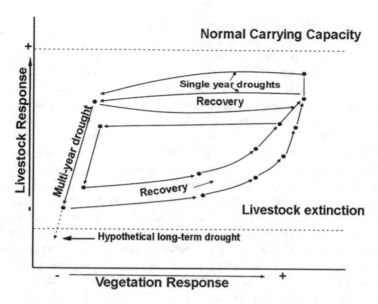

Fig. 6.5 State-and-transition models are a representation of all known or anticipated stable states that may occupy an individual ecological site. States are assumed to be separated by thresholds that are considered to be nonreversible without management intervention. Community phases represent recognizable variations of a state that are readily reversible (from Stringham et al. 2003)

equilibrium states. Consequently, STMs are best interpreted as multiple equilibria rather than disequilibrium models (Petraitis 2013). Rangeland ecologists adopted the "nonequilibrium" terminology utilized by Westoby et al. (1989) because STMs were introduced as an alternative to the range model that was severely criticized for overemphasis of equilibrium dynamics associated with Clementsian succession (Joyce 1993; Briske et al. 2005). In addition, the focus of STMs has moved beyond threshold identification to the management of controlling variables and feedback mechanisms supporting resilience of stable states, but the nonequilibrium terminology has remained. Chapter 9 by Bestelmeyer and coauthors comprehensively describe the development, implementation, and interpretation of STMs.

6.2.6   What Has Been Learned?

Nonequilibrium ecology as described by the NEP model has been reinterpreted to indicate that while livestock may not necessarily be in equilibrium with forage during the wet season, they will always be in equilibrium with key forage resources during the dry season (Illius and O'Connor 1999). The recognition of two categories of forage resources with grazed systems indicates that those with low and highly variable rainfall do not function exclusively as "nonequilibrium systems", because an equilibrial relationship with forage availability and livestock exists during the dry or winter season. Similarly, systems considered to be equilibrial are likely to experience intervals of nonequilibrium between livestock and forage during wet or dry seasons or years when forage production is high or when animal numbers are low. The current status of this rigorous debate is that appropriate qualification of equilibrium ecology, rather than its complete replacement by nonequilibrium models, offers more realistic interpretations for both plant–herbivore interactions and vegetation dynamics than does complete reliance on disturbance-driven events.

Resilience provides a framework to accommodate the occurrence of dynamic equilibria within ecosystems (i.e., engineering resilience) and the potential for ecosystems to transition to alternative stable states (i.e., ecological resilience). Recognition of nonlinear and nonreversible shifts between stable states initially focused attention on identification and characterization of thresholds separating these states. However, thresholds have proven very difficult to identify prior to their occurrence which minimizes their management value. An alternative approach that has greater management value focuses on the trajectory of state resilience relative to the proximity of potential thresholds, with the use of ecological indicators, rather than on thresholds themselves (Watson et al. 1996; Briske et al. 2008). State-and-transition models developed and maintained by the USDA-NRCS and its partners represent the major framework for application of ecological resilience to rangelands in the USA and elsewhere. Greater insight into feedback mechanisms and controlling variables establishing resilience, and ecological indicators to assess the trajectory of state resilience, are needed to promote implementation of resilience-based management (Bestelmeyer and Briske 2012).

6.3 Resilience of Social-Ecological Systems

Resilience theory is currently being explored in coupled social-ecological systems (SESs), in addition to ecological systems previously described. The objective is to provide a "humans-in-nature" perspective that serves to guide natural resource management, policy, and governance. As the name suggests, SESs are integrated systems of ecosystems, humans, institutions, and social organizations that contain feedbacks and interdependences among system components (Folke et al. 2010). This approach has provided common ground for social and natural sciences to interact and evaluate multiple knowledge sources addressing human–environment relationships. SESs are founded on the recognition that ecological information represents necessary but insufficient knowledge to manage ecosystems because they are strongly influenced by human needs, values, and goals (Chapter 8, this volume). Put more simply, "natural resource problems are human problems" (Ludwig 2001).

The concept of SESs emerged from interaction between social and ecological scientists in response to what was seen as failed natural resource management policy from the 1970s to the early 1990s. These policies often resulted in unintended negative consequences because they neglected the complex and often unrecognized interactions that exist between social and ecological components of these systems.

These failures primarily originated from the ineffectiveness of the steady-state management model that emphasized maximum sustainable production of one or a few ecosystem services such as livestock production from rangelands and timber from forests, with little concern for other components of these complex systems. The deficiencies of this management model reside in the narrow interpretation of sustainability, inevitable trade-offs between sustainability and maximum resource yield, and tendency to overestimate resource availability and the consistency of resource yield (Holling and Meffe 1996; Ascher 2001). In contrast, resilience and resilience-based management focus on the variability and uncertainty of SESs and encourage managers to guide the direction of change, rather than to prevent change, to provide diverse ecosystem services to society (Chapin et al. 2010). However, in spite of the recognized deficiencies of the steady-state management model, it still remains the most widely used natural resource management strategy today (Anderies et al. 2006). Its persistence is likely a consequence of the central role that optimal control procedures have played in natural resource economics and the absence of a viable alternative management strategy.

At the same time that C.S. Holling and coworkers were studying the problems caused by the "command-and-control pathology of natural resource management" discussed above, Elinor Ostrom was also questioning the rationality of top-down, command-and-control governance structures from a policy perspective (Ostrom 1990). Her work demonstrated that small groups of people can effectively manage complex natural resource challenges without top-down governance. She further suggested that top-down interventions could have "pathological" effects on social systems by reducing their capacity to solve problems similar to the way in which Holling envisioned the negative impact of top-down control on ecosystems. These two independent assessments of natural resource management failures—one ecological and other sociological—eventually converged to giving rise to the concept of SESs in the 1900s (Berkes et al. 2003) (Chapter 8, this volume).

6.3.1 Resilience Thinking

Resilience thinking provides a general framework for organizing and analyzing information regarding SESs to guide sustainable development and natural resource use (Folke 2006; Cote and Nightingale 2012). In this context, resilience is more appropriately interpreted as an approach and set of assumptions to analyze and guide SESs, rather than a system property as described in the previous sections (Biggs et al. 2015). The application of resilience to SESs implies "general" resilience that considers the potential existence of multiple drivers, disturbances, and thresholds, as opposed to "specified" resilience that emphasizes the impact of a smaller number of impacts on a specific threshold (Anderies et al. 2006). Resilience thinking is widely viewed as being comprised of three broad components (Folke et al. 2010). First, as previously defined for ecological systems, resilience describes the capacity of SESs to continually change and adapt, and yet remain within their current basin of attraction. Second, adaptive capacity describes the ability of humans to guide and direct change by enabling SESs to respond and adapt to internal and external events so that they can maintain their integrity and function. Third, transformation describes the capacity to create an alternative SES when resilience of the previous system can no longer be maintained by incremental adaptation (Folke et al. 2010). Each of these components will be discussed further in the following sections.

6.3.1.1 Social Resilience

Social resilience refers to the ability of human communities to withstand external shocks to their social infrastructure, such as environmental variability or social, economic, and political disruption (Adger 2000). However, attempts to transfer resilience theory from ecological systems to social systems have encountered several major concerns. Central among them is the validity of the assumption of persistent relationships—feedback mechanisms as described for ecosystems—which determine the ability of SESs to absorb natural and social disruption, change, and continue to persist (Cote and Nightingale 2012; Brown 2014). Social resilience emphasizes societal values and human behaviors, including power relations, equity and justice, and function of social organizations, which are central to human–environment relationships and social change, but are absent from ecological systems (Cote and Nightingale 2012; Olsson et al. 2015). Therefore, thresholds may represent changes in institutional procedures, including power sharing in decision making, wealth distribution, and land tenure, rather than irreversible divisions between two stable states (Christensen and Krogman 2012). In addition, neither social nor ecological resilience are normative so they do not provide a basis to distinguish between desirable and undesirable expressions of resilience (Brown 2014). Consequently, the value of resilience as an analytical tool for SESs has proven even more difficult to define and implement than for ecological systems (Benson and Garmestani 2011). The broad integration of knowledge and perspectives that conveys value to resilience as an organizing framework likely represents the same attributes that restrict its application to standardized management protocols and policies (Cote and Nightingale 2012).

6.3.1.2 Adaptive Capacity and Social Learning

Adaptive capacity describes the ability of humans to create, and shape variability and change in SESs, and it represents a central component of social resilience. Adaptive management, social learning, and adaptive governance are all components of adaptive capacity that are necessary for putting resilience into practice. Adaptive management is often described as learning by doing and it operates in an iterative manner acknowledging that our understanding of complex systems is incomplete and constantly changing. Adaptive management was introduced shortly following development of resilience theory to incorporate the inevitable constraints of uncertainty and insufficient knowledge into natural resource management (Holling 1978). Social learning describes the process by which groups of individuals assess social-ecological conditions and respond in ways that meet desired objectives. This represents a central component of the ability of small coordinated groups of natural resource users to develop more effective local governance than top-down policies (Ostrom 1990) (Chapter 11, this volume).

6.3.1.3   Anticipating System Transformation

Transformation becomes necessary when adaptive capacity and available adaptation strategies are no longer sufficient to maintain resilience of SESs. Transformation describes the capacity to create fundamentally new SESs when ecological, economic, or social conditions make the existing system unsustainable (Walker et al. 2004). Ecological change in the form of increased climatic variability or social disruption regarding the availability and allocation of land, labor, and capital can initiate the need for transformation. The intended goal of transformation is to reorganize SESs around alternative and likely novel basins of attraction that can provide ecosystem services to sustainably support human livelihoods when previous SESs have failed (Walker and Salt 2012).

The capacity of SESs to successfully manage transformation has only recently been considered and is not yet well developed. However, successful transformation may be dependent on five key considerations: awareness, incentives, networks, experimentation, and assets. Awareness of the need to implement transformative strategies is dependent upon the ability of members of the SES to recognize and broadly communicate the occurrence of unsustainable conditions and the need for transformation (Carpenter and Folke 2006; Marshall et al. 2011). Incentives may be required to encourage voluntary change because indecision or denial is often an immediate response to the loss of resilience in SESs (Walker and Salt 2012). The strength of social networks, ability of participants to experiment, preferably at local to regional scales until cost-effective strategies have been established, and leadership are important components for implementing transformation (Nelson et al. 2007; Folke et al. 2010). Finally, effective transformation requires flexibility in the assets or resources necessary to implement change. Recognizing and communicating the need for transformation and developing policies, programs, and actions to support determination of when and how to initiate transformational change represent important challenges for the rangeland profession (Joyce et al. 2013) (Chapter 15, this volume).

6.3.2   Resilience-Based Governance and Policy

Institutions and policies governing the behavior of SESs influence resilience by defining the rights and capacities of managers to make decisions regarding social-ecological trade-offs (Horan et al. 2011). A framework capable of supporting resilience thinking in SESs must be able to address the problems of governance, including management and policy decisions in the face of uncertainty, disputed values, and potential shifts to alternative states (Carpenter and Brock 2008). The failure to recognize and manage feedbacks in SESs, including limited monitoring of policy outcomes and insufficient adaptive management, may be among the primary reasons previous institutions have contributed to natural resource management failures (Holling and Meffe 1996; Ascher 2001). Therefore, a key component of the resilience of SESs is the reorganization of institutions to monitor policy outcomes and implement information feedbacks to learn from previous actions. This may be best achieved by focusing on experimentation, adaptation, and social learning within local communities, rather than on implementation of generalized policies originating from static institutions that are assumed to be efficient or "right" (Benson and Garmestani 2011; Anderies and Janssen 2013).

Designing governance for resilient SESs requires understanding of how biophysical conditions, social structure, and institutional policies interact and affect each other and the entire system. Resilient governance regimes are those that achieve a good fit between these different system components so that they continue to function under conditions of uncertainty and disturbance to the system (Anderies and Janssen 2013). In this way, "resilience thinking" departs considerably from traditional policy and resource management which typically takes a narrower view of imposing top-down policies to drive the system toward specific outcomes. Understanding how different components in SESs respond and interact to management strategies and policies and what outcomes will be produced has proven extremely difficult. Consequently, only general principles of governance have been identified to promote resilience in SESs (Anderies et al. 2006). Three of these major principles follow:

(1) Collaboration to build trust and promote dialogue toward a shared understanding of the system among stakeholders is necessary to mobilize action and self-organize SESs (Ostrom 1990).
(2) Multilayered institutions located at various scales within the system improve the fit between knowledge, action, and social-ecological interactions in ways that promote adaptive societal responses at appropriate times and locations (Ostrom 1990; Anderies and Janssen 2013).
(3) Accountable authorities that pursue the just distribution of benefits and involuntary risks among stakeholders to enhance the adaptive capacity of the most vulnerable groups (Lebel et al. 2006).

6.3.3 Resilience Analysis and Management

Resilience management in SESs emphasizes two major, interrelated goals. First is to prevent the system from transitioning to an undesirable stable state, and second is to retain and promote the ability of SESs to reorganize following major change (Walker et al. 2002). The intent is not to direct the trajectory of SESs toward a predetermined endpoint, but rather to strengthen the internal feedbacks to enhance

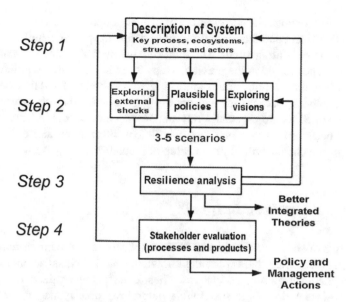

Fig. 6.6 A resilience analysis and management procedure that consists of four broad, interrelated steps that are to be conducted by multiple stakeholder groups as a means to retain and promote resilience of social-ecological systems (Walker et al. 2002)

general resilience to both anticipated and unanticipated future change. A planning approach consisting of four broad, interrelated steps has been developed as a means to retain and promote general resilience of SESs (Walker et al. 2002). The planning process is to be conducted by multiple stakeholders possessing diverse interests and knowledge of the SES under consideration (Fig. 6.6).

Step 1. Develop a Conceptual System Model. The initial step is to establish boundaries for the SESs, and to identify the major management issues, critical stakeholders groups, and primary drivers of change. Identification of major eco-system services, primary controlling variables, and institutional and governance arrangements is also important. Investigation of the historical profile of SESs emphasizing the impact and adaptive responses to previous disturbances will provide valuable baseline information.

Step 2. Create Scenarios of Future System Trajectories. The second step is the development of a limited set (3–5) of future scenarios that address a range of potential SES trajectories in response to the major drivers of change identified in step 1. The scenarios may include a business-as-usual scenario, a more confined or conservative scenario, and one or two more exploratory scenarios. These scenarios are not intended to be predictions, but rather broad plausible visions of potential outcomes that are consistent with existing evidence.

Step 3. Resilience Analysis. The goal is to assess how the system may change within each of the scenarios identified in the previous step. This assessment should emphasize anticipated responses of SESs to the drivers and processes that influence stakeholder interests. The identification of thresholds, alternative states, and other potential surprises are of primary importance. This step is highly context dependent and therefore difficult to define in specific detail.

Step 4. Resilience Management. The final step involves stakeholder evaluation of the knowledge created in the previous steps for management and policy considerations. The implications of this knowledge for assessment and management of critical feedback mechanisms and controlling variables that determine general resilience of SESs are especially relevant. A specific trajectory for SESs is not selected because it is assumed that insufficient information and occurrence of unanticipated change will limit the value of predictions.

6.3.4 What Has Been Learned?

The concepts of social resilience and SESs emphasize a "humans-in-nature" perspective that recognizes that ecological knowledge alone is insufficient to sustainably manage human-dominated systems. These concepts recognize the importance of human values and goals to sustainable natural resource management and they create opportunities for multiple stakeholders to adaptively design management strategies and policies. In the context of social resilience, sustainability is pursued by acknowledging the existence of uncertainty, incomplete knowledge, and the potential for abrupt shifts to alternative states, as opposed to steady state management that attempts to minimize variability and target specific outcomes. The application of resilience to SESs emphasizes "general" resilience to various human and natural disturbances, rather than "specified" resilience that emphasizes the impact of a small number of known impacts on specific thresholds. Adaptive capacity has emerged as a key feature of general resilience in SESs that includes adaptive management, social learning, and adaptive governance. Adaptive capacity may be best achieved by focusing on experimentation, adaptation, and collaboration within local communities, rather than on implementation of generalized policies that are assumed to be efficient or "right." Rangeland SESs and the human livelihoods they support are especially vulnerable to a loss of resilience given that they are often characterized by resource scarcity and variability. In cases where the resilience of SESs has been exceeded, transformational change will be required to reorganize these systems within other basins of attraction to support human livelihoods through the production of alternative ecosystem services with different management strategies.

6.4 Future Perspectives

In this section we summarize several perspectives regarding the future development and implementation of resilience that emerged during the writing of this chapter.

6.4.1   Heterogeneity and Livestock-Vegetation Dynamics

Reinterpretation of the NEP model has indicated that grazed systems often contain two unique resource categories: one in which livestock may not necessarily be in equilibrium during the wet season and the other is key resource areas with which livestock are always in equilibrium (Illius and O'Connor 1999). These resource categories are created by heterogeneity in the relative proportion and spatial arrangements of wet- and dry-season forage resources within the landscape (Hempson et al. 2015). The manner in which these spatial attributes influence coupling of vegetation and livestock in relation to length of the wet and dry seasons would provide greater insight into this component of grazed systems. The importance of functional heterogeneity needs to be more effectively incorporated into management recommendations and policy decisions (Chapter 5, this volume).

6.4.2   Procedures to Implement Resilience-Based Management

Resilience has gained wide acceptance as a framework to guide natural resource management, but the procedures necessary to implement it require additional development. Thresholds and alternative states have received the greatest attention, but they do not necessarily provide the best information to guide resilience-based management because they are often only recognized after their occurrence. Greater emphasis needs to be focused on the identification of ecological indicators of controlling variables and feedback mechanisms to assist managers in assessing the trajectory of ecosystem resilience and to identify appropriate management strategies to modify these trajectories when desired.

The search for procedures to implement resilience has encountered some friction between traditional and contemporary scientific approaches regarding the trade-off between precision and vagueness. It has been argued that vagueness, which is normally viewed as being detrimental to scientific progress, may actually promote creativity and problem solving within the context of resilience-based management (Strunz 2012). This creates considerable uncertainty regarding the extent to which resilience procedures should be standardized and formalized for application. State-and-transition models as a component of Ecological Site Descriptions are currently the primary procedure for implementation of resilience-based management on rangelands. It remains uncertain whether procedures addressing the resilience of SESs should be incorporated into this framework or if a separate framework specifically focusing on SESs is required. Emphasis on SESs will require a reorientation from specified resilience emphasizing specific stable states and thresholds, as widely applied today, to general resilience of entire SESs that exist at larger spatial scales.

6.4.3   Recognizing and Guiding Transformation

Globalized markets, climate change, loss of biodiversity, and species invasion collectively contribute to conditions in which SESs and the human livelihoods they provide may become unsustainable. Rangeland SESs may be especially vulnerable because they are frequently characterized by resource scarcity and variability, limited infrastructure and financial capital, and few viable alternatives to sustain human livelihoods (Sayre et al. 2013). Development of policies and programs to increase awareness and communication of the need for transformational change represents an important challenge. The ability to determine when an SES is becoming unsustainable and how to effectively guide transformation of SESs toward a more sustainable alternative requires greater consideration. This will require an assessment of resilience over multiple timescales (Anderies et al. 2013). Short-term decisions primarily focus on specified resilience to maintain stabilizing feedbacks of a desirable state that will minimize development of an alternative state. Mid- and long-term decisions require emphasis on general resilience by adapting SESs to increasing uncertainty and new conditions, and on transformational change when appropriate. Strategies focused on general resilience will require a greater understanding of the organization and function of SESs than is currently available.

6.4.4   Institutional Reorganization to Promote Resilience

A centralized organizational structure controlling both power and financial resources often supports a command-and-control management strategy which is recognized as an impediment to resilience-based management (Holling and Meffe 1996). Consequently, the traditional management, policy, and institutions responsible for natural resource management present major challenges to the implementation of resilience thinking (Benson and Garmestani 2011). Formal institutional arrangements within existing laws and regulations often ignore ecological complexity and variability, and emphasize a "preservation paradigm" that is focused on minimizing or mitigating human damage to ecosystems. Current natural resource management policies and incentives are often designed to prevent change in the desired "optimal" state and they often represent perverse incentives that may eventually reduce resilience of the system (Anderies et al. 2006). In addition, management agencies often focus exclusively on ecological components of natural resource challenges, but largely overlook the associated social components. This is largely a consequence of legal and policy frameworks that separate decision making regarding these two systems. However, as previously indicated, ecological and social systems are tightly linked through reciprocal feedbacks that require simultaneous consideration for development of effective management and policy.

Barriers that must be overcome to modify governance to enhance resilience of SESs include (1) the tendency for political expediency to modify, rather than change, existing institutions; (2) reliance on traditional procedures and knowledge to address novel, complex problems; and (3) fragmented governance among land ownership patterns and institutional jurisdictions (Brunckhorst 2002). Potential implementation of the Endangered Species Act to address conservation and management of greater sage-grouse (*Centrocercus urophasianus*) and its habitat represents an excellent example of an institutional mismatch with complex natural resource management challenges (Boyd et al. 2014). The regulatory approach is incapable of addressing the ecological and social complexity of these challenges because it is not designed to empower collaborative adaptive management among diverse stakeholders (Chapter 11, this volume).

6.5 Summary

Rangeland ecology and management have undergone a major transformation in the past quarter century as nonequilibrium ecology and resilience theory were adopted to increase consistency with observed ecological dynamics and management outcomes. Equilibrium ecology had long been a guiding ecological principle that emphasized linear and predictable ecosystem dynamics and it supported the steady-state management model that prevailed throughout the twentieth century. Equilibrium ecology was challenged on the basis of both theoretical inconsistencies and its inability to account for observed ecosystem dynamics. Nonequilibrium ecology recognizes that disturbance, spatial heterogeneity, and multiple stable states, in addition to internal biotic regulation, have a major influence on ecosystem dynamics (Wu and Loucks 1995).

In addition to these broad implications of nonequilibrium ecology, rangeland ecology was explicitly challenged by the persistent nonequilibrium model that originated in pastoral systems in East Africa. This model indicated that forage availability and livestock numbers were seldom in equilibrium because vegetation recovered more rapidly than livestock numbers following multiyear drought (Ellis and Swift 1988). It was further assumed that this weak coupling between vegetation and livestock minimized the potential for grazing to degrade rangeland resources. However, the persistent nonequilibrium model has been qualified by recognizing that livestock are always in equilibrium with the key resource areas of a grazed system, even though they may only be loosely coupled to abundant wet-season resources. Reinterpretation of the nonequilibrium persistent model challenges the legitimacy of the "new" rangeland ecology by reaffirming that stocking rate and carrying capacity are valid concepts for grazing management, albeit in the context of larger landscapes and longer time periods.

Two categories of resilience exist to describe unique patterns of ecosystem dynamics. Engineering resilience assumes that systems are confined to a single basin of attraction and it is represented by the time required for a system to return to its original equilibrium point following disturbance (Holling 1973). In contrast, ecological resilience recognizes that ecosystems may not return to their original equilibrium point, but that they may reorganize around alternative equilibrium points (Gunderson 2000). Resilience is often expressed graphically with "basin-of-attraction" or "ball-and-cup" diagrams to further clarify this abstract concept. In this graphic representation, the ball represents the structural variables of the system in relation to the controlling variables and feedback mechanisms that are represented by the shape and number of the basins (Beisner et al. 2003; Walker et al. 2012). Engineering resilience emphasizes the depth of the basin which determines the rate of recovery following disturbance. Ecological resilience emphasizes the width of the basin to represent the amount of disturbance that a system can withstand without the ball rolling beyond the rim of the basin (threshold) into an alternative basin. Ecological resilience is most commonly applied to natural resource management where thresholds and the formation of alternative stable states are of major concern.

State-and-transition models represent a conceptual advance from the traditional range model that was founded on Clementsian succession by recognizing the occurrence of nonlinear vegetation dynamics and the potential existence of alternative stable states on individual ecological sites. These models were originally introduced as a "management language," rather than an ecological theory, to organize and interpret rangeland vegetation dynamics. However, the STM framework has become a major tool for interpreting and integrating both management and ecological information. Subsequently, these models are broadly viewed as being supported by ecological resilience and are most appropriately considered equilibrium models because individual states exist with a single basin of attraction and thresholds represent boundaries between multiple equilibrium states. Ecological indicators of state variables, controlling variables, and to lesser extent feedback mechanisms that underpin resilience are used to determine if a state is trending toward or away from pending thresholds (Briske et al. 2008). This information can inform managers of the need to implement actions to modify state resilience to achieve desired outcomes.

The adoption of resilience theory has had broad consequences for natural resource management by providing an alternative to command-and-control management. Command and control employs a top-down, regulatory strategy that often ignores variation and the complexity of interactions among ecological and social system components (Anderies et al. 2006). The typical response to uncertainty and surprise is to increase control which often further reduces resilience and moves the system toward pending thresholds (Holling 1973; Holling and Meffe 1996). Management strategies to minimize variability and optimize production efficiency in one portion of the system frequently increase vulnerability in another portion of the system. These top-down management strategies have also been founded on unrealistic assumptions regarding the ability of managers to anticipate and implement actions, often as technological solutions, in time to prevent natural resource degradation or ecosystem shifts to alternative states.

In contrast to command and control, resilience-based management recognizes both the inevitability of change and the need for change, such that it seeks to guide change, rather than to control it to maintain a single optimal state. A family of concepts—alternative stable states, thresholds, controlling variables, and feedbacks—has evolved around resilience theory that has both increased its potential conceptual value and introduced considerable ambiguity (Strunz 2012). The vagueness inherent to resilience theory is counter to the clarity and precision normally associated with science, but this trade-off may be necessary to promote creativity, trans-disciplinarity, and cooperation among diverse stakeholder groups that is needed to contend with the complexity of natural resource problems currently confronting society. The broad and often ambiguous elements of resilience that have contributed to its intuitive value and appeal are likely the same attributes that make it challenging to interpret and implement (Cote and Nightingale 2012).

Some of the critical challenges confronting the application of resilience to natural resource management are represented by the following questions. How much disturbance can ecosystems absorb before they cross thresholds and reorganize as alternative stable state (Standish et al. 2014)? Is it possible to identify indicators of resilience within existing stable states so that management actions can be implemented to reduce the probability of a threshold being crossed? How can restoration programs best prioritize efforts to reestablish stable states after they have crossed a threshold and reorganized as an alternative stable state? How can resilience thinking be most effectively incorporated into management recommendations and policy decisions without impairing generality and flexibility inherent to the theory (Strunz 2012)? Is the resilience framework relevant to SESs (Anderies et al. 2006)? These questions are especially challenging given that experimental evidence is very limited and that experiments to address these questions are extremely difficult to conduct in large complex systems. Historical data has been suggested as a means to investigate resilience by retrospective analysis of events in both ecological (Standish et al. 2014) and SESs (Stafford Smith et al. 2007). All indications are that resilience will continue to be a work in progress given both the complexity of the concepts involved and the enormity of natural resource challenges to be addressed.

Resilience has been envisioned as a framework to guide society on a path toward sustainability (Folke 2006). It has even been suggested that resilience should replace sustainability as an organizing framework to support environmental management and ecological governance because it is often impossible to know what should and can be sustained in a world of increasing complexity, uncertainty, and rapid change (Benson and Craig 2014). Resilience-based management emphasizes adaptive management and the development of adaptive capacity to guide change, rather than management actions to reduce variability in an attempt to stabilize desired systems. In this context, resilience may represent a more valuable framework than sustainability for natural resource management because it anticipates uncertainty and emphasizes learning to contend with future challenges (Strunz 2012). However, for resilience to meaningfully contribute to this enormous challenge it must be put it into practice at multiple scales of application or it may encounter the same ambiguous outcome as sustainability.

References

Adger, W.N. 2000. Social and ecological resilience: Are They Related? *Progress in Human Geography* 24: 347–364.

Anderies, J.M., and M.A. Janssen. 2013. Robustness of social–ecological systems: Implications for public policy. *Policy Studies Journal* 41(3): 513–536.

Anderies, J.M., B.H. Walker, and A.P. Kinzig. 2006. Fifteen weddings and a funeral: Case studies and resilience-base management. *Ecology & Society* 11(1): 21.

Anderies, J.M., C. Folke, B. Walker, and E. Ostrom. 2013. Aligning key concepts for global change policy: Robustness, resilience and sustainability. *Ecology & Society* 18(2): 8.

Ascher, W. 2001. Coping with complexity and organizational interests in natural resource management. *Ecosystems* 4: 742–757.

Bagchi, S., D.D. Briske, B.T. Bestelmeyer, and X.B. Wu. 2013. Assessing resilience and state-transition models with historical records of cheatgrass *Bromus tectorum* invasion in North American sagebrush-steppe. *Journal of Applied Ecology* 50: 1131–1141.

Behnke, R.H., I. Scoones, and C. Kerven. 1993. *Range ecology at disequilibrium: New models of natural variability and pastoral adaptation in African savannas*. London: Overseas Development Institute.

Beisner, B.E., D.T. Haydon, and K. Cuddington. 2003. Alternative stable states in ecology. *Frontiers in Ecology and the Environment* 1: 376–382.

Benson, M.H., and R.K. Craig. 2014. The end of sustainability. *Society and Natural Resources* 27: 777–782.

Benson, M.H., and A.S. Garmestani. 2011. Can we manage for resilience? The integration of resilience thinking into natural resource management in the United States. *Environmental Management* 48: 392–399.

Berkes, F., J. Colding, and C. Folke. 2003. *Navigating social-ecological systems: Building resilience for complexity and change*. Cambridge: Cambridge University Press.

Bestelmeyer, B.T., and D.D. Briske. 2012. Grand challenges for resilience-based management of rangelands. *Rangeland Ecology & Management* 65: 654–663.

Bestelmeyer, B.T., J.R. Brown, K.M. Havstad, R. Alexander, G. Chavez, and J.E. Herrick. 2003. Development and use of state-and-transition models for rangelands. *Journal of Range Management* 56: 114–126.

Bestelmeyer, B.T., A.M. Ellison, W.R. Fraser, K.B. Gorman, S.J. Holbrook, C.M. Laney, M.D. Ohman, D.P.C. Peters, F.C. Pillsbury, A. Rassweiler, R.J. Schmitt, and S. Sharma. 2011. Analysis of abrupt transitions in ecological systems. *Ecosphere* 2: 129. doi:10.1890/ES11-00216.1.

Bestelmeyer, B.T., M.C. Duniway, D.K. James, L.M. Burkett, and K.M. Havstad. 2013. A test of critical thresholds and their indicators in a desertification-prone ecosystem: More resilience than we thought. *Ecology Letters* 16: 339–345.

Biggs, R., M. Schluter, and M.L. Schoon. 2015. An introduction to the resilience approach and principles to sustain ecosystem services in social–ecological systems. In *Principles for building resilience: Sustaining ecosystem services in social–ecological systems*, ed. R. Biggs, M. Schluter, and M.L. Schoon, 1–31. Cambridge, UK: Cambridge University Press.

Boyd, C.S., D.D. Johnson, J.D. Kerby, T.J. Svejcar, and K.W. Davies. 2014. Of grouse and golden eggs: Can ecosystems be managed within a species-based regulatory framework? *Rangeland Ecology & Management* 67: 358–368.

Briske, D.D., S.D. Fuhlendorf, and F.E. Smeins. 2003. Vegetation dynamics on rangelands: A critique of the current paradigms. *Journal of Applied Ecology* 40: 601–614.

———. 2005. State-and-transition models, thresholds, and rangeland health: A synthesis of ecological concepts and perspectives. *Rangeland Ecology & Management* 58: 1–10.

———. 2006. A unified framework for assessment and application of ecological thresholds. *Rangeland Ecology & Management* 59: 225–236.

Briske, D.D., B.T. Bestelmeyer, T.K. Stringham, and P.L. Shaver. 2008. Recommendations for development of resilience-based state-and-transition models. *Rangeland Ecology & Management* 61: 359–367.

Brown, K. 2014. Global environmental change I: A social turn for resilience? *Progress in Human Geography* 38: 107–117.

Brunckhorst, D.J. 2002. Institutions to sustain ecological and social systems. *Ecological Management & Restoration* 3: 108–116.

Bureau of Land Management (BLM). 2010. *Rangeland interagency ecological site manual.* Washington, DC: BLM Manual 1734-1.

Carpenter, S.R., and W.A. Brock. 2008. Adaptive capacity traps, *Ecology & Society* 13(2): 40.

Carpenter, S.R., and C. Folke. 2006. Ecology for transformation. *Trends in Ecology and Evolution* 21: 309–315.

Chapin III, F.S., S.R. Carpenter, G.P. Kofinas, C. Folke, N. Abel, W.C. Clark, P. Olsson, D.M. Stafford Smith, B. Walker, O.R. Young, F. Berkes, R. Biggs, J.M. Grove, R.L. Naylor, E. Pinkerton, W. Stephen, and F.J. Swanson. 2010. Ecosystem stewardship: Sustainability strategies for a rapidly changing planet. *Trends in Ecology and Evolution* 25: 241–249.

Christensen, L., and N. Krogman. 2012. Social thresholds and their translation into social–ecological management practices. *Ecology & Society* 17(1): 5.

Contamin, R., and A.M. Ellison. 2009. Indicators of regime shifts in ecological systems: What do we need to know and when do we need to know it? *Ecological Applications* 19(3): 799–816.

Cote, M., and A.J. Nightingale. 2012. Resilience thinking meets social theory: Situating social change in socio-ecological systems. *Progress in Human Geography* 36: 475–489.

Cowling, R.M. 2000. Challenges to the 'new' rangeland science. *Trends in Ecology and Evolution* 15: 303–304.

Curtain, C.G., and J.P. Parker. 2014. Foundations of resilience thinking. *Conservation Biology* 28(4): 912–923.

Dakos, V., S.R. Carpenter, W.A. Brock, A.M. Ellison, V. Guttal, et al. 2012. Methods for detecting early warnings of critical transitions in time series illustrate using simulated ecological data. *PLoS ONE* 7(7), e41010.

Díaz, S., I. Noy-Meir, and M. Cabido. 2001. Can grazing response of herbaceous plants be predicted from simple vegetative traits? *Journal of Applied Ecology* 38: 497–508.

Dyksterhuis, E.J. 1949. Condition and management of rangeland based on quantitative ecology. *Journal of Range Management* 2: 104–105.

Ellis, J.E., and D.M. Swift. 1988. Stability of African pastoral ecosystems: Alternate paradigms and implications for development. *Journal of Range Management* 41: 450–459.

Folke, C. 2006. Resilience: The emergence of a perspective for social–ecological systems analyses. *Global Environmental Change* 16: 253–267.

Folke, C., S.R. Carpenter, B. Walker, M. Scheffer, T. Chapin, and J. Rockstrom. 2010. Resilience thinking: Integrating resilience, adaptability and transformability. *Ecology & Society* 15(4): 20.

Gunderson, L.H. 2000. Ecological resilience—In theory and application. *Annual Review of Ecology and Systematics* 31: 425–439.

Hastings, A., and D.B. Wysham. 2010. Regime shifts in ecological systems can occur with no warning. *Ecology Letters* 13: 464–472.

Hempson, G.P., A.W. Illius, H.H. Hendricks, W.J. Bond, and S. Vetter. 2015. Herbivore population regulation and resource heterogeneity in a stochastic environment. *Ecology* 96: 2170–2180.

Holling, C.S. 1973. Resilience and stability of ecological systems. *Annual Review of Ecology and Systematics* 4: 1–23.

———. 1978. *Adaptive environmental assessment and management.* Chinchester: John Wiley and Sons.

Holling, C.S., and G.K. Meffe. 1996. Command and control and the pathology of natural resource management. *Conservation Biology* 10: 328–337.

Horan, R.D., E.P. Fenichel, K.L.S. Drury, and D.M. Lodge. 2011. Managing ecological thresholds in coupled environmental–human systems. *Proceedings of the National Academy of Science, USA* 108: 7333–7338.

Illius, A.W., and T.G. O'Connor. 1999. On the relevance of non-equilibrium concepts to arid and semi-arid grazing systems. *Ecological Applications* 9: 798–813.

Illius, A.W., and T.G. O'Connor. 2000. Resource heterogeneity and ungulate population dynamics. *Oikos* 89: 283–294.

Jones, N.A., H. Ross, T. Lynam, P. Perez, and A. Leitch. 2011. Mental models: An interdisciplinary synthesis of theory and methods. *Ecology & Society* 16:46 [online]. Retrieved from http://www.ecologyandsociety.org/vol16/iss1/art46/.

Joyce, L.A. 1993. The life cycle of the range condition concept. *Journal of Range Management* 46: 132–138.

Joyce, L.A., D.D. Briske, J.R. Brown, H.W. Polley, B.A. McCarl, and D.W. Bailey. 2013. Climate change and North American rangelands: Assessment of mitigation and adaptation strategies. *Rangeland Ecology & Management* 66: 512–528.

Laycock, W.A. 1991. Stable states and thresholds of range condition on North American rangelands: A viewpoint. *Journal of Range Management* 44: 427–433.

Lebel, L., J.M. Anderies, B. Campbell, C. Folke, S. Hatfield-Dodds, T.P. Hughes, and J. Wilson. 2006. Governance and the capacity to manage resilience in regional social–ecological systems. *Ecology & Society* 11(1): 19.

Lewontin, R.C. 1969. The meaning of stability. In *Diversity and stability of ecological systems*. Brookhaven Symposia in Biology No. 22. Brookhaven, New York.

Limb, R.F., D.M. Engle, A.L. Alford, and E.C. Hellgren. 2014. Plant community response following removal of *Juniperus virginiana* from tallgrass prairie: Testing for restoration limitations. *Rangeland Ecology & Management* 67: 397–405.

llius, A.W., and T.G. O'Connor. 2000. Resource heterogeneity and ungulate population dynamics. *Oikos* 89: 283–294.

Ludwig, D. 2001. The era of management is over. *Ecosystems* 4: 758–764.

Lynam, T.J.P., and M. Stafford Smith. 2004. Monitoring in a complex world—Seeking slow variables, a scaled focus, and speedier learning. *African Journal of Range & Forage Science* 21: 69–78.

Marshall, N.A., I.J. Gordon, and A.J. Ash. 2011. The reluctance of resource-users to adopt seasonal climate forecast to enhance resilience to climate variability on the rangelands. *Climatic Change* 107: 511–529.

May, R.M. 1977. Thresholds and breakpoints in ecosystems with a multiplicity of stable states. *Nature* 269: 471–477.

Muthoni, F.K., T.A. Groen, A.K. Skidmore, and P. van Oel. 2014. Ungulate herbivory overrides rainfall impacts on herbaceous regrowth of residual biomass in a key resource area. *Journal of Arid Environments* 100–101: 9–17.

Nelson, D.R., W.N. Adger, and K. Brown. 2007. Adaptation to environmental change: Contributions of a resilience framework. *Annual Review of Environment and Resources* 32: 395–419.

NRC (National Research Council). 1994. *Rangeland health: New methods to classify, inventory, and monitor rangelands*. Washington, DC: National Academy Press. 180 p.

Olsson, L., A. Jerneck, H. Thoren, J. Persson, and D. O'Byrne. 2015. Why resilience is unappealing to social science: Theoretical and empirical investigations of the scientific use of resilience. *Science Advances* 1(4), e1400217.

Ostrom, E. 1990. *Governing the commons: The evolution of institutions for collective action*. New York: Cambridge University Press.

Petraitis, P. 2013. *Multiple stable states in natural ecosystems*, 188. Oxford, UK: Oxford University Press.

Pickett, S.T.A., and R.S. Ostfeld. 1995. The shifting paradigm in ecology. In *A new century for natural resource management*, ed. R.L. Knight and S.F. Bates, 261–278. Washington, DC: Island Press.

Sayre, N.F., R.R.J. McAllister, B.T. Bestelmeyer, M. Moritz, and M.D. Turner. 2013. Earth stewardship of rangelands: Coping with ecological, economic and political marginality. *Frontiers in Ecology and the Environment* 11: 348–354.

Scheffer, M., and S. Carpenter. 2003. Catastrophic regime shifts in ecosystems: Linking theory to observation. *Trends in Ecology and Evolution* 18: 648–656.

Scheffer, M., S.R. Carpenter, T.M. Lenton, J. Bascompte, W. Brock, et al. 2012. Anticipating critical transitions. *Science* 338: 344–348.

Smith, E.L. 1988. *Successional concepts in relation to range condition assessment*, 113–133. Netherlands: Springer.

SRM Task Group (Society for Range Management Task Group on Unity in Concepts and Terminology Committee, Society for Range Management). 1995. New concepts for assessment of rangeland condition. *Journal of Range Management* 48: 271–282.

Stafford Smith, D.M., G.M. McKeon, I.W. Watson, B.K. Henry, G.S. Stone, W.B. Hall, and S.M. Howden. 2007. Learning from episodes of degradation and recovery in variable Australian rangelands. *Proceedings of the National Academy of Sciences USA* 104: 20690–20695.

Standish, R.J., R.J. Hobbs, M.M. Mayfield, B.T. Bestelmeyer, et al. 2014. Resilience in ecology: Abstraction, distraction, or where the action is? *Biological Conservation* 177: 43–51.

Stringham, T.K., W.C. Krueger, and P.L. Shaver. 2003. State and transition modeling: An ecological process approach. *Journal of Range Management* 56: 106–113.

Strunz, S. 2012. Is conceptual vagueness an asset? Arguments from philosophy of science applied to the concept of resilience. *Ecological Economics* 76: 112–118.

Suding, K.N., and R.J. Hobbs. 2009. Threshold models in restoration and conservation. *Trends in Ecology and Evolution* 24: 271–279.

von Wehrden, H., J. Hanspach, P. Kaczensky, J. Fischer, and K. Wesche. 2012. Global assessment of the non-equilibrium concept in rangelands. *Ecological Applications* 22: 393–399.

Walker, B., and D. Salt. 2012. *Resilience practice: Building capacity to absorb disturbance and maintain function*, 226. Washington, DC: Island Press.

Walker, B., S. Carpenter, J. Anderies, N. Abel, et al. 2002. Resilience management in social-ecological systems: A working hypothesis for a participatory approach. *Conservation Ecology* 6(1):14 [online]. Retrieved from www.consecol.org/vol6/iss1/art14.

Walker, B., C.S. Holling, S.R. Carpenter, and A. Kinzig. 2004. Resilience, adaptability and transformability in social–ecological systems. *Ecology & Society* 9(2):5. Retrieved from http://www.ecologyandsociety.org/vol9/iss2/art5.

Walker, B.H., S.R. Carpenter, J. Rockstrom, A.-S. Crepin, and G.D. Peterson. 2012. Drivers, "slow" variables, "fast" variables, shocks, and resilience. *Ecology & Society* 17(3): 30.

Watson, I.W., D.G. Burnside, and A.M. Holm. 1996. Event-driven or continuous; which is the better model for managers? *Australian Rangeland Journal* 18: 351–369.

Westoby, M., B.H. Walker, and I. Noy-Meir. 1989. Opportunistic management for rangelands not at equilibrium. *Journal of Range Management* 42: 266–274.

Wiens, J.A. 1984. On understanding a nonequilibrium world: Myth and reality in community patterns and processes. In *Ecological communities: Conceptual issues and the evidence*, ed. D.R. Strong, D. Simberloff, L. Abele, and A.B. Thistle, 439–458. New Jersey: Princeton University Press.

Wu, J., and O.L. Loucks. 1995. From balance of nature to hierarchical patch dynamics: A paradigm shift in ecology. *Quarterly Review of Biology* 70: 439–466.

Chapter 7
Ecological Consequences of Climate Change on Rangelands

H. Wayne Polley, Derek W. Bailey, Robert S. Nowak, and Mark Stafford-Smith

Abstract Climate change science predicts warming and greater climatic variability for the foreseeable future. These changes in climate, together with direct effects of increased atmospheric CO_2 concentration on plant growth and transpiration, will influence factors such as soil water and nitrogen availability that regulate the provisioning of plant and animal products from rangelands. Ecological consequences of the major climate change drivers—warming, precipitation modification, and CO_2 enrichment—will vary among rangelands partly because temperature and precipitation shifts will vary regionally, but also because driver effects frequently are nonadditive, contingent on current environment conditions, and interact synergistically with disturbance regimes and human interventions. Consequences of climate change that are of special relevance to rangelands are modification of forage quantity and quality, livestock metabolism, and plant community composition. Warming is anticipated to be accompanied by a decrease in precipitation in already arid to semiarid rangelands in the southwestern USA, Central America, and south and southwestern Australia. Higher

---

Mention of trade names or commercial products does not imply endorsement by the US Department of Agriculture. USDA is an equal opportunity provider and employer.

H.W. Polley (✉)
USDA-Agricultural Research Service, Grassland, Soil & Water Research Laboratory, Temple, TX 76502, USA
e-mail: wayne.polley@ars.usda.gov

D.W. Bailey
Animal and Range Sciences Department, New Mexico State University, Las Cruces, NM 88003, USA
e-mail: dwbailey@ad.nmsu.edu

R.S. Nowak
Department of Natural Resources and Environmental Science, University of Nevada Reno, Reno, NV 89557, USA
e-mail: nowak@cabnr.unr.edu

M. Stafford-Smith
CSIRO Land and Water, Canberra, ACT 2601, Australia
e-mail: Mark.Staffordsmith@csiro.au

© The Author(s) 2017
D.D. Briske (ed.), *Rangeland Systems*, Springer Series on Environmental Management, DOI 10.1007/978-3-319-46709-2_7

temperatures combined with drought will significantly impair livestock production by negatively impacting animal physiological performance, increasing ectoparasite abundances, and reducing forage quality and quantity. Conversely, the warmer, wetter conditions anticipated in the northwestern USA, southern Canada, and northern Asia may increase animal productivity by moderating winter temperatures, lengthening the growing season, and increasing plant productivity. Synergist interactions between climate change drivers and other human impacts, including changes in land-use patterns, intensification of disturbances, and species introductions and movements, may further challenge ecosystem integrity and functionality. Evidence from decades of research in the animal and ecological sciences indicates that continued directional change in climate will substantially modify ecosystem services provisioned by the world's rangelands.

Keywords Atmospheric CO_2 • Atmospheric warming • Forage quality and quantity • Livestock production • Precipitation • Soil water availability

7.1 Introduction

Climate change science predicts warming and greater climatic variability for the foreseeable future, including more frequent and severe droughts and storms, as a consequence of increasing atmospheric concentrations of greenhouse gases (GHGs). These gases, which include carbon dioxide (CO_2), methane (CH_4), nitrous oxide (N_2O), and tropospheric ozone (O_3), reduce cooling of the Earth by partially blocking the emission of long-wave infrared radiation into space to create a "greenhouse effect" that is vital to buffering day-night temperature fluctuations on Earth. GHG concentrations are rising largely as a result of human activities (IPCC 2013) and will continue to rise for the foreseeable future even if emission rates decline because GHGs remain in the atmosphere for hundreds of years (Karl et al. 2009). Increased GHG concentrations will amplify the current greenhouse effect and further warm the Earth and modify precipitation patterns.

These changes in climate, together with the direct effects of increased CO_2 concentration on plant growth and transpiration, will influence factors such as soil water availability and nitrogen (N) cycling that regulate the provisioning of plant and animal products and other services from rangeland ecosystems (Walther 2003, 2010; Joyce et al. 2013; Polley et al. 2013). Climate change alone or in combination with impacts of other human activities, such as intensification of disturbances, may force ecosystems beyond their historical range of variability. This may result in a change in ecosystem structure and function that will be difficult to reverse on the management timescale of decades (Joyce et al. 2013).

We provide a (1) brief review of climatic trends during the twentieth century that provide evidence of a climate change signature, (2) summary of changes in climate anticipated on the world's rangelands during the twenty-first century, (3) synthesis of key "principles" from the ecological and climatological sciences that are foundational to projecting climate change impacts, and (4) assessment of three plausible climate change scenarios for rangelands.

7.2   Recent Climatic Trends: A Climate Change Signature

The Earth's climate changed throughout geological history in response to natural events, but we have entered an era in which human impacts on global fluxes of radiative energy have demonstrable effects on climate (IPCC 2013). This modern climate change signature is evident in the form of atmospheric warming, rapid glacial retreat, accelerated plant phenology, modified precipitation patterns, and increasing wild fires (Parmesan and Yohe 2003; IPCC 2013). Global mean temperature increased during recent decades, particularly in northern latitudes and over land. Each of the last three decades has been successively warmer at the Earth's surface than any preceding decade since 1850 (IPCC 2013). Six of the 10 years from 1998 through 2007 were among the hottest 10 % recorded for much of North America (NOAA National Climatic Data Center 2013).

Precipitation has also increased on average during recent decades, particularly in the Northern Hemisphere and at midlatitudes (IPCC 2013). Importantly, however, precipitation has declined by >5 % in several areas of the world with extensive grazing lands. These areas include southeastern Australia, central Africa, the Mediterranean grasslands, and woodlands in southern Europe, and rangelands of northwestern North America.

A climate change signature also is evident in the form of an increasing frequency of extreme weather events, including greater precipitation variability. The frequency of intense precipitation events has increased in North America and Europe, but there is lesser confidence that this change has occurred for other continents (IPCC 2013).

7.3   Climate Change Projections

Atmospheric temperature increased by 1 °C since industrialization (ca. 1750) largely as a result of increasing concentrations of CO_2 (Keeling et al. 2009) and other GHGs (IPCC 2007). Globally, surface air temperature is predicted to increase by 2–4 °C by the final decades of this century, relative to the 1986–2005 average (IPCC 2013). Warming is expected to be greatest at northern latitudes (Fig. 7.1) and at night.

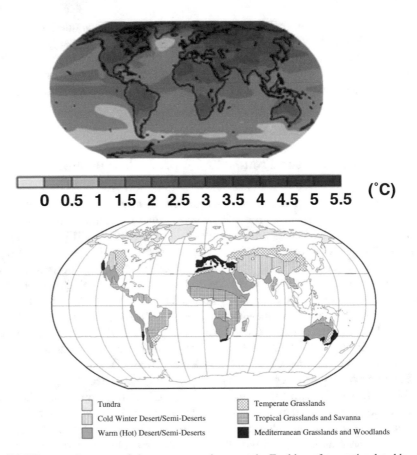

Fig. 7.1 The annual average of air temperature change at the Earth's surface as simulated by climate models for the period 2046–2065 (*upper panel*) and distribution of the world's rangelands (*lower panel*; redrawn from Allen-Diaz 1996). Temperature change is calculated relative to the average for the period 1980–1999. The *upper panel* is adapted from IPCC (2007)

Warming is anticipated to increase weather extremes, including both intra- and interannual precipitation variability and the occurrence of both drought and heat waves, by altering atmospheric circulation patterns (Easterling et al. 2000; McCabe and Clark 2006). Projections of regional and seasonal shifts in precipitation include greater uncertainty than those of CO_2 or temperature modification, but precipitation generally is expected to increase near the poles and decrease elsewhere, with significant seasonal variation (IPCC 2007). Precipitation is anticipated to increase during winter in the northwestern USA, southern Canada, and northern Asia, and during spring in central North America, but decrease by 10–20 % during winter and spring in southern Africa and during spring and summer in central Asia. Average annual precipitation is expected to decrease in southwestern North America.

Reduced precipitation and more frequent droughts are forecast for much of Australia, particularly southern and southwestern regions (IPCC 2007). Climate change also may increase the size or intensity of events when they do occur (Groisman et al. 2005; Karl et al. 2009). Heat waves are anticipated to increase in both frequency and magnitude in proportion to increasing mean temperature.

The issues of global GHG accumulation and its consequences for future climate patterns are necessarily complex (Lindzen 1999), as evidenced by the unanticipated slowing of global warming for more than a decade (Smith 2013). Despite uncertainties and the apparent "pause" in the warming trend, which has been attributed partly to heat uptake by ocean waters (Guemas et al. 2013), it would be irresponsible to ignore the cumulative evidence for a climate change footprint and the well-documented and continuing GHG accumulation that, given current understanding, must eventually lead to additional warming at the Earth's surface (IPCC 2013).

The climate change drivers of warming, precipitation modification, and CO_2 enrichment each will influence rangeland structure and function with impacts on animal production. Temperature regulates rates of chemical reactions, animal metabolism, and water and energy fluxes. However, CO_2 concentration influences rates with which leaves exchange CO_2 and H_2O with the atmosphere. Precipitation regulates plant productivity and associated ecosystem processes by determining soil water availability.

7.4 Key Scientific Principles for Projecting Climate Change Impacts

In this section, we discuss key scientific findings or principles that form the basis for our assessment of climate change impacts on rangelands. We regard each principle as a critical generalization resulting from years of research in the ecological or climatological sciences. Together, these principles inform our evaluation of rangeland-climate interactions.

7.4.1 Magnified Greenhouse Effects Are Irreversible

Climatic consequences of the magnified greenhouse effect are irreversible for a minimum period of decades to centuries given current technologies. GHG concentrations and climate are changing at a rate that is and likely will continue to be exceedingly rapid compared to past changes.

Despite important uncertainties about the magnitude of atmospheric warming, dynamics in the Earth climate system make additional climate change a virtual certainty. Indeed, centuries may be required for climate to equilibrate with current levels of GHGs (Matthews and Weaver 2010). For instance, warmed oceans cause air temperature to increase even if atmospheric CO_2 concentration is stabilized. The decrease in warming that would result from slowly declining CO_2 concentrations would largely be offset by heat loss from warmed oceans (Solomon et al. 2009). Dynamics of C-cycle processes also will lead to warming by causing CO_2 concentration to continue to rise in the absence of anthropogenic emissions if, as anticipated, climate change leads to loss of organic C from terrestrial ecosystems by causing widespread forest loss or thawing of permafrost (Koven et al. 2011).

Climate is changing at an unprecedented rate. For example, rates of change were greater during the period from 1880 to 2005 than during the Little Ice Age and early Holocene (Diffenbaugh and Field 2013) and further acceleration is anticipated. These unprecedented rates of change challenge the coping capacity of social-economic-biophysical systems (Joyce et al. 2013; Chap. 15, this volume) and the ability of many organisms to track favorable climatic conditions across the landscape. Consequences include shifts in vegetation patterns and range distributions, increases in rapidly dispersed "weedy" species, and the potential occurrence of plant communities that have not previously existed (Polley et al. 2013).

7.4.2 Ecological Consequences of Climate Change Will Vary Regionally

Climate change will impact ecological processes differently in different regions because the magnitude, decadal timing, or seasonal patterns of warming and precipitation modification will be expressed differently among regions. Climate change impacts likely will be greatest for rangelands where climate shifts amplify currently positive climatic effects or exacerbate climatic limitations on plant and animal productivity or surface water supplies. For example, intensification of drought (warmer, drier conditions) may elevate the risk of extensive plant mortality and even of biome reorganization in arid ecosystems in which plants already function with limited water availability (Ponce Campos et al. 2013; see Box 7.1).

7.4.3 Climate Drivers Have Unique but Potentially Interactive Effects on Plants and Ecosystem Processes

Climate drivers have unique effects on plants and ecosystems. For example, temperature regulates water and energy fluxes between land surfaces and the atmosphere, whereas CO_2 concentration influences rates of leaf photosynthesis and transpiration, in addition to retaining thermal energy near the Earth's surface.

Text Box 7.1: Climatic drying accelerates degradation of semiarid shrubland ecosystems

A recent trend toward greater aridity intensified stresses in a semiarid shrubland in Spain on a drought-prone soil (Vicente-Serrano et al. 2012). Increased temperature increased evaporative demand causing plant cover to decline during summer and in areas in which water limitation is common. Plant cover increased between 1984 and 2009 as temperature increased on some shrublands (*top panel*), but decreased strongly in more water-limited communities on infertile gypsum soil (*lower panel*). Photographs are reproduced from Vicente-Serrano et al. (2012) with permission from the Ecological Society of America.

(continued)

Text Box 7.1: (continued)

The combined effect of climate change drivers may differ from that anticipated by summing single-driver effects (Fig. 7.2; but see, Dukes et al. 2005). For instance, plant biomass responses to combined warming and CO_2 enrichment treatments often are smaller than anticipated from single-factor experiments (Morgan et al. 2011; Dieleman et al. 2012). Interactions also are evident in driver effects on plant community composition. Responses of foliar cover to experimental warming were species specific in an old field, varied through time, and were contingent on precipitation treatments (Engel et al. 2009). Warming may exacerbate effects of reduced precipitation on rangelands (Hovenden et al. 2008; Sherry et al. 2008), but the ecosystem consequences of the combination of climate change drivers generally are not as extreme as would be anticipated from single-driver effects (Wu et al. 2011; Dieleman et al. 2012).

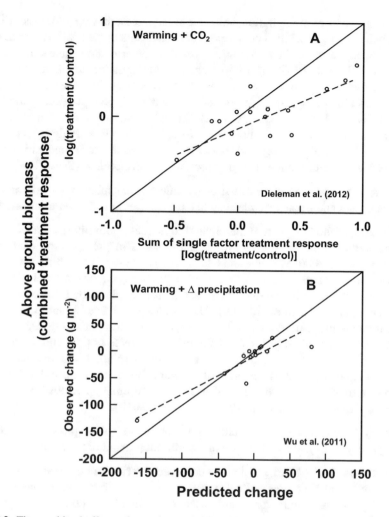

Fig. 7.2 The combined effects of warming and either CO_2 enrichment (**a**) or precipitation treatment (**b**) on aboveground biomass plotted versus biomass values predicted by summing responses from single-factor manipulations. Data are reported as either the logarithm of the ratio of treatment to control biomass (**a**) or as the difference in biomass between the treatment and control (**b**). The *solid line* is the 1:1 line representing the responses expected if warming did not modify the CO_2 or precipitation response. Most observations fall below the 1:1 line, indicating that combined treatment effects are smaller than anticipated by summing effects from single-factor manipulations

7.4.4   Rangelands Will Respond Strongly to Driver Effects on Soil Water Availability

Plant productivity and community composition are regulated by soil water availability on rangelands.

The importance of precipitation, by inference of soil water availability, to plant productivity is evident in positive relationships between plant productivity and various measures of precipitation, both across (e.g., Rosenzweig 1968; Lieth 1973; Sala et al. 2012) and within ecosystem types (e.g., Sala et al. 1988; Huxman et al. 2004). Indeed, water is the most limiting resource on arid and semiarid rangelands (Smith and Nowak 1990). The importance of precipitation to plant community structure and composition is evident in the marked shift in grassland vegetation that occurs along a west-to-east gradient of increasing precipitation in the Great Plains of North America (Risser et al. 1981) and in the major shifts in plant composition that occurred during the 1930s' drought in the Great Plains (Weaver and Albertson 1943). The strong link between precipitation and rangeland function implies four climate change-relevant corollaries to Principle 4 (Sect. 7.4.4).

Corollary 1 Climate change drivers strongly influence precipitation patterns and rates of evapotranspiration, potentially leading to large changes in soil water availability.

Warming of the biosphere is anticipated to increase both interannual and intra-annual precipitation variability with possible shifts in precipitation seasonality (Easterling et al. 2000; McCabe and Clark 2006) and associated shifts in temporal patterns of soil water content. One of the primary ecosystem-level effects of warmer temperatures is to reduce water availability to plants by increasing evaporative demand (McKeon et al. 2009). The resulting increase in plant water stress often reduces plant productivity in the absence of compensating precipitation changes (Parton et al. 2007). One of the primary effects of CO_2 enrichment is to increase plant growth per unit of water transpired (plant water use efficiency; Ainsworth and Long 2005) and, at least temporarily, reduce canopy-level transpiration rate to slow the decline in soil water content during periods between precipitation events (Morgan et al. 2004b; Fay et al. 2012).

Corollary 2 The effect of precipitation variability on plant productivity differs among ecosystems and as a function of the current precipitation regime.

Plant productivity (NPP) of a given ecosystem varies among years in response to interannual precipitation variability, but the relationship of NPP to precipitation variability differs among rangelands as a function of the current precipitation (Fig. 7.3). The NPP-precipitation relationship differs among ecosystems partly

Fig. 7.3 The interannual coefficient of variation (CV) in aboveground net primary productivity (ANPP) for desert, grassland, old-field, and forest sites, respectively, along a gradient of increasing annual precipitation. The figure is adapted from Knapp and Smith (2001)

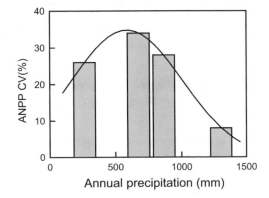

because rain-use efficiency varies among ecosystems as a result of differences in plant life history traits and biogeochemical cycling (Huxman et al. 2004). In general, NPP is more responsive to annual increases in precipitation than to precipitation declines in warm and either mesic or semiarid than in desert rangelands because the NPP response is constrained in deserts by low plant density and leaf area (Knapp and Smith 2001).

The response of rangeland NPP to intra-annual variability in precipitation also differs among ecosystems and depends on current precipitation. The anticipated shift to larger, but less frequent precipitation events (Groisman et al. 2005; Karl et al. 2009) is expected to increase the duration and severity of drought stress and hence reduce aboveground NPP (ANPP) in mesic rangelands (Knapp et al. 2002), but increase ANPP in more arid systems (Heisler-White et al. 2009). The ANPP impact of changing the size and frequency of precipitation events can be envisioned using a conceptual "soil water bucket" model (Knapp et al. 2008; Fig. 7.4). According to this model, the amount of water in the rooting zone of plants (the soil water bucket) has both upper and lower stress thresholds for plant and ecosystem processes. Plant and ecosystem processes approach maximum rates when soil water availability is neither limiting nor excessive. On mesic rangelands with annual precipitation of approximately 600–1000 mm, the soil water bucket usually is moderately full. For these ecosystems, a shift to larger, but less frequent precipitation events increases the frequency and duration of periods during which soil water content falls below the lower stress threshold by increasing water losses to runoff and percolation to groundwater. In more arid ecosystems where soil water content usually is low, precipitation that arrives in fewer, larger events is anticipated to increase the proportion of precipitation that

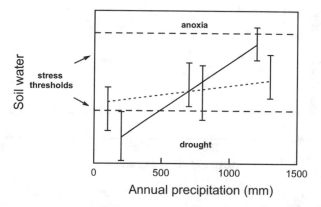

Fig. 7.4 A conceptual depiction of effects of mean annual precipitation on the response of soil water dynamics to precipitation and fewer, larger precipitation events (Knapp et al. 2008). *Vertical bars* represent temporal fluctuations in soil water content as influenced by intra-annual variation in precipitation for each of the three sites along a precipitation gradient (arid; <500 mm, mesic, wet; >1000 mm). A *solid line* connects idealized responses given the current size and frequency of precipitation events. A *dashed line* connects responses envisioned under a precipitation regime characterized by larger, but fewer events

percolates beneath upper soil layers where it is most susceptible to evaporation. Greater percolation to depth should increase ANPP by reducing the frequency or duration of periods during which soil water content remains below the lower stress threshold. Consistent with the soil water bucket model, it was found that delivering the same total of precipitation as fewer, but larger events increased ANPP by 30 % in semiarid shortgrass steppe, but decreased ANPP by 10–18 % in tallgrass prairie (Knapp et al. 2002; Heisler-White et al. 2009).

Corollary 3 Precipitation regulates the response of productivity to CO_2 but not to warming.

The ANPP response to CO_2 varies as a function of precipitation when considered across ecosystems. The ANPP-CO_2 response peaks at "moderate" levels of annual precipitation (300–400 mm) across desert and grassland ecosystems (Fig. 7.5; Nowak et al. 2004; Morgan et al. 2004b). Productivity responds relatively little to CO_2 when precipitation is very low because water stress inhibits growth and may contribute to plant senescence and mortality if it becomes sufficiently severe. The benefits of elevated CO_2 on productivity are reduced when annual precipitation is >400 mm partly because the efficient use of water use is no longer a critical variable (Nowak et al. 2004; Morgan et al. 2004b). Therefore, the average across-ecosystem response of NPP to CO_2 likely will be greatest in systems in which NPP is or will become moderately water limited (Nowak et al. 2004; Morgan et al. 2004b, 2011; Webb et al. 2012).

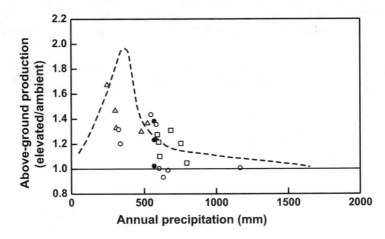

Fig. 7.5 The ratio of ANPP at elevated compared to ambient CO_2 (E/A) varies as a function of precipitation for grassland and desert ecosystems. The *dashed line* in the figure represents a conceptual model of the maximal E/A vs. precipitation relationship developed by Nowak et al. (2004). Symbols denote annual data from field experiments with open-top chambers on tallgrass prairie (*open circles*; Owensby et al. 1999), shortgrass steppe (*open triangles*; Morgan et al. 2004a), and Swiss calcareous grassland (*open squares*; Niklaus and Körner 2004), and the 5-year average of E/A for tallgrass prairie vegetation grown in elongated chambers on each of the three soil types (*closed circles*; Fay et al. 2012; Polley et al. 2012) as a function of growing season precipitation. The figure is adapted from Nowak et al. (2004) and Morgan et al. (2004b)

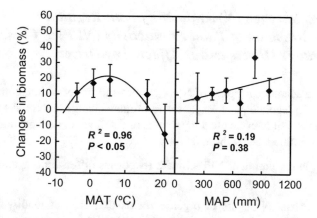

Fig. 7.6 Biomass responses of terrestrial plants to experimental warming plotted versus the mean annual temperature (MAT) and mean annual precipitation (MAP) at the site of origin. Values are means ± 95 % CI. The biomass response to warming varies as a quadratic function of MAT, but is not significantly related to MAP. Figures are adapted from Lin et al. (2010)

By contrast, there is no consistent relationship between biomass response to experimental warming and mean annual precipitation (MAP) across terrestrial plants studied (Fig. 7.6; Lin et al. 2010). Experimental warming has been shown to reduce soil water content by increasing evapotranspiration (Harte and Shaw 1995; Wan et al. 2005), with potentially negative effects on ANPP. Warming also may enhance productivity by alleviating low temperature limits on plant growth (Luo 2007; Lin et al. 2010) and increasing N mineralization (Rustad et al. 2001; Dijkstra et al. 2008). These opposing influences of warming may explain why many rangeland experiments have shown little consistent warming effect (Polley et al. 2013) and why interactive effects of warming and altered precipitation are smaller than expected from single-factor effects (Fig. 7.2b; Wu et al. 2011; Morgan et al. 2011; Xu et al. 2013).

Corollary 4 Precipitation seasonality regulates ecosystem responses to both CO_2 and warming.

Warming and CO_2 enrichment combined should lead to earlier and more rapid plant growth in ecosystems dominated by winter precipitation because warming will reduce temperature limitations on both growth and growth responses to CO_2. On the other hand, warming may reduce CO_2 benefits on arid and semiarid rangelands dominated by summer precipitation by increasing evaporative demand. The evaporative potential of air increases nonlinearly with temperature, such that a given increase in temperature will cause a disproportionately large increase in evaporative demand when temperature is high.

Plant responses to CO_2 enrichment alone also depend on the seasonal distribution of precipitation. CO_2 stimulated ANPP of Australian grassland most during years when summer rainfall exceeded rainfall during spring and autumn (Hovenden et al. 2014). High rainfall during cool spring and autumn seasons may reduce production in this grassland by intensifying N limitation to plant growth.

7.4.5   Soil Nitrogen (N) Availability both Regulates the Response of Plant Productivity (NPP) to Climate Change Drivers and Is Affected by Drivers

Plants require adequate supplies of N for growth and reproduction; consequently N availability will regulate plant responses to climate change drivers. As many of the pathways in the soil N cycle are temperature and soil water dependent, N availability is, in turn, influenced by these drivers.

Corollary 1   N regulation of NPP-climate responses differs among climate change drivers.

Most terrestrial N occurs in organic forms that are not readily available to plants; hence rangeland responses to climate change drivers depend partly on how quickly N is mineralized from organic to inorganic forms that are available to plants. Low N availability frequently limits plant productivity on rangelands (e.g. Seastedt et al. 1991) and may reduce or even eliminate any benefit of CO_2 enrichment for plant growth (Owensby et al. 1994; Reich et al. 2006; Reich and Hobbie 2012). Conversely, N availability does not appear to restrain plant response to experimental warming (Lin et al. 2010). Typically, N limits plant production in arid and semiarid ecosystems only if drought stress is alleviated (Ladwig et al. 2012), implying that N availability will have little effect on ANPP on arid and semiarid rangelands such as those in the southwestern North America and south and southwestern Australia that are predicted to become even drier.

Corollary 2   Climate change drivers may accelerate or slow N cycling with possible feedbacks on NPP.

Experimental warming increases N availability to plants by accelerating N mineralization rates, provided that soil water is available (Rustad et al. 2001; Dijkstra et al. 2008), but warming also may increase N losses (Wu et al. 2012). Water addition to dry soil greatly increases N mineralization rates, but the long-term relationship between precipitation and ecosystem N pools and N cycling is more complicated. Nitrogen pools increase along a gradient of increasing precipitation from shortgrass steppe to tallgrass prairie in the central USA, but rates of litter decomposition and of resin-captured N in soil decline (McCulley et al. 2009).

CO$_2$ enrichment has been hypothesized to create or reinforce N limitations on ANPP by reducing N mineralization. To the extent that CO_2 enrichment increases plant production and ecosystem C accumulation, it also increases the sequestration of N and other elements in long-lived plant material and organic matter leading over time to a decline in N availability to plants [the progressive N limitation (PNL) hypothesis; Luo et al. 2004]. Elevated CO_2 reduces soil N availability in some cases (Reich et al. 2006; Reich and Hobbie 2012), but results from most experiments indicate that the strength of this negative feedback is reduced by processes that delay the onset of N limitation or enhance N accumulation in soil-plant systems (Polley et al. 2011). For example, rates of N input in arid ecosystems are thought to be balanced by similar rates of gaseous loss of N (Peterjohn and Schlesinger 1990), but the addition of labile C (McCalley and Sparks 2008), as may result from CO_2 enrichment (Schaeffer et al. 2003), greatly decreases N emissions. Microbial activity in arid lands is limited by available C (Schaeffer et al. 2007); consequently CO_2 enrichment may increase C inputs into soil, which then accelerate rates of soil organic N mineralization to increase the availability of inorganic N to plants (Billings et al. 2004; Jin and Evans 2007; Jin et al. 2011). CO_2 enrichment may further increase N availability by increasing the biomass and diversity of fungi (Jin and Evans 2010; Nguyen et al. 2011) that utilize recalcitrant soil substrates and by increasing the activities of enzymes involved in N and C cycling (Jin and Evans 2007; Jin et al. 2011). Increased N_2 fixation, increased N use efficiency, and increased root foraging for N at elevated atmospheric CO_2 also may delay potential decreases in soil N (Luo et al. 2006). The preponderance of evidence indicates that gradual sequestration of N in organic matter will not strongly limit the responses of rangeland plants to climate change drivers.

7.4.6 Ecosystem Responses to Climate Change Drivers Vary Because of Differences in Management Practices and Historical Land-Use Patterns

Effects of climate change drivers on ANPP and other ecosystem processes are governed by a set of variables "internal" to ecosystems. These internal controls, which include soil resource supply and characteristics, current or potential biota, and current and historical land-use patterns and accompanying disturbance regimes, all contribute to variation in ecosystem responses to climate change drivers (Lindenmayer et al. 2010; Polley et al. 2013). For example, fire indirectly affects responses to drivers by modifying soil resources and vegetation. Fires volatilize substantial quantities of N (Seastedt et al. 1991); hence frequent burning may constrain ecosystem responses to drivers by reinforcing N limitations on plant growth. Land uses such as grazing also regulate rangeland responses to climate change. Sheep grazing limited CO_2 stimulation of grassland productivity by selectively consuming the two groups of plants (legumes, forbs) with the greatest growth responses to CO_2 (Newton et al. 2014).

7.4.7 Climate Change Drivers Affect Livestock Production both Directly and Indirectly

Corollary 1 Warming causes greatest physiological impairment to livestock in environments that are currently warm.

Livestock performance is optimal when ambient temperatures are within the "thermo-neutral zone" (Ames and Ray 1983) where forage intake and energy requirements of livestock are not affected by temperature. At temperatures below

Table 7.1 Potential consequences of heat stress for cattle performance

Immediate (hours to days)	Medium term (days to weeks)	Long term (weeks to years)
Increased activity to dissipate heat (e.g., increased sweating and time in shade)	Altered metabolism (e.g., decreased feed conversion efficiency)	Increased ectoparasite loads
Increased water intake	Reduced milk production	Reduced reproduction
Reduced forage intake	Reduced growth	Possibly increased mortality

the thermo-neutral zone, animals increase energy production and forage intake to maintain homeostasis. At temperatures above the thermo-neutral zone, animals become stressed and must actively dissipate heat by reducing walking and spending more time in shade (Table 7.1). Heat stress can dramatically reduce milk production and the efficiency of feed conversion (Wayman et al. 1962; McDowell 1968) by reducing forage intake, changing nutrient portioning independent of intake, and reducing capacity to mobilize body fat and employ glucose-sparing mechanisms (Baumgard and Rhoads 2012). Heat stress also reduces reproduction rates by reducing male and female gamete production, embryonic development, and fetal growth (Hansen 2009; Kadokawa et al. 2012). Heat stress can be fatal to livestock, especially those fed large quantities of high-quality feeds, as evidenced by the mass mortality of feedlot cattle during heat waves (Hahn 1999). Values of a temperature humidity index (THI; the weighted product of air temperature and relative humidity) greater than 80 are considered heat stress days, requiring deployment of sprinklers, additional shade, or similar cooling measures to minimize deaths among feedlot cattle (Hahn 1999). The number of days that THI exceeds 80 may increase by 138 % by 2070 (Howden et al. 2008).

Detrimental effects of heat stress will be intensified if climate change results in more frequent and severe droughts, as are forecast for southwestern North America (Seager and Vecchi 2010) and south and southwestern Australia (IPCC 2007). Springs and dugouts on arid and semiarid rangelands often dry during drought, requiring that livestock travel greater distances for water. Walking increases the heat load, and animals must increase sweating and respiration rates to maintain homeostasis (Moran 1973).

Corollary 2 Climate change drivers reduce livestock production by increasing abundances of ectoparasites and reducing forage quality and quantity.

Warmer temperatures may suppress livestock productivity by increasing winter survival of ectoparasites to facilitate larger populations (Karl et al. 2009). Horn flies [*Haematobia irritans* (L.)] are the primary pest of concern for cattle in the USA (Byford et al. 1992). Horn flies can reduce cattle weight gain by 4–14 %, and adversely impact animal physiological functions by increasing cortisol production and decreasing N retention (Byford et al. 1992; Oyarzún et al. 2008). Horn flies also promote livestock-avoidance behaviors, such as walking and tail switching (Harvey and Launchbaugh 1982). These additional activities and energy expenditures directly contribute to reduced livestock performance. Ticks (*Amblyomma americanum* Koch) reduce weight gains of British breeds, such as Hereford, by greater than 30 % (Byford et al. 1992). By contrast, ticks have less effect on Brahman crosses with British and Continental cattle, and little, if any, effect on Brahman cattle (Utech et al. 1978; George et al. 1985). Livestock gains in Australia could decrease by greater than 18 % because of increased tick infestations associated with climate change unless European and British cattle breeds are replaced by Brahmans or other tick-resistant breeds (White 2003).

Climate change drivers also are anticipated to reduce livestock performance on many rangelands by reducing forage quantity, quality, or both (Polley et al. 2013). Drought obviously reduces forage production, whereas elevated CO_2 has consistently been shown to increase plant C:N ratios and decrease tissue N concentrations (Cotrufo et al. 1998; Morgan et al. 2001), resulting in forage with reduced crude protein levels. Forage digestibility often declines at elevated CO_2 (Morgan et al. 2004a; Milchunas et al. 2005). Regional-scale analyses indicate that livestock become more nutrient limited in warmer and drier climates as both dietary crude protein and digestible organic matter content of forages decrease (Craine et al. 2010). A 4-year study in the tallgrass prairie showed that warming reduced N concentration and, correspondingly, crude protein levels of live and dormant forage from five warm-season grasses (An et al. 2005). In the absence of protein supplementation, cattle production will decline if forage quality declines too greatly because forage intake is volume limited. Forage quality may increase in cooler regions, such as the northern USA and southern Canada (Craine et al. 2010), if winters become warmer and wetter as anticipated.

7.4.8 Climate Change Indirectly Affects Vegetation Composition and Structure by Influencing Fire Regimes

Fire affects the composition of vegetation partly by favoring fire tolerant over fire-sensitive species. Fire is an important, although not exclusive, predictor of the relative abundances of comparably fire-tolerant grasses and fire-sensitive woody vegetation (Bond 2008) and of the global distribution of the savanna biome (Staver et al. 2011). By increasing fine fuel loads, exotic grasses can increase wildfire frequency to the detriment of native vegetation. Proliferation of the exotic annual grass *Bromus tectorum* (cheatgrass), for example, has significantly increased fire frequency and even the spatial areas of fires in invaded areas of the Great Basin, USA (Balch et al. 2013).

Shifts in temperature and precipitation are known to modify fire regime. Most wildfires in the western USA occur during the hottest, driest portion of the year (Westerling et al. 2003). Fires are largest in grass- and shrub-dominated ecosystems when unusually wet-growing seasons during which fine fuels accumulate are followed by dry conditions that enhance fuel flammability and ignition (Littell et al. 2009). In contrast, fires are largest in western forested ecosystems of North America when precipitation is low and temperature is high in both the fire year and the preceding year. Fire activity is projected to increase if the climate becomes both warmer and drier (Pechony and Shindell 2010).

Warmer and drier conditions are conducive to more frequent fires, but sufficient fuels will be required to sustain rangeland fires. Increased CO_2 could increase fire frequency or intensity (Sage 1996) and thereby reinforce the effects of fire on ecosystem processes by increasing plant production (fuel load) and fuel flammability or favoring fire-adapted plant species. Plant growth appears to be particularly responsive to CO_2 among several fire-adapted annual grasses, including *Bromus tectorum* and *Avena barbata* (Jackson et al. 1994; Ziska et al. 2005).

7.4.9 Climate Change May Lead to Communities That Are Unlike any Found Today, with Important Consequences for Ecosystem Function and Management

Climate change may alter the composition or relative abundances of species in plant communities. Increased weather variability is anticipated to favor short-lived plants and other organisms (e.g., animals, insects, diseases) that can response rapidly to environmental change. Vegetation shifts are expected to result largely from changes in the amount, seasonal pattern, and vertical distribution of soil water (Knapp et al. 2008; Volder et al. 2010).

Climate change will lead to combinations of seasonal temperature and precipitation that differ from current climatic conditions. Consequently, climate shifts may support plant communities that are compositionally unlike any found today (no-analog or novel communities; Williams and Jackson 2007), perhaps accompanied by changes in the number and functional types of species present. For example, species diversity is lower and the latitudinal shift in the ratio of C_3 to C_4 species is more pronounced in novel than native communities in the tallgrass prairie region of North America (Martin et al. 2014).

Past episodes of climate change drove local extinctions that led to vegetation change as geographic shifts in climatic conditions outpaced the capacity of many plant species to migrate and establish (Blois et al. 2013). Species that successfully colonize following extinctions often are ecological generalists (e.g., many "weedy" species). Increased establishment of generalist species leads, in turn, to the development of progressively more homogenous plant assemblages across spatial scales. Climate change may exacerbate vegetation homogenization by favoring generalist and often exotic species (Everard et al. 2010; Blois et al. 2013), with potentially negative effects on ecosystem function (Isbell et al. 2011).

7.4.10   Increased Climatic Variability Increases Fluctuations in Ecological Systems, Rendering Sustainable Management More Difficult

Human learning to avoid cycles of degradation is hindered by variability in both ecological and social-economic systems (Stafford Smith et al. 2007). Increased variability in ecological systems in future climates, combined with mismatches in the temporal or spatial scales at which human impacts and ecological processes change, can lead to rangeland degradation. Stafford Smith et al. (2007), for example, described a cycle of rangeland degradation that has been repeated multiple times in Australia over the past century. Favorable climatic and economic conditions lead initially to an increase in the number of livestock maintained per unit of land area (stocking rate). Subsequent drought decreases forage production, often leading to overgrazing, especially when drought is coupled with government inducements (e.g., feed subsidies) or reduced livestock prices that lessen incentives for producers to destock (Thurow and Taylor 1999). Excessive defoliation reduces herbaceous cover potentially leading to greater soil erosion. The combined impacts of drought, current and previous stresses, including water limitation and overgrazing, and the occurrence of a warmer, drier climate might be reflected first in directional shifts in the cover or composition of the plant community or as increased variability in ANPP (Fig. 7.7). Rangeland degradation results if plant productivity and diversity decrease to levels that cannot be economically sustained or ecologically reversed (Willms et al. 1985; Milton et al. 1994) (Chaps. 8 and 15, this volume).

7.5   An Assessment of Climate Change Scenarios

Following, we provide an assessment of the implications of climate change for rangelands organized around three plausible, but regionally unique, climate change scenarios. Greater uncertainty is associated with projected changes in precipitation than CO_2 concentration or warming, particularly at regional scales. The following should be viewed as a general assessment of the collective impact of a given climate change scenario (Box 7.2).

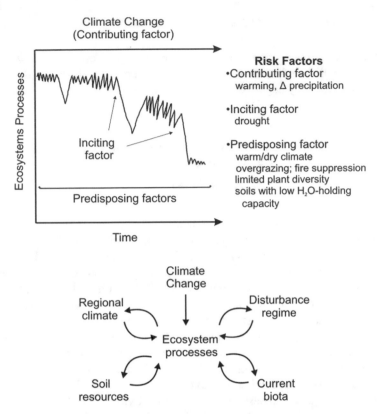

Fig. 7.7 A framework for envisioning how climate change alone or in interaction with other risks may cause a large decline in ecosystem services, including plant and livestock productivity [*upper panel*; adapted from Manion (1981)]. We consider three categories of risks—predisposing, inciting, and contributing. Predisposing factors are current and antecedent stresses that increase an ecosystem's susceptibility to short-duration stresses, such as drought (inciting factor), or to longer term change in climate (contributing factor). Predisposing factors include stresses related to current states or values of variables that are internal to ecosystems and dynamic over ecological timescales (current regional climate, soil resources, biota, and disturbance regime; *lower panel*). In the *upper panel*, we illustrate a case in which climate change (contributing factor) exacerbates negative effects of a recent stress, such as drought (inciting factor), for a rangeland on which processes already are limited by a warm/dry climate, overgrazing, or a sparse, species-poor plant canopy (predisposing factors) to increase process variability sufficiently to change average process levels

Text Box 7.2: Key risks associated with climate change for rangelands
Illustration of representative risks for livestock production on rangelands subject to each of the three general climate change scenarios. Climate-related drivers of risk are indicated by icons. Key risks are those shifts in biotic or abiotic conditions judged likely to lead to ecosystem changes that are of large magnitude or are irreversible at management timescales of decades. An overall potential risk is assessed for each scenario assuming an increase in mean temperature of either 2 or 4 °C given current management strategies and adaptation capacity. The risk associated with any given biotic or abiotic shift will vary among regions depending on current biophysical and socioeconomic factors.

(continued)

Text Box 7.2: (continued)

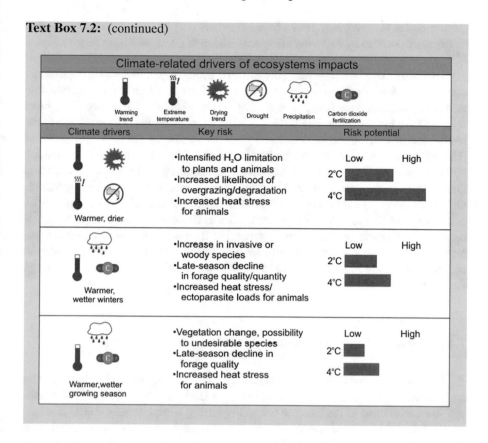

7.5.1   Warmer, Drier Climate Scenario

Regional specificity—Climate models project that warming will be accompanied by a decrease in precipitation in arid to semiarid rangelands in the southwestern North America, Central America, and south and southwestern Australia (IPCC 2007). Precipitation is projected to decrease by 10–20 % during winter and spring in southern Africa and during spring and summer in central Asia.

NPP and forage quality—Warming and drying will reduce soil water availability leading to a decrease in NPP on rangelands that already are warm and dry. NPP is more sensitive to the amount by which precipitation declines than to the amount of warming (Xu et al. 2013). Elevated CO_2 will not greatly alleviate negative effects of drying on NPP because of persistent soil water limitations. Forage quality likely will decline in response to combined effects of CO_2 enrichment, warming, and reduced precipitation.

Livestock production—Warmer and potentially drier conditions likely will reduce cattle production, resulting in fewer cattle operations. Remaining cattle operations will likely switch to the most heat tolerant of the British and European breeds (*Bos taurus*) or from *Bos taurus* breeds to the more heat-tolerant *Bos indicus* breeds, including Brahman and Brahman crosses, Romosinuano and Senepol

(Hammond et al. 1996; Kay 1997). Some ranchers may be forced to change species of livestock, such as by replacing cattle with sheep or goats which are better adapted to warm temperatures and drought (Kay 1997).

Vegetation composition—Intensification of drought (warmer, drier conditions) elevates the risk of extensive plant mortality and even of biome reorganization in arid ecosystems (Ponce Campos et al. 2013). For example, the severe drought of 2002–2003 caused greater than 90 % mortality of pinyon pine (*Pinus edulis*) in the southwestern USA whereas associated trees of *Juniperus monosperma* survived (Breshears et al. 2005, 2009). Warmer and drier conditions also may alter vegetation by favoring more frequent fires. Increasing fire frequency in the Mojave Desert and Great Basin in the past 20 years has converted communities of desert shrublands and shrub steppe to annual grasslands (CCSP 2008; Balch et al. 2013). Some shrub and woodland ecosystems of rangelands, such as those in the western USA, developed under climatic conditions more favorable than those forecast. These ecosystems may be subject to increased fire frequencies if more severe droughts lead to episodes of woody mortality that produce fuel loads sufficient to sustain fire (e.g., Breshears et al. 2005).

7.5.2   Warmer, Wetter Winters Scenario

Regional specificity—Climate models project that warming will be accompanied by an increase in precipitation in the northwestern USA, southern Canada, and northern Asia, with most of the increase occurring during winter months (IPCC 2007).

NPP and forage quality—Warming and increased precipitation, coupled with elevated CO_2 concentration, should increase NPP, especially early in the season. Experimental warming and increased precipitation generally stimulate plant growth (Wu et al. 2011). NPP likely will decline in the latter portion of the growing as warming and reduced precipitation limit soil water availability. In combination, climate change drivers may have little effect on forage quality.

Livestock production—The combination of warmer temperatures, greater precipitation, and CO_2 enrichment should increase livestock productivity, particularly in cooler regions such as the northern USA and Canada, by moderating winter temperatures (Baker et al. 1993; Eckert et al. 1995; Rötter and Geijn 1999), lengthening the growing season, and increasing NPP.

Vegetation composition—Warmer, wetter winters likely will favor plants that grow early in the season or access soil water accumulated early in the growing season. For example, ponderosa pine (*Pinus ponderosa*) established in grassland-forest ecotones in northern Colorado during years when spring and autumn precipitation were high in association with El Niño events (League and Veblen 2006). By increasing soil water content, higher CO_2 and precipitation also favor recruitment of taprooted invasive forbs like leafy spurge (*Euphorbia esula*), diffuse knapweed (*Centaurea diffusa),* and baby's breath *(Gypsophila paniculata),* as well as some subshrubs (Owensby et al. 1999; Morgan et al. 2007; Blumenthal et al. 2008).

7.5.3 Warmer, Wetter Growing Season Scenario

Regional specificity—Climate models project an increase in winter and spring precipitation in northern portions of Great Plains region of central North America (IPCC 2007; Walthall et al. 2012).

NPP and forage quality—Warming combined with wetter conditions during the early growing season and elevated CO_2 should increase NPP, partly by extending the growing season. Grass production in tallgrass prairie in this region is highly responsive to early-season precipitation (Craine et al. 2012). The anticipated shift to larger, but less frequent precipitation events may reduce ANPP in mesic rangelands (Knapp et al. 2002), but increase ANPP in more arid systems in the region (Heisler-White et al. 2009). Forage quality may improve, especially if forage benefits of increased early-season precipitation are not negated by effects of late-season dry periods.

Livestock production—The combination of warmer temperatures, greater precipitation, and CO_2 enrichment should increase livestock productivity by moderating winter temperatures (Baker et al. 1993; Eckert et al. 1995; Rötter and Geijn 1999), lengthening the growing season, and increasing early-season NPP. Wetter growing seasons allow rangeland livestock producers to provide additional watering locations and, perhaps, shift to more heat-tolerant breeds.

Vegetation composition—Warmer temperatures and wetter winter and spring seasons likely will favor plants that grow early in the season.

7.6 Knowledge Gaps

The following are important knowledge gaps that limit our ability to predict rangeland responses to climate change.

- Uncertainty in climate projections, especially in projections of precipitation trends. Uncertainty is greater at regional than global scales.
- Limited understanding of how climate change drivers interact to affect key ecosystem variables and processes, including soil water content and dynamics, NPP, and forage quality.
- Limited capacity to discern impacts of climate change from those of management and disturbances.
- Limited capacity to breed livestock that are adapted to weather and rangeland conditions anticipated as climate changes.

Most livestock currently are selected for growth and carcass traits (Cartwright 1970). Climate change may render the capacity to survive and reproduce under harsh conditions more important than growth rate. Progress in molecular genetics and animal breeding is required to identify and select livestock with improved adaptation to warmer and drier conditions. Because the traits selected to improve performance under harsher environments may be opposite to those traditionally selected (e.g., productivity traits; Prayaga et al. 2009), programs are required to breed animals adapted to harsh environments that also produce red meat of acceptable quality to consumers.

7.7 Summary

Climate is changing at an unprecedented rate as a consequence of increasing atmospheric concentrations of greenhouse gases including carbon dioxide (CO_2), methane (CH_4), and nitrous oxide (N_2O). Dynamics in the Earth climate system make additional climate change a virtual certainty because centuries will be required for climate to equilibrate with current levels of greenhouse gases.

Climate change science predicts warming and greater climatic variability for the foreseeable future. Warming is expected to be greatest at northern latitudes and over land. Precipitation is anticipated to increase at higher latitudes and decrease elsewhere, with significant seasonal variation and greater interannual or intra-annual variability.

The major climate change drivers—warming, precipitation modification, and direct effects of increased CO_2 concentration on plant growth and transpiration—will alter soil water availability and N cycling, processes that regulate the provisioning of plant and animal products and other services from rangelands. For example, increased precipitation variability will be manifest as shifts in precipitation seasonality and in the temporal patterns of soil water availability on some rangelands. Warmer temperatures will increase evaporative demand, thereby reducing soil water in the absence of compensating change in precipitation, whereas CO_2 enrichment could reduce transpiration rates and increase plant growth per unit of water transpired. Climate change drivers also will influence soil N availability to plants, with potential feedbacks on plant responses to drivers. Warming increases N availability by accelerating N mineralization rates when soil water is available, but CO_2 enrichment can create or reinforce N limitations on productivity by increasing the sequestration of N in long-lived plant material and organic matter. Limited N, in turn, can constrain benefits of CO_2 enrichment for plant growth.

The ecological consequences of climate change drivers will vary among rangelands partly because the magnitude, decadal timing, or seasonal patterns of warming and precipitation modification will be expressed differently among regions. Climate change impacts will be greatest for rangelands where climate shifts amplify currently positive climatic effects or exacerbate climatic limitations on plant and animal productivity.

The ecological consequences of climate change drivers also will vary among rangelands because driver effects frequently are nonadditive and contingent on current precipitation regimes. Nonadditive effects are indicated when the combined influence of shifts in climate drivers differs from that anticipated by summing single-driver effects. Plant growth responses to combined warming and CO_2

enrichment treatments often are smaller than anticipated from single-factor experiments, for example. The effects of climate drivers also vary as a function of the current precipitation regime. The responses of plant productivity to both CO_2 and precipitation variability depend on the current precipitation regime, for instance. Plant productivity generally is more responsive to interannual precipitation variability in mesic or semiarid than desert rangelands. The stimulating effect of CO_2 on productivity also depends on precipitation, peaking at moderate levels of annual precipitation (300–400 mm) when considered across desert and grassland ecosystems.

Warming, especially when combined with drought, will cause greatest physiological impairment to livestock in already warm environments by negatively impacting animal physiological performance, increasing abundances of ectoparasites, and reducing forage quality and quantity. Heat stress reduces milk production, forage intake, and reproduction rate, whereas warmer temperatures increase winter survival of ectoparasites and may reduce forage quality.

Climate change will lead to combinations of seasonal temperature and precipitation that differ from current climatic conditions. Warmer, drier conditions also are anticipated to increase fire activity. This shift in climatic conditions and fire activity could lead to plant communities that are compositionally unlike any found today. The rapid rate at which climate is changing, in combination with other drivers and disturbances, will favor ecological generalists such as short-lived plants (e.g., weeds), animals, insects, and disease organisms. Increased establishment of generalist plant species could, in turn, hasten the development of more homogeneous plant assemblages with negative effects on species diversity and animal productivity.

Rangelands increasingly are being transformed by climate change and a variety of human impacts, including change in land-use patterns, intensification of disturbances, and accelerated species introductions and movements. Singly, any of these changes will alter services provisioned by rangelands. Combined, the imprint of human activities may challenge ecosystem integrity and functionality by increasing variability in ecosystem processes and production enterprises beyond their historical range. As a result, livestock production and supply of other ecosystem services could decline to levels that cannot be economically sustained or ecologically reversed. Given this potential outcome, we suggest that an overarching goal of management and monitoring must be to improve our ability to predict the vulnerability of rangelands and rangeland production systems to continued climate change.

References

Ainsworth, E.A., and S.P. Long. 2005. What have we learned from 15 years of free-air CO_2 enrichment (FACE)? A meta-analytic review of the responses of photosynthesis, canopy properties and plant production to rising CO_2. *New Phytologist* 165: 351–372.

Allen-Diaz, B. 1996. Rangelands in a changing climate: Impacts, adaptations, and mitigation. In *Climate change 1995—Impacts, adaptations and mitigation of climate change: Scientific-technical analyses*, ed. R.T. Watson, M.C. Zinyowera, R.H. Moss, and D.J. Dokken. New York: Cambridge University Press.

Ames, D.R., and D.E. Ray. 1983. Environmental manipulation to improve animal productivity. *Journal of Animal Science* 57: 209–220.

An, Y., S. Wan, X. Zhou, A.A. Subedar, L.L. Wallace, et al. 2005. Plant nitrogen concentration, use efficiency, and contents in a tallgrass prairie ecosystem under experimental warming. *Global Change Biology* 11: 1733–1744.

Baker, B., J. Hanson, R. Bourdon, and J. Eckert. 1993. The potential effects of climate change on ecosystem processes and cattle production on US rangelands. *Climatic Change* 25: 97–117.

Balch, J.K., B.A. Bradley, C.M. D'Antonio, and J. Gómez-Dans. 2013. Introduced annual grass increases regional fire activity across the arid western USA (1980–2009). *Global Change Biology* 19: 173–183.

Baumgard, L.H., and R.P. Rhoads. 2012. Ruminant nutrition symposium: Ruminant production and metabolic responses to heat stress. *Journal of Animal Science* 90: 1855–1865.

Billings, S., S.M. Schaeffer, and R.D. Evans. 2004. Soil microbial activity and N availability with elevated CO_2 in Mojave Desert soils. *Global Biogeochemical Cycles* 18: GB1011. doi:10.1029 /2003GB002137.

Blois, J.L., P.L. Zarnetske, M.C. Fitzpatrick, and S. Finnegan. 2013. Climate change and the past, present, and future of biotic interactions. *Science* 341: 499–504.

Blumenthal, D., R.A. Chimner, J.M. Welker, and J.A. Morgan. 2008. Increased snow facilitates plant invasion in mixed-grass prairie. *New Phytologist* 179: 440–448.

Bond, W.J. 2008. What limits trees in C_4 grasslands and savannas? *Annual Review of Ecology, Evolution, and Systematics* 39: 641–659.

Breshears, D.D., N.S. Cobb, P.M. Rich, K.P. Price, C.D. Allen, et al. 2005. Regional vegetation die-off in response to global-change-type drought. *Proceedings of the National Academy of Sciences of the United States of America* 102: 15144–15148.

Breshears, D.D., O.B. Myers, C.W. Meyer, F.J. Barnes, C.B. Zou, et al. 2009. Tree die-off in response to global change-type drought: Mortality insights from a decade of plant water potential measurements. *Frontiers in Ecology and the Environment* 7: 185–189.

Byford, R.L., M.E. Craig, and B.L. Crosby. 1992. A review of ectoparasites and their effect on cattle production. *Journal of Animal Science* 70: 597–602.

Cartwright, T.C. 1970. Selection criteria for beef cattle for the future. *Journal of Animal Science* 30: 706.

Cotrufo, M.F., P. Ineson, and A. Scott. 1998. Elevated CO_2 reduces the nitrogen concentration of plant tissues. *Global Change Biology* 4: 43–54.

Craine, J.M., A.J. Elmore, K.C. Olson, and D. Tolleson. 2010. Climate change and cattle nutritional stress. *Global Change Biology* 16: 2901–2911.

Craine, J.M., J.B. Nippert, A.J. Elmore, A.M. Skibbe, S.L. Hutchinson, et al. 2012. Timing of climate variability and grassland productivity. *Proceedings of the National Academy of Sciences of the United States of America* 109: 3401–3405.

Dieleman, W.I.J., S. Vicca, F.A. Dijkstra, F. Hagedorn, M.J. Hovenden, et al. 2012. Simple additive effects are rare: A quantitative review of plant biomass and soil process responses to combined manipulations of CO_2 and temperature. *Global Change Biology* 18: 2681–2693.

Diffenbaugh, N.S., and C.B. Field. 2013. Changes in ecologically critical terrestrial climate conditions. *Science* 341: 486–492.

Dijkstra, F.A., E. Pendall, A.R. Mosier, J.Y. King, D.G. Milchunas, et al. 2008. Long-term enhancement of N availability and plant growth under elevated CO_2 in a semi-arid grassland. *Functional Ecology* 22: 975–982.

Dukes, J.S., N.R. Chiariello, E.E. Cleland, L.A. Moore, M.R. Shaw, et al. 2005. Responses of grassland production to single and multiple global environmental changes. *PLoS Biology* 3(10): e319.

Easterling, D.R., G.A. Meehl, C. Parmesan, S.A. Changnon, T.R. Karl, et al. 2000. Climate extremes: Observations, modeling, and impacts. *Science* 289: 2068–2074.

Eckert, J.B., B.B. Baker, and J.D. Hanson. 1995. The impact of global warming on local incomes from range livestock systems. *Agricultural Systems* 48: 87–100.

Engel, E.C., J.F. Weltzin, R.J. Norby, and A.T. Classen. 2009. Responses of an old-field plant community to interacting factors of elevated [CO$_2$], warming, and soil moisture. *Journal of Plant Ecology* 2: 1–11.

Everard, K., E.A. Seabloom, W.S. Harpole, and C. de Mazancourt. 2010. Plant water use affects competition for nitrogen: Why drought favors invasive species in California. *American Naturalist* 175: 85–97.

Fay, P.A., V.L. Jin, D.A. Way, K.N. Potter, R.A. Gill, et al. 2012. Soil-mediated effects of subambient to increased carbon dioxide on grassland productivity. *Nature Climate Change* 2: 742–746.

George, J.E., R.L. Osburn, and S.K. Wikel. 1985. Acquisition and expression of resistance by Bos indicus and Bos indicus x Bos taurus calves to Amblyomma americanum infestation. *The Journal of Parasitology* 71: 174–182.

Groisman, P.Y., R.W. Knight, D.R. Easterling, T.R. Karl, G.C. Hegerl, et al. 2005. Trends in intense precipitation in the climate record. *Journal of Climate* 18: 1326–1350.

Guemas, V., F.J. Doblas-Reyes, I. Andreu-Burillo, and M. Asif. 2013. Retrospective prediction of the global warming slowdown in the past decade. *Nature Climate Change* 3: 649–664.

Hahn, G.L. 1999. Dynamic responses of cattle to thermal heat loads. *Journal of Animal Science* 77(Suppl 2): 10–20.

Hammond, A., T. Olson, C. Chase, E. Bowers, R. Randel, et al. 1996. Heat tolerance in two tropically adapted Bos taurus breeds, Senepol and Romosinuano, compared with Brahman, Angus, and Hereford cattle in Florida. *Journal of Animal Science* 74: 295–303.

Hansen, P.J. 2009. Effects of heat stress on mammalian reproduction. *Philosophical Transactions of the Royal Society B* 364: 3341–3350.

Harte, J., and R. Shaw. 1995. Shifting dominance within a montane vegetation community: Results of a climate-warming experiment. *Science* 267: 876–880.

Harvey, T.L., and J.L. Launchbaugh. 1982. Effect of horn flies (Diptera, muscidae) on behavior of cattle. *Journal of Economic Entomology* 75: 25–27.

Heisler-White, J.L., J.M. Blair, E.F. Kelly, K. Harmoney, and A.K. Knapp. 2009. Contingent productivity responses to more extreme rainfall regimes across a grassland biome. *Global Change Biology* 15: 2894–2904.

Hovenden, M.J., P.C.D. Newton, K.E. Willis, J.K. Janes, A.L. Williams, et al. 2008. Influence of warming on soil water potential controls seedling mortality in perennial but not annual species in a temperate grassland. *New Phytologist* 180: 143–152.

Hovenden, M.J., P.C.D. Newton, and K.E. Willis. 2014. Seasonal not annual rainfall determines grassland biomass response to carbon dioxide. *Nature* 511: 583–586.

Howden, S.M., S.J. Crimp, and C.J. Stokes. 2008. Climate change and Australian livestock systems: Impacts, research and policy issues. *Australian Journal of Experimental Agriculture* 48: 780–788.

Huxman, T.E., M.D. Smith, P.A. Fay, A.K. Knapp, M.R. Shaw, et al. 2004. Convergence across biomes to a common rain-use efficiency. *Nature* 429: 651–654.

Intergovernmental Panel on Climate Change [IPCC]. 2007. *Climate change 2007: The physical science basis, contribution of Working Group I to the Fourth Assessment Report of the Intergovernmental Panel on Climate Change* eds. S. Solomon, D. Qin, M. Manning, Z. Chen, M. Marquis, K.B. Averyt, M. Tignor, and H.L. Miller. Cambridge, UK: Cambridge University Press.

———. 2013. Summary for policymakers. In *Climate change 2013: The physical science basis. Contribution of Working Group I to the Fifth Assessment Report of the Intergovernmental Panel on Climate Change*, eds. T.F. Stocker, D. Qin, G.-K. Plattner, M. Tignor, S.K. Allen, J. Boschung, A. Nauels, Y. Xia, V. Bex, and P.M. Midgley. Cambridge, UK: Cambridge University Press.

Isbell, F., V. Calcagno, A. Hector, J. Connolly, W.S. Harpole, et al. 2011. High plant diversity is needed to maintain ecosystem services. *Nature* 477: 199–202.

Jackson, R.B., O.E. Sala, C.B. Field, and H.A. Mooney. 1994. CO$_2$ alters water use, carbon gain, and yield of dominant species in a natural grassland. *Oecologia* 98: 257–262.

Jin, V., and R.D. Evans. 2007. Elevated CO_2 affects microbial carbon substrate use and N cycling in Mojave Desert soils. *Global Change Biology* 13: 452–465.

Jin, V.L., and R.D. Evans. 2010. Microbial ^{13}C utilization patterns via stable isotope probing of phospholipid biomarkers in Mojave Desert soils exposed to ambient and elevated atmospheric CO_2. *Global Change Biology* 16: 2334–2344.

Jin, V.L., S.M. Schaeffer, S.E. Ziegler, and R.D. Evans. 2011. Soil water availability and microsite mediate fungal and bacterial phospholipid fatty acid biomarker abundances in Mojave Desert soils exposed to elevated atmospheric CO_2. *Journal of Geophysical Research: Biogeosciences* 116. 10.1029/2010JG001564.

Joyce, L.A., D.D. Briske, J.R. Brown, H.W. Polley, B.A. McCarl, et al. 2013. Climate change and North American rangelands: Assessment of mitigation and adaptation strategies. *Rangeland Ecology and Management* 66: 512–528.

Kadokawa, H., M. Sakatani, and P.J. Hansen. 2012. Perspectives on improvement of reproduction in cattle during heat stress in a future Japan. *Animal Science Journal* 83: 439–445.

Karl, T.R., J.M. Melillo, and T.C. Peterson (eds.). 2009. *Global climate change impacts in the United States*. New York: Cambridge University Press.

Kay, R.N.B. 1997. Responses of African livestock and wild herbivores to drought. *Journal of Arid Environments* 37: 683–694.

Keeling, R.F., S.C. Piper, A.F. Bollenbacher, and J.S. Walker. 2009. Atmospheric CO_2 records from sites in the SIO air sampling network. In *Trends: A compendium of data on global change*. Oak Ridge, TN: Carbon Dioxide Information Analysis Center, Oak Ridge National Laboratory, U.S. Department of Energy.

Knapp, A.K., and M.D. Smith. 2001. Variation among biomes in temporal dynamics of aboveground primary production. *Science* 291: 481–484.

Knapp, A.K., P.A. Fay, J.M. Blair, S.L. Collins, M.D. Smith, et al. 2002. Rainfall variability, carbon cycling, and plant species diversity in a mesic grassland. *Science* 298: 2202–2205.

Knapp, A.K., C. Beier, D.D. Briske, A.T. Classen, Y. Luo, et al. 2008. Consequences of more extreme precipitation regimes for terrestrial ecosystems. *BioScience* 58: 811–821.

Koven, C.D., B. Ringeval, P. Friedlingstein, P. Ciaia, P. Cadule, et al. 2011. Permafrost carbon-climate feedbacks accelerate global warming. *Proceedings of the National Academy of Sciences of the United States of America* 108: 14769–14774.

Ladwig, L.J., S.L. Collins, A.L. Swann, Y. Xia, M.F. Allen, et al. 2012. Above- and belowground responses to nitrogen addition in a Chihuahuan Desert grassland. *Oecologia* 169: 177–185.

League, K., and T. Veblen. 2006. Climatic variability and episodic *Pinus ponderosa* establishment along the forest-grassland ecotones of Colorado. *Forest Ecology and Management* 228: 98–107.

Lieth, H. 1973. Primary production: Terrestrial ecosystems. *Human Ecology* 1: 303–332.

Lin, D., J. Xia, and S. Wan. 2010. Climate warming and biomass accumulation of terrestrial plants: A meta-analysis. *New Phytologist* 188: 187–198.

Lindenmayer, D.B., W. Steffen, A.A. Burbidge, L. Hughes, R.L. Kitching, et al. 2010. Conservation strategies in response to rapid climate change: Australia as a case study. *Biological Conservation* 143: 1587–1593.

Lindzen, R.S. 1999. The greenhouse effect and its problems. In *Climate policy after Kyoto*, ed. T.R. Gerholm. Brentwood, UK: Multi-Science Publishing Company.

Littell, J.S., D. McKenzie, D.L. Peterson, and A.L. Westerling. 2009. Climate and wildfire area burned in western U.S. ecoprovinces, 1916–2003. *Ecological Applications* 19: 1003–1021.

Luo, Y. 2007. Terrestrial carbon-cycle feedback to climate warming. *Annual Review of Ecology, Evolution, and Systematics* 38: 683–712.

Luo, Y., B. Su, W.S. Currie, J.A. Dukes, A. Finzi, et al. 2004. Progressive nitrogen limitation of ecosystem responses to rising atmospheric carbon dioxide. *BioScience* 54: 731–739.

Luo, Y., C.B. Field, and R.B. Jackson. 2006. Special feature: Does nitrogen constrain carbon cycling, or does carbon input stimulate nitrogen cycling? *Ecology* 87: 3–4.

Manion, P.D. 1981. *Tree disease concepts*. Englewood Cliffs, NJ: Prentice-Hall.

Martin, L.M., H.W. Polley, P.P. Daneshgar, M.A. Harris, and B.J. Wilsey. 2014. Biodiversity, photosynthetic mode, and ecosystem services differ between native and novel ecosystems. *Oecologia* 175: 687–697.

Matthews, H.D., and A.J. Weaver. 2010. Committed climate warming. *Nature Geoscience* 3: 142–143.

McCabe, G.J., and M.P. Clark. 2006. Shifting covariability of North American summer monsoon precipitation with antecedent winter precipitation. *International Journal of Climatology* 26: 991–999.

McCalley, C.K., and J.P. Sparks. 2008. Controls over nitric oxide and ammonia emissions from Mojave Desert soils. *Oecologia* 156: 871–881.

McCulley, R.L., I.C. Burke, and W.K. Lauenroth. 2009. Conservation of nitrogen increases with precipitation across a major grassland gradient in the Central Great Plains of North America. *Oecologia* 159: 571–581.

McDowell, R.E. 1968. Climate versus man and his animals. *Nature* 218: 641–645.

McKeon, G.M., G.S. Stone, J.I. Syktus, J.O. Carter, N.R. Flood, et al. 2009. Climate change impacts on northern Australian rangeland livestock carrying capacity: A review of issues. *Rangeland Journal* 31: 1–29.

Milchunas, D., A. Mosier, J. Morgan, D. LeCain, J. King, et al. 2005. Elevated CO_2 and defoliation effects on a shortgrass steppe: Forage quality versus quantity for ruminants. *Agriculture, Ecosystems & Environment* 111: 166–184.

Milton, S.J., M.A. du Plessis, and W.R. Siegfried. 1994. A conceptual model of arid rangeland degradation. *BioScience* 44: 70–76.

Moran, J. 1973. Heat tolerance of Brahman cross, buffalo, Banteng and Shorthorn steers during exposure to sun and as a result of exercise. *Australian Journal of Agricultural Research* 24: 775–782.

Morgan, J.A., D.R. Lecain, A.R. Mosier, and D.G. Milchunas. 2001. Elevated CO_2 enhances water relations and productivity and affects gas exchange in C_3 and C_4 grasses of the Colorado shortgrass steppe. *Global Change Biology* 7: 451–466.

Morgan, J.A., A.R. Moiser, D.G. Milchunas, D.R. Lecain, J.A. Nelson, et al. 2004a. CO_2 enhances productivity, alters species composition, and reduces digestibility of shortgrass steppe vegetation. *Ecological Applications* 14: 208–219.

Morgan, J.A., D.E. Pataki, C. Körner, H. Clark, S.J. Del Grosso, et al. 2004b. Water relations in grassland and desert ecosystems exposed to elevated atmospheric CO_2. *Oecologia* 140: 11–25.

Morgan, J.A., D.G. Milchunas, D.R. Lecain, M. West, and A.R. Mosier. 2007. Carbon dioxide enrichment alters plant community structure and accelerates shrub growth in the shortgrass steppe. *Proceedings of the National Academy of Sciences of the United States of America* 104: 14724–14729.

Morgan, J.A., D.R. Lecain, E. Pendall, D.M. Blumenthal, B.A. Kimball, et al. 2011. C_4 grasses prosper as carbon dioxide eliminates desiccation in warmed semi-arid grassland. *Nature* 476: 202–206.

Newton, P.C.D., M. Lieffering, A.J. Parsons, S.C. Brock, P.W. Theobald, et al. 2014. Selective grazing modifies previously anticipated responses of plant community composition to elevated CO_2 in a temperate grassland. *Global Change Biology* 20: 158–169.

Nguyen, L.M., M.P. Buttner, P. Cruz, S.D. Smith, and E.A. Robleto. 2011. Effects of elevated atmospheric CO_2 on rhizosphere soil microbial communities in a Mojave Desert ecosystem. *Journal of Arid Environments* 75: 917–925.

Niklaus, P.A., and C. Körner. 2004. Synthesis of a six year study of calcareous grassland responses to in situ CO_2 enrichment. *Ecological Monographs* 74: 491–511.

NOAA National Climatic Data Center. 2013. State of the Climate: National overview for annual 2012. Published online December 2012. http://www.ncdc.noaa.gov/sotc/national/2012/13. Accessed 23 November 2015.

Nowak, R.S., D.S. Ellsworth, and S.D. Smith. 2004. Functional responses of plants to elevated atmospheric CO_2—do photosynthetic and productivity data from FACE experiments support early predictions? *New Phytologist* 162: 253–280.

Owensby, C.E., L.M. Auen, and P.I. Coyne. 1994. Biomass production in a nitrogen-fertilized, tallgrass prairie ecosystem exposed to ambient and elevated levels of CO_2. *Plant and Soil* 165: 105–113.

Owensby, C.E., J.M. Ham, A.K. Knapp, and L.M. Auen. 1999. Biomass production and species composition change in a tallgrass prairie ecosystem after long-term exposure to elevated atmospheric CO_2. *Global Change Biology* 5: 497–506.

Oyarzún, M.P., A. Quiroz, and M.A. Birkett. 2008. Insecticide resistance in the horn fly: Alternative control strategies. *Medical and Veterinary Entomology* 22: 188–202.

Parmesan, C., and G. Yohe. 2003. A globally coherent fingerprint of climate change impacts across natural systems. *Nature* 421: 37–42.

Parton, W.J., J.A. Morgan, G. Wang, and S. Del Grosso. 2007. Projected ecosystem impact of the Prairie Heating and CO_2 Enrichment experiment. *New Phytologist* 174: 23–834.

Pechony, O., and D.T. Shindell. 2010. Driving forces of global wildfires over the past millennium and the forthcoming century. *Proceedings of the National Academy of Sciences of the United States of America* 107: 19167–19170.

Peterjohn, W.T., and W.H. Schlesinger. 1990. Nitrogen loss from deserts in the southwestern United States. *Biogeochemistry* 10: 67–79.

Polley, H.W., J.A. Morgan, and P.A. Fay. 2011. Application of a conceptual framework to interpret variability in rangeland responses to atmospheric CO_2 enrichment. *Journal of Agricultural Science* 149: 1–14.

Polley, H.W., V.L. Jin, and P.A. Fay. 2012. Feedback from plant species change amplifies CO_2 enhancement of grassland productivity. *Global Change Biology* 18: 2813–2823.

Polley, H.W., D.D. Briske, J.A. Morgan, K. Wolter, D.W. Bailey, et al. 2013. Climate change and North American rangelands: Trends, projections, and implications. *Rangeland Ecology and Management* 66: 493–511.

Ponce Campos, G.E., M.S. Moran, A. Huete, Y. Zhang, C. Bresloff, et al. 2013. Ecosystem resilience despite large-scale altered hydroclimatic conditions. *Nature* 494: 349–352.

Prayaga, K., N. Corbet, D. Johnston, M. Wolcott, G. Fordyce, et al. 2009. Genetics of adaptive traits in heifers and their relationship to growth, pubertal and carcass traits in two tropical beef cattle genotypes. *Animal Production Science* 49: 413–425.

Reich, P.B., and S.E. Hobbie. 2012. Decade-long soil nitrogen constraint on the CO_2 fertilization of plant biomass. *Nature Climate Change* 3: 278–282.

Reich, P.B., S.E. Hobbie, T. Lee, D.S. Ellsworth, J.B. West, et al. 2006. Nitrogen limitation constrains sustainability of ecosystem response to CO_2. *Nature* 440: 922–925.

Risser, P.G., E.C. Birney, H.D. Blocker, S.W. May, W.J. Parton, et al. 1981. *The true prairie ecosystem*. Stroudsburg, PA: Hutchinson Ross Publishing.

Rosenzweig, M.L. 1968. Net primary productivity of terrestrial communities: Predictions from climatological data. *American Naturalist* 102: 67–74.

Rötter, R., and V.D. Geijn. 1999. Climate change effects on plant growth, crop yield and livestock. *Climatic Change* 43: 651–681.

Rustad, L.E., J.L. Campbell, G.M. Marion, R.J. Norby, M.J. Mitchell, et al. 2001. A meta-analysis of the response of soil respiration, net nitrogen mineralization, and aboveground plant growth to experimental ecosystem warming. *Oecologia* 126: 543–562.

Sage, R.F. 1996. Modification of fire disturbance by elevated CO_2. In *Carbon dioxide, populations, and communities*, ed. C. Körner and F.A. Bazzaz. San Diego, CA: Academic Press.

Sala, O.E., W.J. Parton, L.A. Joyce, and W.K. Lauenroth. 1988. Primary production of the central grassland region of the United States. *Ecology* 69: 40–45.

Sala, O.E., L.A. Gherardi, L. Reichmann, E. Jobbagy, and D. Peters. 2012. Legacies of precipitation fluctuations on primary production: Theory and data synthesis. *Philosophical Transactions of the Royal Society B* 367: 3135–3144.

Schaeffer, S.M., S.A. Billings, and R.D. Evans. 2003. Responses of soil nitrogen dynamics in a Mojave Desert ecosystem to manipulations in soil carbon and nitrogen availability. *Oecologia* 134: 547–553.

Schaeffer, S.M., S.A. Billings, R.D. Evans. 2007. Laboratory incubations reveal potential responses of soil nitrogen cycling to changes in soil C and N availability in Mojave Desert soils exposed to elevated atmospheric CO_2. *Global Change Biology* 13:854–865.

Seager, R., and G.A. Vecchi. 2010. Climate change and water in southwestern North America special feature: Greenhouse warming and the 21st century hydroclimate of southwestern North America. *Proceedings of the National Academy of Sciences of the United States of America* 107: 21277–21282.

Seastedt, T.R., J.M. Briggs, and D.J. Gibson. 1991. Controls on nitrogen limitation in tallgrass prairie. *Oecologia* 87: 72–79.

Sherry, R.A., E. Weng, J.A. Arnone, D.W. Johnson, D.S. Schimel, P.S. Verburg, L.L. Wallace, and Y. Luo. 2008. Lagged effects of experimental warming and doubled precipitation on annual and seasonal aboveground biomass production in a tallgrass prairie. *Global Change Biology* 14: 2923–2936.

Smith, D. 2013. Has global warming stalled? *Nature Climate Change* 3: 618–619.

Smith, S.D., and R.S. Nowak. 1990. Ecophysiology of plants in the Intermountain lowlands. In *Ecological studies*, eds. C.B. Osmond, L.F. Pitelka, and G.M. Hidy. Vol. 80. Plant biology of the basin and range. Heidelberg: Springer-Verlag.

Solomon, S., G.-K. Plattner, R. Knutti, and P. Friedlingstein. 2009. Irreversible climate change due to carbon dioxide emissions. *Proceedings of the National Academy of Sciences of the United States of America* 106: 1704–1709.

Stafford Smith, D.M., G.M. McKeon, I.W. Watson, B.K. Henry, G.S. Stone, et al. 2007. Learning from episodes of degradation and recovery in variable Australian rangelands. *Proceedings of the National Academy of Sciences of the United States of America* 104: 20690–20695.

Staver, A.C., S. Archibald, and S.A. Levin. 2011. The global extent and determinants of savanna and forest as alternative biome states. *Science* 334: 230–232.

Thurow, T.L., and C.A. Taylor Jr. 1999. Viewpoint: The role of drought in range management. *Journal of Range Management* 52: 413–419.

U.S. Climate Change Science Program [CCSP] . 2008. Synthesis and Assessment Product 4.3 (SAP 4.3): The effects of climate change on agriculture, land resources, water resources, and biodiversity in the United States. Washington, DC.

Utech, K., R. Wharton, and J. Kerr. 1978. Resistance to Boophilus microplus (Canestrini) in different breeds of cattle. *Australian Journal of Agricultural Research* 29: 885–895.

Vicente-Serrano, S.M., A. Zouber, T. Lasanta, and Y. Pueyo. 2012. Dryness is accelerating degradation of vulnerable shrublands in semiarid Mediterranean environments. *Ecological Monographs* 82: 407–428.

Volder, A., M.G. Tjoelker, and D.D. Briske. 2010. Contrasting physiological responsiveness of establishing trees and a C_4 grass to rainfall events, intensified summer drought, and warming in oak savanna. *Global Change Biology* 16: 3349–3362.

Walthall, C.L., J. Hatfield, P. Backlund, L. Lengnick, E. Marshall, et al. 2012. *Climate change and agriculture in the United States: Effects and adaptation*. Washington, DC: USDA Technical Bulletin 1935.

Walther, G.R. 2003. Plants in a warmer world. *Perspectives in Plant Ecology* 6(3): 169–185.

Walther, G.R. 2010. Community and ecosystem responses to recent climate change. *Philosophical Transactions of the Royal Society B* 365:2019–2024.

Wan, S., D. Hui, L.L. Wallace, and Y. Luo. 2005. Direct and indirect effects of experimental warming on ecosystem carbon processes in a tallgrass prairie. *Global Biogeochemical Cycles* 19: 2014. doi:10.1029/2004GB002315.

Wayman, O., H.D. Johnson, I.L. Berry, and C.P. Merilan. 1962. Effect of ad libitum or force-feeding of two rations on lactating dairy cows subject to temperature stress. *Journal of Dairy Science* 45: 1472–1478.

Weaver, J.E., and F.W. Albertson. 1943. Resurvey of grasses, forbs, and underground plant parts at the end of the great drought. *Ecological Monographs* 13: 63–117.

Webb, N.P., C.J. Stokes, and J.C. Scanlan. 2012. Interacting effects of vegetation, soils and management on the sensitivity of Australian savanna rangelands to climate change. *Climatic Change* 112: 925–943.

Westerling, A.L., H.G. Hidalgo, D.R. Cayan, and T.W. Swetnam. 2003. Warming and earlier spring increases western U.S. forest wildfire activity. *Science* 313: 940–943.

White, N. 2003. The vulnerability of the Australian beef industry to impacts of the cattle tick (*Boophilus microplus*) under climate change. *Climatic Change* 61: 157–190.

Williams, J.W., and S.T. Jackson. 2007. Novel climates, no-analog communities, and ecological surprises. *Frontiers in Ecology and the Environment* 5: 475–482.

Willms, W., S. Smoliak, and J. Dormaar. 1985. Effects of stocking rate on a rough fescue grassland vegetation. *Journal of Range Management* 38: 220–225.

Wu, Z., P. Dijksta, G.W. Koch, J. Peñuelas, and B.A. Hungate. 2011. Responses of terrestrial ecosystems to temperature and precipitation change: A meta-analysis of experimental manipulation. *Global Change Biology* 17: 927–942.

Wu, Z., P. Dijkstra, G.W. Koch, and B.A. Hungate. 2012. Biogeochemical and ecological feedbacks in grassland responses to warming. *Nature Climate Change* 2: 458–461.

Xu, X., R.A. Sherry, S. Niu, D. Li, and Y. Luo. 2013. Net primary productivity and rain-use efficiency as affected by warming, altered precipitation, and clipping in a mixed-grass prairie. *Global Change Biology* 19: 2753–2764.

Ziska, L.H., J.B. Reeves, and R.R. Blank. 2005. The impact of recent increases in atmospheric CO_2 on biomass production and vegetative retention of cheatgrass (*Bromus tectorum*): Implications for fire disturbance. *Global Change Biology* 11: 1325–1332.

Section II
Management

Chapter 8
Rangelands as Social–Ecological Systems

Tracy Hruska, Lynn Huntsinger, Mark Brunson, Wenjun Li,
Nadine Marshall, José L. Oviedo, and Hilary Whitcomb

Abstract A social–ecological system (SES) is a combination of social and ecological actors and processes that influence each other in profound ways. The SES framework is not a research methodology or a checklist to identify problems. It is a conceptual framework designed to keep both the social and ecological components of a system in focus so that the interactions between them can be scrutinized for drivers of change and causes of specific outcomes. Resilience, adaptability, and transformability have been identified as the three related attributes of SESs that determine their future trajectories. Identifying feedbacks between social and ecological components of the system at multiple scales is a key to SES-based analysis. This chapter explores the spectrum of different ways the concept has been used and defined, with a focus on its application to rangelands. Five cases of SES analysis are

T. Hruska • L. Huntsinger (✉)
Department of Environmental Science, Policy, and Management, University of California,
Berkeley, 310 Mulford Hall, Berkeley, CA 94720-3110, USA
e-mail: t.hruska@berkeley.edu; huntsinger@berkeley.edu

M. Brunson
Department of Environment & Society, Utah State University,
5215 Old Main Hill, Logan, UT 84322-5215, USA
e-mail: mark.brunson@usu.edu

W. Li
Department of Environmental Management, College of Environmental Sciences and
Engineering, Peking University, Beijing 100871, China
e-mail: wjlee@pku.edu.cn

N. Marshall
CSIRO, Land and Water, James Cook University,
ATSIP Building, Townsville, QLD Q4811, Australia
e-mail: nadine.marshall@csiro.au

J.L. Oviedo
Institute of Public Goods and Policies (IPP), Spanish National Research Council (CSIC),
Albasanz 26-28, E-28037 Madrid, Spain
e-mail: jose.oviedo@csic.es

H. Whitcomb
U.S. Fish & Wildlife Service, Utah Ecological Services Field Office,
2369 Orton Circle, Suite 50, West Valley City, UT 84119, USA
e-mail: Hilary_Whitcomb@fws.gov

© The Author(s) 2017
D.D. Briske (ed.), *Rangeland Systems*, Springer Series on Environmental
Management, DOI 10.1007/978-3-319-46709-2_8

263

presented from Australia, China, Spain, California, and the Great Basin of the USA. In each case, the SES framework facilitates identification of cross-system feedbacks to explain otherwise puzzling outcomes. While information intensive and logistically challenging in the management context, the SES framework can help overcome intractable challenges to working rangelands such as rangeland conversion and climate change. The primary benefit of the SES framework is the improved ability to prevent or correct social policies that cause negative ecological outcomes, and to achieve ecological objectives in ways that support, rather than hurt, rangeland users.

Keywords Endangered species • Ranch economics • Restoration • Resilience • Climate change • Complex adaptive systems

8.1 Introduction: What Is a Social–Ecological System?

The dependence of humans on natural systems makes it essential to understand how human use and management affect the capacity of ecosystems to sustainably support human needs. Yet, too often, social and ecological systems have been studied as if they operate independently. There is a critical need for comprehensive, multidisciplinary approaches to understanding the social and ecological components, interactions, and processes that shape rangeland conditions, including the social, economic, cultural, and political attributes of the people and communities within rangeland systems. Environmental problems arise from failures in social processes as much as from ecological processes, and recognizing this, a common framework is needed for understanding and analyzing the drivers that lead to improvement or deterioration of natural resources (Ostrom 2009). The "social–ecological system" (SES) concept provides a framework for analyzing complex rangeland dynamics and identifying interventions that can increase rangeland sustainability and support the production of desired goods and services. Here we explore a spectrum of different ways the concept has been used and defined in research on rangelands.

Humans alter natural systems in an effort to increase human benefit. Some changes are dramatic, such as cultivation for crop production, but others are less obvious, such as vegetation changes caused over time by extensive livestock grazing. Human systems react to ecosystem changes in many different ways, as with the economic, demographic, and policy responses to drought, wildfire, or deforestation. While range science has developed sound techniques for examining both the ecological and social components of rangelands, there has been little progress so far in seeing them as integrated and interdependent, or in developing techniques to resolve potentially competing goals (e.g., species conservation, open land access, economic benefits) within a rangeland SESs. Range ecology research has traditionally focused on grazing regimes and ecological indicators with less attention paid to the needs and goals of the livestock owner or property manager. This neglects real-world concerns about the finances, labor, limited time and information, and multiple goals of ranchers, pastoralists, and resource managers that influence *why* livestock are grazed in a certain way. Conversely, the social sciences have provided a wealth of information on who uses rangelands and why, but has been less successful at linking social, cultural, political, and ecological factors to ecological outcomes (Brunson 2012).

Purely technical interventions in rangelands often fail when researchers and managers do not consider their impacts on economic, political, cultural, and social well-being. To illustrate, introducing improved livestock to replace local breeds has at times been proposed to improve the livelihoods of pastoralists in developing countries. Some researchers have pointed out that improved livestock in such settings have had unintended consequences including increased financial risk, altered grazing patterns and gender roles, increased labor needs, and decreased income for women (Wangui 2008). Improving livestock breeds may do little to alleviate what might be the overarching problems of inadequate markets, government and industry land grabs, crop encroachment, and even climate change. In addition, the acceptability and practicality of a new technology for the people expected to use it must be considered. The goals of individual ranching enterprises may or may not mesh with those supposed by researchers and agency managers. In addition, drought, government policy, and livestock prices are external drivers affecting any proposed innovation at the ranch enterprise level, while personal beliefs and traditions, and family relationships, have implications for the ability and willingness to cope with change and adopt new technology. Rangeland research and management cannot afford to overlook human dimensions if the expectation is to contribute to the solution of real-world problems.

The social processes that sustain or degrade the ecosystem's current state, and the ecological processes that both drive ecosystem change and shape human use and benefits occur at multiple scales and are fraught with uncertainties. To improve the sustainability of natural resource use, managers need not only better or more complete ecological data, but also a clear understanding of where, when, and how resources are used and who gets to use them, and how and why use varies over time and across the landscape. The SES framework allows managers to treat all these interacting dynamics as part of a single integrated system (Fig. 8.1).

The notion that ecosystems and societies are shaped by one another is not a new idea (Norgaard 1994), but it has not been sufficiently emphasized in rangeland science. The edited book entitled "Linking Social and Ecological Systems" by Berkes and Folke (1998) was groundbreaking because it provided an integrated approach to simultaneously analyze both social and ecological systems for the purpose of natural resource management, and launched the term "social–ecological systems." Foundational work on the SES concept replaced the "view that resources can be treated as discrete entities in isolation from the rest of the ecosystem and the social system" (Berkes and Folke 1998, p. 2). Since the term emerged in the late 1990s, SESs have also, but less commonly, been called "coupled human-natural systems" to reflect the fact that both society and ecosystems have distinct internal dynamics

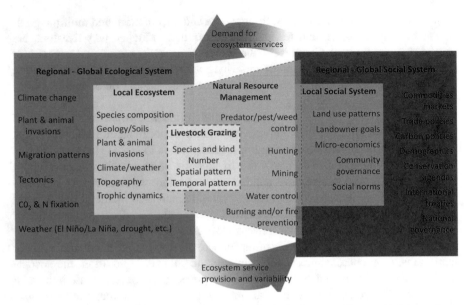

Fig. 8.1 Generalized diagram of a rangeland social–ecological system. Humans and the environment interact in countless ways outside of natural resource management, but the interactions are most directly planned, manipulated, and monitored in natural resource management activities. Local, regional, and global social processes can all shape natural resource use and management activities. While resource policy may be set at large geographical scales (e.g., national), management activities occur within a single ecosystem. Livestock grazing differs from other types of natural resource use in that it is indirect; rather than directly manipulating a rangeland ecosystem, livestock operators devote their primary attention to managing livestock, and the livestock interact directly with the rangeland (adapted from Reid et al. 2014)

but react in response to one another, sometimes in unanticipated ways (Liu et al. 2007a, b; Turner et al. 2003).

Much of SES research has focused on their resilience, describing various characteristics that allow an SES to persist and adapt to changing circumstances (e.g., Gunderson and Holling 2002; Berkes et al. 2003). This vein of SES scholarship is dominated by systems theory and treats SESs as complex adaptive systems that self-organize (e.g., Folke et al. 2005) and operate with feedbacks and thresholds (e.g., Walker et al. 2004). Drawn almost wholly from the natural sciences, this framing of SES has been critiqued by some social scientists on the grounds that such ecological principles cannot be so simply applied to social systems nor, by extension, to SESs (Olsson et al. 2015).

A challenge for applying the SES framework is in analyzing how social and ecological components of the system interact in iterative cycles. Too often, only single cross-system influences are emphasized in SESs, such as how changes in resource or social policy affect rangeland ecosystems, without following up to see how altered ecological processes feed back to affect the social system. While several conceptual models have been created for rangeland SES that might address this shortcoming (e.g., Fox et al. 2009; Walker et al. 2009), it is not always clear how to use them and they have not been widely applied.

8.1.1 Conceptualizing SESs

SESs are typically too large and complex to analyze all their structural and functional components at once. Creating a conceptual framework is one way of thinking through the complexity of SESs. The primary purpose of SES frameworks is the identification of specific components, processes, or feedbacks for analysis, and a metric for assessing their roles and interactions in the system. For example, who are the resource uses, and do they share information about the resource with each other? How far is a population from where policy and management decisions are made, and how valued is their input about resource use? Policy makers or resource managers can create their own frameworks in order to analyze their resource systems of interest. The key is identifying the important variables, the scales on which they operate, figuring out how they interact, and then measuring them over time. When problems arise, trying to solve the "why" will often entail finding unexpected connections between multiple components within the SES. Understanding how the SES has reacted to perturbations in the past can be of great help in this effort. At best, the use of SES frameworks will spotlight where interventions are needed or possible to achieve management goals, and will detect system changes over time in a way that allows for some level of prediction.

One SES framework originating from political and economic science is meant to allow identification of SES components and interactions within systems of resource use such as fish, groundwater, or pastures. This framework divides an SES into seven categories for analysis: resource systems, resource units, governance, users, interactions, outcomes, and related (or adjacent) ecosystems (Ostrom 2007, 2009). Each of the seven categories is then subdivided into a set of components in order to identify causal relationships and drivers, so that different systems can be compared. A different framework focuses on the exposure of an SES to a particular hazard, and then tracks sensitivity and resilience of both social and ecological components with the aim of analyzing vulnerability (Turner et al. 2003). A third, called the Drylands Development Paradigm, aims to synthesize lessons from research on desertification and economic development, and to act as a template whereby each of the five key principles of SESs can be examined and tested in case studies (Reynolds et al. 2007). The Resilience Alliance has created its own framework specifically for assessing the resilience of an SES (Resilience Alliance). Each of these frameworks is intended for a specific set of contexts and types of resource use, but the underlying assumptions about the interdependency of SESs are the same (for a review of ten SES frameworks, see Binder et al. 2013).

Researchers and managers tend to focus on the components of an SES most likely to be influenced by a given change or intervention, or perhaps those most amenable to analysis or management based on their own discipline, as the methods and theories of their own discipline are most familiar to them. Team approaches that include social as well as ecological scientists can help to assure a more comprehensive approach. Three key characteristics critical for analyzing an SES—scale, feedback, and resilience—may be difficult to recognize and measure, and thus may be overlooked. A rangeland SES is profoundly affected by attributes such as the system's geographical location, social context, governance structure, management dynamics, uses of natural resources, and economic relationships. These attributes can all be helpfully analyzed according to their scale, feedback, and resilience in the SES.

8.1.2 Scale

Understanding of ecosystems and their response to use and management has often been hampered by a failure to appreciate the role of *scale* (Cash et al. 2006). Each scale, such as spatial or temporal, may have different dominant patterns and processes at different hierarchical levels (Fig. 8.2). For example, in considering sustainability of rangelands, at the level of a rangeland ecological site, the selectivity and distribution of grazing animals may be critical. At the regional level, the price of real estate and zoning laws may be the critical factors in rangeland sustainability. Although they are not always, hierarchical levels may be nested. For example, if conservation does not occur at the regional level, then there may be no rangeland

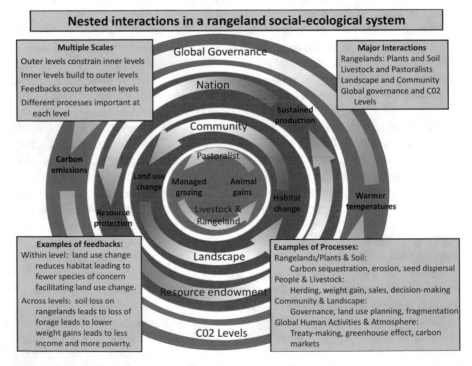

Fig. 8.2 An SES portrayed as a nested hierarchy illustrates how feedbacks occur across and within scales, and that different processes act as important system drivers at different scales. It also illustrates how some factors are largely outside the control of the pastoralist, for example national politics or drought and warming. On the other hand, outer levels are shaped by lower levels: without pastoralists, there is no pastoral community

ecological site left to manage (Chap. 5, this volume). Conversely, if ecological sites are not well managed, negative perceptions by the public may erode support for ranching, a possibility that lends support to the creation of regulations governing grazing use implemented at the regional or even national level. Temporal scale is similarly important, as processes may happen quickly or slowly, last only briefly, have a legacy effect influencing future processes, or persist for a very long time. Interactions among scales and levels are common and may cross social and ecological systems.

Unfortunately, monitoring or evaluating systems at multiple scales is often beyond the budget and knowledge of natural resource managers or scientific researchers. For this reason, some authors have recommended concentrating analysis on interactions of specific subcomponents (Roe et al. 1998), while others have suggested that social–ecological interactions are typically determined by a small number of "controlling variables" that should be the focal point of analysis (Holling 2001). In either case, the spatial or temporal scale of management may not be adequate to address the scale of ecological processes, a social–ecological mismatch leading either to mismanagement, ineffectual management, or an absence of management (Cumming et al. 2006). For example, a land manager might set a single stocking rate for an entire property based on average grass cover. Within that property, however, that stocking rate might result in overgrazing of some pastures and undergrazing in others. In this instance, the geographical level of management—the property—is too large for sustainable management of some individual pastures. On the other hand, some natural resource problems may occur at a larger level than the property, calling for a watershed-level approach that crosses property lines and involves understanding what drives cross-boundary cooperation. The challenge is to integrate and validate social and ecological data from multiple scales and levels when crafting policy and management prescriptions.

8.1.3 Feedbacks

A *feedback*, or feedback loop, is when a variable within a system changes in such a way that increases the likelihood and strength of further change (positive feedback) or decreases the likelihood or strength of future change (negative feedback). Positive feedback loops are self-reinforcing or amplifying, while negative feedback loops are self-regulating, or stabilizing. For example, conservation initiatives directed at ecological systems may alter the living situation or behavior of local social groups who might then increase or reduce their environmental impact as a result (Miller et al. 2012). As an example of a positive feedback loop, the sale of several ranches for residential development in an area formerly dominated by ranching can increase land values and alter community dynamics, causing more ranchers to sell their land to developers (Huntsinger 2009). Negative feedback loops dampen a particular effect or make an action less likely to be repeated. For example, in a natural system an overpopulation of herbivores reduces the forage available to each animal to the point where reproduction slows and mortality increases, lowering the number of herbivores. In SESs, interactions between hierarchical levels may function as feedbacks (Gunderson and Holling 2002).

From the perspective of rangeland managers or policy organizations, feedbacks may act as both vulnerabilities and opportunities. Where positive feedback loops have negative consequences, such as the conversion of ranchland mentioned previously, extra precautions should be taken to prevent those feedbacks from taking effect. Where feedback loops create positive change, short-term expenditures may be justified by long-term benefits. For example, establishing venues for stakeholder meetings and management collaboration may require investing additional time and money but result in steadily increasing participation that reduces management costs and improves outcomes in the long run. Whether positive or negative, recognizing the presence of feedback is a crucial step.

8.1.4 Resilience and Adaptability

Rangeland management and science have increasingly focused on *resilience* of rangeland ecosystems, including the resilience of social actors. Resilience can be defined as the capacity of a system to absorb disturbance and reorganize so as to retain essentially the same function, structure, identity, and feedbacks (Walker et al. 2004) (Chap. 6, this volume). Disturbances may originate in social or ecological subsystems and may occur slowly or rapidly (May 1977). A non-resilient SES may change or lose components and functionality when an unusual change, or perturbation, occurs in either the social or the ecological subsystem; a resilient SES will not only maintain function, but may also benefit from disturbance by reorganizing to further increase resilience (Gunderson and Holling 2002; Berkes et al. 2003). Resilient systems are those that can more readily adapt to new forces without losing functionality or transforming in fundamental ways. It is important to note that resilience is not an inherently good or bad quality. Degraded, unproductive rangelands or impoverished communities might be just as or even more resilient to change (i.e., improvement) than are more desirable and productive states (Cote and Nightingale 2012).

Resilience is not the same thing as stability. Stability is the ability of a system to return to an equilibrium state following a temporary disturbance (Holling 1973). Ecological stability has been challenged by alternative ecological models that reject the notion that ecosystems have a single equilibrium state (Westoby et al. 1989). Managers often attempt to create a stable flow of inputs and outputs from a managed system, because managing more variable systems requires frequent monitoring and the ability to rapidly alter resource-use patterns—both of which are expensive and difficult to carry out. Unfortunately, the resulting simplification of the managed system frequently results in reduced resilience (Holling 1973; Walker et al. 1981). The resilience concept does not preclude small changes or variation within the system, thus providing a better fit with dynamic, multi-equilibrium rangelands. For example, the establishment of stocking rates at a rangeland's perceived carrying capacity may be assumed to foster stable, sustainable livestock production. Such a steady-state view overlooks the impact of variable rainfall and temperature on forage production, which may lead to undesirable grazing outcomes in above- or below-average years and ultimately result in loss of ecosystem functionality (Chap. 6, this volume).

Analyses of integrated rangeland SESs have tended to view rangelands as *complex adaptive systems* that should be managed to enable adaptation to ecological and social change (Walker and Janssen 2002; Walker et al. 2009; Huber-Sannwald et al. 2012). Complex adaptive systems have many components that adapt or learn as they interact (Holland 1992) (Chap. 11, this volume). For example, cheatgrass (*Bromus tectorum*) invasion of the US Great Basin has resulted in ecosystem shifts away from dominance by perennial grasses and shrubs to dominance by cheatgrass, an annual species. This has influenced both the biological and human components of the ecosystem. The monocultural stands now common in North America facilitate the spread of a native generalist fungal pathogen called "black fingers of death" (*Pyrenophora semeniperda*) that colonizes cheatgrass stands across a broad distribution (Meyer et al. 2008). Livestock operators have had to adapt their grazing regimes to fit the timing of cheatgrass productivity and the periodic loss of forage caused by increasingly frequent wildfires. The transition to cheatgrass dominance has altered wildlife habitat and reduced the populations of some species, spurring conservation and restoration efforts. Cheatgrass is thought to be spread by livestock grazing, but grazing also serves to reduce cheatgrass biomass and thus the likelihood of damaging wildfires (Knapp 1996). Grazing, restoration efforts, and wildfires all interact in the production of ecosystem services. Interventions by livestock operators, range managers, and policy makers may have an effect, but the ultimate outcomes are difficult to predict given the complex ecologic and climatic factors involved.

The ability to cope with disturbance and respond to change has been termed *adaptive capacity* (Plummer and Armitage 2010). Within a given SES, adaptive capacity may vary at different scales, for different processes, and for different organizations and individuals. An individual or community with many diverse resources may be better able to adapt to change. A multispecies rangeland is usually better able to maintain productivity despite fluctuations in weather or drought, or the introduction of a plant disease, because some species will thrive better than others in the new conditions. Similarly, some people may have the flexibility of mind to adapt to new life conditions while others may not.

Adaptation does not only occur after singular, discrete perturbations, however. The dynamic nature of both ecosystems and society entails a constant state of change, meaning that adaptation is a continual, iterative process (Rammel et al. 2007). Change can originate in either society or the ecosystem and does not necessarily result in a functional, or successful, adaptation by the other system. People and institutions may not perceive change or the necessity of change, may be unwilling to change, may be unable to adapt successfully, or may change in a way that does not help. In society, the ability to adapt and the options available for adaptation are limited by power dynamics that are often overlooked or wishfully assumed to be less significant than they are. Adaptation should not be assumed to follow change, nor should it be assumed to be beneficial when it does (Watts 2015a). While it is common to hear of the need for society to adapt to climate change or other environmental forces in order to reduce vulnerability, it is rarely pointed out that the reason society is vulnerable and must adapt is because of the way the social–ecological landscape has developed. In many cases, society has created its own vulnerability to climate change and other environmental forces (Taylor 2014).

Ecological diversity and the presence of redundant components have also been highlighted as central for maintaining resilience (Walker 1995; Walker et al. 1999). While some theorists have proposed that a diversity of institutions and stakeholders in governance and management structures can benefit natural resource management, social scientists have questioned the extent to which such ecologically based notions can be extended to social systems. For example, some consolidated authoritarian regimes have proven to be remarkably resilient by monopolizing power and violently crushing any challenges (Agrawal 2005). Resilience was incorporated into ecology decades ago and is now ubiquitous in that field, but it has also emerged as a central feature in SES analyses, including rangeland SESs (Folke 2006; Reid et al. 2014). Despite the concept's recent prevalence in such institutions as the World Bank and the US military, many social scientists are critical about applying the concept in social contexts, including to SESs (Olsson et al. 2015). There are several key reasons for this critique.

First, by placing emphasis on resilience to disturbance rather than on disturbance itself, less attention is paid to the more politically sensitive questions of who is vulnerable and why, and how future disturbances might be avoided (Walker and Cooper 2011; Watts 2015b). Coming as it does largely from the natural sciences, resilience is often viewed as a rather mechanistic cause-and-effect process that does not account for human agency and goal formation (Davidson 2010). Furthermore, the formation of resilient livelihoods in SESs may be promoted by governments or other institutions but perceived by individuals or communities as radical, undesirable cultural change (Crane 2010). This is in part because what constitutes a social or environmental "problem" is highly subjective and frequently politically motivated (Castree 2001). The role of environmental shocks in driving social or SES change must be balanced by an awareness of the political and economic factors that create or allow "natural" disasters such as famines (Watts 1983). Despite these critiques, it is also true that the resilience concept has been adopted by many social movements around the world as a way to frame projects of social adaptation to new challenging circumstances (Brown 2014).

It must also be noted that the centrality of the system concept inherent in SES, resilience, and complex adaptive system frameworks is not without problems. The concept of a system inherently involves thinking about a multitude of components with coordinated actions and potentially even a unitary goal. In both human and ecological settings, who and what constitutes "the system" is by no means clear, and it would be inappropriate to assume that coordinated activity or collective goals are common outcomes of human interactions (Olsson et al. 2015).

8.2 Environmental Governance

The richly interconnected view of resource systems in an SES stands in opposition to strategies that attempt to reduce system complexity by focusing on only a small number of target resources, species, or indicators as is typical of maximum sustained yield and steady-state natural resource management (Holling and Meffe 1996). History has shown that these types of management strategies are often ecologically unsustainable because of unrecognized slow system change, sudden unpredicted disturbance, and/or unknown interconnections. On the social dimension, these approaches often fail as a result of an inability to understand what people want from natural resources, a lack of capacity to govern human resource use, and the broad perception of an accompanying policy or distribution of benefits as unjust.

Given the evolving nature of complex adaptive systems and their lack of predictability, much of the work on SESs in rangelands and elsewhere focuses on developing responsive policy and governance that supports system resilience (e.g., Walker et al. 2004; Armitage et al. 2009) rather than attempting to provide specific and relatively inflexible resource management prescriptions. Social groups do not maintain consistent or uniform relationships with their surrounding environment, but change in either social or ecological patterns cannot necessarily be attributed to a corresponding driver in the other system (Vayda and McCay 1975). Resource management and governance policies must therefore monitor and be responsive to ecological and social processes that may or may not create new drivers of change within the SES. It is the inclusion of both ecological and social variables within the frame of analysis that makes the SES framework useful for management. Changes in the price of beef or altered land tenure policies, for example, have to be considered alongside fluctuations in climate and vegetation composition when planning management actions or policies for a rangeland SES.

Problems that cross scales or levels within SESs can prove challenging for two reasons: *perception* and *communication*. First, the occurrence of a phenomenon at one level must be perceived as having been caused by a driver at another level. Second, that observation has to be communicated—persuasively—to the person or organization capable of solving the problem, and that person or entity has to decide to address the problem. Solutions involving changes in policy need to be effectively communicated to the affected population, ideally with buy-in from the affected populations. The perception problem can be met with a combination of thorough cross-scale monitoring and diverse information networks. Communication problems require adaptations to governance structures and strategies that facilitate information sharing and learning across sectors and hierarchical levels. An increasing number of groups, such as the Sustainable Rangelands Roundtable (http://www.sustainablerangelands.org/) and the California Rangeland Conservation Coalition (http://carangeland.org/) in the USA, are devoted to encouraging this type of communication about rangelands and range management.

Inclusion of various stakeholders in goal-setting, planning, monitoring, research, data interpretation, and decision making is one way that managers can create improved integration of ecosystem management with the social system, and gather more information about the system. Various models and terms have been created for this type of process, including community-based natural resource management (Leach et al. 1999) and adaptive comanagement (Olsson et al. 2004). Through the inclusion of multiple stakeholders, a project can gain access to information about the social needs and dynamics of the SES and to traditional and local knowledge about the ecosystem, which optimally increases the benefits of management to both the social and ecological components of the SES (Olsson et al. 2004). Engaging stakeholders can start to build consensus around an initiative, constructing the social networks needed for implementation and adaptation across the many dimensions of an SES. One model for a participatory approach to increasing SES resilience involves collaboration between many stakeholders to define the bounds of the SES and the trajectory of progress desired, followed by scientific study to determine how resilience can be maximized under such trajectories, and lastly a collaborative assessment of policy and management implications (Walker et al. 2004). In this process, stakeholders can provide information and insights that managers or scientists cannot, while networks and relationships are formed that can foster the iterative learning central to adaptation.

Some social and ecological problems occur at extensive spatial scales which only organizations with broad jurisdiction may be equipped to handle, such as regulating the migration of livestock herds under transhumance (Turner 2011). Furthermore, participatory approaches must be tailored to the specific management context and be flexible in response to social needs and the respective strengths of different stakeholders, which may mean employing different collaboration techniques and reaching out to different stakeholders at different times (Stringer et al. 2006). Increased stakeholder participation can also prevent making timely or difficult management decisions, particularly regarding the curtailment of resource use. For this reason it is best to adopt governance strategies that incorporate stakeholder input without causing decision-making stalemates. Providing a process to sanction the decision-making authority helps to ensure that decision makers who do make nonconsensual management decisions remain accountable for those decisions, hopefully leading to fully participatory negotiations and decisions most of the time (Lebel et al. 2006) (Chap. 11, this volume).

One hurdle in the way of improving the management of rangeland SESs is that regulatory policies are usually enacted on the premise that the problem faced is homogeneous across different times and places and that a single policy applied consistently will solve this problem in all locations. Unfortunately, rangeland problems are rarely so consistent and neither are the agencies tasked with implementing government policy. The single agency, single policy type of policy implementation is an example of *centralized governance*. Centralized systems assume that all information can be routed through a single office and that solutions can come from that same office. In contrast, *polycentric governance* systems have multiple locations for collecting data and issuing and carrying out management actions. Polycentric governance models do not rely on a single solution to a perceived single problem, but rather seek to coordinate activities working toward a common goal.

For example, rather than ordering a single government bureau to apply a herbicide to an invasive weed wherever it is found at the same time every year, a polycentric governance system might rely on some federal offices, some counties, and some local nonprofit groups to eradicate that same weed at the time and in the manner that work best in that area. Polycentric governance systems may be more difficult to coordinate logistically but they are more likely to account for local social and ecological differences in a manner that increases project efficacy (Nagendra and Ostrom 2012). Given their more diverse constituents, polycentric governance systems are also more open to different types of information than centralized systems, and may be more creative in finding solutions (Lebel et al. 2005). Polycentric systems may also find it more difficult to reach consensus among their constituents, which can slow down decision making and delay projects.

The recent turn toward adaptive management and comanagement models tends to feature government agencies, NGOs, and other institutions as the principle actors in SESs. This institutional bias risks excluding individuals or groups that lack the relevant job titles from having a voice in how resources are managed. This is especially problematic for politically marginalized groups, such as most mobile pastoralists, who may not be considered viable rangeland managers by governmental or international entities. The institutional bias in both the resilience and adaptive management frameworks works to overlook power imbalances between various stakeholders, encouraging the false assumption that resulting decisions are consensual. Even in community-based natural resource management models, which have been extensively deployed in sustainable development projects worldwide, the "community" is all too often assumed to be a singular, cohesive group with internally uniform characteristics and goals, when in reality this is seldom, if ever, true (Agrawal and Gibson 1999).

8.3 Case Studies

In the following case studies, each conducted by different authors of this chapter, the SES framework is used to focus on different aspects of the SES as they influence the adaptive capacity and resilience of the system. Some authors focus more on the ecological dynamics shaping the ability to adapt, while others are most attentive to the social components. The focal spatial scale ranges from entire regions in the USA's Great Basin and Australia to a couple of counties in California, down to the village scale in China, and finally to the scale of individual enterprises in Spain. Each study intends to improve understanding and support of the social and ecological drivers of resilient rangeland SESs.

8.3.1 Adaptation to Climate Change by Australian Livestock Managers

In northern Australia, climate change is expected to lead to increasingly dry conditions (Marshall 2010; Marshall and Stokes 2014). These changes are anticipated to be unprecedented—projections suggest that the scale and rate of change driven by increasing concentration of greenhouse gases in the atmosphere will significantly alter the distribution and quality of rangeland resources (IPCC 2014). The most likely climate future for the North based on RCP4.5 and RCP8.5 model projections for 2030 and 2090 suggests that temperatures will be warmer and hotter, respectively, but no "most likely" future with respect to rainfall is suggested (www.climatechangeinaustrala.gov.au). Although Australian rangelands have historically been highly resilient to a range of environmental, economic, and social changes, climate change in northern Australia is expected to reduce forage production, livestock profit margins, and biodiversity. Rangeland livestock operations are already struggling to maintain profitability because of recent drought conditions (Marshall and Stokes 2014). If managers and operators are not able to adapt to changing conditions, the extensive lands currently utilized for grazing might be converted to other, less valuable, states. Should grazing cease, extensive areas may transition to new ecological states that provide fewer or less valued ecosystem services. This means that sustaining rangeland landscapes in Australia is tied to the profitability of rangeland operations compared to the alternative possible uses. This study uses the SES framework to highlight how changes to the ecological system must be matched by adaptive changes in the social system in order to maintain the resilience of these pastoral SESs.

The future of the Australian rangeland SESs depends on the capacity of managers to sustainably manage rangelands, and the employment of managers is dependent on the condition of the rangelands (Marshall et al. 2011, 2014). Occupying some 70 % of the Australian landmass (Stafford Smith et al. 2007), rangelands are sparsely populated and of spatially and temporally variable productivity due to erratic rainfall. High variation in weather and seasons means that droughts are "normal" across the country, and drought declaration can occur more often than 3 years in 10 (McKeon et al. 2000). Livestock managers have had to cope with drought against an existing backdrop of conventional economic, biophysical, institutional, cultural, and political pressures and uncertainties (Howden et al. 2007). It is uncertain whether livestock operators have the adaptive capacity to adjust grazing practices to the altered conditions of a changed climate and remain both ecologically sustainable and economically viable (Marshall and Stokes 2014).

Environmental degradation on Australian rangelands can occur when, in an attempt to minimize the costs of a drought, livestock managers mismanage stocking rates, exacerbating pressures on already stressed grasslands (McKeon et al. 2004). One way for Australian livestock managers to adapt to climate change would be through making the most of good years and avoiding losses and reductions in resource condition in drought years (McKeon et al. 2004). Knowing when to alter stocking rates, when to supplement with outside feed, when to move livestock to other properties, when to burn, and when to alter water supplies, for example, can differentiate between those producers likely to be successful in the long term and those that are not (Hansen 2002). If stocking rates are too high at the onset of drought, for example, soil compaction and erosion will reduce productivity in future years (McKeon et al. 2004). In order to avoid damaging rangelands in bad times but reap rewards in good times, livestock managers have to remain flexible by having backup plans and the ability to quickly adapt grazing plans to match present conditions. They need to balance economic, environmental, and social trade-offs, and manage their system as an SES, rather than attempting to make the system profitable every year. However, not all managers in Australia have the vision or capacity to maintain rangeland resilience (Marshall and Smajgl 2013).

Adaptive capacity in people or organizations is typically associated with creativity and innovation (Holling 2001); testing and experimenting (Folke et al. 2005); effective feedback mechanisms (Adger et al. 2011; Cumming et al. 2005); adaptive management approaches (Briske et al. 2008); flexibility (Cumming et al. 2006); reorganizing given novel information (Marshall et al. 2013); managing risk (Howden et al. 2007); and having the necessary resources at hand (Marshall and Stokes 2014). These characteristics are critical at all scales. On Australian rangelands, the adaptive capacity of individual managers has been conceptualized and operationalized as comprising four main dimensions; (1) how risks and uncertainty are managed; (2) the extent of skills in planning, learning, and reorganizing; (3) financial and psychological flexibility to undertake change; and (4) anticipation of the need and willingness to contemplate and undertake change (Marshall 2010; Marshall et al. 2014). A livestock manager who ranks highly in all four dimensions is thought to be more able to adapt to changing circumstances, in other words possess greater adaptive capacity. These four dimensions have been used to examine the adaptive capacity of managers to sustainably manage rangelands (Marshall and Smajgl 2013). Based on a survey-based evaluation of these dimensions, only 16 % of managers across northern Australia have the capacity to meet the challenges of a changing climate, and the remaining majority may be unable to maintain successful grazing operations into the future (Marshall et al. 2014). Vulnerability was assessed as a function of both adaptive capacity and climate sensitivity, where managers who were assessed as more dependent on the grazing resource were assumed to be affected by smaller changes in local climate. The northern beef industry as a whole was regarded as vulnerable particularly because of poorly managed operational risk, weak support networks, and low strategic skills or interest in changing behavior by managers (Marshall et al. 2014).

The SES concept recognizes the link between the continuation of a specific ecological system and the continued socioeconomic viability of the livestock industry. By 2030, some areas of northern Australia will be experiencing more droughts and lower summer rainfall (Cobon et al. 2009). Livestock managers need support in accepting that they must adapt and in developing and implementing effective adaptations. Possible avenues for intervention might be in teaching managers about climate change, disseminating up-to-date climate and ecological data, determining appropriate stocking rates for new climatic conditions, assisting with financial tools to support rapid sales or purchases of livestock when conditions change, and improving the monitoring strategies or adaptability of grazing plans more generally. By providing knowledge of the different types of vulnerability of resource users, vulnerability assessments can enable decision makers to prioritize their efforts, provide a basis for early engagement, and tailor a range of adaptation approaches to most effectively accommodate and support the divergent requirements of the different categories of resource users. Given the coupling of social and ecological systems, maintaining rangeland resilience across scales by supporting human adaptation processes is likely to be an essential strategy for adapting to the challenges of the future.

8.3.2   Climate Change and Forb Restoration in the Great Basin, USA

The SES framework was used to understand factors that impeded the use of herbaceous broad-leaved plants, or forbs, in restoration of cheatgrass (*Bromus tectorum*) invaded Great Basin sagebrush steppe ecosystems of the western USA. Most of these ecosystems are managed by federal agencies, in particular the Bureau of Land Management. Forbs are an important component of biodiversity in these ecosystems (West 1993) and increasing native forb species richness can enhance resistance to invasive plants (Pokorny et al. 2005) including cheatgrass. By providing fine-textured, combustible fuels, cheatgrass increases susceptibility to wildfires, and wildfires have been growing in frequency and severity across the Great Basin (Brooks et al. 2004), a trend that is expected to continue as a result of climate change (Abatzoglou and Kolden 2011). Yet when rangeland managers choose seed mixes for restoring native plant communities after a wildfire, forbs are often underutilized. While reduced forb abundance after wildfires is a local- to regional-scale ecological issue, applying an SES framework revealed that it stems partly from higher level processes that affect agency budget choices, as well as individual variation in how managers perceive and interpret scientific information about rangelands and climate change. The SES framework accounts for factors, relationships, and feedbacks among scales that influence the relationship between forb restoration, climate change, invasive plants, and manager decision making. An SES-based analysis of key drivers of manager decision making helped to understand the limitations to the adaptive capacity of managers.

Land managers may know which plant species to reseed after wildfire to suit past conditions, but predictions of future climate in these regions suggest more variable and extreme weather events, longer droughts, and increasing summer high temperatures (Ackerly et al. 2010; Polley et al. 2013). Part of this study was to evaluate the effect of summer warming on forbs to test the assumption that forb species choices for postfire rangeland seedings might need to be adapted to suit future climate conditions (Whitcomb 2011). Summer air temperatures at the soil surface are predicted to increase +4.5 to +6 °C in the Great Basin by the year 2100 (Jiang et al. 2013). A field experiment was conducted over 2 years at an experiment station near Logan, Utah, in which selected native and non-native forbs were grown to test their responses to increases in air temperature (Post and Pederson 2008). As hypothesized, the different plant species responded differently to warming, indicating that changes in species fitness and ultimately composition under warming conditions are likely. If managers are to effectively implement postfire seeding practices for these new conditions, they need to have the adaptive capacity to try new seed mixes, despite concerns about costs and uncertainty about propagation (Sheley and Half 2006).

Land management decisions were examined in order to assess the interactions and factors shaping postfire rehabilitation practices. Most Great Basin public land is managed by natural resource professionals who are expected to be responsive to the interests of the public and to use scientific information, admittedly in short supply, to manage sustainably. In this case, information about the prospect of climate change and the response of different species *should* have driven managers to choose rehabilitation methods that anticipate climate change effects on the ecosystem. The available climate and ecological information suggests that forb rehabilitation should be prioritized in management decisions, and that using seed mixes that are adapted for climate change will increase the likelihood of diverse forb communities over the long term. Yet knowing which species are more likely to survive in a warmer climate is only part of the management picture.

Using the SES framework it became apparent that managers' attitudes toward using available scientific information were influenced by broader scale US political debates about the existence, causes, and appropriate response to climate change. Research has shown that the best predictor of viewpoints about climate change is personal value orientation (Leiserowitz 2006). Managers employed by government agencies are partially influenced by the policies and norms of the agencies that employ them, but personal values also can affect management decisions (Richards and Huntsinger 1994). To understand how these social factors influence rangeland rehabilitation decisions in a sagebrush steppe SES, managers employed by various agencies across the region were interviewed regarding their opinions whether local weather events are indicators of larger climate trends; their concerns about the risks associated with climate change in their jurisdictions; current management activities to address future climate predictions; and perceptions about the role of forbs in ecosystem resilience.

Insights from 20 usable interviews conducted in May 2010 found that managers may not use available data about temperature changes or forb responses when choosing species for seed mixes. Thirty-year climate data showed that precipitation had declined at 18 of the 20 locations where the interviewees worked, with an average decrease of 12 %, and maximum temperatures overall had increased. Yet when asked whether the climate was changing locally or not, only about half had noticed changes. Those who thought that the climate was changing typically had spent more time in that location than those who did not think so. This finding may indicate that managers with local experience based their answers on personal experience, while more recent arrivals relied instead on general beliefs about climate change. Managers in both groups stated that while their organizations had policies in place that encouraged consideration of climate change in management, they were hesitant to do so without more specific guidance about how to use climate change information in their decision making. This range of responses demonstrates the complexity of managerial decision making and the unpredictable array of variables that influence adaptive capacity.

Further limitations to adaptive capacity were revealed in the interviews. Some managers were uncertain about the role and status of native forbs in their jurisdictions. Most reported using custom seed mixes that included native forbs as well as grasses, but forb diversity was low with only one or two species included, typically due to the generally high cost of forb seed. As wildfires become more frequent and severe across the region, managers struggle to obtain the resources needed to keep up with postfire rehabilitation needs. Budget shortages also inhibit the ability to take the risks needed to successfully adapt to changing conditions. Together with a lack of firm conviction about the occurrence of climate change, a choice *not* to change practices could be easily made.

Considered in its entirety, analysis of the social context suggests that many rangeland managers were unprepared to adapt to climate change when implementing postfire rehabilitation seedings. Over time, such a failure in adaptability, if it continues, could lead to the transition of more areas of sagebrush steppe to alternative ecological states, which in turn would affect land-use practices by local communities. The SES framework made it possible to examine how local land management practices are affected by large-scale social and ecological forces that do not seem directly related, but are linked and mediated through manager perceptions.

One might conclude that the key to changing seed choices is to influence manager beliefs about the importance of forbs to ecosystems and the reality of climate change. Yet climate change beliefs are highly related to personal values, and value-based attitudes are highly resistant to change (Eagly and Kulesa 1997). Manager beliefs are also shaped partially by prevailing opinions in the local community (Kennedy et al. 2001), and these may be even more resistant to change. A more fruitful intervention might be to provide specific agency-wide guidance for the use of new seed mixes, framing the need not in terms of climate change, but as related to problems managers experience directly such as non-native species invasions, higher fire frequencies, and drought. Facilitating communication between managers who are actively preparing for climate change and those who are not may clarify the benefits of adaptation measures and enhance adaptive capacity. Increasing budgets to increase purchasing power and devoting more resources to identifying new seed sources and seeding technologies would also help to improve manager ability to use native forb seeds effectively in future conditions.

8.3.3 California Black Rail Habitat in the Sierra Nevada Foothills

Concerns about the welfare of a rare bird, the California black rail (*Laterallus jamaicensis coturniculus*), led to a study of the SES that sustains the small wetlands that are its primary habitat in the Sierra Nevada foothills. More than two-thirds of the wetlands in the area are fed primarily by irrigation water, either by irrigation runoff or through leaks in earthen irrigation canals and ditches, and are scattered within grazed annual grasslands that are mostly in private ownership. Wetlands fed by irrigation water are also more consistently wet and had greater bird use than those subject to seasonal water variations (Richmond et al. 2010). Designing the study and analyzing research results using an SES framework revealed that many wetlands are functionally "accidental" and have little impact on land use or productivity from the landowner perspective. They are largely ignored by landowners, and while this benign neglect is to some degree why they have served as black rail habitat for decades, changing environmental and economic conditions could lead to their demise. In this study, the SES framework linked factors outside of the land manager-ecosystem relationship to strong impacts on the potential for conserving rail habitat, and revealed a need for governance that facilitates feedbacks from rail habitat conditions to water districts.

The secretive black rail is a small ground-dwelling marsh bird, and was known only from large marshes in San Francisco Bay and along the lower Colorado River until it was "discovered" in the Sierra foothills of Yuba, Nevada, Placer, and Butte counties in 1994 (Richmond et al. 2008). The SES framework enabled researchers to conceptualize and model the ecosystem service of rail habitat provision as a product of the interaction of humans and environment, rather than a service provided by the ecosystem alone (Huntsinger and Oviedo 2014). Researchers hypothesized that the interaction of landowners and environment is driven mostly by water scarcity, fears of mosquito-related illness, ranching activities, water price, and landowner goals for their land (Fig. 8.3).

To understand how landowner decisions influenced black rail habitat, landowners within the bird's habitat distribution were surveyed about water and land management goals and practices in 2014. Results showed that about half the landowners purchased irrigation water from a water district. Water districts are local government institutions that supply water to farms and homes in a rural area. They typically serve hundreds to thousands of properties. While many respondents reported having a small wetland that could be rail habitat on their property, few survey respondents reported any management of such wetlands, with about 9 % reporting draining a wetland in the last 5 years, and 9 % reporting that they created a wetland during the last 5 years. About half said that they valued wetlands as wildlife habitat, about a quarter thought that the green forage was useful for livestock, but about a quarter reported not doing any management because the wetlands simply did not "bother" them.

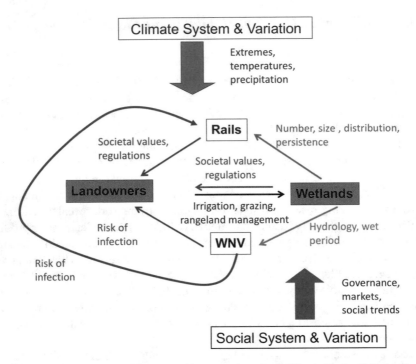

Fig. 8.3 Initial model of the SES proposed by researchers for California black rail in the Sierran foothills (Hruska et al. 2015). Much of the focus was on the relationship between landowners and wetlands. While the focus of this excerpt is on the social element, researchers also studied bird ecology, epidemiology, and hydrology as part of the study

The history of real estate appreciation in the area has also influenced water use. Because of strong competition for water allocations from water districts, it is often difficult to get a new allocation or increase an old one, but once granted, allocations are rarely taken away. As a result, landowners who get an allocation keep purchasing that amount of water every year, whether they can use it or not, to avoid losing their allocation. Having an allocation makes a property more valuable. Under these conditions there is little motivation to conserve water. Despite the fact that California was in the third year of severe drought, in 2013–2014 water district water purchasers were more likely than non-purchasers to respond in the survey that they had plenty of water for their property. Water districts buffer the drought for their customers, maintaining existing flow largely without reductions. Using water district water apparently changes the timing and nature of feedbacks to management from drought—it took 4 years of drought before water districts began cutbacks and landowners felt the impacts in our study area. While the lack of reduced water use during drought is a problem from a water conservation standpoint, it is positive for maintaining rail habitat.

Researchers learned that water districts and their policies have an extraordinarily important role in determining how people use water, especially during droughts.

This finding came despite the fact that, because they are so diverse, numerous, and little understood, water districts were not included in the initially proposed SES. The two most common actions respondents said that they would take if water districts increased prices substantially would be to reduce or cancel water purchases, or to reduce or eliminate irrigated pasture, two actions that would strongly affect rail habitat. Similarly, many respondents reported that they would reduce the size of irrigated pastures or decrease their irrigation frequency if water districts provided less water. Interviews with water districts revealed that while there are feedbacks between landowners and water districts, there seem to be no feedbacks from wetlands to water districts—in general, water districts have no legal or political motivation to consider the impacts on habitat from water conservation or water delivery practices (Fig. 8.4). In addition, state policy is encouraging water districts to conserve water, and districts are now making substantial investments in their infrastructure to prevent leaks and seeps, both water sources that create rail habitat. Water conservation efforts throughout California will be translated to many land-owners and wetlands via the water districts.

The SES framework made crucial "weak links" in sustaining habitat for the rail quite clear: state water policy, local water districts, and landowners are unaware of

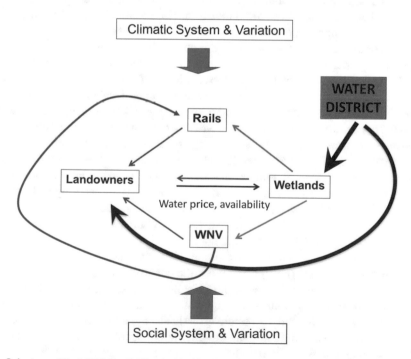

Fig. 8.4 A modified SES for California black rail habitat in the Sierran Foothills. New, critical players were identified as the study progressed and the areas where interventions would be important were located. Landowners only have an indirect effect on water districts, and at a different scale—in the aggregate. Water districts are directly affected by climate and regulations, both larger scale processes

any reason to consider impacts of management decisions on small wetlands. In conclusion, maintaining the resilience of the SES will require finding points of leverage for influencing water district actions, a process that will involve changes in governance. Given that most rail habitat is on private land, improving the resilience of wetlands must also incorporate outreach to landowners and water districts, a process that would require collaboration between multiple organizations throughout the SES. The fact that landowners expressed a strong interest in wildlife will help guide outreach activities.

8.3.4 Nomad Sedentarization Project in Xinjiang, China

Grassland covers 41.7 % of China and is home to some 17 million registered pastoralists and agro-pastoralists. Most pastoralists are ethnic minorities that have traditionally moved mixed herds of livestock up and down an elevation gradient on a seasonal basis, or across large distances to avoid drought and seek good weather and range conditions. Mobile livestock management buffers the spatially and temporally variable conditions in arid rangelands and is deeply intertwined with social and cultural practice and traditions (Roe et al. 1998; Li and Huntsinger 2011). Mobility and opportunistic grazing, common adaptive strategies in arid land pastoralism, are important components of the resilience that has enabled pastoralists to persist in environments with unpredictable forage production. The SES framework can be used to assess the resilience of pastoralist SESs in response to development policies, in this case sedentarization projects that decrease mobility of livestock herds. State-driven nomad sedentarization projects in China are intended to improve household income while decreasing grazing pressure on local grasslands (Harris 2010). By examining sedentarization at multiple spatial scales—village, county, and water catchment—researchers found that these projects have met objectives at some scales and in some locations but not others. An SES approach revealed how new patterns of resource use created by sedentarization policies have had significant environmental consequences, weakening the resilience of pastoralist communities in the study area within Xinjiang Uyghur Autonomous Region (hereafter Xinjiang).

Since 2006, the Chinese Government has enacted a series of Nomad Sedentarization Projects (NSP) throughout the country's six largest pastoral areas. The NSPs in China are designed to provide improved social services to herders, including construction of houses with tap water and electricity, and development of alternative livelihoods, and to restore grasslands by reducing grazing pressure through decreased stocking rates and the planting of supplemental fodder near settlements. The projects are funded directly by the central government, with annual budgets sometimes exceeding the equivalent of 200 million US dollars (Ministry of Finance 2011). In contrast to previous studies of sedentarization that focused only on individual villages, this study sought to examine the effects of sedentarization on the pastoral SES across different spatial scales: economic and rangeland conditions at

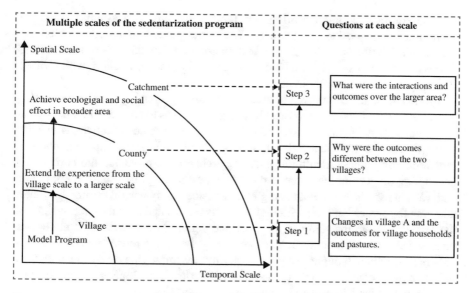

Fig. 8.5 Questions developed based on the SES framework for two sedentarized villages in Xinjiang Province, China (from Fan et al. 2014)

the village scale, social and economic processes at the county scale, and ecological processes at the catchment scale (Fig. 8.5). This allows analysis connecting management impacts at one scale to unexpected consequences or feedbacks to smaller or larger scales (Pelosi et al. 2010).

In Jinghe County of Xinjiang, sedentarization of mobile pastoralists began in the late 1990s and was completed in 2007, by which time 98 % of pastoral households had settled. Jinghe County lies within the Ebinur Lake catchment, and the two study villages are both within the county and within 50 km of the lake's shore. Given the low precipitation at the catchment floor of only 60–80 mm per year, the snowmelt-fed lake and associated wetlands play an important role in sustaining the regional ecology.

Researchers surveyed herder households in two villages, here referred to as Village A and Village B, to document household income and herder opinions on sedentarization's effect on grassland health and livelihoods. Prior to sedentarization, all households had annually moved livestock through four seasonal pastures and had similar standards of living, though households in Village A owned more livestock (406 ± 142) than those in Village B (308 ± 142). As part of sedentarization, all households built permanent homes in their former autumn pastures and were allotted adjacent land for cultivation. Access to traditional pastures was limited, allowing for only a two-season (summer–winter) rotational cycle, with livestock spending more time near residences. Importantly, Village A was allotted 13.4 ha of private land per household by the government whereas Village B received only 5.4 ha per household due to a new protected area nearby. Though livestock husbandry still accounted for approximately three-quarters of total income, new income sources included renting out land, government grassland subsidies, agriculture, and outside employment. While the pattern of sedentarization was quite similar for the two villages, its effect on household livelihoods was not. Village A was a local success story and held up as a model of modernization, while Village B struggled to meet household needs.

In Village A, all 23 surveyed households reported preferring their new sedentary life because they liked the improved housing and access to services, and the overall "more convenient" lifestyle. Nearly all had increased livestock numbers, with a new average herd size of 1002 ± 548. These larger herds were given supplemental feed grown on household agricultural plots or purchased from outside the community, decoupling them from the variable rangeland productivity. Households built barns and warming sheds that maintained greater livestock body weight during the winter, allowing for earlier lambing with higher survival rates and heavier lamb weights at the time of sale. In the summer, livestock were grazed on traditional summer pastures and also on summer pastures rented from neighboring townships. Despite this, only one household thought that pasture quality was improving over time, with most linking pasture condition to rainfall. At the village level, ecological conditions were not believed to have improved because of the policy.

Sedentarization in Village B was significantly different. Due to smaller and more dispersed household allotments, few households were able to irrigate their land or rent it out to professional farmers. Village B was unable to rent additional summer pasture land, as Village A had. Households could not increase their livestock numbers, and average income is now 50 % of that of Village A. Households in Village B were unable to make comparable investments in infrastructure and supplemental feeds. Many households required bank loans just to meet household expenses. Thus, while both villages turned to agriculture and supplemental food and settled in permanent housing, the two villages had significantly different outcomes in terms of household income, herd size, and use of irrigation. At the county level, the sedentarization created inequitable outcomes among villages which could ultimately destabilize the social system.

At the catchment level, there have been dramatically increased rates of groundwater withdrawal for irrigated agriculture, especially cotton. The nearby Ebinur Lake is shrinking rapidly and local river flows are decreasing or disappearing entirely. The lake now has half the surface area that it had in 1950, with steady declines marked since 2003, correlating with the increased area under cultivation (Sun and Gao 2010). Human activity is held responsible, with most of the water used for crops (Qian et al. 2004; Cheng and Hong 2011). Survey respondents remarked that it was becoming increasingly difficult to get drinking water from shallow wells, as the local water table was dropping. Sedentarization has had significant ecological impacts at the catchment scale that may undermine the resilience of the SES of all villages in the watershed, even the more successful Village A.

The SES framework reveals that environmental policy has both social and ecological effects, and that they may be different at different spatial scales. Ecologically, water limitations at the catchment scale seem likely to feed back to the village level, making the current agricultural uses that resulted from sedentarization unsustainable.

Declining pasture conditions may lead to more reliance on irrigated crops, feeding back to increased water demand, and worsening water loss at the catchment level. The disparity in economic impacts apparent at the county level may destabilize the SES socially by creating feelings of inequity at the village level. Sedentarization has undermined the resilience of these pastoral communities by generating social and ecological tensions at multiple spatial scales. By becoming aware of the interconnections of social and ecological systems, and considering outcomes at multiple spatial scales, managers in this case would be better equipped to establish development policies that sustain households and villages, pastures, and watersheds.

8.3.5 Environmental Accounting for Spanish Private Dehesa Properties

Oak woodland *dehesa* is an ancient and extensive agro-sylvo-pastoral system in southwestern Spain's Mediterranean climate zone that produces multiple products, including cork, acorns, and wood from oaks; forage for diverse breeds of cattle, sheep, goats, and bees; habitat for game and mushrooms; recreation and scenery; and acorns for Iberian pigs. The characteristic pattern of well-spaced oaks with a mostly herbaceous understory is shaped by human management. Landowners enjoy many nonmarket ecosystem services (also called private amenities) from the land, including a beautiful setting, recreation, the status of owning a large property, hunting, a traditional lifestyle, the rewards and challenges of stewardship, and the possibility of passing the property on to their heirs. Woodland ecosystems actively managed as dehesa have notably high biodiversity, higher than similar systems under alternate land uses (Bugalho et al. 2011). Dehesa is threatened by abandonment because of the low prices for commodities such as cork and by competition from agricultural intensification and development. The SES framework is applied here to explore the feedbacks between the environment and the individual dehesa enterprise to understand factors shaping the persistence of the dehesa system at the household level.

   Like ranches in the USA (Oviedo et al. 2013), dehesa properties command higher prices than can be explained solely by income from commercial production. Commercial income is often low and governmental subsidies supplement the operations. However, the cultural and ecological nonmarket ecosystem services consumed by dehesa owners partially explain why they chose to pay expensive dehesa land prices, just as they have been used to explain why ranchers in the USA persist in ranching when other investment choices might show greater monetary returns (Smith and Martin 1972; Oviedo et al. 2012) (Chap. 14, this volume). In this study, researchers sought to quantify these nonmarket landowner benefits in order to determine how much dehesa owners "earn" from nonmarket ecosystem services compared to commercial income. This comparison would make it possible to consider these services, as motivators for land ownership, when making land use policy. In this case study, an "agroforestry accounting system" (Campos 2000) is used at the property level to monetize some of the nonmarket benefits to the landowner from dehesa, and to place them into a fuller accounting of income to the landowner. Presented here is a summary from the detailed study of Oviedo et al. (2015a, b).

The Agroforestry Accounting System (AAS) is a framework developed to overcome the limitations of conventional income accounting by incorporating, among other things, ecosystem amenity benefit streams in economic analysis (Campos 2000). Though this approach can be used to estimate the economic value of the ecosystem to society, here we focus on benefits to the landowner as the focal environmental feedback to the SES. The AAS includes capital gains, with land appreciation as its main component. Land appreciation is part of landowner income because the landowner will realize this value when the property is eventually sold. As with land valuation more broadly, land appreciation of dehesa is better explained by nonmarket ecosystem service amenities available to dehesa owners than by commercial income potential (Oviedo et al. 2015b). The study differs from previous economic analyses of dehesa and other agroforestry ecosystems in that both market commodities and nonmarket ecosystem services are calculated together to create a more complete ledger of total income equivalents for dehesa owners.

The "ecosystem services" concept originated with the idea that land-use decision making would be improved if the nonmarket benefits from ecosystems could be quantified in monetary terms (Chap. 14, this volume). For example, a municipality might develop a forest watershed and gain tax dollars, but it might trade off substantial water filtration and provision services, erosion control, and recreation opportunities. Recreation and water are sold in various markets, and the prices can be used to generate an estimate of those monetary values, but erosion risk, for example, is difficult to valuate. "Nonmarket" benefits such as aesthetic beauty or cultural heritage values are even more difficult to monetize. One method often used is "contingent valuation analysis" (CVA), where people who benefit from a particular environmental feature, such as a stand of trees used for walking or communing with nature, are asked how much money they would be willing to pay to maintain the existence of, or receive in compensation for the loss of, that feature (Mitchell and Carson 1989; Campos et al. 2009).

Data from three different sources was integrated in the AAS methodology. One source was a contingent valuation survey of 765 landowners used to obtain estimates of the value of nonmarket ecosystem service benefits to landowners (Oviedo et al. 2015b). A second source was the nominal cumulative land revaluation rate for Spanish dry natural grassland for the period 1994–2010 (MARM 2011), used to roughly approximate land appreciation value. Third, commercial income data for three sample dehesas was gathered from account books, in-depth interviews, and field data (Oviedo et al. 2015a). Government net subsidies were not included here because they are temporary, and depend largely on the varying European economic context.

In all three dehesas, commercial activities alone result in negative operating income. Capital gains from land appreciation are positive in all cases and make up some, but not all, of the difference between expenditures and revenue. The inclusion of the nonmarket ecosystem services consumed by dehesa owners and quantified by CVA makes the income for all three dehesas positive, and in fact makes up a higher share of total income than commercial activities. As the principal drivers of appreciating land values, landowner nonmarket ecosystem services are doubly important in this accounting.

When the low or negative commercial income, positive income from capital gains, and value of owner-consumed ecosystem services are factored together, real total profitability ranges from 3.2 to 5.6 % (Oviedo et al. 2015a). If conventional income accounting were applied, the feedback from the ecosystem to the landowner from nonmarket ecosystem services would be overlooked, and our understanding of the interactions, or feedbacks, between landowners and the dehesa ecosystem would be less complete.

Dehesas would not be "profitable" for landowners without the ecosystem services they provide, making nonmarket ecosystem services consumed by the landowner a positive feedback that should strengthen the landowner's bond to the property. As these are nonmarket benefits and do not produce cash, the landowner must be able to afford, and be willing to pay, property expenses in order to acquire them. This capacity is key to the ecological sustainability of dehesas and implies that dehesas must be owned by people either with substantial savings or with monetary income from other sources. When considered in the larger portfolio of income streams from dehesas, having nonmarket sources diversifies the operation, increasing the resilience of ownership in the face of unpredictable and changing markets for the more tangible products.

The translation of nonmarket benefits to the landowner into monetary terms allows us to better understand when subsidies or other interventions are needed to maintain the integrated "profitability" that motivates landowner choice in this SES. It also highlights the need to maintain the production of landowner-consumable ecosystem services in order to motivate ownership and management investment. This is critical in Spain, for while land use controls inhibit conversion of dehesa to other uses, they do not sustain the active dehesa management necessary to preserve the considerable benefits of the system to Spanish society, including wildlife habitat, carbon sequestration, watershed protection, and scenery. The governmental subsidies that are provided to landowners reflect an appreciation of these values. An open question is whether any of these nonmarket goods and services can eventually find a nongovernmental market (Caparrós et al. 2013). Environmental accounting provides insight into how regulations, social pressures, significant ecological change, or other factors that reduce the nonmarket ecosystem services a landowner can consume from their property may put the dehesa at risk as much or more than low commercial profits.

8.4   What Can Be Learned from These Case Studies?

It has been suggested that resilience, adaptability, and transformability are the three related attributes of SESs that determine their future trajectories (Walker et al. 2004). Resilience and adaptability figure prominently in the case studies, but the third, transformability, or the ability of a system to transition to a new SES, has been less studied and is not found in the case studies. In each, the SES framework was used to locate key interventions needed to maintain or increase the resilience or adaptive capacity of the system (Table 8.1). The SES framework revealed multi-scalar feedbacks and drivers that otherwise likely would have been overlooked, and helped to understand at least some of the relationships that influence the resilience of complex systems. Each of the case studies reveals certain weak points in the SES that are either being neglected or exacerbated by current management strategies, and which may threaten the long-term persistence of the SES.

The focal SES in each case study was found to have vulnerabilities that threaten to cause significant shifts in the landscape and the resident social groups. By analyzing each of the case studies as an SES, it was possible to identify threats as well as points where adaptive capacity and resilience were low, in essence revealing the weakest link in the chain of interactions and the point at which intervention should be made. Unfortunately, the synthetic research that went into each of the case studies is rare in natural resource management, and underscores the need for a large amount of information in order to create sound policies and management prescriptions. Since it would likely be expensive and time consuming for any one agency or actor to gather all the necessary data individually, gathering together people from many different sectors of the SES may be the best way to detect threats and how they could be avoided.

8.5   Future Perspectives

Writing about the need for global action to respond to the profound changes in ecosystems caused by human activity, Carpenter and others state, "The challenge of sustainable development is to … transform social-ecological systems to provide food, water, energy, health and well-being in a manner that is economically, ecologically and socially viable for many generations in the future and for people in all parts of the world" (2012). The Millennium Ecosystem Assessment published in 2005 identified gaps in current knowledge linking ecosystem services and human well-being, including the need to understand how SESs evolve over time and respond to policy interventions, trade-offs among different ecosystem services, and how to integrate the expectation of nonlinear and abrupt changes into policy and planning (MA 2005). The SES concept is still relatively new to range science, and it has yet to be widely applied. It is acknowledged that social factors deserve more attention and more in-depth research in range management, but research to date on the interactions of society and rangelands has rarely been able to escape the bounds of a single discipline (Brunson 2012). Most range research has tended to focus on the ecology, management strategies, or economics of rangelands, but has rarely synthesized these different components.

Table 8.1 Overview of the SES framework and five case studies

Case	Problem	SES highlights	Interventions
Adaptation to climate change by Australian livestock managers	Lack of adaptive capacity in the face of anticipated climate change leads to poor management decisions and reduced economic resilience to climate change Scale: Region and enterprise	Ranching collapse would cause undesirable social and ecological change. Feedbacks between climate change and rangeland productivity are connected to ranch economic welfare	Education and support for rancher adaptive capacity, and mission-oriented research into rancher needs to guide education and outreach
Climate change and forb restoration in the Great Basin	Limited manager adaptive capacity in developing postfire seeding practices that anticipate climate change, leading to decreased ecosystem resilience Scale: Region and administrative units	Individual and community values limit personal adaptive capacity, costs and uncertainty of successful regeneration constrain management adaptive capacity	Support receptivity to learning; problem should be stated in terms other than adapting to climate change; financial support to reduce risk is needed
The California black rail and small wetlands in the Sierran foothills	Small wetland habitats for a rare bird are at risk from climate change and nonadaptive water conservation policy Scale: Individual landowners and water districts	Policy at the state and local scale inadvertently threatens small-scale wetland habitat for a rare bird. A lack of feedback to local or state policy makers about wetlands because they are "invisible"	New governance or policies for water districts are needed; outreach to landowners about maintaining small wetlands as wildlife habitat
Sedentarization of pastoralists in Xinjiang	Sedentarization and irrigated agriculture put the grassland SES at risk ecologically and socially due to economic inequality and overuse of water Scale: Household, village, county, and catchment	Understanding of the multiple scales of an SES and how they interact can be used to assess resiliency; impacts to and feedbacks from processes at the broader scale undermine resilience at the household scale	Programs should be revised using SES assessment of impacts at multiple scales, including equity of outcomes; development plans should be altered to fit environmental constraints
Environmental accounting for dehesas in Spain	Reasons for owning traditional woodlands producing many ecosystem services are not recognized by policy, yet support resilience to fluctuating prices of agricultural products Scale: Individual landowners, ecosystem	The consumption of ecosystem services by landowners acts as a positive feedback on the resilience of regionally valuable dehesa	Understand and support feedbacks that enhance landowner commitment to maintaining dehesa enterprises, including landowner-consumed ecosystem services

Rangelands around the world operate under a broad array of governance systems and property rights regimes. Thus far the SES concept has been used far more effectively to analyze past and present situations than in providing clear steps for future work. Future research should go further in determining how these different components are linked, and further suggest policy improvements that might better support ranchers, pastoralists, and rangeland managers. The effect of large-scale economic and political forces on local environments lends itself well to SES analysis. For example, in the USA, zoning laws discourage conversion of private rangeland to other uses within certain geographical areas, but ranch conversion is not otherwise prohibited and ranches are rarely the most profitable land use. Alternatively, many European countries have national laws that ban the conversion of certain agricultural lands—including dehesa—to nonagricultural uses. In the western USA, there is growing interest in conserving "working rangelands"—rangelands that produce ecosystem services as well as commodities. Yet there has been little research to date comparing the effectiveness of various national land policies on maintaining working rangelands or supporting their active management at the household scale. Comparing the effects of land-use policies on the ecology of working rangelands would provide needed policy-relevant information.

At a recent national workshop on "usable science," participants, including scientists, livestock operators, and land managers, ranked 142 identified issues proposed by five working groups (water, animals, vegetation, soils, and socioeconomics). The number one-ranked issue overall came out of the Socio-Economics Working Group: understanding and managing for variability (climate, drought, fire), adaptation, and recovery (Brunson et al. 2016). This topic is admittedly broad, but the SES approach is a good fit for analyzing key components: ecosystem change, adaptive capacity, and resilience in rangeland systems. Rangelands are subject to high variability of climate, vegetation, and market influences of livestock and feed prices, and dealing with such variability is a constant challenge for livestock operators and land managers. SES frameworks might productively be used to increase the resilience of working rangelands by identifying beneficial ecological traits but also by constructing social and economic support systems for livestock operators and land managers. Research exploring the use of SES frameworks to help practitioners and managers anticipate and manage variability and change in environment and society is a needed contribution.

Trade-offs and synergies among the various goods and services derived from rangelands need more attention in general. On public lands, agency interventions on specific allotments could have impacts on entire landscapes: examining these cross-scale effects, and the trade-offs among them, requires greater attention. Designation of a park or preserve may be of great benefit in meeting conservation and recreation needs, but might have devastating effects on individual livestock operators, and lead to a transformation in nearby communities with various ecological and social outcomes at diverse scales. SES analysis could help anticipate these effects, providing a fuller picture of the opportunities and trade-offs of the change. This type of research is hampered, however, by the difficulties of cross-disciplinary research. SESs are inherently interdisciplinary, but different disciplines use different research methods, and the multiple geographic scales used by SES researchers require multiple researchers to work together in different places. Greater emphasis is necessary to integrate and balance multidisciplinary programs and projects addressing rangelands, including the use of multiple research methods.

Finally, as pointed out earlier, transformability as an SES characteristic has not received enough attention from SES researchers. Yet transformations are occurring in rangelands in many parts of the world. In China, when does a sedentarized nomad community shift to a different SES and what does that mean for the well-being of the people in the community? In the USA, many ranching communities have experienced an influx of new residents working in businesses related to mining and tourism, or seeking a place to retire, vacation, or telecommute. When does a ranching community transform to another type of community or SES altogether? What does this mean for the economy and the environment? The SES framework could be used to assess the impact of such transformations in a comprehensive way, and to analyze the resilience and adaptability of the new SES.

8.6 Summary

An SES is a combination of social and ecological components that shape each other in profound ways. For example, a grassland landscape is radically altered when it is converted to agriculture. The natural components of that system are affected by farming and land management practices, water use, infrastructure, etc. Farming communities are impacted by the productivity of the soil, by precipitation and temperature, and by the multitude of plants and animals they either rely on (for pollination or soil health) or compete with (crop pests or predators). Similarly, livestock operators graze their animals and conduct management activities in ways that shape rangeland ecology, but also respond to changing ecological conditions such as invasive plant species or variable productivity caused by irregular rainfall. Larger scale patterns such as climate change, demographic trends, and global meat prices also affect rangelands both directly and by altering land-use patterns.

The SES concept is not a methodology for research or a checklist to identify problems. It is a conceptual framework designed to keep both the social and ecological components of a system in focus so that the interactions between them can be scrutinized for drivers of change and causes of specific outcomes. Furthermore, change may cross back and forth between the social and ecological subsystems in ongoing feedbacks. Most research and land management policies are based predominantly on either ecological or social phenomena and problems. This type of single-discipline thinking leads to policies which either fail to address the problem or cause unintended consequences. SES analysis requires a great deal of information from multiple disciplines and often at multiple sites, which is logistically challenging and has served as a barrier to widespread use of the SES framework until recently.

SESs exist because human life depends on ecosystems, and human actions perpetually affect ecosystem components and functions. Rangeland managers work at the intersection of human enterprise and rangeland ecosystems. Managers must remain flexible and adaptive enough not only to tailor grazing and management activities to suit unpredictable environmental conditions, but also to respond to changing policy, economics, demands for ecosystem services, and management capacity. Such flexibility and adaptability constitute a serious challenge especially given climate change and decreasing profitability of range-fed livestock. If we as a society want to continue to have working rangelands, policies to promote more cross-disciplinary research and education, flexible land use, and novel economic programs to satisfy multiple objectives for rangelands are sorely needed.

It has been suggested that resilience, adaptability, and transformability are the three related attributes of SESs that determine their future trajectories. Resilience can be defined as the capacity of a system to absorb disturbance and reorganize while undergoing change so as to retain essentially the same function, structure, identity, and feedbacks (Walker et al. 2004). In the face of a disturbance, a resilient SES will not only maintain function but may even use the disturbance as an opportunity to reorganize and further develop resilience (Gunderson and Holling 2002; Berkes et al. 2003). Disturbances can originate in social or ecological systems and can happen rapidly or gradually (May 1977; Chap. 6, this volume). The ability of the SES to adapt to change is a key to resilience. If change overwhelms the resilience and adaptive capacity of an SES, it will transform to a new type of SES. SES analyses should strive to identify the interactions that lead to resilience, adaptation, or transformation.

The application of the resilience concept to social settings and, to a lesser degree, the use of the SES concept itself have been critiqued for overlooking the role of human autonomy, cultural values, social heterogeneity, and power relations among actors in SESs. An overemphasis on institutions as environmental managers and decision makers all too often obscures the role of individuals and loosely affiliated groups in social–environmental relations. In the future, range SES analyses could be improved by better accounting for social difference among stakeholders and their ability to take part in political and decision-making processes.

Components of SESs that figure prominently into their analysis include scale and feedbacks. Understanding of ecosystems and their response to management has often been hampered by a failure to appreciate the role of scale. Different patterns and processes are characteristic of different temporal and spatial scales. Research too often focuses on a single scale, overlooking processes that occur primarily at larger or smaller scales, but which nonetheless critically impact the components of the focal SES. Identifying feedbacks between social and ecological components of the system at multiple scales is a key to SES-based analysis. For example, household economics may be affected by international meat prices or a consolidation of the meat-packing industry. Droughts that occur only once a decade can have lasting effects on rangeland ecology, herd sizes, management strategies, and local poverty.

Five case studies using the SES concept were presented. In each, the SES framework was used to locate key interventions needed to maintain or increase the resilience or adaptive capacity of the system. Analyses identified important processes and interactions at scales from the personal values of an individual to region-wide watershed impacts. The SES framework revealed multi-scalar feedbacks and drivers that otherwise likely would have been overlooked, and helped to understand at least some of the relationships that influence the resilience of complex systems. Each of the case studies revealed weak points in the SES that were either neglected or exacerbated by current management strategies, and which may undermine the long-term persistence of the SES, causing significant shifts in both landscapes and social groups. By analyzing each of the case studies as an SES, it was possible to determine how the ecological and social components of the systems were affecting each other. This allowed for an assessment of how the SES was being threatened, where adaptive capacity was low, and when resilience was breaking down, in essence revealing the weakest link in the chain of interactions and the point at which interventions should be made.

Unfortunately, the synthetic research that went into each of the case studies is rare in natural resource management, and indicates the need for a large amount of information in order to create sound policies and management prescriptions. Since it would likely be expensive and time consuming for any one agency or actor to gather all the necessary data individually, gathering together people from many different sectors of the SES to share information and collaborate on solutions may be the best way to detect threats and how they can be avoided. Research is needed on integrative metrics for cross-disciplinary projects, the on-the-ground impacts of social interactions and processes, the policy interventions that support resilience, and evaluating trade-offs and synergies.

One hurdle in the way of improving the management of rangeland SESs is the fact that regulatory policies are usually enacted on the premise that a problem is consistent across different times and places and that the policy will solve this problem when applied everywhere uniformly. Unfortunately, rangeland problems are rarely so consistent and neither are the agencies tasked with implementing government policy. The single agency, single policy type of policy implementation is an example of centralized governance. Centralized systems assume that all information can be routed through a single office and that solutions can come from that same office. In contrast, polycentric governance systems have multiple locations for collecting data and issuing and carrying out management actions. Polycentric governance models do not rely on a single solution to a perceived single problem, but rather seek to coordinate activities working toward a common goal.

Managers should aim to maximize the resilience of both the ecological and social elements of a desirable SES, which calls for favoring diversity and adaptability over maximizing yield and efficiency (Holling 1973; Holling and Meffe 1996). Given that rangeland managers typically have very limited control over the social components of range SESs, increasing participation and cooperation between managers, other invested actors, and the public to maximize information sharing, cooperation, and adaptive capacity of management activities would likely improve outcomes of rangeland SES management (Walker et al. 2004; Gunderson 2001; Olsson et al. 2006). Adaptive comanagement has become a common prescription for ecosystems and for SESs, and may include collaboration among agencies whose jurisdictions intersect in a particular SES, or participatory efforts with diverse stakeholders

(Stringer et al. 2006). Adaptive management also has relatively high information needs and institutional costs, making it difficult for many agencies to undertake (Jacobson et al. 2006). Increasing the involvement and number of stakeholders may improve monitoring of SESs and generate more viable alternatives and solutions, but it does not itself constitute a solution. Some stakeholders inevitably have more power than others in shaping how an SES functions.

Rangeland managers and policy makers would be well advised to keep the following points in mind when creating management plans for SESs:

1. Ecological diversity and redundancy of components are beneficial for resilience, and should be preserved through management activities. Feedbacks may support or weaken resilience. Undesirable states can also be resilient.
2. Stakeholders in any system are typically stratified throughout several hierarchical levels of geographical scale and legal authority. These hierarchical levels do not have the same motivations nor are they affected by the same processes. Interactions between them are complex.
3. All SESs are complex, and changes within them may be difficult or impossible to predict. Management plans should thus be adaptive, with changes contingent on consistent monitoring to guide both short- and long-term planning. Governance systems should likewise be adaptive, for similar reasons.
4. Uncertainty within the system can be minimized through the inclusion of all relevant stakeholders in the management process. Genuine inclusion implies that stakeholders have a chance to affect outcomes and receive benefits while acknowledging that authority and benefits are rarely shared equally and that action must usually be taken based on incomplete information and a lack of consensus. When successful, stakeholder inclusion increases information gathering and feedback and decreases uncooperative behavior and unpredicted behavioral change (Armitage et al. 2009).
5. SES research teams should include both social and ecological scientists. Unbalanced funding and emphasis can lessen the prospects for a successfully interactive project. Funding agencies need to emphasize this balance in granting programs.

The primary benefit of the SES framework is the improved ability to prevent or correct social policies that cause negative ecological outcomes, and to respond to ecological problems in ways that support, rather than hurt, social actors. By utilizing an SES analysis, rangeland managers and policy makers can create beneficial feedback loops such that society benefits from sustainable utilization of rangelands, and ecological objectives are met in ways that benefit livestock operators and the broader society. Managing solely for social or ecological objectives has a long history of unintended consequences, including ecosystem collapse and social unrest. While information intensive, conceptually complex, and logistically challenging in the management context, the SES framework can help overcome intractable challenges to working rangelands such as climate and land-use change.

References

Abatzoglou, J.T., and C.A. Kolden. 2011. Climate change in western US deserts: Potential for increased wildfire and invasive annual grasses. *Rangeland Ecology and Management* 64: 471–478.

Ackerly, D.D., S.R. Loarie, W.K. Cornwell, et al. 2010. The geography of climate change: Implications for conservation biogeography. *Diversity and Distributions* 16: 476–487.

Adger, W.N., K. Brown, D.R. Nelson, et al. 2011. Resilience implications of policy responses to climate change. *Wiley Interdisciplinary Reviews: Climate Change* 2: 757–766.

Agrawal, A. 2005. *Environmentality*. Durham, NC: Duke University Press.

Agrawal, A., and C.C. Gibson. 1999. Enchantment and disenchantment: The role of community in natural resource conservation. *World Development* 27: 629–649.

Armitage, D.R., R. Plummer, F. Berkes, et al. 2009. Adaptive co-management for social-ecological complexity. *Frontiers in Ecology and the Environment* 7: 95–102.

Berkes, F., and C. Folke (eds.). 1998. *Linking social and ecological systems: Management practices and social mechanisms for building resilience*. New York: Cambridge University Press.

Berkes, F., J. Colding, and C. Folke (eds.). 2003. *Navigating social–ecological systems: Building resilience for complexity and change*. New York: Cambridge University Press.

Binder, C.R., J. Hinkel, P.W.G. Bots, and C. Pahl-Wostl. 2013. Comparison of frameworks for analyzing social-ecological systems. *Ecology and Society* 18: 26.

Briske, D.D., B.T. Bestelmeyer, T.K. Stringham, and P.L. Shaver. 2008. Recommendations for development of resilience-based state-and-transition models. *Rangeland Ecology and Management* 61: 359–367.

Brooks, M.L., C.M. D'Antonio, J.B. Richardson, et al. 2004. Effects of invasive alien plants on fire regimes. *BioScience* 54: 677–688.

Brown, K. 2014. Global environmental change: A social turn for resilience? *Progress in Human Geography* 38: 107–117.

Brunson, M.W. 2012. The elusive promise of social-ecological approaches to rangeland management. *Rangeland Ecology and Management* 65: 632–637.

Brunson, M.W., L. Huntsinger, U. Kreuter, and J. Ritten. 2016. Usable socio-economic science for rangelands. *Rangelands* 38: 85–89.

Bugalho, M.N., M.C. Caldeira, J.S. Pereira, et al. 2011. Mediterranean cork oak savannas require human use to sustain biodiversity and ecosystem services. *Frontiers in Ecology and the Environment* 9: 278–286.

Campos, P. 2000. An agroforestry account system. In *Institutional aspects of managerial and accounting in forestry*, ed. H. Joebstl, M. Merlo, and L. Venzi, 9–19. Viterbo: IUFRO and University of Viterbo.

Campos, P., J.L. Oviedo, A. Caparrós, et al. 2009. Contingent valuation of private amenities from oak woodlands in Spain, Portugal, and California. *Rangeland Ecology and Management* 62: 240–252.

Caparrós, A., L. Huntsinger, J.L. Oviedo, et al. 2013. Economics of ecosystem services, Chapter 12. In *Mediterranean oak woodland working landscapes: Dehesas of Spain and ranchlands of California*, ed. P. Campos, L. Huntsinger, J.L. Oviedo, et al. New York, NY: Springer, Landscape Series.

Carpenter, S.R., C. Folke, A. Norstrom, et al. 2012. Program on ecological change and society: An international research strategy for integrated social-ecological systems. *Current Opinion in Environmental Sustainability* 4: 134–138.

Cash, D.W., W. Adger, F. Berkes, et al. 2006. Scale and cross-scale dynamics: Governance and information in a multilevel world. *Ecology and Society* 11(2): 8.

Castree, N. 2001. Socializing nature: Theory, practice, and politics. In *Social nature: Theory, practice and politics*, ed. N. Castree and B. Braun, 1–21. Oxford: Blackwell.

Cheng, T., and J.J. Hong. 2011. Water resources balancing analysis of Ebinur Lake Area in the next 10 years. *Gansu Water Conservancy and Hydropower Technology* 47: 11–13 (in Chinese).

Cobon, D.H., G.S. Stone, J.O. Carter, et al. 2009. The climate change risk management matrix for the grazing industry of northern Australia. *Rangeland Journal* 31: 31–49.

Cote, M., and A.J. Nightingale. 2012. Resilience thinking meets social theory: Situating social change in socio-ecological systems (SES) research. *Progress in Human Geography* 36: 475–489.

Crane, T.A. 2010. Of models and meanings: Cultural resilience in social-ecological systems. *Ecology and Society* 15: 19.

Cumming, G.S., J. Alcamo, O. Sala, et al. 2005. Are existing global scenarios consistent with ecological feedbacks? *Ecosystems* 8: 143–152.

Cumming, G.S., D.M. Cumming, and L. Redman. 2006. Scale mismatches in social-ecological systems: Causes, consequences, and solutions. *Ecology and Society* 11: 14.

Davidson, D.J. 2010. The applicability of the concept of resilience to social systems: Some sources of optimism and nagging doubts. *Society and Natural Resources* 23: 1135–1149.

Eagly, A.H., and P. Kulesa. 1997. Attitudes, attitude structure, and resistance to change. In *Environment, ethics, and behavior: The psychology of environmental valuation and degradation*, ed. M.H. Bazerman, D.M. Messick, A.E. Tenbrunzel, and K.A. Wade-Bensone. San Francisco: New Lexington Press.

Fan, M., W.J. Li, C. Zhang, and L. Li. 2014. Impacts of nomad sedentarization on social and ecological systems at multiple scales in Xinjiang Uyghur Autonomous Region, China. *Ambio* 43: 673–686.

Folke, C. 2006. Resilience: The emergence of a perspective for social-ecological systems analysis. *Global Environmental Change* 16: 253–267.

Folke, C., T. Hahn, P. Olsson, and J. Norberg. 2005. Adaptive governance of social-ecological systems. *Annual Review of Environment and Resources* 30: 441–473.

Fox, W.E., D.W. McCollum, J.E. Mitchell, L.E. Swanson, U.P. Kreuter, J.A. Tanaka, G.R. Evans, and H.T. Heintz. 2009. An Integrated Social, Economic, and Ecologic Conceptual (ISEEC) framework for considering rangeland sustainability. *Society & Natural Resources* 22: 593–606.

Gunderson, L.H. 2001. Managing surprising ecosystems in southern Florida. *Ecological Economics* 31: 371–378.

Gunderson, L.H., and C.S. Holling (eds.). 2002. *Panarchy: Understanding transformations in human and natural systems*. Washington, DC: Island Press.

Hansen, J.W. 2002. Applying seasonal climate prediction to agricultural production. *Agricultural Systems* 74: 305–307.

Harris, R.B. 2010. Rangeland degradation on the Qinghai-Tibetan plateau: A review of the evidence of its magnitude and causes. *Journal of Arid Environments* 74: 1–12.

Holland, J.H. 1992. Complex adaptive systems. *Daedalus* 121: 17–30.

Holling, C.S. 1973. Resilience and stability of ecological systems. *Annual Review of Ecology and Systematics* 4: 1–23.

———. 2001. Understanding the complexity of economic, ecological, and social systems. *Ecosystems* 4:390–405.

Holling, C.S., and G. Meffe. 1996. Command and control and the pathology of natural resource management. *Conservation Biology* 10: 328–337.

Howden, S.M., J. Soussana, F.N. Tubiello, et al. 2007. Adapting agriculture to climate change. *Proceedings of the National Academy of Sciences of the United States of America* 104: 19691–19696.

Hruska, T.V., L. Huntsinger, and J. Oviedo. 2015. An accidental resource: The social ecological system framework applied to California black rails inhabiting small wetlands in Sierran foothill oak woodlands. In *Proceedings of the seventh California oak symposium: Managing oak woodlands in a dynamic world*, eds. R.B. Standiford, and K. Purcell (tech coords), 231–238. November 3–6, 2014, Visalia, CA. Berkeley, CA: USDA Forest Service General Technical Report PSW-GTR-251.

Huber-Sannwald, E., M. Ribeiro Palacios, J.T. Arredondo Moreno, et al. 2012. Navigating challenges and opportunities of land degradation and sustainable livelihood development in dryland social-ecological systems: A case study from Mexico. *Philosophical Transactions of the Royal Society of London. Series B: Biological Sciences* 367: 3158–3177.

Huntsinger, L. 2009. Into the wild: Vegetation, alien plants, and familiar fire at the exurban frontier. Chapter 8. In *The planner's guide to natural resource conservation: The science of land development beyond the metropolitan fringe*, ed. A. Esparza and G. McPherson. New York, NY: Springer.

Huntsinger, L., and J. Oviedo. 2014. Ecosystem services may be better termed social ecological services in a traditional pastoral system: The case in California Mediterranean rangelands at multiple scales. *Ecology and Society* 19: 8.

IPCC [Intergovernmental Panel on Climate Change]. 2014. *Impacts, adaptation, and vulnerability part A: Global and sectoral aspects working group II contribution to the fifth assessment report of the Intergovernmental Panel on Climate Change*. Cambridge: Cambridge University Press.

Jacobson, S.K., J.K. Morris, J.S. Sanders, et al. 2006. Understanding barriers to implementation of an adaptive land management program. *Conservation Biology* 20: 1516–1527.

Jiang, X., S.A. Rauscher, T.D. Ringler, et al. 2013. Projected future changes in vegetation in western North America in the twenty-first century. *Journal of Climate* 26: 3671–3687.

Kennedy, J.J., T.J. Ward, and P. Glueck. 2001. Evolving forestry and rural development beliefs at midpoint and close of the 20th century. *Forest Policy and Economics* 3: 81–95.

Knapp, P.A. 1996. Cheatgrass (*Bromus tectorum* L) dominance in the Great Basin Desert: History, persistence, and influences to human activities. *Global Environmental Change* 6: 37–52.

Leach, M., R. Mearns, and I. Scoones. 1999. Environmental entitlements: Dynamics and institutions in community-based natural resource management. *World Development* 27: 225–247.

Lebel, L., P. Garden, and M. Imamura. 2005. The politics of scale, position, and place in the governance of water resources in the Mekong region. *Ecology and Society* 10: 18.

Lebel, L., J.M. Anderies, B. Campbell, et al. 2006. Governance and the capacity to manage resilience in regional social-ecological systems. *Ecology and Society* 11: 1.

Leiserowitz, A. 2006. Climate change risk perception and policy preferences: The role of affect, imagery, and values. *Climatic Change* 77: 45–72.

Li, W., and L. Huntsinger. 2011. China's grassland contract policy and its impacts on herder ability to benefit in Inner Mongolia: Tragic feedbacks. *Ecology and Society* 16: 1.

Liu, J., T. Dietz, S.R. Carpenter, et al. 2007a. Coupled human and natural systems. *Ambio* 8: 639–649.

Liu, J., T. Dietz, S.R. Carpenter, M. Alberti, et al. 2007b. Complexity of coupled human and natural systems. *Science* 317: 1513–1516.

MA [Millennium Ecosystem Assessment]. 2005. *Ecosystems and human well-being: Summary for decision makers*. Washington, DC: Millennium Assessment.

MARM. 2011. Encuesta de precios de la tierra 2010. Ministerio de Medio Ambiente y Medio Rural y Marino. http://www.magrama.gob.es/es/estadistica/temas/encuesta-de-precios-de-la--tierra/. Viewed July 2012.

Marshall, N.A. 2010. Understanding social resilience to climate variability in primary enterprises and industries. *Global Environmental Change* 20: 36–43.

Marshall, N.A., and A. Smajgl. 2013. Understanding variability in adaptive capacity on rangelands. *Rangeland Ecology and Management* 66: 88–94.

Marshall, N.A., and C.J. Stokes. 2014. Identifying thresholds and barriers to adaptation through measuring climate sensitivity and capacity to change in an Australian primary industry. *Climatic Change* 126: 399–411.

Marshall, N.A., M. Friedel, R.D. Van Klinken, et al. 2011. Considering the social dimension of contentious species: The case of buffel grass. *Environmental Science & Policy* 14: 327–338.

Marshall, N.A., S. Park, S.M. Howden, et al. 2013. Climate change awareness is associated with enhanced adaptive capacity. *Agricultural Systems* 117: 30–34.

Marshall, N.A., C.J. Stokes, N.P. Webb, et al. 2014. Social vulnerability to climate change in primary producers: A typology approach. *Agriculture Ecosystems and Environment* 186: 86–93.

May, R.M. 1977. Thresholds and breakpoints in ecosystems with a multiplicity of stable states. *Nature* 269: 471–477.

McKeon, G.M., A. Ash, W. Hall, and Smith M. Stafford. 2000. Simulation of grazing strategies for beef production in North-East Queensland. In *Applications of seasonal climate forecasting in agricultural and natural ecosystems*, ed. G.L. Hammer, N. Nicholls, and C. Mitchell. London: Kluwer Academic Publishers.

McKeon, G.M., W.B. Hall, B.K. Henry, et al. 2004. *Pasture degradation and recovery in Australia's rangelands: Learning from history*, 1–255. Brisbane, AU: Queensland Department of Natural Resources, Mines and Energy.

Meyer, S.E., J. Beckstead, P.S. Allen, and D.C. Smith. 2008. A seed bank pathogen causes seed-borne disease: *Pyrenophora semeniperda* on undispersed grass seeds in western North America. *Canadian Journal of Plant Pathology* 30: 525–533.

Miller, B.W., S.C. Caplow, and P.W. Leslie. 2012. Feedbacks between conservation and social-ecological systems. *Conservation Biology* 26: 218–227.

Ministry of Finance (People's Republic of China). 2011. http://jjs.mof.gov.cn/zhengwuxinxi/touruqingkuang/201108/t20110802_583336.html. Accessed 3 August 2011.

Mitchell, R.C., and R.T. Carson. 1989. *Using surveys to value public goods: The contingent valuation method*. Washington, DC: Resources for the Future.

Nagendra, H., and E. Ostrom. 2012. Polycentric governance of multifunctional forested landscapes. *International Journal of Commons* 6: 104–133.

Norgaard, R. 1994. *Progress betrayed: The demise of development and a co-evolutionary revisioning of the future*. London: Routledge.

Olsson, P., C. Folke, and F. Berkes. 2004. Adaptive comanagement for building resilience in social–ecological systems. *Environmental Management* 34: 75–90.

Olsson, P., L.H. Gunderson, S.R. Carpenter, et al. 2006. Shooting the rapids: Navigating transitions to adaptive governance of social-ecological systems. *Ecology and Society* 11: 18.

Olsson, L., A. Jerneck, H. Thoren, et al. 2015. Why resilience is unappealing to social science: Theoretical and empirical investigations of the scientific use of resilience. *Science Advances* 1: e1400217.

Ostrom, E. 2007. A diagnostic approach for going beyond panaceas. *Proceedings of the National Academy of Sciences of the United States of America* 104: 15181–15187.

Ostrom, E. 2009. A general framework for analyzing sustainability of social-ecological systems. *Science* 325: 419–422.

Oviedo, J.L., L. Huntsinger, P. Campos, and A. Caparros. 2012. Income value of private amenities assessed in California oak woodlands. *California Agriculture* 66: 91–96.

Oviedo, J.L., P. Ovando, L. Forero, et al. 2013. The private economy of dehesas and ranches: Case studies, Chapter 13. In *Mediterranean oak woodland working landscapes: Dehesas of Spain and ranchlands of California*, ed. P. Campos, L. Huntsinger, J.L. Oviedo, et al. New York: Springer, Landscape Series.

Oviedo, J.L., L. Huntsinger, and P. Campos. 2015a. Landowner total income from oak woodland working landscapes in Spain and California. In *Proceedings of the Seventh California oak symposium: Managing oak woodlands in a dynamic world*, eds. R.B. Standiford, and K. Purcell. November 3–6, 2014, Visalia, CA. Washington, DC: USDA Forest Service General Technical Report PSW-GTR-XX.

Oviedo, J.L., P. Campos, A. Caparrós. 2015b. Valoración de servicios ambientales privados de propietarios de fincas agroforestales de Andalucía. In *Renta total y capital de las fincas agroforestales de Andalucía*, eds. P. Campos, and P. Ovando. Memorias científicas de RECAMAN 4: 4.1.

Pelosi, C., M. Goulard, and G. Balent. 2010. The spatial scale mismatch between ecological processes and agricultural management: Do difficulties come from underlying theoretical frameworks? *Agriculture Ecosystems and Environment* 139: 455–462.

Plummer, R., and D. Armitage. 2010. Integrating perspectives on adaptive capacity and environmental governance. In *Adaptive capacity and environmental governance*, ed. D. Armitage and R. Plummer. Berlin: Springer.

Pokorny, M.L., R.L. Sheley, C.A. Zabinsky, et al. 2005. Plant functional group diversity as a mechanism for invasion-resistance. *Restoration Ecology* 13: 448–459.

Polley, H.W., D.D. Briske, J.A. Morgan, et al. 2013. Climate change and North American rangelands: Trends, projections and implications. *Rangeland Ecology and Management* 66: 493–511.

Post, E., and C. Pederson. 2008. Opposing plant community responses to warming with and without herbivores. *Proceedings of the National Academy of Sciences of the United States of America* 105: 12353–12358.

Qian, Y.B., Z.N. Wu, J. Jiang, and Q. Yang. 2004. Eco-environmental change and its impact factors in the Ebinur Lake catchments during the past 50 years. *Journal of Glaciology and Geocry* 01: 17–26 (in Chinese, English summary).

Rammel, C., S. Stagl, and H. Wilfing. 2007. Managing complex adaptive systems—a coevolutionary perspective on natural resource management. *Ecological Economics* 63: 9–21.

Reid, R.S., M.E. Fernandez-Gimenez, and K.A. Galvin. 2014. Dynamics and resilience of rangelands and pastoral peoples around the globe. *Annual Review of Environment and Resources* 39: 217–242.

Resilience Alliance. n.d. Assessing resilience in social-ecological systems: Workbook for practitioners 2.0. www.resalliance.org/resilience-assessment. Viewed 9 November 2015.

Reynolds, J.F., D.M.S. Smith, E.F. Lambin, et al. 2007. Global desertification: Building a science for dryland development. *Science* 316: 847–851.

Richards, R., and L. Huntsinger. 1994. BLM employee attitudes toward environmental conditions on rangelands. *Journal of Range Management* 47: 365–368.

Richmond, O.M.W., J. Tecklin, and S.R. Beissinger. 2008. Distribution of California black rails in the Sierra Nevada foothills. *Journal of Field Ornithology* 79: 381–390.

Richmond, O.M.W., S.K. Chen, B.B. Risk, et al. 2010. California black rails depend on irrigation-fed wetlands in the Sierra Nevada foothills. *California Agriculture* 64: 85–93.

Roe, E., L. Huntsinger, and K. Labnow. 1998. High reliability pastoralism. *Journal of Arid Environments* 39: 39–55.

Sheley, R.L., and M.L. Half. 2006. Enhancing native forb establishment and persistence using a rich seed mixture. *Restoration Ecology* 14: 627–635.

Smith, A.H., and W.E. Martin. 1972. Socioeconomic behavior of cattle ranchers, with implications for rural community development in the west. *American Journal of Agricultural Economics* 54: 217–225.

Stafford Smith, D.M., G.M. McKeon, I.W. Watson, et al. 2007. Learning from episodes of degradation and recovery in variable Australian rangelands. *Proceedings of the National Academy of Sciences of the United States of America* 104: 20690–20695.

Stringer, L.C., A.J. Dougill, E. Fraser, et al. 2006. Unpacking "participation" in the adaptive management of social–ecological systems: A critical review. *Ecology and Society* 11: 39.

Sun, L., and Y.Q. Gao. 2010. Influences of changes in cultivated land area on the surface area of Ebinur Lake Valley in Xinjiang. *Guangxi Agricultural Science* 41: 848–852 (in Chinese, English summary).

Taylor, M. 2014. *The political ecology of climate change adaptation: Livelihoods, agrarian change and the conflicts of development*. London: Routledge.

Turner, M.D. 2011. The new pastoral development paradigm: Engaging the realities of property institutions and livestock mobility in dryland Africa. *Society and Natural Resources* 24: 469–484.

Turner, B.L., R.E. Kasperson, P.A. Matson, et al. 2003. A framework for vulnerability analysis in sustainability science. *Proceedings of the National Academy of Sciences of the United States of America* 100: 8074–8079.

Vayda, A.P., and B.J. McCay. 1975. New directions in ecology and ecological anthropology. *Annual Review of Anthropology* 4: 293–306.

Walker, B.H. 1995. Conserving biological diversity through ecosystem resilience. *Conservation Biology* 9: 747–752.

Walker, J., and M. Cooper. 2011. Genealogies of resilience from systems ecology to the political economy of crisis adaptation. *Security Dialogue* 42: 143–160.

Walker, B.H., and M.A. Janssen. 2002. Rangelands, pastoralists and governments: Interlinked systems of people and nature. *Philosophical Transactions of the Royal Society B* 357: 719–725.

Walker, B.H., D. Ludwig, C. Holling, and R. Peterman. 1981. Stability of semi-arid savanna grazing systems. *Journal of Ecology* 69: 473–498.

Walker, B.H., A. Kinzig, and J. Langridge. 1999. Plant attribute diversity, resilience, and ecosystem function: The nature and significance of dominant and minor species. *Ecosystems* 2: 95–113.

Walker, B.H., C.S. Holling, S.R. Carpenter, and A. Kinzig. 2004. Resilience, adaptability, and transformability in social-ecological systems. *Ecology and Society* 9: 5.

Walker, B.H., N. Abel, J.M. Anderies, and P. Ryan. 2009. Resilience, adaptability, and transformability in the Goulburn-Broken Catchment, Australia. *Ecology and Society* 14: 12.

Wangui, E.E. 2008. Development interventions, changing livelihoods, and the making of female Maasai pastoralists. *Agriculture and Human Values* 25: 365–378.

Watts, M.J. 1983. On the poverty of theory: Natural hazards research in context. In *Interpretations of Calamity*, ed. K. Hewitt. London: Allen and Unwin.

———. 2015b. Adapting to the Anthropocene: Some reflections on development and climate in the West African Sahel. *Geographical Research* 53:288–297.

———. 2015a. Now and then: The origins of political ecology and the rebirth of adaptation as form of thought. In *The Routledge handbook of political ecology*, eds. T. Perreaut, G. Bridge, and J. McCarthy. London: Routledge.

West, N.E. 1993. Biodiversity of rangelands. *Journal of Range Management* 46: 2–13.

Westoby, M., B. Walker, and I. Noy-Meir. 1989. Opportunistic management for rangelands not at equilibrium. *Journal of Range Management* 42: 266–274.

Whitcomb, H.L. 2011. *Temperature increase effects on sagebrush ecosystem forbs: Experimental evidence and range manager perspectives (MS thesis)*. Logan, UT: Utah State University.

Chapter 9
State and Transition Models: Theory, Applications, and Challenges

Brandon T. Bestelmeyer, Andrew Ash, Joel R. Brown, Bulgamaa Densambuu,
María Fernández-Giménez, Jamin Johanson, Matthew Levi, Dardo Lopez,
Raul Peinetti, Libby Rumpff, and Patrick Shaver

Abstract State and transition models (STMs) are used to organize and communicate information regarding ecosystem change, especially the implications for management. The fundamental premise that rangelands can exhibit multiple states is now widely accepted and has deeply pervaded management thinking, even in the absence of formal STM development. The current application of STMs for management, however, has been limited by both the science and the ability of institutions to develop and use STMs. In this chapter, we provide a comprehensive and contemporary overview of STM concepts and applications at a global level. We first review the ecological concepts underlying STMs with the goal of bridging STMs to recent theoretical developments in ecology. We then provide a synthesis of the history of

B.T. Bestelmeyer (✉) • M. Levi
USDA-ARS, Jornada Experimental Range, New Mexico State University,
MSC 3JER, Box 30003, Las Cruces, NM 88003, USA
e-mail: Brandon.Bestelmeyer@ars.usda.gov

A. Ash
CSIRO, 306 Carmody Road, St Lucia, QLD 4067, Australia

J.R. Brown
USDA-NRCS, Jornada Experimental Range, New Mexico State University,
Las Cruces, NM 88003, USA

B. Densambuu
Green Gold Mongolia, Sky Plaza Business Centre,
Olympic Street 12, Khoroo 1, Ulaanbaatar, Mongolia

M. Fernández-Giménez
Department of Forest and Rangeland Stewardship, Colorado State University,
Fort Collins, CO 80523, USA

J. Johanson
USDA-NRCS, 42 Engdahl Drive, Dover-Foxcroft, ME 04426, USA

D. Lopez
INTA Estación Forestal Villa Dolores, Unidad de Investigación en Bosque Nativo,
Camino Viejo a San José Villa, Dolores-Traslasierra, Córdoba 5870, Argentina

© The Author(s) 2017   303
D.D. Briske (ed.), *Rangeland Systems*, Springer Series on Environmental
Management, DOI 10.1007/978-3-319-46709-2_9

STM development and current applications in rangelands of Australia, Argentina, the United States, and Mongolia, exploring why STMs have been limited in their application for management. Challenges in expanding the use of STMs for management are addressed and recent advances that may improve STMs, including participatory approaches in model development, the use of STMs within a structured decision-making process, and mapping of ecological states, are described. We conclude with a summary of actions that could increase the utility of STMs for collaborative adaptive management in the face of global change.

Keywords Digital soil mapping • Ecological site description • Resilience • State transition • Structured decision-making • Transient dynamics

9.1   Introduction

State and transition models (STMs) were conceived as a means to organize and communicate information about ecosystem change as a basis for management. While some authors regard "the state and transition model" as a specific theory about how ecosystems respond to disturbance (see review in Pulsford et al. 2014), we take the view that STMs are not a theory per se, but are a flexible way of organizing information about ecosystem change that may draw on a wide range of concepts about ecosystem dynamics (Westoby et al. 1989). The value of STMs for rangeland managers is in fostering a general understanding of how rangelands function and respond to management actions, thereby leading to more efficient and effective allocation of management efforts.

The fundamental idea is simple (see the Caldenal STM at http://jornada.nmsu.edu/esd/international/argentina). Vegetation, a commonly used indicator of ecosystem conditions, is described according to discrete plant communities (such as an open *Prosopis caldenia* forest with grassy understory). In doing so, we develop a logic for distinguishing different communities so that stakeholders can communicate effectively about them. Next, we describe the multiple plant communities that can occur on a particular site. The key problem in this step is to define the characteristics of the "site"—its climate, soils, and topographic position. Otherwise we might conclude erroneously that a set of plant communities are alternative states of a specific site when in fact they exist on different sites. Finally, we identify the

R. Peinetti
Facultad de Agronomía, Universidad Nacional de La Pampa,
Santa Rosa, La Pampa 6300, Argentina

L. Rumpff
School of Biosciences, University of Melbourne, Melbourne, VIC 3010, Australia

P. Shaver
Rangeland Management Services L.L.C., 2510 Meadow Lane, Woodburn, OR 97071, USA

causes of transitions between communities and the constraints to recovery of particular communities, including succession, event-driven change, and persistent transitions to alternative stable states (Briske et al. 2003). The causes and constraints to change are often incompletely understood, but they can be tested by monitoring the effects of management and restoration actions.

These steps allow managers to link information about plant community composition collected during inventory with concepts of ecosystem dynamics to develop management plans aimed at long-term stewardship. For instance, management actions may seek to maintain a desired plant community with high forage quality, to restore native plants and animals that formerly occupied the site, or to create a mosaic of different plant communities favoring wildlife. In this way, STMs can help specify management objectives for a site, and serve as guides to maintain and restore ecosystem services.

The diagrammatic and narrative portions of STMs synthesize various sources of knowledge about ecosystem change, including scientific results, historical anecdotes, and local knowledge. The synthesis is used to develop predictions for how ecosystems respond to natural events and management actions (Bestelmeyer et al. 2009b). Conceptual STMs can be expanded into quantitative models by including estimates of the likelihood of change.

Well-developed STMs can serve as a basis for collaborative adaptive management (i.e., management by iterative hypothesis testing, involving multiple stakeholders; Susskind et al. 2012) (Chap. 1, this volume). These guidelines can be updated based on monitoring and new knowledge. In this way, STMs can facilitate a shift from rigid prescriptions based on a one-way relationship between science and management toward a constantly evolving set of recommendations based on collaborative learning and adaptation. Collaborative adaptive management is likely to be more effective than rigid rules of thumb as a basis for environmental stewardship, especially as global climate continues to change.

Because of the potential for STMs to link science to management, they are being developed with increasing frequency in rangelands and other ecosystems on several continents (Hobbs and Suding 2009). While some STMs were never intended to be used for management, others were developed as a basis for outreach and decision support. The linkage of STMs to on-the-ground decision-making, however, remains limited for a number of reasons, including a lack of adequate detail and specificity in STMs and the inability of institutions to develop and use STMs. Moreover, it is inherently difficult to determine the likelihood of transitions, especially given time lags and long timeframes needed to observe some transitions. Nonetheless, there is continued optimism that STMs can provide useful tools for bridging the science-management divide (Knapp et al. 2011b).

Our approach in this chapter is to (1) review the ecological basis for STMs, (2) outline the fundamental components of STMs, (3) review the experiences in several countries with the development and use of STMs (Australia, Argentina, the United States, and Mongolia), (4) identify and address challenges to the use of STMs for management, and (5) describe recent technical advances that may improve STMs, including participatory approaches in model construction, the use of STMs within a structured decision-making process, and mapping of ecological states. We conclude with a summary of strategies to improve the utility of STMs for collaborative adaptive management.

9.2   Conceptual Advances in the Ecology of State Transitions[1]

The publications of seminal papers on ecosystem resilience and event-driven vegetation dynamics in rangelands catalyzed a significant shift in thought among scientists and managers beginning in the 1970s (Westoby 1980; Walker and Westoby 2011) (Chap. 6, this volume). Prior to this time, the notions of climax vegetation and orderly succession following disturbance, stemming from early American plant ecology, were used to interpret vegetation dynamics, even in systems where vegetation change is now known to be discontinuous and irreversible (e.g., Campbell 1929). It is now widely acknowledged that (1) vegetation change in response to grazing or weather variations may not occur along a single continuum but rather may produce multiple stable plant communities; (2) vegetation change is not necessarily reversible; and (3) vegetation change can be discontinuous and sudden. While recognition of these patterns occurred prior to the development of STMs, the formalization of "state-and-transition" thinking via the models promoted a broadened view of how vegetation can change (Westoby et al. 1989).

In spite of the impact of STMs on general thought, the continuing challenge is to represent accurately the patterns, timescales, and drivers of change among states in particular settings. To this end, it is important to distinguish transient dynamics from persistent transitions between alternative states (Bestelmeyer et al. 2003; Stringham et al. 2003). *Transient dynamics*, driven by disturbance or weather events, produce significant but temporary changes in vegetation composition or production that can be reversed in a few years to several decades (e.g., via moderation of disturbance, succession, or weather events). *State transitions*, on the other hand, involve persistent changes in vegetation such that recovery of the former state is dependent on unacceptably long recovery times, active restoration, extreme events, or a reversal of climatic change that occurs over several decades or never occurs (Suding and Hobbs 2009). Below, we review the conceptual distinction between these types of dynamics, acknowledging that it may be difficult to distinguish them in practice.

9.2.1   Transient Dynamics

Whether a system undergoes transient dynamics or a state transition following a disturbance is influenced by a variety of factors, including plant traits that evolved in response to disturbance, the ability of alternative plant species to colonize a site, and

---

[1] Primary author B. Bestelmeyer.

the resistance of soils to degradation (Seybold et al. 1999; Cingolani et al. 2005). For example, in the Chihuahuan Desert where most historical grasslands have converted to eroding shrublands, grasslands dominated by the perennial grass tobosa (*Pleuraphis mutica*) have been comparatively resilient to drought and overgrazing episodes owing to its low palatability, vegetative reproduction via rhizomes that are protected below ground, and its dominance on landforms that receive water runoff and sediment from upslope positions (Herbel and Gibbens 1989; Yao et al. 2006). While disturbances such as continuous heavy grazing can cause significant change in vegetation cover and composition in many rangelands, recovery can be rapid, taking only a few growing seasons in productive settings (Fig. 9.1a), or occur slowly, taking decades in resource-limited environments (Miriti et al. 2007; Lewis et al. 2010). Species having slow recruitment and growth rates may exhibit significant time lags in recovery. Nonetheless, adjustments to the management strategy or disturbance

Fig. 9.1 Examples of transient dynamics and state transitions in rangelands. (**a**) Transient dynamics featuring a reversible shift between communities dominated by western wheatgrass, *Pascopyrum smithii*, (*1*) and blue grama grass, *Bouteloua gracilis* (*2*), in the northern Great Plains of North Dakota, USA; recovery of the more productive *P. smithi* community can occur in several years with changes to grazing management (courtesy of Jeff Printz). (**b**) Transient dynamics on the Santa Rita Experimental Range in the Sonoran Desert of Arizona, USA, starting with cholla cactus (*Opuntia imbricata*) dominance in 1948 (*1*), then burroweed (*Ambrosia dumosa*) dominance in 1962 (2), and increasing dominance of blue palo verde (*Parkinsonia florida*) from 1988–2007, which might represent a state transition (*3* and *4*; courtesy of Mitch McClaran). (**c**) A state transition from grassland to shrubland on the Jornada Experimental Range in the Chihuahuan Desert of New Mexico, USA, starting with high cover of black grama grass, *Bouteloua eriopoda*, (*1*) that may be reduced (*2*) and subsequently recovered, unless a threshold is crossed (*3*) after which *B. eriopoda* goes extinct and mesquite (*Prosopis glandulosa*) dominates (*4*). This change results in an eroding shrubland state that experiences infrequent co-dominance by another perennial grass, *Sporobolus* spp. (*5*), during periods of high rainfall

regime (e.g., via reduced stocking rates or reestablishment of natural fire disturbance regimes to remove woody plants) can be used to initiate recovery.

Weather variations are especially important causes of transient dynamics in rangelands featuring high interannual rainfall variability. For example, high winter precipitation initiates recruitment of burroweed (*Isocoma tenuisecta*) in the Sonoran Desert and burroweed dominance can be sustained for one to two decades until dry periods and senescence cause declines in density (Fig. 9.1b; McClaran et al. 2010; Bagchi et al. 2012). Such vegetation changes can be abrupt, but they do not necessarily represent a transition between alternative stable states. This is because vegetation change is predictably related to recent environmental conditions and it can be reversed via plant senescence or subsequent, common weather events (Jackson and Bartolome 2002; McClaran et al. 2010).

9.2.2 State Transitions

The hallmark of a state transition (sometimes referred to as a "regime shift"; Scheffer and Carpenter (2003)) is long-term persistence of new plant communities, or a new range of variation among plant communities that differs from that of the previous state. The persistence of new states can be caused by mechanisms that are internal to the ecosystem, such as competitive dominance of invaders or plant-environment feedbacks favoring new species under the same soil and climate conditions. In addition, directional changes in external environmental drivers, such as climate change, can cause the persistence of new states.

State transitions in rangelands have been described with the following sequence in some STMs produced in the United States (Fig. 9.1c). Weather variations or disturbances can cause transient dynamics within a historical or "reference" state resulting in two (or more) distinct communities (Bestelmeyer et al. 2003; Stringham et al. 2003). Certain of these communities may have low resilience and be susceptible to a state transition (called an "at-risk community"; Briske et al. 2008). Recovery to communities less likely to undergo a transition can occur with the return of favorable weather or reduced disturbance frequency or intensity. Alternatively, an intensification of adverse weather or disturbance can cause the plant community to cross a threshold (often called a "tipping point") into a new state. The new state may be stable with respect to the dominance of key plant species, but still exhibit transient dynamics among a set of plant communities that did not exist in the previous state (Friedel 1991).

The persistence of alternative states can be caused by invaders that are superior competitors when given a foothold in a community (Seabloom et al. 2003). Alternatively, the cessation of natural disturbances can lead to the dominance of superior competitors. For example, the cessation of fire in prairie grasslands can lead to increases in woody plant density and size. When the density of woody plants limits grass (and fuel) continuity and fire spread, and when woody plants grow to a size that limits their mortality in response to fire, then reintroduction of fire can no longer recover the grassland state (Twidwell et al. 2013b) (Chap. 2, this volume). These two types of state transitions involve changes in dominant plants, but not necessarily a change in plant production or other ecosystem properties. While production and soil carbon levels may be maintained (or even increased) with such transitions (Barger et al. 2011), the provision of other ecosystem services (e.g., forage for livestock production) is often reduced (Eldridge et al. 2011) (Chap. 14, this volume).

Plant production can be reduced when the loss of dominant perennial plants leads to a reduction in soil water infiltration, accelerated erosion that reduces soil fertility, or rising water tables resulting in salinization (D'Odorico et al. 2013). In arid and semiarid rangelands, there may be thresholds in plant patch organization below which positive feedbacks between plant patches, resource acquisition, and plant survival and reproduction break down, resulting in a persistent low-productivity/high bare ground state (Kéfi et al. 2011). In other words, if larger plant patches become fragmented too much, the plants occupying those patches suffer due to increased soil erosion and decreased resource availability (Svejcar et al. 2015). State transitions associated with soil degradation are often called "desertification."

State transitions often have multiple, interacting causes (Fig 9.2; Walker and Salt 2012). *Drivers* that are external to the system can cause a gradual or abrupt change in *controlling* (or "slow") variables. The controlling variables directly determine the *state* variables of interest. An example would be a change in the intensity and duration of grazing periods (the driver) that gradually reduces grass root mass, basal cover, and soil organic matter (controlling variables) to affect plant foliar cover, production, and composition (state variables). *Triggering events* occurring over relatively short time

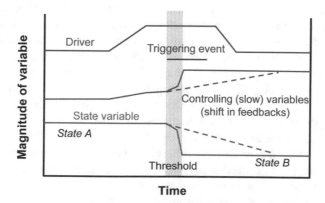

Fig. 9.2 A schematic illustrating the pattern and interaction of variables over time involved in state transitions. Elements include external drivers that are returned to pre-transition levels, discrete triggering events occurring over short periods that exacerbate the effects of changing drivers, responses of internal controlling (a.k.a slow) variables that may exhibit feedbacks with state variables, and transitions in state variables. The position of different states and the threshold between states are noted. *Dashed lines* indicate that changes in controlling and state variables need not be abrupt

periods, such as an extreme drought, can amplify the rate and magnitude of change in controlling variables and may produce abrupt changes in state variables.

Abrupt changes can occur in both transient dynamics and state transitions. In state transitions, however, *feedbacks* among controlling variables and state variables can lead to persistence of the new state. For example, reduced plant cover leads to increased soil erosion and reduced litter inputs, accelerating the loss of soil organic matter and the ability of the soil to store moisture. Modifications to one or more feedbacks can produce an abrupt change in state variables (i.e., community structure and composition) to create an alternative state, even after the driver has returned to previous levels. The *threshold* between states is the period in time when changes in controlling variables, and possibly feedbacks, lead to persistent changes in state variables.

State transitions need not always be abrupt, however. Abrupt changes in controlling variables can cause strongly lagged, nearly linear responses in state variables. For example, long-lived plants can persist long after the environment required for their establishment has disappeared, leading to a gradual transition after the threshold is crossed. Alternatively, controlling variables may change gradually and be tracked by gradual changes in plant composition, such as with climate change (see the dashed lines in Fig. 9.2; Hughes et al. 2013). Even irreversible state transitions can occur gradually.

9.2.3   Distinguishing Transient Dynamics from State Transitions

The criteria used to distinguish transient dynamics from state transitions depend on the length of time needed for recovery and the implications of these timelines for management. Recovery that does not occur within an acceptable management time-frame without intensive effort is often categorized as a state transition (Watson and Novelly 2012). What is deemed "acceptable" varies among users and contexts, but should ideally be based on measurable recovery criteria. For example, recovery that takes longer than 3 years following a change in grazing management is treated as a state transition in Mongolia by the Mongolian government (National Agency for Meteorology and Environmental Monitoring and Ministry of Environment 2015). For the US government, changes are called state transitions when they are irreversible or take "several decades" for recovery of the former state (Caudle et al. 2013).

It is also important to realize that the type of dynamics recognized might depend on the specific plant functional groups considered. For example, in the Calden (*Prosopis caldenia*) forests of central Argentina, herbaceous plants can exhibit transient dynamics and be managed over multi-year timescales (Llorens 1995), even as gradual shrub and tree encroachment over decades increasingly constrains herbaceous cover and composition, representing a state transition (Dussart et al. 1998). Unlike the simpler models of the past, transient dynamics and state transitions can be represented simultaneously in STMs.

9.3   Development of State and Transition Models

STMs should be designed to serve land managers and policymakers by: (1) communicating locally relevant indicators of transient dynamics and state transitions and their consequences for ecosystem services; (2) describing the drivers and environmental conditions affecting susceptibility to transitions; (3) recommending management to avoid undesirable transitions (i.e., resilience management) and to obtain desired ecosystem services; and (4) identifying realistic restoration or adaptation options for alternative states (Bestelmeyer et al. 2009a).

Assembling the evidence to support an STM can be accomplished in most cases using a combination of sources. Monitoring data, historical records, comparisons of plant communities and surface soil characteristics among sites with different management histories, published experiments, and local knowledge can be combined to infer vegetation dynamics (Bestelmeyer et al. 2009b). In any case, it is important to recognize that the dynamics represented in STMs are hypotheses that should be tested through the outcomes of management decisions.

The structure of STMs represented in the literature to date is highly diverse. Different authors have used different conventions to develop model diagrams and narratives. Models can be entirely qualitative/descriptive (Knapp and Fernandez-Gimenez 2009; Kachergis et al. 2013), quantify only properties of states (Bestelmeyer et al. 2010; Miller et al. 2011), or quantify states and/or transitions (Jackson and Bartolome 2002; Czembor and Vesk 2009; Rumpff et al. 2011). Across all model types, however, there are a set of common elements that define an STM.

9.3.1   Define the "Site"

An STM should focus on the alternative states and dynamics of an environmentally uniform area (Peterson 1984). STMs focus on temporal dynamics, so inclusion of significant ecosystem differences due to inherent differences in soil or climate confuses space and time and may lead to flawed interpretations. In rangelands and forests, terrestrial land units such as ecological sites or potential vegetation types approximate areas of environmental uniformity and can define the spatial extent of individual STMs (Bestelmeyer et al. 2003; Yospin et al. 2014). Attempts to define STMs at too fine a spatial scale, however, may result in an unwieldy number of STMs and make comparisons among environmental contexts difficult. For this reason, STMs can be developed at a relatively broad spatial extent, such as a landscape, and the effects of varying soil and climate context within the landscape can be described as a narrative for transitions. Grouping land areas according to "disturbance response groups" in the northwestern USA similarly seeks to produce more general STMs that sacrifice spatial precision for greater efficiency of development and use (T. K. Stringham, pers. comm.).

9.3.2 Define the Alternative States

Each state that is possible for a site is described. In some instances, plant communities linked via transient dynamics are represented as "states" in the broad sense (Jackson and Bartolome 2002; Bagchi et al. 2012) and in other cases, alternative stable states in the narrow sense are emphasized and transient dynamics within states are described separately or ignored (Miller et al. 2011).

Descriptions of transient dynamics have been based on differences in species composition of plant communities that are relevant to management, such as grazing use or wildlife habitat value. Descriptions of alternative states tend to focus on the relationships of vegetation structure to the processes maintaining that structure, such as erosion, fire frequency, or nitrogen fixation (Petersen et al. 2009; Kachergis et al. 2011). Some STMs depict both alternative states and transient dynamics within states by using boxes for plant communities and separating certain communities using irreversible transitions across a threshold boundary, signifying a state transition (Oliva et al. 1998). US agencies developing STMs for Ecological Site Descriptions (see Sect. 9.4.3; Fig. 9.3) identify

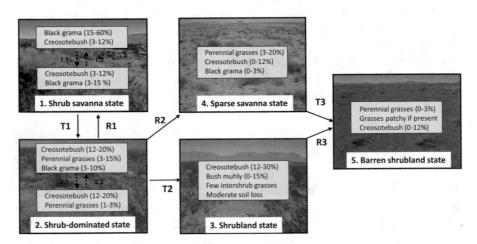

T1. Continuous heavy grazing, thinning and patchy loss of black grama, lack of fire, shrub proliferation, patchy erosion.
R1. Shrub control associated with grazing deferment or prescribed grazing and climate permitting black grama recovery
R2. Shrub control after grass is sparse or erosion advanced, or followed by poorly planned grazing
T2. Loss of remaining interspace grasses, gradual loss of soil organic matter, infill of shrubs, and soil erosion
R3. Shrub control when soil loss or other constraints preclude grass establishment
T3. Poorly planned grazing that causes collapse of grass population

Fig. 9.3 An example of an STM developed for the Gravelly ecological site, including soils that are loamy-skeletal Haplocalcids and non-carbonatic Petrocalcids in the 200–250 mm precipitation zone of the Southern Desertic Basins, Plains, Mountains Major Land Resource Area (MLRA 42) of New Mexico and west Texas, USA. Following conventions used by US federal land management agencies, rapidly reversible community phases are small boxes whereas states defined by important management and ecological thresholds are defined by large boxes. Each phase is characterized by foliar cover values for dominant or key plant species or functional groups that distinguish it from other phases. In the abbreviated narrative, *T* signifies an unintentional transition whereas *R* signifies a transition caused by restoration action (that can have unintended consequences)

transient dynamics among communities within a state (called "community phases") as smaller boxes connected by reversible arrows, that are nested within larger boxes representing alternative states (USDA Natural Resources Conservation Service 2014).

Each community or state is typically given a narrative to describe its characteristics and, in some cases, the important ecosystem services it provides. Numerical values allow quantitative distinction of states (Fig. 9.3). It is useful to describe the management actions or natural processes that maintain or weaken the resilience of each state and the conditions characterizing low resilience (Standish et al. 2014). Alternative states may exhibit variations in resilience, such that undesirable shifts can be avoided (Briske et al. 2008) and opportunities for restoration toward desirable states can be exploited (Holmgren and Scheffer 2001).

9.3.3   Describe Transitions

Each transition, represented by arrows, is given a narrative. Transient dynamics are typically attributed to perturbations such as grazing or fire, rainy periods or droughts, or to succession. As described in Sect. 9.2.2 and Fig. 9.2, state transitions can be described using four basic elements. First, the mechanisms causing a shift among states are described, including external drivers or triggering events, changes in controlling variables and feedbacks, and indicators of change based on controlling variables (e.g., evidence of soil erosion) or state variables (changes in plant composition). Timelines for transitions can be described, such as whether they are gradual or abrupt relative to management timeframes. Second, the constraints to recovery of the former state can be described (sometimes referred to as a threshold), including how altered feedbacks or environmental conditions preclude the appearance of some plant communities. Third, strategies for the reversal of transitions through restoration actions can be described. Fourth, context dependence in space (such as soils or climate) or time (such as weather conditions) that affects the likelihood of undesirable transitions or restoration success can be described.

9.4   Development and Applications of STMs in Rangeland Management

Although many STMs have been created, four countries have produced groups of STMs to support rangeland management. How these efforts originated and progressed (or didn't progress) provide important lessons for future efforts. Below, authors familiar with the history of STM development in Australia, Argentina, the United States, and Mongolia offer accounts representing a variety of global contexts.

9.4.1   Australia[2]

9.4.1.1   History

Australia was an early adopter of STMs, particularly in their application to range-land management. This early interest stems from two developments. First, Australian rangeland ecologists were at the forefront of considering how concepts of nonequilibrium dynamics and thresholds were applicable to the management of arid range-lands (Westoby 1980; Friedel 1991). Second, unlike the United States where formal monitoring of rangelands had been instituted based on the range condition and trend concept (Dyksterhuis 1949; Shiflet 1973), Australia had no single or dominant institutionalized model for rangeland monitoring and, consequently, a number of approaches were developed (e.g., Watson et al. 2007).

The absence of a widely accepted framework for describing plant community dynamics in Australia, coupled with the appeal of the state-and-transition format, led to keen interest from the rangeland research community. Adoption was particularly rapid in tropical Australia where the research and management of tropical grazing lands was moving away from a long phase of pasture agronomy associated with the use of introduced species to one based on sustainable utilization of the largely intact, native savannas (Ash et al. 1994; Brown and Ash 1996). STMs provided an effective approach for describing the dynamics of many plant communities in tropical rangelands. This resulted in a special edition of the journal Tropical Grasslands on STMs (Taylor et al. 1994).

In addition to providing qualitative STMs for the major plant communities used for livestock production across northern Australia, the journal issue raised a number of concerns about the broader use of STMs in rangeland management. Major concerns included strategies for communication using models and their role in management; the ability (or inability) to define quantitatively both states and transitions for specific plant communities; and incorporation of spatial processes, such as water flow (Brown 1994; Grice and MacLeod 1994; Scanlan 1994). Shortly afterward, Watson et al. (1996) questioned the strong focus on event-driven processes and abrupt change and suggested that a model of more continuous, cumulative change was just as appropriate to describe vegetation dynamics in many systems. Further, they suggested an emphasis on the management of vegetation within an ecological state to either prime it for a desired transition or protect it from an undesired transition.

Acceptance of STMs was also evident in southern Australia, such as the original bladder saltbush model used by Westoby et al. (1989), as well as in arid rangelands, particularly where piosphere effects can lead to alternative vegetation states within a management unit (Hunt 1992). A strong interest from ecologists in the fragmented and remnant temperate woodlands drove further conceptual development of STMs,

---

[2] Primary authors are A. Ash and J. Brown.

primarily in the context of restoration (Price and Morgan 2008; Hobbs and Suding 2009; Rumpff et al. 2011).

9.4.1.2 Current Applications

The early interest in developing and applying STMs was not followed by a well-resourced or formal approach to embedding STMs in management of rangelands used for livestock grazing. STM development was carried out via research projects, or by informal approaches in land management or extension agencies, often driven by enterprising individuals, but rarely through systematic institutional initiatives.

One of the limitations in using STMs has been a robust approach to defining states and the thresholds between states. There is a lack of quantitative data for the majority of plant communities and descriptions of dynamics have tended to be qualitative. Moreover, defining and applying threshold concepts in practical management can be problematic because of the potential misinterpretation of management needs (Bestelmeyer 2006). Thus, a quantitative basis for distinguishing state transitions from transient dynamics is immensely important.

There were early efforts in Australia to describe transitions quantitatively using Markov models (Scanlan 1994). The use of Bayesian belief networks to better incorporate uncertainty and expert knowledge has provided an improved conceptual basis for defining states (Bashari et al. 2009) but to date has had limited application. Other approaches included simulation/scenario modeling based on historical rainfall, understory grassland growth, and utilization rates by livestock (Hill et al. 2005). While the simulation modeling approach has proved useful in research for understanding system dynamics, it has not translated well to practical application. Another approach to testing the applicability of the state transition concept is to monitor how frequently transitions are occurring. Watson and Novelly (2012) used an extensive, long-term monitoring dataset from Western Australia to determine how often predefined thresholds were crossed. During a 17-year evaluation period 11 % of grassland sites and 1 % of shrubland sites were judged to have undergone a transition. More recently, a study in semiarid wetlands in Australia provided a robust approach for quantifying the causes of state transitions and using logistic models to generate future transition scenarios (Bino et al. 2015). While there has been a range of quantitative approaches tested, a consistent, structured approach to defining and testing for state transitions is still lacking.

Following the initial interest in the STM approach and continued, sporadic development of models for different plant communities (e.g., Phelps and Bosch 2002), there is little evidence that STMs have been formally incorporated into pastoral management in Australia, by either individual producers or by land management agencies (Watson and Novelly 2012). There is, however, anecdotal evidence that STMs have influenced how rangeland professionals communicate with land managers. One argument is that this "mindset change" is sufficient and that institutionalizing a highly proscribed approach to STMs will stifle flexibility. However, this may be outweighed by the risk of not having a consistent, institutionalized approach to vegetation management in an environment where there are declining resources and capacity in management agencies to proactively assist land managers.

Why has the development of STMs slowed in Australia while it has gained momentum in other countries, most notably the USA? Australia lacks the critical mass of research and extension personnel to develop a comprehensive catalogue of STMs for plant communities at a spatial scale relevant to management. In addition, there is a paucity of robust information on the management-scale distribution of soil properties and accompanying plant community dynamics, exacerbated by the absence of a well-supported and consistent national approach to field-based rangeland monitoring. While that deficiency is being overcome to some extent through a more coordinated national approach to synthesizing information on rangeland condition and trend (Bastin et al. 2009), Australia still lacks a widely applied, ecologically based site classification system such as "ecological sites" in the USA (Brown 2010) which underpins the development of spatially specific STMs.

The lack of formalized STMs does not mean that rangeland management is occurring in the absence of general principles and locally explicit guidelines. Many rangeland professionals working in land management agencies across Australia have been exposed to STMs and either implicitly or explicitly use STM concepts when engaging with producers. In addition, considerable effort has been expended on developing grazing land management education courses for producers, with the most visible example being in northern and central Australia (Quirk and McIvor 2003). However, in an effort to simplify concepts of land condition and its interaction with grazing management, STMs within these educational courses have been replaced by a simple four-level (A[best], B, C, D[worst]) land condition class scheme (e.g., Bartley et al. 2014). While this has been effective as a communication tool, it has tended to de-emphasize the importance of processes responsible for long-term vegetation change. For example, Bartley et al. (2014) showed that even with recommended grazing management practices over a 10 year period, the improvement from class "C" to "B" was proceeding very slowly. This might indicate a state transition related to soil degradation and/or the presence of an exotic grass that was limiting native perennial grass re-establishment. The land condition classes cannot distinguish transient from state transition dynamics or capture the mechanisms involved.

Having been the leaders in the initial development of STMs, rangeland ecologists and land administrators in Australia should consider how development of STMs has progressed elsewhere in the world to see what innovations in application might be relevant to Australia. Recent approaches provide useful frameworks for incorporating STMs into practical management (Bestelmeyer et al. 2009b; Suding and Hobbs 2009). These frameworks go well beyond the development of STMs themselves to include aspects of empirical data to support development of STMs, monitoring protocols, and adaptive management. Ultimately, success will be judged by the utility and relevance of STMs to rangeland managers.

9.4.2   Argentina[3]

9.4.2.1   History

Interest in STMs began in the early 1990s following the publication of the seminal paper by Westoby et al. (1989). STMs were motivated in large part by the need for a new framework to describe plant community dynamics. A series of STMs developed for the arid Patagonian region were the earliest examples (Paruelo et al. 1993). Models for arid environments usually involved the effects of grazing, initially causing a loss of palatable grass species but eventually causing a reduction in total grass cover associated with increasing bare soil and erosion rates. Following these models, a decline in plant cover results in a reduction of soil water holding capacity and plant production, causing a feedback to water and wind erosion that further inhibits reestablishment of grass species (Cesa and Paruelo 2011) (Chap. 3, this volume). State transitions were regarded as irreversible or difficult to reverse. This sequence corresponds to most STMs developed for the Patagonian steppe in Paruelo et al. (1993).

STMs were developed for more humid environments later in the 1990s, including montane grasslands (Barrera and Frangi 1997; Pucheta et al. 1997), Pampean grasslands (Aguilera et al. 1998; Laterra et al. 1998; León and Burkart 1998), and herbaceous vegetation of the Caldenal/Espinal ecoregion (Llorens 1995). These STMs emphasized changes in species composition rather than large decreases in total plant cover. In these models, grazing did not produce noticeable changes in soil physical properties through erosion as observed in the Patagonian region because total plant cover is usually not greatly reduced by grazing.

A third type of STM described state transitions in "mallines," a local name for meadows with high productivity and biodiversity within the Patagonian steppe, and which are an important source of forage for livestock (Paruelo et al. 1993). Overgrazing and trampling by livestock in mallines produces a transition to an alternative state due to the loss of plant cover that promotes increased runoff and/or soil salinization. Increased runoff and erosion result in gully formation. Consequently, altered hydrology causes a shift in plant communities. Similar hydrologically based state transitions are observed in alluvial floodplains of the Chaco region (Menghi and Herrera 1998).

In contrast to early expectations, these STMs had little impact on science and management in Argentina. Exploring the reasons why interest and activity waned may provide insights for improving the usefulness of the STM framework in Argentina and elsewhere. First, STMs developed in the 1990s did not feature adequate detail. STMs described drivers associated with transitions but provided little description on processes and mechanisms controlled by the drivers. Narratives did not contain information on thresholds and processes controlling the functions of alternative states (i.e., feedbacks). Transitions identified in these models were rarely

---

[3] Primary authors are D. Lopez and R. Peinetti.

experimentally tested (López 2011). Most models were superficial representations of community dynamics that did not provide useful predictions.

Second, STMs were used to synthesize general regional information on ecosystem dynamics but lacked the site specificity needed for practical applications. They contained few recommendations on management practices or restoration actions to reverse undesirable transitions. Similar to Australia, the lack of a land classification system tied to STMs such as ecological sites (Bestelmeyer et al. 2009b) led to confusion about the spatial domain to which a particular model applied.

Third, important types of state change were simply not addressed by existing models. Tree and shrub encroachment and "thicketization" of woody plants represents one of the most important kinds of state change occurring in several ecosystems in the central and northern parts of the country (Brown et al. 2006). The thicketization of forests and grasslands has received a great deal of attention in basic and applied sciences (e.g., Dussart et al. 1998), including information about management practices, but there have been very few cases in which this understanding was incorporated in STMs.

Finally, there have been few incentives for scientists to expand development of STMs. Modern Argentinean ecological science, as directed by funding and reward systems over the last few decades, has been focused on short-term studies that yield rapid publication and career advancement (Farji-Brener and Ruggiero 2010). In this environment, there was little incentive for integration across different case studies at a regional level or long-term studies to support STM development.

9.4.2.2 Current Applications

There is a substantial demand from society for responsible natural resource management, in part due to the alarming deforestation rates of the last 10–15 years in the semiarid and humid forests of Argentina (Gasparri et al. 2013). Societal demand for rational management forced the establishment of new federal regulations on the use of natural resources. To apply these regulations, policymakers have recognized that a new suite of management decision tools and a basis for assessment and monitoring are required, leading to a renewed interest in STMs manifest in the recent Argentinean Rangeland Congress in 2013 (http://inta.gob.ar/documentos/jornadas-taller-post-congreso-argentino-mercosur-de-pastizales-cap2013). At this meeting, there was general consensus among participants that STMs associated with ecological site concepts should be explored as an option to organize the available information under a common framework for both rangelands and forests. A research network was proposed. It is hoped that this network will serve as a platform for interactions between different research groups and thereby stimulate the production of systematically structured STMs and ecological site classifications across the nation. As yet, funding the network and motivating coordination among researchers via a network remains a significant challenge.

9.4.3   United States[4]

9.4.3.1   History

The official adoption of STMs in 1997 as a component of land evaluation can be considered a paradigm shift in US rangeland science. Clementsian, or succession-based, concepts of community dynamics originating in the early twentieth century provided acceptable explanations for observed vegetation changes in rangelands, particularly in response to livestock grazing. Succession concepts embodied in the "range succession" or "range condition" model (Westoby 1980; Joyce 1993) worked fairly well in highly resilient prairie ecosystems where much of the grazing livestock and conservation efforts were concentrated. Even leading proponents of the range condition model (Dyksterhuis 1958; Passey and Hugie 1962), however, noted that the scope of this model was limited to forage for domestic livestock and climax plant communities dominated by perennial, herbaceous species.

In spite of these caveats, use of the range condition model spread throughout US rangelands and was linked to evaluation procedures and financial and technical assistance from federal land management agencies. A relatively well-trained and mature workforce able to detect discrepancies between model predictions and actual conditions, and make ad hoc adjustments to management prescriptions (Shiflet 1973), created a sense of complacency among adherents (Joyce 1993). Strong connections among universities, agencies, and managers strengthened the ability of the rangeland profession to adapt to these inconsistencies (Svejcar and Brown 1991). However, as the application of the range condition concepts spread into diverse rangeland settings, such as those experiencing long-term shrub encroachment, significant limitations in model application became apparent.

As the applicability of the range condition model began to be questioned, theoretical ecologists were developing alternatives to the Clementsian model to explain how ecosystems, and specifically rangelands, behave (Holling 1973; May 1977). The multiple stable state model was less deterministic than the range condition model and multiple trajectories were possible, better matching observations of rangeland change. Soon afterward in the 1980s, concern about the appropriateness of range condition as a universal metric of rangeland function surfaced within US land management agencies. The inability to link non-forage values to the range condition model was now recognized as a major limitation of assessment procedures (Society for Range Management 1983). By the end of the decade, there was widespread dissatisfaction with the application of the range condition model to all rangeland ecosystems (Lauenroth and Laycock 1989; Pieper and Beck 1990).

In this context, the impact of the first publication on STMs (Westoby et al. 1989) was rapid and substantial. Following this paper, there was a flurry of experimental and review papers exploring the application of STMs to particular rangeland ecosystems, both within and outside of the USA. Federal land management agencies

---

[4]Primary authors are J. Brown and P. Shaver.

undertook extensive reviews of the use of the range condition model as a basis for technical and financial assistance versus implementation of an STM-based approach, culminating in publications by the US National Research Council (National Research Council 1994) and the Society for Range Management (Task Group on Unity in Concepts and Terminology Committee Members 1995). The two reviews called for standardization of rangeland evaluation approaches and replacement of the range condition model with a model that could account for multiple stable states. Different plant communities could have distinct values to society and call for different management approaches, but a primary focus was to preserve "site potential"— the option to sustain desired plant communities and services—by avoiding accelerated soil erosion.

These two reviews were catalysts for adoption of STM concepts by natural resource agencies. Beginning in late 1990s, STMs began to be developed and used by rangeland specialists, primarily those associated with USDA Natural Resource Conservation Service (NRCS), for communication with ranchers about management needs and to provide guidance in administering federal financial assistance. The policy implications of the latter led to a systematic approach to STM development within NRCS. Widespread development of STMs, however, was delayed because they had to be linked to "ecological site descriptions" (ESDs; formerly called "range site descriptions"). ESDs are documents that had long served as the site-specific basis for management recommendations by federal land management agencies. ESDs are linked to soil survey databases through the connection of ecological sites to soil maps maintained by the NRCS. Application of the range condition model via ESDs involved the calculation of plant community similarity between an observed and a single, historical climax plant community identified for each ecological site (Dyksterhuis 1949). In order for the rangeland condition model to be replaced, thousands of STMs would have to be developed for ecological sites across the USA, each requiring the description of multiple plant communities.

9.4.3.2 Current Applications

Acceleration of STM development represents a major logistical challenge because of the large number of STMs needed, particularly in the eastern half of the USA. Added to these logistical concerns, there has been a lack of clear institutional guidance on how to structure STMs that were developed in the late 1990s and early 2000s. Few agency employees have been dedicated to ESD and STM development, and in some locations, contracts were awarded to private enterprises to work on STMs. In most locations, however, STM development was an added duty for existing federal agency staff. Much of this work, although creative, lacked coordination. In some cases, transitions featuring overwhelming indicators of persistence were presented as transient dynamics following the range condition model. In other systems that feature transient behavior, community variations were presented as alternative stable states. The resulting inaccuracies in some STMs have elicited criticism of how they are produced (Twidwell et al. 2013a).

In spite of these problems, STMs have gained greater visibility and are increasingly viewed as useful tools for communicating research and management recommendations. New definitions of STM components, scale considerations, and a greater variety of ecosystem attributes linked to STMs (Briske et al. 2008; Bestelmeyer et al. 2010; Holmes and Miller 2010) have emerged. Systematic approaches to the development, evaluation, and refinement of STMs (Bestelmeyer et al. 2009b; Bestelmeyer and Brown 2010), informed by the successes and limitations of early model development efforts, have been incorporated in recent US government guidelines (Caudle et al. 2013; USDA Natural Resources Conservation Service 2014). These comprehensive guidelines address priority setting, resource allocation, and progress reporting. They also incorporate recent scientific literature, diverse agency policies, and user needs. Nonetheless, significant challenges remain, particularly (1) funding and expertise required to accelerate STM development and deliver STMs to the public, (2) inclusion of information pertaining to ecosystem services other than livestock production, such as climate change mitigation and adaptation, hydrology, and species of conservation concern, (3) how to make STM development more participatory and inclusive to support adaptive management, and (4) how to address the impending effects of climate change in models developed with a high degree of spatial specificity (Knapp et al. 2011b; Twidwell et al. 2013a). Current NRCS and interagency efforts are focused on these concerns.

9.4.4   Mongolia[5]

9.4.4.1   History

Mongolia is dominated by rangelands, and livestock production is a critical component of the national economy and cultural traditions. Nonetheless, Mongolia never adopted well-defined or universally accepted rangeland evaluation concepts or procedures. The shift from a nomadic or transhumant, subsistence herding system into a market economy in 1993 led to dramatic increases in livestock numbers and loss of herder mobility (Fernández-Giménez 2002). The perception of widespread rangeland degradation associated with overgrazing (Bruegger et al. 2014; Hilker et al. 2014) motivated interest in rangeland evaluation and monitoring procedures. A systematic approach was needed because assertions about rangeland degradation have been challenged within the Mongolian government and the broader academic community (Addison et al. 2012), creating conflict about the need for interventions to reduce stocking rates versus calls by some officials to encourage larger livestock numbers. Beginning in 2004, Green Gold Mongolia (GG), a project funded by the Swiss Agency for Development and Cooperation, initiated efforts to build a national capacity for reporting on the present state and future trend of Mongolian rangelands. In addition, GG sought to develop tools to facilitate rangeland management

---

[5] Primary authors are B. Densambuu and B. Bestelmeyer.

at local, regional, and national levels. Following exposure to the concept of ESD-based STMs from US scientists and land managers in the mid-2000s, GG and its government partners undertook an effort to develop ESDs for Mongolia.

9.4.4.2 Current Applications

In many ways, the relatively recent Mongolian experience with STMs takes advantage of what was learned in the early development efforts of Australia, Argentina, and the USA. STM development began in concert in 2008 with the development of a standard methodology for vegetation measurement, based on procedures used by US government agencies (Herrick et al. 2005). These procedures were officially adopted by the Mongolian government in 2011. In addition to providing a sound basis for reporting trends in rangeland vegetation, adoption of a unified measurement method ensured that cover and production values reported in STMs were comparable to monitoring data produced by the National Agency for Meteorology and Environmental Monitoring (NAMEM). Training for a GG research team on methods to develop STMs and database management began in the USA in early 2009, followed by data collection co-occurring with training in Mongolia from 2009 to 2014. Following recommendations adopted by US agencies (Bestelmeyer et al. 2009b; Knapp et al. 2011a; USDA Natural Resources Conservation Service 2014), inventory of vegetation and soils was conducted at over 600 sites across Mongolia, coupled to workshops aimed at eliciting local knowledge about reference conditions, the presumed causes of vegetation change, and to identify informative sites for inventory. These data are the basis for STMs that were included as a report for the Mongolian government in 2015 (https://www.eda.admin.ch/content/dam/countries/countries-content/mongolia/en/Mongolia-Rangeland-health-Report_EN.pdf).

A National Ecological Site Core Group was established in 2011 composed of experienced plant community ecologists representing different ecoregions across Mongolia as well as decision-makers of key institutes able to develop shared interpretations of inventory data. The National Core group (1) reviews published materials to establish reference conditions and causes of state change, (2) works in close collaboration with the GG research team in developing STMs, and (3) performs outreach to encourage adoption of materials by local government and herder cooperatives.

Because of the magnitude of the project, the limited budget, and the need for landscape-level information matched to herding and transhumance patterns, the decision was made to produce broad-level concepts for ecological sites based primarily on landforms and large differences in soil texture or hydrology. These classes, called Ecological Site Groups (ESG), combine finer-level soil classes that are equivalent to ecological sites in the USA (e.g., Moseley et al. 2010). STMs are developed for each ESG, resulting in 3–5 STMs per ecoregion and 25 total STMs for Mongolia (http://jornada.nmsu.edu/files/STM_Mongolian-catalogue-revised_2015.pdf). Because vegetation dynamics do not differ strongly across ecological sites within an ESG, the general models are deemed adequate for evaluation and management recommendations.

The specification of rangeland management strategies to maintain or recover perennial grasses is a primary objective of the STM development effort. In most of the sites sampled, the presence of well-distributed, remnant perennial grasses suggests that plant community recovery could occur in a few years to several decades with changes to grazing management (Khishigbayar et al. 2015). Thus, STMs are being designed to contain detailed information about recommended stocking rates and grazing deferment periods, tailored to the objectives of either maintaining a state or recovering a former state. Recommendations and expectations are linked to specific vegetation cover indicators that can be monitored.

In addition to their use as rangeland management guides by local governments and herder groups, STMs are being embedded in the activities of two government agencies. NAMEM has responsibility for monitoring 1550 plots across Mongolia to report on national rangeland trends. A lack of information about reference conditions and trends in monitoring data has precluded clear statements about rangeland health. Based on STMs drafted for most common rangeland communities in different ecoregions of Mongolia, NAMEM was able to conclude, preliminarily, that Mongolian rangeland communities are in general altered from historical reference states but that relatively rapid recovery was possible in the majority of cases.

STMs can provide a link between monitoring interpretations and management recommendations at the local level. The Agency for Land Affairs, Geodesy and Cartography (ALAGAC) is responsible for land management planning and its implementation nationally. STM concepts are being integrated into participatory rangeland management plans in several pilot areas. These pilot programs will provide a test of the value of the information content of STMs and therefore how they should be refined. As of 2015, expectations are high. Herder groups are using maps based on STMs (including information about recent forage availability and desired community change) to plan grazing and resting periods. It is encouraging that STMs are being used as a basis for such specific management actions.

9.4.5   Summary of STM Applications

The cases described above suggest that major efforts to develop STMs have taken different trajectories following the introduction of the concept in 1989. In Australia and Argentina, initial enthusiasm and progress was not sustained due to limitations in the data available to develop STMs, the dearth of land classification systems as a basis for STMs, and lack of resources and incentives for scientists and managers. In the USA, these limitations were overcome to varying degrees by the linkage of STMs to rangeland evaluation systems and financial assistance programs supported by government agencies. The vast scientific and administrative infrastructure provided by well-funded US government agencies has supported the nationwide development of numerous STMs. While this strategy has dramatically accelerated STM development compared to Australia and Argentina, it also introduced logistical difficulties associated with managing such a large number of STMs.

The Mongolian effort takes advantage of recent advances and lessons learned. STM development there was motivated by national concerns over rangeland degradation that attracted international development support. A dedicated team of scientists worked with government agencies to develop a relatively simple land classification system as a basis for STMs and employed a broadly collaborative approach to develop STMs. Furthermore, the STMs and related educational materials were purpose-built for collaborative rangeland management at broad spatial scales characteristic of transhumant and nomadic grazing systems of the country. The Mongolian experience may provide a useful model for STM development efforts for many parts of the world.

9.5   Knowledge Gaps[6]

The limitations to STM use highlighted above and recent evaluations of STMs in the USA (Knapp et al. 2011b; Twidwell et al. 2013a) suggest several overarching challenges that must be addressed in order to develop more useful STMs and better employ them for management. Below, we describe the main challenges and strategies for responding to them.

9.5.1   Reference States, History, and Novel Ecosystems

STMs, such as those used in the USA and Mongolia, often define a reference state that represents historical or a "healthy" set of ecosystem conditions for society, such that a primary goal of management is to maintain the reference state or to restore it (Fulé et al. 1997; Stoddard et al. 2006). Reference states are usually ascertained using historical information or measurements gathered in areas that have not been transformed relative to historical conditions. In many ecosystems, the societal significance and desirability of the reference state is straightforward when that state is well known and when it supports a set of ecosystem services valued by stakeholders.

In other cases, however, there can be difficulties in identifying a meaningful reference state. Historical conditions may be poorly understood, such that there is controversy about the plant communities present and the nature of disturbance regimes (Whipple et al. 2011; Lanner 2012). This may be especially problematic for plant and animal species that rely on a variety of states (Fuhlendorf et al. 2012). For example, a persistent, low plant cover state associated with prairie dog disturbance is necessary to support some native bird species in shortgrass steppe ecosystems (Augustine and Derner 2012). Thus, areas that may appear degraded to some

---

[6] Primary author is B. Bestelmeyer.

observers, and with respect to some ecosystem functions, may support biodiversity and valued species.

Furthermore, the recent concept of "novel ecosystems" acknowledges that it may not be practical to target a historical state as a management goal if the likelihood for restoration success is low or the costs high (Hobbs et al. 2009) (Chap. 13, this volume). In such cases, the costs of restoration should be evaluated relative to the ecosystem services provided by different states (Belnap et al. 2012). In some cases, it may be preferable to manage for alternative states. For some scientists, however, evaluations based on ecosystem services rather than historical fidelity are controversial (Doak et al. 2014).

The designation of reference conditions should be based on a broadly collaborative process and take into consideration several factors including history (both recent and evolutionary), the physical processes affecting potential plant communities (climate, soils, and topography), a recognition of specific time scales for disturbance and other processes, practicality of use, and the variety of ecosystem services of interest in particular ecosystems. Similarly, management objectives should be defined in a circumspect and collaborative manner. Managing toward reference conditions may be preferred in some locations, while managing for alternative states may be useful in others.

9.5.2   Broader Representation of Ecosystem Services

Given that STMs are principally used for communication with particular sets of managers, grazing managers for example, they often emphasize a relatively narrow set of ecosystem services (Twidwell et al. 2013a) (Chap. 14, this volume). Minimal recognition of other ecosystem services, including biodiversity and the regulation of water supply, will limit the utility of such STMs for other users. Quantitative interpretations about the different ecosystem services provided by ecological states could be added to STMs (Brown and MacLeod 2011; Koniak et al. 2011). Such information could be used to evaluate the financial costs of restoring a historical state against the change in benefits relative to the current state. Similarly, trade-offs among ecosystem services associated with transitions between states can be communicated in terms of specific variables such as forage provision, species losses, and changes to groundwater recharge rates. As noted above, such exercises may reveal that states considered to be degraded by some observers offer important ecosystem services to others (Mascaro et al. 2012). They may also clarify the trade-offs between specific services, such as forage production vs. biodiversity (Fuhlendorf et al. 2012).

Although it is useful to communicate about states in terms of ecosystem services, it is prudent to acknowledge our limited ability to comprehensively measure all of them effectively. Certain attributes of reference states will be overlooked if they are not adequately measured, especially the biodiversity of organisms that are not the focus of management (Bullock et al. 2011; Reyers et al. 2012). Historical states will continue to be valued for this reason.

9.5.3   Climate Change

STMs often implicitly assume that long-term climate properties and potential vegetation are stable (i.e., stationarity). This assumption leads to an emphasis on recent history in designating alternative states (Twidwell et al. 2013a) (Chap. 7, this volume). Given that climate change is likely to cause directional changes in environmental conditions, plant community responses to management observed in the recent past may become less informative in the future. At present, however, forecasts of climate change effects on vegetation, especially at the resolution of STMs, are not well developed (Settele et al. 2014). STMs could benefit from linkages to species distribution models (Bradley 2010) and models examining the role of soil profile properties in mediating water availability (Zhang 2005). Narratives highlighting the consequences of recent extreme events, such as the tree die-off during an extreme drought in the southwestern USA (Breshears et al. 2005), could be readily included in STMs. Particularly in arid rangelands, management strategies aimed at promoting resilience to known extreme events (especially water deficits) would be similar to strategies implemented to adapt to climate change, at least over the next decade or two (Ash et al. 2012).

9.5.4   Testable Mechanisms

The inclusion of sufficient detail on mechanisms of vegetation change has been a primary limitation of STMs (Knapp et al. 2011b; Svejcar et al. 2014, Sect. 9.4). For example, transitions in some grassland STMs are sometimes ascribed only to the driver (e.g., continuous heavy grazing) without more detailed analysis of the mechanisms by which transitions occur. Information on plant demography (plant death, lack of recruitment), the timeframe for transitions (1 year or several decades), specific indicators of the risk of transition (reduced reproduction rates, indications of erosion), and the management strategies used to prevent transitions given the processes (proper timing of defoliation to permit successful reproduction during favorable years) are often not described in STMs. Richness of detail may be lacking because (1) the information is believed to be too complicated to include and therefore best left to direct interactions between managers and extension specialists; (2) simple lack of effort on the part of model developers; or (3) a lack of detailed knowledge.

These reasons notwithstanding, model developers should strive to include details in a systematic way (e.g., Sect. 9.3; the Caldenal STM at http://jornada.nmsu.edu/esd/international/argentina) in order for STMs to be used and, more importantly, be tested and improved via adaptive management (Briske et al. 2008; Bestelmeyer et al. 2010) (Chap. 9, this volume). Even when the specific mechanisms of state transitions (or resilience of a state) are not well understood, they can be postulated by blending local knowledge with the rich body of work in ecological science (Kachergis et al. 2013). This can be aided by the development of general STMs at the level of broad ecosystem types that can be refined, if needed, to finer-grained land units such as ecological sites. Analysis of historical treatments and new monitoring data can then be used to revisit the hypotheses. For example, shrub-dominated coppice dune states of sandy soils in the Chihuahuan Desert were believed to resist widespread perennial grass recovery based on historical observations and the notion that high erosion rates precluded grass establishment. An unusual sequence of years with high precipitation, and other poorly understood factors, led to a flush of grass recruitment that was unexpected (Peters et al. 2012). The STM for this system has been modified to include this new information. In this way, STMs can be regarded as theoretical constructs that synthesize what is known, use that knowledge to generate management hypotheses, and are updated as new knowledge is acquired.

9.5.5 Information Delivery and Use

If STMs are to be used as tools for long-term environmental stewardship, then the information presented in STMs must be accessible to land managers and/or become integrated in outreach and management activities. Developing and conveying the information in STMs to users such that they can guide management decisions is a multifaceted problem that should be carefully considered by the institutions developing STMs (and see Sect. 9.4). General approaches to information transfer include (1) collaborative development of STMs that include the managers who will use them (see Sect. 9.6.1; Knapp et al. 2011a), (2) initiation of collaborative adaptive management projects at the scale of landscapes that include STM development and use as key components (Bestelmeyer and Briske 2012), (3) the use of web-based technologies and mobile devices to link users to STMs pertaining to specific localities (Herrick et al. 2013), and (4) the distillation of STM information into simple presentation materials (such as pictorial field guides, web-based materials) and the use of field-based workshops to enable understanding of these materials. The use of STMs for management will require concerted efforts by scientists, government agencies, educators, and technical experts and cannot be limited to the production of reports, publications, and associated databases by a handful of managers and ecologists.

9.6 Future Perspectives

Three emerging approaches are currently transforming how STMs are developed and used, including participatory development of STMs with stakeholders as part of community-based management approaches, structured decision-making via STMs, and the use of digital mapping approaches to provide spatially explicit information on ecological states. Here we summarize the current status and future goals of these three approaches.

9.6.1 Participatory Approaches to Model Development[7]

Participatory and collaborative STM development approaches emerged for two practical reasons. First, available field data rarely cover the landscape adequately at a sufficiently fine resolution, or over timescales sufficient to detect transitions and calculate their probabilities. Key types and combinations of management and environmental drivers often are not represented in the available data. Second, models based solely on the knowledge of individual scientists or land management professionals may rely too heavily on a single person's observations and experiences, which can result in biases similar to using monitoring data from only a few locations on a landscape or points in time. These limitations suggest that a more inclusive and participatory approach that integrates multiple knowledge sources may be a pragmatic solution to the challenges inherent in STM development (Kachergis et al. 2013) (Chap. 11, this volume).

Perhaps even more important, participatory approaches will increase the utility, credibility, and use of STMs by managers. Recent surveys have shown that many ranchers and natural resource professionals have little knowledge or experience with STMs when they are available (Kelley 2010). Engaging these potential "end-users" of STMs in the process of developing the models increases STM awareness and acceptance, and thus the likelihood that the models will be used to guide and refine management. An acknowledged limitation of many existing STMs is a focus on a narrow set of ecological attributes and management practices to characterize states and transitions, and a limited suite of management interpretations emphasizing livestock production (Sect. 9.5.2; Knapp et al. 2011b). If STMs are to represent multiple ecosystem values and services, and not just changes in vegetation composition and production for a single or narrow range of uses (e.g., forage production), then multiple disciplines and perspectives are needed.

Participatory or collaborative STM development has taken a variety of forms. The most familiar in the USA is the "technical team," an interdisciplinary collaboration of specialists (e.g., rangeland ecology, soils, hydrology, fire, wildlife, geographic information systems, and cultural resources), often involving several natural resources agencies and academic experts, convened to develop STMs for a particular area. In some areas, such technical teams have been expanded to include landowners or ranchers (Johanson and Fernandez-Gimenez 2015). Collaborative STM development usually takes place over a period of months to a few years and may involve multiple meetings and field trips. The "model development workshop" is another type of participatory approach in which a multi-stakeholder group with diverse knowledge and interests in a particular ecological site or set of sites is

---

[7] Primary authors are M. Fernandez-Gimenez and J. Johanson.

brought together for a single workshop or series of workshops to develop or refine STMs (Knapp et al. 2011a). Such workshops often have an explicit aim to include the local knowledge of long-time residents in an area as well as professional and scientific knowledge. Kachergis et al. (2013) proposed a hybrid approach that involves a diverse set of stakeholders and a combination of literature review, workshops, and field sampling. When it is not possible or practical to bring diverse stakeholders together in one location, or when knowledge documentation is an objective, interviews or surveys with stakeholders can provide a means of recording valuable information that can inform model development (Knapp et al. 2010; Runge 2011).

There is no one best way to facilitate a collaborative or participatory STM development process, but several groups with experience using different collaborative approaches have described the processes that have worked for them (Knapp et al. 2011a; Kachergis et al. 2013; T. K. Stringham, pers. comm.). The process outlined by T. K. Stringham (pers. comm.), which follows the expanded technical team model, focuses on assembling a core team of highly experienced and committed disciplinary experts and inviting participation from a broader group of agency specialists. The workshop model (Knapp et al. 2011a) and integrated literature, workshop, and field sampling approach (Kachergis et al. 2013) draw from a wider array of stakeholders and emphasize the value of including long-term residents and those whose knowledge is derived from land-based livelihoods. All three of these processes begin with a draft graphical model that serves as the basis for initial discussions and feedback from the group.

Johanson and Fernandez-Gimenez (2015) drew on these experiences together with those of participants in 16 collaborative ESD and STM development projects in the USA to identify common outcomes, challenges, and keys to success. Most efforts were successful in producing an STM or portion of an ESD. Additional outputs included publications, applications of the models to management, workshops, and databases. Many benefits beyond these tangible outputs were also identified, such as improved working relationships and communication among participants from different organizations, decreased conflict, increased efficiency of STM development, greater use of STMs, and improved data credibility.

Participatory processes are never without challenges. The most frequently cited concerns were related to the quality, diversity, management, and analysis of available data. Reconciling different concepts for classifying ecosystems and their dynamics and agreeing on goals for STM development efforts were common challenges in expanded collaborations. Time and funding constraints and recruitment/retention of participants were additional obstacles. Because many natural resource professionals are unfamiliar with ESDs and STMs, key concepts must be taught to all participants and reinforced with additional teaching throughout the process. Similarly, when working with nontechnical stakeholders, care must be taken to define key terms in a clear and accessible manner and to provide an introduction to STM concepts and applications. Although some professionals express skepticism about the accessibility of STMs to nonprofessionals (Knapp et al. 2011b), we have found that most people readily grasp these concepts, especially once they are engaged in the process of model development.

The keys to successful participatory STM development are similar to those for any participatory natural resource management effort (Wondolleck and Yaffee 2000; Daniels and Walker 2001). First, involve the right people at the right time. Make sure that the needed expertise is present, particularly experienced specialists in soils and rangeland ecology, but also hydrology, fire, wildlife, geographic information systems, and cultural resources. When integrating local knowledge is an important objective, seek diversity and depth of experience in local knowledge holders. Community referrals are often an effective way to identify knowledgeable residents (Knapp and Fernandez-Gimenez 2009; Knapp et al. 2010).

Second, it is important to maintain clear and open communication, a willingness to learn from others, and focus on mutually beneficial outcomes. In multiagency collaborations, conflicts can arise over the differing mandates and procedures of different agencies. When multiple stakeholders are involved, careful facilitation is required to balance power dynamics and ensure that the contributions of all participants are respected. Clear ground rules should be established regarding the criteria for including states and transitions and how potentially conflicting views of ecosystem dynamics will be handled and represented in the model. In multi-stakeholder STM workshops, the level of agreement among participants about each state and transition can be explicitly documented and used to identify uncertainties to test through targeted field sampling or adaptive management experiments (Knapp et al. 2011a; Kachergis et al. 2013). This leads to more efficient use of limited field sampling resources.

Third, support from management within participating agencies is critical. If administrators do not value collaboration and support their staff in participating in such efforts, it is very difficult to sustain the level of participation and commitment needed for success. Fourth, many participants reported that joint field visits were key to successful collaborative STM development. Discussing conditions observed in specific areas can help resolve misunderstandings and elicit new sources of information. Fifth, because many of the challenges identified relate to data collection, management, and analysis, it is important to discuss and agree upon responsibilities and protocols for these activities up front. Often the university or research partners in STM collaborations take the lead on data analysis. However, we strongly encourage groups to invite broad participation in data analysis and especially in data interpretation. We also recommend formal data sharing and use agreements to facilitate information sharing and protect confidentiality where needed.

Reported participant experiences suggest that collaboration is a good investment that increases the efficiency of STM development. It requires significant human, financial, and time resources, but yields both tangible and intangible benefits that participants perceive to increase the quality, credibility, and utility of STMs.

9.6.2 Structured Decision-Making via State and Transition Models[8]

In this section, we ask: can STMs be used in a more systematic way to prioritize management objectives and to efficiently allocate management funds? Below we discuss why managers may benefit from integrating STMs into a structured decision-making process, and developing STMs such that they enable quantitative predictions of management outcomes.

Ecosystem management decisions are invariably complex. There may be a lack of understanding about the processes underlying a specific problem. Alternatively, there may be multiple and potentially competing objectives for management, which may not be readily apparent, but which should be determined before developing the model. For instance, when faced with an imperative to both manage for a certain plant community and protect a threatened species, it may be that the habitat for that species does not correspond to the desired vegetation state. In addition, it may be that an objective to minimize costs is at odds with the funds required to restore a community to the desired state. Stakeholders will not value all of these objectives in the same way, but it is the role of the decision-maker to evaluate these trade-offs. Last, there may be multiple potential alternative management strategies, but high uncertainty and disagreement about ecosystem responses to management. For the decision-maker, choosing the best course of action to help achieve the specified objectives can be extremely difficult (Runge 2011; Gregory et al. 2012).

Many of these problems can be addressed by using a systematic approach to the decision-making process. The term "structured decision-making" broadly refers to a framework that incorporates a logical sequence of steps to help decision-makers (1) define their decision context; (2) identify measurable objectives; (3) formulate alternative management strategies; (4) explore the consequences of those alternatives in relation to the specified objectives; and, if necessary (5) make trade-offs among objectives (Gregory et al. 2012). The framework utilizes a broad suite of decision-analysis tools that can aid transparent and logical decision-making (Addison et al. 2013). Despite the multitude of tools and methods that may be applied, the basic premise is a framework that is driven by the objectives, or values, of those involved in the decision-making process (Keeney 1996; Runge 2011).

STMs are typically developed as conceptual models, informed by expert knowledge and existing data. Such models may quantify the characteristics of states but lack a quantification of transition probabilities given particular values of controlling variables and management actions (i.e., they are qualitative or semiquantitative STMs). Within the structured decision-making framework (Fig. 9.4), a qualitative STM can be used to clarify the decision context among stakeholders, the desired direction of change and key attributes of interest (objectives), and the different management interventions that might be employed to achieve this change (alternatives). In addition, qualitative STMs could be used to begin exploring the consequences of

---

[8] Primary author is L. Rumpff.

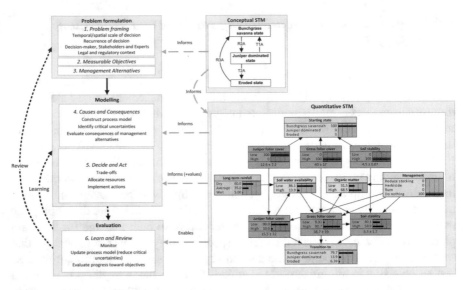

Fig. 9.4 The structured decision-making framework adapted from Wintle et al. (2011). A conceptual STM (adapted from Bestelmeyer et al. 2010) is commonly used to frame the problem, whereas the quantitative version of the STM (structured here as a Bayesian network) is useful to identify, explore and resolve critical uncertainty

the alternatives in relation to the objectives. As a decision-support tool, a qualitative STM is often all that is required to guide a good management decision within the structured decision-making process. For instance, an STM (based on Bestelmeyer et al. 2010) can be used to identify the interventions required to achieve the ecological conditions for a reference state (Bunchgrass savannah; Fig. 9.4). In this instance there is one objective (the reference state), and clearly defined interventions. However, recognized uncertainty about the effects of climate change may result in different models of cause-and-effect, uncertainty about the most effective interventions, or even uncertainty about whether the goal state is attainable. In cases where there are numerous alternatives to choose from, multiple and competing objectives, conflicting values among stakeholders, differing stories of cause-and-effect, or "critical uncertainty" (i.e., uncertainty that bears on key decisions), decision-making based on quantitative STMs can help select the best decision.

Quantitative (or process-based) models are useful for identifying and exploring the uncertainties that impact management decisions (Duncan and Wintle 2008; Rumpff et al. 2011). A process-based model represents the current state of knowledge and assumptions about the dynamics of the system, and allows predictions to be made about the efficacy of the different management strategies in relation to the objectives of interest. For instance, in Fig. 9.4, the assumptions behind the STM have been quantified and converted into a probabilistic model of cause-and-effect (a Bayesian network). Probabilistic transition estimates now include uncertainty about the efficacy of management interventions under various climatic scenarios.

A management decision will often involve multiple objectives, with no one management strategy that maximizes all objectives. For example, there may be a trade-off between achieving the reference state and maximizing agricultural productivity. The quantitative model should first be expanded to enable predictions for both objectives. The predictions can then be combined with value judgments (or preferences) that specify which objective should benefit over the other, given the range of possible outcomes (Gregory et al. 2012). The true value of an alternative management strategy is a combination of the consequences (including uncertainty), and the weight or value attributed to the objectives (step 5, Fig. 9.4). At this point, the decision may be obvious, or uncertainty may be obscuring the preferred management strategy.

Uncertainty is inevitable, but decision-makers should pay particular attention to resolving critical uncertainties, as this can result in modified and potentially more effective management decisions. Monitoring is used to resolve this uncertainty, by iteratively updating the knowledge within the process-based model (step 6, Fig. 9.4). This is known as adaptive management, which is a form of structured decision-making, required when decisions are recurrent and hampered by critical uncertainty (Runge 2011). Thus, adaptive management requires extra steps in the structured decision-making framework to provide a plan for motivating, designing, and interpreting the results of monitoring.

Although the development of quantitative state-and-transition models has increased (Bashari et al. 2009; Nicholson and Flores 2011; Rumpff et al. 2011), to date their application in a management context is rare. Thus, it can be concluded that STMs have yet to reach their full potential as decision-support tools for the implementation of natural resource management and the evaluation of its outcomes. Both quantitative and qualitative models can be used to capture our current understanding about system dynamics, and to identify and explore uncertainty surrounding the response to management (Rumpff et al. 2011; Runge 2011). The choice of decision support tool should be dictated by the availability and form of knowledge, whether qualitative or quantitative predictions are required to make a decision, and whether quantitative skills are accessible given the timeframe available for decision-making.

Whether the model is quantitative or qualitative, structured decision-making can help to provide a systematic and transparent framework for identifying objectives, collate existing knowledge, explore the consequences of management alternatives and identify and evaluate uncertainty. The value of qualitative STMs to help frame and guide vegetation management decisions in rangelands is not in question. Rather, managers and researchers should acknowledge the complexities of their particular problem context, and assess whether structured decision-making approaches are useful.

9.6.3 Mapping State-and-Transition Model Information[9]

Managers currently represent ecosystem variations across a wide range of scales for various uses. In rangelands, potential natural vegetation is mapped via land unit classifications such as habitat types (Jensen et al. 2001), range units and range sites (Kunst et al. 2006), and ecological sites (Bestelmeyer et al. 2003). More recently,

---

[9] Primary author is M. Levi.

attention has focused on the delineation of the current states of a set of land units based on its STM (Steele et al. 2012). The product is called a "state map" that can make the information within STMs spatially explicit for its use in management.

STMs are typically linked to land units that define the spatial extent to which information in STMs should be extrapolated. Soil survey is often used to map land units bearing distinct STMs (such as ecological sites), particularly in the USA. Hence, STMs can be linked to maps of soil types or landforms. Soil maps thus provide a template for mapping ecological states across multiple STMs. One constraint in linking soil maps to STMs is that any errors in existing soil spatial data are transferred to the state map. Many soil maps in rangelands consist of "soil map units" that represent multiple soil types either due to a limitation of mapping scale or landscape heterogeneity (Duniway et al. 2010). In some cases, soil types are similar and grouped to the same ecological site; however, soil types with contrasting properties combined within the same soil map unit may belong to different ecological sites and STMs. In the USA, it has been a priority to resolve these discrepancies in order to improve the utility of STMs (Steele et al. 2012).

Ecological sites and states can be mapped simultaneously using environmental variables, such as from remote sensing products (Browning and Steele 2013; Hernandez and Ramsey 2013). One benefit of utilizing remotely sensed data to characterize ecological sites and states is the ability to produce scalable information that can be tailored to particular needs (Kunst et al. 2006). For example, West et al. (2005) outlined a strategy for producing a hierarchical map of ecological units for 4.5 million hectares area in western Utah based on a variety of data sources. The finest level was a "vegetation stand" that is similar to ecological states represented in STMs.

Mapping of ecological states can be difficult in rangelands because spectral data from conventional sources, such as MODIS or LANDSAT satellites, is often not of sufficient resolution or quality to distinguish states. Blanco et al. (2014) integrated hyperspectral and multi-spectral remote sensing data to identify ecological sites in rangelands of Argentina. This approach could be extended to map states. Steele et al. (2012) outlined a framework for mapping ecological sites and states in rangelands of southern New Mexico using a combination of soil survey spatial data combined with image interpretation of aerial photography to manually delineate ecological site and state polygons (i.e., line maps).

Digital soil mapping (DSM) is an emerging technique that can improve estimates of soil property and ecological state information at fine spatial scales in rangelands by predicting the properties of pixels of varying resolution (e.g., to 5 m) (Levi and Rasmussen 2014; Nauman et al. 2014). Although DSM has not yet been applied to state mapping, it could fill a much needed gap by increasing automation, using a greater range of data sources, and allowing for rapid updating of state maps when new data become available. Data-driven classification algorithms can greatly reduce the time needed to produce state maps because they provide a means of grouping pixels into similar units, thereby reducing the burden of hand digitizing (Laliberte 2007; MacMillan et al. 2007). DSM approaches can also be scaled up or down to meet desired management objectives, which is currently difficult to do with polygon-

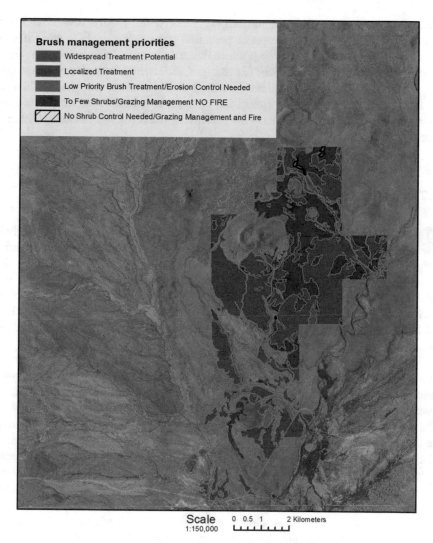

Fig. 9.5 An example of a product based on an ecological state map for a single ecological site type. The map illustrates interpretations of an STM according to brush management treatment options (courtesy of Eldon Ayers)

based maps. In turn, DSM could be used to identify vegetation responses to soil properties that may improve STMs (Browning and Duniway 2011).

State mapping can extend the utility of STMs for management. In landscapes with a mix of ecological sites and states, state mapping distills information across multiple STMs into a simpler classification scheme that can be used for communication among stakeholders and to develop action plans (Fig. 9.5). For example, a state map was used in the southwestern USA in planning for brush control treatments to identify areas that were (1) near a desired reference condition where no treatment was needed, (2) areas that had experienced soil erosion where treatment would likely not produce increases in perennial grass cover, and (3) areas where treatment would be most likely to produce desired changes. In a similar way, state mapping can be used to plan for land use changes, such as by prioritizing development away from desirable reference states (Stoms et al. 2013). State mapping could also be used to visualize or model spatial interactions in a landscape, such as where increases in grass cover would have the greatest impact on water retention within a watershed.

9.7 Summary

STMs evolved from the recognition that vegetation change was more complex than could be accounted for by succession alone, and could occur along numerous pathways, be discontinuous, and result in multiple stable states in the same environment. Conceptualizing vegetation as discrete states also provides a useful platform for tailoring management actions to the properties and possibilities associated with each state. For rangeland managers, the value of STMs resides both in their flexibility for organizing information and in their ability to foster a general understanding about how rangelands function.

Progress toward developing rangeland STMs at a global level has been uneven due to several factors, including limitations of data and fiscal and personnel resources. As strategies to overcome these limitations are developed, the ultimate success of STMs as management tools will require careful attention to several topics. First, there should be a clear understanding of the characteristics of alternative states, including a reference state where such a concept is meaningful. Field sampling, synthesis of experimental results and long-term vegetation records, and participatory approaches are important resources for defining states. State characterization should ideally represent information on a variety of ecosystem services. In most cases, this will require coordinated sampling efforts to link variations in plant community states to empirical or model-based evaluations of habitat quality, soil carbon storage potential, and value for livestock, for example.

Second, STMs should attempt to distinguish transient dynamics from state transitions. Evidence-based approaches necessitate clear statements not only about drivers of transition but also about the controlling variables and processes constraining recovery and timelines for ecosystem change. STMs should feature logical and testable statements about how states will respond to management, such that STMs can support experimentation, quantitative models, and eventual revision. Even where data are scarce, local knowledge can be framed as testable propositions. Predictions regarding the effects of climate change on ecosystems may best be addressed at a regional scale, but information on the impact of past extreme events can be highlighted. Strategies to manage alternative states, such as through novel uses of states invaded by woody plants, may help with climate adaptation over the longer term.

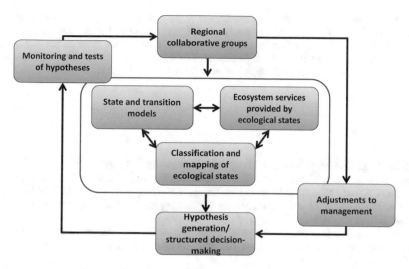

Fig. 9.6 A schematic of how STMs can be used in collaborative adaptive management, adapted from Bestelmeyer and Briske (2012)

Third, STM development programs should consider how to make information available, useful, and believable to users. Participatory approaches can promote understanding and acceptance of STMs. There should be a clear link between STMs and specific management actions, which can facilitate the inclusion of STMs into collaborative adaptive management programs supported by local communities, non-governmental organizations, or governmental agencies (Fig. 9.6). Regional or landscape collaborative groups can develop STMs and identify ecosystem services of interest from different states. The linkage of STMs to maps of ecological states can facilitate management application and testing. Hypotheses for management responses can be developed for specific land units (Fig. 9.5) and structured decision-making approaches can be used for cases when multiple management options are possible, trade-offs make decisions difficult, and the preferred decision is unclear or controversial. Tests of hypotheses via monitoring can be used to either revise the STM or make minor management adjustments.

In order to facilitate their use in collaborative adaptive management, STMs should be presented and used in a variety of ways, including simple extension materials, formal hypotheses for ecological research and tests of management efficacy, rangeland evaluation criteria, maps, or Bayesian models. Policymakers, technical assistance personnel, regulators, scientists, land managers, and stakeholders should be working from the same general understanding of how a rangeland ecosystem functions, even if those parties differ in their preferred states or ecosystem services. STMs should link understanding across different organizational levels as a basis for collaborative adaptive management. Our hope is that the recommendations presented here will promote development of STMs that are indispensable for the management of global rangelands.

Acknowledgements We thank David Briske, Ian Watson, and an anonymous reviewer for comments that substantially improved this chapter. B.T.B. was supported by appropriated funds to the USDA Agricultural Research Service and by the Jornada Basin LTER program (DEB-1235828). M.E.F.-G. was supported by Colorado Agricultural Experiment Station Project COL00698A and NRCS Conservation Innovation Grant Agreement No. 69-3A75-12-213.

References

Addison, J., M. Friedel, C. Brown, J. Davies, and S. Waldron. 2012. A critical review of degradation assumptions applied to Mongolia's Gobi Desert. *Rangeland Journal* 34: 125–137.

Addison, P.F.E., L. Rumpff, S.S. Bau, J.M. Carey, Y.E. Chee, F.C. Jarrad, M.F. McBride, and M.A. Burgman. 2013. Practical solutions for making models indispensable in conservation decision-making. *Diversity and Distributions* 19: 490–502.

Aguilera, M.O., D.F. Steinaker, M.R. Demaría, and A.O. Ávila. 1998. Estados y transiciones de los pastizales de Sorghastrum pellitum del área medanosa central de San Luis, Argentina. *Ecotropicos* 11: 107–120.

Ash, A.J., J.A. Bellamy, and T.G.H. Stockwell. 1994. State and transition models for rangelands. 4. Application of state and transition models to rangelands in northern Australia. *Tropical Grasslands* 28: 223–228.

Ash, A., P. Thornton, C. Stokes, and C. Togtohyn. 2012. Is proactive adaptation to climate change necessary in grazed rangelands? *Rangeland Ecology & Management* 65: 563–568.

Augustine, D.J., and J.D. Derner. 2012. Disturbance regimes and mountain plover habitat in shortgrass steppe: Large herbivore grazing does not substitute for prairie dog grazing or fire. *Journal of Wildlife Management* 76: 721–728.

Bagchi, S., D.D. Briske, X.B. Wu, M.P. McClaran, B.T. Bestelmeyer, and M.E. Fernandez-Gimenez. 2012. Empirical assessment of state-and-transition models with a long-term vegetation record from the Sonoran Desert. *Ecological Applications* 22: 400–411.

Barger, N.N., S. Archer, J. Campbell, C. Huang, J. Morton, and A. Knapp. 2011. Woody plant proliferation in North American drylands: A synthesis of impacts on ecosystem carbon balance. *Journal of Geophysical Research – Biogeosciences* 116: G00K07. doi:10.1029/201 0JG001506.

Barrera, M.D., and J.L. Frangi. 1997. Modelo de estados y transiciones de la arbustificación de pastizales de Sierra de la Ventana, Argentina. *Ecotropicos* 10: 161–166.

Bartley, R., J.P. Corfield, A.A. Hawdon, A.E. Kinsey-Henderson, B.N. Abbott, S.N. Wilkinson, and R.J. Keen. 2014. Can changes to pasture management reduce runoff and sediment loss to the Great Barrier Reef? The results of a 10-year study in the Burdekin catchment, Australia. *Rangeland Journal* 36: 67–84.

Bashari, H., C. Smith, and O.J.H. Bosch. 2009. Developing decision support tools for rangeland management by combining state and transition models and Bayesian belief networks. *Agricultural Systems* 99: 23–34.

Bastin, G.N., D.M.S. Smith, I.W. Watson, and A. Fisher. 2009. The Australian Collaborative Rangelands Information System: Preparing for a climate of change. *Rangeland Journal* 31: 111–125.

Belnap, J., J.A. Ludwig, B.P. Wilcox, J.L. Betancourt, W.R.J. Dean, B.D. Hoffmann, and S.J. Milton. 2012. Introduced and invasive species in novel rangeland ecosystems: Friends or foes? *Rangeland Ecology & Management* 65: 569–578.

Bestelmeyer, B.T. 2006. Threshold concepts and their use in rangeland management and restoration: The good, the bad, and the insidious. *Restoration Ecology* 14: 325–329.

Bestelmeyer, B.T., and D.D. Briske. 2012. Grand challenges for resilience-based management of rangelands. *Rangeland Ecology & Management* 65: 654–663.

Bestelmeyer, B.T., and J.R. Brown. 2010. An introduction to the special issue on ecological sites. *Rangelands* 32: 3–4.

Bestelmeyer, B.T., J.R. Brown, K.M. Havstad, R. Alexander, G. Chavez, and J. Herrick. 2003. Development and use of state-and-transition models for rangelands. *Journal of Range Management* 56: 114–126.

Bestelmeyer, B., K. Havstad, B. Damindsuren, G. Han, J. Brown, J. Herrick, C. Steele, and D. Peters. 2009a. Resilience theory in models of rangeland ecology and restoration: The evolution and application of a paradigm. In *New models for ecosystem dynamics and restoration*, ed. R.J. Hobbs and K.N. Suding, 78–96. Washington, DC: Island Press.

Bestelmeyer, B.T., A.J. Tugel, G.L. Peacock, D.G. Robinett, P.L. Shaver, J.R. Brown, J.E. Herrick, H. Sanchez, and K.M. Havstad. 2009b. State-and-transition models for heterogeneous landscapes: A strategy for development and application. *Rangeland Ecology & Management* 62: 1–15.

Bestelmeyer, B.T., K. Moseley, P.L. Shaver, H. Sanchez, D.D. Briske, and M.E. Fernandez-Gimenez. 2010. Practical guidance for developing state-and-transition models. *Rangelands* 32: 23–30.

Bino, G., S.A. Sisson, R.T. Kingsford, R.F. Thomas, and S. Bowen. 2015. Developing state and transition models of floodplain vegetation dynamics as a tool for conservation decision-making: A case study of the Macquarie Marshes Ramsar wetland. *Journal of Applied Ecology* 52: 654–664.

Blanco, P.D., H.F. del Valle, P.J. Bouza, G.I. Metternicht, and L.A. Hardtke. 2014. Ecological site classification of semiarid rangelands: Synergistic use of Landsat and Hyperion imagery. *International Journal of Applied Earth Observation and Geoinformation* 29: 11–21.

Bradley, B.A. 2010. Assessing ecosystem threats from global and regional change: Hierarchical modeling of risk to sagebrush ecosystems from climate change, land use and invasive species in Nevada, USA. *Ecography* 33: 198–208.

Breshears, D.D., N.S. Cobb, P.M. Rich, K.P. Price, C.D. Allen, R.G. Balice, W.H. Romme, J.H. Kastens, M.L. Floyd, J. Belnap, J.J. Anderson, O.B. Myers, and C.W. Meyer. 2005. Regional vegetation die-off in response to global-change-type drought. *Proceedings of the National Academy of Sciences of the United States of America* 102: 15144–15148.

Briske, D.D., S.D. Fuhlendorf, and F.E. Smeins. 2003. Vegetation dynamics on rangelands: A critique of the current paradigms. *Journal of Applied Ecology* 40: 601–614.

Briske, D.D., B.T. Bestelmeyer, T.K. Stringham, and P.L. Shaver. 2008. Recommendations for development of resilience-based state-and-transition models. *Rangeland Ecology & Management* 61: 359–367.

Brown, J.R. 1994. State and transition models for rangelands. 2. Ecology as a basis for rangeland management: Performance criteria for testing models. *Tropical Grasslands* 28: 206–213.

———. 2010. Ecological sites: Their history, status, and future. *Rangelands* 32:5–8.

Brown, J.R., and A.J. Ash. 1996. Managing resources: Moving from sustainable yield to sustainability in tropical rangelands. *Tropical Grasslands* 30: 47–57.

Brown, J., and N. MacLeod. 2011. A site-based approach to delivering rangeland ecosystem services. *Rangeland Journal* 33: 99–108.

Brown, A., O.U. Martinez, M. Acerbi, and J. Corcuera (eds.). 2006. *La situación ambiental Argentina 2005*. Buenos Aires, Argentina: Fundación Vida Silvestre Argentina.

Browning, D.M., and M.C. Duniway. 2011. Digital soil mapping in the absence of field training data: A case study using terrain attributes and semiautomated soil signature derivation to distinguish ecological potential. *Applied and Environmental Soil Science*. 421904, 12, doi:10.1155/2011/421904.

Browning, D.M., and C.M. Steele. 2013. Vegetation index differencing for broad-scale assessment of productivity under prolonged drought and sequential high rainfall conditions. *Remote Sensing* 5: 327–341.

Bruegger, R.A., O. Jigjsuren, and M.E. Fernandez-Gimenez. 2014. Herder observations of rangeland change in Mongolia: Indicators, causes, and application to community-based management. *Rangeland Ecology & Management* 67: 119–131.

Bullock, J.M., J. Aronson, A.C. Newton, R.F. Pywell, and J.M. Rey-Benayas. 2011. Restoration of ecosystem services and biodiversity: Conflicts and opportunities. *Trends in Ecology & Evolution* 26: 541–549.

Campbell, R.S. 1929. Vegetative succession in the Prosopis sand dunes of southern New Mexico. *Ecology* 10: 392–398.

Caudle, D., J. Dibenedetto, M. Karl, H. Sanchez, and C. Talbot. 2013. *Interagency ecological site handbook for rangelands*. United States Government.

Cesa, A., and J.M. Paruelo. 2011. Changes in vegetation structure induced by domestic grazing in Patagonia (Southern Argentina). *Journal of Arid Environments* 75: 1129–1135.

Cingolani, A.M., I. Noy-Meir, and S. Díaz. 2005. Grazing effects on rangeland diversity: A synthesis of contemporary models. *Ecological Applications* 15: 757–773.

Czembor, C.A., and P.A. Vesk. 2009. Incorporating between-expert uncertainty into state-and-transition simulation models for forest restoration. *Forest Ecology and Management* 259: 165–175.

D'Odorico, P., A. Bhattachan, K.F. Davis, S. Ravi, and C.W. Runyan. 2013. Global desertification: Drivers and feedbacks. *Advances in Water Resources* 51: 326–344.

Daniels, S.E., and G.B. Walker. 2001. *Working through environmental conflict: The collaborative learning approach*. Westport, CT: Praeger Publishers.

Doak, D.F., V.J. Bakker, B.E. Goldstein, and B. Hale. 2014. What is the future of conservation? *Trends in Ecology & Evolution* 29: 77–81.

Duncan, D., and B. Wintle. 2008. Towards adaptive management of native vegetation in regional landscapes. In *Landscape analysis and visualisation*, ed. C. Pettit, W. Cartwright, I. Bishop, K. Lowell, D. Pullar, and D. Duncan, 159–182. Berlin, Heidelberg: Springer.

Duniway, M.C., B.T. Bestelmeyer, and A. Tugel. 2010. Soil processes and properties that distinguish ecological sites and states. *Rangelands* 32: 9–15.

Dussart, E., P. Lerner, and R. Peinetti. 1998. Long term dynamics of 2 populations of Prosopis caldenia Burkart. *Journal of Range Management* 51: 685–691.

Dyksterhuis, E.J. 1949. Condition and management of range land based on quantitative ecology. *Journal of Range Management* 2: 104–115.

———. 1958. Ecological principles in range evaluation. *The Botanical Review* 24:253–272.

Eldridge, D.J., M.A. Bowker, F.T. Maestre, E. Roger, J.F. Reynolds, and W.G. Whitford. 2011. Impacts of shrub encroachment on ecosystem structure and functioning: Towards a global synthesis. *Ecology Letters* 14: 709–722.

Farji-Brener, A.G., and A. Ruggiero. 2010. ¿Impulsividad o paciencia? Qué estimula y qué selecciona el sistema científico argentino. *Ecología Austral* 20: 307–314.

Fernández-Giménez, M.E. 2002. Spatial and social boundaries and the paradox of pastoral land tenure: A case study from postsocialist Mongolia. *Human Ecology* 30: 49–78.

Friedel, M.H. 1991. Range condition assessment and the concept of thresholds: A viewpoint. *Journal of Range Management* 44: 422–426.

Fuhlendorf, S.D., D.M. Engle, R.D. Elmore, R.F. Limb, and T.G. Bidwell. 2012. Conservation of pattern and process: Developing an alternative paradigm of rangeland management. *Rangeland Ecology & Management* 65: 579–589.

Fulé, P.Z., W.W. Covington, and M.M. Moore. 1997. Determining reference conditions for ecosystem management of southwestern Ponderosa Pine forests. *Ecological Applications* 7: 895–908.

Gasparri, N.I., H.R. Grau, and J. Gutiérrez Angonese. 2013. Linkages between soybean and neotropical deforestation: Coupling and transient decoupling dynamics in a multi-decadal analysis. *Global Environmental Change* 23: 1605–1614.

Gregory, R., L. Failing, M. Harstone, G. Long, T. McDaniels, and D. Ohlson. 2012. *Structured decision making: A practical guide to environmental management choices*. West Sussex: Wiley-Blackwell.

Grice, A.C., and N.D. MacLeod. 1994. State and transition models for rangelands. 6. State and transition models as aids to communication between scientists and land managers. *Tropical Grasslands* 28: 241–246.

Herbel, C.H., and R.P. Gibbens. 1989. Matric potential of clay loam soils on arid rangelands in southern New Mexico. *Journal of Range Management* 42: 386–392.

Hernandez, A.J., and R.D. Ramsey. 2013. A landscape similarity index: Multitemporal remote sensing to track changes in big sagebrush ecological sites. *Rangeland Ecology & Management* 66: 71–81.

Herrick, J.E., J.W. Van Zee, K.M. Havstad, L.M. Burkett, and W.G. Whitford. 2005. *Monitoring manual for grassland, shrubland and savanna ecosystems*. Volume I: Quick Start. Volume II: Design, supplementary methods and interpretation. Las Cruces, NM: USDA-ARS Jornada Experimental Range.

Herrick, J.E., K.C. Urama, J.W. Karl, J. Boos, M.V.V. Johnson, K.D. Shepherd, J. Hempel, B.T. Bestelmeyer, J. Davies, J.L. Guerra, C. Kosnik, D.W. Kimiti, A.L. Ekai, K. Muller, L. Norfleet, N. Ozor, T. Reinsch, J. Sarukhan, and L.T. West. 2013. The global Land-Potential Knowledge System (LandPKS): Supporting evidence-based, site-specific land use and management through cloud computing, mobile applications, and crowdsourcing. *Journal of Soil and Water Conservation* 68: 5A–12A.

Hilker, T., E. Natsagdorj, R.H. Waring, A. Lyapustin, and Y.J. Wang. 2014. Satellite observed widespread decline in Mongolian grasslands largely due to overgrazing. *Global Change Biology* 20: 418–428.

Hill, M.J., S.H. Roxburgh, J.O. Carter, and G.M. McKeon. 2005. Vegetation state change and consequent carbon dynamics in savanna woodlands of Australia in response to grazing, drought and fire: A scenario approach using 113 years of synthetic annual fire and grassland growth. *Australian Journal of Botany* 53: 715–739.

Hobbs, R.J., and K.N. Suding. 2009. *New models for ecosystem dynamics and restoration*. Washington, DC: Island Press.

Hobbs, R.J., E. Higgs, and J.A. Harris. 2009. Novel ecosystems: Implications for conservation and restoration. *Trends in Ecology & Evolution* 24: 599–605.

Holling, C.S. 1973. Resilience and stability of ecological systems. *Annual Review of Ecology and Systematics* 4: 1–23.

Holmes, A.L., and R.F. Miller. 2010. State-and-transition models for assessing grasshopper sparrow habitat use. *Journal of Wildlife Management* 74: 1834–1840.

Holmgren, M., and M. Scheffer. 2001. El Niño as a window of opportunity for the restoration of degraded arid ecosystems. *Ecosystems* 4: 151–159.

Hughes, T.P., C. Linares, V. Dakos, I.A. van de Leemput, and E.H. van Nes. 2013. Living dangerously on borrowed time during slow, unrecognized regime shifts. *Trends in Ecology & Evolution* 28: 149–155.

Hunt, L. 1992. Piospheres and the state-and-transition model of vegetation change in chenopod shrublands. *Proceedings of the 7th Biennial Conference of the Australian Rangeland Society*, Cobar, Australia, 37–45.

Jackson, R., and J. Bartolome. 2002. A state-transition approach to understanding nonequilibrium plant community dynamics in Californian grasslands. *Plant Ecology* 162: 49–65.

Jensen, M.E., J.P. Dibenedetto, J.A. Barber, C. Montagne, and P.S. Bourgeron. 2001. Spatial modeling of rangeland potential vegetation environments. *Journal of Range Management* 54: 528–536.

Johanson, J., and M. Fernandez-Gimenez. 2015. Developers of ecological site descriptions find benefits in diverse collaborations. *Rangelands* 37: 14–19.

Joyce, L.A. 1993. The life cycle of the range condition concept. *Journal of Range Management* 46: 132–138.

Kachergis, E., M.E. Rocca, and M.E. Fernandez-Gimenez. 2011. Indicators of ecosystem function identify alternate states in the sagebrush steppe. *Ecological Applications* 21: 2781–2792.

Kachergis, E.J., C.N. Knapp, M.E. Fernandez-Gimenez, J.P. Ritten, J.G. Pritchett, J. Parsons, W. Hibbs, and R. Roath. 2013. Tools for resilience management: Multidisciplinary development of state-and-transition models for northwest Colorado. *Ecology and Society* 18.

Keeney, R.L. 1996. *Value-focused thinking: A path to creative decisionmaking*. Cambridge, MA: Harvard University Press.

Kéfi, S., M. Rietkerk, M. Roy, A. Franc, P.C. de Ruiter, and M. Pascual. 2011. Robust scaling in ecosystems and the meltdown of patch size distributions before extinction. *Ecology Letters* 14: 29–35.

Kelley, W. 2010. *Rangeland managers' adoption of innovations, awareness of state and transition models, and management of Bromus tectorum*. MS Thesis, Colorado State University.

Khishigbayar, J., M.E. Fernández-Giménez, J.P. Angerer, R.S. Reid, J. Chantsallkham, Y. Baasandorj, and D. Zumberelmaa. 2015. Mongolian rangelands at a tipping point? Biomass and cover are stable but composition shifts and richness declines after 20 years of grazing and increasing temperatures. *Journal of Arid Environments* 115: 100–112.

Knapp, C.N., and M.E. Fernandez-Gimenez. 2009. Understanding change: Integrating rancher knowledge into state-and-transition models. *Rangeland Ecology & Management* 62: 510–521.

Knapp, C.N., M.E. Fernandez-Gimenez, and E. Kachergis. 2010. The role of local knowledge in state-and-transition model development. *Rangelands* 32: 31–36.

Knapp, C.N., M. Fernandez-Gimenez, E. Kachergis, and A. Rudeen. 2011a. Using participatory workshops to integrate state-and-transition models created with local knowledge and ecological data. *Rangeland Ecology & Management* 64: 158–170.

Knapp, C.N., M.E. Fernandez-Gimenez, D.D. Briske, B.T. Bestelmeyer, and X. Ben Wu. 2011b. An assessment of state-and-transition models: Perceptions following two decades of development and implementation. *Rangeland Ecology & Management* 64: 598–606.

Koniak, G., I. Noy-Meir, and A. Perevolotsky. 2011. Modelling dynamics of ecosystem services basket in Mediterranean landscapes: A tool for rational management. *Landscape Ecology* 26: 109–124.

Kunst, C., E. Monti, H. Perez, and J. Godoy. 2006. Assessment of the rangelands of southwestern Santiago del Estero, Argentina, for grazing management and research. *Journal of Environmental Management* 80: 248–265.

Laliberte, A.S. 2007. Combining decision trees with hierarchical object-oriented image analysis for mapping arid rangelands. *Photogrammetric Engineering and Remote Sensing* 73: 197.

Lanner, R.M. 2012. How did we get it so wrong? *Journal of Forestry* 110: 404.

Laterra, P., O.R. Vignolio, L.G. Hidalgo, O.N. Fernández, M.A. Cauhépé, and N.O. Maceira. 1998. Dinámica de pajonales de paja colorada (Paspalum spp) manejados con fuego y pastoreo en la Pampa Deprimida Argentina. *Ecotropicos* 11: 141–149.

Lauenroth, W.K., and W.A. Laycock. 1989. *Secondary succession and the evaluation of rangeland condition*. Boulder, CO.: Westview Press Inc.

León, R.J.C., and S.E. Burkart. 1998. El pastizal de la Pampa Deprimida: estados alternativos. *Ecotropicos* 11: 121–130.

Levi, M.R., and C. Rasmussen. 2014. Covariate selection with iterative principal component analysis for predicting physical soil properties. *Geoderma* 219–220: 46–57.

Lewis, T., N. Reid, P.J. Clarke, and R.D.B. Whalley. 2010. Resilience of a high-conservation-value, semi-arid grassland on fertile clay soils to burning, mowing and ploughing. *Austral Ecology* 35: 464–481.

Llorens, E.M. 1995. Viewpoint: The state and transition model applied to the herbaceous layer of Argentina's Calden forest. *Journal of Range Management* 48: 442–447.

López, D.R. 2011. *Una aproximación estructural–funcional del modelo de estados y transiciones para el estudio de la dinámica de la vegetación en estepas de Patagonia norte*. San Carlos de Bariloche: Universidad Nacional del Comahue.

MacMillan, R.A., D.E. Moon, and R.A. Coupe. 2007. Automated predictive ecological mapping in a forest region of BC, Canada, 2001–2005. *Geoderma* 140: 353–373.

Mascaro, J., R.F. Hughes, and S.A. Schnitzer. 2012. Novel forests maintain ecosystem processes after the decline of native tree species. *Ecological Monographs* 82: 221–238.

May, R.M. 1977. Thresholds and breakpoints in ecosystems with a multiplicity of stable states. *Nature* 269: 471–477.

McClaran, M.P., D.M. Browning, and C.y. Huang. 2010. Temporal dynamics and spatial variability in desert grassland vegetation. In *Repeat photography: Methods and application in natural sciences*, 145–166. Washington, DC: Island Press.

Menghi, M., and M. Herrera. 1998. Modelo de estados y transiciones para pastizales del valle de inundación del río Dulce (Depresión de Mar Chiquita, Córdoba, Argentina). *Ecotropicos* 11: 131–140.

Miller, M.E., R.T. Belote, M.A. Bowker, and S.L. Garman. 2011. Alternative states of a semiarid grassland ecosystem: Implications for ecosystem services. *Ecosphere* 2:art55.

Miriti, M.N., S. Rodriguez-Buritica, S.J. Wright, and H.F. Howe. 2007. Episodic death across species of desert shrubs. *Ecology* 88: 32–36.

Moseley, K., P.L. Shaver, H. Sanchez, and B.T. Bestelmeyer. 2010. Ecological site development: A gentle introduction. *Rangelands* 32: 16–22.

National Agency for Meteorology and Environmental Monitoring and Ministry of Environment, G. D. a. T. 2015. *National Report on the Rangeland Health of Mongolia*, 66. Ulaanbaatar, Mongolia: Government of Mongolia.

National Research Council. 1994. *Rangeland health: New methods to classify, inventory, and monitor rangelands*. Washington, DC: National Academies Press.

Nauman, T.W., J.A. Thompson, and C. Rasmussen. 2014. Semi-automated disaggregation of a conventional soil map using knowledge driven data mining and random forests in the Sonoran Desert, USA. *Photogrammetric Engineering and Remote Sensing* 80: 353–366.

Nicholson, A.E., and M.J. Flores. 2011. Combining state and transition models with dynamic Bayesian networks. *Ecological Modelling* 222: 555–566.

Oliva, G., A. Cibils, P. Borrelli, and G. Humano. 1998. Stable states in relation to grazing in Patagonia: A 10-year experimental trial. *Journal of Arid Environments* 40: 113–131.

Paruelo, J.M., M.B. Bertiller, T.M. Schlichter, and F.R. Coronato. 1993. Secuencias de deterioro en distintos ambientes Patagónicos: Su caracterización mediante el modelo de estados y transiciones. In *Convenio Argentino-Alemán, Cooperación técnica INTA-GTZ. Lucha contra la Desertificación en la Patagonia a través de un sistema de monitoreo ecológico (LUDEPA– SME)*, 104. San Carlos de Bariloche, Argentina.

Passey, H.B., and V.K. Hugie. 1962. Application of soil-climate-vegetation relations to soil survey interpretations for rangelands. *Journal of Range Management* 15: 162–166.

Peters, D.P.C., J. Yao, O.E. Sala, and J.P. Anderson. 2012. Directional climate change and potential reversal of desertification in arid and semiarid ecosystems. *Global Change Biology* 18: 151–163.

Petersen, S.L., T.K. Stringham, and B.A. Roundy. 2009. A process-based application of state-and-transition models: A case study of western Juniper (Juniperus occidentalis) encroachment. *Rangeland Ecology & Management* 62: 186–192.

Peterson, C.H. 1984. Does a rigorous criterion for environmental identity preclude the existence of multiple stable points? *American Naturalist* 124: 127–133.

Phelps, D.G., and O.J.H. Bosch. 2002. A quantitative state and transition model for the Mitchell grasslands of central western Queensland. *The Rangeland Journal* 24: 242–267.

Pieper, R.D., and R.F. Beck. 1990. Range condition from an ecological perspective: Modifications to recognize multiple use objectives. *Journal of Range Management* 43: 550–552.

Price, J.N., and J.W. Morgan. 2008. Woody plant encroachment reduces species richness of herb-rich woodlands in southern Australia. *Austral Ecology* 33: 278–289.

Pucheta, E., M. Cabido, and S. Diaz. 1997. Modelo de estados y transiciones para los pastizales de altura de las Sierras de Córdoba, Argentina. *Ecotropicos* 10: 151–160.

Pulsford, S.A., D.B. Lindenmayer, and D.A. Driscoll. 2014. A succession of theories: Purging redundancy from disturbance theory. *Biological Reviews*.

Quirk, M., and J. McIvor. 2003. *Grazing land management: Technical manual*. Australia: Meat and Livestock Australia Sydney.

Reyers, B., S. Polasky, H. Tallis, H.A. Mooney, and A. Larigauderie. 2012. Finding common ground for biodiversity and ecosystem services. *Bioscience* 62: 503–507.

Rumpff, L., D.H. Duncan, P.A. Vesk, D.A. Keith, and B.A. Wintle. 2011. State-and-transition modelling for Adaptive Management of native woodlands. *Biological Conservation* 144: 1224–1236.

Runge, M.C. 2011. An introduction to adaptive management for threatened and endangered species. *Journal of Fish and Wildlife Management* 2: 220–233.

Scanlan, J.C. 1994. State and transition models for rangelands. 5. The use of state and transition models for predicting vegetation change in rangelands. *Tropical Grasslands* 28: 229–240.

Scheffer, M., and S.R. Carpenter. 2003. Catastrophic regime shifts in ecosystems: Linking theory to observation. *Trends in Ecology & Evolution* 18: 648–656.

Seabloom, E.W., W.S. Harpole, O.J. Reichman, and D. Tilman. 2003. Invasion, competitive dominance, and resource use by exotic and native California grassland species. *Proceedings of the National Academy of Sciences* 100: 13384–13389.

Settele, J., R. Scholes, R. Betts, S.E. Bunn, P. Leadley, D. Nepstad, J.T. Overpeck, and M.A. Taboada. 2014. Terrestrial and inland water systems. In *Climate change 2014: Impacts, adaptation, and vulnerability. Part A: Global and sectoral aspects. Contribution of Working Group II to the Fifth Assessment Report of the Intergovernmental Panel of Climate Change*, eds. C.B. Field, V.R. Barros, D.J. Dokken, K.J. Mach, M.D. Mastrandrea, T.E. Bilir, M. Chatterjee, K.L. Ebi, Y.O. Estrada, R.C. Genova, B. Girma, E.S. Kissel, A.N. Levy, S. MacCracken, P.R. Mastrandrea, and L.L. White, 271–359. Cambridge, UK and New York, NY: Cambridge University Press.

Seybold, C.A., J.E. Herrick, and J.J. Brejda. 1999. Soil resilience: A fundamental component of soil quality. *Soil Science* 164: 224–234.

Shiflet, T.N. 1973. Range sites and soils in the United States. *Arid Shrublands: Proceedings of the Third Workshop of the US/Australia Rangeland Panel*, 33. Denver, CO: Society for Range Management.

Society for Range Management. 1983. *Guidelines and terminology for range inventories and monitoring*. Report of the Range Inventory Standardization Committee, Denver, CO.

Standish, R.J., R.J. Hobbs, M.M. Mayfield, B.T. Bestelmeyer, K.N. Suding, L.L. Battaglia, V. Eviner, C.V. Hawkes, V.M. Temperton, V.A. Cramer, J.A. Harris, J.L. Funk, and P.A. Thomas. 2014. Resilience in ecology: Abstraction, distraction, or where the action is? *Biological Conservation* 177: 43–51.

Steele, C.M., B.T. Bestelmeyer, L.M. Burkett, P.L. Smith, and S. Yanoff. 2012. Spatially explicit representation of state-and-transition models. *Rangeland Ecology & Management* 65: 213–222.

Stoddard, J.L., D.P. Larsen, C.P. Hawkins, R.K. Johnson, and R.H. Norris. 2006. Setting expectations for the ecological condition of streams: The concept of reference condition. *Ecological Applications* 16: 1267–1276.

Stoms, D.M., S.L. Dashiell, and F.W. Davis. 2013. Siting solar energy development to minimize biological impacts. *Renewable Energy* 57: 289–298.

Stringham, T.K., W.C. Krueger, and P.L. Shaver. 2003. State and transition modeling: An ecological process approach. *Journal of Range Management* 56: 106–113.

Suding, K.N., and R.J. Hobbs. 2009. Threshold models in restoration and conservation: A developing framework. *Trends in Ecology & Evolution* 24: 271–279.

Susskind, L., A.E. Camacho, and T. Schenk. 2012. A critical assessment of collaborative adaptive management in practice. *Journal of Applied Ecology* 49: 47–51.

Svejcar, T., and J.R. Brown. 1991. Failures in the assumptions of the condition and trend concept for management of natural ecosystems. *Rangelands* 13: 165–167.

Svejcar, T., J. James, S. Hardegree, and R. Sheley. 2014. Incorporating plant mortality and recruitment into rangeland management and assessment. *Rangeland Ecology & Management* 67: 603–613.

Svejcar, L., B. Bestelmeyer, M. Duniway, and D. James. 2015. Scale-dependent feedbacks between patch size and plant reproduction in desert grassland. *Ecosystems* 18: 146–153.

Task Group on Unity in Concepts and Terminology Committee Members. 1995. New concepts for assessment of rangeland condition. *Journal of Range Management* 48: 271–282.

Taylor, J., N. MacLeod, and A. Ash. 1994. State and transition models: Bringing research, extension and management together. Proceedings of a workshop held at the Forestry Training Centre, Gympie, Queensland, September 13–14, 1993. *Tropical Grasslands* 28: 193–194.

Twidwell, D., B.W. Allred, and S.D. Fuhlendorf. 2013a. National-scale assessment of ecological content in the world's largest land management framework. *Ecosphere* 4:art94.

Twidwell, D., S.D. Fuhlendorf, C.A. Taylor, and W.E. Rogers. 2013a. Refining thresholds in coupled fire-vegetation models to improve management of encroaching woody plants in grasslands. *Journal of Applied Ecology* 50: 603–613.

USDA Natural Resources Conservation Service. 2014. *National ecological site handbook*. Washington, DC: United States Department of Agriculture.

Walker, B., and D. Salt. 2012. *Resilience practice: Building capacity to absorb disturbance and maintain function*. Washington, DC: Island Press.

Walker, B., and M. Westoby. 2011. States and transitions: The trajectory of an idea, 1970–2010. *Israel Journal of Ecology & Evolution* 57: 17–22.

Watson, I.W., and P.E. Novelly. 2012. Transitions across thresholds of vegetation states in the grazed rangelands of Western Australia. *Rangeland Journal* 34: 231–238.

Watson, I.W., D.G. Burnside, and A.M. Holm. 1996. Event-driven or continuous; which is the better model for managers? *The Rangeland Journal* 18: 351–369.

Watson, I.W., P.E. Novelly, and P.W.E. Thomas. 2007. Monitoring changes in pastoral rangelands – the Western Australian Rangeland Monitoring System (WARMS). *The Rangeland Journal* 29: 191–205.

West, N.E., F.L. Dougher, G.S. Manis, and R.D. Ramsey. 2005. A comprehensive ecological land classification for Utah's West Desert. *Western North American naturalist* 65: 281–309.

Westoby, M. 1980. Elements of a theory of vegetation dynamics in arid rangelands. *Israel Journal of Botany* 28: 169–194.

Westoby, M., B. Walker, and I. Noy-Meir. 1989. Opportunistic management for rangelands not at equilibrium. *Journal of Range Management* 42: 266–274.

Whipple, A.A., R.M. Grossinger, and F.W. Davis. 2011. Shifting baselines in a California oak savanna: Nineteenth century data to inform restoration scenarios. *Restoration Ecology* 19: 88–101.

Wintle, B.A., S.A. Bekessy, D.A. Keith, B.W. van Wilgen, M. Cabeza, B. Schroder, S.B. Carvalho, A. Falcucci, L. Maiorano, T.J. Regan, C. Rondinini, L. Boitani, and H.P. Possingham. 2011. Ecological-economic optimization of biodiversity conservation under climate change. *Nature Climate Change* 1: 355–359.

Wondolleck, J.M., and S.L. Yaffee. 2000. *Making collaboration work: Lessons from innovation in natural resource management*. Washington, DC: Island Press.

Yao, J., D.C. Peters, K. Havstad, R. Gibbens, and J. Herrick. 2006. Multi-scale factors and long-term responses of Chihuahuan Desert grasses to drought. *Landscape Ecology* 21: 1217–1231.

Yospin, G.I., S.D. Bridgham, R.P. Neilson, J.P. Bolte, D.M. Bachelet, P.J. Gould, C.A. Harrington, J.A. Kertis, C. Evers, and B.R. Johnson. 2014. A new model to simulate climate change impacts on forest succession for local land management. *Ecological Applications* 25: 226–242.

Zhang, X.C. 2005. Spatial downscaling of global climate model output for site-specific assessment of crop production and soil erosion. *Agricultural and Forest Meteorology* 135: 215–229.

Chapter 10
Livestock Production Systems

Justin D. Derner, Leigh Hunt, Kepler Euclides Filho, John Ritten,
Judith Capper, and Guodong Han

Abstract Rangelands, 50 % of the earth's land surface, produce a renewable resource of cellulose in plant biomass that is uniquely converted by ruminant livestock into animal protein for human consumption. Sustainably increasing global animal production for human consumption by 2050 is needed while reducing the environmental footprint of livestock production. To accomplish this, livestock producers can interseed legumes and use bioenergy protein by-products for increased dietary protein, develop forage "hot spots" on the landscape, use adaptive grazing management in response to a changing climate, incorporate integrated livestock-crop production systems, improve fertility to increase birth rates, and reduce livestock losses due to disease and pest pressure. Conceptual advances in livestock production systems have expanded the utility of livestock in conservation-oriented approaches that include (1) efforts to "engineer ecosystems" by altering vegetation structure for increased habitat and species diversity, and structural heterogeneity; (2) use of targeted grazing to reduce invasive annual grasses and invasive weeds, and fuel reduction to decrease wildfires; and (3) improvement of the distribution of livestock grazing across the landscape. Livestock production systems need to increase output of animal protein by implementation of knowledge and technology, but this production must

J.D. Derner (✉)
USDA-ARS, High Plains Grassland Research Station, Cheyenne, WY, USA
e-mail: Justin.Derner@ars.usda.gov

L. Hunt
CSIRO Ecosystem Sciences, Darwin, NT, Australia

K.E. Filho
Department of Research and Development, Brazilian Agricultural Research Corporation
(EMBRAPA), Brasilia, Brazil

J. Ritten
Department of Agricultural and Applied Economics, University of Wyoming,
Laramie, WY, USA

J. Capper
Livestock Sustainability Consultancy, Oxford, United Kingdom

G. Han
Department of Grassland Science, College of Ecology and Environmental Science,
Inner Mongolia Agricultural University, Hohhot, China

© The Author(s) 2017
D.D. Briske (ed.), *Rangeland Systems*, Springer Series on Environmental
Management, DOI 10.1007/978-3-319-46709-2_10

347

be sustainable and society needs to have confidence that animals were raised in a humane and environmentally acceptable manner such that the quality and safety of the animal protein are acceptable for consumers.

Keywords Adaptations for increasing climatic variability • Adaptive grazing management • Flexible stocking rate strategies • Forage hot spots on landscape • Ruminant livestock • Sustainable intensification

10.1 Introduction

10.1.1 Goals and Objectives

Livestock production systems utilizing global rangelands provide the ability for humans to effectively harvest animal protein from plants. These systems, which occur on six of the seven continents (Antarctica is the exception), are highly diverse, ranging from low-input, pastoral production systems located in arid and semiarid environments on communally owned lands to highly intensive production systems in more mesic environments which can integrate livestock-crop-forage systems to improve feed efficiency and reduce time from birth to harvest. The goals of this chapter are to review the important conceptual and technological advances in livestock production systems of the past 25 years, and look forward to the key opportunities that will influence global livestock production in the next 25 years.

For the retrospective look back 25 years, our objectives are to (1) showcase global trends that have occurred for ruminant livestock; (2) demonstrate, using the USA as an example, the economics of livestock production; (3) review the contributions of technological advances; (4) exhibit the shifts in management strategy toward increased emphasis on livestock in conservation-oriented approaches for land managers; (5) address environmental considerations of livestock production with an emphasis on greenhouse gas emissions; and (6) examine the changes in livestock production systems of South America's largest beef producer, Brazil.

For the prospective look forward 25 years, our overall objective is to demonstrate the significant influence of increasing global population and a rising middle class on increasing demand for animal protein. Here, we address (1) sustainable intensification of livestock production systems, including options in Australia; (2) adaptations to climatic variability; (3) customer influence on livestock production systems; (4) the need for increased feed efficiency and fertility; (5) reducing losses to disease; and (6) the importance of genetics and genomics, including rapidly emerging DNA and RNA genetic tools, and conclude with (7) an emerging integrated livestock-crop production system involving pasture, crops, and forestry.

10.1.2   Global Significance of Ruminant Livestock

Ruminant livestock, by hosting specialized microbes in their digestive system, serve as energy brokers between cellulose in plant biomass and energy and protein available for human consumption. Worldwide, rangelands provide 70 % of the forage for ruminant livestock (Holechek 2013) as domesticated livestock graze about 50 % of the world's land surface (Holechek et al. 2011), primarily occurring on lands which are ill suited for crop production (Steinfeld et al. 2006; Fig. 10.1). Furthermore, livestock can enhance efficiency of crop production through consumption of plant residues and increased rates of nutrient cycling.

10.1.3   Global Livestock Production

Globally, numbers of cattle, goats, and sheep increased from 1979 to 2009, with percentage increases of 14 % (cattle), 93 % (goats), and 1 % (sheep) observed (Fig. 10.2, FAO 2011). The continents of Africa and Asia experienced the largest percentage increases in cattle numbers, whereas Europe had the largest percentage decrease. In Africa, Asia, North America, and Oceania the numbers of goats at least doubled; only Europe had a decline in goat numbers. Sheep numbers were about 50 % higher in 2009 for Africa, Asia, and Central America/Caribbean, but about 50 % lower for

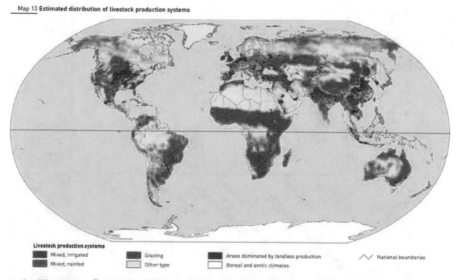

Fig. 10.1 Distribution of livestock production systems (FAO)

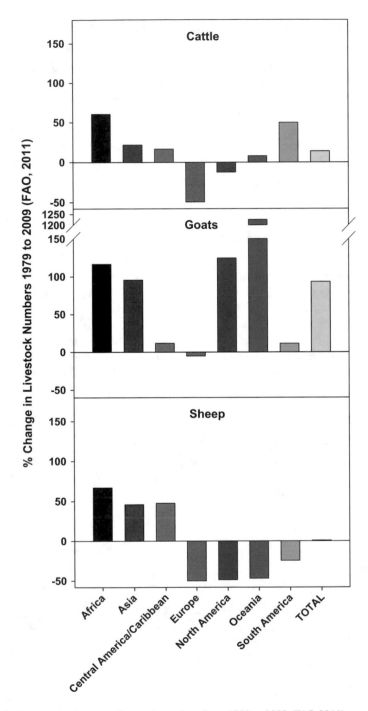

Fig. 10.2 Percentage change in livestock numbers from 1979 to 2009 (FAO 2011)

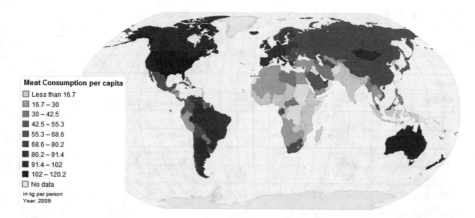

Meat Consumption per capita
- ☐ Less than 16.7
- ☐ 16.7 – 30
- ■ 30 – 42.5
- ■ 42.5 – 55.3
- ■ 55.3 – 68.6
- ■ 68.6 – 80.2
- ■ 80.2 – 91.4
- ■ 91.4 – 102
- ■ 102 – 120.2
- ☐ No data

in kg per person
Year: 2009

Fig. 10.3 Worldwide annual meat consumption per capita (FAO 2013). Current Worldwide Annual Meat Consumption per capita, Livestock and Fish Primary Equivalent, Food and Agriculture Organization of the United Nations, viewed 31st March, 2013

Europe, North America, and Oceania and 25 % lower for South America (Fig. 10.2). Increasing incomes in emerging economies of Africa and Asia, as well as increased urbanization on these continents, are driving these changes in global patterns of livestock production due to greater amounts of animal protein in human diets.

Annual growth rates in meat production over the past three decades for beef, sheep, and goats were 3–4 % in developing countries compared to slightly negative rates in developed countries[1]. Beef production is expanding globally. For example, breeding cattle (primarily the Angus breed) are being purchased by Russia[2] as well as former Soviet Republics (e.g., Kazakhstan[3]), and Brazil (see below), through direct purchases from the USA and Australia. Worldwide meat consumption per capita is highest in North America and Australia, and lowest in Africa (Fig. 10.3, FAO 2013). Beef and African buffalo contribute about one-fourth of the worldwide meat production (Table 10.1). Areas of the world with more than half of their total meat production from beef and buffalo are eastern Africa, central Asia, and Australia and New Zealand. Sheep and goat production contributes about 5 % of the worldwide meat production, with western Africa, central Asia, and Australia and New Zealand having 20 % or more of their total meat production from sheep and goats. Total meat imports across the world increased by about 60 % from 2000 to 2010, with percentage increases highest for southern and central Asia, and western Africa (Table 10.2). Cattle density is highest in India, the eastern Great Plains of the USA, western Europe, and southeastern South America (Fig. 10.4). Sheep density is highest in the Mediterranean region, and southwestern and southeastern Australia (Fig. 10.4), while density of goats is highest in India and the tropical region of Africa (Fig. 10.4).

---

[1] http://faostat.fao.org/

[2] http://www.businessweek.com/articles/2012-06-21/beef-the-new-opiate-of-the-russian-masses

[3] http://www.themeatsite.com/meatnews/22375/kazakhstan-to-grow-beef-herd

Table 10.1 Global meat production (1000 tons) in 2010 (FAO 2013 Stats)

Groups of countries	Total	Beef and buffalo Production	% of total	Sheep and goat Production	% of total	Pig	Poultry
World	296,107	67,776	22.9	13,459	4.5	109,370	99,050
Africa	17,309	6684	38.6	2872	16.6	1239	4769
Eastern	3595	1808	50.3	522	14.5	408	503
Middle	1189	413	34.7	143	12.0	153	118
Northern	5977	2504	41.9	1098	18.4	1	2059
Southern	3075	955	31.1	208	6.8	322	1507
Western	3473	1004	28.9	901	25.9	354	582
Latin America/ Caribbean	46,253	17,386	37.6	438	0.9	6553	21,310
North America	46,626	13,318	28.6	92	0.2	12,112	20,800
Asia	123,501	16,623	13.5	7716	6.2	62,054	34,858
Central	2323	1346	57.9	472	20.3	246	155
Eastern	86,904	7386	8.5	4136	4.8	54,194	19,447
South-eastern	15,948	1737	10.9	221	1.4	7164	6760
Southern	12,342	4853	39.3	1941	15.7	352	4951
Western	5984	1300	21.7	947	15.8	98	3545
Europe	56,628	11,001	19.4	1287	2.3	26,939	16,222
Eastern	16,825	3166	18.8	302	1.8	6639	6222
Northern	8099	1978	24.4	369	4.6	3399	2305
Southern	11,881	2091	17.6	417	3.5	6004	2964
Western	19,823	3766	19.0	200	1.0	10,897	4731
Oceania	5789	2764	47.7	1053	18.2	474	1092
Australia and New Zealand	5297	2744	51.8	1053	19.9	383	1065
Melanesia	478	18	3.8	0	0.0	80	24
Micronesia	3	0	0.0	0	0.0	2	1
Polynesia	12	2	16.7	0	0.0	8	2

10.1.4 Economics of Livestock Production: The US Cattle Example

The cattle production industry in the USA has become increasingly specialized, with individual sectors focusing on calf production, yearlings (mostly background-ing on forages—e.g., wheat or grasslands/rangelands/pasture), feedlots for finish-ing, processors, packers, and retail marketers. For all sectors, risk management is key. For example, cow-calf producers in highly variable environments often employ conservative stocking rates as a strategy across years to reduce risk (Torell et al. 2010). Generally, low-input producers utilize conservative stocking, minimization of debt, and enterprise and income diversification to maintain economic sustain-ability (Kachergis et al. 2014; Roche et al. 2015). Producers having access to

Table 10.2 Volume (1000 tons) of total meat imports and exports in 2000 and 2010 and net importer or exporter status in 2000 and 2010 (FAO 2013 Stats)

Groups of countries	Imports		Exports		Net importer/exporter	
	2000	2010	2000	2010	2000	2010
World	23,441	37,239	24,359	39,530	Exporter	Exporter
Africa	778	1753	118	189	Importer	Importer
Eastern	29	74	23	19	Importer	Importer
Middle	181	533	0	0	Importer	Importer
Northern	235	428	10	8	Importer	Importer
Southern	217	322	84	105	Importer	Importer
Western	117	396	2	57	Importer	Importer
Latin America/ Caribbean	1858	3266	2424	7840	Exporter	Exporter
North America	2320	2197	5881	8029	Exporter	Exporter
Asia	7650	11,820	2568	3736	Importer	Importer
Central	89	347	1	1	Importer	Importer
Eastern	5856	6823	1701	1898	Importer	Importer
South-eastern	598	1581	501	753	Importer	Importer
Southern	37	343	312	742	Exporter	Exporter
Western	1070	2725	52	341	Importer	Importer
Europe	10,642	17,849	10,909	17,212	Exporter	Importer
Eastern	1680	4519	794	2187	Importer	Importer
Northern	2090	3455	2872	3416	Exporter	Importer
Southern	2875	3453	1072	2187	Importer	Importer
Western	3998	6422	6172	9422	Exporter	Exporter
Oceania	193	354	2459	2523	Exporter	Exporter
Australia and New Zealand	80	229	2456	2521	Exporter	Exporter
Melanesia	63	70	2	2	Importer	Importer
Micronesia	7	1	0	0	Importer	Importer
Polynesia	43	55	0	0	Importer	Importer

additional forage (e.g., irrigated pastures), crop residues, and grazing of cover crops can utilize higher input strategies to optimize economic returns. Relatively cheap grains and growth of the feedlot industry since the 1960s have encouraged many cow-calf producers to sell calves rather than carry yearlings as costs of gain in feedlots have historically been cheaper than cost of gain on forage. Feedlots sell finished animals into the packer-processor market where animals are harvested and moved into the wholesale market. From here, beef is dispersed into the retail market. Cow herd numbers in the USA have decreased over 30% since the 1970s, and are at their lowest levels since the end of World War II (Fig. 10.5), leading to record high cattle prices in 2014 and 2015. Production outputs such as weaning weights have increased due to implementation of multiple technologies (Ash et al. 2015).

In 1970, there were over three feeder cattle outside feedlots for every animal in a feedlot; currently this value is less than two animals as comparatively cheap grains

Fig. 10.4 World density maps in 2005 (numbers of animal km⁻²) for cattle (*top*), sheep (*middle*), and goats (*bottom*) (FAO)

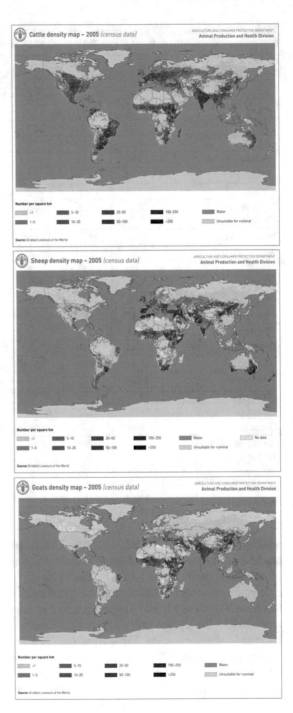

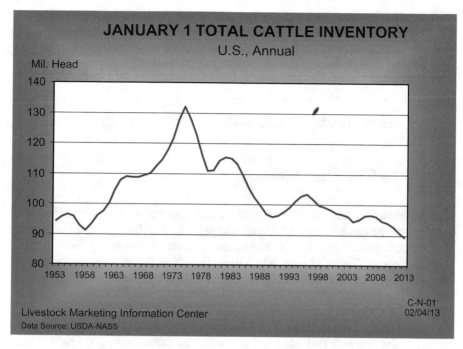

Fig. 10.5 US total cattle inventory (1953–2013) (USDA-NASS)

have increased the reliance on feedlot gains for cattle. The vast majority of feedlots in the USA are located in the Great Plains where most of the corn is produced. In addition, most of the biofuel (e.g., ethanol) refineries are located in this same region, which provides a large supply of by-products for use in feedlots. Feedlots having capacity for greater than 32,000 head market around 40% of fed cattle, and greater feedlot operation size, have been shown to significantly reduce costs of production. Genetic selection for animals with greater gains kg of feed^{-1} has increased carcass weights. Residual feed intake is garnering attention as a tool that allows producers to select for efficiency while accounting for body weight and a wide range of factors above and beyond simply total feed consumption and total animal gain (Herd et al. 2003; Herd and Arthur 2009). Successful technological advances that increase feed efficiency to enhance productivity (e.g., improved feeding and management), and structural shifts in the livestock sector to reduce adverse environmental impacts, can increase profitability and sustainability of livestock production systems (Herrero et al. 2013).

Feedlot operators are increasingly turning to formula and grid markets, where prices are determined on actual carcass qualities instead of the traditional markets that base price on live or dressed weights. Grid-based prices rely on USDA quality and yield grades to impact prices through premiums for those carcasses with desirable grades, while many formulas also include premiums for branded beef (e.g., Certified Angus Beef) and discounts for characteristics such as injection-site blemishes and hide damage. Cattle price on grids has increased from 14% in 1996 to

50 % in 2001, and to 75 % by 2012. The grid system can be a benefit to the cattle industry as it sends clear signals to feeders of changing consumer preferences. For example, premiums occur for branded products that meet customer preferences for reduced used of antibiotics and hormones (e.g., natural beef).

10.2 Looking Back: Livestock Production—The Previous 25 Years

10.2.1 Technological Advances

Technological advances and implementation of management practices derived from experimental research in the past 25 years have focused on increasing the efficiency of livestock production and reducing environmental impacts. Technological advances include (1) use of artificial insemination for breeding livestock with superior genetic traits, (2) crossbreeding to achieve heterosis (hybrid vigor), (3) emerging use of DNA and RNA technology for advances in genetic trait selection, (4) use of grains for improved gain efficiency to finish livestock for harvest, and (5) use of growth hormones to shorten the time from birth to harvest leading to increased efficiency. Associated with this increased production efficiency has been a reduction in greenhouse gas emissions kg beef^{-1} (Capper and Hayes 2012). Other technological advances have included additional water developments—water systems, pipelines, spring developments, and installation of ponds—which improve livestock distribution and utilization of available forage, and dietary supplements (e.g., urea, phosphorus, minerals) to overcome seasonal deficiencies in forage quality, especially in tropical and subtropical rangelands. Grazing strategies across the globe are quite diverse from low-intensity management of pastoral and communal approaches to high-intensity management facilitated by infrastructure developments (Roche et al. 2015). Key management practices that have been developed include (1) the application of sustainable stocking rates to maintain or improve the health of rangelands, including riparian habitats (Briske et al. 2011), as well as associated optimization of net income (Holechek 2013; Kemp et al. 2013), and (2) matching of calving season to the prevailing environment to reduce associated harvested feed costs (Grings et al. 2005; Griffin et al. 2012).

10.2.2 Shifts in Production Strategies

Within the first decade of the twenty-first century, livestock production strategies have increased emphasis on livestock in conservation-oriented approaches to include (1) efforts to "engineer ecosystems" by altering vegetation structure for increased habitat and species diversity, and structural heterogeneity to achieve desired contemporary outcomes (Derner et al. 2009, 2013); (2) use of targeted grazing involving application of a specific kind of livestock at a determined season, duration, and intensity to accomplish defined vegetation or landscape goals (Launchbaugh and Walker 2006), to reduce invasive annual grasses (Diamond et al. 2010) and invasive weeds (Goehring et al. 2010), as well as fuel reduction efforts (Davison 1996; Clark et al. 2013); and (3) improvement of the distribution of livestock grazing across the landscape through the use of low-stress stockmanship methods using herding, strategic location of low-moisture supplement blocks (Bailey et al. 2008), patch burn grazing in mesic (Fuhlendorf and Engle 2004) and semiarid (Augustine and Derner 2014) ecosystems, and foraging and learning through past experiences that increase the likelihood of animals learning to eat different plants (e.g., Provenza et al. 2003). These ecological benefits from conservation-management applications have been attained without negatively impacting livestock production (Limb et al. 2011; Augustine and Derner 2014).

10.2.3   Environmental Considerations: Beef Production and Greenhouse Gas Emissions

Greenhouse gas emissions kg beef^{-1} produced between 9.9 kg CO_2-eq (carbon dioxide equivalent) and 36.4 CO_2-eq in intensive systems, and from 12.0 to 44.0 kg CO_2-eq kg^{-1} beef in extensive systems (Pelletier et al. 2010; Capper 2012). These differences in GHG emissions may be attributed to differences in assessment methodology in addition to the direct consequence of different production systems. Assessment methodology differences include variation in the system boundaries and underlying assumptions and model complexity that may have a considerable effect upon the results of environmental impact assessments (Bertrand and Barnett 2011). Until a global analysis of environmental impact from beef production is conducted, with similar methodology for each region, it is difficult to draw firm conclusions as to the variation within and between global regions. The study by Herrero et al. (2013) is a first approximation to spatially disaggregate a global livestock dataset to assist in determinations of greenhouse gas emissions at regional to local scales. Most of the developed countries have low greenhouse gas emission intensities due to improved and intensive feeding practices, and higher feed quality; conversely, sub-Saharan Africa is a hot spot for high greenhouse gas emission intensities as a result of low-quality feeds and animals with low productive potential animal^{-1} (Herrero et al. 2013).

Adverse environmental effects are minimized by improving productivity in the metrics of carcass weight and growth rate (Capper and Hayes 2012; White and Capper 2013). As productivity increases, the proportion of daily energy allocated to maintenance decreases and maintenance requirements of the total animal population decreases. Improvement in growth rate reduces time from birth to harvest and increases the total production of meat yield in a shorter time frame without affecting herd numbers (Capper 2011a; Capper and Hayes 2012; White and Capper 2013). With increasing intensity of a production system, corresponding improvements in productivity and efficiency are usually exhibited resulting in "more intensive"-type production systems tending to use fewer resources and having lower greenhouse gas emissions than "less intensive" or "extensive" systems (Capper 2011b).

Efficiency in the US beef industry markedly increased between 1977 and 2007 with growth rates (kg head^{-1} day^{-1}) increasing by 64%, harvest weights increasing by 30%, and days from birth to harvest decreasing by 20% (Capper 2011a). These advances over the 30 years were attributed to improvements in genetics, nutrition, and management, as well as use of fossil fuels and irrigation water development. As a result, 12–33% less land, water, and feed were needed, and greenhouse gas emissions kg beef^{-1} decreased by 16% (Capper 2011a). Greenhouse gas emissions unit beef^{-1} were 67% greater in pasture-finished systems than in feedlot systems (Capper 2012). To maintain current US beef production (11.8 billion kg) from an entirely pasture-based system would require an additional 52 million ha of land (Capper 2012). If such conversion did occur, the increase in carbon emissions from these extensive systems would be equal to adding 25.2 million automobiles to the road year^{-1} (Capper 2012). Questions remain for feedlot systems with respect to long-term availability and cost of quantity and quality of water, as well as fossil fuels for associated crop production to maintain the current advantage in greenhouse gas emissions for these systems.

10.2.4   Beef Production: A Brazilian Example

Beef production in South America is primarily confined to pastures compromised of introduced forage species that vary with regional environmental characteristics. For example, temperate species of grasses and legumes dominate pastures in southern Brazil, while the remainder of Brazil has mostly tropical species. Brazil has approximately 200 million head of cattle. Increases in the numbers of cattle in Brazil are occurring in spite of a decrease in total pasture area (Martha et al. 2012). This is a result of production intensification and increased efficiency that supports higher stocking rates. In addition, age to harvest has decreased, as has the age at the beginning of reproduction. The quality of carcasses produced has markedly improved as a result of better animal and forage genetics, better management practices—financial, nutrition, reproduction, and health—and organization of the beef supply chain. For example, beef production increased only 0.3% year^{-1} from 1950 to 1975, but increased to 3.6% year^{-1} from 1975 to 1996, and then further increases to 6.6% year^{-1} were observed from 1996 to 2006 (Martha et al. 2012). Key beef production metrics for Brazil increased from 1994 to 2007 (Table 10.3). For example, over these 13 years, harvest rate (13%), numbers of cattle harvested (73%), beef production (77%), consumption per capita (13%), domestic consumption (37%), and the amount (540%) and value (694%) of beef exports increased, whereas the amount of beef imported decreased by almost half.

Table 10.3 Key metrics of beef production in Brazil (1994–2007)

	1994	1998	2002	2006	2007
Harvest rate (%)	16.4	19.1	19.8	21.7	21.7
Harvest (millions of head)	26.0	30.2	35.5	44.4	45.0
Beef production (1000 tons of carcass)	5200	6040	7300	8950	9200
Consumption per capita (kg of carcass)	32.6	35.8	36.6	36.6	36.7
Domestic consumption (1000 tons of carcass)	5017	5797	6395	6780	6880
Export (1000 tons of carcass)	378.4	377.6	1006.0	2200.0	2420.0
Import (1000 tons of carcass)	195.9	135.1	100.7	30.0	100.0
Export (US$ million)	573	589	1107	3800	4552
Import (US$ million)	230.5	220.0	84.0	63.0	210.0

Source: Adapted from CNPC. Available at http://www.cnpc.org.br/site/Balanco.xls

For livestock production in Brazil, improvements in forage management involving combinations of new grass and legume cultivars developed by breeding programs, and associated grazing systems adapted to regional-specific conditions can increase livestock production, with the caution that mineral supplementation is necessary due to the weathered soils (Ferraz and de Felicio 2010). Greater understanding of soil-plant-animal interactions in both temperate and tropical areas will result in better pasture management, as well as improved livestock production efficiency. The seasonality of forage production in tropical regions can limit sustainability of livestock production as forage limitations in dry seasons reduce animal gains. Intensifying management in these regions remains a challenge, but opportunities exist for (1) adapting stocking rates for seasonal differences in forage production, (2) irrigating pastures where water is available and it is recommended, (3) stockpiling of forage for dry season, (4) use of diet supplementation, (5) increased use of feedlots as a strategy to finish animals, (6) improved animal breeding, and (7) a combination of two or more of these strategies. Intensifying management provides important co-benefits such as reducing the pressure for clearing new areas for pasture and lessening deforestation impacts, decreasing costs of beef production by keeping the production systems closer to the infrastructure system of roads and industries already in place, and contributing to decreases in total greenhouse gas emissions since intensification shortens the time to harvest.

10.3 Looking Ahead: Livestock Production in the Next 25 Years

The global livestock production industry faces a significant challenge in producing sufficient animal protein to supply an increasing population, including an expanding middle class. Predictions are that animal production will need to increase by 70% by 2050 to accommodate an additional two billion humans (from seven to nine billion

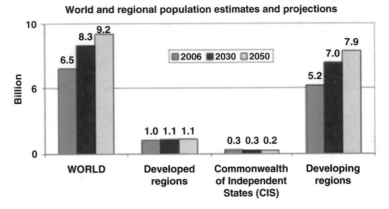

Fig. 10.6 World and regional population estimates and projections. Source: UN Population Division, 2006 Revision, World Population Prospects

Growth in Demand for Beef 2000 - 2030

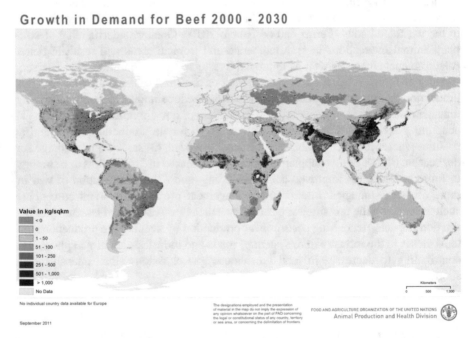

Fig. 10.7 Growth in demand for beef (2000–2030) (FAO)

plus) and a rise in the global middle class (FAO 2009, 2011; Fig. 10.6). Growth in demand for animal protein through 2030 will be largely concentrated in the tropical regions of Asia (e.g., India and China), Africa, and Latin America (Fig. 10.7). As a result, livestock production will need to employ sustainable intensification in terms of management to increase production while having neutral environmental effects.

The feasibility of sustainable intensification will largely depend on the ability of managers to adaptively match forage production, forage demand, and forage quality with increasing weather and climatic variability. The flexibility of operational structure such as cows and calves with yearlings provides substantial economic outcomes with adaptive management, but is dependent on high quality and accurate seasonal climate forecasts that are not currently available (Torell et al. 2010). Forage quality concerns of reduced crude protein concentrations and forage digestibility are predicted to offset greater forage production associated with increases in atmospheric CO_2 concentration (e.g., Milchunas et al. 2005; Craine et al. 2010).

Intensification of livestock production will largely occur in mesic, rather than xeric, environments. In drier environments, management emphasis will encompass resiliency, risk reduction, avoidance of debt and degradation of natural resources, and low input for sustainability. Livestock production intensification in more mesic environments will be determined by four variables and external forces (Euclides Filho 1996). First, improvement of efficiency and economic viability of livestock production enterprises is dependent on effectively managing available natural resources and efficiently utilizing available technologies, including the potential risks and adoption rates of best management practices by producers. Second, market-driven competitiveness and achieving consumer expectations will require a capacity to consistently provide high-quality products with reliable taste and tenderness within price ranges affordable for society. For example, value-added products with unique niche markets could be associated with traceability and certification systems to document origin of the livestock for the consumer. Third, recognition of constraints in land-use decisions related to coexistence of food, feed, fiber, and energy production from lands will require increasing integration of these production systems. Fourth, emphasis on economic, social, and environmental benefits, including greater improved distribution of livestock profits among segments of the supply chain, and an increased concern about the collective and individual well-being of both humans and animals. In summary, livestock production systems must intensify their activities by introducing knowledge and technology that not only assures sustainable production, but also demonstrates transparency for the general public that will increase confidence in the quality of animal protein for consumers, as well as humane treatment of animals and environmental impacts of livestock production.

10.3.1 Sustainable Intensification: An Australian Example

Sustainable intensification options in Australia include oversowing native perennial grass pastures with legumes, the development of "mosaic" irrigation for forage "hot spots" on the landscape where limited areas of suitable soil on large properties are developed with irrigation to produce high-value forages to enable animals to be finished for market, and potential on-farm production of low-cost supplements such as high-protein algal-based supplements using technology from the biofuel industry. While the technology may not yet be in place to support all these developments, and the economics may not be positive in all cases, these options do indicate that there is potential to increase livestock production by 2% annually over a 20-year time frame (Ash et al. 2015). Other developments such as genetic improvements for feed conversion efficiency, improvement in veterinary care, and increased use of supplements can also have an important role in boosting productivity (Ash et al. 2015).

However, most intensification options require greater managerial commitment and capital infrastructure costs, and are often associated with increased risk. The economic risk is generally high, because of the large capital costs associated with intensification options and frequency and duration of drought which reduces the value of livestock due to excess market supply (Coppock et al. 2009; Ash et al. 2015). Livestock enterprises will also need to adapt to a low-carbon operating environment as societal pressure increases to reduce carbon emissions while enhancing carbon sequestration. The potential benefits of various intensification options for livestock production and environmental sustainability require greater consideration on rangelands.

10.3.2 Adaptations for Increasing Climactic Variability

Climatic variability is an area of considerable concern for livestock production (Polley et al. 2013; Reeves et al. 2013). A high degree of climatic variability is a key feature of arid and semiarid rangelands worldwide as droughts are a major cause of land degradation and economic loss (Stafford Smith et al. 2007; Coppock 2011). Preparedness of land managers for such events and their responses are crucial to minimizing negative effects on natural resources as well as the financial consequences for the production enterprise (Kachergis et al. 2014). A high degree of reliance on emergency government financial assistance to support drought-affected properties has encouraged land managers to maintain current stocking rates with consequential land degradation. Adverse effects of poor drought management on other rangeland values such as biodiversity and water resources are also of increasing concern.

The use of diverse management strategies is often necessary to manage risk associated with climate variability (McAllister 2012). Options include adaptation strategies associated with flexible herd management, alternative livestock types and breeds, modified enterprise structures, and geographic relocation (Joyce et al. 2013). The main livestock management options range from the use of conservative stocking rates (Hunt 2008) to adopting a flexible stocking rate strategy in which livestock numbers are varied in response to changing and seasonal forage availability (Ritten et al. 2010; Torell et al. 2010). Incorporation of a yearling enterprise in addition to cow-calf operations can increase flexibility by providing (1) extra grazing animals during periods of high forage availability, (2) readily marketable animals during drought periods, and (3) ability to preserve herd genetics by selling yearlings, rather than the base cow herd. Challenges to adding this second enterprise include increased managerial effort and skills, contacts in the industry for the supply and sale of yearling animals or other classes of animals, and additional financial outlay, cash flow, and marketing. For example, obtaining yearling animals during favorable times and having somewhere to send these animals as forage conditions deteriorate can present challenges for land managers under a flexible strategy. For more remote areas distant from markets, a flexible strategy can present challenges due to logistical difficulties and costs of transporting animals. Transporting livestock from a property with a forage deficit to another property with a forage surplus (agistment), where the livestock owner leases the land for grazing the livestock, is an important management option in Australia (McAllister et al. 2006). Benefits to the livestock owner can include (1) an avoided forced sale of livestock with depressed prices, (2) a more rapid vegetation recovery following the end of drought by reducing degradation due to drought, and (3) maintaining a core breeding herd. A risk is prolonged drought and the leased property also running out of forage, potentially forcing the eventual sale of the livestock. Agistment is less common outside of Australia, although in the USA there was a substantial movement of cattle from Texas and California to the Northern Great Plains during recent droughts. Nomadism and transhumance are practiced in some African and Asian rangelands as a means of buffering the effects of climate variability, but this mobility is rapidly being lost (Chap. 17, this volume).

Provision of supplementary feed as a drought management strategy is problematic in many rangelands because of the conflict between drought policies and economics of purchasing supplemental feed. The cost of purchase and transport of supplemental feed can be considerable, particularly for remote regions and extensive enterprises with large herds or flocks typical of many rangelands, and in the case of multiyear droughts (Kachergis et al. 2014). Maintaining livestock on rangelands by means of supplementary forage can degrade natural resources, including vegetation and soils. Providing supplementary feed can only be justified for maintaining a limited number of animals, such as valuable core breeding stock. For example, placing cows in drylots for calving and breeding may be advantageous for both cow and subsequent calf performance compared to supplementing cows on pastures during these periods, due to controlling rations for protein, energy, and fat content (Wilson et al. 2015). In shrub-dominated rangelands experiencing drought, livestock will increase use of shrubs (Estell et al. 2012).

10.3.3   Customer-Driven Demand for Livestock Quality and Products

Consumers of animal protein and fiber products are becoming increasingly concerned about the nature of livestock production systems, the welfare of animals, and the effects of livestock production on the environment. Consumers expect these products to be produced using humane methods in largely natural environments and without adverse environmental consequences, and this affects their buying habits (Grandin 2007). There has also been a trend toward the development of "low-stress" stockmanship methods of livestock handling and production, and land managers are increasingly adopting such techniques. Adoption of these methods can have an added benefit as there is increasing evidence that productivity and profitability of livestock enterprises are improved (Cote 2004; Grandin 2007). Although survey data overwhelmingly concludes that price, taste, convenience, and nutrition are the major factors affecting purchasing decisions (Vermeir and Verbeke 2006; Simmons 2009), these are still dependent on the product being morally or ethically acceptable to the consumer. Furthermore, legislative efforts could potentially dictate production practices for livestock production.

10.3.4 Improving Feed Efficiency

The poultry and swine industries have made significant gains in improving feed efficiency over the past several decades, and considerable interest currently exists within the beef industry to select cattle for improved feed efficiency. This may be achieved through an improvement in residual feed intake (RFI), defined as reduced feed consumption required to support maintenance and production compared to the predicted or average quantity (Archer et al. 1999). Steers selected for high efficiency (low RFI) consumed less feed over the finishing period compared to low-efficiency (high RFI) cohorts in a large-scale feedlot study while maintaining harvest weight and exhibiting a greater dressing percentage (Herd et al. 2009). Furthermore, Angus steers selected for low RFI had reduced methane emissions consistent with reduced dry matter intake (Hegarty et al. 2007). If productivity can be maintained with reduced dry matter intake, then resource use, greenhouse gas emissions, and feed costs would also be predicted to decrease unit production output^{-1}. National research programs[4] are evaluating genetic improvements for feed efficiency in beef cattle to reduce feed resources, increase production of animal protein without additional feed inputs, and reduce greenhouse gas emissions of beef production systems.

10.3.5 Improving Fertility

Livestock fertility is arguably the major factor by which global livestock producers could improve the sustainability of animal protein. For example, within the USA, 89 % of cows bear a live calf each year (USDA 2009), but this number declines to 50–60 % in South American countries (e.g., Brazil, Argentina, and Chile). Cow-calf operations contribute up to 80 % of greenhouse gas emissions unit beef^{-1} (Beauchemin et al. 2010) and productivity improvements post-calving cannot compensate for the resource use and greenhouse gas emissions associated with maintaining a nonproductive cow. Management practices and technologies that improve birth rates offer significant opportunities to reduce land and water use, greenhouse gas emissions, and feed costs (Capper 2013a).

---

[4] http://www.beefefficiency.org/

10.3.6   Reducing Losses to Disease

Globally averaged livestock losses due to disease are more than 20%; thus considerable gains could be made through treating diseases or conditions that have a negative impact upon livestock performance. For example, prudent use of parasiticides in beef cattle improves performance, with associated positive environmental and economic impacts (Lawrence and Ibarburu 2007; Capper 2013c). To date, effects of many of the less tangible productivity losses within livestock systems such as male fertility, clinical and subclinical morbidity, and growth of replacement animals have yet to be quantified. Development of new vaccines provides opportunities for targeted protection from losses associated with livestock viruses, including bovine viral diarrhea, BVD; bovine respiratory syncytial virus, BRSV; infectious bovine rhinotracheitis, IBR; and parainfluenza 3 virus, PI3[5].

10.3.7   Hormone Use and Sustainability

Consumer aversion to the use of chemicals in food production is often cited as a retailer rationale for removing technologies such as hormones, beta agonists, or antibiotics from the food supply chain. The market share for organic, natural, or local foods is small, but growing. Technologies such as ionophores, steroid implants, and beta agonists have had significant roles in reducing environmental impacts of ruminant production in those regions where they are registered for use (Capper 2013b). For example, steroid implants and beta agonists improve growth rates and harvest weights which reduce land and water use, as well as greenhouse gas emissions unit beef^{-1} (Capper 2013b). Removing these production enhancement technologies from US beef production would increase land and water use, and global greenhouse gas emissions, while also resulting in increased imports from countries will less efficient production systems (Capper and Hayes 2012).

10.3.8   Genetics and Genomics

Current selection efforts in livestock production, especially beef, have resulted in animals with increased growth and carcass qualities. As a result of these selection efforts, animals now have a larger mature size with greater maintenance requirements, which increase production costs (Williams et al. 2009). However, genetic improvement in reproductive performance has not occurred (Garrick 2011). This may be attributable to low selection accuracy in traits such as longevity, lifetime

---

[5] http://purduephil.wordpress.com/2013/10/10/usda-approves-first-combination-mlv-vaccine-to-provide-targeted-protection-against-bvd-1b/

reproductive performance, and fertility (Garrick 2011). Advances in breeding programs and best management agricultural practices have produced linear gains in global food production equal to 32 million metric tons year^{-1} (Tester and Langridge 2010). However, this rate will need to be increased to 44 million metric tons year^{-1} to accommodate predicted global food production needs by 2050. Completion of the Genome Sequencing for cattle (Zimin et al. 2010), goats (Dong et al. 2013), and sheep (The International Sheep Genomics Consortium 2009) provides the genetic template for using DNA and RNA technologies to improve production efficiency of these species. Use of genetic markers for determination of parentage has been commercialized with costs of DNA sequencing precipitously decreasing from 1990 to 2012 (Eggen 2012). Genomic selection has been put forward as a new breeding paradigm (Meuwissen et al. 2001; Eggen 2012), with use of molecular breeding values being combined with traditional expected progeny differences as indexes for traits. Fundamental to livestock producers will be the combination of new genomic information with traditional pedigree and performance data, but the genomic information needs to be cost effective and have a high accuracy (Johnston et al. 2012). It is anticipated that rapidly emerging DNA and RNA genetic tools (Johnston et al. 2012) will permit the advancement of genomic selection programs for individual and multiple traits simultaneously. New genetic tools may also allow selection for traits that are difficult to measure, such as adaptability and grazing distribution (Bailey et al. 2015). In addition, newly emergent genetic field such as epigenetics, which is the study of cellular and physiological traits that are heritable and not caused by changes in the DNA sequence where inheritance patterns differ even when DNA sequences are the same, provides opportunities to turn on or off different sets of genes (e.g., Rada-Iglesias and Wysocka 2011). Here for example, hair coat color could be modified to adapt to seasonal environmental stresses, with hair color darker during the colder months and lighter during the hotter months.

10.3.9 Integrated Livestock-Crop Production Systems

Integrated livestock-crop production systems (Sulc and Franzluebbers 2014) are an emergent management strategy. Integrated livestock-crop production systems can reduce enterprise risk, restore degraded land, increase productivity, diversify production, and enhance resiliency of the land (Palmer 2014). In addition, by integrating livestock with crops as well as with forests, manure from livestock can be used as fertilizer to improve soil nutrient status and soil organic matter (Sulc and Franzluebbers 2014). Combining crops, livestock, and forestry might be done in rotation to capitalize on synergies among the ecosystem components by improving physical and chemical characteristics of soils which results in decreasing the need of new areas for increased production.

10.4   Future Perspectives

Livestock production enters a period of opportunity to address increasing efficiencies for the provision of animal protein as well as reducing environmental footprints. First, the incorporation of adaptive grazing management, with monitoring-informed decision making, to optimize forage demand with available forage will increase efficiency of livestock weight gain on rangelands while providing additional economic benefits for producers. This will concurrently reduce negative environmental impacts associated with improper stocking rate during dry periods and drought, and capture additional livestock gain during wet periods for increased net economic returns. Second, increasing availability and reliability of transportation infrastructure in developed and developing countries provide capacity to more efficiently move animals to feed which (1) reduces GHG emissions with transportation of feed to animals, (2) reduces land degradation associated with drought, and (3) increases animal weight gains due to removing constraints in systems that are characterized by extended dry seasons where animals typically lose weight. Third, continued advances in genetics for feed efficiency and carcass quality provide opportunities to capture additional economic income from value-added niche markets for the delivery of animal protein products in highly transparent manner to society with concomitant reductions in the GHG footprint of livestock production due to shortened times from birth to harvest. Fourth, improvements in birth and weaning rates and reductions in losses to disease and pest provide inherent efficiencies to global livestock production to increase provision of animal protein without any increase in land area, thereby again reducing environmental footprints.

With an additional two billion humans and a rise in the middle class by 2050, animal protein production will need to increase by 70% to meet global demand. This growth, largely concentrated in the tropical regions of Asia, Africa, and Latin America, will require sustainable intensification of livestock production systems. Two possible strategies to meet this demand are (1) further increases in "more intensive" production systems that incorporate grains in countries like Brazil to increase gain efficiency, reduce time from birth to harvest, and reduce greenhouse gas emissions, and (2) intensification of rangeland systems by interseeding legumes and use of bioenergy protein by-products for increased dietary protein, and development of forage "hot spots" on the landscape. Further, livestock production efficiencies can be increased through (1) adopting a flexible stocking rate strategy to vary livestock numbers on rangelands in response to changing seasonal forage availability, (2) selecting animals for improved feed efficiency, (3) improving fertility to increase birth rates, and (4) reducing livestock losses due to disease and pest pressure. Rapidly emerging DNA and RNA genetic tools, combined with completion of genomic sequencing for livestock, provide capacity for advancement of genomic selection programs for individual and multiple traits simultaneously as well as mitigation and adaptation to a changing climate and environmental stresses. In summary, livestock production systems must intensify their activities by introducing knowledge and technology that not only assures sustainable production, but also improves transparency for increased confidence in the quality of animal protein.

10.5 Summary

Ruminant livestock uniquely convert the high cellulose biomass of grasses, forbs, and woody plants produced on rangelands, which occupy about 50 % of the world's land surface, as a renewable dietary source of energy and animal protein for human consumption. Globally, increasing incomes in emerging economies of Africa and Asia, as well as increased urbanization on these continents, are driving increasing livestock numbers and changes in global patterns of livestock production due to greater levels of animal protein in human diets. Although livestock production systems have already benefited from fossil fuel inputs and many technological and conceptual advances, sustainably increasing global animal protein production for human consumption by 2050 is needed while also reducing the environmental footprint of livestock production. Genetic improvement technology involving artificial insemination for breeding to increase superior genetic traits, crossbreeding to achieve heterosis or hybrid vigor, and emerging use of DNA and RNA technologies combined with the completion of genomic sequencing for livestock provide capacity for advancement of genomic selection programs through transparent efforts that increase societal confidence in the quality of animal protein for consumers, as well as increasing production efficiency and reducing land and water use, and greenhouse gas emissions per unit beef. Further, increasing abundance of "more intensive"-type production systems that incorporate grains, growth hormones, and dietary supplements for improved gain efficiency to finish livestock for harvest use fewer resources and have lower greenhouse gas emissions per unit animal protein than "less intensive" or "extensive" systems. Livestock producers can interseed legumes and use bioenergy protein by-products for increased dietary protein, develop forage "hot spots" on the landscape, use adaptive grazing management in response to a changing climate, incorporate integrated livestock-crop production systems, improve fertility to increase birth rates, and reduce livestock losses due to disease and pest pressure. These management strategies can be used in an effort to (1) improve gain efficiency facilitating finishing livestock for harvest using fewer resources, (2) reduce time from birth to harvest, and (3) have lower greenhouse gas emissions per unit animal protein. Economics of livestock sales for harvest are increasingly turning to formula and grid markets, where prices are determined on actual carcass qualities instead of the traditional markets that base price on live or dressed weights; premiums are paid for carcasses that meet customer demands, whereas discounts are associated with less desirable carcasses. Conceptual advances in livestock production systems have expanded the utility of livestock in conservation-oriented approaches to include (1) efforts to "engineer ecosystems" by altering vegetation structure for increased habitat and species diversity, and structural heterogeneity); (2) use of targeted grazing to reduce invasive annual grasses and invasive weeds, and fuel reduction to decrease wildfires; and (3) improvement of the distribution of livestock grazing across the landscape. Livestock production systems need to increase their output of animal protein by implementation of knowledge and technology, but this production must be sustainable and society needs to have confidence that animals were raised in a humane and environmentally acceptable manner such that the quality and safety of the animal protein are acceptable for consumers.

References

Archer, J.A., E.C. Richardson, R.M. Herd, and P.F. Arthur. 1999. Potential for selection to improve efficiency of feed use in beef cattle: A review. *Australian Journal of Agricultural Research* 50: 247–261.

Ash, A., L. Hunt, C. McDonald, J. Scanlan, L. Bell, R. Cowley, I. Watson, J. McIvor, and N. MacLeod. 2015. Boosting the productivity and profitability of northern Australian beef enterprises: Exploring innovation options using simulation modelling and systems analysis. *Agricultural Systems* 139: 50–65.

Augustine, D.J., and J.D. Derner. 2014. Controls over the strength and timing of the fire-grazing interactions in a semi-arid rangeland. *Journal of Applied Ecology* 51: 242–250.

Bailey, D.W., C. VanWagoner, R. Weinmeister, and D. Jensen. 2008. Evaluation of low-stress herding and supplement placement for managing cattle grazing in riparian and upland areas. *Rangeland Ecology and Management* 61: 26–37.

Bailey, D.W., S. Lunt, A. Lipka, M.G. Thomas, J.F. Medrano, A. Canovas, G. Rincon, M.B. Stephenson, and D. Jensen. 2015. Genetic influences on cattle grazing distribution: Associated of genetic markers with terrain use in cattle. *Rangeland Ecology and Management* 68: 142–149.

Beauchemin, H.H., S.M. Janzen, T.A. Little, McAllister, and S.M. McGinn. 2010. Life cycle assessment of greenhouse gas emissions from beef production in western Canada: A case study. *Agricultural Systems* 103: 371–379.

Bertrand, S., and J. Barnett. 2011. Standard method for determining the carbon footprint of dairy products reduces confusion. *Animal Frontiers* 1: 14–18.

Briske, D.D., J.D. Derner, D.G. Milchunas, and K.W. Tate. 2011. An evidence-based assessment of prescribed grazing practices. In *Conservation benefits of rangeland practices: Assessment, recommendations, and knowledge gaps*, ed. D.D. Briske. Washington, DC: USDA-NRCS.

Capper, J.L. 2011. The environmental impact of beef production in the United States: 1977 compared with 2007. *Journal of Animal Science* 89: 4249–4261.

———. 2011b. Replacing rose-tinted spectacles with a high-powered microscope: The historical vs. modern carbon footprint of animal agriculture. *Animal Frontiers* 1:26–32.

———. 2012. Is the grass always greener? Comparing resource use and carbon footprints of conventional, natural and grass-fed beef production systems. *Animals* 2:127–143.

———. 2013a. The environmental and economic impact of calving rate within U.S. beef production. In *Proceedings of the ADSA-ASAS Joint Annual Meeting*. Indianapolis, IN.

———. 2013b. The environmental and economic impact of steroid implant and beta-adrenergic agonist use within U.S. beef production. In *Proceedings of the ADSA-ASAS Joint Annual Meeting*. Indianapolis, IN.

———. 2013c. The environmental and economic impact of withdrawing parasite control (Fenbendazole) from U.S. beef production. In *Proceedings of the ADSA-ASAS Joint Annual Meeting*. Indianapolis, IN.

Capper, J.L., and D.J. Hayes. 2012. The environmental and economic impact of removing growth-enhancing technologies from United States beef production. *Journal of Animal Science* 90: 3527–3537.

Clark, A., T. DelCurto, M. Vavra, and B.L. Dick. 2013. Stocking rate and fuels reduction effects on beef cattle diet composition and quality. *Rangeland Ecology and Management* 66: 714–720.

Coppock, D.L. 2011. Ranching and multiyear droughts in Utah: Production, impacts, risk perceptions, and changes in preparedness. *Rangeland Ecology and Management* 64: 607–618.

Coppock, D.L., D.L. Snyder, L.D. Sainsbury, M. Amin, and T.D. McNiven. 2009. Intensifying beef production on Utah private land: Productivity, profitability, and risk. *Rangeland Ecology and Management* 62: 253–267.

Cote, S. 2004. *Stockmanship—A powerful tool for grazing land management*. Arco, ID: USDA Natural Resources Conservation Service.

Craine, J.M., A.J. Elmore, K.C. Olson, and D. Tollenson. 2010. Climate change and cattle nutritional stress. *Global Change Biology* 16: 2901–2911.

Derner, J.D., W.K. Lauenroth, P. Stapp, and D.J. Augustine. 2009. Livestock as ecosystem engineers for grassland bird habitat in the western Great Plains of North America. Rangeland Ecology & Management 62: 111–118.

Davison, J. 1996. Livestock grazing in wildland fuel management programs. *Rangelands* 18(6): 242–245.

Diamond, J.M., C.A. Call, and N. Devoe. 2010. Effects of targeted cattle grazing on fire behavior of cheatgrass-dominated rangeland in the northern Great Basin, USA. *International Journal of Wildland Fire* 18(8): 944–950.

Dong, Y., M. Xie, Y. Jiang, N. Xiao, X. Du, et al. 2013. Sequencing and automated whole-genome optical mapping of the genome of a domestic goat (Capra hircus). *Nature Biotechnology* 31: 135–143.

Eggen, A. 2012. The development and application of genomic selection as a new breeding paradigm. *Animal Frontiers* 2: 10–15.

Estell, R.E., K.M. Havstad, A.F. Cibils, E.L. Fredrickson, D.M. Anderson, et al. 2012. Increasing shrub use by livestock in a world with less grass. *Rangeland Ecology and Management* 65: 553–562.

Euclides, F.K. 1996. A pecuária de corte brasileira no terceiro milênio. In SIMPÓSIO SOBRE O CERRADO, 8.; INTERNATIONAL SYMPOSIUM ON TROPICAL SAVANNAS, 1. Brasília. Anais. Planaltina: EMBRAPA CPAC, 118–120.

FAO. 2011. FAOSTAT statistical database. http://faostat.fao.org/site/573/default.aspx#ancor. Accessed 29 June 2011.

———. 2013. *World agriculture towards 2030/2050: The 2012 revision.* Rome, Italy: United Nations Food and Agriculture Organization.

Ferraz, J.B.S., and P.E. de Felicio. 2010. Production systems—An example from Brazil. *Meat Science* 84: 238–243.

Food and Agriculture Organization of the United Nations. 2009. *How to feed the world in 2050.* Rome, Italy: FAO.

———. 2011. *The state of the world's land and water resources for food and agriculture.* Rome, Italy: Food and Agriculture Organization of the United Nations.

Fuhlendorf, S.D., and D.M. Engle. 2004. Application of the fire-grazing interaction to restore a shifting mosaic on tallgrass prairie. *Journal of Applied Ecology* 41: 604–614.

Garrick, D.J. 2011. The nature, scope and impact of genomic prediction in beef cattle in the United States. *Genetics Selection Evolution* 43: 17.

Goehring, B.J., K.L. Launchbaugh, and L.M. Wilson. 2010. Late-season targeted grazing of yellow starthistle (*Centaurea solstitialis*) with goats in Idaho. *Invasive Plant Science and Management* 3(2): 148–154.

Grandin, T. 2007. Introduction: Effect of customer requirements, international standards and marketing structure on the handling and transport of livestock and poultry. In *Livestock handling and transport*, ed. T. Grandin. Wallingford, UK: CAB International.

Griffin, W.A., L.A. Stalker, D.C. Adams, R.N. Funstron, and T.J. Klopfenstein. 2012. Calving date and wintering system effects on cow and calf performance I: A systems approach to beef production in the Nebraska Sandhills. *The Professional Animal Scientists* 28(3): 249–259.

Grings, E.E., R.E. Short, K.D. Klement, T.W. Geary, M.D. MacNeil, et al. 2005. Calving system and weaning age effects on cow and preweaning calf performance in the Northern Great Plains. *Journal of Animal Science* 83: 2671–2683.

Hegarty, R.S., J.P. Goopy, R.M. Herd, and B. McCorkell. 2007. Cattle selected for lower residual feed intake have reduced daily methane production. *Journal of Animal Science* 85: 1479–1486.

Herd, R.M., and P.F. Arthur. 2009. Physiological basis for residual feed intake. *Journal of Animal Science* 87(14): E64–E71.

Herd, R.M., J.A. Archer, and P.F. Arthur. 2003. Reducing the cost of beef production through genetic improvement in residual feed intake: Opportunity and challenges to application. *Journal of Animal Science* 81(13): E9–E17.

Herd, R.M., S. Piper, J.M. Thompson, P.F. Arthur, B. McCorkell, et al. 2009. Benefits of genetic superiority in residual feed intake in a large commercial feedlot. *Proceedings of the Association for the Advancement of Animal Breeding and Genetics* 18: 476–479.

Herrero, M., P. Havlik, H. Valin, A. Notenbaert, M.C. Rufino, et al. 2013. Biomass use, production, feed efficiencies, and greenhouse gas emissions from global livestock systems. *PNAS* 110: 20888–20893.

Holechek, J.L. 2013. Global trends in population, energy use and climate: Implications for policy development, rangeland management, and rangeland users. *The Rangeland Journal* 35: 117–129.

Holechek, J.L., R.D. Piper, and C.H. Herbel. 2011. *Range management: Principles and practices*, 6th ed. Upper Saddle River, NJ: Prentice-Hall/Pearson, Inc.

Hunt, L.P. 2008. Safe pasture utilization rates as a grazing management tool in extensively grazed tropical savannas of northern Australia. *The Rangeland Journal* 30: 305–315.

Johnston, D.J., B. Tier, and H.U. Graser. 2012. Beef cattle breeding in Australia with genomics: Opportunities and needs. *Animal Production Science* 52(3): 100–106.

Joyce, L.A., D.D. Briske, J.R. Brown, H.W. Polley, B.A. McCarl, et al. 2013. Climate change and North America Rangelands: Assessment of mitigation and adaptation strategies. *Rangeland Ecology and Management* 66: 512–528.

Kachergis, E.J., J.D. Derner, B.B. Cutts, L.M. Roche, V.T. Eviner, M.N. Lubell, and K.W. Tate. 2014. Increasing flexibility in rangeland management during drought. *Ecosphere* 5: 1–14.

Kemp, D.R., G.D. Han, X.Y. Hou, D.L. Michalk, F.J. Hou, et al. 2013. Innovative grassland management systems for environmental and livelihood benefits. *PNAS* 110: 8369–8374.

Kigas, D.T., B.P. Dalrymple, M.P. Heaton, J.F. Maddox, A. McGrath, P. Wilson, R.G. Ingersoll, R. McCulloch, S. McWilliam, D. Tang, J. McEwan, N. Cockett, V.H. Oddy, F.W. Nicholas, H. 2009. Raadsma for the International Sheep Genomics Consortium. A genome wide survey of SNP variation reveals the genetic structure of sheep breeds. PLOS One http://dx.doi.org/10.1371/journal/pone/0004668.

Launchbaugh, K., and J. Walker. 2006. Targeted grazing: A new paradigm for livestock management. In *Targeted grazing: A natural approach to vegetation management and landscape enhancement—A handbook on grazing as a new ecological service*, ed. K. Launchbaugh. www.sheepusa.org/targetedgrazing. Accessed 4 October 2013.

Lawrence, J.D., and M.A. Ibarburu. 2007. Economic analysis of pharmaceutical technologies in modern beef production. *Proceedings of the NCCC-134 Conference on Applied Commodity Price Analysis, Forecasting, and Market Risk Management*. Chicago, IL. http://www.farmdoc.uiuc.edu/nccc134.

Limb, R.F., S.D. Fuhlendorf, D.M. Engle, J.R. Weir, R.D. Elmore, et al. 2011. Pyric–herbivory and cattle performance in grassland ecosystems. *Rangeland Ecology and Management* 64(6): 659–663.

Martha Jr., G.B., E. Alves, and E. Contini. 2012. Land-saving approaches and beef production growth in Brazil. *Agricultural Systems* 110: 173–177.

McAllister, R.R.J. 2012. Livestock mobility in arid and semi-arid Australia: Escaping variability in space. *The Rangeland Journal* 34: 139–147.

McAllister, R.R.J., I.J. Gordon, M.A. Janssen, and N. Abel. 2006. Pastoralists' responses to variation of rangeland resources in time and space. *Ecological Applications* 16: 572–583.

Meuwissen, T.H., B.J. Hayes, and M.E. Goddard. 2001. Prediction of total genetic value using genomewide dense marker maps. *Genetics* 157: 1819–1829.

Milchunas, D.G., A.R. Mosier, J.A. Morgan, D.R. LeCain, J.Y. King, et al. 2005. Elevated CO_2 and defoliation effects on a shortgrass steppe: Forage quality versus quantity for ruminants. *Agriculture, Ecosystems & Environment* 111: 166–184.

Palmer, L. 2014. A new climate for grazing livestock. *Nature Climate Change* 4: 321–323.

Pelletier, N., R. Pirog, and R. Rasmussen. 2010. Comparative life cycle environmental impacts of three beef production strategies in the Upper Midwestern United States. *Agricultural Systems* 103: 380–389.

Polley, H.W., D.D. Briske, J.A. Morgan, K. Wolter, D. Bellarby, et al. 2013. Climate change and North American rangelands: Trends, projections, and implications. *Rangeland Ecology and Management* 66: 493–511.

Provenza, F.D., J.J. Villalba, L.E. Dziba, S.B. Atwood, and R.E. Banner. 2003. Linking herbivore experience, varied diets, and plant biochemical diversity. *Small Ruminant Research* 49: 257–274.

Rada-Iglesias, A., and J. Wysocka. 2011. Epigenomics of human embryonic stem cells and induced pluripotent stem cells: Insights into pluripotency and implications for disease. *Genome Medicine* 3: 36.

Reeves, J.L., J.D. Derner, M.A. Sanderson, M.K. Petersen, L.T. Vermeire, et al. 2013. Temperature and precipitation affect steer weight gains differentially by stocking rate in northern mixed-grass prairie. *Rangeland Ecology and Management* 66: 438–444.

Ritten, J.P., W.M. Frasier, C.T. Bastian, S.T. Gray. 2010. Optimal rangeland stocking decisions under stochastic and climate-impacted weather. *Ameriican Journal of Agricultural Economics* 92: 1242–1255.

Roche, L.M., B.B. Cutts, J.D. Derner, M.N. Lubell, and K.W. Tate. 2015. On-ranch grazing strategies: Context for the rotational grazing dilemma. *Rangeland Ecology and Management* 68: 248–256.

Simmons, J. 2009. *Food economics and consumer choice*. Greenfield, IN: Elanco Animal Health.

Stafford Smith, D.M., G.M. McKeon, I.W. Watson, B.K. Henry, G.S. Stone, et al. 2007. Learning from episodes of degradation and recovery in variable Australian rangelands. *PNAS* 104: 20690–20695.

Steinfeld, H., P. Gerber, T. Wassenaar, V. Castel, M. Rosales, et al. 2006. *Livestock's long shadow. Environmental issues and options*. Rome, Italy: FAO.

Sulc, R.M., and A.J. Franzluebbers. 2014. Exploring integrated crop-livestock systems in different ecoregions of the United States. *European Journal of Agronomy* 57: 21–30.

Tester, M., and P. Langridge. 2010. Breeding technologies to increase crop production in a changing world. *Science* 327: 818–822.

Torell, L.A., S. Murugan, and O.A. Ramirez. 2010. Economics of flexible versus conservative stocking strategies to manage climate variability risk. *Rangeland Ecology and Management* 63: 415–425.

USDA. 2009. *Beef 2007–08 Part II: Reference of beef cow-calf management practices in the United States, 2007–08*. Fort Collins, CO: USDA:APHIS:VS, CEAH, National Animal Health Monitoring System.

Vermeir, I., and W. Verbeke. 2006. Sustainable food consumption: Exploring the consumer "attitude - behavioral intention" gap. *Journal of Agricultural and Environmental Ethics* 19: 169–194.

White, R.R., and J.L. Capper. 2013. An environmental, economic, and social assessment of improving cattle finishing. *Journal of Animal Science* 91(12): 5801–5812.

Williams, J.L., D.J. Garrick, and S.E. Speidel. 2009. Reducing bias in maintenance energy expected progeny difference by accounting for selection on weaning and yearling weights. *Journal of Animal Science* 87: 1628–1637.

Wilson, T.B., D.B. Faulkner, and D.W. Shike. 2015. Influence of late gestation drylot rations differing in protein degradability and fat content on beef cow and subsequent calf performance. *Journal of Animal Science* 93(12): 5819–5828.

Zimin, A.V., A.L. Delcher, L. Florea, D.R. Kelley, M.C. Schatz, et al. 2009. A whole-genome assembly of the domestic cow, *Bos Taurus*. *Genome Biology* 10: R42.

Chapter 11
Adaptive Management of Rangeland Systems

Craig R. Allen, David G. Angeler, Joseph J. Fontaine, Ahjond S. Garmestani, Noelle M. Hart, Kevin L. Pope, and Dirac Twidwell

Abstract Adaptive management is an approach to natural resource management that uses structured learning to reduce uncertainties for the improvement of management over time. The origins of adaptive management are linked to ideas of resilience theory and complex systems. Rangeland management is particularly well suited for the application of adaptive management, having sufficient controllability and reducible uncertainties. Adaptive management applies the tools of structured decision making and requires monitoring, evaluation, and adjustment of management. Adaptive governance, involving sharing of power and knowledge among relevant stakeholders, is often required to address conflict situations. Natural resource laws and regulations can present a barrier to adaptive management when requirements for legal certainty are met with environmental uncertainty. However, adaptive management is possible, as illustrated by two cases presented in this chapter. Despite challenges and limitations, when applied appropriately adaptive management leads to improved management through structured learning, and rangeland management is an area in which adaptive management shows promise and should be further explored.

C.R. Allen (✉) • J.J. Fontaine • K.L. Pope
U.S. Geological Survey, Nebraska Cooperative Fish and Wildlife Research Unit,
School of Natural Resources, University of Nebraska, Lincoln, NE, USA
e-mail: callen3@unl.edu; jfontaine2@unl.edu; kpope2@unl.edu

D.G. Angeler
Department of Aquatic Sciences and Assessment, Swedish University
of Agricultural Sciences, Uppsala, Sweden
e-mail: david.angeler@slu.se

A.S. Garmestani
National Risk Management Research Laboratory, U.S. Environmental
Protection Agency, Cincinnati, OH, USA
e-mail: garmestani.ahjond@epa.gov

N.M. Hart
Nebraska Cooperative Fish and Wildlife Research Unit, School
of Natural Resources, University of Nebraska, Lincoln, NE, USA
e-mail: nchaine@huskers.unl.edu

D. Twidwell
Department of Agronomy and Horticulture, University of Nebraska, Lincoln, NE, USA
e-mail: dirac.twidwell@unl.edu

© The Author(s) 2017
D.D. Briske (ed.), *Rangeland Systems*, Springer Series on Environmental
Management, DOI 10.1007/978-3-319-46709-2_11

Keywords Collaborative adaptive management • Structured decision making • Resilience • Adaptive governance • Natural resource law • Uncertainty

11.1 Introduction

Adaptive management (AM) is an approach to management that emphasizes structured learning through decision making for situations where knowledge is incomplete and managers must act despite uncertainty regarding management outcomes (Walters 1986). Adaptive management produces iterative decisions based on information resulting from management, and builds knowledge and improves management over time (Allen and Garmestani 2015). Natural resource management contains numerous uncertainties and ecosystem managers can make better decisions in the future if they can learn, and these ideas underlying AM hold intuitive appeal and should be common sense. Indeed, C.S. Holling, the describer of AM, recognized, "Adaptive management is not really much more than common sense. But common sense is not always in common use" (Holling 1978).

Although AM may be "common sense," there continues to be confusion regarding what actually constitutes AM. This misunderstanding is largely based upon the belief that AM is what management has always been—trial-and-error attempts to improve management. However, unlike a trial-and-error approach, AM has explicit structure, including a careful description of objectives, hypotheses of problem causation, alternative management approaches, predicted consequences of implementing management alternatives, procedures for collection and analysis of monitoring data, and a mechanism for updating management as learning occurs.

From its inception until the 1960s, fish, wildlife, and range management in many nations focused primarily on the management of game and commercially important species, including domestic livestock. Game management included such activities as the control of predators, the establishment of hunting and fishing regulations, and the direct manipulation and creation of habitat considered suitable for target species. This focus has gradually broadened and during the last two decades a convergence of the formerly discrete fields of fish and wildlife biology, ecology, rangeland ecology, and conservation biology has occurred, reflecting a shift in dominant stakeholder groups from consumptive to nonconsumptive users (van Heezik and Seddon 2005). Range management has started to embrace a broader view of management that includes non-game species, and management no longer is exclusively focused on providing harvestable resources, but increasingly deals with conservation of threatened species, invasive species control, and the regulation of populations that are perceived as overabundant. Globally, fish, wildlife, and range management has followed similar patterns in different countries over the last few decades as international boundaries have become more open and communication and travel easier and faster. However, attitudes toward "management" and "conservation" still bear the stamp of historical contingency and reflect the norms of the cultures and governments of the countries within which managers reside.

A relatively recent trend in fish and wildlife biology is a more explicit focus on biodiversity conservation, monitoring, and the protection of endangered species and their critical habitat (Baxter et al. 1999). During the 1980s and 1990s, there was an increase in awareness of the social issues and uncertainties surrounding fish, wildlife, and range management (Cutler 1982). Increasingly, managers have implicitly or explicitly recognized that managing natural resources includes managing people, an area where adaptive management can provide a useful approach.

Adaptive management has been attempted in a variety of settings, including in river and watershed management (Habron 2003; Allan et al. 2008; Smith 2011), park management (Agrawal 2000; Varley and Boyce 2006; Moore et al. 2011), and wildlife harvest management (Williams and Johnson 1995; Johnson 2011), with varying success. Varied success is in part because AM is not a panacea for the navigation of "wicked problems" (Rittel and Webber 1973; Ludwig 2001) and does not produce easy answers. Adaptive management is only appropriate for a subset of natural resource management problems in which both uncertainty and controllability are high (Fig. 11.1) (Peterson et al. 2003). It is a poor fit for solving problems of high complexity, high external influences, long temporal extent, high structural uncertainty, and where there is low confidence in assessments—climate change for example (Gregory et al. 2006). Although even in these situations, concepts of AM are useful because they emphasize the need for clear objectives, flexibility, and learning.

Rangeland management in particular shows promise for application of AM (Bashari et al. 2009; Boyd and Svejcar 2009), having a tradition of modeling system dynamics (e.g., state-and-transition models) (Westoby et al. 1989; Anderies et al.

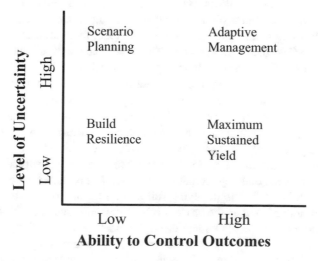

Fig. 11.1 Adaptive management and scenarios are complementary approaches to understanding complex systems. Adaptive management functions best when both uncertainty and controllability are high, which means the potential for learning is high, and the system can be manipulated (adapted from Peterson et al. 2003)

2002), identifiable spatial management units (e.g., pastures), clear management objectives (e.g., maintain forage production), and reducible uncertainties related to management impacts. In this chapter we discuss the techniques and challenges of AM and apply them to rangeland systems with two case examples.

11.2 Development of Adaptive Management

Adaptive management was founded in decision approaches of other fields (Williams 2011a), including business (Senge 1990), experimental science (Popper 1968), systems theory (Ashworth 1982), and industrial ecology (Allenby and Richards 1994). Adaptive management philosophies in natural resource management may be traced back to Beverton and Holt (1957) in fisheries management, although the term AM was not used (reviewed in Williams (2011a)). Adaptive management did not come into common usage until C.S. Holling, building upon his own work on resilience theory (Holling 1973), edited the volume *Adaptive Environmental Assessment and Management* in 1978. The concept of resilience, predicated upon the occurrence of more than one ecological state for complex systems such as ecosystems, had several ramifications. First, it meant that managers should be careful not to exceed thresholds that might change the state of the system being managed, and that the location of those thresholds is largely unknown. Second, for ecological systems in a desired state, management should focus on maintaining that regime, and enhancing its resilience, and management should not inadvertently erode the resilience of the system being managed. Adaptive management was developed as a method to continue management while probing the dynamics and resilience of systems using "management experiments" to enhance learning and reduce uncertainty (Chap. 6, this volume).

Carl Walters (1986) built upon Holling's foundational contribution (1973) and further developed AM ideas, especially in regard to modeling. Whereas Holling's original emphasis was in bridging the gap between science and practice, Walters emphasized treating management activities as experiments designed to reduce uncertainty. Both scientists sought an approach that allowed resource management to continue while explicitly acknowledging and reducing uncertainties. Walters (1986) described the process of AM as beginning "with the central tenet that management involves a continual learning process that cannot conveniently be separated into functions like research and ongoing regulatory activities, and probably never converges to a state of blissful equilibrium involving full knowledge and optimum productivity." Walters characterized AM as the process of defining and bounding the management problem, representing what is known through models, and identifying: (1) assumptions and predictions, (2) sources of uncertainty, (3) alternate hypotheses, and (4) policies that allow continued resource management while enhancing learning.

Adaptive management has been referenced either implicitly (Beverton and Holt 1957) or explicitly (Holling 1978; Walters and Hilborn 1978) for more than 50 years, but despite a relatively long theoretical history, AM has been difficult to implement in natural resource management. The limited implementation of AM stems from four fundamental problems: (1) a lack of clarity in definition and approach, with multiple interpretations of AM falling upon a continuum of complexity and design from simple "learning by doing" to more complex processes with planning and design linked with evaluation and monitoring (Holling 1978; Wilhere 2002; Aldridge et al. 2004); (2) a limited number of successful examples (Lee 1993, 1999; McLain and Lee 1996; Moir and Block 2001; Walters 2007); (3) management, policy, and funding that favor reactive approaches (Ascher 2001; Schreiber et al. 2004); and (4) laws, policy, and management plans built upon equilibrium-based conceptions of nature (Garmestani and Allen 2014). These challenges have slowed the development of AM and resulted in incomplete and inappropriate implementation of AM.

Despite implementation issues, momentum and interest in the subject continues to grow. An indication of the growing movement toward taking a more proactive approach to natural resource management is the publication by the United States Department of Interior of an AM technical guide (Williams et al. 2009) and applications guide (Williams and Brown 2012), and the policies developed around these manuals to:

"Incorporate adaptive management principles, as appropriate, into policies, plans, guidance, agreements, and other instruments for the management of resources under the Department's jurisdiction."—Department of Interior Manual (522 DM 1)

11.3 Process of Adaptive Management

Deciding on the objectives and management options is critical to any management approach. This is challenging for natural resource management because social–ecological systems are complex, including multiple objectives and stakeholders, overlapping jurisdictions, and short- and long-term effects, and they are characterized by multiple sources of uncertainty, both social and ecological (Chap. 8, this volume). Decision makers are presented with challenging decisions—predicted consequences of proposed alternatives, value-based judgments about priorities, preferences, and risk tolerances—often under enormous pressure (economic, environmental, social, and political) and with limited resources. This can result in management paralysis, or continuation of the status quo, as managers and policymakers become overwhelmed by the decision-making process and lose track of the desired social–ecological conditions they are charged with achieving. Resource management can be arduous and controversial, particularly with diverse stakeholders. Fortunately, there are methods to overcome these pitfalls and maximize the potential for success.

Structured decision making is one method of overcoming management paralysis and mediating stakeholder conflicts. Borrowed from the sociological fields, structured decision making is an approach to identify and evaluate alternative resource management options by engaging stakeholders, experts, and decision makers in the decision process and addressing the complexity and uncertainty inherent in resource

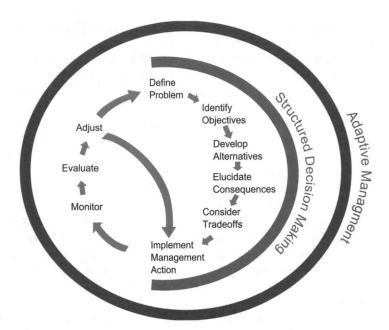

Fig. 11.2 The minimum steps necessary to implement a structured decision-making process; more complex integration of individual steps may be necessary if future steps clarify the process or if the decision is iterative over time

management in a proactive and transparent manner. Structured decision making uses a set of steps to evaluate a problem and integrates planning, analysis, and management into a transparent process focused on achieving the fundamental objectives (Fig. 11.2). Central to the success of the structured decision-making process are clearly articulated objectives, explicit acknowledgement of uncertainty, and transparent incorporation of stakeholder interests into the decision process. The conceptual simplicity inherent in structured decision making makes the process useful for a variety of decisions.

In addition to structured decision making, AM requires the potential for learning through monitoring, evaluation, and adjustment of decisions based on what is learned. Combining the essential steps of structured decision making—monitoring, evaluation, and adjustment—creates the cycle of AM. Therefore, AM can be seen as a special case of structured decision making (Walters 1986; Williams et al. 2002).

11.3.1 Steps of Structured Decision Making

11.3.1.1 Define the Problem

The first step in a structured decision-making process is a clear and concise description of the problem and the motivation underlying the need to address the problem. Although identifying the problem may seem self-evident, failure to clearly define the problem to stakeholders and subsequent agreement by stakeholders as to the nature of the problem are often cited as the primary reason management and policy actions fail. To facilitate this process, decision makers need to ask:

(a) What decision(s) have to be made?
(b) What is the scope of the decision?
(c) Will the decision be iterated over time?
(d) What are the constraints within which the decision will be made?
(e) What stakeholders should be involved in the decision process and what are their respective roles?

11.3.1.2 Identify Objectives

The centerpiece of the structured decision-making process is a set of clearly elucidated objectives. They define the "why do we care" about the decision by describing stakeholder values. Objectives also facilitate the search for alternatives and become the metric for comparing and evaluating management outcomes. Ideally, objectives are stated with the desired direction of change and in quantitative terms that relate to parameters that can be measured and evaluated. Objectives are meant to focus efforts on the importance of the decision in a consistent and transparent manner that exposes key trade-offs and uncertainties so decision makers can generate creative and proactive alternatives. Objectives should be complete, controllable, concise, measurable, and understandable (McDaniels 2000). To achieve this requires working closely with stakeholders to identify what is important about the decision. The outcome of such efforts may produce a variety of objectives that will need to be simplified.

Objectives can be separated into fundamental objectives (the ultimate goals) and means objectives (ways of achieving the ends) to ensure that management actions really affect the defined problem. For example, "maximize forage" may be an important objective for the management of a ranch, but if the property is being managed for multiple objectives including wildlife, forage is primarily important because it increases the diversity of wildlife supported. "Maximize forage" is thus a means objective for reaching the fundamental objective of "maximizing wildlife populations." Clearly there are other means objectives that would also facilitate this fundamental objective (e.g., minimize mortality, maximize forage diversity). The benefit of distinguishing the two types of objectives is that the identification of means objectives can help lead to alternative management actions, while the identification of fundamental objectives gives a basis for evaluating and comparing alternatives. The status of fundamental or means is not an innate quality of an objective, but rather is context dependent. Consequently, what was a means objective for one decision may be a fundamental objective for another.

After developing a careful list of objectives, it can be useful to develop a hierarchy, or means-ends diagram, to group similar objectives and clarify the links and relationships between means and fundamental objectives. An objectives hierarchy can help clarify the context of each fundamental objective by identifying all the important elements that are affected by the decision process and demonstrate to stakeholders the importance of all objectives even those that are not "fundamental objectives."

11.3.1.3 Develop Alternatives

Management success is only as likely as the creativity and diversity of management alternatives. Unfortunately management paralysis and status quo too often limit managers to few options and thereby impede management success. The process of identifying management alternatives, like the process of identifying objectives, starts with brainstorming. Identifying alternative management actions is a process that should be addressed iteratively, as knowledge of best practices and the creativity to develop novel ideas should not be expected to develop instantaneously. It is important to have a group with a set of interdisciplinary skills that represent the larger decision to ensure that the needs of stakeholders are not overlooked. This is not to say that the stakeholders involved in identifying alternative management actions are the same as the larger stakeholder group, usually they are not. This is primarily due to the technical knowledge necessary to present plausible alternatives. Still there are opportunities where the benefit of being naive may present novel actions that might not otherwise be considered.

The brainstorming process should begin by identifying alternatives for individual objectives, but should also be looking for opportunities when one action may fulfill the needs of multiple objectives. Identifying alternatives also means acknowledging those actions that must be done (e.g., policy) as well as constraints and potential trade-offs between objectives and management actions. It is important that the "brainstorming" process focus on developing management actions that are (1) designed to address the outlined objectives, (2) built on the best known practices, (3) comprehensive enough to include the technical understanding for implementation, (4) expose trade-offs between the decision process by having mutually exclusive strategies, and (5) achieve the maximum benefit for the stakeholders involved.

Once an extensive list of alternatives has been identified, it can be useful to group them into strategies based on general similarities in what they aim to achieve. Sometimes these represent the needs of specific stakeholder groups or specific conditions that could be achieved. For example, management actions on a rangeland may be grouped into those addressing the needs of cattle, wildlife, or diversity; alternatively, they may be grouped based on their ability to restore the landscape to 50, 75, or 95 % of historical heterogeneity. Both methods have merit; the first method makes it clear to the stakeholders what objectives are being met and where trade-offs must occur, and the second minimizes the inherent interests of any particular group to make the process less contentious.

11.3.1.4   Exploring Consequences

The list of alternative management actions is only effective if it creates an opportunity to evaluate and compare actions in light of the objectives before implementation. It is important to realize that the process of identifying management consequences is not a value judgment, but an assessment of the likely outcomes of the action(s). Using the best knowledge available, this process is an exercise focused on predicting the likely outcomes of each alternative and the likelihood that each achieves the desired objective. Depending upon knowledge of the system this process can be highly quantitative where extensive data are modeled and probabilities assigned, or as is often the case, it can depend heavily on expert opinion or comparisons to similar systems. In both cases, there is a degree of uncertainty associated with predicted outcomes as well as the parameters included in the modeling process. Decisions are almost always made in the face of uncertainty because system function is rarely precisely understood and the effect of management actions is never certain. Uncertainty can make differentiating among alternatives difficult. It is important that uncertainty be confronted throughout the decision process and that the uncertainties are identified and the possible impacts on the system and the ability to achieve stated objectives documented. These uncertainties may be reduced through the addition of monitoring, evaluation, and adjustment steps as part of an AM cycle (discussed in detail shortly).

Once the modeling process has predicted the likely outcomes of each management action, the next step is to develop a consequence table. The purpose of a consequence table is to produce a summary of the anticipated consequences of each potential management action on each of the objectives in a table or matrix. A consequence table can take a variety of forms, from a simple rating system (e.g., consumer report five-star rating) to a complex table with specific probabilities of outcomes and subsequent likelihoods of achieving each objective. Independent of the complexity of the underlying models that populate the matrix, the purpose of the consequence table is to ease and facilitate direct comparison of each management actions' ability to achieve each objective.

11.3.1.5   Consider Trade-Offs

Ideally the structured decision-making process would lead to a clear management alternative that achieves the objectives of all interested parties; unfortunately, this is rarely the case. Generally, the process of developing a consequence table will make clear which options are the least likely to be effective, but if there are multiple stakeholders and multiple objectives most decisions will require a trade-off between the ability of the remaining options to achieve each objective. The process of identifying where these trade-offs arise is analytical, but the decision process itself is highly value laden and dependent upon stakeholders. In most complex decisions, this will involve stakeholders choosing between less-than-perfect alternatives. There are a variety of methods to facilitate highly value laden decisions by weighting options based on the values of the stakeholders and then comparing alternatives to find the "best" compromise solutions. However, trade-offs are real and it is unlikely that all parties will be totally satisfied with the eventual outcome. Indeed, the benefit of the structured decision-making process is that even if there is disagreement, the process makes the disagreement transparent and enables stakeholders to re-evaluate using new knowledge and perspectives.

11.3.1.6 Implement Management Action

The final step in the structured decision-making process is implementation. Although this may always seem to be the desired outcome of a decision process, social and political pressures to reach "perfection" often impede implementation and leave decisions in a continuous state of inaction. To ensure success, managers, policymakers, and stakeholders must collaborate to move through the decision process in a timely manner to ensure action can be taken. Failure to take action is a decision, whether it is made passively or actively.

11.3.2 Monitoring, Evaluation, and Adjustment

The steps of structured decision making are a useful way to begin the planning and management process by allowing for transparent decisions, but structured decision making alone is not sufficient for AM. In order for a project to be truly AM, there must be (1) potential for learning through monitoring and evaluation of results and (2) adjustment of decisions following learning. As such, monitoring and evaluation are key components. Ongoing monitoring can be resource demanding and seen as an unnecessary expense; budgets often do not incorporate funds and personnel to support monitoring. Even when monitoring does occur, it is only as useful as its use in evaluation. Monitoring must be conducted rigorously, following a structured protocol, and designed such that learning about system dynamics and the impact of management can occur. The learning from evaluation must be used to adjust future management.

Monitoring, evaluation, and adjustment are key steps of AM and create an ongoing cycle of managing and learning. The cycle of managing and learning can be divided into two phases, a setup and an iterative phase (Williams 2011a). The setup phase is made up of the structured decision-making steps, while the iterative phase is a cycle from decision making to monitoring to evaluation and back again. Learning occurs during the iterative phase, but re-evaluation of the structured decision-making process should also happen periodically to examine how the context has changed.

11.4 Types of Adaptive Management

There are two prevailing schools of thought emerging from different traditions of AM, the resilience-experimentalist school and the decision-theoretic school (McFadden et al. 2011). The resilience-experimentalist school emphasizes inclusion of stakeholders throughout the process, active learning about ecosystem resilience through experimentation, acknowledgement of cross-scale linkages, and potential for surprises in complex systems. Adaptive management involving a large degree of stakeholder collaboration has also been called collaborative AM or adaptive co-management. The decision-theoretic school includes stakeholders to properly identify the problem, objectives, and alternatives. Relatively simple decision-focused models are then developed following principles of decision theory to identify the appropriate management action.

The resilience-experimentalist school recognizes a need for bridging organizations to address cross-scale linkages found in nested, complex social–ecological systems. Bridging organizations connect stakeholders and policymakers at different levels (Olsson et al. 2007). To do so successfully, bridging organizations must formulate strategies, coordinate joint action, address uncertainty, and link diverse stakeholders in a world of increasing complexity. Brown (1993) investigated bridging organizations from across the world, and a variety of applications—regional economic policy in the USA; small-scale irrigation projects in Indonesia; and agricultural productivity in Zimbabwe—found that bridging organizations are independent of stakeholders in a social–ecological system, which allows them to negotiate with stakeholders and advocate multiple positions. This unique role in the management of social–ecological systems affords bridging organizations the capacity to catalyze the formation of policies that are flexible and reflective of the realities of ecosystems and institutions (Brown 1993). In addition, bridging organizations have the capacity to reduce transaction costs and provide a mechanism to enforce adherence to desired policies, despite their lack of regulatory authority (Hahn et al. 2006).

Examples of bridging organizations include (1) assessment teams, which are made up of actors across sectors in a social–ecological system; (2) nongovernmental organizations, which create an arena for trust-building, learning, conflict resolution, and adaptive co-management; and (3) the scientific community, which acts as a watchdog as well as a facilitator for AM.

The decision-theoretical school applies the tools of decision science to select optimal management choices under conditions of uncertainty. A distinction is made between passive and active AM. In either case learning occurs, but in passive management the emphasis is on achievement of the management objective with learning a by-product, and in active AM, reducing uncertainty is an objective and management actions are selected based on the potential for learning (Williams 2011a, b).

11.5 Adaptive Management in Rangelands

In this section, we outline the implementation of AM in two case examples that reference both the decision-theoretical and resilience-experimentalist approach. The first example is the US Fish and Wildlife Service (USFWS) Adaptive Harvest Management

Plan (AHM), which is often heralded as the most successful case of AM and it provides an example of passive AM following the decision-theoretical school. The second case is directly tied to the management of rangelands and describes AM in prescribed burn associations (PBAs), a private citizen-led management effort in rangelands that follows the resilience-experimentalist AM approach.

11.5.1   USFWS Adaptive Harvest Management

Adaptive harvest management (AHM) is one of the most successful efforts to apply the principles of AM and demonstrate how to successfully manage natural resources through improving our understanding of natural systems through management actions. The AM processes of AHM have greatly improved our understanding of the harvest potential of waterfowl populations, the ability of managers to regulate harvest, and the importance of monitoring and assessment programs to support the decision-making process.

Why has AHM succeeded while so many other attempts to implement AM have stalled? First, AHM developed a clear and concise objective: maximize long-term waterfowl harvest while ensuring long-term viability of waterfowl populations. The development and agreement by stakeholders to a concise set of fundamental objectives is paramount to ensuring the success of any AM program. Failure to agree upon fundamental objectives will ensure management will fail. The second key to AHM success was the simultaneous support for management, research, and monitoring. Waterfowl research and management in North America are nearly unequaled by any other natural resources management program in terms of history, scope, and investment (Hawkins et al. 1984). The combination of well-supported management, research, and monitoring programs has resulted in a reduction in the uncertainty of how waterfowl populations respond to management and enabled managers and policymakers to meet stated objectives. Unfortunately, attempts to implement AM often fail to address all of these requirements. In particular, resources for monitoring and research are often undervalued with the outcome being a series of management actions with no capacity for understanding the implications of those actions.

The final key to the success of AHM has been the ability to implement management and policy decisions based on the best information available. One reason for this is that the model predictions have dictated liberal harvest as the supported management action, meaning tough trade-offs have not needed to be made between hunter satisfaction and sustaining waterfowl populations. In many attempts to implement AM, the regulatory body charged with implementation of management recommendations is unable, or worse unwilling, to implement actions proposed by the outcome of the AM process; the body in charge of regulatory control is too often a stakeholder in the process of AM with an agenda independent of regulating the resource alone. In contrast to AHM, which is regulated by the USFWS with support from various stakeholders with parallel interests, several regulatory agencies often control resources for a given program, each an independent stakeholder with an independent agenda. Such a situation can make implementation of management recommendations challenging, especially if it contradicts long-standing dogma. Consider for example the management of Glen Canyon Dam and the waters of the Colorado River. Heralded by Congress as an AM success story, the Colorado River Adaptive Management Program has fallen short of success despite years of work, and the ecological status of the Colorado River and the conflict inherent to the development of an AM program continue to worsen (Susskind et al. 2010). The regulatory agency that controls the flow of water throughout the Colorado River Basin, the Bureau of Reclamation, is also one of the major stakeholders in the AM process with an agenda (water storage) that conflicts with several other stakeholders and regulatory agencies that manage people and wildlife along the Colorado River (e.g., California Department of Water Resources, Mexican National Water Commission, USFWS). From these examples one might conclude that AM is difficult to implement for management of resources where various stakeholders and regulators are at odds. Actually, implementation of AM is appropriate in both examples. The Colorado River Adaptive Management Program example highlights the importance of collaboration, the benefits of a single regulatory body, and the need to agree upon a priori objectives that guide long-term management decisions despite short-term political, societal, economic, or even environmental impacts.

11.5.2   Adaptive Management in PBAs

Prescribed burn associations (PBAs; also referred to as prescribed burn cooperatives) have risen to the forefront of prescribed fire management in central North America private lands (Twidwell et al. 2013a) and provide an example of the implementation of the resilience-experimentalist AM approach in rangelands. Prescribed fire associations are neighbor-help-neighbor partnerships where members pool knowledge, training, and resources to implement prescribed fires for rangeland management (Taylor 2005). These associations have emerged as a bottom-up response to broad-scale encroachment of *Juniperus* species and its negative impact on multiple grassland services important to rangeland managers (Twidwell et al. 2013a). We outline how PBAs are now operating under the resilience-experimentalist AM framework; however, it is interesting to note that AM was not explicitly considered during the early formulation of PBAs. Instead, the use of AM has emerged as a need to provide solutions to a biome-level threat to rangeland resources. As burn associations have matured over the past 20 years, so has their ability to integrate the full scope of AM principles outlined previously.

At the heart of PBAs exists a tight coupling among stakeholders, scientists, and agency personnel engaged in bridging organizations and shaping decision making. University scientists and outreach professionals host workshops regularly, providing training, scientific outreach, and an open forum that targets adaptive learning outcomes among participants. State and federal natural resource agencies recognize the joint mission among agencies and landowners, and as a result have started funding prescribed burns associations to help local groups buy equipment and conduct prescribed burns. The management objectives of PBAs have triggered applied research experiments conducted through a resilience lens, resulting in the identification of fire thresholds across alternative states that fire practitioners can target and learn from (Twidwell et al. 2013b). Such management-research linkages have contributed to the increasing use of high intensity fires in areas that have undergone a shift to an alternative state dominated by non-resprouting *Juniperus* trees, while providing a cautionary learning experience for land managers in other regions also susceptible to this type of transformation (Twidwell et al. 2013a).

The implementation of AM among PBAs reveals how the resilience-experimentalist approach can lead to more flexible policies and legislation. Over the last century, controlling and limiting variability in fire behavior has been a central priority of natural resource management across the globe. Yet, more flexible policies are consistently called for in fire management to more closely mimic variation in natural fire regimes to manage species dependent upon such variability (Hutto 2008; Conway and Kirkpatrick 2007; van Wilgen 2013; Odion et al. 2014). In local areas, some PBAs, through AM, have successfully shifted regulatory constraints governing the use of prescribed fire in the private sector. Special legal exemptions have been granted to a small proportion of PBAs to provide flexibility to conduct fires during periods of government-mandated outdoor burning bans for restoration purposes (Twidwell et al. 2013a). While this has allowed some associations to conduct prescribed fires in conditions capable of overcoming woody plant mortality objectives, members recognize that legislation can shift to their disadvantage (Toledo et al. 2013). As a response, burn associations are moving beyond local affiliations of landowners and developing a formal hierarchical structure with existing alliances in the state (e.g., Prescribed Burn Alliance of Texas; http://pbatexas.org) and region (e.g., Alliance of Prescribed Burn Associations). Ongoing discussions are addressing the creation of a national alliance.

Clear recommendations have now been developed for cross-organizational and cross-scale monitoring and evaluation of PBA management actions. Such recommendations were provided, in part, to maintain engagement among stakeholders, university personnel, and agency professionals throughout both phases of AM—structured decision making, and monitoring and evaluation—with the intent of learning and informing future decisions (Table 11.1).

11.6 Adaptive Governance

Administrative agencies typically change incrementally (Lindblom 1959), and as such changes in policy are small because there is not enough information to make large overhauls of organization policy. Standard operating procedures often contribute to organizational inertia, as they slow the bureaucratic process (Allison 1969). Further, the lack of organizations matched to the appropriate scale is a significant barrier for sound environmental management (Dietz et al. 2003). Within this

Table 11.1 Example of how the two-phase process of adaptive management is being implemented to foster learning and adjust decision making of prescribed burn cooperatives dealing with *Juniperus* encroachment (adapted from Twidwell et al. 2013a)

Phase I. Structured decision making	Example of formalizing AM in PBAs
1. Define the problem	Juniper encroachment, loss of grasslands, and the services they provide
2. Identify objectives	Use fire to prevent juniper encroachment, reduce juniper abundance, and restore grassland services
3. Identify management alternatives	Use mechanical and chemical treatments to supplement fire activities; alter grazing practices to increase fuel loading and fire intensity
4. Explore consequences	Develop a consequence table summarizing potential consequences of each management action and likelihood of achieving objectives
5. Identify and evaluate trade-offs	Assess successional trajectory of vegetation following fire, potential to trigger invasion of exotic plant or animal species, negative responses from neighbors or urban residents impacted by smoke
6. Implement management action(s)	Conduct prescribed burns in conditions capable of meeting management objectives
Phase II. Monitoring and evaluation	
7. Monitoring and evaluation	Track fire effects on juniper and changes in juniper abundance, the reestablishment of grassland vegetation, potential livestock stocking rates, and biodiversity and conservation values. Recognize long-term monitoring is needed for accurate evaluation of management actions in many rangelands (Herrick et al. 2006)
8. Adjustment	Adjust management actions and targeted burning conditions based on monitoring programs; assess need to adjust structured decision-making steps

context, adaptive governance can help overcome this scale mismatch via collaboration of a diverse set of stakeholders at multiple scales (Hughes et al. 2005). Adaptive governance is a form of governance that incorporates formal organizations, informal groups and networks, and individuals at multiple scales for purposes of collaborative environmental management (Folke et al. 2005). Bridging organizations, enabling legislation, and government policies can also contribute to the success of an adaptive governance framework; governance creates a vision and management actualizes the vision (Folke et al. 2005).

Adaptive governance works via sharing of management power and responsibilities and promotes a collaborative, participatory process. It is dependent upon adaptive co-management, and adaptive co-management in turn is dependent upon social networks for success. Social networks have the capacity to allow for development of new ideas, to facilitate communication between entities, and create the flexibility necessary for the interplay of the fluid (ecological systems) and the rigid (organizations) to be successful for environmental management (Folke et al. 2005). Leadership has been well established as a critical factor in facilitating good environmental management. Leaders develop and facilitate a vision for environmental management, incorporating local knowledge and information from social networks (Folke et al. 2005).

Studies of adaptive co-management in Sweden and Canada have concluded that this form of management of ecological systems was most effective when there was: leadership with vision for the system of interest; legislation that created the environment for AM; funds for AM; monitoring of the ecological system; information flow (i.e., cross-scale linkages); combination of a variety of sources of knowledge; and venue for collaboration (Olsson et al. 2004). These factors are critical to manage for resilience in social–ecological systems, as they help to protect the system from the failure of management decisions under uncertainty (i.e., imperfect information). Adaptive governance is facilitated by informal networks and leadership, which creates the capacity for development of novel ideas for environmental management (Folke et al. 2005). These informal networks have the capacity to generate political, financial, and legal support for novel environmental management (Folke et al. 2005). Further, adaptive governance is dependent upon polycentric institutions that are redundant (e.g., scale-specific) and are quasi-autonomous (Olsson et al. 2006). A comparison of five case studies from around the world concluded that in order for a social–ecological system to transition to adaptive governance, it must undergo a preparation and a transformation phase, linked by a window of opportunity (Olsson et al. 2006).

11.7   Adaptive Management and Law

One of the most significant barriers for managing social–ecological systems is that aspects of society, especially the certainty of law and institutional rigidity, are not in concert with ecological realities, including multiregimes and nonlinear systems and responses (Garmestani et al. 2013; Garmestani and Allen 2014). The certainty of law and institutional rigidity often limit experimentation that is necessary for AM (Garmestani et al. 2009). This is critical, and some scholars contend that environmental governance can only succeed if rules evolve with the system of interest (Dietz et al. 2003).

Ecosystem management has been applied within the outdated framework of the Endangered Species Act (ESA), but would be better suited for an AM framework (Ruhl 2004). In its current form, the ESA does not have the flexibility in its regulatory language to effectively implement adaptive responses to changing environmental conditions (Boyd et al. 2014). The fundamental constraint to AM is the current state of administrative law (Ruhl 1998). As the law now stands, the procedural rules require a vast amount of work before an agency promulgates a rule or issues a permit (Ruhl 2008). This "pre-decision" activity allows for public input and prepares agencies for judicial review. Ruhl (2008) contends that "agencies will find that interest groups and courts relentlessly will erode adaptive agency behavior, using all the tools conventional administrative law puts at their disposal." Having to operate in an atmosphere where each policy is evaluated on the "front-end," in anticipation of public and legal scrutiny, has squelched agencies' appetite for AM.

US administrative law is a two-step process, in which the first allows for public comment on draft documents and alternative options (Ruhl and Fischman 2010). The second step is final agency action, which creates "certainty" to the process and makes the decision subject to judicial review. This process is based on the assumption that agencies have the capacity to predict the consequences of a "final agency action" (Ruhl and Fischman 2010). This establishes a fundamental conflict between linear legal processes (i.e., administrative law) based on "stationarity" and environmental management frameworks (i.e., AM) based on the realization of dynamic systems characterized by nonstationarity (Ruhl and Fischman 2010).

Only in rare cases, such as AHM and PBAs, has AM been successful within the current regulatory framework. In effect, administrative agencies in the United States largely do not conduct AM as it was originally conceived (Ruhl and Fischman 2010). Rather, agencies conduct AM "lite," a form of partially flexible management, because the courts have provided some leeway to AM projects provided they have requirements that are legally enforceable (Ruhl and Fischman 2010). The primary problem with AM "lite" is that it does not measure up to the standards of AM theory, nor does it hold up under the scrutiny of substantive and procedural law.

11.8 Future Perspectives

Considerable confusion exists regarding what constitutes adaptive management. The methods and theory behind AM have been described (here, and citations herein), as have the barriers to successful implementation (Allen and Gunderson 2011). However, implementation remains frequently problematic, with trial-and-error approaches described as adaptive management, and frequent application of adaptive management to intractable large-scale management problems that are inappropriate for adaptive management, largely because controllability is not present. A simple process of structured decision making can be applied in such situations. However, many of the challenges found in range management are appropriate for adaptive management, because grazing unit replication is possible and most management interventions applied are controlled by managers or landowners. Integration of adaptive management and range management should increase the rate of learning, necessary in a rapidly changing world.

In order for AM to move past AM "lite" and realize its true potential for rangeland management, administrative law will likely need to be reformed. Administrative law would then proceed on two trajectories: (1) a fixed-rule track that would apply unless an agency can justify otherwise; and (2) an AM track, where a new set of administrative law standards specific to AM would hold precedence, in order to actualize AM as a tool for rangeland management (Karkkainen 2005). A recent law review article heeded this call for an AM track and provided a model law for administrative procedures and AM (Craig and Ruhl 2014). In particular, the model law highlights the categories of agency decision making that are amenable to AM (see Craig and Ruhl 2014).

11.9 Summary

The conceptual underpinnings for AM are simple; there will always be inherent uncertainty and unpredictability in the dynamics and behavior of complex ecological systems, yet management decisions must still be made. The strength of AM is in the recognition and confrontation of such uncertainty. Rather than ignore uncertainty, or use it to preclude management actions, AM can foster resilience and flexibility to cope with an uncertain future, and develop management approaches that acknowledge inevitable changes and surprises. Since its initial introduction, AM has been hailed as a solution to endless trial-and-error approaches to complex management challenges. However, it does not produce easy answers, and it is appropriate in only a subset of management problems.

Clearly AM has great potential when applied appropriately. A prime example in rangeland management is PBAs, now established throughout the Great Plains. Rangelands in general are appropriate for application of AM because of the ability to model system dynamics (state-and-transition models), identifiable management units across large areas, clear management objectives (e.g., maintain native grasses), and reducible uncertainties related to management impacts. Adaptive management may be the best way forward for improving how we approach rangeland management, but will require more than most current applications of the strategy. In particular, in order to account for coupled human and natural systems, AM will require (at a minimum) legal reform (Craig and Ruhl 2014), integration with adaptive governance (Folke et al. 2005), and accounting for scale and cross-scale interactions (Garmestani et al. 2013). If these steps are taken, perhaps then AM will fulfill its promise for rangeland management.

Acknowledgements This research was supported in part by an NSF IGERT grant, DGE-0903469, and the August T. Larsson Foundation (NJ Faculty, SLU). Any opinions, findings, and conclusions or recommendations expressed in this material are those of the authors and do not necessarily reflect the views of the NSF. The Nebraska Cooperative Fish and Wildlife Research Unit is jointly supported by a cooperative agreement between the United States Geological Survey, the Nebraska Game and Parks Commission, the University of Nebraska–Lincoln, the United States Fish and Wildlife Service, and the Wildlife Management Institute. Any use of trade, firm, or product names is for descriptive purposes only and does not imply endorsement by the US Government. The views expressed in this paper are those of the authors and do not represent the views or policies of the US Environmental Protection Agency.

References

Agrawal, A. 2000. Adaptive management in transboundary protected areas: The Bialowieza National Park and Biosphere Reserve as a case study. *Environmental Conservation* 27: 326–333.

Aldridge, C.L., M.S. Boyce, and R.K. Baydack. 2004. Adaptive management of prairie grouse: How do we get there? *Wildlife Society Bulletin* 32: 92–103.

Allan, C., A. Curtis, G. Stankey, and B. Shindler. 2008. Adaptive management and watersheds: A social science perspective. *Journal of the American Water Resources Association* 44: 166–174.

Allen, C., and A.S. Garmestani. 2015. *Adaptive management of social-ecological systems.* Dordrecht, The Netherlands: Springer.

Allen, C.R., and L.H. Gunderson. 2011. Pathology and failure in the design and implementation of adaptive management. *Journal of Environmental Management* 92: 1379–1384.

Allenby, B.R., and D.J. Richards. 1994. *The greening of industrial ecosystems.* Washington, DC: National Academy Press.

Allison, G. 1969. Conceptual models and the Cuban missile crisis. *The American Political Science Review* 63: 689–718.

Anderies, J.M., M.A. Janssen, and B.H. Walker. 2002. Grazing management, resilience, and the dynamics of a fire-driven rangeland system. *Ecosystems* 5: 23–44.

Ascher, W. 2001. Coping with complexity and organizational interests in natural resource management. *Ecosystems* 4: 742–757.

Ashworth, M.J. 1982. *Feedback design of systems with significant uncertainty.* Chichester, UK: Research Studies Press.

Bashari, H., C. Smith, and O.J.H. Bosch. 2009. Developing decision support tools for rangeland management by combining state and transition models and Bayesian belief networks. *Agricultural Systems* 99: 23–34.

Baxter, G.S., M. Hockings, R.W. Carter, and R.J.S. Beeton. 1999. Trends in wildlife management and the appropriateness of Australian university training. *Conservation Biology* 13: 842–849.

Beverton, R.J.H., and S.J. Holt. 1957. *On the dynamics of exploited fish populations.* London, UK: Her Majesty's Stationery Office.

Boyd, C.S., and T.J. Svejcar. 2009. Managing complex problems in rangeland ecosystems. *Rangeland Ecology and Management* 62: 491–499.

Boyd, C.S., D.D. Johnson, J.D. Kerby, T.J. Svejcar, and K.W. Davies. 2014. Of grouse and golden eggs: Can ecosystems be managed within a species-based regulatory framework? *Rangeland Ecology and Management* 67: 358–368.

Brown, L.D. 1993. Development bridging organizations and strategic management for social change. *Advances in Strategic Management* 9: 381–405.

Conway, C.J., and C. Kirkpatrick. 2007. Effect of forest fire suppression on buff-breasted flycatchers. *Journal of Wildlife Management* 71: 445–457.

Craig, R.K., and J.B. Ruhl. 2014. Designing administrative law for adaptive management. *Vanderbilt Law Review* 67: 1–87.

Cutler, M.R. 1982. What kind of wildlifers will be needed in the 1980s? *Wildlife Society Bulletin* 10: 75–79.

Dietz, T., E. Ostrom, and P.C. Stern. 2003. The struggle to govern the commons. *Science* 302: 1907–1912.

Folke, C., T. Hahn, P. Olsson, and J. Norberg. 2005. Adaptive governance of social-ecological systems. *Annual Review of Environment and Resources* 30: 441–473.

Garmestani, A.S., and C.R. Allen. 2014. *Social-ecological resilience and law.* New York: Columbia University Press.

Garmestani, A.S., C.R. Allen, and H. Cabezas. 2009. Panarchy, adaptive management and governance: Policy options for building resilience. *Nebraska Law Review* 87: 1036–1054.

Garmestani, A.S., C.R. Allen, and M.H. Benson. 2013. Can law foster social-ecological resilience? *Ecology and Society* 18: 37. http://www.ecologyandsociety.org/vol18/iss2/art37/.

Gregory, R., D. Ohlson, and J. Arvai. 2006. Deconstructing adaptive management: Criteria for applications to environmental management. *Ecological Applications* 16: 2411–2425.

Habron, G. 2003. Role of adaptive management for watershed councils. *Environmental Management* 31: 29–41.

Hahn, T., P. Olsson, C. Folke, and K. Johansson. 2006. Trust-building, knowledge generation and organizational innovations: The role of a bridging organization for adaptive co-management of a wetland landscape around Kristianstad, Sweden. *Human Ecology* 34: 573–592.

Hawkins, A.S., R.C. Hanson, H.K. Nelson, and H.M. Reeves (eds.). 1984. *Flyways: Pioneering waterfowl management in North America*. Washington, DC: U.S. Government Printing Office.

Herrick, J.E., G.E. Schuman, and A. Rango. 2006. Monitoring ecological processes for restoration projects. *Journal of Nature Conservation* 14: 161–171.

Holling, C.S. 1973. Resilience and stability of ecological systems. *Annual Review of Ecology and Systematics* 4: 1–23.

———. 1978. *Adaptive environmental assessment and management*. Chichester, UK: Wiley.

Hughes, T.P., D.R. Bellwood, C. Folke, R.S. Steneck, and J. Wilson. 2005. New paradigms for supporting the resilience of marine ecosystems. *Trends in Ecology & Evolution* 20: 380–386.

Hutto, R.L. 2008. The ecological importance of severe wildfires: Some like it hot. *Ecological Applications* 18: 1827–1834.

Johnson, F.A. 2011. Learning and adaptation in the management of waterfowl harvests. *Journal of Environmental Management* 92: 1385–1394.

Karkkainen, B.C. 2005. Panarchy and adaptive change: Around the loop and back again. *Minnesota Journal of Law, Science & Technology* 7: 59–77.

Lee, K.N. 1993. *Compass and gyroscope: Integrating science and politics for the environment*. Washington, DC: Island Press.

———. 1999. Appraising adaptive management. *Conservation Ecology* 3:3.

Lindblom, C. 1959. The science of muddling through. *Public Administration Review* 19: 79–88.

Ludwig, D. 2001. The era of management is over. *Ecosystems* 4: 758–764.

McDaniels, T. 2000. Creating and using objectives for ecological risk assessment and management. *Environmental Science & Policy* 3: 299–304.

McFadden, J.E., T.L. Hiller, and A.J. Tyre. 2011. Evaluating the efficacy of adaptive management approaches: Is there a formula for success? *Journal of Environmental Management* 92: 1354–1359.

McLain, R.J., and R.G. Lee. 1996. Adaptive management: Promises and pitfalls. *Environmental Management* 20: 437–448.

Moir, W.H., and W.M. Block. 2001. Adaptive management on public lands in the United States: Commitment or rhetoric? *Environmental Management* 28: 141–148.

Moore, C.T., E.V. Lonsdorf, M.G. Knutson, H.P. Laskowski, and S.K. Lor. 2011. Adaptive management in the U.S. National Wildlife Refuge System: Science-management partnerships for conservation delivery. *Journal of Environmental Management* 92: 1395–1402.

Odion, D.C., C.T. Hanson, A. Arsenault, W.L. Baker, D.A. DellaSala, et al. 2014. Examining historical and current mixed-severity fire regimes in ponderosa pine and mixed-conifer forests of western North America. *PLoS One* 9: e87852.

Olsson, P., C. Folke, and F. Berkes. 2004. Adaptive co-management for building resilience in social-ecological systems. *Environmental Management* 34: 75–90.

Olsson, P., L.H. Gunderson, S.R. Carpenter, P. Ryan, L. Lebel, et al. 2006. Shooting the rapids: Navigating transitions to adaptive governance of social-ecological systems. *Ecology and Society* 11: 18.

Olsson, P., C. Folke, V. Galaz, T. Hahn, and L. Schultz. 2007. Enhancing the fit through adaptive co-management: Creating and maintaining bridging functions for matching scales in the Kristianstads Vattenrike Biosphere Reserve Sweden. *Ecology and Society* 12: 28.

Peterson, G.D., G.S. Cumming, and S.R. Carpenter. 2003. Scenario planning: A tool for conservation in an uncertain world. *Conservation Biology* 17: 358–366.

Popper, K.R. 1968. *The logic of scientific discovery*, 2nd ed. New York, NY: Harper and Row.

Rittel, H.W.J., and M.M. Webber. 1973. Dilemmas in a general theory of planning. *Policy Sciences* 4: 155–173.

Ruhl, J.B. 1998. The Endangered Species Act and private property: A matter of timing and location. *Cornell Journal of Law and Public Policy* 8: 37–54.

———. 2004. Taking adaptive management seriously: A case study of the Endangered Species Act. *University of Kansas Law Review* 52:1249–1284.

———. 2008. Adaptive management for natural resources – inevitable, impossible, or both? *Rocky Mountain Mineral Law Institute* 54:11-1.

Ruhl, J.B., and R.L. Fischman. 2010. Adaptive management in the courts. *Minnesota Law Review* 95: 424–484.

Schreiber, E.S.G., A.R. Bearlsin, J. Nicol, and C.R. Todd. 2004. Adaptive management: A synthesis of current understanding and effective application. *Ecological Management and Restoration* 5: 177–182.

Senge, P.M. 1990. *The fifth discipline: The art and practice of the learning organization*. New York, NY: Currency Doubleday.

Smith, C.B. 2011. Adaptive management on the central Platte River – science, engineering, and decision analysis to assist in the recovery of four species. *Journal of Environmental Management* 92: 1414–1419.

Susskind, L., A.E. Camacho, and T. Schenk. 2010. Collaborative planning and adaptive management in Glen Canyon: A cautionary tale. *Columbia Journal of Environmental Law* 35: 1–54.

Taylor Jr., C.A. 2005. Prescribed burning cooperatives: Empowering and equipping ranchers to manage rangelands. *Rangelands* 27: 18–23.

Toledo, D., M.G. Sorice, and U.P. Kreuter. 2013. Social and ecological factors influencing attitudes toward the application of high-intensity prescribed burns to restore fire adapted grassland ecosystems. *Ecology and Society* 18: 9.

Twidwell, D., W.E. Rogers, S.D. Fuhlendorf, C.L. Wonkka, D.M. Engle, et al. 2013a. The rising Great Plains fire campaign: Citizens' response to woody plant encroachment. *Frontiers in Ecology and the Environment* 11: e64–e71.

Twidwell, D., S.D. Fuhlendorf, C.A. Taylor Jr., and W.E. Rogers. 2013b. Refining fire thresholds in coupled fire-vegetation models to improve management of encroaching woody plants in grassland. *Journal of Applied Ecology* 50: 603–613.

van Heezik, Y., and P.J. Seddon. 2005. Structure and content of graduate wildlife management and conservation biology programs: An international perspective. *Conservation Biology* 19: 7–14.

van Wilgen, B.W. 2013. Fire management in species-rich Cape fynbos shrublands. *Frontiers in Ecology and the Environment* 11: e35–e44.

Varley, N., and M.S. Boyce. 2006. Adaptive management for reintroductions: Updating a wolf recovery model for Yellowstone National Park. *Ecological Modeling* 193: 315–339.

Walters, C.J. 1986. *Adaptive management of renewable resources*. New York, NY: McMillan.

———. 2007. Is adaptive management helping to solve fisheries problems? *Ambio* 36:304–307.

Walters, C.J., and R. Hilborn. 1978. Ecological optimization and adaptive management. *Annual Review of Ecology and Systematics* 9: 157–188.

Westoby, M., B. Walker, and I. Noy-Meir. 1989. Opportunistic management for rangelands not at equilibrium. *Journal of Range Management* 42: 266–274.

Wilhere, G.F. 2002. Adaptive management in habitat conservation plans. *Conservation Biology* 16: 20–29.

Williams, B.K. 2011. Adaptive management of natural resources—framework and issues. *Journal of Environmental Management* 92: 1346–1353.

———. 2011b. Passive and active adaptive management: Approaches and an example. *Journal of Environmental Management* 92:1371–1378.

Williams, B.K., and E.D. Brown (eds.). 2012. *Adaptive management: The U.S. Department of the Interior applications guide*. Washington, DC: Adaptive Management Working Group, US Department of the Interior.

Williams, B.K., and F.A. Johnson. 1995. Adaptive management and the regulation of waterfowl harvests. *Wildlife Society Bulletin* 23: 430–436.

Williams, B.K., J.D. Nichols, and M.J. Conroy. 2002. *Analysis and management of animal populations*. San Diego, CA: Academic Press.

Williams, B.K., R.C. Szaro, and C.D. Shapiro. 2009. *Adaptive management: The U.S. Department of the Interior technical guide*. Washington, DC: Adaptive Management Working Group, US Department of the Interior.

Chapter 12
Managing the Livestock–Wildlife Interface on Rangelands

Johan T. du Toit, Paul C. Cross, and Marion Valeix

Abstract On rangelands the livestock–wildlife interface is mostly characterized by management actions aimed at controlling problems associated with competition, disease, and depredation. Wildlife communities (especially the large vertebrate species) are typically incompatible with agricultural development because the opportunity costs of wildlife conservation are unaffordable except in arid and semi-arid regions. Ecological factors including the provision of supplementary food and water for livestock, together with the persecution of large predators, result in livestock replacing wildlife at biomass densities far exceeding those of indigenous ungulates. Diseases are difficult to eradicate from free-ranging wildlife populations and so veterinary controls usually focus on separating commercial livestock herds from wildlife. Persecution of large carnivores due to their depredation of livestock has caused the virtual eradication of apex predators from most rangelands. However, recent research points to a broad range of solutions to reduce conflict at the livestock–wildlife interface. Conserving wildlife bolsters the adaptive capacity of a rangeland by providing stakeholders with options for dealing with environmental change. This is contingent upon local communities being empowered to benefit directly from their wildlife resources within a management framework that integrates land-use sectors at the landscape scale. As rangelands undergo irreversible changes caused by species invasions and climate forcings, the future perspective favors a proactive shift in attitude towards the livestock–wildlife interface, from problem control to asset management.

J.T. du Toit (✉)
Department of Wildland Resources, Utah State University,
5230 Old Main Hill, Logan, UT 84322, USA
e-mail: johan.dutoit@usu.edu

P.C. Cross
U.S. Geological Survey, Northern Rocky Mountain Science Center, Bozeman, MT, USA
e-mail: pcross@usgs.gov

M. Valeix
Laboratoire de Biométrie et Biologie Evolutive, CNRS, Lyon, France
e-mail: mvaleix@yahoo.fr

© The Author(s) 2017
D.D. Briske (ed.), *Rangeland Systems*, Springer Series on Environmental Management, DOI 10.1007/978-3-319-46709-2_12

Keywords Livestock–wildlife conflict • Disease ecology • Livestock depredation •
Rangeland resilience • Social–ecological systems • Novel ecosystems

12.1   Introduction

We write this chapter in a time of increasing recognition of the value of rangelands
as providers of ecosystem services, broadening the traditionally focused view of
rangelands as areas for the production of commodities from free-ranging livestock
(Havstad et al. 2007). It is also a time in which ecologists are calling for conventional production-maximizing management approaches to be transformed into a
"resilience framework" for the stewardship of social–ecological systems (Chapin
et al. 2009). We are living in the Anthropocene epoch, and have been for quite a
long time already (Balter 2013), during which we have unwittingly changed just
about everything on our planet, including the climate. In this time of self-awareness
of our environmental responsibility, rangelands provide a stage on which new
approaches to natural resource management can be developed and implemented. On
a global scale, rangelands are particularly important for wildlife conservation and
there is urgency in the search for effective ways of reconciling conservation with
livestock production (du Toit et al. 2010). This is a vast topic with many facets that
we cannot comprehensively cover and so we focus on what we consider to be the
key issues, with our treatment of wildlife applying mainly to large (>5 kg) free-
living mammals. We have structured this chapter to first review some of the main
conceptual advances over the past 25 years or so, as we see them, at the livestock–
wildlife interface on rangelands. We then suggest the social implications of translating those concepts—when and where possible—into applications that could
contribute to rangeland resilience.

12.2   Conceptual Advances at the Livestock–Wildlife
Interface

Reconciling wildlife conservation with livestock production on rangelands
requires a departure from the conventional "either/or" model in which conservation and agriculture are represented by separate, competing sectors of society
and governmental administration. Integrated approaches are required at the
landscape scale (Sayer et al. 2013) with local communities empowered to benefit
from wildlife and livestock together, and with management agencies geared for
enhancing the resilience of entire social–ecological systems (Biggs et al. 2012).
Resilient systems can maintain their function, structure, identity, and feedbacks
by absorbing disturbances and reorganizing within continually changing environments (Walker et al. 2004; Chap. 6, this volume). But achieving an integrated

approach requires an understanding of the reasons why the sectoral management approach persists across most rangelands, together with an understanding of how it can be mitigated.

Fundamentally, the practice of livestock production stems from the domestication of wild species, which has been accomplished through the removal of "wildness" over centuries. That process has built a culture of livestock husbandry that sequesters and protects livestock from the ever-present forces of wildness, which is obviously necessary in times and places where wildness is overwhelmingly intractable. Now, however, the wildness-to-tameness ratio has long since flipped on most of the world's rangelands and conceptual advances point towards a new and very different model for rangeland management, or stewardship (Chapin et al. 2009; Walker 2010). Here we review those advances and consider how science and management might interact to better understand and work with, rather than against, the main features that define the livestock–wildlife interface on rangelands.

12.2.1   Competition

Livestock–wildlife competition operates through two sets of processes within the social–ecological systems we call rangelands: economic processes influence agricultural and wildlife-based enterprises as sources of income for producer communities; ecological processes influence the relative efficiencies of livestock and wildlife species in utilizing the food and water resources occurring in their shared range.

At the global scale, economic processes generally result in agricultural returns outcompeting wildlife returns and the patchwork of land use within rangelands intensifying towards croplands and fragmented rangelands (Hobbs et al. 2008). As markets, technology, and infrastructure develop, the position of a rangeland on its production possibility frontier (PPF) changes (Bastian et al. 1991; Smith et al. 2012) with agricultural production becoming specialized, driving down the possibilities for wildlife production (Fig. 12.1). The transition begins with a fully intact wildlife community (Point A) as still occurs in wildlife reserves, game ranches, and areas where diseases (e.g., trypanosomiasis in Africa) exclude livestock. Eventually livestock production is so specialized (Point C; irrigated and fertilized pastures, winter supplementation, fenced paddocks, etc.) that wildlife production is impossible. In some cases, well-regulated hunting for trophies and meat can add to the rangeland's production potential from livestock (Point B), and infrastructure (waterpoints, access roads, etc.) provided for livestock production can also be beneficial to the sustainable utilization of wildlife. However, in most rangelands the transition has proceeded directly to maximizing livestock production (Point C) and restoring the wildlife community would necessitate a disproportionate pull-back in agricultural production, with unaffordable opportunity costs. Avoiding or overcoming those costs—incurred by foregoing land-use opportunities that are incompatible with wildlife—requires innovative policies to enable competitive and sustainable returns from wildlife to local communities and private landowners (Norton-Griffiths and

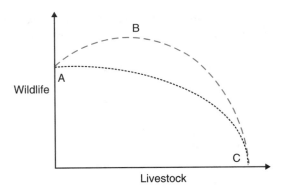

Fig. 12.1 Production possibility frontiers (PPFs) are represented by hypothetical curves describing the maximum possible output of goods from wildlife relative to each possible level of output of livestock from a shared rangeland. The curve between points A and C is typical, with intensifying livestock production forcing a decline in wildlife resources due to persecution and competition for habitats and food. Raising the PPF to include point B is an option where the back-and-forth transmission of diseases between livestock and wildlife is not a major concern

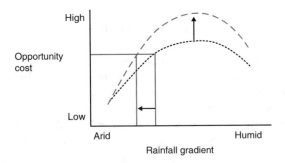

Fig. 12.2 The opportunity cost of conserving wildlife on a rangeland increases across the rainfall gradient, with mesic areas being valuable for agriculture and therefore unavailable as habitat for wildlife. In very humid areas the agricultural potential declines but various land-use options (e.g., logging followed by cattle production) can still be more lucrative than conserving wildlife for ecotourism. Developments in agricultural technology and changing market forces can raise the relative value of non-wildlife options and thus inflate the opportunity cost of conserving wildlife (from *dotted* to *dashed line, vertical arrow*). Then, without a commensurate increase in the value of wildlife, the areas in which any particular opportunity cost can be met by wildlife are reduced towards the arid end of the gradient (*horizontal arrow*)

Said 2010). Wildlife resources on rangelands are especially vulnerable to increasing opportunity costs driven by technological advances in exploration for, and exploitation of, groundwater, natural gas, and oil (e.g., Altchenko and Villholth 2013; Copeland et al. 2013; Northrup and Wittemyer 2013). Increasingly, arid rangelands are being converted to supply agricultural commodity markets (Fig. 12.2) at the expense of wildlife habitat. In addition, even the most remote rangelands are influenced by globalization, which can drive price anomalies in certain livestock products. Examples include cashmere wool from goats in Central Asia (Berger et al.

2013) and cattle dung used as fertilizer for organically grown coffee in India (Madhusudan 2005). In such cases the producer communities overstock, resulting in indigenous herbivores being outcompeted and predators persecuted with added intensity because the opportunity costs of tolerating them are hyper-inflated.

Despite the global trend towards degradation and fragmentation of wildlife habitats on rangelands, wildlife and mixed wildlife–livestock enterprises can be successful in semi-arid rangelands where opportunity costs are low (du Toit 2010). Commercial wildlife management ("game ranching") now predominates over cattle ranching on private land in the dry "bushveld" regions of South Africa where climatic, political, and economic changes have increased the relative profitability of wildlife enterprises (Carruthers 2008). Elsewhere in the world, some socioeconomic transitions have caused sharp declines in agricultural production resulting in land abandonment, most notably in post-Soviet Eastern Europe (Prishchepov et al. 2012) and also in Oceania and parts of North America (Queiroz et al. 2014). Where land abandonment occurs, low opportunity costs to wildlife conservation allow the resilience of social–ecological systems to be rebuilt with projects that restore biodiversity and ecological processes.

In the Great Plains of North America, for example, shrinking rural economies in the 1980s prompted a proposal that the USA's federal government should buy back failed farmlands and create a vast "buffalo commons" (Popper and Popper 1987). Since then, despite political and cultural resistance to that proposal from local communities, and without government buy-backs, the agrarian downtrend has opened new opportunities for wildlife restoration. Coordinated initiatives involving the nonprofit sector, private landowners, Native American tribes, and federal and state agencies are currently using American plains bison (*Bison bison bison*) as the flagship species for restoring ecologically significant expanses of prairie (Freese et al. 2007). Similar opportunities exist in Eastern Europe, where abundant habitat is now available for European bison (*Bison bonasus*) in depopulated areas of former farmland and rangeland in which livestock numbers have dropped by 60–75 % since the early 1990s (Kuemmerle et al. 2011).

The outlook is mixed, however, because in the western USA there are some areas that were once rangeland but are being—or already have been—transformed into exurban and suburban housing developments. This has been facilitated by the rise of the Internet, which enables executives to live in comparatively remote areas where they can work from home while enjoying environmental recreation opportunities. The situation is exacerbated in areas with access to both public land and a regional airport (Rasker et al. 2013; Fig. 12.3). Thus, whereas livestock ranching can be somewhat detrimental to wildlife it does still conserve habitats that are lost when converted into housing developments (Hansen et al. 2005).

As for ecological processes, the widespread competitive success of livestock *vs.* wildlife depends on the facilitation of favorable conditions for livestock by humans. That is how a small number of livestock species—usually fewer than five in any one area—has come to overwhelmingly dominate the herbivore biomass of rangelands that were formerly the natural habitats of diverse assemblages of coevolved wild species (du Toit and Cumming 1999). The global trend towards sedentary pastoralism on

Fig. 12.3 Examples of
urban (*above*) and exurban
(*below*) encroachment on
rangelands in the western
United States. Aerial
photographs show the
boundary between the
town of Jackson,
Wyoming, and the USFWS
National Elk Refuge
(*above*), and a recent
development outside of
Pinedale, Wyoming
(*below*). Photo credits:
USGS

rangelands (Hobbs et al. 2008; Western et al. 2009; Groom and Western 2013) is
linked as both cause and effect with habitat fragmentation, supplementary fodder
production, predator eradication, and water provision. In combination those factors
typically maintain artificially high livestock densities and chronic overutilization of
the remaining rangeland, outcompeting indigenous herbivores (e.g., Mishra et al.
2004; Ogutu et al. 2010). Wild ungulates can only coexist with cattle in the few subsistence pastoral systems in which watering points are widely distributed and wildlife
species are not intensively persecuted (Georgiadis et al. 2007). Wild ungulates generally do not share watering points with livestock (Sitters et al. 2009) probably because
of the activities of herders (and their dogs) around watering points. Overall, therefore,
the combined effect of all aspects of livestock production across most rangelands
worldwide is that indigenous large herbivores and their predators get ratcheted back
to isolated habitat refugia and protected reserves. That process is, however, being at
least partially reversed (large predators excluded) in some areas such as in the western
USA, where valued game populations are increasingly being allowed to recover on
rangelands under mixed livestock–wildlife management.

There is growing recognition that some plant–herbivore interactions specific
to non-livestock large herbivores are potentially valuable and worth conserving, restoring, or introducing to rangelands. For example, where cattle graze

together with wild ungulates in East Africa during the wet season they achieve higher weight gains than in foraging areas from which wild ungulates are excluded (Odadi et al. 2011). That is probably because zebras (*Equus quagga*), being comparatively large hind-gut fermenters, reduce the stem:leaf ratio in the sward and thereby facilitate conditions for grazing ruminants such as cattle. Even though competition occurs in the dry season, weight gains from the facilitative effect in the wet season are sufficient that a net benefit might be achievable with management. In theory, managers could impose seasonal shifts in both the foraging areas used by livestock (currently common) and the mixes of livestock and wildlife species foraging together in each season and area (currently uncommon). In practice, however, there are few areas of rangeland in which livestock and wildlife can be actively managed together in a commercial operation at the landscape scale. Yet there are mounting scientific arguments in favor of selectively exploiting the diverse functional properties of wildlife species (du Toit 2011), despite the barriers of command-and-control laws and societal inertia.

The megafaunal extinctions that occurred on all continents other than Africa in the late Quaternary Period are attributed to a suite of factors that include overhunting by humans (Brook and Bowman 2004). Now, introducing morphologically and ecologically similar surrogates—mostly from Africa—has been proposed to restore the ecosystem processes once driven by those now-extinct megaherbivores in North America (Donlan et al. 2005) and Australia (Bowman 2012). Those "rewilding" proposals are contentious because benchmark conditions of archaic ecosystems are unknown, the global climate is continually changing, and the practicalities are prohibitive (Rubenstein et al. 2006). Also, sociopolitical acceptance of the rewilding argument requires a leap of faith in the net benefits of quasi-Pleistocene assemblages replete with large predators and megaherbivores (Soulé and Noss 1998). Nevertheless, "hybrid" or "novel" ecosystems (Hobbs et al. 2009) are emerging all the time mainly as a result of unintended species invasions and climate change, as well as the intended engineering activities of humans. For example, in the Netherlands 6000 ha of grazing land was created from seabed less than 50 years ago and is now a popular wildlife refuge called Oostvaardersplassen (Marris 2009). The bold approach adopted there was to stock the area with relict breeds of cattle and horses, together with indigenous cervids, lagomorphs, and waterfowl that moved in, and allow competitive and facilitative interactions to play out through time. With minimal intervention from managers, Oostvaardersplassen now supports one of Western Europe's richest terrestrial faunas occupying a patchwork of grassland, woodland, and wetland habitats. Those plant communities were established naturally, mainly from seeds brought in by birds, and the landscape's heterogeneity is maintained by interactions among top-down and bottom-up processes operating through herbivory, the climate, and soil.

The ongoing global changes that define the Anthropocene epoch (Zalasiewicz et al. 2010) mean that the reference state of an ecosystem cannot be an historic condition, but is the most desired of the potential alternative states at some future stage along that ecosystem's trajectory of change. Some rangelands (e.g., in the Great Basin of the western USA) are now so radically altered by invasive plants and the fire regimes they support, on top of past overgrazing, changing land use, and ongoing climatic changes, that managers cannot restore them to their former states (Davies et al. 2012). They are either drifting through hybrid states towards thresholds of irreversible change or are reorganizing as novel ecosystems and so historic benchmark conditions are irrelevant and misleading. Within these novel ecosystems the basis for classifying species as "native," "exotic," or "feral" has become unclear, apart from legal mandates to conserve native threatened and endangered species.

Conserving ecological resilience involves conserving the full suite of functional types within a species assemblage, which entails developing a functional typology (cataloging the key functional types and quantifying their equilibrium biomass densities) for each target ecosystem. This is an emerging challenge for wildlife ecologists, many (perhaps most) of whom are reluctant to shift from their professionally trained focus on the taxonomic typology of the pristine state of the ecosystem at some benchmark stage in history. Nevertheless, while conservation biologists debate whether it is right or wrong (e.g., Doak et al. 2014; Marvier and Kareiva 2014), biodiversity is inevitably managed more for what it does than for what it is. Building resilience in transformed rangelands would involve finding the "right" combination of large herbivores to represent the particular mix of functional types (e.g., large-, medium-, and small-bodied grazers and browsers) needed to reach a feasibly desired state of the rangeland in question. This would enhance heterogeneity at the plant–herbivore interface, facilitate processes such as nutrient cycling and seed dispersal, and diversify the portfolio of options for stakeholders in the social–ecological system. If the need for a functional type could not be filled by species native to the geographical area, or by local livestock breeds, then it would be pragmatic to trial selected exotic species within an adaptive management framework. A key management consideration would, however, be the feasibility of controlling introduced species where they are desired and preventing them from invading where they are not, which could be difficult or impossible.

12.2.2 *Disease*

The impacts of pathogens and parasites were historically ignored by wildlife ecologists who mostly considered disease as a compensatory form of mortality. That was probably because, with the exception of acute disease outbreaks, infectious diseases were not easily observed and those disease outbreaks that did result in large-scale die-offs were perceived as random one-time events. Only by knowing the disease status of individuals, and following them through time, do some of the underlying processes become apparent. Due to these observational challenges the importance of disease to the functioning of ecosystems was undervalued.

12.2.2.1   Development of Disease Ecology as a Discipline

One of the key developments in disease ecology was when May and Anderson (1979) explored how disease dynamics might change when the host population is considered to be dynamic rather than assumed to be constant. Then, depending on the specifics of how the parasite is transmitted, disease can emerge as a strong factor regulating the host population. This development probably sounds obvious to many ecologists, but at that time much of disease ecology was borrowed from studies on humans, in less variable populations. Many principles of wildlife disease ecology continue to be derived from human systems, where the datasets tend to be richer (Grenfell et al. 2002). However, there are several important ways in which wildlife populations are likely to differ from human or livestock populations: wildlife populations fluctuate more; reproduction is more closely tied to resources and population density; movements can be more localized; and predation can interact with disease. These differences, independently or in combination, can have important management implications. For example, "critical community size" is the population size required for a disease to persist. The concept is implicit in the rationale of those managers who are inclined towards reducing a wildlife population's size as a disease control strategy. It originated from analyses of measles, which dies out in cities smaller than about 200,000–300,000 people (Bartlett 1957; Grenfell et al. 2002). With humans the number of new susceptible individuals recruited into a population tends to be highly correlated with population size. This is also to be expected with livestock, but wildlife populations near carrying capacity have much-reduced recruitment and so the number of new susceptibles required for the disease to persist might not be correlated with total population size (Lloyd-Smith et al. 2005).

Even when episodic, diseases can have long-term effects on ecosystems (Dobson and Hudson 1986). For example, rinderpest, a morbillivirus of artiodactyls (Plowright 1962), was introduced to Africa in the late nineteenth century and caused massive die-offs across sub-Saharan Africa before being eradicated in the early part of the twenty-first century. This outbreak and subsequent die-off resulted in a large-scale release of herbivory on trees and shrubs by ungulates, triggering long-term disturbances across the African savanna biome, such as in northern Botswana (Vandewalle and Alexander 2014). Similar ecosystem-level effects, this time acting through release of predation on insects, might occur as a result of the ongoing epidemic of a fungal pathogen (*Geomyces destructans*) causing white-nose syndrome in bats in eastern North America (Frick et al. 2010).

Developments in disease ecology have followed a similar progression as those in general ecology. First there was a focus on population dynamics of the host and/or pathogen and issues of density dependence for each. This was followed by work on spatial structure and metapopulations (Hess 1994, 1996a, b). More recently, disease ecology has branched out to multi-host/multi-pathogen interactions (Jolles et al. 2008; Viana et al. 2014), community-level interaction networks (Lafferty et al. 2006), and the effects of biodiversity on disease dynamics (Johnson and Thieltges 2010; Johnson et al. 2013).

Research on the effects of biodiversity on disease dynamics has been motivated, in part, by the example of Lyme disease (Ostfeld and Keesing 2000). The causative bacterium *Borrelia burgdorferi* is hosted by white-footed mice (*Peromyscus leucopus*) that reach higher densities in less diverse ecosystems where mammalian mesopredators and avian raptors are rare or absent. An ensuing debate has concentrated on whether this represents a general relationship between higher biodiversity and lower disease or if this is idiosyncratic and system-specific (Lafferty 2012). Yet for parasites with complex lifestyles—such as trematodes, cestodes, nematodes, acanthocephalans, chytridiomycetes, oomycetes, and myxosporeans that move between multiple host species to complete their life cycles—there are multiple different mechanisms by which "dilution" or "decoy" effects could moderate disease prevalence (Johnson and Thieltges 2010). The debate over the impact of biodiversity on disease is similar to the debate over the mechanisms by which biodiversity may affect ecosystem processes (Loreau et al. 2001; Hooper et al. 2005).

Traditionally parasites and pathogens were, and mostly still are, viewed as things to be controlled or eradicated where possible. More recently, however, researchers have been investigating the effects of parasites on ecosystem function and whether a stable and resilient ecosystem has a rich parasite assemblage (Hudson et al. 2006; Vannier-Santos and Lenzi 2011; Hatcher et al. 2012). Similarly, there is an explosion of research on the microbiome, or the microbial community within an individual, and how that community composition affects nutrition and obesity, immune function, disease risk, cancer, and so forth (Kau et al. 2011; Vannier-Santos and Lenzi 2011). Healthy wildlife populations are not necessarily devoid of pathogens (Stephen 2014) but have a mix they coevolved with and which protect them from invasions of novel types.

12.2.2.2 Management of Wildlife Diseases

In rangelands, various zoonoses (rabies, bovine tuberculosis, brucellosis, etc.) can be hosted by wildlife species but are most commonly transmitted to humans through their domesticated animals. For example, brucellosis, which affects most pastoral societies, is one of the most common zoonotic infections worldwide with more than 500,000 new cases annually (Pappas et al. 2006). Also, because human populations in rangelands depend to greater or lesser degrees on domesticated animals for their livelihoods, nonzoonotic diseases with wildlife reservoirs, such as foot-and-mouth disease, are also of concern. Political and social controversies at the interface between domesticated and wild animals in rangelands are thus likely to include disease issues of some type (Kock et al. 2010).

Such controversies are heightened by the poor success rate of campaigns to eradicate diseases in wildlife reservoirs, with the notable exception of fox rabies in Europe where a safe and effective vaccine bait was available (Brochier et al. 1991). In systems where diseases co-circulate in livestock and wildlife, control efforts can be successful when they are targeted in livestock, such as with rinderpest for example (Mariner et al. 2012; Roeder et al. 2013). Similarly, in Spain, brucellosis in red deer (*Cervus elaphus*) declined as a result of control efforts in livestock (Serrano et al. 2011). In these cases the wildlife populations were not competent disease reservoirs, failing to sustain the infection without co-circulation through livestock. In cases where wildlife populations are competent reservoir hosts, control efforts are complicated by social, logistical, and ecological factors (e.g., Donnelly et al. 2006).

Culling and increased hunting of wildlife are often proposed to control wildlife diseases but they are seldom effective. First, it might be difficult to achieve the necessary hunter participation to create large reductions in host density (e.g., Heberlein 2004). Many hunters will want to maintain high densities to maximize their future hunting opportunities. Second, the cost of culling wildlife, particularly in a test and cull strategy, can be very high and so is only applicable to localized operations (Wolfe et al. 2004). One of the rationales for culling wildlife is that disease transmission might be positively correlated with host density (Lloyd-Smith et al. 2005). Although plausible, this assumption is not always supported by data and even if transmission is density-dependent, culling can have adverse consequences. For example, culling badgers (*Meles meles*) to control bovine tuberculosis in Britain causes the disease to spread even further due to social disruption, with surviving badgers roaming more widely to find mates (Bielby et al. 2014). In addition, increased hunting in regions of easy access might locally aggregate wildlife in areas of limited hunter access even if the regional density declines.

Part of the problem, at least for diseases that can be transmitted back and forth between livestock and wildlife, is that disease control measures used on wildlife are based on those developed for livestock. Therein lies a mismatch in spatial scale, because livestock can be intensively managed within fences at the ranch scale but wildlife populations are free-ranging at the landscape scale (Bienen and Tabor 2006). There is also a mismatch in temporal scale because disease eradication campaigns are typically continuous for livestock but in wildlife they can be tactically scheduled to take advantage of episodic natural disturbances. For example, in the Kruger ecosystem of South Africa, buffalo (*Syncerus caffer*) in herds with a higher prevalence of bovine tuberculosis lose condition faster during dry seasons (Caron et al. 2003). It follows that any attempts to minimize the disease in the buffalo population should be reserved for immediately after droughts, which occur about once per decade. That is when die-offs of 50 % or more leave a smaller population size to contend with, and presumably a much reduced proportion of infected animals (du Toit 2010).

For rangelands in particular, progress in reducing—or at least accommodating—the risks of wildlife-borne diseases to livestock will depend upon the better integration of veterinary practice and epidemiology with disease ecology. As a case in point, before brucellosis was virtually eradicated in cattle in the USA it was transmitted to elk (*Cervus canadensis*) and bison in the Greater Yellowstone Area, where it has repeatedly been transmitted back to cattle from elk (Rhyan et al. 2013). Free-ranging elk are now increasing in the area in which brucellosis is endemic and so the veterinary achievements in fighting the disease in livestock over the past 75 years must now be followed by ecologically based adaptive management approaches to control the spread of the disease in elk (Rhyan et al. 2013). Devising such approaches for any wildlife host of a livestock disease will depend upon studies at the landscape scale and population level to understand such issues as: seasonal movements between feeding areas; joining-and-leaving behavior of contagious individuals moving between social groups; use of habitats at key times for disease transmission; effects of climate on the dynamics of populations and the aggregation and dispersal of groups and individuals; and responses to fear induced by hunting and predation. In addition, there are calls for trade policy reforms to relax stringent veterinary regulations such as those responsible for veterinary cordon fences in southern Africa (Fynn and Bonyongo 2011; McGahey 2011; Thomson et al. 2013). There, the fences have decimated migratory wildlife populations that, in the longer term, could be more valuable to local communities than export-quality beef. At a minimum, certain wildlife species should be destigmatized as disease vectors, such as bison in North America and buffalo in South Africa, because disease-free herds can be established and used to restock areas where they can safely comingle with cattle.

12.2.3   Predation

Considering the obvious risks that large predators impose on humans and the animals they depend upon for their welfare, it is not surprising that large predators are now threatened worldwide (Ripple et al. 2014). The geographic expansion of human activities results in large predator populations becoming increasingly fragmented (Woodroffe 2000), for example the Iberian lynx *Lynx pardinus* (Rodriguez and Delibes 1992) and the African lion *Panthera leo* (Riggio et al. 2013). Most species have become extirpated from parts of their range with, for example, the African lion having lost ~85 % of its range in the last 500 years (Morrison et al. 2007). This global erosion of large predator guilds raises concerns about the loss of natural predation as an ecosystem process. In a visionary article, Hairston et al. (1960) suggested "the world is green" because herbivore populations are limited by their predators and so major ecosystem-level effects should result from the dwindling abundance and distribution of apex predators. Fifty years later, evidence is mounting that apex predators represent a functionally important component of stable ecosystems (Sergio et al. 2008; Ritchie and Johnson 2009; Estes et al. 2011; Ripple et al. 2014). Their effects flow through trophic cascades, in which lethal and nonlethal interactions between predators and prey drive sequential responses down the food chain. Artificially replicating predation is problematic because sport hunting or culling is selective and episodic, being a pulse disturbance instead of the press disturbance imposed by an intact, coevolved, predator guild. As is characteristic of a press disturbance, the ecosystem-level effects of predators are most apparent when they are either removed or reintroduced. For instance, in some regions of the world the eradication of large predators has resulted in problematically high densities of native and feral herbivores, with associated impacts on biodiversity (Terborgh et al.

2001). Examples include feral horses *Equus caballus* in the western USA (Garrott 1991), wild boar *Sus scrofa* in western Europe (Sáez-Royuela and Tellería 1986), and white-tailed deer *Odocoileus virginianus* in eastern North America (Côté et al. 2004), for which the ecological effects of their population irruptions are similar to those following the introduction of large herbivores to predator-free islands (Allombert et al. 2005a, b; Martin et al. 2010). Conversely, some evidence suggests that the reintroduction of wolves *Canis lupus* to Yellowstone National Park in 1995 triggered a trophic cascade that is ultimately contributing to changes in plant communities as different as riparian thickets and montane grasslands (White et al. 2013). The interactions are, however, complex and cannot be predicted from trophic dynamics alone (Marshall et al. 2014).

Prey populations are directly influenced by the consumptive effects of predation but there are also indirect, nonlethal influences arising from fear-driven behavioral responses of prey to their risk of predation. The "ecology of fear" (Lima 1998; Brown et al. 1999; Laundré 2010) is now an accepted subdiscipline focusing on how prey populations respond to predation risk across a suite of response variables including spatial movement and habitat use (Creel et al. 2005; Valeix et al. 2009a; Courbin et al. 2015), temporal niche (Valeix et al. 2009b), vigilance level (Laundré et al. 2001), group size (Creel and Winnie 2005), and so forth. Spatial heterogeneity in predation risk and corresponding behavioral adjustments of prey give rise to "a landscape of fear" (Laundré et al. 2001, 2014), which might then influence ecosystem structure and function (Ripple et al. 2001; Ripple and Beschta 2006; Kuijper et al. 2013). Predator-induced behavioral adjustments by prey might involve energetic costs and physiological responses (Creel et al. 2007; Barnier et al. 2014) that can ultimately affect prey demography (Creel and Christianson 2008; Christianson and Creel 2010; Zanette et al. 2011). Indeed, predators could have a greater effect on prey demography through fear than through direct consumption of individual prey (Preisser et al. 2005). For livestock on rangelands the indirect effects of predators are likely to include decreased conception rates and weaning weights (Howery and DeLiberto 2004; Steele et al. 2013). On the other hand, apex predators dominate intraguild relations (Palomares and Caro 1999; Caro and Stoner 2003) such that a collapse in an apex predator population typically results in the phenomenon of mesopredator release (Prugh et al. 2009; Ritchie and Johnson 2009), which can have negative implications for biodiversity conservation (Johnson et al. 2006).

Overall, by promoting biodiversity and trophic web integrity, apex predators contribute to the resilience of ecosystems challenged by biological invasions (Wallach et al. 2010), disease outbreaks (Pongsiri et al. 2009), and climate change (Wilmers et al. 2006). Conservation of apex predators is now considered a worldwide priority and, as a result, large carnivores have been reintroduced to several ecosystems (Hayward and Somers 2009). Because large carnivores roam in the matrix outside protected areas (Woodroffe and Ginsberg 1998; Elliot et al. 2014a), people in rangelands have an especially important role to play in their conservation. The livestock production that defines these social–ecological systems is inevitably associated with an entrenched antipathy towards large predators across all continents. Evidence nevertheless indicates, at least in the USA, that feasible management options exist for

Fig. 12.4 Apex predators can influence rangeland ecosystems in ways that are far more complex than their best-known role as problem animals responsible for livestock depredation. Interactions can be direct (*solid arrows*) and indirect (*dashed*)

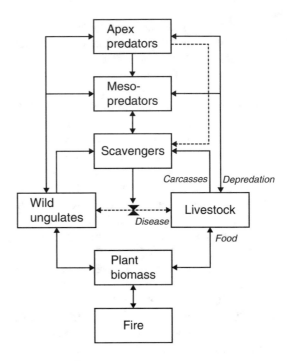

conserving large predators on rangelands without compromising the economic viability of the ranching lifestyle (Shivik 2014). Solutions require an understanding of how predators directly and indirectly influence livestock production either positively or negatively (Fig. 12.4). First, overabundant wild ungulate populations are associated with an increased risk of disease transmission to domesticated animals and an increased risk of zoonosis. Predators regulate prey populations and also comb out the sick and weak individuals, thereby contributing to animal health in rangelands. Additionally, they provide carrion for scavengers such as vultures (Wilmers et al. 2003), which have a controlling influence on the spread of diseases (Sekercioglu 2006). Some apex predators are facultative scavengers and so they interact indirectly with obligate scavengers by competing not only at their own kills, but also at carcasses they have stolen or found. Second, the dynamics of herbivory and fire are tightly coupled (Gill et al. 2009; Holdo et al. 2009) and fire is fundamental to rangeland ecology (Bond et al., this volume), so predators and diseases can influence a rangeland's structure and function indirectly via the fire regime. Third, abundant wild ungulate populations can create political tensions and socioeconomic costs on rangelands, as demonstrated by elk and feral horses in the western USA, which could be reduced by apex predators (Beschta et al. 2013). Finally, numerous studies have highlighted the perverse consequences of apex predator persecution, which include mesopredator release. Sheep ranchers in the Rocky Mountains of the USA, for example, might come to appreciate the drop in coyote density caused by the return of wolves (Berger and Gese 2007), given that depredation by coyotes is commonly perceived as an important factor in the declining sheep industry (Berger 2006).

Creativity is needed in employing effective tools and management practices to mitigate the negative effects of predators because the problem is globally urgent. Despite drastically declining populations in most large carnivore species, human–carnivore conflicts are steadily increasing (Treves and Karanth 2003; Woodroffe et al. 2005) with the most common reason being depredation on livestock (Sillero-Zubiri and Laurenson 2001; Thirgood et al. 2005). Examples of "problem" predators include lions, leopards *Panthera pardus*, spotted hyaenas *Crocuta crocuta*, cheetahs *Acinonyx jubatus* and wild dogs *Lycaon pictus* in Africa (e.g., Ogada et al. 2003), wolves and lynx in Europe (e.g., Sunde et al. 1998), wolves, coyotes *Canis latrans* and grizzly bears *Ursus arctos* in North America (e.g., Knowlton et al. 1999), pumas *Puma concolor* and jaguars *Panthera onca* in South America (e.g., Palmeira et al. 2008), and tigers *Panthera tigris* and snow leopards *Panthera uncia* in Asia (e.g., Bagchi and Mishra 2006).

Lethal control of predator populations has been the common rule for centuries but applied research over recent decades has developed a diverse "toolbox" for effective conflict mitigation (Breitenmoser et al. 2005; Thirgood et al. 2005; Shivik 2014). Technological advances have brought new methods of nonlethal deterrence (Shivik 2006) and predator-proof fencing, which can be used with or without changes in husbandry practices and guarding (Ogada et al. 2003, Woodroffe et al. 2005). Additionally, financial instruments can offset the costs of the conflicts and ameliorate human–carnivore coexistence (Dickman et al. 2011). With human–wildlife conflict, and especially depredation, being one of the most widespread and urgent issues facing conservation biologists today (Inskip and Zimmermann 2009), the publication rate in this field has steadily increased over the past 25 years (Dickman 2010). Emerging from this burgeoning literature are three branches of research into the prevention or mitigation of human–carnivore conflicts:

12.2.3.1 Consequences of Lethal Control

Lethal control, whether nonselective population reduction, illegal persecution of the species (snaring, poisoning), or retaliatory killing in the context of problem-animal-control policy, has been practiced for centuries. However, in territorial species (as most carnivores are), removal of a territory holder creates a vacuum that is rapidly filled by neighboring individuals or dispersers. This has been demonstrated for lions in Zimbabwe, where some territories outside the protected area of Hwange National Park were successively filled as its occupants were, one after another, removed by sport hunters (Loveridge et al. 2007; see also van de Meer et al. 2014 for territorial drift in wild dogs). Because of reduced levels of intraspecific competition, vacant territories are particularly attractive to dispersing subadults, which are often less efficient hunters than adults and less able to compete for occupied territories. However, compared with residents, dispersers are more daring and thus more likely to use human-dominated landscapes (Elliot et al. 2014a) and kill livestock (Patterson et al. 2004). Hence, the vacuum effect caused by indiscriminate retaliatory killing might not only compromise the viability of some carnivore populations (Woodroffe

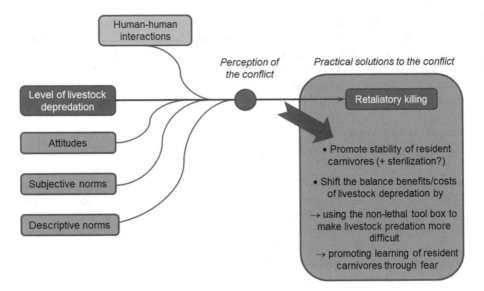

Fig. 12.5 In its most simple form, conflict between humans and large predators on rangelands is driven mainly by some level of depredation on livestock resulting in a negative perception that leads to retaliatory killing (*blue boxes*). However, large predators have important functional properties at the ecosystem level and so there is a mounting need to address the conflict. Recent research reveals that local people's perceptions of the conflict are influenced by a variety of factors that can be channeled into finding smarter solutions (*green boxes*)

and Ginsberg 1998) and disrupt their social stability (e.g., increased infanticide, Swenson et al. 1997; Loveridge et al. 2010—and delayed dispersal, Elliot et al. 2014b), but also be counterproductive for the mitigation of the human–carnivore conflicts. Hence, maintaining resident carnivores without disrupting their social structure is likely the most effective way forward to efficiently control livestock depredation (Fig. 12.5). Indeed, Bromley and Gese (2001a, b) showed that sterilization successfully reduced coyote depredation on sheep by removing the seasonal spike in nutritional demands imposed by provisioning pups, while territorial and social behaviors were maintained. Hence, resident ("better-behaved") sterile coyotes can be used as a management tool to exclude itinerant sheep-killing coyotes.

12.2.3.2 Carnivore Behavioral Ecology in Human-Dominated Landscapes

It is now recognized that carnivores create a landscape of fear for their prey (Laundré et al. 2001). What is less recognized is that carnivores themselves live in landscapes of fear too. People kill carnivores when they are perceived as a threat and so the spatial ecology of carnivores is influenced by their fear of humans. Studies on wolves (Theuerkauf et al. 2003) and lions (Valeix et al. 2012; Oriol-Cotterill et al. 2015a) have shown changes in the behavior of carnivores in the vicinity of human-dominated areas, as indicated by spatial avoidance, temporal shift, change of travel speed, use of suboptimal habitats, and so on. Carnivores appear to balance the benefits of accessing livestock with the costs associated with livestock raiding, supporting the hypothesis that fear influences the behavioral ecology of carnivores in human-dominated landscapes (Oriol-Cotterill et al. 2015b). This raises new possibilities for the tactical use of fear in managing human–carnivore conflicts (Fig. 12.4) just as hunting for fear has been proposed to induce strong behavioral responses in ungulates as a way of diverting them from areas where their impacts are undesired (Cromsigt et al. 2013).

12.2.3.3 Human Attitudes

Traditionally, research on human–carnivore conflicts has focused on the numerically tractable aspects of ecology and economy. However, the killing of predators is not always simply retaliatory and so the social drivers of human attitudes have to be identified (Lindsey et al. 2005; Dickman 2010; Macdonald et al. 2010; Guerbois et al. 2012). The difference between expected and observed outcomes of mitigation measures is likely to lie in the mismatch between common assumptions made by conservation practitioners and the perceptions and attitudes held by local people (Calvacanti et al. 2010; Dickman 2010; Marchini and Macdonald 2012). This can explain why one conflict mitigation effort might be successful in one area but fail elsewhere. In a study inspired by social sciences, Marchini and Macdonald (2012) revealed that the intention of ranchers to kill jaguars in Brazil is influenced not only by the perceived impact of jaguars on livestock but also by attitudes (an individual's personal feelings about killing a jaguar), subjective norms (an individual's perception of whether important people would approve), and descriptive norms (an individual's perception of whether other people would do the same thing). Human–human antagonisms about response options can lie at the core of a human–wildlife conflict and so the relevant social issues have to be teased out (Fig. 12.5) before workable mitigation measures can be devised (Marshall et al. 2007; Peterson et al. 2010).

12.3 Societal Implications of Integrating the Management of Wildlife and Livestock

Globally, rangelands are crucial for wildlife conservation and much has been written about why and how to conserve wildlife while maintaining livestock in semiarid ecosystems (see du Toit et al. 2010). Yet implementation is the crux of the problem because rangelands, which comprise about 40 % of global land cover, include countries with the lowest standards of governance that are home to the most impoverished and war-torn segments of humanity. Responsibility thus rests with societies occupying rangelands in those other parts of the world where reconciling agriculture with wildlife conservation is an achievable goal. Conventional approaches to natural resource management place social–ecological systems on disappointing trajectories because the economically (and therefore politically) dominant sectors of agriculture and mining outcompete the environmental sector, commonly with negative implications for the sustainable livelihoods of local communities. As an alternative to this "sectoral approach," a review conducted for the UN's Convention on Biological Diversity distilled out a "landscape approach" (Sayer et al. 2013). The principles of that approach strive for the integration of sectoral priorities for the long-term benefit of human livelihoods within defined landscapes. In a western USA rangeland, for example, a ranching landscape might comprise cattle production on private and public land, mineral extraction, wind and solar energy production, irrigated cultivation, hunting and other recreational activities, and small-scale commerce and industry in rural towns. The landscape approach would aim to muddle through land-use conflicts by fostering an integrated long-term perspective by, and for, the diverse stakeholders across all active sectors in that ranching landscape. In some cases this is beginning to happen where agencies are realizing that building local support is the key to cooperative efforts that minimize the potential for legal challenges.

Competition between wildlife and livestock can be both actual and perceived, with perceived competition prevailing especially where livestock owners—and sometimes land owners—derive no tangible benefits from wildlife (Ranglack et al. 2015). The absence of benefits is likely because they are not allowed by command-and-control laws administered by government agencies. For example, the North American model of wildlife management rests on the doctrine that wildlife is held in trust by the government for the public and therefore cannot be owned or marketed. That model has resulted in many conservation successes but is at odds with global evidence of bottom-up models being more effective for ecosystem conservation, especially in rangelands (Victurine and Curtin 2010). Now, because ranchers in the USA still control access to their land, they can sell temporarily exclusive access to outfitter operations whose clients have acquired hunting licenses from the state. This inclusion of outfitting operations into a mixed income system does not happen easily, because it represents a new way of doing business on ranchland and a break from the historic community service provided by ranchers to local hunters (Haggerty and Travis 2006).

In the USA, state fish and wildlife management agencies depend on fishing and hunting license revenue and so they are most responsive to the demands of their hunter constituency. This commodity orientation discounts ecosystem services and brings wildlife management into direct competition with livestock production for three main reasons: (1) high-value game species include ungulates (such as elk) that share food and diseases with livestock; (2) revenue from wildlife goes to the relevant management agency whereas livestock supports local ranching livelihoods; (3) ranchers with permits to graze on public land are required (by a federal agency) to immediately destock if range quality deteriorates, but without commensurate reductions (by a state agency) of wild ungulate populations. Even where wild ungulate population densities are not high enough to measurably reduce range quality for livestock, negative effects are perceived and such beliefs are entrenched in ranching communities that have railed against officialdom for generations (Ranglack et al. 2015). Now, the natural resource management paradigm of the twenty-first century is all about "resilience thinking" (Walker and Salt 2006; Allen et al. 2011; Curtin and Parker 2014) within a landscape approach (Sayer et al. 2013). For the USA this implies the empowerment of local institutions (counties, landowner associations, grazing associations, etc.) and landowners to economically benefit from, and actively participate in, the management of the local wildlife resource (threatened and endangered species aside), requiring substantial changes to the prevailing North American model of wildlife management.

An example to illustrate the motivation for the above changes can be found in the intermountain rangelands of the western USA. There, quaking aspen (*Populus tremuloides*) forests are vulnerable to chronic browsing of sprouts from the rootstock and when aspen stands die off they are typically replaced by coniferous woodland. This aspen-to-conifer shift and associated changes to ecosystem function exert negative impacts on biodiversity (Seager et al. 2013). There is also a catchment-wide decline in water yield for runoff and groundwater recharge (LaMalfa and Ryel 2008). Major browsers of aspen sprouts are elk and deer, for which state wildlife agencies set population objectives to satisfy the hunting lobby, which in turn generates important revenue for the agencies through license fees. To achieve their population objectives for elk and deer, state wildlife agencies typically suppress predator populations (Beschta et al. 2013). Hunting "tags" are a valuable commodity but local communities earn no share of the hunting revenue from public lands. Furthermore, around the Greater Yellowstone Area, expanding elk herds occurring at higher densities are boosting the spread of elk-borne brucellosis that is now being transmitted back to cattle (Rhyan et al. 2013). The costs of elevated elk and deer densities in upland catchments thus ultimately settle on local communities that incur negative impacts on their ranching enterprises and on the downstream availability of water for agricultural and urban needs. Finally, integrated management of herbivory (by wildlife and livestock), forest dynamics, hydrology, and animal health is bedeviled by the fact that each falls within the domain of a separate state or federal government agency. Meanwhile, in the absence of an integrated approach, the biodiversity of these rangelands is being eroded by exurban housing developments (Hansen et al. 2005).

Wildlife has to contribute meaningfully to community-level economies and become integrated with livestock and other land uses to enhance adaptive capacity at the landscape scale. Although still anathema to many in the North American wildlife profession, such concepts have been tested elsewhere and lessons can be learned. For example, since the latter decades of the twentieth century, integrated conservation and development projects (ICDPs) have been the cornerstone of most schemes to alleviate concurrent crises in human welfare and biodiversity conservation in developing countries. Success is contingent upon government hierarchies passing down appropriate authority to lower levels, or at least allowing revenue sharing, so that rural communities may develop a proprietary interest in their local wildlife resources. One such concept emerged in Zimbabwe in the late 1980s in the form of CAMPFIRE, a program empowering peasant farmers in communal lands to benefit directly from the wildlife around them. The rapid success of CAMPFIRE in strengthening rural communities and fostering bottom-up conservation has made it an object lesson in the international development sector (Frost and Bond 2008). Unfortunately, ICDP schemes are vulnerable to autocracy and institutionalized corruption (Garnett et al. 2007) yet that does not diminish the concept's potential in countries with better governance. It offers an example of how a tightly legislated top-down wildlife management model can be transformed into a bottom-up model, enabling wildlife to contribute to the resilience of social–ecological systems, with particular application in rangelands (Ranglack and du Toit 2016).

12.4   Future Perspectives

Achieving the effective integration of livestock and wildlife management at the landscape scale requires the negotiation of multiple social and political barriers, and for many rangelands the opportunity might be lost before adequate change can occur. Nevertheless, there is mounting evidence that rangelands can be managed for both livestock and wildlife where the costs of competition, predation, and disease can be offset by mixed revenue streams and facilitated grazing. In affluent countries there are additional possibilities including marketing strategies, such as labeling livestock products as "wildlife-friendly," which might allow for additional premiums to be charged to help offset costs. In developing countries, integrated conservation and development projects (ICDPs) are advocated, funded, and technically supported by foreign aid agencies and international conservation organizations (Garnett et al. 2007). Corruption, poverty, and weak institutions hamper conservation on the rangelands of developing countries, but the flip side has problems too. Paradoxically, the developed countries that support ICDPs in developing countries have poor records of bringing back and adopting the best practices that have emerged. Tightly legislated and compartmentalized government agencies that exert centralized top-down control are not conducive to the emergence of an integrated landscape-scale approach. Local communities on rangelands have to muddle through the process of building their own management frameworks, for which they need technical support from extension services and cooperation from government agencies.

Completely overcoming a "wicked" problem such as livestock–wildlife conflict is virtually impossible because it is too complex to be clearly defined and so there is no clear solution. Nevertheless, progress towards at least a partial solution should be possible with the coordination of efforts in an integrated approach at the landscape scale. The key change for rangelands will be a shift in policies and incentives to sustain ecosystem services despite the pressure for commodity production (Havstad et al. 2007; Norton-Griffiths and Said 2010). Wildlife communities are integral to such services and so wildlife and livestock have to be, and can be, woven together into an integrated management framework if resilient rangelands are to be sustained for the long-term benefit of the people who live off them. But how might these changes come about and what forces will drive them?

In the context of social–ecological systems, it appears that transformations—whether intended or unintended, desirable or undesirable—are most likely to occur when crises force stakeholders to find new ways of doing business (Chapin et al. 2009). The future perspective for rangelands, already affected by the increasing frequency and duration of drought conditions associated with global climate change, is not lacking in crisis scenarios. In some areas, such as privately owned commercial ranchland in southern Africa, drought crises have already driven transitions from cattle-based to wildlife-based enterprises that have transformed the local ranching culture (Carruthers 2008). Yet a transformation of that type is only possible where indigenous wildlife communities and their habitats, and therefore the adaptive capacity of the system, remain sufficiently conserved. The lesson for global rangelands is twofold: (1) drought-driven transformation is to be expected; (2) the outcome could be more desirable than the alternatives if the management objective for the wildlife–livestock interface is proactively shifted from problem control to asset management.

12.5 Summary

Integrating wildlife conservation with livestock production is implicit in the paradigm shift from production maximizing to resilience building in the social–ecological systems known as rangelands. Globally, rangelands are especially important areas for wildlife conservation in return for which wildlife can provide benefits to local communities, both directly through consumptive and nonconsumptive uses and indirectly through the facilitation of ecosystem services. The main issues to accommodate are human–wildlife conflicts arising from competition, disease, and predation. Economic competition tends to relegate wildlife conservation to the arid side of the rainfall gradient for rangelands, where opportunity costs are low. Elsewhere, wild ungulates are generally outcompeted because humans facilitate conditions for livestock, persecute wildlife, and fragment and transform natural habitats.

Diseases that co-circulate through livestock and wildlife populations are difficult to control because veterinary protocols developed for livestock at the ranch scale are seldom effective on free-ranging wildlife populations at the landscape scale. Consequently there have been few successes in effectively controlling "wildlife diseases," many of which were transmitted to wildlife from livestock in the first place. Discussions about controlling diseases that affect livestock on rangelands tend to focus on imposing a spatiotemporal separation at the livestock–wildlife interface. That is simply necessary in certain circumstances but a pragmatic view is needed of the trade-offs associated with blanket animal health policies that enforce such separation. Innovative trade policy reforms are needed before intact wildlife communities can share rangelands with commercially produced livestock, adding adaptive capacity to their social–ecological systems.

Just as with diseases, pragmatic approaches to conserving large predators on rangelands require that societal stigmas be overcome. Apex predators have been extirpated from most of their former ranges but a growing literature demonstrates their diverse and previously underestimated effects at the ecosystem level. This is despite the inevitable association between large carnivores and livestock depredation, so achieving their effective conservation is an immense challenge. Recent methodological advances do nevertheless enable some innovative approaches to mitigating human–carnivore conflicts. These include the use of fear as a tactic to promote learning among resident carnivores without disrupting their social structures. Additionally, the development of a nonlethal toolbox of deterrence methods and changes in livestock husbandry practices should be helpful in shifting the balance between persecution and acceptance of large predators in rangeland ecosystems. Retaliatory killing can be counterproductive albeit necessary in certain circumstances as a short-term response to placate affected livestock owners. Sterilization is a management-intensive option to stabilize localized populations of territorial predators and is likely to have longer-term effectiveness than lethal control. Finally, carnivore conservation cannot be successful without the support of local communities and so the key to coexistence is an understanding of the drivers of human attitudes to large carnivores. Conflict mitigation requires a balance of practical solutions, outreach, and the best available information on both the ecology of the carnivore species concerned and the human dimensions of the problem.

Mounting evidence confirms that functionally intact wildlife assemblages have properties of importance at the ecosystem level. To conserve and restore such functional properties, policy changes and extension programs are needed for local communities to become proprietors of at least a segment of the local wildlife resource. The global experience is that livelihoods on rangelands are most likely sustained in the face of externally driven challenges if communities can self-organize within resilient social–ecological systems. Resilience can be enhanced by weaving wildlife into the frameworks of those systems, which requires that centralized wildlife management agencies adapt to "resilience thinking." Proactive management of the livestock–wildlife interface is integral to the process of strategizing for climate-driven transformations of global rangelands.

References

Allen, C.R., S. Cumming, A.S. Garmestani, P.D. Taylor, and B.H. Walker. 2011. Managing for resilience. *Wildlife Biology* 17: 337–349.
Allombert, S., A.J. Gaston, and J.L. Martin. 2005a. A natural experiment on the impact of overabundant deer on songbird populations. *Biological Conservation* 126: 1–13.
Allombert, S., S. Stockton, and J.L. Martin. 2005b. A natural experiment on the impact of overabundant deer on forest invertebrates. *Conservation Biology* 19: 1917–1929.
Altchenko, Y., and K.G. Villholth. 2013. Transboundary aquifer mapping and management in Africa: A harmonized approach. *Hydrogeology Journal* 21: 1497–1517.

Bagchi, S., and C. Mishra. 2006. Living with large carnivores: Predation on livestock by the snow leopard (*Uncia uncia*). *Journal of Zoology* 268: 217–224.

Balter, M. 2013. Archeologists say the 'Anthropocene' is here – but it began long ago. *Science* 340: 261–262.

Barnier, F., M. Valeix, P. Duncan, S. Chamaillllé-Jammes, P. Barre, et al. 2014. Diet quality in a wild grazer declines under the threat of an ambush predator. *Proceedings of the Royal Society B-Biological Sciences* 281: 20140446.

Bartlett, M.S. 1957. Measles periodicity and community size. *Journal of the Royal Statistical Society: Series A (Statistics in Society)* 120: 48–71.

Bastian, C.T., J.J. Jacobs, L.J. Held, and M.A. Smith. 1991. Multiple use of public rangeland: Antelope and stocker cattle in Wyoming. *Journal of Range Management* 44: 390–394.

Berger, K.M. 2006. Carnivore-livestock conflicts: Effects of subsidized predator control and economic correlates on the sheep industry. *Conservation Biology* 20: 751–761.

Berger, K.M., and E.M. Gese. 2007. Does interference competition with wolves limit the distribution and abundance of coyotes? *Journal of Animal Ecology* 76: 1075–1085.

Berger, J., B. Buuveibaatar, and C. Mishra. 2013. Globalization of the cashmere market and the decline of large mammals in Central Asia. *Conservation Biology* 27: 679–689.

Beschta, R.L., D.L. Donahue, D.A. DellaSala, J.J. Rhodes, J.R. Karr, et al. 2013. Adapting to climate change on Western public lands: Addressing the ecological effects of domestic, wild, and feral ungulates. *Environmental Management* 51: 474–491.

Bielby, J., C.A. Donnelly, L.C. Pope, T. Burke, and R. Woodroffe. 2014. Badger response to small-scale culling may compromise targeted control of bovine tuberculosis. *Proceedings of the National Academy of Sciences of the United States of America* 111: 9193–9198.

Bienen, L., and G. Tabor. 2006. Applying an ecosystem approach to brucellosis control: Can an old conflict between wildlife and agriculture be successfully managed? *Frontiers in Ecology and the Environment* 4: 319–327.

Biggs, R., M. Schlüter, D. Biggs, E. Bohensky, S. BurnSilver, et al. 2012. Toward principles for enhancing the resilience of ecosystem services. *Annual Review of Environment and Resources* 37: 421–448.

Bowman, D. 2012. Bring elephants to Australia? *Nature* 482: 30.

Breitenmoser, U., C. Angst, J.-M. Landry, C. Breitenmoser-Würsten, J.D.C. Linnell, et al. 2005. Non-lethal techniques for reducing depredation. In *People and wildlife: Conflict or coexistence?* ed. R. Woodroffe, S. Thirgood, and A. Rabinowitz. Cambridge, UK: Cambridge University Press.

Brochier, B., M.P. Kieny, F. Costy, P. Coppens, B. Bauduin, et al. 1991. Large-scale eradication of rabies using recombinant vaccinia-rabies vaccine. *Nature* 354: 520–522.

Bromley, C., and E.M. Gese. 2001. Surgical sterilization as a method of reducing coyote predation on domestic sheep. *Journal of Wildlife Management* 65: 510–519.

———. 2001b. Effects of sterilization on territory fidelity and maintenance, pair bonds, and survival rates of free-ranging coyotes. *Canadian Journal of Zoology* 79: 386–392.

Brook, B.W., and D.M.J.S. Bowman. 2004. The uncertain blitzkrieg of Pleistocene megafauna. *Journal of Biogeography* 31: 517–523.

Brown, J.S., J.W. Laundré, and M. Gurung. 1999. The ecology of fear: Optimal foraging, game theory, and trophic interactions. *Journal of Mammalogy* 80: 385–399.

Calvacanti, S.M., S. Marchini, A. Zimmermann, E.M. Gese, and D.W. Macdonald. 2010. Jaguars, livestock and people: Reality and perceptions behind the conflicts in Brazil. In *The biology and conservation of wild felids*, ed. D.W. Macdonald and A.J. Loveridge. Oxford, UK: Oxford University Press.

Caro, T., and C. Stoner. 2003. The potential for interspecific competition among African carnivores. *Biological Conservation* 110: 67–75.

Caron, A., P.C. Cross, and J.T. du Toit. 2003. Ecological implications of bovine tuberculosis in African buffalo herds. *Ecological Applications* 13: 1338–1345.

Carruthers, J. 2008. Wilding the farm or farming the wild? The evolution of scientific game ranching in South Africa from the 1960s to the present. *Transactions of the Royal Society of South
Africa* 63: 160–181.

Chapin III, F.S., G.P. Kofinas, and C. Folke (eds.). 2009. *Principles of ecosystem stewardship:
Resilience-based natural resource management in a changing world.* New York, USA:
Springer.

Christianson, D., and S. Creel. 2010. A nutritionally mediated risk effect of wolves on elk. *Ecology*
91: 1184–1191.

Copeland, H.E., A. Pocewicz, D.E. Naugle, T. Griffiths, D. Keinath, et al. 2013. Measuring the
effectiveness of conservation: A novel framework to quantify the benefits of sage-grouse conservation policy and easements in Wyoming. *PLoS One* 8(6): e67261. doi:10.1371/journal.
pone.0067261.

Côté, S.D., T.P. Rooney, J.P. Tremblay, C. Dussault, and D.M. Waller. 2004. Ecological impacts of
deer overabundance. *Annual Review of Ecology Evolution and Systematics* 35: 113–147.

Courbin, N., A.J. Loveridge, D.W. Macdonald, H. Fritz, M. Valeix, et al. 2015. Reactive responses
of zebras to lion encounters shape their predator-prey space game at large scale. *Oikos.*
doi:10.1111/oik.02555.

Creel, S., and D. Christianson. 2008. Relationships between direct predation and risk effects.
Trends in Ecology & Evolution 23: 194–201.

Creel, S., and J. Winnie. 2005. Responses of elk herd size to fine-scale spatial and temporal variation in the risk of predation by wolves. *Animal Behaviour* 69: 1181–1189.

Creel, S., J. Winnie Jr., B. Maxwell, K. Hamlin, and M. Creel. 2005. Elk alter habitat selection as
an antipredator response to wolves. *Ecology* 86: 3387–3397.

Creel, S., D. Christianson, S. Liley, and J.A. Winnie Jr. 2007. Predation risk affects reproductive
physiology and demography of elk. *Science* 315: 960.

Cromsigt, J.P.G.M., D.P.J. Kuijper, M. Adam, R.L. Beschta, M. Churski, et al. 2013. Hunting for
fear: Innovating management of human-wildlife conflicts. *Journal of Applied Ecology* 50:
544–549.

Curtin, C.G., and J.P. Parker. 2014. Foundations of resilience thinking. *Conservation Biology* 28:
912–923.

Davies, K.W., J.D. Bates, and A.M. Nafus. 2012. Mowing Wyoming big sagebrush communities
with degraded herbaceous understories: Has a threshold been crossed? *Rangeland Ecology &
Management* 65: 498–505.

Dickman, A.J. 2010. Complexities of conflict: The importance of considering social factors for
effectively resolving human-wildlife conflict. *Animal Conservation* 13: 458–466.

Dickman, A.J., E.A. Macdonald, and D.W. Macdonald. 2011. A review of financial instruments to
pay for predator conservation and encourage human-carnivore coexistence. *Proceedings of the
National Academy of Sciences of the United States of America* 108: 13937–13944.

Doak, D.F., V.J. Bakker, B.E. Goldstein, and B. Hale. 2014. What is the future of conservation?
Trends in Ecology & Evolution 29: 77–81.

Dobson, A.P., and P.J. Hudson. 1986. Parasites, disease and the structure of ecological communities. *Trends in Ecology & Evolution* 1: 11–15.

Donlan, C.J., H.W. Green, J. Berger, C.E. Bock, and J.H. Bock. 2005. Re-wilding North America.
Nature 436: 913–914.

Donnelly, C.A., R. Woodroffe, D.R. Cox, F.J. Bourne, C.L. Cheeseman, et al. 2006. Positive and
negative effects of widespread badger culling on tuberculosis in cattle. *Nature* 439: 843–846.

du Toit, J.T. 2010. Considerations of scale in biodiversity conservation. *Animal Conservation* 13:
229–236.

———. 2011. Coexisting with cattle. *Science* 333:1710–1711.

du Toit, J.T., and D.H.M. Cumming. 1999. Functional significance of ungulate diversity in African
savannas and the ecological implications of the spread of pastoralism. *Biodiversity and
Conservation* 8: 1643–1661.

du Toit, J.T., R. Kock, and J.C. Deutsch (eds.). 2010. *Wild rangelands: Conserving wildlife while
maintaining livestock in semi-arid ecosystems.* Chichester, UK: Blackwell Publishing.

Elliot, N.B., S.A. Cushman, D.W. Macdonald, and A.J. Loveridge. 2014a. The devil is in the dispersers: Predictions of landscape connectivity change with demography. *Journal of Applied Ecology* 51: 1169–1178. doi:10.1111/1365-2644.12282.

Elliot, N.B., M. Valeix, D.W. Macdonald, and A.J. Loveridge. 2014b. Social relationships affect dispersal timing revealing a delayed infanticide in African lions. *Oikos* 123: 1049–1056.

Estes, J.A., J. Terborgh, J.S. Brashares, M.E. Power, J. Berger, et al. 2011. Trophic downgrading of Planet Earth. *Science* 333: 301–306.

Freese, C.H., K.E. Aune, D.P. Boyd, J.N. Derr, S.C. Forrest, et al. 2007. Second chance for the plains bison. *Biological Conservation* 136: 175–184.

Frick, W.F., J.F. Pollock, A.C. Hicks, K.E. Langwig, D.S. Reynolds, et al. 2010. An emerging disease causes regional population collapse of a common North American bat species. *Science* 329: 679–682.

Frost, P.G.H., and I. Bond. 2008. The CAMPFIRE programme in Zimbabwe: Payments for wildlife services. *Ecological Economics* 65: 776–787.

Fynn, R.W.S., and M.C. Bonyongo. 2011. Functional conservation areas and the future of Africa's wildlife. *African Journal of Ecology* 49: 175–188.

Garnett, S.T., J. Sayer, and J. du Toit. 2007. Improving the effectiveness of interventions to balance conservation and development: A conceptual framework. *Ecology and Society* 12(1): 2. http://www.ecologyandsociety.org/vol12/iss1/art2/.

Garrott, R.A. 1991. Feral horse fertility control: Potential and limitations. *Wildlife Society Bulletin* 19: 52–58.

Georgiadis, N.J., J.G.N. Olwero, G. Ojwang, and S.R. Romañach. 2007. Savanna herbivore dynamics in a livestock-dominated landscape: I. Dependence on land use, rainfall, density, and time. *Biological Conservation* 137: 461–472.

Gill, J.L., J.W. Williams, S.T. Jackson, K.B. Lininger, and G.S. Robinson. 2009. Pleistocene megafaunal collapse, novel plant communities, and enhanced fire regimes in North America. *Science* 326: 1100–1103.

Grenfell, B.T., O.N. Bjørnstad, and B.F. Finkenstädt. 2002. Dynamics of measles epidemics: Scaling noise, determinism, and predictability with the TSIR model. *Ecological Monographs* 72: 185–202.

Groom, R.J., and D. Western. 2013. Impact of land subdivision and sedentarization on wildlife in Kenya's southern rangelands. *Rangeland Ecology and Management* 66: 1–9.

Guerbois, C., A.-B. Dufour, G. Mtare, and H. Fritz. 2012. Insights for integrated conservation from attitudes of people toward protected areas near Hwange National Park, Zimbabwe. *Conservation Biology* 27: 844–855.

Haggerty, J.H., and W.R. Travis. 2006. Out of administrative control: Absentee owners, resident elk and the shifting nature of wildlife management in southwestern Montana. *Geoforum* 37: 816–830.

Hairston, N.G., F.E. Smith, and L.B. Slobodkin. 1960. Community structure, population control, and competition. *The American Naturalist* 94: 421–425.

Hansen, A.J., R.L. Knight, J.M. Marzluff, S. Powell, K. Brown, et al. 2005. Effects of exurban development on biodiversity: Patterns, mechanisms, and research needs. *Ecological Applications* 15: 1893–1905.

Hatcher, M.J., J.T. Dick, and A.M. Dunn. 2012. Diverse effects of parasites in ecosystems: Linking interdependent processes. *Frontiers in Ecology and the Environment* 10: 186–194.

Havstad, K.M., D.P.C. Peters, R. Skaggs, J. Brown, B. Bestelmeyer, et al. 2007. Ecological services to and from rangelands of the United States. *Ecological Economics* 64: 261–268.

Hayward, M.W., and M.J. Somers (eds.). 2009. *Reintroduction of top-order predators*. Chichester, UK: Blackwell Publishing.

Heberlein, T.A. 2004. 'Fire in the Sistine Chapel': How Wisconsin responded to chronic wasting disease. *Human Dimensions of Wildlife* 9: 165–179.

Hess, G.R. 1994. Conservation corridors and contagious disease: A cautionary note. *Conservation Biology* 8: 256–262.

————. 1996a. Disease in metapopulation models: Implications for conservation. *Ecology* 77:1617–1632.

————. 1996b. Linking extinction to connectivity and habitat destruction in metapopulation models. *The American Naturalist* 148:226–236.

Hobbs, N.T., K.A. Galvin, and C.J. Stokes. 2008. Fragmentation of rangelands: Implications for humans, animals, and landscapes. *Global Environmental Change* 18: 776–785.

Hobbs, R.J., E. Higgs, and J.A. Harris. 2009. Novel ecosystems: Implications for conservation and restoration. *Trends in Ecology & Evolution* 24: 599–605.

Holdo, R.M., A.R.E. Sinclair, A.P. Dobson, K.L. Metzger, B.M. Bolker, et al. 2009. A disease-mediated trophic cascade in the Serengeti and its implications for ecosystem C. *PLoS Biology* 7(9): e1000210.

Hooper, D.U., F.S. Chapin, J.J. Ewel, A. Hector, P. Inchausti, et al. 2005. Effects of biodiversity on ecosystem functioning: A consensus of current knowledge. *Ecological Monographs* 75: 3–35.

Howery, L.D., and T.J. DeLiberto. 2004. Indirect effects of carnivores on livestock foraging behavior and production. *Sheep and Goat Research Journal* 19: 53–57.

Hudson, P.J., A.P. Dobson, and K.D. Lafferty. 2006. Is a healthy ecosystem one that is rich in parasites? *Trends in Ecology & Evolution* 21: 381–385.

Inskip, C., and A. Zimmermann. 2009. Human-felid conflict: A review of patterns and priorities worldwide. *Oryx* 43: 18–34.

Johnson, P.T.J., and D.W. Thieltges. 2010. Diversity, decoys and the dilution effect: How ecological communities affect disease risk. *Journal of Experimental Biology* 213: 961–970.

Johnson, C.N., J.L. Isaac, and D.O. Fischer. 2006. Rarity of a top predator triggers continent-wide collapse of mammalian prey: Dingoes and marsupials in Australia. *Proceedings of the Royal Society B: Biological Sciences* 274: 341–346.

Johnson, P.T., D.L. Preston, J.T. Hoverman, and K.L. Richgels. 2013. Biodiversity decreases disease through predictable changes in host community competence. *Nature* 494: 230–233.

Jolles, A.E., V.O. Ezenwa, R.S. Etienne, W.C. Turner, and H. Olff. 2008. Interactions between macroparasites and microparasites drive infection patterns in free-ranging African buffalo. *Ecology* 89: 2239–2250.

Kau, A.L., P.P. Ahern, N.W. Griffin, A.L. Goodman, and J.I. Gordon. 2011. Human nutrition, the gut microbiome and the immune system. *Nature* 474: 327–336.

Knowlton, F.F., E.M. Gese, and M.M. Jaeger. 1999. Coyote depredation control: An interface between biology and management. *Journal of Range Management* 52: 398–412.

Kock, R., M. Kock, S. Cleaveland, and G. Thomson. 2010. Health and disease in wild rangelands. In *Wild rangelands: Conserving wildlife while maintaining livestock in semi-arid ecosystems*, ed. J.T. du Toit, R. Kock, and J.C. Deutsch. Chichester, UK: Blackwell Publishing.

Kuemmerle, T., V.C. Radeloff, K. Perzanowski, P. Kozlo, T. Sipko, et al. 2011. Predicting potential European bison habitat across its former range. *Ecological Applications* 21: 830–843.

Kuijper, D.P.J., C. de Kleine, M. Churski, P. van Hooft, J. Bubnicki, et al. 2013. Landscape of fear in Europe: Wolves affect spatial patterns of ungulate browsing in Białowieża Primeval Forest, Poland. *Ecography* 36: 1263–1275.

Lafferty, K.D. 2012. Biodiversity loss decreases parasite diversity: Theory and patterns. *Philosophical Transactions of the Royal Society B Biological Sciences* 367: 2814–2827.

Lafferty, K.D., A.P. Dobson, and A.M. Kuris. 2006. Parasites dominate food web links. *Proceedings of the National Academy of Sciences of the United States of America* 103: 11211–11216.

LaMalfa, E.M., and R. Ryel. 2008. Differential snowpack accumulation and water dynamics in aspen and conifer communities: Implications for water yield and ecosystem function. *Ecosystems* 11: 569–589.

Laundré, J.W. 2010. Behavioral response races, predator-prey shell games, ecology of fear, and patch use of pumas and their ungulate prey. *Ecology* 91: 2995–3007.

Laundré, J.W., L. Hernández, and K.B. Altendorf. 2001. Wolves, elk, and bison: Reestablishing the 'landscape of fear' in Yellowstone National Park, USA. *Canadian Journal of Zoology* 79: 1401–1409.

Laundré, J.W., L. Hernández, P.L. Medina, A. Campanella, J. López-Portillo, et al. 2014. The landscape of fear: The missing link to understand top-down and bottom-up controls of prey abundance? *Ecology* 95: 1141–1152.

Lima, S.L. 1998. Nonlethal effects in the ecology of predator-prey interactions. *BioScience* 48: 25–34.

Lindsey, P.A., J.T. du Toit, and M.G.L. Mills. 2005. Attitudes of ranchers towards African wild dogs *Lycaon pictus*: Conservation implications on private land. *Biological Conservation* 125: 113–121.

Lloyd-Smith, J.O., P.C. Cross, C.J. Briggs, M. Daugherty, W.M. Getz, et al. 2005. Should we expect population thresholds for wildlife disease? *Trends in Ecology & Evolution* 20: 511–519.

Loreau, M., S. Naeem, P. Inchausti, J. Bengtsson, J.P. Grime, et al. 2001. Biodiversity and ecosystem functioning: Current knowledge and future challenges. *Science* 294: 804–808.

Loveridge, A.J., A.W. Searle, F. Murindagomo, and D.W. Macdonald. 2007. The impact of sport-hunting on the population dynamics of an African lion population in a protected area. *Biological Conservation* 134: 548–558.

Loveridge, A.J., G. Hemson, Z. Davidson, and D.W. Macdonald. 2010. African lions on the edge: Reserve boundaries as 'attractive sinks'. In *Biology and conservation of wild felids*, ed. D.W. Macdonald and A.J. Loveridge. Oxford, UK: Oxford University Press.

Macdonald, D.W., A.J. Loveridge, and A. Rabinowitz. 2010. Felid futures: Crossing disciplines, borders, and generations. In *Biology and conservation of wild felids*, ed. D.W. Macdonald and A.J. Loveridge. Oxford, UK: Oxford University Press.

Madhusudan, M.D. 2005. The global village: Linkages between international coffee markets and grazing by livestock in a south Indian wildlife reserve. *Conservation Biology* 19: 411–420.

Marchini, S., and D.W. Macdonald. 2012. Predicting ranchers' intention to kill jaguars: Case studies in Amazonia and Pantanal. *Biological Conservation* 147: 213–221.

Mariner, J.C., House JA, C.A. Mebus, A.E. Sollod, D. Chibeu, et al. 2012. Rinderpest eradication: Appropriate technology and social innovations. *Science* 337: 1309–1312.

Marris, E. 2009. Reflecting the past. *Nature* 462: 30–32.

Marshall, K., R. White, and A. Fischer. 2007. Conflicts between humans over wildlife management: On the diversity of stakeholder attitudes and implications for conflict management. *Biodiversity and Conservation* 16: 3129–3146.

Marshall, K.N., D.J. Cooper, and N.T. Hobbs. 2014. Interactions among herbivory, climate, topography and plant age shape riparian willow dynamics in northern Yellowstone National Park, USA. *Journal of Ecology* 102: 667–677.

Martin, J.L., S.A. Stockton, S. Allombert, and A.J. Gaston. 2010. Top-down and bottom-up consequences of unchecked ungulate browsing on plant and animal diversity in temperate forests: Lessons from a deer introduction. *Biological Invasions* 12: 353–371.

Marvier, M., and P. Kareiva. 2014. The evidence and values underlying 'new conservation'. *Trends in Ecology & Evolution* 29: 131–132.

May, R.M., and R.M. Anderson. 1979. Population biology of infectious diseases: Part II. *Nature* 280: 455–461.

McGahey, D.J. 2011. Livestock mobility and animal health policy in southern Africa: The impact of veterinary cordon fences on pastoralists. *Pastoralism* 1: 14.

Mishra, C., S.E. van Wieren, P. Ketner, I.M.A. Heitkönig, and H.H.T. Prins. 2004. Competition between domestic livestock and wild bharal *Pseudois nayaur* in the Indian Trans-Himalaya. *Journal of Applied Ecology* 41: 344–354.

Morrison, J.C., W. Sechrest, E. Dinerstein, D.S. Wilcover, and J.F. Lamoreux. 2007. Persistence of large mammal faunas as indicators of global human impacts. *Journal of Mammalogy* 88: 1363–1380.

Northrup, J.M., and G. Wittemyer. 2013. Characterizing the impacts of emerging energy development on wildlife, with an eye towards mitigation. *Ecology Letters* 16: 112–125.

Norton-Griffiths, M., and M.Y. Said. 2010. The future of wildlife on Kenya's rangelands: An economic perspective. In *Wild rangelands: Conserving wildlife while maintaining livestock in semi-arid ecosystems*, ed. J.T. du Toit, R. Kock, and J.C. Deutsch. Chichester, UK: Blackwell Publishing.

Odadi, W.O., M.K. Karachi, S.A. Abdulrazak, and T.P. Young. 2011. African wild ungulates compete with or facilitate cattle depending on season. *Science* 333: 1753–1755.

Ogada, M.O., R. Woodroffe, N.O. Oguge, and L.G. Frank. 2003. Limiting depredation by African carnivores: The role of livestock husbandry. *Conservation Biology* 17: 1521–1530.

Ogutu, J.O., H.-P. Piepho, R.S. Reid, M.E. Rainy, R.L. Kruska, et al. 2010. Large herbivore responses to water and settlements in savannas. *Ecological Monographs* 80: 241–266.

Oriol-Cotterill, A., D.W. Macdonald, M. Valeix, S. Ekwanga, and L.G. Frank. 2015a. Spatiotemporal patterns of lion space use in a human-dominated landscape. *Animal Behaviour* 101: 27–39.

Oriol-Cotterill, A., M. Valeix, L.G. Frank, C. Riginos, and D.W. Macdonald. 2015b. Landscapes of coexistence for terrestrial carnivores: The ecological consequences of being downgraded from ultimate to penultimate predator by humans. *Oikos* 124: 1263–1273.

Ostfeld, R.S., and F. Keesing. 2000. Biodiversity and disease risk: The case of Lyme disease. *Conservation Biology* 14: 722–728.

Palmeira, F.B.L., P.G. Crawshaw, C.M. Haddad, K.M.P.M.B. Ferraz, and L.M. Verdade. 2008. Cattle depredation by puma (*Puma concolor*) and jaguar (*Panthera onca*) in central-western Brazil. *Biological Conservation* 141: 118–125.

Palomares, F., and T. Caro. 1999. Interspecific killing among mammalian carnivores. *The American Naturalist* 153: 492–508.

Pappas, G., P. Papadimitriou, N. Akritidis, L. Christou, and E.V. Tsianos. 2006. The new global map of brucellosis. *Lancet Infectious Diseases* 6: 91–99.

Patterson, B.D., S.M. Kasiki, E. Selempo, and R.W. Kays. 2004. Livestock predation by lions (*Panthera leo*) and other carnivores on ranches neighbouring Tsavo National Park, Kenya. *Biological Conservation* 119: 507–516.

Peterson, M.N., J.L. Birckhead, K. Leong, M.J. Peterson, and T.R. Peterson. 2010. Rearticulating the myth of human-wildlife conflict. *Conservation Letters* 3: 74–82.

Plowright, W. 1962. Rinderpest virus. *Annals of the New York Academy of Sciences* 101: 548–563.

Pongsiri, M.J., J. Roman, V.O. Ezenwa, T.L. Goldberg, H.S. Koren, et al. 2009. Biodiversity loss affects global disease ecology. *BioScience* 59: 945–954.

Popper, D.E., and F.J. Popper. 1987. The Great Plains: From dust to dust. A daring proposal for dealing with an inevitable disaster. *Planning* 53: 12–18.

Preisser, E.L., D.I. Bolnick, and M.F. Benard. 2005. Scared to death? The effects of intimidation and consumption in predator-prey interactions. *Ecology* 86: 501–509.

Prishchepov, A.V., V.C. Radeloff, M. Baumann, T. Kuemmerle, and D. Müller. 2012. Effects of institutional changes on land use: Agricultural land abandonment during the transition from state-command to market-driven economies in post-Soviet Eastern Europe. *Environmental Research Letters* 7: 024021.

Prugh, L.R., C.J. Stoner, C.W. Epps, W.T. Bean, W.J. Ripple, et al. 2009. The rise of the mesopredator. *BioScience* 59: 779–791.

Queiroz, C., R. Beilin, C. Folke, and R. Lindborg. 2014. Farmland abandonment: Threat or opportunity for biodiversity conservation? A global review. *Frontiers in Ecology and the Environment* 15: 288–296.

Ranglack, D.H., and J.T. du Toit. 2016. Bison with benefits: Towards integrating wildlife and ranching sectors on a public rangeland in the western USA. *Oryx* 50(3): 549–554.

Ranglack, D.H., S. Durham, and J.T. du Toit. 2015. Competition on the range: Science vs. perception in a bison-cattle conflict in the western USA. *Journal of Applied Ecology* 52: 467–474.

Rasker, R., P.H. Gude, and M. Delorey. 2013. The effect of protected federal lands on economic prosperity in the neo-metropolitan West. *Journal of Regional Analysis and Policy* 43: 110–122.

Rhyan, J.C., P. Nol, C. Quance, A. Gertonson, J. Belfrage, et al. 2013. Transmission of brucellosis from elk to cattle in the Greater Yellowstone Area, USA, 2002–2012. *Emerging Infectious Diseases* 19: 1992–1995.

Riggio, J., A. Jacobson, L. Dollar, H. Bauer, M. Becker, et al. 2013. The size of savannah Africa: A lion's (*Panthera leo*) view. *Biodiversity and Conservation* 22: 17–35.

Ripple, W.J., and R.L. Beschta. 2006. Linking a cougar decline, trophic cascade, and catastrophic regime shift in Zion National Park. *Biological Conservation* 133: 397–408.

Ripple, W.J., E.J. Larsen, R.A. Renkin, and D.W. Smith. 2001. Trophic cascades among wolves, elk and aspen on Yellowstone National Park's northern range. *Biological Conservation* 102: 227–234.

Ripple, W.J., J.A. Estes, R.L. Beschta, C.C. Wilmers, E.G. Ritchie, et al. 2014. Status and ecological effects of the world's largest carnivores. *Science* 343: 1241484.

Ritchie, E.G., and C.N. Johnson. 2009. Predator interactions, mesopredator release and biodiversity conservation. *Ecology Letters* 12: 982–998.

Rodriguez, A., and M. Delibes. 1992. Current range and status of the Iberian lynx *Felis pardina* Temminck, 1824 in Spain. *Biological Conservation* 61: 189–196.

Roeder, P., J. Mariner, and R. Kock. 2013. Rinderpest: The veterinary perspective on eradication. *Philosophical Transactions of the Royal Society of London. Series B: Biological Sciences* 368: 20120139.

Rubenstein, D.R., D.I. Rubenstein, P.W. Sherman, and T.A. Gavin. 2006. Pleistocene Park: Does re-wilding North America represent sound conservation for the 21st century? *Biological Conservation* 132: 232–238.

Sáez-Royuela, C., and J.L. Tellería. 1986. The increased population of the wild boar (*Sus scrofa* L) in Europe. *Mammal Review* 16: 97–101.

Sayer, J., T. Sunderland, J. Ghazoul, J.-L. Pfund, D. Sheil, et al. 2013. Ten principles for a landscape approach to reconciling agriculture, conservation, and other competing land uses. *Proceedings of the National Academy of Sciences of the United States of America* 110: 8349–8356.

Seager, S.T., C. Eisenberg, and S.B. St. Clair. 2013. Patterns and consequences of ungulate herbivory on aspen in western North America. *Forest Ecology and Management* 299: 81–90.

Sekercioglu, C.H. 2006. Increasing awareness of avian ecological function. *Trends in Ecology & Evolution* 21: 464–471.

Sergio, F., T. Caro, D. Brown, B. Clucas, J. Hunter, et al. 2008. Top predators as conservation tools: Ecological rationale, assumptions, and efficacy. *Annual Review of Ecology and Systematics* 39: 1–19.

Serrano, E., P.C. Cross, M. Beneria, A. Ficapal, J. Curia, et al. 2011. Decreasing prevalence of brucellosis in red deer through efforts to control disease in livestock. *Epidemiology and Infection* 139: 1626–1630.

Shivik, J.A. 2006. Tools for the edge: What's new for conserving carnivores. *BioScience* 56: 253–259.

———. 2014. *The predator paradox: Ending the war with wolves, bears, cougars, and coyotes.* Boston, MA: Beacon Press.

Sillero-Zubiri, C., and M.K. Laurenson. 2001. Interactions between carnivores and local communities: Conflict or coexistence? In *Carnivore conservation*, ed. J.L. Gittleman, S.M. Funk, D.W. Macdonald, and R.K. Wayne. Cambridge, UK: Cambridge University Press.

Sitters, J., I.M.A. Heitkonig, M. Holmgren, and G.S.O. Ojwang. 2009. Herded cattle and wild grazers partition water but share forage resources during dry years in East African savannas. *Biological Conservation* 142: 738–750.

Smith, P.F., R. Gorddard, A.P.N. House, S. McIntyre, and S.M. Prober. 2012. Biodiversity and agriculture: Production frontiers as a framework for exploring trade-offs and evaluating policy. *Environmental Science & Policy* 23: 85–94.

Soulé, M., and R. Noss. 1998. Rewilding and biodiversity: Complimentary goals for continental conservation. *Wild Earth* 8: 18–28.

Steele, J.R., B.S. Rashford, T.K. Foulke, J.A. Tanaka, and D.T. Taylor. 2013. Wolf (*Canis lupus*) predation impacts in livestock production: Direct effects, indirect effects, and implications for compensation ratios. *Rangeland Ecology and Management* 66: 539–544.

Stephen, C. 2014. Toward a modernized definition of wildlife health. *Journal of Wildlife Diseases* 50: 427–430.

Sunde, P., K. Overskaug, and T. Kvam. 1998. Culling of lynxes *Lynx lynx* related to livestock predation in a heterogeneous landscape. *Wildlife Biology* 4: 169–175.

Swenson, J.E., F. Sandegren, A. Soderberg, A. Bjarvall, R. Franzen, and P. Wabakken. 1997. Infanticide caused by hunting of male bears. *Nature* 386: 450–451.

Terborgh, J., L. Lopez, P. Nuñez, M. Rao, G. Shahabuddin, et al. 2001. Ecological meltdown in predator-free forest fragments. *Science* 294: 1923–1926.

Theuerkauf, J., W. Jędrzejewski, K. Schmidt, and R. Gula. 2003. Spatiotemporal segregation of wolves from humans in the Białowieża Forest (Poland). *Journal of Wildlife Management* 67: 706–716.

Thirgood, S., R. Woodroffe, and A. Rabinowitz. 2005. The impact of human-wildlife conflict on human lives and livelihoods. In *People and wildlife: Conflict or coexistence?* ed. R. Woodroffe, S. Thirgood, and A. Rabinowitz. Cambridge, UK: Cambridge University Press.

Thomson, G.R., M.-L. Penrith, M.W. Atkinson, S. Thalwitzer, A. Mancuso, S.J. Atkinson, and S.A. Osofsky. 2013. International trade standards for commodities and products derived from animals: The need for a system that integrates food safety and animal disease risk management. *Transboundary and Emerging Diseases* 60: 507–515.

Treves, A., and U. Karanth. 2003. Human-carnivore conflict and perspectives on carnivore management worldwide. *Conservation Biology* 17: 1491–1499.

Valeix, M., A.J. Loveridge, S. Chamaillé-Jammes, Z. Davidson, F. Murindagomo, et al. 2009a. Behavioral adjustments of African herbivores to predation risk by lions: Spatiotemporal variations influence habitat use. *Ecology* 90: 23–30.

Valeix, M., H. Fritz, A.J. Loveridge, Z. Davidson, J.E. Hunt, et al. 2009b. Does the risk of encountering lions influence African herbivore behaviour at waterholes? *Behavioral Ecology and Sociobiology* 63: 1483–1494.

Valeix, M., G. Hemson, A.J. Loveridge, G. Mills, and D.W. Macdonald. 2012. Behavioural adjustments of a large carnivore to access secondary prey in a human-dominated landscape. *Journal of Applied Ecology* 49: 73–81.

van de Meer, E., H. Fritz, P. Blinston, and G.S.A. Rasmussen. 2014. Ecological trap in the buffer zone of a protected area: Effects of indirect anthropogenic mortality on the African wild dog *Lycaon pictus*. *Oryx* 48: 285–293.

Vandewalle, M.E., and K.A. Alexander. 2014. Guns, ivory and disease: Past influences on the present status of Botswana's elephants and their habitats. In *Elephants and savanna woodland ecosystems: A study from Chobe National Park, Botswana*, ed. C. Skarpe, J.T. du Toit, and S.R. Moe. Chichester, UK: John Wiley & Sons, Ltd.

Vannier-Santos, M.A., and H.L. Lenzi. 2011. Parasites or cohabitants: Cruel omnipresent usurpers or creative "éminences grises"? *Journal of Parasitology Research* 2011: 214174.

Viana, M., R. Mancy, R. Biek, S. Cleaveland, P.C. Cross, et al. 2014. Assembling evidence for identifying reservoirs of infection. *Trends in Ecology & Evolution* 29: 270–279.

Victurine, R., and C. Curtin. 2010. Financial incentives for rangeland conservation: Addressing the 'show-us-the-money' challenge. In *Wild rangelands: Conserving wildlife while maintaining livestock in semi-arid ecosystems*, ed. J.T. du Toit, R. Kock, and J.C. Deutsch. Chichester, UK: Blackwell Publishing.

Walker, B. 2010. Riding the rangelands piggyback: A resilience approach to conservation management. In *Wild rangelands: Conserving wildlife while maintaining livestock in semi-arid ecosystems*, ed. J.T. du Toit, R. Kock, and J.C. Deutsch. Chichester, UK: Blackwell Publishing.

Walker, B., and D. Salt. 2006. *Resilience thinking: Sustaining ecosystems and people in a changing world*. Washington, DC: Island Press.

Walker, B., C.S. Holling, S.R. Carpenter, and A. Kinzig. 2004. Resilience, adaptability and transformability in social-ecological systems. *Ecology and Society* 9(2): 5. http://www.ecologyandsociety.org/vol9/iss2/art5.

Wallach, A.D., C.N. Johnson, E.G. Ritchie, and A.J. O'Neill. 2010. Predator control promotes invasive dominated ecological states. *Ecology Letters* 13: 1008–1018.

Western, D., R.J. Groom, and J. Worden. 2009. The impact of subdivision and sedentarization of pastoral lands on wildlife in an African savanna ecosystem. *Biological Conservation* 142: 2538–2546.

White, P.J., R.A. Garrott, and G.E. Plumb (eds.). 2013. *Yellowstone's wildlife in transition.* Cambridge, MA: Harvard University Press.

Wilmers, C.C., R.L. Crabtree, D.W. Smith, K.M. Murphy, and W.M. Getz. 2003. Trophic facilitation by introduced top predators: Grey wolf subsidies to scavengers in Yellowstone National Park. *Journal of Animal Ecology* 72: 909–916.

Wilmers, C.C., E. Post, R.O. Peterson, and J.A. Vucetich. 2006. Predator disease out-break modulates top-down, bottom-up and climatic effects on herbivore population dynamics. *Ecology Letters* 9: 383–389.

Wolfe, L.L., M.W. Miller, and E.S. Williams. 2004. Feasibility of "test-and-cull" for managing chronic wasting disease in urban mule deer. *Wildlife Society Bulletin* 32: 500–505.

Woodroffe, R. 2000. Predators and people: Using human densities to interpret declines of large carnivores. *Animal Conservation* 3: 165–173.

Woodroffe, R., and J. Ginsberg. 1998. Edge effects and the extinction of populations inside protected areas. *Science* 280: 2126–2128.

Woodroffe, R., S. Thirgood, and A. Rabinowitz (eds.). 2005. *People and wildlife: Conflict or coexistence?* Cambridge, UK: Cambridge University Press.

Zalasiewicz, J., M. Williams, W. Steffen, and P. Crutzen. 2010. The new world of the Anthropocene. *Environmental Science & Technology* 44: 2228–2231.

Zanette, L.Y., A.F. White, M.C. Allen, and M. Clinchy. 2011. Perceived predation risk reduces the number of offspring songbirds produce per year. *Science* 334: 1398–1401.

Section III
Challenges

Chapter 13
Invasive Plant Species and Novel Rangeland Systems

Joseph M. DiTomaso, Thomas A. Monaco, Jeremy J. James, and Jennifer Firn

Abstract Rangelands around the world provide economic benefits, and ecological services are critical to the cultural and social fabric of societies. However, the proliferation of invasive non-native plants have altered rangelands and led to numerous economic impacts on livestock production, quality, and health. They have resulted in broad-scale changes in plant and animal communities and alter the abiotic conditions of systems. The most significant of these invasive plants can lead to ecosystem instability, and sometimes irreversible transformational changes. However, in many situations invasive plants provide benefits to the ecosystem. Such changes can result in novel ecosystems where the focus of restoration efforts has shifted from preserving the historic species assemblages to conserving and maintaining a resilient, functional system that provides diverse ecosystem service, while supporting human livelihoods. Thus, the concept of novel ecosystems should consider other tools, such as state-and-transition models and adaptive management, which provide holistic and flexible approaches for controlling invasive plants, favor more desirable plant species, and lead to ecosystem resilience. Explicitly defining reclamation, rehabilitation, and restoration goals is an important consideration regarding novel ecosystems and it allows for better identification of simple, realistic targets and goals. Over the past two decades invasive plant management in rangelands has adopted an ecosystem

J.M. DiTomaso (✉)
Department of Plant Sciences, University of California, Davis, CA, USA
e-mail: jmditomaso@ucdavis.edu

T.A. Monaco
USDA-ARS Forage and Range Research Laboratory, Logan, UT, USA
e-mail: tom.monaco@ars.usda.gov

J.J. James
Sierra Foothill Research and Extension Center, University of California, Browns Valley, CA, USA
e-mail: jjjames@ucanr.edu

J. Firn
School of Earth, Environmental, and Biological Sciences, Queensland University of Technology, Brisbane, QLD, Australia
e-mail: Jennifer.firn@qut.edu.au

© The Author(s) 2017
D.D. Briske (ed.), *Rangeland Systems*, Springer Series on Environmental Management, DOI 10.1007/978-3-319-46709-2_13

429

perspective that focuses on identification, management, and monitoring ecological processes that lead to invasion, and to incorporating proactive prevention programs and integrated management strategies that broaden the ecosystem perspective. Such programs often include rehabilitation concepts that increase the success of long-term management, ecosystem function, and greater invasion resistance.

Keywords Ecosystem resilience • Successional management • Novel ecosystems • Management • Rehabilitation • Resilience

13.1   Introduction

The economic impact of invasive plants on livestock production includes interfering with grazing practices, lowering yield and quality of forage, increasing costs of managing and producing livestock, slowing animal weight gain, reducing the quality of meat, milk, wool, and hides, and poisoning livestock (DiTomaso 2000). In rangelands, these noxious invasive plants were estimated to cause $2 billion (USD) in annual losses in the USA (Bovey 1987), which is more than all other pests combined (Quimby et al. 1991).

In the USA, the most prevalent invasive plants have been estimated to occupy between 41 and 51 million hectares of public and private land (Duncan et al. 2004) and they continue to spread at a rate of about 14 % per year (Westbrooks 1998) with no expectation that this rate will decline. Moreover, invasive plants have invaded over half of the non-Federal rangelands and they comprise more than 50 % of the plant cover in 6.6 % of these lands (USDA 2010). In Australia, more than 600 exotic plant species are recorded within rangelands, with 160 of these identified as threats to biodiversity (Grice and Martin 2006; Firn and Buckley 2010), and 20 considered Weeds of National Significance. Therefore, the challenges posed by invasive plants in rangelands are of serious concern and it is expected to increase in the next several decades.

Although it is difficult to assign a monetary value to the adverse consequences of invasive plants, they also adversely impact all categories of ecosystem services that are provided by healthy functional rangelands. Healthy rangelands not only provide economic importance around the world, but also multiple ecosystem services that benefit millions of people in both rural and urban areas. These services include food, fiber, clean water, recreational opportunities, and open space, minerals, religious sites, aesthetics, and natural medicines (Havstad et al. 2007; Rudzitis 1999). Furthermore, rangelands provide important nontraditional ecological services, including biodiversity, wildlife habitat, and carbon sequestration (Havstad et al. 2007).

Many invasive plant-infested areas have experienced drastic changes in vegetative structure and function, including plant community composition and forage quantity and quality. Plant invasion can reduce biological diversity, threaten rare and endangered species, reduce wildlife habitat and forage, alter fire frequency, increase erosion, and deplete soil moisture and nutrient levels (DiTomaso 2000). In various quantitative assessments, invasive plants have been estimated to decrease range productivity by 23–75 % (Eviner et al. 2010), native plant diversity by 44 %, and abundance of animal species by 18 % (Vilà et al. 2010). In addition, they have increased fire frequency and intensity and the amount of area burned, as well as the prevalence of other invasive species (Mack et al. 2000). For example, the dominance of *Bromus tectorum* dramatically shortened the fire frequency intervals throughout the Great Basin region of the USA, leading to near elimination of much of the native shrub vegetation (Whisenant 1990). Even in tropical areas of Hawaii the invasion of non-native warm- and cool-season grasses has provided an abundance of fine fuels, which have increased fire frequencies (D'Antonio and Vitousek 1992). This has subsequently led to dominance by more fire-tolerant non-native species (Fig. 13.1).

The management of invasive plants on rangelands can be more complicated and difficult than weed control in agricultural systems. While control often refers to population reduction of a target weed or invasive plant species, the term management is more inclusive and encompasses control efforts within the crop or rangeland ecosystem. In agricultural areas, for example, the goal of weed management is to eliminate all vegetation to enhance the yield of the desired crop. In contrast, on rangeland systems the goal of a management program is to preserve or enhance all desired species, yet remove one or a few undesirable species. In addition, unlike agricultural

Fig. 13.1 *Bromus tectorum* (downy brome or cheatgrass) infested area within the western United States. The invasive European grass has converted millions of hectare from sagebrush steppe to annual grasslands

production that occurs on private lands, approximately 50 % of rangelands in the 14 western US states are in public ownership that are managed by several federal, state, or local government agencies (Havstad et al. 2007). However, both public and privately owned rangelands confront similar challenges regarding invasive plant management. In many other areas of the world, particularly in less-developed countries, the motivation to manage invasive plants is more a function of human subsistence and survival, than it is about increased profitability or a return to a traditional historic community (Hobbs et al. 2009). In these situations, it would be expected that invasive plant management would be seldom attempted. These considerations can impact decisions or approaches to managing invasive plants on rangelands.

The overarching goal of this chapter is to summarize proactive strategies, and their corresponding conceptual frameworks that offer the greatest success in achieving desired outcomes in invasive plant management programs on rangelands. Our specific objectives are threefold. First, our goal is to clearly define relationships between invasive plants and ecosystem services, and to identify management systems that yield the greatest overall return of ecosystem services. To achieve this goal requires greater emphasis on important ecosystem services that can be realized through a more integrated management program, as well as to recognize under what conditions invasive species removal is possible. The concept of novel ecosystems may be more realistic for severely invaded and modified ecosystems. Second, our objective is to discuss the societal implications of novel ecosystems, the services they provide, and the consequences of short-, medium-, and long-term management activities. Integrated within a management program is the need for prevention strategies, including predictive models for assessing invasion risk and understanding the biological causes of succession and invasion, and flexible restoration strategies to maximize the success of converting degraded communities into functional systems. Our final objective is to provide a theoretical framework for recovery of degraded communities that contrasts reclamation, restoration, and rehabilitation and the expected outcomes and costs for each approach.

13.2 Scope of the Invasive Plant Problem

The widespread invasion and undesirable impacts of rangeland invasive plants have been recognized for well over 100 years. Much of the private rangeland in the western USA is now occupied by a variety of invasive plant species (Fig. 13.2). In the USA alone, it is estimated that there are over 3000 non-native plant species that have become naturalized and are able to maintain self-sustaining populations within rangelands (Kartesz 2010). However, only 37–60 non-native species are considered of major economic and ecologic importance (DiTomaso 2000). Many of these invasive plants were introduced with the genuine intention to improve ecosystems for a specific land use objective. Some of these introductions have proven successful, but many have not (Cook and Dias 2006).

In the western USA, several annual grasses (*Bromus hordeaceus* (soft brome), *Avena barbata* (slender oat), and *Lolium perenne* ssp. *multiflorum* (Italian ryegrass))

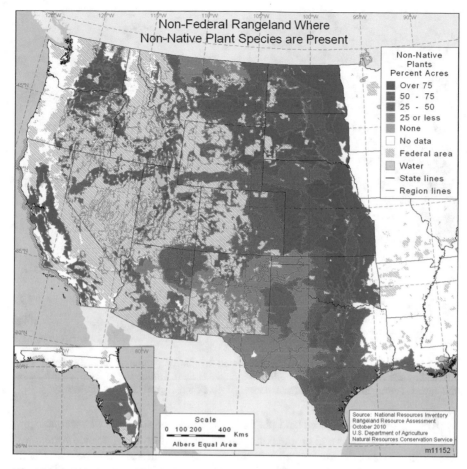

Fig. 13.2   Non-federal rangeland were invasive plants are present (USDA 2010)

and perennial grasses (*Phalaris aquatica* (hardinggrass), *Pennisetum ciliare* (buffel-grass), and *Eragrostis lehmanniana* (Lehmann lovegrass)) were intentionally introduced as forage species or as potential crops (e.g., *Isatis tinctoria* (dyer's woad)). However, the majority of invasive species, particularly thistles, were accidentally introduced as contaminants in seed, transported on equipment and vehicles, or in fur and clothing (DiTomaso 2000). By comparison, the Australian Commonwealth Plant Introduction Scheme was initiated in 1929, and over-time introduced more than 5000 species of grasses, legumes, and other forage and browse plants, including woody species (Cook and Dias 2006). In tropical Australia, 13 % of introductions have become a problem, with only 5 % being considered useful for agriculture (Lonsdale 1994). Low (1997) suggested that 5 out of 18 of Australia's worst tropical environmental invasive plants were intentionally introduced as pasture grasses. The degradation of these natural ecosystems, including rangelands, has generally occurred despite the best intentions of improving an ecosystem to provide ecosystem services people desire (Fig. 13.3).

Fig. 13.3 Mitchell grassland normally dominated by native perennial grass (*Astrebla* spp.) invaded by the African shrub *Acacia nilotica* (prickly acacia). Unlike the western United States, this invasive shrub has converted large expanses of the Mitchell grasslands to scrubland

Among the invasive rangeland plants in the USA, annual bromes (*Bromus* spp.) are the most pervasive and problematic. Of the annual bromes species, *Bromus tectorum* (cheatgrass, downy brome) is by far the most problematic and now infests 23 million hectares (Duncan et al. 2004) and 28 % of all non-Federal rangelands (USDA 2010). Downy brome was first introduced to the western USA in 1861, and by the early 1900s was widely distributed in many rangelands, particularly *Artemisia* spp.-dominated (sagebrush) ecosystems that were overgrazed (Billings 1994). In these areas, it altered the natural fire regime to replace native, perennial species as the dominant vegetation (Whisenant 1990). Similarly, *Centaurea solstitialis* (yellow starthistle) was introduced from Chile to California around 1850 and to other South American countries even earlier (Gerlach 1997). By the early 1900s it was a common invasive plant of rangelands, roadsides, grain fields, and alfalfa fields in northern California and Argentina, where it has outcompeted most native annual species. Today it is estimated to infest nearly six million hectares of rangeland in the USA (Duncan et al. 2004). Some of the other widely distributed and problematic rangeland invasive species in the western USA include medusahead (*Taeniatherum caput-medusae*), other *Centaurea* species, especially diffuse knapweed (*C. diffusa*) and spotted knapweed (*C. stoebe*), musk thistle (*Carduus nutans*), Canada thistle (*Cirsium arvense*), and leafy spurge (*Euphorbia esula*). Combined they are estimated to infest about 16 million hectares in the USA (Duncan et al. 2004). One of the most important rangeland invasive species in Australia, *Echium plantagineum* (Paterson's curse, salvation Jane) was introduced from the Mediterranean region in 1843, and by 1900 it was well established in rangelands of southeast Australia (Parsons and Cuthbertson 2001). Like *Centaurea solstitialis*, it has impacted native plant diversity, as well as pasture legumes, in Australian grazing lands (Cullen and Delfosse 1985). While invasive plants can cause broad-scale changes in plant communities, historical cultural practices, particularly overgrazing, can increase invasive plant establishment and proliferation. For example, by 1895 overgrazing of rangelands in several Canadian provinces and 16 western states of the United States (US) led to dense infestations of *Salsola tragus* (Russian-thistle) (Young and Evans 1979).

13.2.1 Ecosystem Services

While the detrimental effects of invasive plants in rangelands and other plant communities are well documented, there are many instances where they provide benefits to the ecosystem (Eviner et al. 2012). Typically, however, there are trade-offs between positive and negative impacts of invasive plants. This is most apparent in highly altered and degraded landscapes where abiotic conditions are so degraded that native species are unable to naturally recover and recovery may not be possible even when mediated by restoration efforts. In these systems, invasive plants may provide a number of beneficial services, including reduced soil erosion, regulation of pests and disturbance regimes, purification of air and water, increasing habitat for pollinators and other species, providing nurse sites for native plant establishment, and facilitating phytoremediation (Diaz et al. 2007; Richardson and Gaertner 2013).

Invasive plant species were even intentionally introduced to restore key ecosystem services in some degraded systems (Eviner et al. 2012). These services included livestock or wildlife forage, wildlife habitat, erosion control, honey source plants, and medicinal or ornamental value (Duncan et al. 2004). For example, in California the largely unintentional introduction of non-native European winter annual grasses, such as *Bromus hordeaceus* (soft brome), *Avena barbata* (slender oat), and *Lolium perenne* ssp. *multiflorum* (Italian ryegrass) have greatly altered the survival of grazing-intolerant native perennial grassland communities to only a fraction of their original composition (Murphy and Ehrlich 1989). Today, however, these annual grasses are considered desirable and productive forage species, particularly in the Central Valley and foothill grasslands of California. In other parts of the world, including the USA and Australia, non-native perennial grasses were also intentionally introduced for increased forage production, drought tolerance, and soil stabilization (D'Antonio and Vitousek 1992; Cook and Dias 2006; Lonsdale 1994). Among the more widely planted perennial grasses include, *Agropyron cristatum* (crested wheatgrass), *Pennisetum ciliare* (buffelgrass), *Eragrostis lehmanniana* (Lehmann lovegrass), and *Eragrostis curvula* (African lovegrass; Australia). All these species present significant trade-offs, depending on the region in which they had been introduced.

Despite the use of some invasive plants to provide ecosystem services, there is a general lack of understanding of how to predict and manage, or even measure, the effects of invasive species on ecosystem services (Eviner et al. 2012; Jeschke et al. 2014). This can limit the decision-making ability of land managers, yet ecosystem services are increasingly being used as criteria for prioritizing efforts to remove or manage invasive plants. In many situations, the focus has shifted from preserving the historic species assemblages within a particular site, to conserving the functionality and services provided by the existing plant community (Hobbs et al. 2011).

13.2.2   Novel Ecosystems and Restoration

While there are a number of factors that contribute to the severity of impacts of non-native plants on ecosystems, ecologists have recently recognized that these impacts may be a symptom of shifting environmental conditions that will no longer support the native community (Eviner et al. 2012; Hobbs et al. 2009). These "novel ecosystems" occur because the species composition and function of greatly altered ecosystems have been completely transformed from the historic system (Hobbs et al. 2009) (Fig. 13.4). In these situations, invasive species may not be dramatically disrupting ecological processes, but rather, they may be sustaining or restoring important ecosystem services under a different set of environment conditions.

Novel ecosystems can be the consequence of abiotic changes brought about through impacts of climate, land use, pollution, CO_2 and atmospheric nitrogen

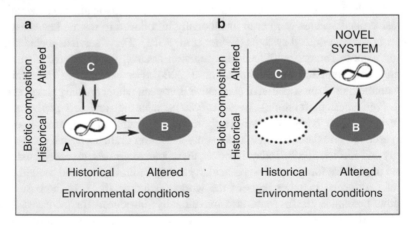

Fig. 13.4 Creation of novel ecosystems via biotic or abiotic change (modified from Suding et al. 2004). The "range of variability" and the adaptive four-phase cycle of a natural ecosystem are collapsed into the range of values found in zone A. (**a**) An ecosystem is altered by directional environmental drivers ($A \rightarrow B$) or the addition or loss of an important species ($A \rightarrow C$). (**b**) Once in the new state (either B or C), internal restructuring due to new biotic and abiotic interactions further alters community composition through changes in abundances or species losses, and through changes in biogeochemical interactions (from Seastedt et al. 2008)

enrichment, altered disturbance regimes, and urbanization (Steffen et al. 2004). These abiotic changes will almost certainly lead to subsequent changes in species composition and biogeochemical cycling that can irreversibly alter the system from its historic condition (Seastedt et al. 2008). In addition to abiotic changes, biotic modifications may also contribute to the development of novel ecosystems. These biotic changes may include new species invasions (plants or animals, including livestock), local extinction of keystone species or ecosystem engineers.

While some ecosystems remain intact, many, if not most, are novel and have an altered structure and function with unprecedented combinations of species under new abiotic conditions, compared to historic systems. This is an important starting point for the development and prioritization of invasive species management programs (Seastedt et al. 2008; Perring et al. 2013). However, it is likely that these novel systems are capable of further transformation and may not necessarily represent a resilient community. This is an important consideration in identifying desired outcomes and long-term management strategies that are designed to maximize ecosystem function and services, yet build and maintain ecological resilience (Eviner et al. 2012). Innovation, adaptation, and social flexibility will be required to attain these goals (Seastedt et al. 2008).

Many restoration efforts have the objective of recreating the historic landscape, despite the uncertainty associated with what is actually the "natural" ecosystem (Hobbs et al. 2009; Jackson and Hobbs 2009). In most cases outside of Europe and Asia, for example, the historic or "natural" ecosystem is defined as what was present before Europeans exerted their widespread influence on landscapes (Black et al. 2006; Bowman 1998). The goal of restoring ecosystems back to their "natural" state can only be effective when the historic range of variability in abiotic and biotic feedback mechanisms are still present (Seastedt et al. 2008). This may be feasible in some systems, for example, high elevation meadows, forests and riparian areas or remote and isolated wildlands, where invasive species have had only a limited impact. Alternatively, if the definition of a historic system is broadened to include a certain amount of modification and addition of new species, it may not be possible to conserve or restore the site to near-historic conditions, yet it may represent a functionally similar system (Hobbs et al. 2009). With an even broader definition of a historic system, it is possible to restore the key features and functions of the ecosystem, without the constraint of eliminating all nonindigenous species. Land managers need to take these considerations into account when assessing the feasibility of success, economic realities, and even intrinsic cultural values.

While restoration to a historic or near-historic ecosystem is possible in some cases, the abiotic and biotic feedbacks may have been so dramatically altered for many rangelands that they now represent novel ecosystems with unique assemblages of species and functions that have no analog to historic systems (Hobbs et al. 2009). Novel ecosystem recovery to conditions resembling historic conditions through restoration is considered very unlikely or impossible (Jackson and Hobbs 2009; Seastedt et al. 2008). Consequently, restoration programs should focus on managing for future change that emphasizes ecosystem function, goods, and services (Hobbs et al. 2011), maintaining genetic and species diversity, and encouraging biogeochemical processes that favor desirable species (Seastedt et al. 2008). The challenge for land managers in the future will be to determine what extent or type of changes are considered beneficial, while avoiding actions that create further ecosystem degradation (Jackson and Hobbs 2009). Restoration options that remove the requirement of attaining a historic ecosystem may reduce both effort and costs, yet may still achieve a desired outcome (Hobbs et al. 2009).

13.3 Major Conceptual Advances

In the current era of anthropogenic change, native species losses have been exceeded by gains in exotic species (Ellis et al. 2012). It is now clear that few native plant communities remain undisturbed and that most are accompanied by exotic species (Ellis and Ramankutty 2008). While novel ecosystems are not ubiquitous (Murcia et al. 2014), the global pattern of species reshuffling due to human disturbances creates the need to improve our understanding of proactive management possibilities. Proactive approaches are warranted because by definition, novel ecosystems have experienced simultaneous biotic and abiotic changes with no historical analog system or clear understanding of how to restore them (Hobbs et al. 2006; Williams and Jackson 2007). Given this high level of uncertainty, we sought to identify some of the major conceptual advances in invasive species management that have occurred during the past 25 years and explore ways to proactively apply them to rangelands that meet the criteria of novel ecosystems.

13.3.1 Integrated Invasive Plant Management

Although more prevalent in past years, invasive plant management has often focused on the control of a single species without regard to the unintended consequences of the control method. This approach typically relied on a single control technology, such as grazing, herbicides, or prescribed burning, and it has generally proven inadequate to keep pace with ecological threats emerging worldwide (Hobbs and Humphries 1995). This strategy has proven unsuccessful in the long-term (Masters and Sheley 2001), but frequently the resident native species or even desirable nonnative species do not benefit from the management strategy and can actually deteriorate further (Seastedt et al. 2008). This occurs because removal of the invasive species does not necessarily restore the ecosystem to a functional system, but may lower the abundance of important desirable plants, and cause further losses in ecological functioning (Pokorny et al. 2005). Instead of recovering rangeland function, this control method may degrade the abiotic environment or open niches for reinvasion or invasion by other undesirable species (Masters and Sheley 2001). For example, the use of the herbicide aminopyralid to control *Centaurea solstitialis* in California can lead to the subsequent invasion and proliferation of even less palatable and equally noxious long-awned annual grasses, such as *Aegilops triun-cialis* (barb goatgrass) or *Taeniatherum caput-medusae* (medusahead), that are not susceptible to the herbicide (DiTomaso et al. 2006).

13.3.1.1 Identifying Causes of Invasion

Greater emphasis needs to be focused on the management of invaded systems and identification of the underlying causal factors responsible for the invasion (Hobbs and Humphries 1995). This would contribute to a more appropriate approach to invasive plant management and to development of a broader ecological understanding of the mechanisms and processes that contribute to invasive plant success and develop management strategies that promote functional systems, provision of ecosystem services, and resilience to reinvasion (Hulme 2006; Seastedt et al. 2008; Sheley and Krueger-Mangold 2003). In some cases, this may require a compromise that is logistically practical and cost effective. Consequently, a broader view of invasive plant management emerged as integrated invasive plant management was combined with other aspects of ecosystem function.

13.3.1.2 Applying Multiple Control Tactics

Integrated pest management (IPM) of invasive plants stemmed from the realization that the dominance and spread of invasive plants indicates an underlying management problem that should be addressed before control can be successful. The basic elements of invasive plant management include the use of multiple control tactics and the careful integration of knowledge regarding the invasive species into the management effort (Buhler et al. 2000). For example, invasive plant management emphasized that recovery of degraded rangelands require more than control of the invasive plant. It is founded on a systematic, sequential application of multiple, combined tactics such as chemical, biological, cultural, and mechanical control measures to remediate ecosystem functions and reduce the negative impacts of invasive species below an economic threshold (Masters and Sheley 2001). This new management approach signified a change in inquiry from "what is that invasive plant and how do I remove it?" to "why is that invasive plant present and how can I manage the system to suppress it, prevent its spread, and remediate its impacts?" The adoption of invasive plant management was also spurred by the need to broaden typical control efforts that relied too heavily on herbicides and tillage (Holt 1994) and lessen the occurrence of herbicide-resistant weeds due to repetitive use of herbicides (Beckie and Reboud 2009). It is also important to recognize that an invasive plant management program is very often closely tied to restoration, mitigation, and rehabilitation efforts. As will be discussed in more detail, the goal of rehabilitation emphasizes both the short- and long-term effects on biodiversity and socioeconomic values.

Recognizing some of the key challenges associated with IPM can enhance the application of invasive plant management on rangelands. For example, biological, implementation, and research challenges must be addressed when developing invasive plant management systems (Buhler et al. 2000). Some of these challenges include failure to account for fecundity and survival of invasive species and excessive emphasis on individual populations in a single year as opposed to adopting a holistic approach based on analysis, theory, and implementation within an ecosystem. Research needs to develop practices capable of directly impacting propagule production, plant survival, and the transition from propagules to seedlings (Buhler et al. 2000). Another challenge of invasive plant management may be the most obvious; that is, prioritizing one control practice at the expense of an overall invasive plant management strategy that is environmentally and economically viable (Buhler 2002).

13.3.1.3   Successional and Process-Based Management

Integrated plant management has also benefited from the adoption of successional theory to understand the causes of succession, and adapting this theory to manage rangeland invasive plants. For example, Sheley et al. (1996) suggested that rangeland invasive plant managers need principles and concepts to guide their decisions as opposed to prescriptions for invasive plant control. They outlined a theoretical framework based on a successional model that emphasized influencing the primary causes of succession (i.e., disturbance, colonization, and the performance of species) to alter the plant community from an undesired state to a desired state (e.g., Pickett et al. 1987). This model was also closely aligned with specific ecological processes that should be influenced in order to affect underlying causes of succession, such as modifying *disturbance* to address site availability, *propagule dispersal* and *reproduction* to influence species availability, and altering *resource availability* or applying *stress* to impact performance of both invasive and desirable species (Sheley and Krueger-Mangold 2003; James et al. 2010). For example, the coordinated control of two invasive species—*Centaurea stoebe* (spotted knapweed or formerly *Centaurea maculosa*) and *Potentilla recta* (sulphur cinquefoil)—and perennial grass habitat restoration was accomplished by successively modifying invasive plant performance with herbicides, disturbance with variable seeding techniques, colonization with different seeding rates, and soil resource availability with cover crops (Sheley et al. 2006). The link between invasive plant management and successional theory has also been augmented by the realization that plant communities exist in alternative ecological states that may shift in nonlinear ways in response to disturbance (Westoby et al. 1989).

Process-based management was not only a core aspect of successional invasive plant management, but it also became a central theme of the emerging field of applied ecological restoration. Akin to invasive plant management, ecological restoration emphasized the importance of developing methodologies for landscape application while recognizing the need to target the specific processes responsible for degradation and recovery (Hobbs and Norton 1996). This process-oriented approach, based on the goals of renewing and maintaining ecosystem health, became an early working definition for the Society for Ecological Restoration (Higgs 1997). As the framework developed, further emphasis was placed on process-oriented restoration principles and practices to repair damaged landscapes (Whisenant 1999). This concept of managing processes at large scales expanded in the 1990s with the emergence of ecosystem-based management in federal US agencies (Koontz and Bodine 2008). Although the four largest land-management agencies in the USA (i.e., Forest Service, Fish and Wildlife Service, National Park Service, and the Bureau of Land Management) had formally adopted ecosystem management by 1994, implementing the preservation of ecological processes was identified as a primary challenge (Koontz and Bodine 2008). In recent years, process-based management has also become a central component of resilience-based management, which continues to explore how the role of ecological variables and processes influence rangeland dynamics at various temporal and spatial scales (Briske et al. 2008a; Bestelmeyer and Briske 2012).

13.3.1.4   Ecosystem Resilience

Resilience-based management should provide physical and ecological conditions that allow the system to be self-sustainable and return to pre-disturbance conditions, or reasonably close, within a fairly short time frame following removal, stress, or disturbance (Walker et al. 2002). In addition, a resilient system should be able to resist successful establishment, spread, and ecosystem change from invasive plants following the introduction of propagules (D'Antonio and Chambers 2006). The "whole-ecosystem approach" is now considered an essential aspect of managing invasive species. This is important because the secondary effects of invasive species removal can result in unexpected changes in other ecosystem components, such as (1) trophic cascades on food-web interactions among producers, consumers, and predators, (2) plant-herbivore interactions, and (3) native species reliance on exotic-species habitats (Zavaleta et al. 2001). An ecosystem perspective also provides linkages between the four common management responses of prevention, rapid response and eradication, control/containment, and restoration/mitigation to mirror the invasion processes of introduction, establishment, spread and impact, respectively (Table 13.1). Linking stages of invasion to specific management actions has since

Table 13.1 Relationships among stages of invasion, management strategy, management efficiency, and management costs (from Hulme (2006) and Simberloff et al. (2013))

Invasion stage	Management strategy	Management efficiency	Management cost
Introduction	Prevention	High	Low
Establishment	Rapid response and prevention	Moderate	Moderate
Spread	Control	Low	High
Impact	Restoration/mitigation	Very low	Very high

Invasion stage refers to sequential degradation of rangeland ecosystems over time from introduction of the invasive species to when its presence impacts ecological processes

been proposed as a unified framework for biological invasions, wherein barriers to individual plants, populations, key processes, or entire species must be overcome in order for invasive species to pass to the next stage (Blackburn et al. 2011). Another theoretical framework similarly built upon invasion stages or processes reduced the redundancy among 29 leading invasion hypotheses. This was accomplished by documenting how propagule pressure, abiotic site characteristics, and biotic characteristics of the invasive species can be utilized to narrow the number of mechanisms and processes involved in invasion, as well as identify sequential steps needed to improve invasive plant management (Catford et al. 2009). One such framework for predicting the suite of traits that confer invasion success identified three primary factors, namely, prevailing environmental conditions, traits of the resident species, and traits of the invading species (Moles et al. 2008).

13.3.2   Managing for Ecosystem Function, Functional Species Groups, and Functional Species Traits

Because novel ecosystems have experienced extreme species reshuffling (e.g., Ellis et al. 2012), irreversible restoration thresholds become a defining characteristic due to species extinctions, invasion by exotic species, and highly modified ecological composition and structure. When abiotic and biotic characteristics are severely modified such that irreversible restoration thresholds are recognized, managing the novel ecosystem pursuant of desired functioning and ecological services may take precedence over futile endeavors to reconstruct historical biotic and abiotic composition and functioning (Hallett et al. 2013). This paradigm shifts attention away from managing for species composition and toward identification of key obstacles to maximizing ecosystem functioning and provisioning of ecosystem services.

Ecosystem function has been studied at multiple levels, including plant communities, species functional groups, species, and species traits. For example, as the importance of biodiversity gained prominence in the early 1990s (e.g., Wilson 1992), its functional role within plant communities became a theoretical arena to explore alternative hypotheses regarding its importance (Johnson et al. 1996). One such hypothesis—the redundancy hypothesis—was introduced by Walker (1992), who proposed that when several species regulate ecosystem processes in similar ways they can be considered a functional group and redundancy among species performing similar function enables the ecosystem to compensate for the loss of one or more species. This new interpretation drew attention away from individual species and emphasized identification of functional groups and their role in sustaining ecosystem processes and functions, including invasion resistance. Accordingly, it is now recognized that species and groups of species can have strong effects on their environment and on specific ecosystem functions. The presence of specific functional groups may be more important than species richness or a specific species, as was shown for novel forests that maintained basic ecosystem processes after widespread loss of native species and replacement by introduced species (Mascaro et al. 2012).

The research topic of functional species traits was suggested as a means to make comparisons across regions and scales and allow researchers to assess relationships between traits and ecological processes (Craine et al. 2002). The importance of functional species traits is also based on strong evidence that key ecosystem processes can be predicted by structural or functional traits (Diaz et al. 2007). Species functional traits have also been used to match traits of invasive species to those of native species in an effort to assemble plant communities with greater invasion resistance (Funk et al. 2008; Drenovsky et al. 2012). When plant communities are composed of, or are created using native species that have functional traits similar to invaders, greater invasion resistance is theoretically possible (Funk et al. 2008). This is particularly true when native and invasive species are functionally similar in phenology, which has been shown to strengthen invasion filters (Cleland et al. 2013). A flipside of this theory is that functionally dissimilar invasive species may be more likely to invade and become abundant due to limited competitive exclusion provided by the resident community (Strayer et al. 2006). Given that invasion resistance has been linked to functional species traits, using these traits as a restoration tool is a promising research field with far-reaching management potential. In particular, as the merits of functional traits and species selection for restoration are pursued in the future, it will be important to improve our understanding of which traits are needed to overcome abiotic and biotic thresholds and promote restoration (Jones et al. 2010).

13.3.3   Rationale for Preventive Measures

Accumulation of non-native species in many regions of the global is accelerating in response to human activities, necessitating the need for predictive models to assess invasion risk (Lockwood et al. 2005). Because novel systems are often infested with exotic species, some of which may be highly invasive, preventative measures to assess risk is needed to monitor the status of invasive species. Although, exotic species arrival to a new region is not necessarily a basis for invasive species status (Valéry et al. 2008), preventative measures must still be pursued because multiple invasive species concurrently exist within novel ecosystems. When dealing with multiple invasive species, which vary in their potential to invade, it becomes important to devise strategies to screen species (Pyšek et al. 2004) and set priorities for control efforts (Hobbs and Humphries 1995).

It is widely established that invasive species prevention is far more cost effective than allocating limited resources to control efforts (Finnoff et al. 2007; Panetta 2009) (Table 13.1). Two potential prevention approaches with very different consequences have been identified: (1) prevention of invasion in the present with low target invasive plant specificity and no damage costs in the future and (2) no costs for prevention in the present but possibly high costs for specific target invasive plant control in the future (Naylor 2000). Although both options come with risk and trade-offs between expected benefits and cumulative damage, early detection via proactive assessment of invasive plant flora within novel ecosystems makes good sense to detect problematic invasive plants that may exist in low densities before they proliferate and require large investments to control them (Hobbs and Humphries 1995; Simberloff et al. 2013). This is particularly true for invasive plants that possess a lag phase following initial invasion, but are predicted to rapidly spread and cause significant damage when barriers to invasion are removed (Cunningham et al. 2004). Preventative measures and control priority should also be given to invasive species that dramatically alter the ecosystem—often referred to as engineer and transformer species—and that are known to impact ecosystem processes (Hastings et al. 2007).

13.3.4 Negative Impacts of Invasive Species Management

Control of invasive species is often followed by increases in native species abundance (e.g., Flory and Clay 2009). However, this is most often observed when invasive species are recognized as ecosystem "drivers," and their removal leads to ecosystem recovery (Bauer 2012). In contrast, when an invasive species is considered an ecosystem "passenger" or "back-seat driver," models suggest the invasive species removal will not promote recovery of the native plant community or will require both removal and ecosystem restoration to promote recovery (Bauer 2012). This establishes that careful consideration should be given to how control measures impact entire ecosystems.

In some cases invasive plant management can have negative impacts on ecosystem properties, such that control efforts exacerbate invasion or damage desirable species (Kettenring and Adams 2011). For example, when control practices disturb soils or release resources, invasive species can often gain a competitive advantage over native species (Davis et al. 2000; Cleland et al. 2013). In some cases, invasive plant treatment can be worse than the cure, as is the case when nontarget species are injured following herbicide treatment (Rinella et al. 2009) or when eradication of target species contributes to invasion by other invasive species (Courchamp et al. 2011). According to this scenario, disturbance created by invasive plant management may cause an open site where the target invasive plant or another undesired species from the community readily reinvades (Buckley et al. 2007). Invasive species control can also exacerbate problems if negative abiotic effects persist after their removal and additional restoration steps are not taken to remediate these effects (Corbin and D'Antonio 2012). Lastly, while invasive species are primarily attributed to negative impacts on ecosystem processes and services, in some cases they may enhance rangeland functioning (Eviner et al. 2012). Consequently, the decision to control an invasive plant or restore a novel ecosystem to a natural system, with corresponding ecological services, is not always clear. Only in certain cases is this decision straightforward and certain, for instance, when the goods and services provided by attaining the natural system outweigh the costs of control and when the value of the restored systems is low, but restoration is inexpensive and easy (Belnap et al. 2012). Policy-makers and managers should rank invasive species and whether to pursue eradication or prevention measures based on two criteria: (1) species impact and (2) feasibility of removal or restoration (Parker et al. 1999). High emphasis should be given to eradication when both impact and feasibility of removal is high. In contrast, high priority should be given to research efforts preceding the prevention of species when their potential future impact is high and feasibility of removal is limited by lack of clear management strategies.

13.4   Implications of Conceptual Advances

Over the last two decades as invasive plant management has shifted away from a focus on tools and technology for short-term invasive plant control and toward an emphasis on identifying, managing, and monitoring ecological processes that drive invasion, an array of conceptual advances have developed. These advances are detailed in the previous section and include a refined understanding of invasive plant impact on ecosystem services as well as insight on potential negative impacts of invasive plant control efforts on ecosystems. These advances also included development of invasive plant management tools and strategies, prevention programs, and a better understanding of plant functional trait attributes and how they can be used effectively to design site-specific invasive plant control and desired plant restoration programs. These conceptual advances have major implications for understanding constraints and opportunities for rangeland invasive plant management now and in the future. Broadly, implications of these key conceptual advances can be organized under four themes, including (1) supporting incentives for ecosystem services, (2) deploying long-term invasive plant management programs, (3) addressing socio-economic dynamics of invasive plant management, and (4) refining how management strategies and goals are identified and developed. Each of these implications is discussed below.

13.4.1   Incentives for Ecosystem Services

One of the most salient implications of recent conceptual advances is centered on the links between rangelands, invasive species, and ecosystem services. While on an area basis rangelands represent one of the largest land cover type in the world, marginal rates of return range from low to subsistence level (Tanaka et al. 2011). However, these systems provide a large array of nonmarket ecosystem services to society across the globe including carbon sequestration, biodiversity conservation and water capture and storage and invasive plants, through various mechanisms, can seriously impact these critical services (Eviner et al. 2012; Plieninger et al. 2012). Broad recognition of ecosystem services has driven development of markets and incentives to maintain or enhance rangeland ecosystem services (Chap. 14, this volume). Development of markets and incentives vary substantially globally, and can include payments for services (e.g., carbon sequestration, targeted grazing for invasive species, or fire protection), cost-share programs, technical assistance, tax incentives, and conservation easements (Lubell et al. 2013). In the majority of situations, costs for invasive plant control on rangeland greatly exceed market benefits and incentive programs have traditionally been central to maintaining invasive plant management programs on private rangeland. A recent assessment initiated by the USDA Natural Resources Conservation Services (NRCS) (Briske 2011) recognized that the conservation benefits of invasive plant control are poorly described (Sheley et al. 2011a), making it difficult to identify the return on taxpayer support for these management practices . This assessment also found that invasive plant control can have pronounced negative effects on ecosystem services (Sheley et al. 2011a). On private rangelands, there is generally less support available from incentive programs than there is demand for incentive support (Aslan et al. 2009). The minimal efficacy of incentive programs question the long-term support that could be allocated to these programs. A similar situation exists on public lands where current management resources for invasive species only cover a fraction of the potential area where management is needed (US Bureau of Land Management 1999). A critical emerging opportunity is to more strongly define the relationship between invasive rangeland plants and ecosystem services (e.g., Vilà et al. 2010), and identify species and management scenarios where management inputs yield the greatest aggregate return on ecosystem services.

13.4.2 Deploying Long-Term Invasive Plant Management Programs

Developing integrated invasive plant management principles and decision tools that provide the foundation for long-term and sustainable invasive plant management programs is essential as society begins to closely examine costs and benefits of invasive plant management on private and public rangeland.

As highlighted in the NRCS assessment (Sheley et al. 2011a), decision support systems are a central component of invasive species management because costs of control are high, risk of practice failure and nontarget effects are substantial, and the response of ecosystems to control efforts are often difficult to predict (Epanchin-Niell and Hastings 2010; Januchowski-Hartley et al. 2011). Stakeholders largely agree that invasive plant management is a complex, iterative, and a long-term process (Aslan et al. 2009; Brunson and Tanaka 2011) and there are a number of examples describing how general principles and decision tools can be applied to guide long-term invasive plant management efforts (e.g., James et al. 2010; Sheley et al. 2011b). Despite these conceptual advances there has been little evaluation of the adoption or impact of this information in management decisions and almost no data available on the long-term efficacy of these alternative strategies (Sheley et al. 2011a). Variables associated with individual rangeland enterprises often constrain deployment of invasive plant management tools and strategies below the optimum scenario. How these constraints influence the efficacy of long-term invasive plant management efforts and the conservation or enhancement of ecosystem services is generally not known. The evaluation of adoption barriers and assessment of decision tools in actual management scenarios provides a major opportunity to bridge the science and practice of invasive plant management. Greater management-science linkages may improve the ability of managers to maintain or enhance ecosystem services with conservation incentives programs and publicly funded invasive plant management programs.

13.4.3   Socioeconomic Dynamics of Invasive Plant Management

A number of key socioeconomic factors that influence development and deployment of long-term integrated invasive plant management programs have been identified over the past two decades. The human dimensions of invasive plant management contribute as much as, or more to the success or failure of invasive plant management as do the heterogeneous and stochastic environmental conditions in which management decisions are made (Chap. 8, this volume). Therefore, identification and mitigation of some of these factors represents key opportunities to advance the adoption and impact of invasive plant management programs on rangeland (Sutherland et al. 2004; Briske et al. 2008b). While invasive plant control failures have commonly been linked to factors such as insufficient policy, funding, or scientific knowledge, the influence of diverse and complex socioeconomic conditions is becoming increasingly recognized (Hershdorfer et al. 2007; Epanchin-Niell et al. 2010). The concept of management mosaics has been used to describe the increasing extent of rangeland fragmentation in the western United States and the increased diversity of land use objectives within these fragmented landscapes (Epanchin-Niell et al. 2010). As fragmentation and diversity of land use goals increases, invasive species control becomes more difficult because it alters the costs and incentives for invasive plant control and prevention. These fragmented management mosaics pose a collective action problem because there is little economic incentive for a landowner to control invasive plants unless surrounding neighbors control invasive plants as well (Hershdorfer et al. 2007; Epanchin-Niell et al. 2010). As fragmentation increases, the number of adjacent landowners also increases. As each manager becomes responsible for managing a smaller portion of the landscape, their optimal invasive plant management actions are increasingly dependent on the invasive plant management decisions of neighbors. A greater number of landowners also mean a greater likelihood that different landowners will have different incentives, policies, or practices for invasive plant management. Landowners with higher control incentives than adjacent neighbors bare a larger proportion of control costs than landowners with neighbors having similar or lower control incentives (Wilen 2007). Land owner interaction can be addressed in one of three possible approaches:

top down regulation by centralized government, bottom-up self-governing efforts, and middle out civic environmentalism efforts (DeWitt et al. 2006). Each approach can have a critical role in determining the amount and impact of invasive species management. However, these approaches do not produce the same outcomes in all management situations, and in some cases certain approaches can have negative effects on invasive plant management adoption, coordination, and impact (Hershdorfer et al. 2007; Tanaka et al. 2011).

While numerous opportunities exist to evaluate how these different socioeconomic approaches may enhance long-term progress of integrated invasive plant management programs, actual and perceived costs-benefit ratios and how these change with changing land ownership are not the only socioeconomic dimensions influencing adoption. Adoption is also highly tied to belief systems, perceived ease and risk of implementation, compatibility with the agricultural enterprise and time period in which results can be observed (Didier and Brunson 2004; Tanaka et al. 2011). In many cases producers adopt non-cost effective conservation practices because they have strong lifestyle and conservation values (Didier and Brunson 2004; Brunson and Huntsinger 2008). In other cases, adoption occurs because producers or managers believe the practice has economical or natural resource value, even if the bulk of existing data is to the contrary (Sutherland et al. 2004; Briske et al. 2008b). In some instances, adoption is less influenced by perceived practice outcomes than the ease in which the practice can be learned or applied and the impacts of the practice observed (Didier and Brunson 2004; Tanaka et al. 2011). Therefore, as the impacts of increasingly complex socioeconomic landscapes on rangeland invasive plant management are quantified and major adoption barriers are identified, there are large opportunities to link research and extension efforts that overcome these socioeconomic barriers to deploying invasive plant management programs.

13.4.4   Identifying and Developing Management Strategies and Goals

Ecological restoration is predominantly focused on the recovery of functional plant communities as plants have a controlling influence on energy flows, hydrology, soil stability, and habitat quality (Young et al. 2005; Kulmatiski et al. 2006; Pocock et al. 2012). Consequently, invasive plant management is often tied directly to broader goals and paradigms of ecosystem restoration. Recent conceptual advances have argued for a more deliberate thought process to determine how ecosystem restoration targets, including invasive plant management efforts, are identified (Hobbs et al. 2011; Monaco et al. 2012; Jones 2013). Systems targeted for invasive plant management often are highly modified and a number of ecological and socioeconomic variables constrain realistic and practical restoration targets. Collectively, this line of thinking has argued that restoration ecology, invasive plant management, and conservation biology are all subsets of the broader field of intervention ecology (Hobbs et al. 2011). One major implication of this inclusive perspective is that it allows managers to select among a number of potential reference states and outcomes on which to base their management goals and strategies (Monaco et al. 2012).

State-and-transition models (STMs) have emerged as the leading conceptual framework to describe vegetation dynamics and to assess management scenarios on rangelands (Quetier et al. 2007). STMs are qualitative flowcharts that describe potential alternative stable vegetation states on individual ecological sites established by different combinations of soil and climate (Chap. 9, this volume). These models also identify potential thresholds between states and the existence of restoration pathways that may potentially reverse transitions between states. In most cases, alternative states, transitions, and restoration pathways are based on management experience and expert opinion. STM models provide managers a general framework to identify potential management goals and some basic understanding and tools to pursue those goals. This provides an opportunity for managers to consider ecosystem states that may represent more realistic goals than the historical reference site given current socioeconomic and ecological constraints. On public lands, a broad set of stakeholders are vested in invasive plant management and restoration decision-making and different stakeholder groups can have different values and goals (Brunson and Huntsinger 2008; Brunson and Tanaka 2011). Framing management decision-making under the general concepts of intervention ecology and multiple alternative states, as well as emphasizing ecosystem function and services instead of historical benchmarks, has provided managers a greatly improved foundation for setting management goals and selecting appropriate tools to achieve identified goals.

A major practical implication of recognizing more than one reference state to set management targets and select appropriate tools is an opportunity to incorporate plant material genetics and selection of desired species into invasive plant management strategies. In many management situations controlling invasive plants does not result in long-term reduction in invasive plant abundance unless desired plant species are sown following invasive plant control (Sheley et al. 2011a). Historically, two competing paradigms have emerged which included a "local is best paradigm" that argued for use of plant material from local provenances to preserve genetic, cultural, or social values. This paradigm assumes that locally collected material would be best adapted to local conditions (Jones 2013). The alternative paradigm argues for the development and use of plant material that could excel in a particular function (typically productivity) across a range of environmental conditions. Over the last two decades the plant materials emphasis has shifted from a taxonomic focus of how plant material is developed and incorporated into restoration towards a greater understanding of how functional trait variation in potential plant material may contribute to management goals, given existing ecological and socioeconomic constraints (Drenovsky et al. 2012; Jones 2013). By considering the possibility of multiple alternative states and that different functional traits may play different roles in each of these states, managers have an improved ability to make decisions about what types of plant material to use given their identified reference state and associated management goals. Managers have a framework to identify when local genetic material is appropriate and when enhancing or altering genetic variation among selected species may be appropriate for various management applications.

13.5   Novel Ecosystems: Are They Useful for Rangeland Application?

Impacts of invasive plants are arguably greatest on "new world" continents that were settled by European immigrants and their traditional agricultural and land management strategies. These immigrants aimed to change "foreign and unique ecosystems" to deliver services that their European culture was more accustomed too (Crosby 1986). For example, both Australia and New Zealand had Commonwealth plant introduction societies, whose aims were to replace the "inferior" plant species of these foreign ecosystems with "superior" plant species providing more for the needs of European culture (Cook and Dias 2006). European landscapes have been managed in this manner for centuries. This establishes that novel ecosystems are not a new concept (Perring et al. 2013). Not only have Europeans dramatically altered their landscapes, but Australian aboriginals and indigenous North America's had also altered landscapes prior to European arrival with use of fire for cultivating crops, as well as hunting practices (Gammage 2012). People are dependent on natural ecosystems for food, water, and their overall livelihoods. Ecosystem conversion becomes an issue when land use is changed, often leading to instability and degradation, or when management actions, such as the introduction of exotic species for forage production or erosion control, did not succeed.

The concept of novel ecosystems is useful in that it provides new terminology for describing complicated ecosystem whose original function, structure, and use are no longer accessible in the short or medium-term. Furthermore, it provides an explicit model for why managers may choose an alternative stable state, rather than attempting to return to the historical state. The concept of novel ecosystems needs to be packaged with other tools, such as alternative states models and adaptive management, which extend beyond new terminology to create a strategy with the goal of controlling invasive plants and favoring more desirable plant species.

13.5.1   Invasive Plant Control Strategies Are Dependent on Goals

As mentioned above, restoration is considered a science and practice that is driven by clearly defined goals to explicitly define realistic short-, medium-, and long-term goals (Hobbs 2007). Goals change depending on land use needs and the state within which an ecosystem resides. The length of time that a site should be managed under each restoration goal will depend on the extent to which biotic and abiotic factors of the ecosystem are degraded (e.g., nutrient cycling, hydrology, energy flows, and native species richness). For these reasons, restoration goals should be flexible so that changes can be made to management strategies when systems do not react according to predictions.

Ecological evidence from more than 20 years of research indicate that increasing number of species within a plant community can produce beneficial effects to ecological functions such as increased production and nutrient cycling (Tilman et al. 1997; Hector et al. 1999). There is agreement that biodiversity matters intrinsically, but also for other values (Naeem and Wright 2003), although how and why it matters remains contested (Kaiser 2000; Naeem and Wright 2003).

The debate has largely concentrated around two hypotheses. One is the "niche complementarity" hypothesis, which proposes that species-rich communities are able to access and utilize limiting resources more efficiently because they contain species with a diverse set of ecological traits. The ecosystem is thought to be more functionally "complete" because species complement each other, allowing them to optimize resource use (Tilman et al. 1997; Hector 1998). An alternative is the "sampling effect" hypothesis, which proposes that more biologically diverse communities have increased productivity because they are likely to contain at least one species that is particularly efficient in how it uses resources. That is, only one or two species within the community may be largely responsible for most of the production. In this case, the subordinate and transient species present may not immediately contribute to functioning of the system, but their presence could provide an ecological buffer by responding to changes in environmental conditions or disturbance regimes (Grime 1998). Recent evidence suggests that when the multiple services provided by grasslands are considered, even higher levels of biodiversity are needed to maintain stability (Isbell et al. 2011).

The benefits of diversity are not limited to ecosystem functions such as production and nutrient cycling, but also increased resilience—defined as the ability to recover after disturbance (Hautier et al. 2014). This idea is based on the insurance hypothesis of ecosystem stability (McCann 2000): an ecosystem with more species contains a diverse set of traits, and therefore, has a higher likelihood of recovering from disturbance (McCann 2000; Naeem and Wright 2003; Suding et al. 2008). Because of the role biodiversity plays in maintaining key functions and building resilience to change, there is increasing recognition that encouraging diversity in rangelands is not counter to production. Instead, focusing management activities on both biodiversity and production could, in the long-run, ensure the sustainability of production and environmental integrity (Firn 2007). These issues of sustainable production and resilience are highly topical, considering predictions for an increase in extreme rainfall variability with global climate change (Hellmann et al. 2008).

However, ecological evidence also suggests that very high biodiversity may reduce ecosystem stability (Pfisterer and Schmid 2002). For example, grassland plots with the lowest species diversity recovered more quickly from drought conditions (Pfisterer and Schmid 2002). These results along with those of other studies (Wardle et al. 1997; Firn et al. 2007) suggest that what matters may not be just the number of species, but the "quality" of the biodiversity; more specifically, the collective traits of the species present and how they respond to perturbations, and in turn, how these responses effect functionality (Lavorel and Garnier 2002; Suding et al. 2008). An increase in "quality" species within a plant community may contribute considerably to function and resilience, although choosing the optimal set of plant traits may lead to altered abiotic and biotic conditions such as fire frequency, grazing intensity, and nitrogen deposition. We infer from this work that with a greater number of species there is a higher likelihood that the "right" species will be present to maintain ecosystem function.

13.5.2 Adaptable Theoretical Framework for Recovery of Degraded Communities

Restoration objectives for degraded ecosystems should be clearly defined for the success of management actions, monitored to determine restoration success and then adapted when necessary (McCarthy and Possingham 2007). These objectives should be more holistic than simply targeting the invasive species, adopting the whole-ecosystem approach discussed above (Zavaleta et al. 2001). Three broad objectives are suggested for the restoration of degraded rainforests, each representing different levels of trade-offs between biodiversity and socioeconomic values (Lamb et al. 2005) (Fig. 13.5). The benefits of adapting a simple framework to underpin the design of restoration goals for rangelands are that trade-offs between multiple objectives can be clearly defined and progress monitored and evaluated in a systematic manner, and most importantly, that strategies and perhaps even broad objectives changed as knowledge of the system increases.

Reclamation involves the complete conversion of an ecosystem to a monoculture of a highly productive species, with the prime objective to recover production (Lamb and Gilmour 2003; Firn et al. 2013). In a degraded rangeland dominated by an invasive plant, the forage species chosen for reclamation should be palatable, high in nutritional content, and competitively superior to the invasive species, so as to gain and maintain dominance. If the establishment of another species is successful this approach could act to increase the livelihood of enterprise operators once the high cost of establishment is recovered (Fig. 13.5; Table 13.2). It could also act to increase income in the long-term; however, this will depend on the consistency of the management practices, environmental conditions, and the disturbance regime over time. Because the prime objective is the recovery of productivity, this approach may reduce biodiversity values even further from the degraded state (Fig. 13.5).

In contrast to reclamation, the goal of restoration involves returning the assemblage of species that were present prior to the dominance of invasive plant, and here the main goal may be to increase biodiversity values (Fig. 13.5). Biodiversity values could be defined as native species richness and abundance, conservation of threatened species, and/or properties of an ecosystem. To apply this approach, production related activities such as grazing livestock may need to be excluded to provide the plant community with an opportunity to recover over time. Sites that have been severely degraded may also need costly strategies to reduce soil nutrients, and return species that are no longer present in the landscape and seed bank. If an area is protected then social and economic value, in terms of income, may need to cease.

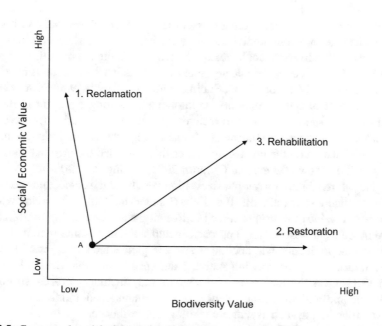

Fig. 13.5 Conceptual model of the trade-offs between approaches for the recovery of a degraded rainforest (adapted from Lamb and Gilmour (2003), Lamb et al. (2005)) including reclamation, rehabilitation, and restoration. Point *A* represents an ecosystem that is degraded and each vector a different theoretical approach to ameliorate both the biodiversity value and the social/economic value from low to high. Reclamation and restoration represent the social/economic and the biodiversity value extremes respectively; while, rehabilitation presents a one to one trade-off between biodiversity and social/economic values

Table 13.2 Potential short- and long-term outcomes for the application of the different approaches to recovery social/economic and biodiversity values in degraded ecosystems

| | Potential | | Potential | |
| | Short-term outcomes | | Long-term outcomes | |
Approaches	Benefits	Costs	Benefits	Costs
Reclamation	Income	Management biodiversity	Income	Biodiversity resilience management
Rehabilitation	Income	Lost income management	Income biodiversity resilience	Lost income management
Restoration	Biodiversity	Lost income management	Biodiversity resilience	Lost income management

Income is defined as the profit received from products derived from the land and costs is defined as the expenditure to manage and produce these products or loss of values such as biodiversity or income

Management costs may then increase from the original degraded state in both the short and long-term, while biodiversity may increase over the short and long-term. An increase in biodiversity could restore functional resilience over the long-term (Table 13.2). There is considerable argument as to whether restoration is an achievable goal (Lamb and Gilmour 2003; Suding et al. 2004; Hobbs et al. 2006) because the original suite of species may not be known and no longer present in the seed bank if the ecosystem is intended to recover by natural regeneration (passive management). Setting targets and milestones for assessing progress may also be difficult because information on the precise dynamics that governed the original ecosystem may also not be known (Lamb and Gilmour 2003; Suding et al. 2004).

A goal of rehabilitation emphasizes biodiversity and socioeconomic values equally, in a one-to-one trade-off (Fig. 13.5) (Lamb et al. 2005). This acknowledges the short-term value of returning some biodiversity, while continuing to utilize productive output. The costs of this approach includes a loss of some income because it is likely that production will decline as some implemented strategies will encourage the return of slower growing native plant species or more desirable exotics (Table 13.1). Over the long-term, the benefits may include more consistent and reliable income and, if management practices maintained biodiversity values, improved functioning such as nutrient cycling and resilience.

Whether the main goal of a resource manager is production, biodiversity conservation, or both, a rehabilitation approach is an effective option for rangeland improvement in the short-term. Investing time and money into the intensive actions needed for either reclamation or restoration will not necessarily deliver their respective and often singularly focused outcomes, particularly in the short-term.

What is clear is there are short-term benefits for the socioeconomic welfare of managers in slowly managing the transition of an invasive plant state to a more diverse state containing a greater proportion of native species. The intensive actions needed to instigate reclamation or restoration does not necessarily provide the socioeconomic values or the biodiversity values desired, without significant additional investments of time and money. Controlling invasive species with intensive strategies can detrimentally effect the remaining native species and lead to further degradation (Rinella et al. 2009), and the method of plant control has a strong effect on ecosystem response (Flory and Clay 2009).

13.5.3   A Revised Rehabilitation-Novel Ecosystem Model

The application of invasive plant control in rangelands is complex because the management practices used can also detrimentally impact functional integrity, including production and nutrient cycling. In this regard, invasive plant control practices themselves can result in further short- and long-term income loss, social distress, and environmental degradation. Common control strategies for invasive plants have generally been designed to eradicate them, but in an enterprise that is dependent on forage availability for livestock consumption, removing even low quality vegetation could prove more detrimental to the livelihood of enterprise operators than the impacts of the invasive plants. Removing available plant cover, especially in arid and semiarid systems, can accelerate soil erosion, nutrient loss, and can lead to the dominance of other invasive plant species. In this case, maintaining a small population of undesirable invasive grass species may be better, in the short-term, then reducing the majority of plant cover, including available forage, and waiting for the system to recover. Instead, control strategies are needed that balance multiple objectives, including short-term income for managers, reasonable costs for establishment and maintenance of productive species, and increased biodiversity to improve ecosystem functioning and resilience.

A revised model for the socioeconomic and biodiversity values in the rehabilitation of rangelands is needed based on the justification provided above (Fig. 13.6).

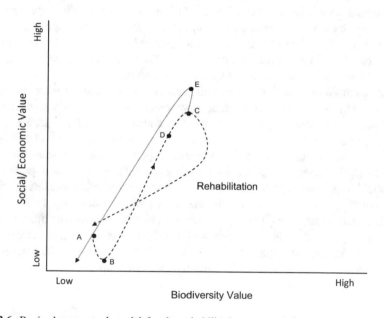

Fig. 13.6 Revised conceptual model for the rehabilitation approach for pastoral improvement. Point *A* represents a degraded state; point *B* the initial cost for the rehabilitation efforts and loss of income; point *C* represents the maximum amount of social/economic value that can be gained from a diverse pasture community; Point *D* represents the maximum social/economic value if environmental conditions become unfavorable; and point *E* represents the initial increase in income associated with over-utilization of the resource and shows the projected decrease in both social/economic and biodiversity value. Management efforts should aim to maintain the ecosystem at point *C* where both biodiversity and social/economic values can be optimized (species that transform an ecosystem), or fragmentation of existing habitats (Seastedt et al. 2008). Theses biotic changes have accelerated in the past few decades due to increased human activities arising from the breakdown of biogeographic barriers and the global human-mediated transport of non-native species (Seastedt et al. 2008). This, in combination with abiotic changes, have increased the rate of appearance of new, non-historical, novel environments, unique species combinations, and altered ecosystem functioning (Hobbs et al. 2006, 2009)

Point *A* represents the socioeconomic and biodiversity values of a degraded state. Point *B* represents the starting point of a rehabilitation approach and illustrates the initial drop in social and economic value associated with the financial costs, time, and effort to increase biodiversity value. Depending on the time since invasion and extent of ecosystem degradation, this drop in socioeconomic value could be higher or lower. Point *C* represents a maximum level of socioeconomic value that could be gained from a rehabilitated ecosystem where biodiversity and socioeconomic goals are both key. Grazing can increase the diversity of species present within ecosystems, particularly if the dominant species are palatable (Fensham et al. 2014; Lunt et al. 2007) and depending on the climatic region and evolutionary history of grazing (Milchunas et al. 1988). In contrast to the model suggested by Lamb et al. (2005) for rainforest communities, there is a higher gain in socioeconomic value with every unit of increase in biodiversity value in pasture communities. We have represented this with a steeper curve (Fig. 13.6) then the model proposed in Fig. 13.5. The grazing pressure should be maintained at a level that encourages diversity within the community, but does not reduce the abundance of key dominant species to a level that opens an opportunity for less desirable species to become established. For this reason, the shape of the curve after point C will likely vary widely depending on the biotic and abiotic characteristics of specific ecosystems.

The maximum socioeconomic benefit gain can be reduced if the favorability of environmental conditions decreases, as represented by point *D* (Fig. 13.6). Because rangelands are generally quick to respond to environmental fluctuations, grazing pressure, and therefore the socioeconomic value would need to be scaled back if conditions became unfavorable for an extended period. This represents the principles of adaptive management because grazing pressure and its impact on biodiversity would need to be regularly assessed to ensure it was at a suitable level for social welfare and to not jeopardize benefits associated with biodiversity. Point *E* represents the system response if grazing pressure is not matched with the qualities of the grazed ecosystem and the environmental conditions. In this case, there may be short-term gains in socioeconomic value due to over-utilization of the resource, but in the long-term any gains in biodiversity and socioeconomic values may be lost.

13.6   Future Perspectives

While the concept of novel ecosystem is not new, it is a useful construct for controlling invasive plant species as it provides explicit terminology for why managers may choose not to return an ecosystem to its historical state. The concept of novel ecosystems needs to be packaged with other tools, such state-and-transition models and adaptive management, which provide holistic and flexible approaches for controlling invasive plants and also considering how abiotic and biotic factors are altered so as to favor more desirable plant species. Explicitly defining reclamation, rehabilitation, and restoration goals is an important addition to the novel ecosystems concept and allows for a more detailed definition and the identification of simple, realistic targets and goals. Most importantly, reclamation, rehabilitation, and restoration goals can be changed as ecosystems and societal needs change, and thus, provide the flexibility and practicality needed for adaptive management practices.

13.7 Summary

Global rangelands provide many important ecosystem services, including food, fiber, clean water, recreational and open space, minerals, religious sites, aesthetics, plant and animal biodiversity, wildlife habitat, and carbon sequestration. Many traditional management practices, coupled with the introduction, establishment and proliferation of invasive non-native plants have led to broad-scale changes in plant communities. The most important of these invasive plants can economically impact numerous aspects of livestock production, including forage yield and quality, animal health and weight gain, and the quality of meat, milk, wool, and hides. In addition, they can reduce the ecological integrity of rangeland communities by altering fire frequency, increasing erosion, depleting soil moisture and nutrients, and reducing plant biodiversity and wildlife habitat and forage. The impacts of invasive plants on community diversity and structure can lead to ecosystem instability, and often irreversible transformational changes within the system. However, there are also many instances where invasive plants can provide benefits to the ecosystem, and thus there is often a trade-off between negative and positive impacts. As a result, the focus of many restoration efforts need to shift from preserving the historic species assemblages to conserving and maintaining a resilient, functional system that provides diverse ecosystem service, in addition to supporting human livelihoods.

Abiotic and biotic feedback mechanisms can be modified by invasive plants to completely and irreversibly transform historic communities to novel ecosystems with different species composition, ecosystem services and function. This can occur though abiotic changes in climate, land use, pollution and nutrient enrichment, altered disturbance regimes, urbanization, or biotic changes associated with local extinction of keystone species, introduction of invasive ecosystem engineers, or habitat fragmentation. In these novel ecosystems, land managers will need to consider how to feasibly and economically approach long-term management that maximizes a different set of ecosystem functions and services, yet maintains ecological and ecosystem resilience. The primary challenge will be to determine the types of changes that are intrinsically desirable and beneficial, without creating other serious problems or further degrading the system.

As many rangelands have experienced anthropogenic changes characteristic of novel ecosystem functioning, managing invasive species in this new era will require proactive strategies and conceptual frameworks that offer greater success in achieving desired outcomes. In the last 25 years, invasive species management has evolved to incorporate integrated strategies that are guided by both successional theory and process-based manipulations of abiotic and biotic factors. During this time frame, invasive species management has also recognized the need to take a broader ecosystem perspective, which has been facilitated by the development of numerous theoretical frameworks that illustrate how management inputs can best be applied to modify specific ecological processes, at specific stages along the invasion continuum. It has also become clear that proactive invasive species management must adopt preventative control strategies that are known to be more economically feasible for rangeland application.

The role of functional species groups and functional species traits have emerged as valuable predictors of ecosystem functioning and as a way to adopt management strategies that foster greater invasion resistance. In particular, research on functional species groups and functional species traits hold tremendous promise to support assessment of ecosystem susceptibility to invasion and the selection species that offer the best trait-matching to compete with invasive species under specific abiotic stresses.

As invasive rangeland plant management has shifted away from a focus on tools and technology for short-term invasive plant control and toward an emphasis on identifying, managing, and monitoring ecological processes that drive invasion, many conceptual advances have developed that have major implications for understanding constraints and opportunities for rangeland invasive plant management. An emerging opportunity that requires greater attention is a clear definition of the relationship between invasive rangeland plants and ecosystem services, and to identify species and management scenarios where management inputs yield the greatest aggregate return on ecosystem services. A second opportunity centers on bridging the science and practice of invasive plant management by evaluating adoption barriers and assessing impacts of invasive plant management decision tools under actual management scenarios. In addition, as the impacts of invasive plant management on socioeconomic systems are quantified and major adoption barriers are identified, there are large opportunities to link research and extension efforts designed to overcome these socioeconomic barriers to deploying sustainable rangeland invasive plant management programs. Third, while a broad set of stakeholders are vested in invasive plant management and restoration decision-making, different stakeholder groups can have different values and goals, which makes management decision-making challenging. Framing management decision-making under the general concepts of intervention ecology and multiple alternative states, as well as emphasizing ecosystem function and services instead of historical benchmarks, provides managers a greatly improved foundation for developing consensus toward management goals and selecting appropriate tools to achieve desired outcomes.

References

Aslan, C., M. Hufford, R. Epanchin-Niell, J. Port, J. Sexton, et al. 2009. Practical challenges in private stewardship of rangeland ecosystems: Yellow starthistle control in Sierra Nevadan Foothills. *Rangeland Ecology and Management* 62: 28–37.

Bauer, J.T. 2012. Invasive species: "back-seat drivers" of ecosystem change? *Biological Invasions* 14: 1295–1304.

Beckie, H.J., and X. Reboud. 2009. Selecting for weed resistance: Herbicide rotation and mixture. *Weed Technology* 23: 363–370.

Belnap, J., J.A. Ludwig, B.P. Wilcox, J.L. Betancourt, W.R.J. Dean, et al. 2012. Introduced and invasive species in novel rangeland ecosystems: Friends or foes? *Rangeland Ecology and Management* 65: 569–578.

Bestelmeyer, B.T., and D.D. Briske. 2012. Grand challenges for resilience-based management of rangelands. *Rangeland Ecology and Management* 65: 654–663.

Billings, W.D. 1994. Ecological affects of cheatgrass and resultant fire on ecosystems in the western Great Basin. In *Proceedings of ecology and management of annual rangelands*, ed. S.B. Monsen and S.G. Kitchen. Ogden, UT: USDA Forest Service Intermountain Research Station Gen. Tech. Rep. INT-GTR-313.

Black, B.A., C.M. Ruffner, and M.D. Abrams. 2006. Native American influences on the forest composition of the Allegheny Plateau, northwest Pennsylvania. *Canadian Journal of Forest Research* 36: 1266–1275.

Blackburn, T.M., P. Pyšek, S. Bacher, J.T. Carlton, R.P. Duncan, et al. 2011. A proposed unified framework for biological invasions. *Trends in Ecology and Evolution* 26: 333–339.

Bovey, R.W. 1987. Weed control problems, approaches, and opportunities in rangeland. *Review Weed Science* 3: 57–91.

Bowman, D.M.J.S. 1998. The impact of Aboriginal landscape burning on the Australian biota. *New Phytologist* 140: 385–410.

Briske DD. ed. 2011. *Conservation benefits of rangeland practices: Assessment, recommendations and knowledge gaps. USDA NRCS*, 429 pages. Lawrence, Kansas: Allen Press.

Briske, D.D., B.T. Bestelmeyer, T.K. Stringham, and P.L. Shaver. 2008a. Recommendations for development of resilience-based state-and-transition models. *Rangeland Ecology and Management* 61: 359–367.

Briske, D.D., J.D. Derner, J.R. Brown, S.D. Fuhlendorf, W.R. Teague, et al. 2008b. Rotational grazing on rangelands: Reconciliation of perception and experimental evidence. *Rangeland Ecology and Management* 61: 3–17.

Brunson, M.W., and L. Huntsinger. 2008. Ranching as a conservation strategy: Can old ranchers save the new west? *Rangeland Ecology and Management* 61: 137–147.

Brunson, M.W., and J. Tanaka. 2011. Economic and social impacts of wildfires and invasive plants in American deserts: Lessons from the Great Basin. *Rangeland Ecology and Management* 64: 463–470.

Buckley, Y.M., B.M. Bolker, and M. Rees. 2007. Disturbance, invasion and re-invasion: Managing the weed-shaped hole in disturbed ecosystems. *Ecology Letters* 10: 809–817.

Buhler, D.D. 2002. Challenges and opportunities for integrated weed management. *Weed Science* 50: 273–280.

Buhler, D.D., M. Liebman, and J.J. Obrycki. 2000. Theoretical and practical challenges to an IPM approach to weed management. *Weed Science* 48: 274–280.

Catford, J.A., R. Jansson, and C. Nilsson. 2009. Reducing redundancy in invasion ecology by integrating hypotheses into a single theoretical framework. *Diversity and Distributions* 15: 22–40.

Cleland, E.E., L. Larios, and K.N. Suding. 2013. Strengthening invasion filters to reassemble native plant communities: Soil resources and phenological overlap. *Restoration Ecology* 21: 390–398.

Cook, G.D., and L. Dias. 2006. It was no accident: Deliberate plant introductions by Australian government agencies during the 20th century. *Australian Journal of Botany* 54: 601–625.

Corbin, J.D., and C.M. D'Antonio. 2012. Gone but not forgotten? Invasive plants' legacies on community and ecosystem properties. *Invasive Plant Science and Management* 5: 117–124.

Courchamp, F., S. Caut, E. Bonnaud, K. Bourgeois, E. Angulo, et al. 2011. Eradication of alien invasive species: Surprise effects and conservation successes. In *Island invasives: Eradication and management*, ed. C.R. Veitch, M.N. Clout, and D.R. Towns. Gland, Switzerland: IUCN.

Craine, J.M., D. Tilman, D. Wedin, P. Reich, M. Tjoelker, et al. 2002. Functional traits, productivity and effects on nitrogen cycling of 33 grassland species. *Functional Ecology* 16: 563–574.

Crosby, A.W. 1986. *Ecological Imperialism: The biological expansion of Europe, 900–1900*. Cambridge: Cambridge University Press.

Cullen, J.M., and Delfosse, E.S. 1985. *Echium plantagineum*: Catalyst for conflict and change in Australia. In *Proceedings VI international symposium on the biological control of weeds*, Vancouver, Canada, 1984, ed. Delfosse, E.S. Ottawa: Agriculture Canada.

Cunningham, D.C., S.C. Barry, G. Woldendorp, and M.B. Burgess. 2004. A framework for prioritizing sleeper weeds for eradication. *Weed Technology* 18: 1189–1193.

D'Antonio, C.M., and J.C. Chambers. 2006. Using ecological theory to manage or restore ecosystems affected by invasive plant species. In *Foundations of restoration ecology*, ed. D.A. Falk, M.A. Palmer, and J.B. Zedler. Washington D.C.: Island Press.

D'Antonio, C.M., and P.M. Vitousek. 1992. Biological invasions by exotic grasses, the grass/fir cycle, and global change. *Annual Review of Ecology and Systematics* 23: 63–87.

Davis, M.A., J.P. Grime, and K. Thompson. 2000. Fluctuating resources in plant communities: A general theory of invasibility. *Journal of Ecology* 88: 528–534.

DeWitt, J., C.H. Wu, K. Horta, R.G. Bell, and A. Schuren. 2006. Top-down, grassroots, and civic environmentalism: Three ways to protect ecosystems. *Frontiers in Ecology and the Environment* 4: 45–51.

Diaz, S., S. Lavorel, F.D. Bello, F. Quetier, K. Grigulis, et al. 2007. Incorporating plant functional diversity effects in ecosystem service assessments. *Proceedings of the National Academy of Sciences of the United States of America* 104: 20684–20689.

Didier, E.A., and M.W. Brunson. 2004. Adoption of range management innovations by Utah ranchers. *Journal of Range Management* 57: 330–336.

DiTomaso, J.M. 2000. Invasive weeds in rangelands: Species, impacts and management. *Weed Science* 48: 255–265.

DiTomaso, J.M., G.B. Kyser, J.R. Miller, S. Garcia, R.F. Smith, et al. 2006. Integrating prescribed burning and clopyralid for the management of yellow starthistle (*Centaurea solstitialis*). *Weed Science* 54: 757–782.

Drenovsky, R.E., B.J. Grewell, C.M. D'Antonio, J.L. Funk, J.J. James, et al. 2012. A functional trait perspective on plant invasion. *Annals of Botany* 110: 141–153.

Duncan, C.A., J.J. Jachetta, M.L. Brown, V.F. Carrithers, J.K. Clark, et al. 2004. Assessing economic, environmental and societal losses from invasive plants on rangeland and wildlands. *Weed Technology* 18: 1411–1416.

Ellis, E.C., and N. Ramankutty. 2008. Putting people in the map: Anthropogenic biomes of the world. *Frontiers in Ecology and the Environment* 6: 439–447.

Ellis, E.C., E.C. Antill, and H. Kreft. 2012. All is not loss: Plant biodiversity in the Anthropocene. *PLoS One* 7: e30535.

Epanchin-Niell, R.S., and A. Hastings. 2010. Controlling established invaders: Integrating economics and spread dynamics to determine optimal management. *Ecology Letters* 13: 528–541.

Epanchin-Niell, R.S., M.B. Hufford, C.E. Aslan, J.P. Sexton, J.D. Port, et al. 2010. Controlling invasive species in complex social landscapes. *Frontiers in Ecology and the Environment* 8: 210–216.

Eviner, V.T., S.A. Hoskinson, and C.V. Hawkes. 2010. Ecosystem impacts of exotic plants can feed back to increase invasion in western U.S. rangelands. *Rangelands* 31: 21–31.

Eviner, V.T., K. Garbach, J.H. Baty, and S.A. Hoskinson. 2012. Measuring the effects of invasive plants on ecosystem services: Challenges and prospects. *Invasive Plant Science and Management* 5: 125–136.

Fensham, R., J.L. Silcock, and J. Firn. 2014. Managed livestock grazing is compatible with the maintenance of plant diversity in semidesert grasslans. *Ecological Applications* 24: 503–517.

Finnoff, D., J.F. Shogren, B. Leung, and D. Lodge. 2007. Take a risk: Preferring prevention over control of biological invaders. *Ecological Economics* 62: 216–222.

Firn, J. 2007. Developing strategies and methods for rehabilitating degraded pastures using native grasses. *Ecological Management and Restoration* 8: 182–186.

Firn, J., and Y.M. Buckley. 2010. Impacts of invasive plants on Australian rangelands. *Rangelands* 32: 48–51.

Firn, J., P.D. Erskine, and D. Lamb. 2007. Woody species diversity influences productivity and soil nutrient availability in tropical plantations. *Oecologia* 154: 521–533.

Firn, J., J.N. Price, and R.D.B. Whalley. 2013. Using strategically applied grazing to manage invasive alien plants in novel grasslands. *Ecological Processes* 2: 26. http://www.ecologicalprocesses.com/content/2/1/26.

Flory, S.L., and K. Clay. 2009. Invasive plant removal method determines native plant community responses. *Journal of Applied Ecology* 46: 434–442.

Funk, J.L., E.E. Cleland, K.N. Suding, and R.S. Zavaleta. 2008. Restoration through reassembly: Plant traits and invasion resistance. *Trends in Ecology and Evolution* 23: 695–703.

Gammage, B. 2012. *The biggest estate on earth*. Crows Nest, NSW: Allen and Unwin.

Gerlach, J.D. 1997. How the west was lost: Reconstructing the invasion dynamics of yellow starthistle and other plant invaders of western rangelands and natural areas. In *Proc California Exotic Pest Plant Council*.

Grice, A.C., and T.G. Martin. 2006. Rangelands, weeds and biodiversity. *Rangeland Journal* 28: 1–2.

Grime, J.P. 1998. Benefits of plant diversity to ecosystems: Immediate, filter and founder effects. *Journal of Ecology* 86: 902–910.

Hallett, L.M., Standish, R.J., Hulvey, K.B., Gardener, M.R., Suding, K.N., et al. 2013. Towards a conceptual framework for novel ecosystems. In *Novel ecosystems: Intervening in the new ecological world order*, ed. R.J. Hobbs, E.S. Higgs, and C.M. Hall. Chichester, West Sussex: Wiley.

Hastings, A., J.E. Byers, J.A. Crooks, K. Cuddington, C.G. Jones, et al. 2007. Ecosystem engineering in space and time. *Ecology Letters* 10: 153–164.

Hautier, Y., E.W. Seabloom, E.T. Borer, P.B. Adler, W.S. Harpole, et al. 2014. Eutrophication weakens stabilizing effects of diversity in natural grasslands. *Nature* 508: 521–525.

Havstad, K.M., D.P.C. Peters, R. Skaggs, J. Brown, B. Bestelmeyer, et al. 2007. Ecological services to and from rangelands of the United States. *Ecological Economics* 64: 261–268.

Hector, A. 1998. The effect of diversity on productivity: Detecting the role of species complementarity. *Oikos* 82: 597–599.

Hector, A., B. Schmid, C. Beierkuhnlein, M.C. Caldeira, M. Diemer, et al. 1999. Plant diversity and productivity experiments in European grasslands. *Science* 286: 1123–1127.

Hellmann, J.J., J.E. Byers, B.G. Bierwagen, and J.S. Dukes. 2008. Five potential consequences of climate change for invasive species. *Conservation Biology* 22: 534–543.

Hershdorfer, M.E., M.E. Fernandez-Gimenez, and L.D. Howery. 2007. Key attributes influence the performance of local weed management programs in the southwest United States. *Rangeland Ecology and Management* 60: 225–234.

Higgs, E.S. 1997. What is good ecological restoration? *Conservation Biology* 11: 338–348.

Hobbs, R.J. 2007. Setting effective and realistic restoration goals: Key directions for future research. *Restoration Ecology* 15: 354–357.

Hobbs, R.J., and S.E. Humphries. 1995. An integrated approach to the ecology and management of plant invasions. *Conservation Biology* 9: 761–770.

Hobbs, R.J., and D.A. Norton. 1996. Towards a conceptual framework for restoration ecology. *Restoration Ecology* 4: 93–110.

Hobbs, R.J., S. Arico, J. Aronson, J.S. Baron, P. Bridgewater, et al. 2006. Novel ecosystems: Theoretical and management aspects of the new ecological world order. *Global Ecology and Biogeography* 15: 1–7.

Hobbs, R.J., E. Higgs, and J.A. Harris. 2009. Novel ecosystems: Implications for conservation and restoration. *Trends in Ecology and Evolution* 24: 559–605.

Hobbs, R.J., L.M. Hallett, P.R. Ehrlich, and H.A. Mooney. 2011. Intervention ecology: Applying ecological science in the twenty-first century. *BioScience* 61: 442–450.

Holt, J.S. 1994. Impact of weed-control on weeds—new problems and research needs. *Weed Technology* 8: 400–402.

Hulme, P.E. 2006. Beyond control: Wider implications for the management of biological invasions. *Journal of Applied Ecology* 43: 835–847.

Isbell, F., V. Calcagno, A. Hector, J. Connolly, W.S. Harpole, et al. 2011. High plant diversity is needed to maintain ecosystem services. *Nature* 477: 199–202.

Jackson, S.T., and R.J. Hobbs. 2009. Ecological restoration in the light of ecological history. *Science* 325: 567–569.

James, J.J., B.S. Smith, E.A. Vasquez, and R.L. Sheley. 2010. Principles for ecologically based invasive plant management. *Invasive Plant Science and Management* 3: 229–239.

Januchowski-Hartley, S.R., P. Visconti, and R.L. Pressey. 2011. A systematic approach for prioritizing multiple management actions for invasive species. *Biological Invasions* 13: 1241–1253.

Jeschke, J.M., S. Bacher, T.M. Blackburn, J.T.A. Dick, F. Essl, et al. 2014. Defining the impact of non-native species. *Conservation Biology* 28: 1188–1194.

Johnson, K.H., K.A. Vogt, H.J. Clark, O.J. Schmitz, and D.J. Vogt. 1996. Biodiversity and the productivity and stability of ecosystems. *Trends in Ecology and Evolution* 11: 372–377.

Jones, T.A. 2013. Ecologically appropriate plant materials for restoration applications. *BioScience* 63: 211–219.

Jones, T.A., T.A. Monaco, and J.J. James. 2010. Launching the counterattack: Interdisciplinary deployment of native-plant functional traits for repair of rangelands dominated by invasive annual grasses. *Rangelands* 32: 38–42.

Kaiser, J. 2000. Rift over biodiversity divides ecologists. *Science* 289: 1282–1283.

Kartesz, J. 2010. A synonymized checklist and atlas with biological attributes for the vascular flora of the United States, Canada, and Greenland. In *Floristic synthesis of North America. CD-ROM version 2.0.*, 2nd edn. Chapel Hill: North Carolina Botanical Garden.

Kettenring, K.M., and C.R. Adams. 2011. Lessons learned from invasive plant control experiments: A systematic review and meta-analysis. *Journal of Applied Ecology* 48: 970–979.

Koontz, T.M., and J. Bodine. 2008. Implementing ecosystem management in public agencies: Lessons from the US Bureau of Land Management and the Forest Service. *Conservation Biology* 22: 60–69.

Kulmatiski, A., K.H. Beard, and J.M. Stark. 2006. Exotic plant communities shift water-use timing in a shrub-steppe ecosystem. *Plant and Soil* 288: 271–284.

Lamb, D., and D. Gilmour. 2003. *Rehabilitation and restoration of degraded forests*. Gland Switzerland and Cambridge, UK: IUCN and WWF.

Lamb, D., P. Erskine, and J. Parotta. 2005. Restoration of degraded tropical forest landscapes. *Science* 310: 1628–1632.

Lavorel, S., and E. Garnier. 2002. Predicting changes in community composition and ecosystem functioning from plant traits: Revisiting the Holy Grail. *Functional Ecology* 16: 545–556.

Lockwood, J.L., P. Cassey, and T. Blackburn. 2005. The role of propagule pressure in explaining species invasions. *Trends in Ecology and Evolution* 20: 223–228.

Lonsdale, W.M. 1994. Inviting trouble: Introduced pasture species in northern Australia. *Australian Journal of Ecology* 19: 345–354.

Low, T. 1997. Tropical pasture plants as weeds. *Tropical Grasslands* 31: 337–343.

Lubell, M.N., B.B. Cutts, L.M. Roche, M. Hamilton, J.D. Derner, et al. 2013. Conservation program participation and adaptive rangeland decision-making. *Rangeland Ecology and Management* 66: 609–620.

Lunt, I.D., D.J. Eldridge, J.W. Morgan, and G.B. Witt. 2007. A framework to predict the effects of livestock grazing and grazing exclusion on conservation values in natural ecosystems in Australia. *Australian Journal of Botany* 55: 401–415.

Mack, R.N., D. Simberloff, W.M. Lonsdale, H. Evans, M. Clout, et al. 2000. Biotic invasions: Causes, epidemiology, global consequences, and control. *Ecological Applications* 3: 689–710.

Mascaro, J., R.F. Hughes, and S.A. Schnitzer. 2012. Novel forests maintain ecosystem processes after the decline of native tree species. *Ecological Monographs* 82: 221–238.

Masters, R.A., and R.L. Sheley. 2001. Principles and practices for managing rangeland invasive plants. *Journal of Range Management* 54: 502–517.

McCann, K.S. 2000. The diversity-stability debate. *Nature* 405: 228–233.

McCarthy, M.A., and H.P. Possingham. 2007. Active adaptive management for conservation. *Conservation Biology* 21: 956–963.

Milchunas, D.G., O.E. Sala, and W.K. Lauenroth. 1988. A generalized-model of the effects of grazing by large herbivores on grassland community structure. *The American Naturalist* 132: 87–106.

Moles, A.T., M.A.M. Gruber, and S.P. Bonser. 2008. A new framework for predicting invasive plant species. *Journal of Ecology* 96: 13–17.

Monaco, T.A., T.A. Jones, and T.L. Thurow. 2012. Identifying rangeland restoration targets: An appraisal of challenges and opportunities. *Rangeland Ecology and Management* 65: 599–605.

Murcia, C., J. Aronson, G.H. Kattan, D. Moreno-Mateos, K. Dixon, and D. Simberloff. 2014. A critique of the 'novel ecosystem' concept. *Trends in Ecology and Evolution* 29: 548–553.

Murphy, D.D., and P.R. Ehrlich. 1989. Conservation biology of California's remnant native grasslands. In *Grassland structure and function: The California annual grassland*, ed. L.F. Huenneke and H.A. Mooney. Dordrecht, The Netherlands: Kluwer.

Naeem, S., and J.P. Wright. 2003. Disentangling biodiversity effects on ecosystem functioning: Deriving solutions to a seemingly insurmountable problem. *Ecology Letters* 6: 567–579.

Naylor, R.L. 2000. The economics of alien species invasions. In *Invasive species in a changing world*, ed. H.A. Mooney and R.J. Hobbs. Washington, D.C.: Island Press.

Panetta, F.D. 2009. Weed eradication—an economic perspective. *Invasive Plant Science and Management* 2: 360–368.

Parker, I.M., D. Simberloff, W.M. Lonsdale, K. Goodell, M. Wonham, et al. 1999. Impact: Toward a framework for understanding the ecological effects of invaders. *Biological Invasions* 1: 3–19.

Parsons, W.T., and E.G. Cuthbertson. 2001. *Noxious weeds of Australia*, 2nd ed. Collingwood, VIC, Australia: CSIRO Publ.

Perring, M.P., R.L. Standish, and R.J. Hobbs. 2013. Incorporating novelty and novel ecosystems into restoration planning and practice in the 21st century. *Ecological Processes* 2: 18.

Pfisterer, A.B., and B. Schmid. 2002. Diversity-dependent production can decrease the stability of ecosystem functioning. *Nature* 416: 84–86.

Pickett, S.T.A., S.L. Collins, and J.J. Armesto. 1987. Models, mechanisms and pathways of duccession. *The Botanical Review* 53: 335–371.

Plieninger, T., S. Ferranto, L. Huntsinger, M. Kelly, and C. Getz. 2012. Appreciation, use, and management of biodiversity and ecosystem services in California's working landscapes. *Journal Environmental Management* 50: 427–440.

Pocock, M.J.O., D.M. Evans, and J. Memmott. 2012. The robustness and restoration of a network of ecological networks. *Science* 335: 973–977.

Pokorny, M.L., R.L. Sheley, C.A. Zabinski, R.E. Engel, T.J. Svejcar, et al. 2005. Plant functional diversity as a mechanism for invasion resistance. *Restoration Ecology* 13: 448–459.

Pyšek, P., D.M. Richardson, and M. Williamson. 2004. Predicting and explaining plant invasions through analysis of source area floras: Some critical considerations. *Diversity and Distributions* 10: 179–197.

Quetier, F., A. Thebault, and S. Lavorel. 2007. Plant traits in a state and transition framework as markers of ecosystem response to land-use change. *Ecological Monographs* 77: 33–52.

Quimby Jr., P.C., W.L. Bruckart, C.J. DeLoach, L. Knutson, and M.H. Ralphs. 1991. Biological control of rangeland weeds. In *Noxious range weeds*, ed. L.F. James, J.O. Evans, M.H. Ralphs, and R.D. Child. San Francisco: Westview Press.

Richardson, D.M., and M. Gaertner. 2013. Plant invasions as builders and shapers of novel ecosystems. In *Novel ecosystems: Intervening in the new ecological world order*, ed. R.J. Hobbs, E.C. Higgs, and C.M. Hall. Oxford: Wiley.

Rinella, M.J., B.D. Maxwell, P.K. Fay, T. Weaver, and R.L. Sheley. 2009. Control effort exacerbates invasive-species problem. *Ecological Applications* 19: 155–162.

Rudzitis, G. 1999. Amenities increasingly draw people to the rural west. *Rural Development Perspectives* 14: 9–13.

Seastedt, T.R., R.J. Hobbs, and K.N. Suding. 2008. Management of novel ecosystems: Are novel approaches required? *Frontiers in Ecology and the Environment* 6: 547–553.

Sheley, R.L., and J. Krueger-Mangold. 2003. Principles for restoring invasive plant-infested rangeland. *Weed Science* 51: 260–265.

Sheley, R.L., T.J. Svejcar, and B.D. Maxwell. 1996. A theoretical framework for developing successional weed management strategies on rangeland. *Weed Technology* 10: 766–773.

Sheley, R.L., J.M. Mangold, and J.L. Anderson. 2006. Potential for successional theory to guide restoration of invasive-plant-dominated rangeland. *Ecological Monographs* 76: 365–379.

Sheley, R., J.J. James, M.J. Rinella, D.M. Blumenthal, and J.M. DiTomaso. 2011a. A Scientific assessment of invasive plant management on anticipated conservation benefits. In *Conservation Benefits of Rangeland Practices: Assessment, Recommendations, and Knowledge Gaps. USDA NRCS*, ed. D.D. Briske. Lawrence, Kansas: Allen Press.

Sheley, R.L., J.J. James, E.A. Vasquez, and T.J. Svejcar. 2011b. Using rangeland health assessment to inform successional management. *Invasive Plant Science and Management* 4: 356–367.

Simberloff, D., J.L. Martin, P. Genovesi, V. Maris, D.A. Wardle, et al. 2013. Impacts of biological invasions: What's what and the way forward. *Trends in Ecology and Evolution* 28: 58–66.

Steffen, W., A. Sanderson, P.D. Tyson, J. Jager, P.A. Matson, et al. 2004. *Global change and the earth system: A planet under pressure*. New York: Springer.

Strayer, D.L., V.T. Eviner, J.M. Jeschke, and M.L. Pace. 2006. Understanding the long-term effects of species invasions. *Trends in Ecology and Evolution* 21: 645–651.

Suding, K.N., K.L. Gross, and G.R. Houseman. 2004. Alternative states and positive feedbacks in restoration ecology. *Trends in Ecology and Evolution* 19: 46–53.

Suding, K.N., S. Lavorel, F.S. Chapin, J.H.C. Cornelissen, S. Diaz, et al. 2008. Scaling environmental change through the community-level: A trait-based response-and-effect framework for plants. *Global Change Biology* 14: 1125–1140.

Sutherland, W.J., A.S. Pullin, P.M. Dolman, and T.M. Knight. 2004. The need for evidence-based conservation. *Trends in Ecology and Evolution* 19: 305–308.

Tanaka, J.A., M.W. Brunson, and L.A. Torell. 2011. A social and economic assessment of rangeland conservation practices. In *Conservation benefits of rangeland practices: Assessment, recommendations, and knowledge gaps. USDA NRCS*, ed. D.D. Briske. Lawrence, Kansas: Allen Press.

Tilman, D., C. Lehman, and K. Thomson. 1997. Plant diversity and ecosystem productivity: Theoretical considerations. *Proceedings of the National Academy of Sciences of the United States of America* 94: 1857–1861.

US Bureau of Land Management. 1999. *Out of ashes, an opportunity*. Boise, ID: National Interagency Fire Center.

US Department of Agriculture (USDA). 2010. *National resources inventory rangeland resource assessment, natural resources conservation service*. Washington, DC: US Department of Agriculture (USDA). http://www.nrcs.usda.gov/Internet/FSE_DOCUMENTS/stelprdb1041751.pdf.

Valéry, L., H. Fritz, J.C. Lefeuvre, and D. Simberloff. 2008. In search of a real definition of the biological invasion phenomenon itself. *Biological Invasions* 10: 1345–1351.

Vilà, M., C. Basnou, P. Pyšek, M. Josefsson, P. Genovesi, et al. 2010. How well do we understand the impacts of alien species on ecosystem services? A pan-European, cross-taxa assessment. *Frontiers in Ecology and the Environment* 8: 135–144.

Walker, B.H. 1992. Biodiversity and ecological redundancy. *Conservation Biology* 6: 18–23.

Walker, B., S. Carpenter, J. Anderies, N. Abel, C.S. Cumming, et al. 2002. Resilience management in social-ecological systems: A working hypothesis for a participatory approach. *Conservation Ecology* 6: 14.

Wardle, D.A., O. Zackrisson, G. Hornberg, and G. Christiane. 1997. The influence of island area on ecosystem properties. *Science* 277: 1296–1299.

Westbrooks, R. 1998. *Invasive plants, changing the landscape of America: Fact book*. Washington, DC: Federal Interagency Committee for the Management of Noxious and Exotic Weeds (FICMNEW).

Westoby, M., B. Walker, and I. Noy-Meir. 1989. Opportunistic management for rangelands not at equilibrium. *Journal of Range Management* 42: 266–274.

Whisenant, S.G. 1990. Changing fire frequencies on Idaho's Snake River Plains: Ecological and management implications. In *Proceedings symposium on cheatgrass invasion, shrub die-off, and other aspects of Shrub Biology and Management*, eds. E.D. McArthur, E.M. Romney, S.D. Smith, and P.T. Tueller. Ogden, UT: USDA Forest Service Intermountain Research Station Gen. Tech. Rept. INT-GTR-313.

Whisenant, S.G. 1999. *Repairing damaged wildlands: A process-orientated, landscape-scale approach Cambridge*. UK: Cambridge University Press.

Wilen, J.E. 2007. Economics of spatial-dynamic processes. *American Journal of Agricultural Economics* 89: 1134–1144.

Williams, J.W., and S.T. Jackson. 2007. Novel climates, no-analog communities, and ecological surprises. *Frontiers in Ecology and the Environment* 5: 475–482.

Wilson, E.O. 1992. *The diversity of life*. Cambridge, Massachusetts: Belknap Press of Harvard University Press.

Young, J.A., and R.A. Evans. 1979. Barbwire Russian thistle seed germination. *Journal of Range Management* 32: 390–394.

Young, T.P., D.A. Petersen, and J.J. Clary. 2005. The ecology of restoration: Historical links, emerging issues and unexplored realms. *Ecology Letters* 8: 662–673.

Zavaleta, E.S., R.J. Hobbs, and H.A. Mooney. 2001. Viewing invasive species removal in a whole-ecosystem context. *Trends in Ecology and Evolution* 16: 454–459.

Chapter 14
Rangeland Ecosystem Services: Nature's Supply and Humans' Demand

Osvaldo E. Sala, Laura Yahdjian, Kris Havstad, and Martín R. Aguiar

Abstract Ecosystem services are the benefits that society receives from nature, including the regulation of climate, the pollination of crops, the provisioning of intellectual inspiration and recreational environment, as well as many essential goods such as food, fiber, and wood. Rangeland ecosystem services are often valued differently by different stakeholders interested in livestock production, water quality and quantity, biodiversity conservation, or carbon sequestration. The supply of ecosystem services depends on biophysical conditions and land-use history, and their availability is assessed using surveys of soils, plants, and animals. The demand for ecosystem services depends on educational level, income, and location of residence of social beneficiaries. The demand can be assessed through stakeholder interviews, questionnaires, and surveys. Rangeland management affects the supply of different ecosystem services by producing interactions among them. Trade-offs result when an increase in one service is associated with a decline in another, and win–win situations occur when an increase in one service is associated with an increase in other services. This chapter provides a conceptual framework in which range management decisions are seen as a challenge of reconciling supply and demand of ecosystem services.

Keywords Provisioning • Regulation • Cultural • Trade-offs • Win–win • Stakeholders • Human well-being

O.E. Sala (✉)
School of Life Sciences and School of Sustainability, Arizona State University,
Tempe, AZ, USA
e-mail: Osvaldo.Sala@asu.edu

L. Yahdjian • M.R. Aguiar
Faculty of Agronomy, Institute for Agricultural Plant Physiology and Ecology (IFEVA),
University of Buenos Aires and CONICET, Buenos Aires, Argentina

K. Havstad
USDA-ARS, Jornada Experimental Range, New Mexico State University,
Las Cruces, NM, USA

© The Author(s) 2017 467
D.D. Briske (ed.), *Rangeland Systems*, Springer Series on Environmental
Management, DOI 10.1007/978-3-319-46709-2_14

14.1   Introduction

Ecosystem services are the benefits that society receives from nature (Daily 1997; MA 2005). They include the provisioning of food, wood and medicinal resources, and services that contribute to climate stability, control of agricultural pests, and purification of air and water (Fig. 14.1). Ecosystem services are broadly classified in four different categories: provisioning, regulating, cultural, and supporting (MA 2005). Provisioning ecosystem services include the contribution of essential goods such as food, fiber, and medicinal. Regulating ecosystem services include carbon sequestration, prevention of soil erosion, and natural flood control. Cultural ecosystem services include intellectual, inspirational, and recreational activities. The fourth category is supporting ecosystem services, which include services that are dependent on ecological processes such as primary production and nutrient cycling and that are intimately related to biological diversity.

Since its conceptualization, the focus of ecosystem services has changed from the description of the processes involved in delivery of a single service at a point in time (Daily 1997) to approaches for analyzing the capacity of nature to produce multiple ecosystem services. The next steps have been assessing multiple ecosystem services

Ecosystem Services

Supporting

-Soil formation
-Biodiversity
-Primary production
-Habitat

Provisioning
-Food and fiber
-Wood
-Clean Water
-Medicinals

Regulating
-Climate Regulation
-Pollination of crops
-Store carbon
-Control flooding

Cultural
-Inspiration
-Recreation
-Education
-Aesthetic

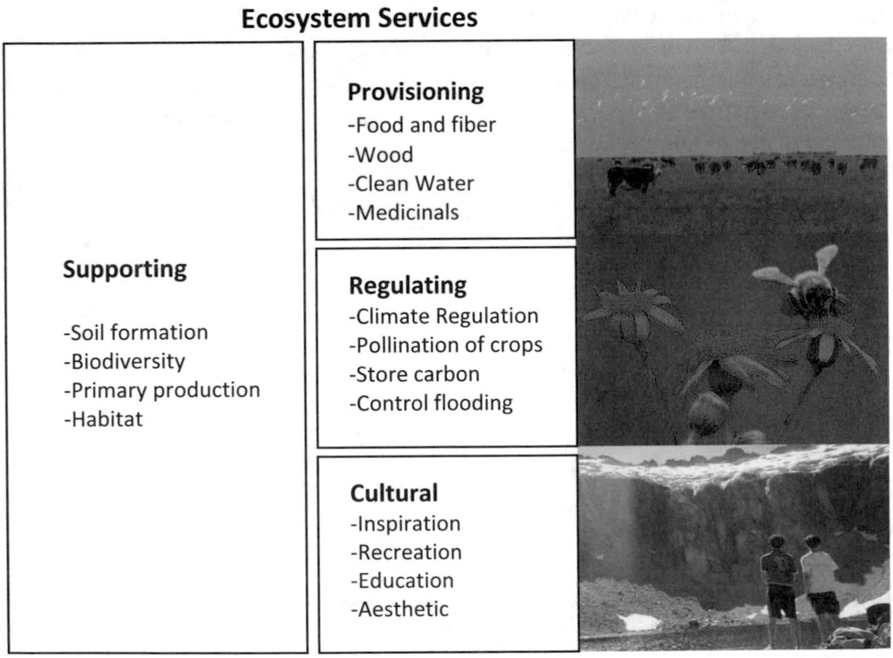

Fig. 14.1 Four categories of ecosystem services as classified by the Millennium Ecosystem Assessment (MA 2005). Photo credits (from *top*): Laura Yahdjian, Magdalena Druille, Felipe Cabrera

under alternative land-use regimes (Foley et al. 2005). Management aimed at increasing the supply of one specific ecosystem service may increase or decrease the supply of others creating trade-offs and win–win situations, respectively.

Rangelands, the land on which the potential native vegetation is predominately grasses, grasslike plants, forbs, or shrubs (Kauffman and Pyke 2001; Chap. 1, this volume), encompass hot and cold deserts, grasslands, savannas, and woodlands. They occupy approximately 54 % of terrestrial ecosystems, and they sustain 30 % of world population, including a myriad of stakeholders (Reynolds et al. 2007; Estell et al. 2012). Rangelands produce a great variety of ecosystem services but only few of them have market value (Sala and Paruelo 1997). For example, commodities produced by rangelands such as meat or wool have market value but other ecosystem services such as regulating, cultural, and supporting services mostly do not have a market value although it is possible to estimate it indirectly.

The science of ecosystem services has grown exponentially as indicated by the number of academic publications per year on this topic, from a few per year in the 1980s to a wealth of papers in the last decade (Fig. 14.2). The total number of peer-reviewed publications addressing ecosystem services exceeded 1200 and entire books and journals have been devoted to this topic. The Millennium Ecosystem Assessment was developed around this concept showing the enormous impact of the ecosystem

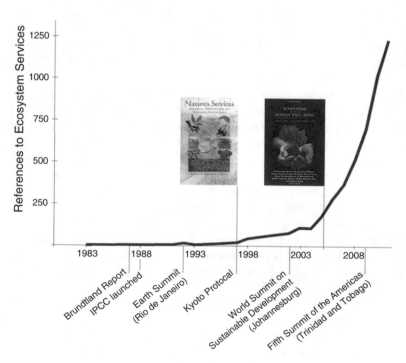

Fig. 14.2 Number of scientific publications emphasizing ecosystem services during the period 1963–2012 (modified from Rositano et al. 2012)

service conceptual framework (MA 2005). The most common approach to study eco-system services has been related to the assessment of the capacity of ecosystems to deliver services, i.e., the supply of ecosystem services. Ecological production functions define how the spatial distribution of ecosystem structure and functioning determine the delivery of ecosystem services (Daily 1997). Even when ecosystem services have been defined as the outcome of ecosystem processes desired by people, the main focus of ecosystem service research has been to identify the potential of a region to produce ecosystem services independently of whether people are demanding them or not. In addition, the economic valuation of ecosystem services has also been a frequent target of research (Costanza et al. 1997; Gomez-Baggethun et al. 2010). In summary, the science of ecosystem services has developed rapidly in the past decades, but it has focused primarily on the supply of services and has largely overlooked the human demand for ecosystem services only until recently (Yahdjian et al. 2015).

Human demand represents the other side of an ecosystem service equation of supply and demand, which is related to the social beneficiaries. Human consumption of resources and utilization of services that are supplied by ecosystems depend upon both their capacity to produce them and the societal value and need placed on those resources and services (Tallis and Polasky 2011). Demand for ecosystem services changes among stakeholders or social beneficiaries, who are the individuals or groups of individuals who have an interest in ecosystem services because they get a profit from them and could have an active or passive influence on their delivery (Lamarque et al. 2011). Stakeholders not only exhibit different demands, but they also have different valuations of various ecosystem services. Indeed, an ecosystem service is not a universally applicable physical phenomenon, but one whose value is shaped by its users. Sustainable land management depends on reconciling supply and demand for ecosystem services by different stakeholders.

Rangelands are ideal for analyzing the balance between supply and demand for different types of services because of the variety of ecosystem services that they provide and the diverse suite of stakeholders interested in different services. In contrast, hyperarid ecosystems provide supporting, cultural, and regulating services but few provisioning services. Similarly humid ecosystems are generally transformed into crop- and wood-production systems, or are subject to human commercial, residential, and industrial development at the expense of cultural, provisioning, and regulating services. In addition, rangelands are broadly threatened by land degradation and climate change (Herrick et al. 2013). In general, rangelands produce abundant ecosystem services in quantity and variety, but the large value and threat of degradation contrast with the fact that humans usually assign small value to them, particularly when compared with tropical or temperate forests (Martin-Lopez et al. 2012).

The transformation of rangeland ecosystems into croplands is constrained by biophysical conditions and economic feasibility (Havstad et al. 2007). For example, mesic rangelands have been converted to agriculture land, while arid and semiarid rangelands continue to be used as grazing lands, with investments in domestic animals, veterinary and reproductive management, fences, and water points that in combination result in a significant increase in livestock production (Oesterheld et al. 1992).

In this chapter, we describe the (1) main ecosystem services provided by rangelands and the major categories of social beneficiaries and (2) most common methods used to estimate supply and demand of these services, and (3) analyze the determinants of supply and demand of ecosystem services and discuss the existence of trade-offs and win–win conditions in the provision of services. Finally, we provide a new conceptual framework for the management of rangelands that is based on reconciling supply and demand of ecosystem services. This framework is dependent on place, time, and the specific valuation that each stakeholder has of specific ecosystem service. The framework recognizes that both supply and demand of ecosystem services change in space and time and are strongly influenced by land management decisions.

14.2   Categories of Rangeland Ecosystem Services

In this section, we describe the main ecosystem services provided by rangelands in each of the four categories of ecosystem services as defined by the Millennium Assessment (MA 2005). We then analyze the balance between supply and demands for each type of ecosystem service.

Provisioning services are the products obtained from ecosystems that can be directly harvested, and, in general, have a market value such as food, fiber, fuel, and freshwater. The main goods produced in rangelands are freshwater for drinking and irrigation; forage to produce meat, milk, wool, and leather; and medicinal products (Sala and Paruelo 1997). What frequently drives the demand for provisioning services is the immediate need of humans for particular plant or animal species, including production of desirable forage species and the harvest of wild game (Perrings et al. 2011). The relationship between supply and demand for these products changes among regions and among the specific provisioning services. At a global scale, the demand for provisioning services in rangelands is higher than the supply, but at local scales, supply may exceed demands (Yahdjian et al. 2015). In the case of water for irrigation, water is required during specific periods when water is scarce, so supply and demand may be spatially or temporally disconnected, which is particularly important since most rangelands are water limited. As such, for provisioning services, the demand surpasses the supply, which is particularly evident for freshwater and food (Yahdjian et al. 2015). The supply of provisioning ecosystem services changes in time at different scales. The supply of meat and wool fluctuates with seasons and production systems, but also changes at decadal timescales as a result of land degradation and market fluctuations (Texeira and Paruelo 2006). Supply of provisioning ecosystem services changes in space over multiple scales, from differences among locations within a specific community to variation along regional precipitation gradients (Adler et al. 2005). Finally, the demand for provisioning services changes among beneficiaries depending on their income, education, and urban versus rural residence (Yahdjian et al. 2015).

Regulating services are the benefits humans receive from regulating ecosystem processes, such as climate regulation, air quality maintenance, water purification, and erosion control. Rangelands sequester large quantities of carbon, principally into the soil, and avoid carbon losses to the atmosphere that would occur if rangelands were to be transformed into croplands or severely degraded (Sala and Paruelo 1997). In the case of carbon sequestration, demand is higher than the supply because this process cannot offset actual carbon emissions from human activities (Tallis and Polasky 2011). Every unit of sequestered greenhouse gas emitted will allow us to minimize environmental and economic damage that would have occurred otherwise. The whole world benefits from a unit of carbon sequestration regardless of where it occurs because greenhouse gases thoroughly mix in the global atmosphere. Carbon sequestration in rangelands is important because of the area that rangelands occupy although per unit area carbon storage is lower than other ecosystems, such as wetlands and forests (Reynolds et al. 2007). Not only do rangelands account for a significant fraction of the global carbon cycle, but they also account for most of the interannual variability in the global carbon sink (Ahlström et al. 2015).

Cultural services are the nonmaterial benefits that humans obtain from ecosystems and they include cultural diversity, spiritual, and religious values, knowledge systems, and recreation. They involve consumptive and nonconsumptive services. Cultural services in rangelands are related to human experiences associated with activities such as wild game hunting, traditional lifestyles, and tourist ranching experiences. The demand for cultural services changes according to the region analyzed (Tallis and Polasky 2011) and has changed over time. For example, in the southwestern USA, the Bureau of Land Management, who administers a large fraction of federal lands in the region, reported an increase in the number of visitors to their lands from 20 to 45 M per year for the 2000–2010 period (Yahdjian et al. 2015). Similarly, the National Park Service reported for the same period an annual increase of 15 M visitors from 35 to 50 M per year.

Supporting services are those that are necessary for the production of all other ecosystem services such as processes that maintain biodiversity to produce goods or cycle nutrients (MA 2005). In rangelands, supporting services are primary production, nutrient cycling, conservation of soils, and biodiversity, which represent a large storehouse of genetic, species, and functional diversity. Rangelands represent the natural ecosystem where annual grasses and legumes are most abundant and from where a large fraction of domesticated species originated (Sala and Paruelo 1997). The key to sustaining biodiversity is harmonizing its protection with the delivery of as many other ecosystem services as possible. Land degradation, which in most cases results from overgrazing, weed invasions, energy extraction, and exurban development, directly affects the provision of supporting services. Arguably, rangeland degradation has a larger and more imminent impact than climate change on the ability of these systems to fulfill human needs (Herrick et al. 2013). At the global scale, the supply of supporting services is higher than the demand, but human use does not directly apply since, by definition, supporting services are not directly used by people, even when they influence the supply of provisioning, regulating, and cultural services.

14.3 Social Beneficiaries of Rangeland Ecosystem Services

In the same way that ecosystem services are classified, the social beneficiaries of services may be classified in categories according to the particular ecosystem services they use. Beneficiaries of ecosystem services are individuals, commercial entities, and the public sector and they may be distributed across local, regional, national, and global scales (Table 14.1). The demand for ecosystem services is complex and the classification of service beneficiaries, who often vary in their ecosystem-service preferences, can be a useful tool for identifying potential trade-offs and for balancing multiple, often conflicting, demands for services. If people's preferences for two or more services are known and they can be expressed accurately in the same units of value, then making the trade-off decision is (at least conceptually) straightforward and involves a simple cost–benefit calculation (Carpenter et al. 2009). Rangeland managers face the need to manage multiple ecosystem services and their interactions (Raudsepp-Hearne et al. 2010) and the demands of multiple beneficiaries (Yahdjian et al. 2015).

The supply and demand of ecosystem services occur at different spatial scales. Some ecosystem services are very local (pollination service, cultural services) whereas others are global (sequestration of greenhouse gases, air and water purification). The different scales involved in the provision of ecosystem services raise the possibility of a mismatch between supply and demand. Mismatches may also occur between those who control the provision of ecosystem services (supply) and those who benefit from them (users).

The main beneficiaries in rangelands are ranchers, land-owner producers, land tenants, service providers, recreational hunters, conservationists, landscape planners, passive and active nature tourists, and government and nongovernmental organizations (Scheffer et al. 2000; Castro et al. 2011; Yahdjian et al. 2015). While ranchers historically have demanded mainly provisioning services, their demands have broadened (Brown and McDonald 1995) while tourists and conservationists classically have demanded more supporting and cultural services. It is important here to further highlight that ranchers vary enormously in their demand

Table 14.1 Potential beneficiaries of ecosystem services across different spatial scales

Spatial scale			
Stakeholders	Local	National/regional	Global
Individual	Hunter/gatherer, subsistence farmers, and tourists	Tourists, consumers, educators, and students	Tourists, consumers, educators, and students
Commercial entity	Local entrepreneurs, farmers, traders, and artisans	Regional economic organizations	International enterprise including fishery and forestry industries
Public sector	Local government	National and regional government	International community

Rows represent stakeholders and *columns* represent various spatial scales at which stakeholders interact with rangeland ecosystems (modified from Newcome et al. (2005))

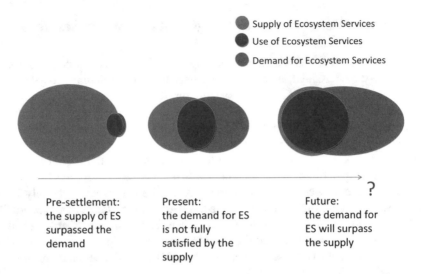

Fig. 14.3 Supply and demand for ecosystem services following European settlement (modified from data in Carpenter et al. (2009))

for ecosystem services. Similarly, people living in urban centers demand clean air and water that are provided by adjacent rural areas. The contrasting demands of different beneficiaries influence the analysis of land-management actions and their consequences on different ecosystem services.

The demand for ecosystem services in rangelands has diversified in the past decades, from mainly provisioning services to an increasing demand for more diverse services including regulating and cultural services (Yahdjian et al. 2015). The balance between supply and demand has also changed greatly from the time of European settlement in North America (Fig. 14.3). The ability of ecosystems to produce services is declining and the demand for them is increasing, with serious implications for both people and the environment. However, we have not developed sufficient knowledge to quantify and model the demand for ecosystem services as we have for their supply. So, the question remains, which category of ecosystem services will have greatest demand in the future? How provision of and demand for ecosystem services will be balanced in rangelands? Which trade-offs among ecosystems services will be most important in the future? Who will be responsible for making these decisions?

14.4 Methods for Estimating Supply and Demand

Different tools and models have been developed to assess the production of ecosystem services, including the valuation of market and nonmarket services, in both economic and noneconomic terms. The combination of ecological production functions and economic valuation describes the monetary value of ecosystem services. Recently, the Natural Capital Project has developed a tool to integrate biophysical and economic information on ecosystem services (Tallis et al. 2011).

The demand for ecosystem services is related to the social beneficiaries and is usually described by the location, type, and intensity of people's demand for services. The demand has been evaluated focusing on the perception of ecosystem services by different stakeholders (De Chazal et al. 2008; Quétier et al. 2010; Martin-Lopez et al. 2012). Preferences have been assessed by compiling responses to questionnaires and interviews (Lamarque et al. 2011; Martin-Lopez et al. 2012). During social surveys, ecosystem services are identified spontaneously, and the more "visible" services, such as recreation, aesthetic, and natural hazard regulation, are commonly described. Other questionnaires request that people rank ecosystem services according to their preferences. During the ranking exercises more "invisible" services, such as pollination and soil fertility, often emerge (Lamarque et al. 2011; Martin-Lopez et al. 2012). Finally, the traditional surveys formally used to value nonmarket ecosystem services, such as the willingness to pay for conservation of certain resources or the existence value, may also be included in studies of demands.

The main drivers associated with people preferences for ecosystem services were monthly income, level of education (from traditional ecological knowledge to formal education), and place of residence (the rural–urban continuum; Yahdjian et al. 2015). In addition, other social variables like age, gender, culture, and geographical location were also associated with the interest that people have in ecosystem services (MA 2005).

The relationship between the supply of ecosystem services and the demand for them determines the actual use of ecosystem services by society (Tallis and Polasky 2011). Food production per hectare or the amount of clean water used for irrigation are examples of estimates of the use of provisioning ecosystem services. When global analyses are implemented, remote drivers and teleconnections, such as international trade practices and agreements, have to be taken into account. Trade patterns, which can be dynamic and quite nuanced, show how demand for certain services in one country leads to changes in the provisioning services in other countries.

14.5   Trade-Offs and Win–Win Interactions

There are cases of synergistic and antagonistic interactions among different types of ecosystem services. Synergistic interactions, or win–win conditions, indicate that management leading to the increase of one type of ecosystem service may result in the increase of other ecosystem services. For example, some ecosystem services respond similarly to specific management practices and ecological conditions, such as those that may lead to increased carbon sequestration then resulting in increased water holding capacity, and, many of them, such as cultural services and biodiversity conservation, produce multiple intertwined values (Bennett et al. 2009).

Antagonistic interactions, or trade-offs, indicate that management practices or events that increase one type of ecosystem service may negatively affect other ecosystem services (Oñatibia et al. 2015). For example, land management practices that lead to increases in the provisioning of food may result in a reduction of clean water purification, creating trade-offs in the provisioning of ecosystem services (Raudsepp-Hearne et al. 2010). Planting trees to increase carbon sequestration or timber production may decrease stream flow in arid areas and represents a trade-off (Nosetto et al. 2008). In summary, ecosystem service research has advanced to identify nature as a complex provider of human benefits (MA 2005).

The rangelands of Patagonia provide an example of trade-offs and win–win relationships among ecosystem services depending on management. An example of win–win is the maximization of carbon, nitrogen, and forage availability at intermediate grazing intensities (Fig. 14.4). A critical provision service, such as forage biomass, is

Fig. 14.4 Example of a win–win interaction between a supportive ecosystem service, carbon and nitrogen stocks, and a provisioning ecosystem service forage production as depicted by the complementary relationships between carbon (C) (**a**) and nitrogen (N) (**b**) in forage of a Patagonian rangeland. Paddocks are used with different stocking rates. Exclosure (Exc) includes fields without domestic animals for at least 27 years. Moderately (Mod) and intensively (Int) grazed paddocks had 0.2 and 0.4 sheep ha⁻¹ (redrawn from Oñatibia et al. (2015))

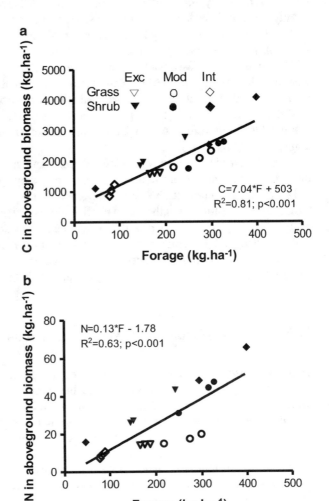

positively related to regulation services such as carbon and nitrogen sequestration at intermediate grazing intensities. Carbon and nitrogen stocks (C) in vegetation (above and belowground) were significantly higher in moderately grazed paddocks than in exclosure and intensively grazed ones (Oñatibia et al. 2015; Fig. 14.4). In this example, the relationship between forage biomass and carbon and nitrogen stocks had a positive linear relationship indicating that a trade-off did not occur (Fig. 14.4).

A trade-off between a supporting and a provisioning ecosystem service occurred in Patagonian rangelands. Grazing intensity shows a unimodal relationship species richness with a maximum value at moderate grazing intensities (Perelman et al. 1997; Fig. 14.5a). A decrease in richness is associated with intensive grazing because local extinction of forage species was not compensated by remaining non-palatable or weedy species. Patch diversity is another critical component of biodiversity in rangelands (Chap. 5, this volume). Abundance of different patch types shows a response similar to that of species richness. Under moderate grazing conditions, high-cover patches decrease whereas low-cover patches increase.

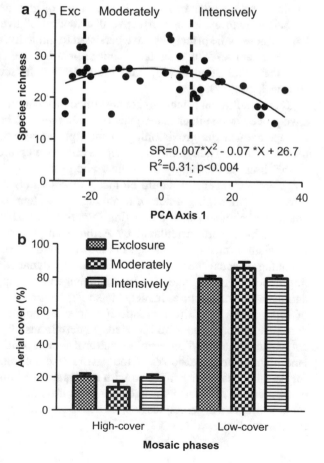

Fig. 14.5 Example of a trade-off between a supportive ecosystem service, biodiversity, and grazing intensity, which is an indicator of a provisioning ecosystem service, livestock production in the Patagonian rangelands, Argentina, under different management strategies. **a** Species richness along three grazing histories (*Exc* exclosures; moderate and intense grazing, separated by *vertical dashed lines*; redrawn from Perelman et al. (1997)), and **b** aerial percent cover of two patch types in low- and high-cover patches under three grazing regimes (mean ± ES), based on sampling the same paddocks as in Fig. 14.4 (redrawn from Cipriotti and Aguiar (2010))

$SR = 0.007 \cdot X^2 - 0.07 \cdot X + 26.7$
$R^2 = 0.31; p < 0.004$

14.6   Rangeland Management and Ecosystem Services: A Historical Perspective

For nearly a century the basic principles of rangeland management have been described and re-explained as the nature of goods and services derived from rangelands. Early in the twentieth century, Sampson (1923) outlined basic management principles and practices to support the continued provision of food and fiber via livestock grazing from rangelands. The need for these principles grew out of an era of resource overexploitation all over the world (Texeira and Paruelo 2006; Sayre et al. 2012). In that era, the provisioning services of food and fiber from rangelands were a central focus. There was either a lack of interest or a general unawareness of other goods and services from these "waste" lands at that time. This would be true for most rangeland environments on all continents in their early stages of settlement and development (Chap. 1, this volume). Management principles of the early twentieth century classically focused on requirements to control overgrazing and erosion through establishment of proper limits of the numbers of livestock, avoidance of grazing forage plants too early in their growth cycle, and effective distribution of livestock use across rangelands. These same principles have persisted to guide livestock grazing as detailed in subsequent texts on rangeland management into the early twenty-first century (Stoddart and Smith 1943; Vallentine 1989; Heitschmidt and Stuth 1991; Holechek et al. 2011).

These traditional principles for sustained provisioning of food and fiber goods have proven to be either inadequate or unappreciated, however, in guiding range management in the provisioning of multiple ecosystem goods and services in recent decades. Though these principles still have application in terms of recognizing limits to the supply of services and extraction of goods, the principles of rangeland management would be more appropriately portrayed as those of the science of managing trade-offs among ecosystem services and negotiating among stakeholders with competing interests. In reality, the management principles, which were articulated by Arthur Sampson nearly a century ago, are insufficient to manage landscapes in the twenty-first century. Currently, the provisioning of ecosystem services is dictated by dynamics of land-use fragmentation, ecological legacies of past management, oppressive constraints of antiquated infrastructure, inadequate social institutions, heterogeneous nuances of topography, fragilities of specific species, economic pressures of global demands for local goods and services, uncertainties of changing climates, political expediencies, and an array of cultural factors seldom acknowledged in the land management textbooks of the past. These complexities have been in play for decades and increasingly so with an expanding human population living on or adjacent to the world's rangelands. The articulation of a more sophisticated set of management principles has not kept pace with these newer landscape realities (Chap. 1, this volume).

14.7   Rangeland Management and Ecosystem Services: Landscape, Time, and Human Interactions

Given these current realities, the provisioning of goods and services from rangelands is now more appropriately perceived as a function of landscape, time, and human gradients (Fig. 14.6). The landscape gradient is shaped by ecological constraints resulting from an array of ecological sites and their existing conditions. It is this landscape gradient that was classically addressed through the basic management principles developed during the twentieth century when rangelands were viewed to provide a more narrow set of goods and services than expected in the twenty-first century.

Governance and socioeconomic conditions represent a critical component of this new conceptual framework. Complex patterns of land ownership (both public and private) and multiple administrative jurisdictions underscore the importance of governance and stakeholder engagement in supplying ecosystem services (Petz et al. 2014). The importance of viewing rangeland landscapes as socio-ecological systems with diverse governing institutions engaged in planning and management has been well described for some landscapes (Huntsinger and Oviedo 2014). Of additional importance are the social and cultural characteristics of resident populations. For

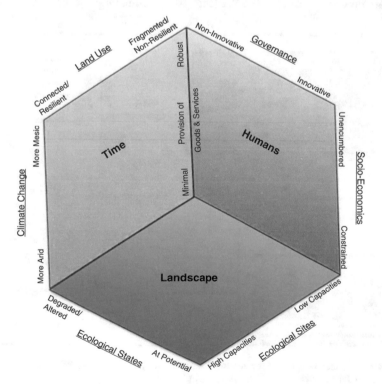

Fig. 14.6 Conceptual diagram of landscape, time, and human gradients that influence the provisioning of ecological goods and services from rangelands (adapted from Sayre et al. (2013))

example, level of education, household income, and place of residence were described as the main human aspects driving demand for ecosystem services (Yahdjian et al. 2015). The need to consider and incorporate these human-related drivers into our management models far exceeds the utilities of the basic principles of rangeland management as articulated frequently throughout the twentieth century.

Land use is a major driver of the array of goods and services that can be supplied, and it is a dynamic feature of landscapes around the world (Foley et al. 2005). Changes in land use can dramatically shift provisions of goods and services, and shifts may signify persistent alterations. Land-use gradients are further complicated by the uncertainties of climate variability and climate change. Though climate models are increasingly sophisticated and generating "near-term" projections (Taylor et al. 2012), their limitations and complexities are still problematic. However, recent statistical methods have provided tools to downscale global climate models to spatial scales that have application to land management (Abatzoglou 2013). What is now emerging is a more focused and applicable set of projections of climatic variables and their probabilities that would improve the ability of designing rangeland management that optimizes the provisioning of different ecosystem services under a changing climate. For example, interactive maps for various climate variables for the 2040–2060 period based on global climate models scaled down to the county level are now available for the USA (Fig. 14.7). Climatic projections at fine scales

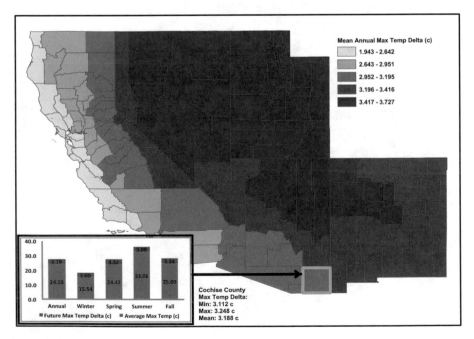

Fig. 14.7 Projected changes in mean annual temperature for the 2040–2060 period scaled to the county level for the five states: California, Nevada, Arizona, Utah, and New Mexico, calculated from Abatzoglou (2013) and adapted from Taylor et al. (2012). Details for Cochise Details for Cochise County, AZ, are included

of time and space will be effective in creating more quantitative understandings of pending changes that will have direct bearing on either the benefits these landscapes can provide or the conditions or processes that lead to desired benefits, either of which are definitions of ecosystem goods and services (Bommarco et al. 2013; see also Chap. 7, this volume).

The utility of the conceptual framework including landscape, time, and human gradients for deriving ecosystem services is evident in recent case studies from specific landscapes around the world (Fig. 14.6). For example, in the Little Karoo of South Africa, four different scenarios were created to evaluate the impact of one service, biodiversity, on the resulting supply of livestock forage, carbon storage, and water recharge from this 19,730 km² landscape (Egoh et al. 2010). Mapped habitats (landscape gradient) were evaluated for capacities to supply these different services. Different realistic land-use potentials (time gradient) were evaluated, such as development of tourism or of carbon markets, in terms of resulting impacts on other ecosystem services. Resulting impacts on revenues and opportunity costs for landowners (human gradient) were also calculated. Scenarios where biodiversity conservation could be achieved with provision of other services were realistic if human factors, including incentives to develop markets for services and institutions to encourage the supply of nontraditional ecosystem services, were emphasized. The importance of understanding landscape capacities to supply various ecosystem services was essential, but consideration of time and human gradients was paramount to managing for ecosystem services over time. Other recent case studies further illustrate the importance of inherent ecological capacities and existing ecological conditions, linkages of ecosystem services to the presence or absence of adequate infrastructure, such as roads and management institutions, subsistence requirements for resident human populations, rates of land-use and land-cover changes that can enhance one set of services at the expense of others, and importance placed on biodiversity within landscapes (Zhao et al. 2004; Muñoz et al. 2013; Pan et al. 2014).

Increasingly, the guidelines for managing ecosystem goods and services materialize for any specific landscape out of quantitative efforts to describe and map their occurrence or potential at spatial scales of relevance to landscape, human, and time gradients (Crossman et al. 2013). Though specific principles guiding processes of description and mapping of ecosystem goods and services are incomplete and methods are diverse, the resulting benefits are tangible. For example, a study both mapped and valued ecosystem services occurring across the Ewaso Ng'iro watershed of northern Kenya (Ericksen et al. 2011). Ecosystem services included medicinal plants, crops, livestock, wildlife, tourism, marketed carbon, wood and fiber, drinking water, flood regulation, cultural identification, and open space. The supply of these services was highly dependent on land use, inherent landscape capacities, existence of supply and their market values, proximity to infrastructure (i.e., accessibility), temporal dynamics of supply, social and cultural values of specific services, and spatial arrangement of the watershed and its sub-catchments. In the end, the final services that could be valued for uses within the watershed were livestock, tourism, and crops, and they were heterogeneously distributed and mapped across the watershed. In this fashion, recommendations could be developed based on specific knowledge of landscapes, demands, and values. For example, the market value of tourism within the watershed was minimal, but the value was highly dependent on infrastructure (national parks, roads). Decisions to further develop tourism could be then made with knowledge of costs both to develop the necessary infrastructure as well as evaluate resulting impacts on other benefits, such as livestock production.

In south-eastern Australia, one study evaluated and ranked existing land-cover types for six ecosystem services of forage production, biodiversity, water regulation, provision of water, carbon stock, and timber production (Baral et al. 2013). In this fashion, the provisioning of these services could be evaluated both spatially for the different cover types and land use and temporally given the known changes in land use and land cover over the past two centuries. In general, their analyses concluded that less modified landscapes resulted in a great supply and diversity of these services. However, specific land uses that modified cover, such as conversion of pasture to plantation, could result in a greater array of services, such as plantations providing timber production from regions within the basin. Their mapping processes provided a basis for evaluating possible services and the resulting trade-offs created by land-use decisions. Characterization and quantification of ecosystem services provide a bridge that can link our knowledge of ecological processes across landscapes, time, and human processes (Fu and Forsius 2015). Continued efforts to clearly establish the principles for these characterizations and quantifications are critical to both understanding linkages among human and ecological processes for a landscape and sustaining output of a demanded supply of goods and services.

Exploitation and utilization of natural resources pose the questions of property ownership and legitimate stakeholders (Latour 2013). Are natural resources the property of individuals, and if so which individuals, or are they the property of humankind? In some regions, provisioning ecosystem services belong to the owners of the land. At the same time, all other ecological services (supporting, regulating, and cultural) in general are not marketable and usually are not claimed or owned by a single individual (single or organized with economic purposes). The number, nature, and diversity of stakeholders have increased in the recent past as the importance of regulating and cultural services has increased relative to provisioning services (Yahdjian et al. 2015). Demand for cultural ecosystem services results in the creation of national parks or natural reserves as in the example for South American rangelands (Murdoch et al. 2010). This strategy raises ethics issue since inhabitants—mostly small ranchers and peasants—are forced to migrate and adopt alternative lifestyles. Additionally migration may also initiate other social conflicts in urban centers (Easdale et al. 2009). It is important to keep in mind as we analyze human-induced degradation of ecological services that there is a network of stakeholders, in addition to ranchers, and sociopolitical processes that are involved (Easdale and Domptail 2014). Many component processes have nonlinear dynamics that may potentially exhibit thresholds (Walker et al. 1981). More than seven decades since the rise of rangeland science the challenge of grazing management remains unresolved because of the changing demands for ecosystem services. Integration of human and biophysical dimensions through the study of ecological services may be a rewarding path.

14.8 Conceptual Framework for Ecosystem Services and Range Management

The concept of ecosystem services has emerged as a powerful tool for guiding management of rangelands in the twenty-first century. Ecosystem services serve as a way of clarifying what is that different stakeholders want from rangelands, ranging in scale from paddocks and counties to regions across national boundaries. Ecosystem services also serve to clarify what goods and services that land is able to supply. The optimal management strategy results from reconciling supply and demand of ecosystem services (Yahdjian et al. 2015) as described in the following equation. Land use is a function of:

$$\text{Land use} = \int \sum_{j=1}^{n} \left[\left(\text{ES}_j \text{ supply} \right), \left(\sum_{i=1}^{m} \left(\text{ES}_j \text{ Demand}_{\text{stakeholder } i} * \text{Political Power}_{\text{stakeholder } i} \right) \right) \right]$$

(14.1)

Here, land use or, in our specific case, rangeland management practices depend on the sum of the supply of all the ecosystem services ES_j from 1 to n, and the sum of the demand for each ES_j from each stakeholder from i to m. Finally, the demand of each stakeholder is weighed by their political power.

For example, rangelands all over the world are being invaded by woody plants (Estell et al. 2012). This transformation from grasslands into shrublands and savannas affects the provisioning of ecosystem services. Woody-plant encroachment affects the provisioning of different ecosystem services from livestock production to maintenance of biodiversity and yielding of clean water (MacLeod and Johnston 1990; Turpie et al. 2008; Anadón et al. 2014). Equation (14.1) can be applied to the rangeland management issue of whether to remove or not woody plants. Rangelands can supply different services including clean water and livestock production, which are enhanced by woody-plant control. On the contrary, the ecosystem service erosion control may be diminished by removal of woody plants. Different stakeholders with different political power value these different services differently. The final solution to the management question of whether to remove woody plants will depend on both (1) the effect that woody plants have at each specific location on ecosystem services and (2) the valuation that each stakeholder has on each ecosystem service.

Similarly, Anadón et al. (2014) analyzed the impact of woody-plant encroachment on livestock production in US and Argentinean rangelands. These are two rangelands, which are similar from the biophysical standpoint and in their ability to supply ecosystem services. However, these rangelands have contrasting socioeconomic conditions that affect the demand for ecosystem services. In Argentina, livestock production is the most valued and primary ecosystem service of interest. In the USA, on the contrary, rangelands face multiple demands for ecosystem services, including biodiversity conservation and recreation, in addition to livestock production. Anadón et al. (2014) showed that the different demand for ecosystem services modified the impact of woody-plant invasions on livestock production. In Argentina, woody-plant cover accounted for 50 % of the livestock production but, in the USA, it explained only half of the variability in livestock production.

14.9 Future Directions

Ecosystem services represent an important conceptual link between the biophysical constraints and human demand. However, our understanding of the supply and demand of ecosystem services is unbalanced. We know much more about the supply of ecosystem services than we know about the demand for different ecosystem services from different beneficiaries. It will be important to enhance our understanding of the demand from different groups of beneficiaries for specific ecosystem services within specific landscapes. In addition, it will be necessary to understand and quantify the determinants of the demand for ecosystem services.

In order to predict the future of rangelands and develop appropriate management strategies, we need to understand the future supply and the demand for ecosystem services. Our strategy to tackle this daunting task is to separate the effect of drivers to the response of supply and demand to their drivers. For example, the future supply of the ecosystem service forage production depends on climate change (the driver) and sensitivity of ecosystems to climate (the response to driver). Similarly, the demand for the ecosystem service recreation depends on the proportion of urban population (driver) and the sensitivity of recreation demand to urbanization (response to driver). This requires an interdisciplinary approach where land owners, land managers (public and/or private), ecologists, climatologists, and social scientists work in close collaboration.

14.10 Summary

Rangeland ecosystem services are the benefits that society receives from rangelands. They include the provisioning of food, wood and medicinal resources, and services that contribute to climate stability, control of agricultural pests, and purification of air and water. Rangeland ecosystem services are classified in four categories: provisioning, regulating, cultural, and supporting. Provisioning ecosystem services include the contribution of essential goods such as food, fiber, and medicinal. Regulating ecosystem services include carbon sequestration, prevention of soil erosion, and natural flood control. Cultural ecosystem services include intellectual, inspirational, and recreational activities. The fourth category is supporting ecosystem services, which include services that are dependent on ecological processes such as primary production and nutrient cycling and that are intimately related to biological diversity.

The supply of ecosystem services is mostly determined by biophysical factors such as climate, soils, as well as historical land use. Human demand represents the other side of the rangeland ecosystem service equation, which is related to the social beneficiaries. Human consumption of resources and utilization of services that are supplied by rangelands depend upon both their capacity to produce them and the societal value and need placed on those resources and services. Demand for ecosystem services changes among social beneficiaries, who are the individuals or groups of individuals who have an interest in ecosystem services. Different methods have been developed to assess the demand for specific ecosystem services, including collation of responses to questionnaires and interviews and classical economic tools such as willingness to pay.

There are cases of trade-offs and win–win interactions among different types of ecosystem services. Win–win conditions occur when management aimed at increasing one type of ecosystem service results in the increase of other ecosystem services. For example, grazing management that results in increased forage supply also increases carbon and nitrogen stocks in rangeland soils. Trade-offs occur when management results in the increase of one ecosystem service and the decrease in other. For example, increasing grazing intensity and livestock production may, in certain cases, decrease biodiversity.

Principles of range management developed in the twentieth century focused primarily on the biophysical components of the rangeland ecosystems and the requirements to control overgrazing and erosion through the establishment of general management principles, including proper limits on numbers of livestock, avoiding grazing of forage plants too early in their growth cycle, and effectively distributing livestock use across rangelands. In the twenty-first century, principles for managing rangelands need to be much broader. Currently, the provisioning of ecosystem services is dictated by dynamics of land-use fragmentation, ecological legacies of past management, infrastructure, fragilities of specific species, economic pressures of global demands for local goods and services, uncertainties of changing climates, political expediencies, and an array of cultural factors seldom acknowledged in the land management textbooks of the past.

Here, we propose a novel conceptual framework for rangeland management based on the premise that management always aims at reconciling supply and demand of ecosystem services. The supply of each ecosystem service is based mostly on biophysical characteristics and the use history that could have affected its potential. The demand for each ecosystem service is different for each group of beneficiaries or stakeholders. Finally, the demands of each group of beneficiaries do not have the same impact because of their differential capacity to influence decision making. Therefore, demands for ecosystem services here are weighed by the political power of each group of beneficiaries or stakeholders. In conclusion, there is not a universally optimal management strategy for a rangeland because demands for ecosystem services, power of each group of beneficiaries, and supply of ecosystem services change through time.

Acknowledgements Authors want to thank Courtney Currier for her support during the course of this project, which received financial support from the US National Science Foundation LTER DEB 1235828.

References

Abatzoglou, J.T. 2013. Development of gridded surface meteorological data for ecological applications and modelling. *International Journal of Climatology* 33: 121–131.

Adler, P.B., D.G. Milchunas, O.E. Sala, I.C. Burke, and W.K. Lauenroth. 2005. Plant traits and ecosystem grazing effects: Comparison of U.S. sagebrush steppe and Patagonian steppe. *Ecological Applications* 15: 774–792.

Ahlström, A., M.R. Raupach, G. Schurgers, B. Smith, A. Arneth, M. Jung, M. Reichstein, J.G. Canadell, P. Friedlingstein, and A.K. Jain. 2015. The dominant role of semi-arid ecosystems in the trend and variability of the land CO_2 sink. *Science* 348: 895–899.

Anadón, J.D., O.E. Sala, B.L. Turner, and E.M. Bennett. 2014. The effect of woody-plant encroachment on livestock production in North and South America. *Proceedings of National Academy of Sciences* 111: 12948–12953.

Baral, H., R.J. Keenan, J.C. Fox, N.E. Stork, and S. Kasel. 2013. Spatial assessment of ecosystem goods and services in complex production landscapes: A case study from south-eastern Australia. *Ecological Complexity* 13: 35–45.

Bennett, E.M., G.D. Peterson, and L.J. Gordon. 2009. Understanding relationships among multiple ecosystem services. *Ecology Letters* 12: 1394–1404.

Bommarco, R., D. Kleijn, and S.G. Potts. 2013. Ecological intensification: Harnessing ecosystem services for food security. *Trends in Ecology and Evolution* 28: 230–238.

Brown, J.H., and W. McDonald. 1995. Livestock grazing and conservation of southwestern rangelands. *Conservation Biology* 9: 1644–1647.

Carpenter, S.R., H.A. Mooney, J. Agard, D. Capistrano, R.S. DeFries, S. Diaz, T. Dietz, A.K. Duraiappah, A. Oteng-Yeboah, H.M. Pereira, C. Perrings, W.V. Reid, J. Sarukhan, R.J. Scholes, and A. Whyte. 2009. Science for managing ecosystem services: Beyond the Millennium Ecosystem Assessment. *Proceedings of the National Academy of Sciences of the United States of America* 106: 1305–1312.

Castro, A., B. Martín-López, M. García-Llorente, P. Aguilera, E. López, and J. Cabello. 2011. Social preferences regarding the delivery of ecosystem services in a semiarid Mediterranean region. *Journal of Arid Environments* 75: 1201–1208.

Cipriotti, P., and M. Aguiar. 2010. Resource partitioning and interactions enable coexistence in a grass-shrub steppe. *Journal of Arid Environments* 74: 1111–1120.

Costanza, R., R. d'Arge, R. de Groot, S. Farber, M. Grasso, B. Hannon, K. Limburg, S. Naeem, R. O'Neill, J. Paruelo, R.G. Raskin, P. Sutton, and M. van den Belt. 1997. The value of the world's ecosystem services and natural capital. *Nature* 387: 253–260.

Crossman, N.D., B. Burkhard, S. Nedkov, L. Willemen, K. Petz, I. Palomo, E.G. Drakou, B. Martín-Lopez, T. McPhearson, and K. Boyanova. 2013. A blueprint for mapping and modelling ecosystem services. *Ecosystem Services* 4: 4–14.

Daily, G.C. (ed.). 1997. *Nature's services. Societal dependence on natural ecosystems.* Washington DC: Island Press.

De Chazal, J., F. Quetier, S. Lavorel, and A. Van Doorn. 2008. Including multiple differing stakeholder values into vulnerability assessments of socio-ecological systems. *Global Environmental Change* 18: 508–520.

Easdale, M., and S. Domptail. 2014. Fate can be changed! Arid rangelands in a globalizing world— a complementary co-evolutionary perspective on the current 'desert syndrome'. *Journal of Arid Environments* 100: 52–62.

Easdale, M.H., M.R. Aguiar, M. Roman, and S. Villagra. 2009. Socio-economic comparison of two biophysical regions: Livestock production systems from Río Negro Province, Argentina. *Cuadernos De Desarrollo Rural* 6: 173–198.

Egoh, B.N., B. Reyers, J. Carwardine, M. Bode, P.J. O'farrell, K.A. Wilson, H.P. Possingham, M. Rouget, W. De Lange, and D.M. Richardson. 2010. Safeguarding biodiversity and ecosystem services in the Little Karoo, South Africa. *Conservation Biology* 24: 1021–1030.

Ericksen, P., M. Said, J.d. Leeuw, S. Silvestri, L. Zaibet, S. Kifugo, K. Sijmons, J. Kinoti, L. Nganga, and F. Landsberg. 2011. *Mapping and valuing ecosystem services in the Ewaso Ng'iro watershed. ILRI–WRI–Danida report*. Nairobi.

Estell, R., K.M. Havstad, A. Cibils, D. Anderson, T. Schrader, and K. James. 2012. Increasing shrub use by livestock in a world with less grass. *Rangeland Ecology and Management* 65: 327–414.

Foley, J.A., R. DeFries, G.P. Asner, C. Barford, G. Bonan, S.R. Carpenter, F.S. Chapin, M.T. Coe, G.C. Daily, H.K. Gibbs, J.H. Helkowski, T. Holloway, E.A. Howard, C.J. Kucharik, C. Monfreda, J.A. Patz, I.C. Prentice, N. Ramankutty, and P.K. Snyder. 2005. Global consequences of land use. *Science* 309: 570–574.

Fu, B., and M. Forsius. 2015. Ecosystem services modeling in contrasting landscapes. *Landscape Ecology* 30: 375–379.

Gomez-Baggethun, E., R. de Groot, P.L. Lomas, and C. Montes. 2010. The history of ecosystem services in economic theory and practice: From early notions to markets and payment schemes. *Ecological Economics* 69: 1209–1218.

Havstad, K.M., D.P.C. Peters, R. Skaggs, J. Brown, B.T. Bestelmeyer, E. Fedrickson, J.E. Herrick, and J. Wright. 2007. Ecological services to and from rangelands of the United States. *Ecological Economics* 64: 261–268.

Heitschmidt, R., and J.P. Stuth. 1991. *Grazing management: An ecological perspective*. Portland: Timber Press.

Herrick, J.E., O.E. Sala, and J.W. Karl. 2013. Land degradation and climate change: A sin of omission? *Frontiers in Ecology and the Environment* 11: 283–283.

Holechek, J.L., R.D. Pieper, and C.H. Herbel. 2011. *Range management: Principles and practices*, 6th ed. New York: Pearson Education, Inc.

Huntsinger, L., and J.L. Oviedo. 2014. Ecosystem services are social-ecological services in a traditional pastoral system: The case of California's Mediterranean rangelands. *Ecology and Society* 19: 8.

Kauffman, J., and D. Pyke. 2001. Range ecology, global livestock influences. In *Encyclopedia of biodiversity*, ed. S. Levin, 33–52. San Diego: Academic.

Lamarque, P., U. Tappeiner, C. Turner, M. Steinbacher, R.D. Bardgett, U. Szukics, M. Schermer, and S. Lavorel. 2011. Stakeholder perceptions of grassland ecosystem services in relation to knowledge on soil fertility and biodiversity. *Regional Environmental Change* 11: 791–804.

Latour, B. 2013. *An inquiry into modes of existence*. Cambridge: Harvard University Press.

MA. 2005. *Millennium Ecosystem Assessment (MA) synthesis report*. Washington, DC: Millennium Ecosystem Assessment.

MacLeod, N., and B. Johnston. 1990. An economic framework for the evaluation of rangeland restoration projects. *The Rangeland Journal* 12: 40–53.

Martin-Lopez, B., I. Iniesta-Arandia, M. Garcia-Llorente, I. Palomo, I. Casado-Arzuaga, D.G. Del Amo, E. Gomez-Baggethun, E. Oteros-Rozas, I. Palacios-Agundez, B. Willaarts, J.A. Gonzalez, F. Santos-Martin, M. Onaindia, C. Lopez-Santiago, and C. Montes. 2012. Uncovering ecosystem service bundles through social preferences. *PLoS One* 7: e38970.

Muñoz, J.C., R. Aerts, K.W. Thijs, P.R. Stevenson, B. Muys, and C.H. Sekercioglu. 2013. Contribution of woody habitat islands to the conservation of birds and their potential ecosystem services in an extensive Colombian rangeland. *Agriculture, Ecosystems and Environment* 173: 13–19.

Murdoch, W., J. Ranganathan, S. Polasky, and J. Regetz. 2010. Using return on investment to maximize conservation effectiveness in Argentine grasslands. *Proceedings of the National Academy of Sciences* 107: 20855–20862.

Newcome, J., A. Provins, H. Johns, E. Ozdemiroglu, J. Ghazoul, D. Burgess, and K. Turner. 2005. *The economic, social and ecological value of ecosystem services: A literature review*. London: Economics for the Environment Consultancy (eftec).

Nosetto, M., E. Jobbágy, T. Tóth, and R. Jackson. 2008. Regional patterns and controls of ecosystem salinization with grassland afforestation along a rainfall gradient. *Global Biogeochemical Cycles* 22: GB2015.

Oesterheld, M., O.E. Sala, and S.J. McNaughton. 1992. Effect of animal husbandry on herbivore-carrying capacity at a regional scale. *Nature* 356: 234–236.

Oñatibia, G.R., M.R. Aguiar, and M. Semmartin. 2015. Are there any trade-offs between forage provision and the ecosystem service of C and N storage in arid rangelands? *Ecological Engineering* 77: 26–32.

Pan, Y., J. Wu, and Z. Xu. 2014. Analysis of the tradeoffs between provisioning and regulating services from the perspective of varied share of net primary production in an alpine grassland ecosystem. *Ecological Complexity* 17: 79–86.

Perelman, S., R. León, and J. Bussaca. 1997. Floristic changes related to grazing intensity in a Patagonian shrub steppe. *Ecography* 20: 400–406.

Perrings, C., S. Naeem, F.S. Ahrestani, D.E. Bunker, P. Burkill, G. Canziani, T. Elmqvist, J.A. Fuhrman, F.M. Jaksic, Z. Kawabata, A. Kinzig, G.M. Mace, H. Mooney, A.H. Prieur-Richard, J. Tschirhart, and W. Weisser. 2011. Ecosystem services, targets, and indicators for the conservation and sustainable use of biodiversity. *Frontiers in Ecology and the Environment* 9: 512–520.

Petz, K., J. Glenday, and R. Alkemade. 2014. Land management implications for ecosystem services in a South African rangeland. *Ecological Indicators* 45: 692–703.

Quétier, F., F. Rivoal, P. Marty, J. de Chazal, W. Thuiller, and S. Lavorel. 2010. Social representations of an alpine grassland landscape and socio-political discourses on rural development. *Regional Environmental Change* 10: 119–130.

Raudsepp-Hearne, C., G.D. Peterson, and E.M. Bennett. 2010. Ecosystem service bundles for analyzing tradeoffs in diverse landscapes. *Proceedings of the National Academy of Sciences of the United States of America* 107: 5242–5247.

Reynolds, J.F., D.M.S. Smith, E.F. Lambin, B. Turner, M. Mortimore, S.P. Batterbury, T.E. Downing, H. Dowlatabadi, R.J. Fernández, and J.E. Herrick. 2007. Global desertification: Building a science for dryland development. *Science* 316: 847–851.

Rositano, F., M. López, P. Benzi, and D.O. Ferraro. 2012. Servicios de los ecosistemas. un recorrido por los beneficios de la naturaleza. Ecosystem services. a travel through natural benefits. Agronomía y ambiente. *Revista de la Facultad de Agronomía de la Universidad de Buenos Aires* 32: 49–60.

Sala, O., and J. Paruelo. 1997. Ecosystem services in grasslands. In *Nature's services: Societal dependence on natural ecosystems*, ed. G.C. Daily, 237–251. Washington, D.C.: Island Press.

Sampson, A.W. 1923. *Range and pasture management*. New York: Wiley.

Sayre, N.F., W. deBuys, B.T. Bestelmeyer, and K.M. Havstad. 2012. "The Range Problem" after a century of rangeland science: New research themes for altered landscapes. *Rangeland Ecology and Management* 65: 545–552.

Sayre, N.F., R.R. McAllister, B.T. Bestelmeyer, M. Moritz, and M.D. Turner. 2013. Earth Stewardship of rangelands: Coping with ecological, economic, and political marginality. *Frontiers in Ecology and the Environment* 11: 348–354.

Scheffer, M., W. Brock, and F. Westley. 2000. Socioeconomic mechanisms preventing optimum use of ecosystem services: An interdisciplinary theoretical analysis. *Ecosystems* 3: 451–471.

Stoddart, L.A., and A.D. Smith. 1943. *Range management*, 547. NY: McGraw-Hill Book Co. Inc.

Tallis, H., and S. Polasky. 2011. Assessing multiple ecosystem services: An integrated tool for the real world. *Natural capital. Theory and practice of mapping ecosystem services*, 34–52. Oxford: Oxford University Press.

Tallis, H., T. Ricketts, A. Guerry, E. Nelson, D. Ennaanay, S. Wolny, N. Olwero, K. Vigerstol, D. Pennington, and G. Mendoza. 2011. *InVEST 2.1 beta user's guide*. Stanford: The Natural Capital Project.

Taylor, K.E., R.J. Stouffer, and G.A. Meehl. 2012. An overview of CMIP5 and the experiment design. *Bulletin of the American Meteorological Society* 93: 485–498.

Texeira, M., and J.M. Paruelo. 2006. Demography, population dynamics and sustainability of the Patagonian sheep flocks. *Agricultural Systems* 87: 123–146.

Turpie, J.K., C. Marais, and J.N. Blignaut. 2008. The working for water programme: Evolution of a payments for ecosystem services mechanism that addresses both poverty and ecosystem service delivery in South Africa. *Ecological Economics* 65: 788–798.

Vallentine, J.F. 1989. *Range development and improvements*. San Diego: Academic.

Walker, B.H., D. Ludwig, C.S. Holling, and R.M. Peterman. 1981. Stability of semi-arid savanna grazing systems. *Journal of Ecology* 69: 473–498.

Yahdjian, L., O.E. Sala, and K.M. Havstad. 2015. Rangeland ecosystem services: Shifting focus from supply to reconciling supply and demand. *Frontiers in Ecology and the Environment* 13: 44–51.

Zhao, B., U. Kreuter, B. Li, Z. Ma, J. Chen, and N. Nakagoshi. 2004. An ecosystem service value assessment of land-use change on Chongming Island, China. *Land Use Policy* 21: 139–148.

Chapter 15
Managing Climate Change Risks in Rangeland Systems

Linda A. Joyce and Nadine A. Marshall

Abstract The management of rangelands has long involved adapting to climate variability to ensure that economic enterprises remain viable and ecosystems sustainable; climate change brings the potential for change that surpasses the experience of humans within rangeland systems. Adaptation will require an intentionality to address the effects of climate change. Knowledge of vulnerability in these systems provides the foundation upon which to base adaptation strategies; however, few vulnerability assessments have examined and integrated the climate vulnerability of the ecological, economic, and social components of rangeland systems. The capacity of ecosystems, humans, and institutions to adjust to potential damage and to take advantage of opportunities is termed adaptive capacity. Given past attempts to cope with drought, current adaptive capacity is not sufficient to sustain rangeland enterprises under increasing climatic variability. Just as ecosystem development is affected by past events, historical studies suggest that past events in human communities influence future choices in response to day-to-day as well as abrupt events. All adaptation is local and no single adaptation approach works in all settings. A risk framework for adaptation could integrate key vulnerabilities, risk, and hazards, and facilitate development of adaptation actions that address the entire socio-ecological system. Adaptation plans will need to be developed and implemented with recognition of future uncertainty that necessitates an iterative implementation process as new experience and information accumulate. Developing the skills to manage with uncertainty may be a singularly important strategy that landowners, managers, and scientists require to develop adaptive capacity.

Keywords Adaptive capacity • Socio-ecological systems • Risk • Vulnerability • Adaptation planning

L.A. Joyce (✉)
USDA-FS Rocky Mountain Research Station, Fort Collins, CO 80526, USA
e-mail: ljoyce@fs.fed.us

N.A. Marshall
CSIRO, Land and Water, James Cook University, Townsville, QLD, Australia
e-mail: nadine.marshall@csiro.au

© The Author(s) 2017
D.D. Briske (ed.), *Rangeland Systems*, Springer Series on Environmental Management, DOI 10.1007/978-3-319-46709-2_15

15.1 Introduction

The management of rangelands has long involved adapting to climate variability in order that economic enterprises remain viable and ecosystems sustainable (Marshall and Stokes 2014). Rangeland management has never been just about the land; "managers have sought to maintain a relationship between rangelands and the people who hoped to benefit from the land, and to do it in such a way that those benefits were realized while the land retained its capacity to provide what society valued" (Brunson 2012). This relationship and the corresponding benefits will be challenged under climate change (IPCC 2014a; Crimp et al. 2010; Chap. 7, this volume).

Climate change brings to this relationship the potential for large-scale modifications, including those that surpass the experience of humans currently living on rangelands. Since the early 1990s, the global scientific community has been studying and reporting on the nature of these global changes in climate, the human and natural activities contributing to these global changes, and the associated impacts to land and water (IPCC 2014a). Warming temperatures are projected as well as changes in seasonal precipitation patterns, total annual precipitation, and the potential for increased drought (Chap. 7, this volume). While rangeland managers and enterprise owners have incorporated strategies to address variability in climate, these future changes may be beyond the variability they have experienced in their lifetime. The enterprise owner and their family, the manager, and employees are embedded within social and economic networks and institutions that are interdependent with the ecological system which includes soil, plants, animals, and ecosystem processes. This interdependent system is formally called a socio-ecological system (Berkes and Folke 1998; Brunson 2012). We view the socio-ecological rangeland system as a collective of economic enterprises (livestock and other market outputs) and the ecological system (Fig. 15.1) (Chap. 8, this volume). We use this framework to explore adaptation to climate change.

The global conversation about adaptation has expanded from an initial focus on ecological and economic impacts and adaptation strategies to a broader vision of ecological, economic, and social impacts and adaptation strategies. Adaptive capacity is the ability of plants, animals, and humans, as well as the systems and institutions to adjust to potential damage or to take advantage of opportunities under climate change (Table 15.1). Social values of the enterprise owner influence management goals while at the same time community values, local and regional economics, and government policy are influencing the owner's values and decisions. Thus understanding the interdependent nature of the socio-ecological rangeland system is key to understanding and facilitating adaptation in rangeland systems (Fig. 15.1).

This chapter explores adaptation to climate change in the context of socio-ecological systems. We review the evolving concept of adaptation and the development of strategies for adaptation to current and future climate change. We explore what we might learn about past attempts to cope with climatic events and how a historical perspective could frame future adaptation strategies on rangelands. Four case studies from around the world are summarized to describe

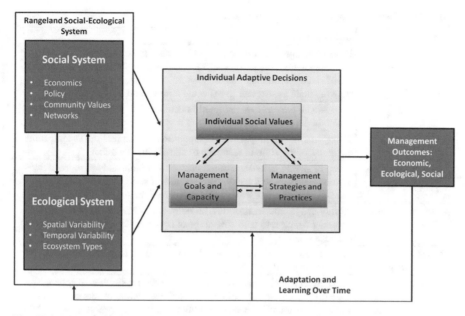

Fig. 15.1 This adaptive decision-making framework emphasizes how the sustainability of individual ranching operations depends on their capacity to adapt to the changing parameters of the social-ecological systems in which their enterprises are embedded (Lubell et al. 2013)

past and future adaptation strategies. We examine what adaptation management on rangelands might look like in the future. The biophysical consequences of climate change on rangelands are described in Chap. 7 of this volume.

15.2   Evolution of Climate Change Adaptation

Our understanding of what adaptation means in response to a changing climate has evolved through the last 25 years and likely will continue to evolve. This evolution is most visible in the five assessment reports of the Intergovernmental Panel on Climate Change (IPCC) where each report (1992, 1995, 2001, 2007, 2014) synthesized the most current published scientific literature on climate change. Two threads in these reports that highlight an evolution in our scientific understanding are of specific interest. First, the definition of adaptation as applied to climate change and related topics such as adaptive capacity has evolved. Second, the discussion of rangelands has shifted from an emphasis on the mismanagement of rangelands to an exploration of the effects of climate change and potential adaptation responses.

Increasingly, the assessment of climate change impacts, vulnerability, and adaptation in the IPCC reports has come to include the economic and social impacts of climate change, and the role of humans in managing natural systems. In the first three IPCC reports, chapters related to rangelands or ecosystems focused on ecological

Table 15.1 Definitions of terms used in this chapter

Term	Definition
Adaptation	Process of adjustment to actual or expected climate and its effects. In human systems, adaptation seeks to moderate or avoid harm or exploit beneficial opportunities. In some natural systems, human intervention may facilitate adjustment to expected climate and its effects
	Incremental adaptation—adaptation actions where the central aim is to maintain the essence and integrity of a system or process at a given scale
	Transformational adaptation—adaptation that changes the fundamental attributes of a system in response to climate and its effects
Adaptation options	Array of strategies and measures that are available and appropriate for addressing adaptation needs. They include a wide range of actions that can be categorized as structural, institutional, or social
Adaptive capacity	Ability of systems, institutions, humans, and other organisms to adjust to potential damage, to take advantage of opportunities, or to respond to consequences
Mitigation (of climate change)	A human intervention to reduce the sources or enhance the sinks of greenhouse gases
Resilience	Capacity of social, economic, and environmental systems to cope with a hazardous event or trend or disturbance, responding or reorganizing in ways that maintain their essential function, identity, and structure, while also maintaining the capacity for adaptation, learning, and transformation
Risk	Potential for consequences where something of value is at stake and where the outcome is uncertain, recognizing the diversity of values. Risk is often represented as probability of occurrence of hazardous events or trends multiplied by the impacts if these events or trends occur. Risk results from the interaction of vulnerability, exposure, and hazard
Sustainable development	Development that meets the needs of the present without compromising the ability of future generations to meet their own needs
Vulnerability	Propensity or predisposition to be adversely affected. Vulnerability encompasses a variety of concepts and elements including sensitivity or susceptibility to harm and lack of capacity to cope and adapt

All definitions from Agard et al. (2014) unless otherwise noted

effects of climate change with very brief discussions of adaptation options (IPCC 1990; Allen-Diaz et al. 1996; Gitay et al. 2001). In the first IPCC report, rangeland adaptation responses were broadly identified as developing emergency and disaster preparedness policies, improving the efficiency of natural resource use and needed research on control measures for desertification, and enhancing adaptability of crops to saline conditions (IPCC 1990). In the Third Assessment Report, the overuse of rangeland resources and the associated rangeland degradation were seen as more impactful than the future effects of climate change (Gitay et al. 2001). Consequently, adaptation options such as selection of plants (legume-based systems) and improved livestock management were identified as a means to address current rangeland degradation as well as the potential effects of climate change.

As the focus expanded to include societal impacts and responses, the structure of the IPCC assessment reports included a more in-depth discussion of adaptation limits and transformation in social and natural systems. The Third Assessment provided a definition of adaptation that specifically included humans: "the adjustment in natural or human systems in response to actual or expected climatic stimuli or their effects, which moderates harm or exploits beneficial opportunities" (IPCC 2001). In this report, adapting to climate change was seen not only as reducing vulnerability to climate change but also as promoting sustainable development, development that meets the needs of the present without compromising the ability of future generations to meet their own needs. Adaptation was characterized in terms of purposefulness (autonomous versus planned), timing (anticipatory, proactive, reactive), temporal scope (short versus long term), spatial scope, form (e.g., structural, legal, institutional), and criteria to evaluate its performance. Although the Third Assessment report did not discuss adaptive capacity with respect to rangelands, adaptive capacity was defined and that definition has been retained by subsequent reports (Table 15.1). In the Fourth Assessment report, adaptive capacity was recognized as being influenced by social variables, in addition to biophysical and economic resources (Adger et al. 2007).

By the Fourth Assessment Report, the scientific and management communities had contributed an extensive literature that could be reviewed in chapters focused on the assessment of adaptation practices, options, constraints and capacity, and interrelationships between adaptation and mitigation. Adaptation rarely was implemented in response to climate change alone and high adaptive capacity did not, in general, lead to actions to reduce vulnerability to climate change. The report identified significant barriers to implementing adaptation that spanned the inability of natural systems to adapt to the rate and magnitude of climate change, but also constraints in technology, financing, cognitive and behavioral components, and social and cultural settings. With respect to ecosystems, adaptation options focused only on altering the context in which ecosystems developed and little attention was given to the human systems component. It was acknowledged that identifying adaptation responses and options for ecosystems was a rapidly developing field (Fischlin et al. 2007). However, it would take a reframing of adaptation in the context of risk to bring the ecological, economic, and social components into a more integrated framework.

In the Fifth Assessment, the definition of adaptation became "the process of adjustment to actual or expected climate and its effects." The definition expands on the human role. "In human systems, adaptation seeks to moderate harm or exploit beneficial opportunities. In natural systems, human intervention may facilitate adjustment to expected climate and its effects" (Table 15.1). Though subtle, this definition is different from previous IPCC definitions in that there is *intentionality* to the adaptation action. It is not just the restoration of a rangeland ecosystem; the adaptation action includes specific consideration of climate change objectives in management. The definition of adaptation was further nuanced. Moving beyond adaptation categories of anticipatory and reactive, private and public, and autonomous and planned, only two types of adaptation were defined in the Fifth Assessment: incremental and transformational (Table 15.1). The report notes that adaptation options to date have been mainly incremental and stresses that adaptation may require transformational changes, in which potentially impacted systems move to fundamentally new patterns, dynamics, and/or locations.

The concept of risk is used in this most recent IPCC Assessment to frame decision making in a changing world, with continuing uncertainty about the severity and timing of climate change impacts and with limits to the effectiveness of adaptation (IPCC 2014b). Risk is defined as "the potential for consequences where something of value is at stake and where the outcome is uncertain, recognizing the diversity of values" (Table 15.1). This introduction of risk allows the discussion of adaptation to integrate the risk of climate-related impacts, climate-related hazards, and vulnerability and exposure of human and natural systems as these risks, hazards, and vulnerabilities interact and are impacted by socioeconomic and climate drivers (Fig. 15.2). When climate change factors from more than one economic sector or geographic region are included in a risk assessment, risks that were not previously assessed or recognized emerge. An example of such interaction is the policy to encourage the use of bioenergy to mitigate climate change by reducing fossil fuel emissions, but which has led to shifting cropland acreage from food production to bioenergy crop production and consequently raising prices for food crops, resulting in a reduction in food security and increasing human vulnerability to climate change

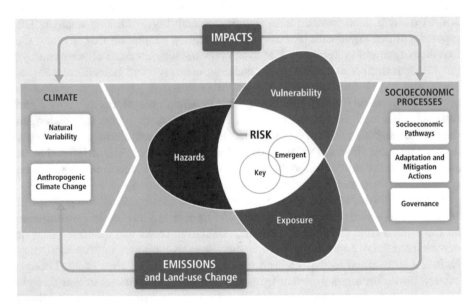

Fig. 15.2 Schematic of the interaction among the physical climate system, exposure, and vulnerability to produce risk. Risk of climate-related impacts results from the interaction of climate-related hazards (including hazardous events and trends) with the vulnerability and exposure of human and natural systems. Vulnerability and exposure are largely the result of socioeconomic pathways and societal conditions (although changing hazard patterns also play a role). Changes in both the climate system (*left side*) and socioeconomic processes (*right side*) are central drivers of the different core components (vulnerability, exposure, and hazards) that constitute risk (Oppenheimer et al. 2014; Fig. 19.1)

(Oppenheimer et al. 2014). This Fifth Assessment report summarizes the key risks globally (Oppenheimer et al. 2014); we will explore the key risks identified for rangelands later in this chapter.

15.3 Assessing Vulnerabilities to Implement Adaptation Actions

Knowledge of vulnerability provides the foundation upon which to develop and select specific adaptations and strategies. However, assessing vulnerability has been challenging as the definition of vulnerability has varied across the ecological-socio-economic spectrum and there has been no standard methodology to assess vulnerability of climate change (Fussel and Klein 2006; Glick et al. 2011; USGCRP 2011). Social characteristics of individuals and communities have been incorporated into vulnerability assessments with respect to disasters; however, most existing climate vulnerability assessments of plants, animals, or ecosystems have limited information on the related social and economic effects of climate (USGCRP 2011). Further, most approaches to assessing vulnerability in natural resource settings have not directly addressed risk.

Within ecological systems, the commonly used framework has focused on quantifying exposure, sensitivity and adaptive capacity of individual plant or animal species, or the ecosystem (Fig. 15.3) (Glick et al. 2011; Furniss et al. 2013). In some cases, the sensitivity of plants, animals, and ecosystems to changes in climate has been documented in the scientific literature or observed in long-term resource inventories (Peterson et al. 2011); additional sources of information include traditional knowledge (Laidler et al. 2009) and expert knowledge (Alessa et al. 2008; McDaniels et al. 2010; Moyle et al. 2013). Tools have also been developed to quantify ecological responses to future climate scenarios (Joyce and Millar 2014), although natural resource vulnerability assessments have been qualitative as well as quantitative.

Vulnerability of economic enterprises on rangelands has not been widely addressed. Few studies have explored the intersection of environmental variability and risk with economic variability and risk in livestock operations (see Torell et al. 2010). Few adaptation strategies identified in ecological or economic vulnerability assessments address social vulnerabilities. However, often the need for understanding the social

Fig. 15.3 Key components of vulnerability, illustrating the relationship among exposure, sensitivity, and adaptive capacity for ecological systems (modified from Glick et al. 2011)

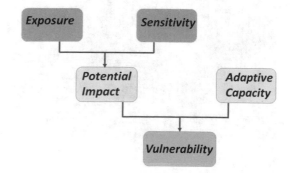

component is identified: for example recognizing that rangeland manager perceptions about climate change inhibit their receptivity to adaptation options (Briske et al. 2015). Vulnerability of the agricultural economic sector has been quantified using market models to identify likely shifts in livestock and crop management strategies based on economic return (Heyhoe et al. 2007; McCarl 2011; Mu et al. 2013).

Box 15.1: Adaptively Managing Environmental and Economic Risks: Pawnee National Grassland, Colorado, USA

East of the Rocky Mountain Front Range, the Pawnee National Grassland, managed by the USDA—Forest Service, sits within a mosaic of private and State of Colorado land, and the USDA Central Plains Experimental Range. The Pawnee National Grassland is managed for multiple ecosystem goods and services— domestic livestock grazing, wildlife, threatened and endangered plants and animals, recreational opportunities, and oil and gas development. These multiple goods and services interconnect the interests of public land managers with private land ranchers (Fig. 15.6). Drought can occur at any time in the region and multiyear droughts of 8–14 years occurred in the 1930s and the early 1950s (Lauenroth and Burke 2008; Evans et al. 2011). Facilitating ecosystem resilience and reducing risk of resource degradation are important to these grassland managers. Reducing economic risks when drought reduces forage availability from public and private land is important to the private land livestock owners. Drought often brings conflict between environmental and economic interests because it directly involves environmental and socioeconomic systems.

Domestic livestock grazing on Pawnee National Grasslands (photo courtesy of David Augustine)

(continued)

(continued)

Adaptive grazing management has been used to create and maintain diverse vegetation structure—a combination of short and mid-tall vegetation patches—that is needed to meet habitat needs for wildlife. The desired objective of rangeland management is to provide available forage for both wildlife and domestic livestock in a manner that is consistent with other resource objectives. Grazing management is accomplished on a total of 162 active allotments in partnership with two grazing associations. The majority of livestock grazing occurs May through October and most allotments are continuously grazed for this period. Annual grazing allocations are cooperatively determined at spring meetings with the FS Range Staff and Grazing Association Boards.

The 2002 drought and high temperatures severely impacted many economic sectors in Colorado (Pielke et al. 2005). In the years preceding 2009, very dry conditions forced grazing allotments on Pawnee to be vacated earlier than initially planned. Grazing association boards and permittees desired more notice about stocking adjustments in order to make more informed decisions about their overall operation. To address these concerns, an annual stocking strategy was developed that employs resource and management information to allocate livestock at the start of the season with a mid-season reevaluation of stocking levels. Resource and management information includes (1) precipitation over the previous year and the last 15 years both annually and for the growing season; (2) stocking rates for the previous year and the last 15 years; and (3) management and objectives including the current management, desired condition of the rangeland, current trends, and priority natural resources to be managed. Using this information, initial stocking recommendations for each allotment reflect condition assessments (poor, moderate, good). Midseason grazing adjustments are based on soil water availability and midseason allotment condition. This strategy is designed to be adaptive, as well as lay out possible scenarios so that the permittees are able to better anticipate their grazing on federal lands and make appropriate adjustments in their overall operations.

Allotment strategy development recognizes the risks of the federal managers and the risks of the private landowners and permittees. USFS personnel have the responsibility to manage the environmental risk as weather and other environmental stressors affect ecosystem services produced from the Pawnee National Grassland. The individual livestock owner has the responsibility to manage the economic risk as influenced by the supply and quality of forage from both federal and private land and livestock market fluctuations. When a drought is widespread, increased demand and high cost of forage may be coincident with volatile and declining cattle prices. The adaptive process gives the federal grassland managers and the permittees information and a timeline in which to make decisions relative to the risks they manage.

Vulnerability has been characterized as a function of both people's sensitivity to a change event and their capacity to adapt to it (Marshall et al. 2013). Consequently, people can be vulnerable because they are highly sensitive to change, or have insufficient adaptive capacity to accommodate change, or both. Importantly, people that are highly sensitive to change are not necessarily vulnerable if they have correspondingly high levels of adaptive capacity. According to this characterization, it is possible to identify who is more vulnerable than whom, and why. Climate sensitivity within the social subsystem is typically measured as a function of resource dependency (Marshall 2011). That is, the more dependent landowners are on the current rangeland enterprise, the more sensitive they are likely to be to climate change. Dependency can be described in economic terms, such as the goods and services produced, income sources, and alternative employment opportunities, and in social terms, including occupational identity, place attachment, employability, networks, environmental knowledge, and awareness (Marshall 2011).

Few assessments have contributed directly to implementation of adaptation actions on the ground (Noble et al. 2014). This lack of action can be ascribed to several factors (Yuen et al. 2013; Joyce and Millar 2014). The assessment could lack clear definitions of vulnerability and adaptive capacity, or have too narrow of a focus, such as natural resources that cannot be managed or changed. Weak quantitative components could include incomplete data, or inadequate descriptions of the interactions between climate change and other environmental stressors. The assessment may have no connection to management decisions such as insufficient information or a method to successfully prioritize among sensitive resources or to evaluate adaptation management. Lastly, the assessment may have failed to engage decision makers and/or the public. Further, few adaptation actions incorporate incentives to encourage human behavior toward management to sustain resilient ecosystems (for example, sustained drought management, Marshall 2010; Marshall and Smajgl 2013). Vulnerability assessments may often fail to implement adaptation because opportunities for collective learning by managers, the public, and decision makers are minimized or overlooked (Yuen et al. 2013). Collective learning arises when various goals, values, knowledge, and points of view are made explicit and questioned to accommodate conflict and reach common agreement. Collective learning represents the basis for identifying the collective action to tackle a shared problem (Yuen et al. 2013) and may be particularly important in a vulnerability assessment of a socio-ecological system.

Codependency between ecological and socioeconomic subsystems suggests that vulnerabilities are intrinsically linked. Further these systems operate in larger societal institutional systems. Using the vulnerability components of exposure, sensitivity, and adaptive capacity, linkages between the ecological and the social subsystems can be conceptualized (Marshall et al. 2014). In this portrayal of a linked socio-ecological system, ecological vulnerability to climate change can be seen as the exposure to climate change in the social subsystem (Fig. 15.4). Vulnerability in the socioeconomic subsystem is a function of the sensitivity of the social subsystem (dependency on natural resources), the adaptive capacity of the socioeconomic subsystem, and the vulnerability of the ecological subsystem (Marshall et al. 2014). There is feedback

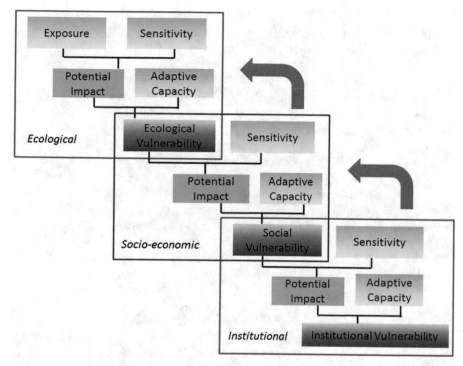

Fig. 15.4 A framework for conceptualizing vulnerability across ecological, socioeconomic, and institutional domains (modified from Marshall et al. 2014). Social vulnerability consists of sensitivity to change, the degree of resource dependency (ecological vulnerability), and adaptive capacity. Similarly, institutional vulnerability consists of institutional sensitivity to change, exposure (social vulnerability), and institutional adaptive capacity

from the socioeconomic system to the ecological system—this feedback may positively or negatively affect ecological vulnerability. We have added an institutional component to the model where the exposure term for the institutional subsystem is the vulnerability of the social subsystem. Vulnerability of the institutional subsystem feeds back to the socioeconomic and the ecological subsystems. Institutional components can be market structures, as when collapse of livestock enterprises led to restructuring of the regional livestock market in the 2012 drought in the USA. Institutional components involve government intervention as in the case with drought relief programs in the USA or national relief following dzud (severe winter weather disaster) in Mongolia. These interventions often do not reflect collective learning or desired collection action across the social and the institutional subsystems and thus may not facilitate resilient decision making at the enterprise or individual level (Thurow and Taylor 1999; Fernández-Giménez et al. 2012). As new markets, new government regulations, and climate change introduce new learning opportunities, vulnerabilities can arise at the household unit or livestock enterprise.

Box 15.2: Collaborative Management as a Means to Minimize Climate Risk: Mongolian Plateau, China

Inner Mongolian rangelands can be environmentally challenging for people and the primary industry of raising livestock. The climate of Inner Mongolia is cold; average annual temperatures vary between 1 and 2.5 °C. Precipitation is low and erratic with the majority occurring during three summer months. Winter storms can be harsh and frequent drought is associated with wind erosion. The culture had adapted to this harsh climate using strategies based on three core components: mobility, cooperation, and reciprocity (Dalintai et al. 2012). These strategies helped to sustain the resilience of this tightly connected socio-ecological system. For drought, these strategies included *otor* and *surug*. In *otor*, herds, through kinship relations, were moved to better grasslands in response to drought conditions; *surug* was a system in which herder leased a core number of their young female animals to herders in areas not as affected by drought. When the conditions in the original herders' area improved, they took back their livestock—this short-term leasing provided a way to maintain the core of their herd by matching forage production with animal demand.

The social setting in this area of China has seen a continual change since the 1950s with collectivization between 1950s and mid-1980s and then market reforms in the early 1980s. These changes affected many of the tight connections in the socio-ecological system. Collectivization strengthened production but weakened the mobility component by encouraging semi-nomadism. Market-oriented reforms emphasized rights of ownership by individual households, attempting to incentivize herders to use their grasslands rationally and sustainably. However, these changes eroded the strategies to adapt to the harsh climate. Mobility of herders was limited; consequently grasslands were overgrazed. Further, government structures assumed the role of providing services, weakening the kinship-based social structure of cooperation and exchange. At the start of the twenty-first century, a top-down effort by the government was initiated to address the degradation of grasslands. Grazing management strategies included restrictions on where and when grazing could occur, thus making it difficult for herders to migrate herds. Grass planting occurred and the government instituted a supplemental feed program. Herders were also moved away from the grasslands. These changes exacerbated the herders' poverty. However, degradation continued; Dalintai et al. (2012) suggest that the policy aimed at protecting the grasslands and improving herder's living standards proved ineffective because these most recent changes were implemented in a top-down manner. Essentially, over all of these social and economic changes, the vulnerability of the socio-ecological system to drought and winter storms increased.

A project was implemented to address issues of poverty and environmental degradation and to help preserve the traditional culture in the Sonid Left Banner area of Inner Mongolia, located on the south-eastern part of the

(continued)

(continued)

Mongolian Plateau (Dalintai et al. 2012). The climatic risks and uncertainties remained as in the past; however the overgrazing had heightened environmental risks which were compounded with greater economic risks. This project attempted to nuance the land tenure structure to bring back the traditional ways but with a new structure that meshed with the land tenure and market systems. Cooperatives were created to facilitate a collective approach for the use of the grassland and a cooperative division of labor. The boundaries of land units from several households were merged, with the households still retaining ownership of the land. All animals were herded together. Herders were encouraged to take on the responsibility to restore the grassland. Collective decisions on grazing methods drew on traditional ways and information provided by research ecologists collecting data and working with the herders. The collective purchasing and marketing of products was an attempt to reduce the economic risks as well as improve market negotiation skills. Herders recognized that reducing the costs of a disaster is in its own way a kind of income. Project scientists realized that lowering the risks to herder production operations was more practical than attempting to increase their incomes. The cooperative's main problem is learning how to adjust to the government's grassland protection policies to better meet the local needs of the herders (Dalintai et al. 2012). The breadth of decisions that the local herders can take on directly affects the final performance of the community-based grassland management projects. Eventually, the restoration of the arid ecosystem and incomes both increase with greater local decision making.

The grand challenge for vulnerability assessments in rangeland systems and in getting adaptation on the ground is to connect awareness of vulnerability with the potential for adaptation across ecological, socioeconomic, and institutional systems. An assessment could identify the level of risk, urgency of action, efficacy, and cost-effectiveness of adaptation options, and engage and empower stakeholders, including vulnerable populations, in adaptation planning (Salinger 2005; Marshall and Johnson 2007; Joyce and Millar 2014). Processes that facilitate collective learning in the vulnerability assessment could help to identify adaptation approaches that most effectively accommodate and support rangeland managers and enterprise owners.

15.4 Resilience in Heterogeneous Systems

15.4.1 Resilience in Socio-Ecological Systems

Resilience has emerged as an important concept to guide and support more inclusive and effective approaches to the management of combined social and ecological systems (Ludwig et al. 1997; Berkes and Folke 1998; Levin et al. 1998). Resilience was first characterized as persistence of ecological systems and described as their ability to absorb change and disturbance and still maintain the same relationships among component parts (Holling 1973; see Chap. 6, this volume). At this time, human and natural systems were treated independently and it was implicitly assumed that ecosystems responded to human use in linear, predictable, and controllable ways (Folke et al. 2002). The concept of resilience has since gained substantial momentum through recognizing the complexity and variability of natural and social systems (Gunderson 1999; Walker and Janssen 2002; Davidson-Hunt and Berkes 2004). For example, social and natural resource systems are intrinsically linked through intricate and dynamic cycles that are, by their very nature, adaptive (Holling and Meffe 1996; Holling 2001). These linked systems continually face interventions or disruptions that "reset" the natural cycles of recovery, growth, and adaptation (Holling 1996). Adaptive systems are flexible, continually change, and can cope and reorganize. Change and adaptation are now understood to be integral features of the dynamics of socio-ecological systems and have replaced the previous concept of ecological stability (Holling 1973; Folke et al. 2002).

The concept of resilience is especially apt when rangelands are envisioned in a socio-ecological context. Rangelands will have to continually adapt to climate-induced changes, including drought, heat waves, wildfires, flooding, greater weather variability, and shifts in rainfall and seasonal patterns (Walker and Schulze 2008), and increased pests and diseases (Volney and Fleming 2000). In addition, rangeland systems will have to address cultural change, including the acceptance and adoption of new best practices, and technology that enhances adaptation to climate change and reduces greenhouse gas emissions (Darnhofer et al. 2010; Fleming and Vanclay 2010; Marshall et al. 2016). In some instances, these climate-induced changes may be so severe or adaptive capacity very limited, that climate-related regulatory change through governance or social institutions will also need to be addressed (Cabrera et al. 2006).

Management of socio-ecological systems involves the maintenance of system properties and feedbacks that confer resilience without compromising the ability to cope and adapt to future change (Holling and Meffe 1996). Specifically, successful adaptation on the rangelands means that landowners and their enterprises will remain viable through time despite an increasing volatility within social and ecological subsystems (Fig. 15.1). Remaining viable depends not only on maximizing productivity during any one season, but also on minimizing negative consequences to future productivity (McKeon et al. 2004). Climate change requires that landholders make the most of good years and avoid losses and reductions in resource condition in drought years to an extent as yet unprecedented (Hobbs et al. 2008). If stocking rates are too high at the onset of drought, for example, soil erosion will be accelerated and the productivity of future years will be diminished (Watson 2004; 2008). These decisions involve trade-offs between short-term profit maximization and risk avoidance (Hammer et al. 1996; Hammer 2000; Hansen 2002; Hertzler 2007).

15.4.2   Variations in Adaptive Capacity of Landowners

Natural systems have an inherent adaptive capacity that has evolved from responding to past disturbances including climate. Adaptation has previously focused on the manipulation of natural resources, and economic resources. However understanding social heterogeneity among enterprise owners is important for effective management of rangelands and climate adaptation planning (Emtage et al. 2007). Adaptive capacity is the ability to convert existing resources—natural, financial, human, social, or physical resources—into a successful adaptation strategy (Marshall et al. 2014). Characteristics that contribute to adaptive capacity include creativity and innovation for identifying adaptations; testing and experimenting with various adaptations; recognizing and responding to effective feedback mechanisms; employing adaptive management approaches; possessing flexibility; being able to reorganize given novel information; managing risk; and having necessary resources at hand (Marshall et al. 2010).

We emphasize that adaptive capacity is not solely dependent on having financial or ecological resources. On rangelands, and at the landowner scale, adaptive capacity has been more comprehensively operationalized according to four measurable attributes reflecting landowners' and managers' skills, circumstances, perceptions, and willingness to change (Marshall et al. 2012). These have been described as (1) how risks and uncertainty are managed; (2) the extent of skills in planning, learning, and reorganizing; (3) the level of financial and psychological flexibility; and (4) the anticipation of the need and willingness to contemplate and undertake change (Marshall 2011; Park et al. 2012). While other measures have been developed in other contexts (e.g., Cinner et al. 2009) these four dimensions have served as the basis from which several studies on rangelands have examined adaptation processes (Marshall et al. 2011; Webb et al. 2013; Marshall and Stokes 2014).

Australian enterprise owners, as a group, exhibited highly heterogeneous levels of adaptive capacity (Marshall and Smajgl 2013). In fact, of the 16 possible combinations describing adaptive capacity on rangelands, all combinations were represented to some extent. Only some individuals had the capacity to respond successfully to policies and practices that enhance climate adaptation. This suggests that the current social heterogeneity in adaptive capacity will profoundly limit the extent to which landowners in Australia can respond to lower summer rainfall and increasing drought projected to occur in 2030 (Cobon et al. 2009).

Landowners that can anticipate or effectively react to the effects of climate change are more likely to adapt to new climate conditions. Landowners with a higher adaptive capacity tend to display consistent characteristics that have enabled researchers to more clearly define or describe what makes for a higher adaptive capacity (Marshall 2010). While management actions cannot eliminate risks of impacts from climate change, management can increase the inherent capacity of ecosystems to adapt to a changing climate (Settele et al. 2014). For example, humans can select adaptation actions that guide the transition or transformation of a socio-ecological system toward an alternative system that may be more resilient to novel climatic conditions (Hobbs et al. 2013). Understanding social heterogeneity across enterprise owners could help tailor climate adaptation planning.

15.5 Management Responses to Past Change

15.5.1 Drought

Human activities can fundamentally alter the social-ecological interactions within rangeland systems (Stafford Smith et al. 2007), particularly as enterprise owners and managers respond to biophysical drivers such as climate, or socioeconomic drivers such as local, regional, and international markets (Reynolds and Stafford Smith 2002; Reynolds et al. 2007; Chap. 8, this volume). Semiarid and arid rangeland systems may be among the most tightly coupled socio-ecological systems because of the high degree of climate variability and dependency among system components (Stafford Smith et al. 2007). We look to studies of past management response to change for insights that could benefit climate change adaptation.

Drought is a normal part of climate and, although common in arid and semiarid rangelands, drought can occur in all types of climate (Thurow and Taylor 1999; Wilhite and Buchanan-Smith 2005). Drought is referred to as a slow-onset natural hazard, where effects of drought accumulate slowly over time. This slow onset, and the temporary nature of drought, often leads to a lagged response by landowners and managers. Drought impacts can be costly, with reductions in water supplies, forage, and livestock productivity. Herd liquidations, one response to drought, often occur on the downside of the price curve for livestock and restocking on the upside of the price curve (Bastian et al. 2006), resulting in financial challenges for the landowner (Torell et al. 2010).

Drought has been a learning experience at the scale of an individual livestock enterprise; however, some enterprises may still be underprepared for subsequent droughts. Over 500 cattle ranchers in the state of Utah were surveyed after the 1999–2004 drought, described as the sixth most severe drought since 1898 in Utah (Coppock 2011). Herd size varied from less than 5 brood cows to over 300 head and grazed ecosystems included desert, grassland, and high-elevation grasslands. Only 14 % of cattle enterprises were prepared for the 1999–2004 drought. The experience of this drought increased the number of ranchers that self-identified as being better prepared for subsequent drought to 29 % when they were surveyed in 2006. A negative experience in the 1999–2004 drought and the perception that another drought was inevitable were primary motivations for increasing drought preparedness. The most common risk management actions put in place by ranchers after the 1999–2004 drought included improving water for livestock, and diversifying family income (Table 15.2). While adaptive capacity for drought improved with this experience, still greater than 50 % of the livestock operators were only somewhat prepared or not prepared for the next drought. This lack of adaptive capacity ensures that crisis drought management will begin again when the next drought occurs.

Table 15.2 Risk-management actions used by Utah ranchers in 2009 for drought preparedness (Coppock 2011)

Tactic	Percentage of respondents saying "Yes, I am doing this" (%)
Improving water for livestock	76±4.0
Diversifying family income	68±4.4
Improving irrigation for hay production	67±4.3
Improving land management	57±4.5
Reducing stocking rates	56±4.9
Enrolling in government disaster compensation programs	55±5.0
Increasing capacity for hay production	53±4.4
Purchasing feed insurance	38±3.9
Seeking extension information	37±3.8
Using Internet drought forecasts	31±4.2
Using forward contracting for livestock sales	30±4.2
Increasing capacity for hay storage	29±3.8
Planning to use grassbanks	26±3.7
Renegotiating bank loans	17±3.5
Other (19 tactics). Most common: (1) expanding grazing land and investing in improved grazing systems (seven), (2) researching drought and drought management (two)	9±2.3
Using forward contracting for hay purchases	8±2.5

Total survey response was 96.7%, resulting in 509 responses

Managers of rangelands enterprises also confront volatility in the meat and fiber market, in addition to climatic variability, and these two events are often interrelated. Using a series of historical drought episodes in Australia (Fig. 15.5), Stafford Smith et al. (2007) identified the linkages between operator decisions and broader social and economic developments. During every drought, a common set of events occurred: (1) good climatic and economic conditions for a period, leading to local and regional social responses of increasing stocking rates, setting the preconditions for rapid environmental collapse, followed by (2) a major drought coupled with a market decline making destocking financially unattractive, further exacerbating grazing impacts, and then (3) permanent or temporary declines in grazing productivity, depending on follow-up seasons coupled again with market and social conditions. One conclusion authors drew from this study is that learning from climate and economic events is often temporally mismatched. Decisions were driven by short-term economic cycles. However, the return times of some climatic events were outside of the life spans of enterprise operators, which limited this information from influencing the short-term decisions. In addition, institutional and government responses, including monitoring and post-drought surveys, were too slow to stop the degradation. Drought management responses require information sharing among managers, industry,

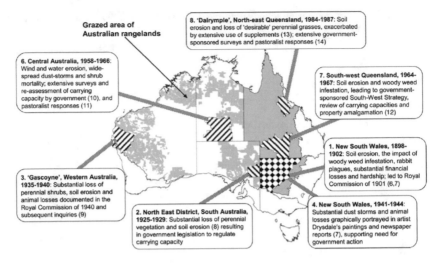

D. Mark Stafford Smith et al. PNAS 2007;104:20690-20695

Fig. 15.5 Drought episodes in Australia were used to describe the interactions between the ecological and the social processes in socio-ecological rangeland systems (from Stafford-Smith et al. 2007). Shading indicates pastoral areas (sheep or cattle), forward hatching indicates episode regions with longer droughts, and back hatching indicates shorter droughts (diamonds indicate that New South Wales had one of each)

science, and institutions at regional and multigenerational scales. Drought-monitoring systems put in place after widespread drought, such as the national monitoring in Australia, or the National Integrated Drought Information System (NIDIS) and the Drought Portal in the USA, offer real-time information about drought as well as local and regional climate.

Drought planning for enterprise operators on rangelands is readily available through government and academic institutions and long-term management strategies for climate variability have been developed. Yet few enterprise operators employ risk management strategies such as conservative stocking or flexible stocking. The use of flexible grazing management that fluctuates with favorable and unfavorable years can produce greater economic return than a set conservative strategy (Torell et al. 2010). However, the return on this approach was highly dependent on the accuracy of seasonal climate forecasts and a careful trade-off analysis of stocking and destocking. Enterprise operators were hesitant to use near-term climate forecasts unless they saw economic and environmental benefits associated with seasonal forecasts (Marshall et al. 2011) or forecast tools were tailored specifically to users' needs (Dilling and Lemos 2011).

Management actions to address proactively as well as during drought are widely available through extension agents, consultants, or professional organizations. Further, governments, industries, and communities have introduced a range of economic and policy initiatives. These have included regulatory instruments, educational instruments, and voluntary and market-based instruments (Moon and Cocklin 2011). However, these efforts have been variable in their success (Sankey et al. 2009; Briske et al. 2011). Research suggests that a significant part of the reason that sustainable practices are not adopted by rangeland enterprise owners is that policies and practices are typically founded on the "average" or "typical" resource user and do not appreciate the extent of diversity among these populations (Marshall and Smajgl 2013; Briske et al. 2015; Roche et al. 2015). Implementing resource-protection strategies without sufficient knowledge of the capacity of people to cope and adapt to them may impose levels of stress upon individuals and communities to such an extent that their ability to adapt, tolerate, or prosper under the new conditions is compromised. Strategies that generate stress and conflict are also likely to result in poor compliance and leave the natural resource unprotected (May 2004).

The greatest challenge under a changing climate may not necessarily be the identification of specific management options, but rather the need to encourage human behavioral changes to sustain the socio-ecological rangeland system. Drought intensity and duration are likely to increase under climate change (Dai 2011, 2013). Few past strategies incentivized human behavior toward management to sustain resilient ecosystems (Marshall and Smajgl 2013). Future management may need to be responsive to the decision-making processes that rangeland owners and operators use. Four distinct patterns of decision making in drought were identified in surveys of cattle ranchers in western USA. The patterns encompassed using a long-term strategy dealing with climatic uncertainty, facing drought by building efficiency into the operation and relying on strong local ranching networks, second-career perspective with reliance on outside income and conservative stocking, and, last, an experimental approach to ranching using evidence-based adaptation to drought (Wilmer and Fernández-Giménez 2015). Outreach by extension or academia to these different patterns of decision making cannot rely on one approach, but rather needs to reflect on decision-maker ways of knowing.

15.5.2 The Influence of History in the Human Response to Change

We close this section with four case studies that identified motivations and adaptive responses to different types of socio-ecological disturbances: environmental change caused by human activity in the Solomon Islands; economic change resulting from closure of a timber mill in Canada; political, social, and economic change in a multiethnic rural village in Romania; and responses to policies for adapting to sea-level change in Australia (Fazey et al. 2015).

Change in each of these case studies was occurring daily; however, the adaptive responses to sudden change were influenced by historical legacies. In the mill closure case study, one community had previously experienced several economic changes (fur trading to mining to timber) and, given this experience, could cultivate a new economy with the mill closure, a contrast to the community without this historical legacy. In the sea-level policy development, the past practice of compensating for damage and current favorable attitudes toward private property rights facilitated the influence of a minority group on resisting change.

Change and the response to change can accelerate further change. On the Solomon Islands, population pressure resulted in subsistence resources dwindling, and the initiation of cash cropping practices, which reduced or moved food gardens from the more fertile areas, increasing pressure on the ecological systems. The increasing numbers of people attempting to respond to this need also resulted in an acceleration of change in the community.

Connections among ecological, economic, and social processes constrained and enhanced the likelihood of success in the multiethnic community. Power in this community was intertwined with conforming to the social norm of a combination of subsistence agriculture and cash-making activities. Political power and education allowed one ethnic group to work outside the village accumulating cash. Newly arrived immigrants creatively adapted; however, their ways did not conform to society's expectations, particularly in subsistence agriculture. Consequently, they were unable to gain important political capital and integrate into village life according to the prevailing social norm.

The development of future adaptation approaches/strategies will need to consider underlying socio-ecological assumptions, values, and principles, and how understanding past change can provide inspiration for new and transformative futures (Fazey et al. 2015). It is recognized that past disturbances influence the response of ecological systems to future disturbances. Equally important, adaptation planning for climate change must recognize that the legacy from historical events influences how individuals and a community will respond to current events and plans for the future.

15.6   Developing Adaptation Options

As we have previously noted, the emphasis on adaptation planning has shifted from a narrow focus on biophysical vulnerability to a broader vision of social, economic, and biophysical vulnerability, including the capacity of humans to respond. Broadly, adaptation needs have been defined by the Intergovernmental IPCC as those circumstances requiring action to ensure safety of populations and security of assets in response to climate impacts (Agard et al. 2014). Effective adaptation planning requires an assessment of the risk of climate-related impacts and hazards, and the vulnerability and exposure of human and natural systems as impacted by socioeconomic and climate drivers (Fig. 15.4). Ecosystem services such as food security, clean water, biodiversity, and disease and flood control are dependent upon ecological processes within the socio-ecological system. Consequently, biophysical needs include sustaining these systems and resources under climate change. Social needs include sustaining financial, human, social, and cultural assets (Noble et al. 2014). In the socio-ecological rangeland system, social needs can involve addressing financial flexibility in a livestock enterprise, risk perceptions of rangeland managers, cultural patterns of grazing, or psychological stresses related to extreme events such as wildfire or drought.

Adaptation options can be classified as structural and physical, social, and institutional (Noble et al. 2014). Within structural and physical, engineering options for drought management could include new or enlarged reservoirs to store water, more efficient water delivery systems, and communications technology as cell phones and drought or flash flood warning systems. Physical adaptation also includes management of ecosystems and watersheds such as enhanced invasive species management, minimized soil erosion, and restoration of ecosystems after natural disturbances (Millar et al. 2007). Social adaptation options could include changes in the enterprise operation such as supplemental feed, conservative stocking, and changing type of livestock (Joyce et al. 2013) as well as options to improve the adaptive capacity of enterprise owner (Marshall and Smajgl 2013).

The private sector and local institutions will bear the greatest responsibility for developing and implementing adaptation strategies and practices (Noble et al. 2014). Livestock enterprise owners and industries associated with these enterprises will be motivated to protect their financial investments under a changing climate— productivity of their land, value of their genetic stock, infrastructure supporting markets, as well as the markets themselves. Local institutions will be key actors in adaptation, as they attempt to implement the top-down flow of policy, such as programs to address responses to extreme climatic events. However, limited availability of funding and resources, especially in developing countries, and the lack of national government support will challenge the ability of the private sector to implement adaptation options. Goals of private sector adaptation actions may not be consistent with local and national adaptation policies (Noble et al. 2014); similarly governmental actions in response to extreme climate events could further exacerbate local adaptation efforts (Fernández-Giménez et al. 2012).

All adaptation is local and no single adaptation approach works in all settings (Noble et al. 2014). Management actions rarely have been motivated by a single objective; consequently, adaptation options have been identified for managing plants, animals, and ecosystem processes along the lines of no-regret, low-regret, and win-win strategies. The motivation here is that these strategies may make ecological and economic sense locally in the current climate and may provide a means of protection as climate continues to change (Millar et al. 2007, 2012; Joyce et al. 2013). For example, in the face of an impending stress such as heat waves or drought, management would focus on actions that protect the existing assets and maintain what humans currently value in ecosystems. Protecting existing animals during increasing temperatures might imply implementing some type of heat stress management. Maintaining what humans value might imply off-enterprise employment to supplement expenses incurred by drought and other weather extremes. These strategies might be considered no-regret or low-regret strategies as heat waves and drought are frequent challenges on rangelands. A more intensive response to impending climate stresses could focus on ensuring that current ecosystems can regenerate or recover after disturbances such as drought or wildfire. These options could involve aggressive invasive species management, alternative feed strategies during drought, or planting after disturbance events—all focused on keeping the ecosystem resilient and sustaining ecosystem services and the current enterprise structure.

As climate continues to change, incremental adaptation actions may not suffice (Kates et al. 2012; Joyce et al. 2013) and, in some cases, may institutionalize management practices that are maladaptive under the continually changing climate (Dilling et al. 2015). Enabling socio-ecological rangeland systems to adapt may be a desired strategy. This approach would assist climatically driven transitions to future novel states while mitigating and minimizing undesired and disruptive outcomes, such as loss of ecosystem productivity, or socioeconomic welfare in the community. Given that shifts in climatic trends and variability will continue into the future, adaptation planning represents an iterative process where climate-related risks and hazards must be continually reevaluated.

Where socio-ecological systems have been resilient in challenging environments, collective learning is likely at the core of that resilience. This collective learning occurs as societies deal with the variability across the biological and socioeconomic environments. Strategies in rangeland communities range from diversifying use of plants and habitats and income opportunities, migration of herds and households, flexibility in social organizations and livelihood strategies, grass banking or grazing reserves, and institutions of reciprocity and exchange (Fernández-Giménez et al. 2012). In Mongolia, the "otor" is one such strategy that has developed over time. Here herders and a portion of the household migrate to fatten animals in the fall, to seek better pastures in a drought or to flee bad weather and poor forage in a dzud. The mobility of herders is somewhat restricted by government policy but not always monitored or checked. However, Fernández-Giménez et al. (2012) concluded that while household units were well prepared for a dzud through the use of otor, these households became vulnerable when in-migration of livestock from other communities occurred. Further, short-term government relief aid in response to these extreme events minimizes loss of life and impoverishment, but it may contribute to social vulnerabilities in the long term, such as lack of individual initiative (Fernández-Giménez et al. 2012; Chap. 17, this volume). Under climate change, adaptation will be a continual process, as individuals and communities seek to adapt to new environmental conditions that arise gradually or through abrupt change.

Collective learning can also occur where local and diverse groups come to realize the challenges that they face, such as the threat of development and rangeland fragmentation (Case Example 15.3) or concern that regulation or legislation will be put in place to protect wildlife or habitat on which the private sector depends (Case Example 15.4). In some cases these groups can self-organize to begin the process of addressing their concerns. In other cases, the group can be motivated by a third party who has little or no stake in the environmental or economic concerns.

Box 15.3: Self-Organizing Community Linking Management and Science: Malpai Borderlands Group in Southwest, USA

A group of private landowners identified the threat of fragmentation from subdivision and development on their landscape in the southwestern corner of New Mexico and the southeastern corner of Arizona. Residential development expanding from urban areas had already resulted in subdivision of some ranches. Additional landscape fragmentation and woody species encroachment of grasslands could permanently limit future options for sustaining rural livelihoods as well as affect ecosystem productivity and biodiversity. The Malpai Borderlands Group (MBG), formally organized in 1994, is a collaborative effort with environmental groups and state and federal agencies, built around goals shared by neighbors within the community and directed at protecting and restoring ecological diversity and productivity of around 324,000 ha. The Board of Directors includes local ranchers, scientists, and other stakeholders. The landscape includes about 57 % private land, 20 % state trust lands, 11 % National Forest, and 7 % Bureau of Land Management-administered land.

The Group's goal is "To preserve and maintain the natural processes that create and protect a healthy, unfragmented landscape to support a diverse, flourishing community of human, plant, and animal life in the borderlands region (http://www.malpaiborderlandsgroup.org/)." To help facilitate this goal, the MBG incorporated as a 501(c)(3) nonprofit organization, and was therefore capable of accepting tax-deductible contributions and holding conservation easements. The MBG has protected 32,000 ha of private land through conservation easements. These easements have had the indirect effort of easing management challenges by enabling ranchers to consolidate properties through purchase of additional properties and for other ranchers, the opportunity to avoid defaulting a mortgage or avoid the need to take a mortgage (Rissman and Sayre 2012). These easements have strengthened the social networks among landowners with easements and the MBG that holds the easement. The resources available for easement owners, such as financial incentives, have promoted increased management on these protected lands.

One of the more innovative projects devised by MBG is the concept of a "grassbank." Originated by the Animas Foundation, owner of the Gray Ranch and a partner in MBG, a grassbank is a concept in which grass on one ranch is made available to another rancher's cattle in return for the conveyance of land-use easements prohibiting subdivision. Grassbanking experiences of three ranchers changed their perceptions of grazing effects and resulted in 30–65 % reductions in their stocking rates on their ranches, a reduction not stipulated in the grassbanking arrangement (Rissman and Sayre 2012).

From the beginning MBG has been strongly committed to using the best available science and technology to achieve their objectives. The Group draws upon the input of a Science Advisory Committee to establish priorities and seek resources. This collaborative effort has resulted in a number of conservation treatments, enabling 28,000 ha of prescribed burn. The collaboration among the MBG and scientists from a wide array of disciplines and affiliations has resulted in enhanced science and management to support adaptation on the ground.

**Box 15.4: Adapting to Climate Change with Social Learning: Gunnison
Basin, Colorado, USA**

The Nature Conservancy, a global nongovernmental organization that emphasizes conservation, began to engage community members in the Gunnison
Basin, Colorado, USA, about climate change. A workshop was held in 2009 to
explore the potential effects of climate change, using climate scenarios and a
structured vulnerability assessment. Many questions about the potential effects
remained within the community. The Nature Conservancy and several other
organizations formed a working group which began the process of exploring
the potential impacts of climate change (http://www.conservationgateway.org/
ConservationByGeography/NorthAmerica/UnitedStates/Colorado/science/
climate/gunnison/Pages/default.aspx). Building on this interest, in 2011, the
Gunnison Climate Working Group was officially formed as a chartered partnership of 14 public and private organizations in Colorado's Upper Gunnison
Basin (http://southernrockieslcc.org/project/gunnison-climate/). Goals of the
Gunnison Climate Working Group are to (1) increase understanding and
awareness of threats posed by climate change to species, ecosystems, and the
benefits they provide to the people of Gunnison Basin; (2) identify and prioritize strategies and techniques for helping people and nature cope with climate
change; and (3) promote coordination, collaboration, and effective implementation of strategies.

The US Fish and Wildlife Service's Southern Rockies Landscape
Conservation Cooperative (LCC) provided funding to the Nature Conservancy,
on behalf of the Working Group, to (1) complete a comprehensive vulnerability assessment to identify species and ecosystems most at risk to climate
change; (2) develop a set of adaptation strategies for priority species and ecosystems; (3) design and begin implementation of a local adaptation demonstration project; and (4) document tools, methods, and lessons learned to
share with others across the Southern Rockies LCC through a climate adaptation learning network.

The Climate Change Vulnerability Assessment for the Gunnison Basin is
a first attempt at identifying ecosystems and species of the Gunnison Basin,
Colorado, that are likely to be affected by climate change and why they are at
risk. Climate projections suggest that the natural environment, ecosystems,
and species of the Gunnison Basin will change significantly over the coming
decades (Neely et al. 2011). The results indicate that many of the natural
features of the Basin (50 % of ecosystems and 74 % of species of conservation concern) are susceptible to loss, degradation, or other changes associated
with warming temperatures. This report provides a foundation for the
Gunnison Climate Working Group's next step: developing social-ecological
adaptation strategies to support resilience of social-ecological systems,
including species, ecosystems, and human livelihoods in the Gunnison Basin.

(continued)

(continued)

The tools, methods, and findings of the Gunnison Basin vulnerability assessment go beyond habitat adaptation strategies applied to support populations of Gunnison sage-grouse. The new tools build ecosystem resilience and support the Gunnison Basin agricultural and recreational economies. The vulnerability assessment provides a scientific foundation for a robust decision-making process which can be carried out over a larger landscape to inform and direct conservation delivery mechanisms for use by multiple partners.

Following completion of the vulnerability assessment, the Gunnison Climate Change Working Group applied for Wildlife Conservation Society funding to design and implement an on-the-ground climate adaptation demonstration project. Wildlife Conservation Society funds, matched by the US Fish and Wildlife Service grant, enabled the Working Group to complete the first phase of a priority strategy. This strategy was considered to be one of no-regrets because it is considered to be effective in the face of a range of future climate change projections. The goal is to enhance ecosystem resilience of wetland and riparian habitats to increase the adaptive capacity to manage for Gunnison sage-grouse. After completing a spatial analysis to identify sites for treatment, the team selected 12 potential sites based on local expertise and conducted rapid field evaluations to determine the top two private land sites for work in 2012. Simple restoration treatments, including one-rock dams, were designed to help retain water in impaired drainages. Partners designed and completed construction of over 100 rock structures on private lands to improve or restore wet meadows—which function as brooding habitat for the Gunnison sage-grouse. This vulnerability assessment and field project demonstrates an approach that facilitated collective learning by a group of diverse users who then went on to implement an adaptation project on the ground.

Many adaptation options have been suggested for the management of ecosystems and socioeconomic systems. Often, these options have been broad, such as drought management planning. In other cases, the options focus on ecosystem management and not the corresponding needs of the resource manager or enterprise owner. In most cases, adaptation options have not been specific enough in terms of the how, the who, and under what conditions these actions can be implemented (Heller and Zavaleta 2009). Further adaptive capacity is influenced not only by physical and economic resources, but also by social factors, factors missing in many of the early papers on this topic. Perhaps the most serious drawback in many of the currently proposed adaptation options is the lack of a linkage across the socio-ecological system. Adaptation in one part of the socio-ecological system, such as energy policy encouraging bioenergy crops, can reduce the adaptive capacity of another part of the system, shifting cropland to bioenergy and reducing food security. What is needed is an organizing framework that can identify key vulnerabilities and risks and integrate adaptation actions across the socio-ecological system. In this manner, the adaptation strategies can be constructed to develop a response to the social, economic, and ecological vulnerabilities. The next section discusses development and application of a risk-management framework for adaptation planning.

15.7 Managing Climate Change Risks Through Adaptation

Rangeland systems have coped or adapted to a wide range of past extreme events, lowering the risk of these events in the future. Risk is quantified as the product of the probability of occurrence of hazardous events and the impact of these events. Climate change could increase the risk of loss of rangeland ecosystem functions such as regeneration and recovery, soil development, and nutrient cycling, and the risk of loss of biodiversity including domestic as well as native plants and animal species. Climate change, coincident with resource management, could increase the risk of degradation or desertification. Future extreme weather events could enhance the risk of loss of infrastructure (buildings, fences, equipment, water systems), enterprise assets (livestock including genetic stock, resource productivity through soil erosion and degradation), and social networks (transportation, informational, and financial). While many risks can be identified, the challenge is to identify those risks that are most important to the sustainability of the socio-ecological system in the future.

Key risks to rangelands are those that portend potentially adverse consequences for humans and social-ecological systems resulting from the interaction of climate-related hazards with the vulnerability of societies and systems exposed. Identifying these types of climate-related risks involves framing the risk as resulting from the interaction of vulnerability, exposure, and hazard (Fig. 15.4). Risks in this climate-related context are considered "key" due to high hazard or high vulnerability of societies and systems exposed, or both. In this framework, emergent risks, not previously considered, can arise from indirect impacts of climate change. For example, encouraging the production of bioenergy crops may decrease food security by reducing the land area producing food crops. The following are identified as key risks (Oppenheimer et al. 2014): risk of food insecurity and the breakdown of food systems linked to warming, drought, flooding, and precipitation variability and extremes, particularly for poorer populations in urban and rural settings; risk of loss of rural livelihoods and income due to insufficient access to drinking and irrigation water and reduced agricultural productivity, particularly for farmers and pastoralists with minimal capital in semiarid regions; and risk of loss of terrestrial and inland water ecosystems, biodiversity, and ecosystem goods, functions, and services they provide for human livelihoods.

Risk perception influences human behavior. Risk perception of owners and managers of rangeland systems builds from past management of drought, wildfire, and extreme heat; however, social considerations, rather than physical vulnerability to climate change (e.g., availability of water), are known to determine managers' perception of the risk of climate change (Marx et al. 2007; Moser and Ekstrom 2010;

Safi et al. 2012). For ranchers and farmers in Nevada, their risk perception of climate change was greater the more dependent their enterprises were on agriculture as their primary income. However, general beliefs about the causes of climate change and linking locally observed impacts to climate change were found to be the factors that most influenced their risk perception of climate change (Safi et al. 2012). Risk perception, informed by collective learning, can be motivation for action (see Case Example 15.4).

Managing climate change risks through adaptation is about planning for the future even though the future is uncertain. In fact, there is a strong link between managing future risks and managing for uncertainty. The extent to which landowners can manage for uncertainty and the associated risks of dealing with uncertainty is one of the more important determinants of the adaptive capacity that landowners possess (Marshall 2010). While some landowners will be unable to develop plans for their enterprise without solid knowledge of what the future may hold, other landowners will be able to develop plans that take into account that the future is unknowable. An important premise in managing uncertainty on rangelands is that they represent complex nonlinear systems which do not always have a definite or repeatable cause-and-effect relationship. Developing the skills to manage under such conditions may be a singularly important strategy for landowners to develop if rangeland systems are to be sustained. Inaction has been shown to be more detrimental than assessing risk and making decisions based on that risk calculation (Howden et al. 2007).

A constructive approach for climate adaptation planning is to plan for a range of plausible climate scenarios, and take the path of "least regrets," which accounts for a range of uncertainties about the future. Uncertainty in the future climate of a region can be ascribed to several sources. We identify six here, each of which need to be explicitly addressed if the risks of climate change are to be effectively managed (www.adaptnrm.org):

1. Natural variability—uncertainty will exist around the ecological conditions, and the spatial and temporal variation in these conditions within a period of time and geographical area.
2. Observation/data error—observation error is the failure to properly observe, measure, or estimate processes and quantities. It results both from imperfect methods of observation, or overlooking key factors, and from sampling error.
3. System uncertainty—system understanding is limited by the understanding of all the links—thus, even with complex models, any projections (qualitative or quantitative) will have uncertainty.
4. Inadequate communication—inadequate communication relates to the difficulty of effectively conveying information between scientists, managers, and stakeholders. When communication is ineffective, information is lost, which can manifest itself as uncertainty.
5. Unclear objectives or lack of goal setting—unclear management objectives are ones that are expressed vaguely, not fully conceived, scaled improperly, or difficult to quantify, and enhance uncertainty within the system.
6. Outcome uncertainty—when actions are not implemented properly and it is not clear whether the model was incorrect or the practices themselves.

The achievement of greater adaptation action will require integration of climate change-related issues with other risk factors, such as climate variability and market risk, and with other policy domains, such as sustainable development (Howden et al. 2007). Dealing with the uncertainties among all aspects of rangeland life will require a comprehensive and dynamic approach covering a range of scales and issues. For example, landowners and managers will need to work with policy makers, practitioners, scientists, and others in their social networks to better assess the climate-related risks and hazards and to establish efficient means to respond to them.

15.8 Knowledge Gaps

This chapter has identified a number of areas where scientific knowledge is limited, quantitative methods are needed to capture ecological and social processes to bound uncertainty, and interdisciplinary research is needed to integrate the ecological and the social components of rangeland systems. Vulnerability assessments and adaptation planning must recognize the variation in the adaptive capacity of both ecological systems and the adaptive capacity of human systems. The previous section identified areas where uncertainty needs to be quantified and bound in order for risks associated with climate change to be identified and prioritized. This is a key area for knowledge development.

One area where very little research is ongoing is the experimentation of proposed adaptation management actions. Adaptation strategies are built on current understanding and practice, but they must recognize and attempt to incorporate future change. Field experimentation testing different proposed adaptation actions would provide a greater understanding of the likely success as well as offer comparisons of how natural systems might respond to the changing climate without adaptation treatments.

15.9 Summary

Climate change adaptation has evolved since the early 1990s and will continue to evolve as the scientific, management, and policy communities grapple with key vulnerabilities, risks, and strategies to adapt to climate change. The greatest learning may take place within the private sector and in local institutions where the greatest responsibility for adaptation may reside. The private sector will be highly motivated to protect their assets and maintain their positions in markets. Local institutions will likely be required to implement top-down adaptation policy developed by regional or national government institutions that may not be consistent with adaptations implemented by the private sector in response to local climatic extremes.

All adaptation is local and no single adaptation approach works in all settings. Understanding the key vulnerabilities and climate risks within the local setting is critical and the base on which adaptation strategies are developed. These vulnerability assessments must connect the understanding of ecological, economic, and social vulnerability with the potential for adaptation. The assessment must provide insights that can assist in the development of land management strategies for rangeland resilience; engage vulnerable populations early in the process; and guide development of strategies that enable decision and policy makers to tailor a range of adaptation approaches that most effectively accommodate the divergent requirements of various resource users. As part of this process, vulnerability assessments must recognize that just as the adaptive capacity of rangeland ecosystems varies across geographic regions, the adaptive capacity among resource managers and owners also varies greatly.

Collective learning is the basis for development of collective action among diverse resource users to tackle shared problems. This learning occurs when information emerges from experience and human interaction such that different goals, values, and knowledge are made explicit and questioned to accommodate conflicts. The challenge for developing and implementing adaptation actions is how to incorporate these learning opportunities into public processes so that underlying ecological and social assumptions about management of the socio-ecological system can be collectively visualized. Getting adaptation options on the ground may be closely tied to the success of such opportunities.

Adaptation requires an intentionality to address the effect of climate change. Specific consideration of how management actions need to respond to projected climate change is a part of the adaptation management strategy. The lack of drought planning in the past suggests that the current adaptive capacity is insufficient to sustain livestock enterprises under more frequent and intense drought in the future. Adapting to future change will require a different strategy than coping with past climatic events; the greatest challenge may be to encourage human behavioral changes to sustain the socio-ecological rangeland system. Just as past events influence future ecosystem development, past events in human communities influence future choices in response to day-to-day activities as well as to sudden and drastic events. The diverse history, experiences, and goals of individual managers represent a heterogeneous adaptive capacity that will greatly affect adaptation planning and the strategies selected and implemented. Adaptation strategies and policies need to reflect this heterogeneity, rather than managing for the average enterprise. Landowners may have different perceptions of risk, administrative or financial skills, access to trusted social networks, dependency to sense of place, or willingness to experiment with novel management practices. Landowners may have different ways of knowing and different past experiences that influence current and future decisions.

Adaptation to climate change will be a continual and iterative process. Landowner enterprises must remain viable as the productivity of the land varies through time. Managing for socio-ecological resilience on rangelands is related to the maintenance of system properties that confer resilience without compromising the ability to cope and adapt to future change. Near-term responses may be incremental; however as climate continues to change, these actions may not suffice and transformative changes in the socio-ecological system may be needed.

Adaptation will need to occur within a system that is complex and where there are not always definite or repeatable cause-and-effect relationships. In addition, adaptation plans will need to occur under significant uncertainty of the future. Uncertainty exists not only within the natural system, but also within the models and modes of understanding of how the system works. Different sources of uncertainty, including uncertainty in other aspects of managing rangelands such as market risk, all need to be managed such that efficient responses can be identified and pathways of "least regrets" can be realized.

References

Adger, W.N., S. Agrawala, M.M.Q. Mirza, C. Conde, K. O'Brien, J. Pulhin, R. Pulwarty, B. Smit, and K. Takahashi. 2007. Assessment of adaptation practices, options, constraints and capacity. In *Climate change 2007: Impacts, adaptation and vulnerability. Contribution of working group II to the fourth assessment report of the Intergovernmental Panel on Climate Change*, ed. M.L. Parry, O.F. Canziani, J.P. Palutikof, P.J. van der Linden, and C.E. Hanson, 717–743. Cambridge, UK: Cambridge University Press.

Agard, J., E.L.F Schipper. 2014. Glossary. In IPCC, 2014: Climate change 2014: impacts, adaptation, and vulnerability. Contribution of working group II to the fifth assessment report of the Intergovernmental Panel on Climate Change, ed. V.R. Barros, C.B. Field, D.J. Dokken, M.D. Mastrandrea, K.J. Mach, T.E. Bilir, M. Chatterjee, K.L. Ebi, Y.O. Estrada, R.C. Genova, B. Girma, E.S. Kissel, A.N. Levy, S. MacCracken, P.R. Mastrandrea, and L.L. White, 1757–1776. Cambridge University Press, Cambridge, United Kingdom.

Alessa, L., A. Kliskey, R. Lammers, C. Arp, D. White, L. Hinzman, and R. Busey. 2008. The Arctic water resource vulnerability index: An integrated assessment tool for community resilience and vulnerability with respect to freshwater. *Climatic Change* 42: 523–541.

Allen-Diaz, B., F.S. Chapin, S. Diaz, M. Howden, J. Puigdefábregas, M. Stafford Smith, T. Benning, F. Bryant, B. Campbell, J. duToit, K. Galvin, E. Holland, L. Joyce, A.K. Knapp, P. Matson, R. Miller, D. Ojima, W. Polley, T. Seastedt, A. Suarez, T. Svejcar, and C. Wessman. 1996. Rangeland in a changing climate: Impacts, adaptations, and mitigation. In *Climate change 1995*, ed. R.T. Watson, M.C. Zinyowera, and R.H. Moss, 133–158. New York, NY, USA: Cambridge University Press (Chapter 2).

Bastian, C.T., S. Mooney, A.M. Nagler, J.P. Hewlett, S.I. Paisley, M.A. Smith, W. Marshall Frasier, and W.J. Umberger. 2006. Ranchers diverse in their drought management strategies. *Western Economics Forum* 5: 1–8.

Berkes, F., and C. Folke. 1998. Linking social and ecological systems for resilience and sustainability. In *Linking social and ecological systems: Management practices and social mechanisms for building resilience*, ed. F. Berkes and C. Folke, 1–25. Cambridge, UK: Cambridge University Press.

Briske, D.D., N.F. Sayre, L. Huntsinger, M. Fernández-Giménez, B. Budd, and J.D. Derner. 2011. Origin, persistence, and resolution of the rotational grazing debate: Integrating human dimensions into rangeland research. *Rangeland Ecology and Management* 64: 325–334.

Briske, D.D., L.A. Joyce, H.W. Polley, J.R. Brown, K. Wolter, J.A. Morgan, B.A. McCarl, and D.W. Bailey. 2015. Climate-change adaptation on rangelands: Linking regional exposure with diverse adaptive capacity. *Frontiers in Ecology and the Environment* 13: 249–256.

Brunson, M.W. 2012. The elusive promise of social-ecological approaches to rangeland management. *Rangeland Ecology and Management* 65: 632–637.

Cabrera, V.E., P. Hildebrand, J.W. Jones, D. Letson, and A. de Vries. 2006. An integrated North Florida dairy farm model to reduce environmental impacts under seasonal climate variability agriculture. *Ecosystems and Environment* 113: 82–97.

Cinner, J., M.M.P.B. Fuentes, and H. Randriamahazo. 2009. Exploring social resilience in Madagascar's marine protected areas. *Ecology and Society* 14: 41.

Cobon, D.H., G.S. Stone, J.O. Carter, J.C. Scanlan, N.R. Toombs, X. Zhang, J. Willcocks, and G.M. McKeon. 2009. The climate change risk management matrix for the grazing industry of northern Australia. *The Rangeland Journal* 31: 31–49.

Coppock, D.L. 2011. Ranching and multiyear droughts in Utah: Production impacts, risk perceptions, and changes in preparedness. *Rangeland Ecology and Management* 64: 607–618.

Crimp, S.J., C.J. Stokes, S.M. Howden, A.D. Moore, B. Jacobs, P.R. Brown, A.I. Ash, P. Kokic, and P. Leith. 2010. Managing Murray–Darling Basin livestock systems in a variable and changing climate: Challenges and opportunities. *The Rangeland Journal* 32: 293–304.

Dai, A. 2011. Drought under global warming: A review. *WIREs Climate Change* 2: 45–65. doi:10.1002/wcc.81 (See also Dai, A. 2012. Erratum 2012. *WIREs Climate Change* 3: 617.).

Dai, A. 2013. Increasing drought under global warming in observations and models. *Nature Climate Change* 3: 52–58.

Dalintai, N.G., N. Gauwau, L. Yanbo, J. Enkhee, and L. Shurun. 2012. The new Otor: Risk management in a desert grassland. In *Restoring community connections to the land*, ed. M.E. Fernández-Giménez, X. Wang, B. Batkhishig, J.A. Klein, and R.S. Reid, 93–112. Wallingford, UK: CAB International.

Darnhofer, I., S. Bellon, B. Dedieu, and R. Milestad. 2010. Adaptiveness to enhance the sustainability of farming systems. A review. *Agronomy for Sustainable Development* 30: 545–555.

Davidson-Hunt, I.J., and F. Berkes. 2004. Nature and society through the lens of resilience: Toward a human-in-ecosystem perspective. In *Navigating social-ecological systems. Building resilience for complexity and change*, ed. F. Berkes, J. Colding, and C. Folke, 53–81. Cambridge, UK: Cambridge University Press.

Dilling, L., and M.C. Lemos. 2011. Creating usable science: Opportunities and constraints for climate knowledge use and their implications for science policy. *Global Environmental Change* 21: 680–689.

Dilling, L., E.D. Meaghan, W.R. Travis, O.V. Wilhelmi, and R.A. Klein. 2015. The dynamics of vulnerability: Why adapting to climate variability will not always prepare us for climate change. *WIREs Climate Change* 6: 413–425.

Emtage, N., J. Herbohn, and S. Harrison. 2007. Landholder profiling and typologies for natural resource-management policy and program support: Potential and constraints. *Environmental Management* 40: 481–492.

Evans, S.E., Byrne, K.M., Lauenroth, W.K., Burke, I.C. 2011. Defining the limit to resistance in a drought-tolerant grassland: long-term severe drought significantly reduces the dominant species and increases ruderals. Journal of Ecology 99: 1500–1507.

Fazey, I., R.M. Wise, C. Lyon, C. Câmpeanu, P. Moug, T.E. Davies. 2015. Past and future adaptation pathways. Climate and Development. doi:10.1080/17565529.2014.989192.

Fernández-Giménez, M.E., B. Batkhishig, and B. Batbuyan. 2012. Cross-boundary and cross-level dynamics increase vulnerability to severe winter disasters (dzud) in Mongolia. *Global Environmental Change* 22: 836–851.

Fischlin, A., G.F. Midgley, J.T. Price, R. Leemans, B. Gopal, C. Turley, M.D.A. Rounsevell, O.P. Dube, J. Tarazona, and A.A. Velichko. 2007. Ecosystems, their properties, goods, and services. In *Climate change 2007: Impacts, adaptation and vulnerability. Contribution of working group II to the fourth assessment report of the Intergovernmental Panel on Climate Change*, 211–272, ed. M.L. Parry, O.F. Canziani, J.P. Palutikof, P.J. van der Linden, and C.E. Hanson. Cambridge, UK: Cambridge University Press.

Fleming, A., and F. Vanclay. 2010. Farmer responses to climate change and sustainable agriculture. A review. *Agronomy for Sustainable Development* 30: 11–19.

Folke, C., S. Carpenter, T. Elmqvist, L. Gunderson, C.S. Holling, and B. Walker. 2002. Resilience and sustainable development: Building adaptive capacity in a world of transformations. *Ambio* 31: 437–440.

Furniss, M.J., K.B. Roby, D. Cenderelli, J. Chatel, C.F. Clifton, A. Clingenpeel, P.E. Hays, D. Higgins, K. Hodges, C. Howe, L. Jungst, J. Louie, C. Mai, R. Martinez, K. Overton, B.P. Staab, R. Steinke, and M. Weinhold. 2013. *Assessing the vulnerability of watersheds to climate change: Results of national forest watershed vulnerability pilot assessments. Gen. Tech. Rep. PNW-GTR-884*, 32 p. Portland, OR: U.S. Department of Agriculture, Forest Service, Pacific Northwest Research Station. (Plus appendix).

Fussel, H.-M., and R.J.T. Klein. 2006. Climate change vulnerability assessments: An evolution of conceptual thinking. *Climatic Change* 75: 301–329.

Gitay, H., S. Brown, W. Easterling, B. Jallow, et al. 2001. Ecosystems and their goods and services. In *Climate change 2001: Impacts, adaptation and vulnerability*, ed. J.J. McCarthy, O.F. Canziani, N.A. Leary, D.J. Dokken, and K.S. White, 237–341. New York, NY: Cambridge University Press.

Glick, P., B.A. Stein, and N.A. Edelson (eds.). 2011. *Scanning the conservation horizon: A guide to climate change vulnerability assessment*. Washington, DC: National Wildlife Federation.

Gunderson, L. 1999. Resilience, flexibility and adaptive management—antidotes for spurious certitude? *Conservation Ecology* 3. www.consecol.org/vol13/iss11/art17.

Hammer, G. 2000. Applying seasonal climate forecasts in agricultural and natural ecosystems—a synthesis. In *Applications of seasonal climate forecasting in agricultural and natural ecosystems*, Atmospheric and Oceanographic Sciences Library, vol. 21, ed. G.L. Hammer, N. Nicholls, and C. Mitchell, 453–462. The Netherlands: Springer.

Hammer, G.L., D.P. Holzworth, and R. Stone. 1996. The value of skill in seasonal climate forecasting to wheat crop management in a region with high climatic variability. *Australian Journal of Agricultural Research* 47: 717–737.

Hansen, J.W. 2002. Realizing the potential benefits of climate prediction to agriculture: Issues, approaches, challenges. *Agricultural Systems* 74: 309–330.

Heller, N. E., Zavaleta, E. S. 2009. Biodiversity management in the face of climate change: A review of 22 years of recommendations. Biological Conservation 142: 14–32.

Hertzler, G. 2007. Adapting to climate change and managing climate risks by using real options. *Australian Journal of Agricultural Research* 58: 985–992.

Heyhoe, E., Y. Kim, P. Kokic, C. Levantis, H. Ahammad, K. Schneider, S. Crimp, R. Nelson, N. Flood, and J. Carter. 2007. Adapting to climate change—issues and challenges in the agriculture sector [online]. *Australian Commodities: Forecasts and Issues* 14(1): 167–178. ISSN: 1321-7844. <http://search.informit.com.au/documentSummary;dn=967460540415764;res=IELBUS>.

Hobbs, N.T., K.A. Galvin, C.J. Stokes, J.M. Lackett, A.J. Ash, R.B. Boone, R.S. Reid, and P.K. Thornton. 2008. Fragmentation of rangelands: Implications for humans, animals, and landscapes. *Global Environmental Change* 18: 776–785.

Hobbs, R.J., E.S. Higgs, and C. Hall. 2013. *Novel ecosystems: Intervening in the new ecological world order*. Chichester: Wiley.

Holling, C.S. 1973. Resilience and stability of ecological systems. *Annual Review in Ecology and Systematics* 4: 1–23.

Holling, C.S. 1996. Surprise for science, resilience for ecosystems, and incentives for people. *Ecological Applications* 6: 733–735.

Holling, C.S. 2001. Understanding the complexity of economic, ecological, and social systems. *Ecosystems* 4: 390–405.

Holling, C.S., and G.K. Meffe. 1996. Command and control and the pathology of natural resource management. *Conservation Biology* 10: 328–337.

Howden, S.M., J. Soussana, F.N. Tubiello, N. Chhetri, M. Dunlop, and H. Meinke. 2007. Adapting agriculture to climate change. *Proceedings of the National Academy of Sciences* 104: 19691–19696.

Intergovernmental Panel on Climate Change [IPCC]. 1990. Climate change the IPCC response strategies. In *Contribution of working group III to the first assessment report of the Intergovernmental Panel on Climate Change*, ed. F. Bernthal, E. Dowdeswell, J. Luo, D. Attard, P. Vellinga, and R. Karimanzira. Cambridge, UK: Cambridge University Press.

Intergovernmental Panel on Climate Change [IPCC]. 2001. Glossary. In Climate change 2001: Impacts, adaptation, and vulnerability. In Contribution of working group II to the third assessment report of the Intergovernmental Panel on Climate Change, ed. J.J. McCarthy, O.F. Canziani, N.A. Leary, D.J. Dokken, and K.S. White. New York, NY, USA: Cambridge University Press.

Intergovernmental Panel on Climate Change [IPCC]. 2014a. Climate change 2014: Impacts, adaptation, and vulnerability. Part A: Global and sectoral aspects. In Contribution of working group II to the fifth assessment report of the Intergovernmental Panel on Climate Change, ed. C.B. Field, V.R. Barros, D.J. Dokken, K.J. Mach, M.D. Mastrandrea, T.E. Bilir, M. Chatterjee, K.L. Ebi, Y.O. Estrada, R.C. Genova, B. Girma, E.S. Kissel, A.N. Levy, S. MacCracken, P.R. Mastrandrea, and L.L. White. Cambridge, UK: Cambridge University Press.

IPCC. 2014b. Summary for policymakers. In Climate change 2014: Impacts, adaptation, and vulnerability. Part A: Global and sectoral aspects. Contribution of working group II to the fifth assessment report of the Intergovernmental Panel on Climate Change, ed. C.B. Field, V.R. Barros, D.J. Dokken, K.J. Mach, M.D. Mastrandrea, T.E. Bilir, M. Chatterjee, K.L. Ebi, Y.O. Estrada, R.C. Genova, B. Girma, E.S. Kissel, A.N. Levy, S. MacCracken, P.R. Mastrandrea, and L.L. White, 1–32. Cambridge, UK: Cambridge University Press.

Joyce, L.A., and C.I. Millar. 2014. Improving the role of vulnerability assessments in decision support for effective climate adaptation. In Forest conservation and management in the Anthropocene: conference proceedings. Proceedings. RMRS-P-71, 494 p., ed. V.A. Sample and R.P. Bixler. Fort Collins, CO: US Department of Agriculture, Forest Service, Rocky Mountain Research Station.

Joyce, L.A., D.D. Briske, J.R. Brown, H.W. Polley, B.A. McCarl, and D.W. Bailey. 2013. Climate change and North American rangelands: Assessment of mitigation and adaptation strategies. Rangeland and Ecology Management 66: 512–528.

Kates, R.W., W.R. Travis, and T.J. Wilbanks. 2012. Transformational adaptation when incremental adaptations to climate change are insufficient. Proceedings of the National Academy of Sciences U.S. A. 109: 7156–7161.

Laidler, G.J., J.D. Ford, W.A. Gough, T. Ikummaq, A.S. Gagnon, S. Kowal, K. Qrunnut, and C. Irngaut. 2009. Travelling and hunting in a changing Arctic: Assessing Inuit vulnerability to sea ice change in Igloolik, Nunavut. Climatic Change 94: 363–397.

Lauenroth, W.K., and I.C. Burke. 2008. Ecology of the shortgrass steppe. New York, NY: Oxford University Press.

Levin, S., S. Barrett, S. Aniyar, W. Baumol, C. Bliss, B. Bolin, P. Dasgupta, P.R. Ehrich, C. Folke, I. Gren, C.S. Holling, A. Jansson, B. Jansson, K. Maler, D. Martin, C. Perrings, and E. Sheshinski. 1998. Resilience in natural and socioeconomic systems. Environment and Development Economics 3: 222–235.

Lubell, M.N., B.B. Cutts, L.M. Roche, M. Hamilton, J.D. Derner, E. Kachergis, and K.W. Tate. 2013. Conservation program participation and adaptive rangeland decision-making. Rangeland Ecology and Management 66: 609–620.

Ludwig, D., B. Walker, C.S. Holling. 1997. Sustainability, stability and resilience. Conservation Ecology 1. www.consecol.org/vol1/iss1/art7.

Marshall, N.A. 2010. Understanding social resilience to climate variability in primary enterprises and industries. Global Environmental Change—Human and Policy Dimensions 20: 36–43.

Marshall, N.A. 2011. Assessing resource dependency on the rangelands as a measure of climate sensitivity. Society and Natural Resources 24: 1105–1115.

Marshall, P.A., and J. Johnson. 2007. The Great Barrier Reef and climate change: Vulnerability and management implications. Australia: Great Barrier Reef Marine Park Authority and Australian Greenhouse Office.

Marshall, N.A., and A. Smajgl. 2013. Understanding variability in adaptive capacity on rangelands. Rangeland Ecology and Management 66: 88–94.

Marshall, N.A., C.J. Stokes, S.M. Howden, and R.N. Nelson. 2010. Enhancing adaptive capacity. In Adapting agriculture to climate change: Preparing Australian agriculture, forestry and fisheries for the future, ed. C. Stokes and S.M. Howden, 245–256. Canberra, Australia: CSIRO Publishing.

Marshall, N.A., I.J. Gordon, and A.J. Ash. 2011. The reluctance of resource-users to adopt seasonal climate forecasts to enhance resilience to climate variability on the rangelands. Climatic Change 107: 511–529.

Marshall, N.A., S.E. Park, W.N. Adger, K. Brown, and S.M. Howden. 2012. Transformational capacity and the influence of place and identity. *Environmental Research Letters* 7: 034022. doi:10.1088/1748-9326/7/3/034022.

Marshall, N.A., S.M. Howden, A.B. Dowd, and E.S. Jakku. 2013. Climate change awareness is associated with enhanced adaptive capacity. *Agricultural Systems* 117: 30–34.

Marshall, N.A., and C.J. Stokes. 2014. Influencing adaptation processes on the Australian rangelands for social and ecological resilience. *Ecology and Society* 19: 14. http://www.ecologyandsociety.org/vol19/iss12/art14.

Marshall, N.A., C.J. Stokes, N.P. Webb, P.A. Marshall, and A.J. Lankester. 2014. Social vulnerability to climate change in primary producers: A typology approach. *Agriculture, Ecosystems and Environment* 186: 86–93.

Marshall, N.A., S. Crimp, M. Curnock, M. Greenhill, G. Kuehne, Z. Leviston, and J. Ouzman. 2016. Some primary producers are more likely to transform their agricultural practices in response to climate change than others. *Agriculture, Ecosystems and Environment* 222: 38–47.

Marx, S.M., E.U. Weber, B.S. Orlove, et al. 2007. Communication and mental processes: Experiential and analytic processing of uncertain climate information. *Global Environmental Change* 17: 47–58.

McCarl, B.A. 2011. Agriculture. In *The impact of global warming on Texas*, 2nd ed, ed. J. Schmandt, G.R. North, and J. Clarkson, 157–171. Austin, TX, USA: University of Texas Press.

McDaniels, T., S. Wilmot, M. Healey, and S. Hinch. 2010. Vulnerability of Fraser River sockeye salmon to climate change: A life cycle perspective using expert judgments. *Journal of Environmental Management* 91: 2771–2780.

McKeon, G.M., W.B. Hall, B.K. Henry, G.S. Stone, and I.W. Watson. 2004. *Pasture degradation and recovery in Australia's rangelands: Learning from history*. Brisbane, Australia: Queensland Department of Natural Resources, Mines and Energy.

Millar, C.I., N.L. Stephenson, and S.L. Stephens. 2007. Climate change and forests of the future: Managing in the face of uncertainty. *Ecological Applications* 17: 2145–2151.

Millar, C.I., K.E. Skog, D.C. McKinley, R.A. Birdsey, C.W. Swanston, S.J. Hines, C.W. Woodall, E.D. Reinhardt, D.L. Peterson, and J.M. Vose. 2012. Adaptation and mitigation. In *Effects of climatic variability and change on forest ecosystems: A comprehensive science synthesis for the U.S. forest sector. Gen. Tech. Rep. PNW-GTR-870*, ed. J.M. Vose, D.L. Peterson, T. Patel-Weynand, 125–192. Portland, OR: U.S. Department of Agriculture, Forest Service, Pacific Northwest Research Station.

Moon, K., and C. Cocklin. 2011. A landholder-based approach to the design of private-land conservation programs. *Conservation Biology* 25: 493–509.

Moser, S.C., and J.A. Ekstrom. 2010. A framework to diagnose barriers to climate change adaptation. *Proceedings of the National Academy of Sciences* 107: 22026–31.

Moyle, P.B., J.D. Kiernan, P.K. Crain, and R.M. Quinones. 2013. Climate change vulnerability of native and alien freshwater fishes of California: A systematic assessment approach. *PLoS One* 8: e63883. doi:10.1371/journal.pone.0063883.

Mu, J.E., B.A. McCarl, and A.M. Wein. 2013. Adaptation to climate change: Changes in farmland use and stocking rate in the U.S. *Mitigation and Adaptation of Strategies for Global Change* 18: 713–730.

Neely, B., R. Rondeau, J. Sanderson, C. Ague, B. Kuhn, J. Siemers, L. Grunau, J. Robertson, P. McCarthy, J. Barsugli, T. Schulz, C. Knapp, eds. 2011. *Gunnison Basin: Climate change vulnerability assessment for the Gunnison climate working group by The Nature Conservancy, Colorado Natural Heritage Program, Western Water Assessment*. Boulder: University of Colorado and Fairbanks: University of Alaska. Project of the Southwest Climate Change Initiative. http://www.conservationgateway.org/ConservationByGeography/NorthAmerica/UnitedStates/Colorado/science/Pages/gunnison-basin-climate-ch.aspx#sthash.gJghvpRH.dpuf.

Noble, I.R., S. Huq, Y.A. Anokhin, J. Carmin, D. Goudou, F.P. Lansigan, B. Osman-Elasha, and A. Villamizar. 2014. Adaptation needs and options. In Climate change 2014: Impacts, adaptation, and vulnerability. Contribution of working group II to the fifth assessment report of the Intergovernmental Panel on Climate Change, ed. C.B. Field, V.R. Barros, D.J. Dokken, K.J. Mach,

M.D. Mastrandrea, T.E. Bilir, M. Chatterjee, K.L. Ebi, Y.O. Estrada, R.C. Genova, B. Girma, E.S. Kissel, A.N. Levy, S. MacCracken, P.R. Mastrandrea, L.L. White, 833–868. Cambridge, UK: Cambridge University Press. Chapter 14.

Oppenheimer, M., M. Campos, R. Warren, J. Birkmann, G. Luber, B. O'Neill, and K. Takahashi. 2014. Emergent risks and key vulnerabilities. In Climate change 2014: Impacts, adaptation, and vulnerability. Part A: Global and sectoral aspects. Contribution of working group II to the fifth assessment report of the Intergovernmental Panel on Climate Change, ed. C.B. Field, V.R. Barros, D.J. Dokken, K.J. Mach, M.D. Mastrandrea, T.E. Bilir, M. Chatterjee, K.L. Ebi, Y.O. Estrada, R.C. Genova, B. Girma, E.S. Kissel, A.N. Levy, S. MacCracken, P.R. Mastrandrea, and L.L. White, 1039–1099. Cambridge, UK: Cambridge University Press.

Park, S.E., N.A. Marshall, E. Jakku, A.M. Dowd, S.M. Howden, E. Mendham, and A. Fleming. 2012. Informing adaptation responses to climate change through theories of transformation. *Global Environmental Change—Human and Policy Dimensions* 22: 115–126.

Peterson, D.L., C.I. Millar, L.A. Joyce, M.J. Furniss, J.E. Halofsky, R.P. Neilson, and T.L. Morelli. 2011. *Responding to climate change in national forests: A guidebook for developing adaptation options. Gen. Tech. Rep. PNW-GTR-855*, 109 p. Portland, OR: U.S. Department of Agriculture, Forest Service, Pacific Northwest Research Station.

Pielke Sr., R.A., N. Doesken, O. Bliss, T. Green, C. Chaffin, J.D. Salas, C.A. Woodhouse, J.J. Lukas, and K. Wolter. 2005. Drought 2002 in Colorado: An unprecedented drought or a routine drought? *Pure and Applied Geophysics* 162: 1455–1479.

Reynolds, J.F., and D.M. Stafford Smith. 2002. Do humans cause deserts? In *Dahlem workshop report 88*, ed. J.F. Reynolds, and D.M. Stafford Smith, 23–40. Berlin, Germany: Dahlem University Press.

Reynolds, J.F., D.M. Stafford Smith, E.F. Lambin, B.L. Turner II, M. Mortimore, S.P. Batterbury, T.E. Downing, H. Dowlatabadi, R.J. Fernandez, J.E. Herrick, E. Hubersannwald, H. Jiang, R. Leemans, T. Lynam, F.T. Maestre, M. Ayarza, and B. Walker. 2007. Global desertification: Building a science for drylands. *Science* 316: 847–851.

Rissman, A.R., and N.F. Sayre. 2012. Conservation outcomes and social relations: A comparative study of private ranchland conservation easements. *Society and Natural Resources* 25: 523–538.

Roche, L.M., T.K. Schohr, J.D. Derner, M.N. Lubell, B.B. Butts, E. Kachergis, V.T. Eviner, and K.W. Tate. 2015. Sustaining working rangelands: Insights from rancher decision making. *Rangeland Ecology and Management* 68: 383–389.

Safi, A.S., W.J. Smith, and Z. Liu. 2012. Rural Nevada and climate change: Vulnerability, beliefs, and risk perception. *Risk Analysis* 32: 1041–1059.

Salinger, M.J. 2005. Increasing climate variability and change: Reducing the vulnerability. *Climatic Change* 70: 1–3.

Sankey, T.T., J.B. Sankey, K.T. Weber, and C. Montagne. 2009. Geospatial assessment of grazing regime shifts and sociopolitical changes in a Mongolian rangeland. *Rangeland Ecology and Management* 62: 522–530.

Settele, J., R. Scholes, R. Betts, S. Bunn, P. Leadley, D. Nepstad, J.T. Overpeck, and M.A. Taboada. 2014. Terrestrial and inland water systems. In Climate change 2014: Impacts, adaptation, and vulnerability. Part A: Global and sectoral aspects. Contribution of working group II to the fifth assessment report of the Intergovernmental Panel on Climate Change, ed. C.B. Field, V.R. Barros, D.J. Dokken, K.J. Mach, M.D. Mastrandrea, T.E. Bilir, M. Chatterjee, K.L. Ebi, Y.O. Estrada, R.C. Genova, B. Girma, E.S. Kissel, A.N. Levy, S. MacCracken, P.R. Mastrandrea, and L.L. White, 833–868. Cambridge, UK: Cambridge University Press.

Stafford Smith, D.M., G.M. McKeon, I.W. Watson, B.K. Henry, G.S. Stone, W.B. Hall, and S.M. Howden. 2007. Learning from episodes of degradation and recovery in variable Australian rangelands. *Proceedings of the National Academy of Sciences* 104: 20690–20695.

Thurow, T.L., and C.A. Taylor. 1999. Viewpoint: The role of drought in range management. *Journal of Range Management* 52: 413–419.

Torell, L.A., S. Murugan, and O.A. Ramirez. 2010. Economics of flexible versus conservative stocking strategies to management climate variability risk. *Rangeland Ecology and Management* 63: 415–425.

US Global Change Research Program [USGCRP]. 2011. Uses of vulnerability assessments for the national climate assessment. *NCA Report Series*, vol. 9. http://assessment.globalchange.gov.

Volney, W.J.A., and R.A. Fleming. 2000. Climate change and impacts of boreal forest insects. *Agriculture, Ecosystems and Environment* 82: 283–294.

Walker, B.H., and M.A. Janssen. 2002. Rangelands, pastoralists and governments: Interlinked systems of people and nature. *Philosophical Transactions of the Royal Society of London Series B—Biological Sciences* 357: 719–725.

Walker, N.J., and R.E. Schulze. 2008. Climate change impacts on agro-ecosystem sustainability across three climate regions in the maize belt of South Africa. *Agriculture, Ecosystems and Environment* 124: 114–124.

Watson, I.W. 2004. I had the right number of sheep, but the wrong amount of rain. In *Conference papers, Australian Rangeland Society 13th Biennial Conference*, Alice Springs, Northern Territory, 5–8 July, ed. G. Bastin, D. Walsh, and C. Nicholson, 291–292.

Watson, I.W; Norton, B.E., Novelly, P.E., Russell, P.J. 2008. Integration of regulation, extension, science, policy and monitoring improves land management in the rangelands of western Australia. In: The Future of Drylands, International Scientific Conference on Desertification and Drylands Research, Tunis, Tunisia, 19-21 June 2006. pp 599 -614. Ed. Lee, C.; Schaaf, T. UNESCO and Springer (http://www.springer.com/us/book/9781402069697)

Webb, N.P., C.J. Stokes, and N.A. Marshall. 2013. Integrating biophysical and socio-economic evaluations to improve the efficacy of adaptation assessments for agriculture. *Global Environmental Change—Human and Policy Dimensions* 23: 1164–1177.

Wilhite, D., and M. Buchanan-Smith. 2005. Drought as hazard: Understanding the natural and social context. In *Drought and Water Crises*, ed. D.A. Wilhite, 3–32. Boca Raton, FL: CRC Press.

Wilmer, H., Fernández-Giménez, M. E. 2015. Rethinking rancher decision-making: a grounded theory of ranching approaches to drought and succession management. The Rangeland Journal 37(5): 517–528.

Yuen, E., S.S. Jovicich, and B.L. Preston. 2013. Climate change vulnerability assessments as catalysts for social learning: Four case studies in south-eastern Australia. *Mitigation and Adaptation Strategies for Global Change* 18: 567–590.

Chapter 16
Monitoring Protocols: Options, Approaches, Implementation, Benefits

Jason W. Karl, Jeffrey E. Herrick, and David A. Pyke

Abstract Monitoring and adaptive management are fundamental concepts to rangeland management across land management agencies and embodied as best management practices for private landowners. Historically, rangeland monitoring was limited to determining impacts or maximizing the potential of specific land uses—typically grazing. Over the past several decades, though, the uses of and disturbances to rangelands have increased dramatically against a backdrop of global climate change that adds uncertainty to predictions of future rangeland conditions. Thus, today's monitoring needs are more complex (or multidimensional) and yet still must be reconciled with the realities of costs to collect requisite data. However, conceptual advances in rangeland ecology and management and changes in natural resource policies and societal values over the past 25 years have facilitated new approaches to monitoring that can support rangeland management's diverse information needs. Additionally, advances in sensor technologies and remote-sensing techniques have broadened the suite of rangeland attributes that can be monitored and the temporal and spatial scales at which they can be monitored. We review some of the conceptual and technological advancements and provide examples of how they have influenced rangeland monitoring. We then discuss implications of these developments for rangeland management and highlight what we see as challenges and opportunities for implementing effective rangeland monitoring. We conclude with a vision for how monitoring can contribute to rangeland information needs in the future.

Keywords Rangeland monitoring • Management objectives • Remote sensing • Sampling design • Core indicators • Land health

J.W. Karl (✉) • J.E. Herrick
USDA-ARS, Jornada Experimental Range, Las Cruces, NM 88003, USA
e-mail: jason.karl@ars.usda.gov; jeff.herrick@ars.usda.gov

D.A. Pyke
U.S. Geological Survey, Forest and Rangeland Ecosystem Science Center, Corvallis, OR 97331, USA
e-mail: david_a_pyke@usgs.gov

© The Author(s) 2017
D.D. Briske (ed.), *Rangeland Systems*, Springer Series on Environmental Management, DOI 10.1007/978-3-319-46709-2_16

16.1   Changing Needs for Rangeland Monitoring

Monitoring has long been recognized as a critical tool for rangeland management. The collection and use of monitoring data to protect and improve rangelands (i.e., principles of adaptive management) have been promoted since the early twentieth century (West 2003a). In 1915, just 3 years after the creation of the Jornada Experimental Range in southern New Mexico, a large network of monitoring plots was established to better understand and address the rapid degradation that was already occurring from excessive livestock grazing (Gibbens et al. 2005). In the first textbook on range management, Sampson (1923) promoted the idea of "a system-ized study, designed to secure the data that will lead to permanent improvement in management and to increased profits from the lands." The basic concepts of moni-toring and adaptive management are now fundamental to rangeland management across land management agencies and embodied as best management practices for private landowners (West 2003a; Boyd and Svejcar 2009).

Historically, rangeland monitoring was limited to determining impacts or maxi-mizing the potential of specific land uses—typically grazing. Because of the recog-nized heterogeneity of rangelands, uneven distribution of land uses (e.g., grazing), and expense associated with obtaining measurements across large areas, monitoring activities were focused on areas where impacts were either observed or expected (Dyksterhuis 1949; Bureau of Land Management 1996).

Over the past several decades, though, the uses of and disturbances to rangelands have increased dramatically. Historically, grazing, has been a dominant land use on rangelands worldwide. In the United States, though, it is increasingly a minor com-ponent in some rangelands compared to other uses such as energy development, recreation, and conservation. Diffuse and widespread disturbances that alter the character and potential of rangelands like non-native plant invasion, woody plant encroachment, and altered wildland fire regimes are prevalent. Additionally, range-lands are increasingly being valued for providing ecosystem services including clean and reliable water supplies, clean air, recreational opportunities, and habitat for many plants and animals, as well as numerous, diverse soil microorganisms (Sala and Paruelo 1997). These changes have altered the need for and character of rangeland monitoring.

The diverse uses and disturbances to rangeland ecosystems are also occurring against a backdrop of global climate change that adds more uncertainty to predic-tions of future rangeland conditions. The effects of changing climatic conditions on plant community composition and production are expected to be variable regionally (Briske et al. 2015), and increasing inter-annual variability of precipitation (IPCC 2007) and temperature may make detecting management-related changes more challenging (Fuhlendorf et al. 2001). Accordingly there is a need for monitoring data to establish baselines of rangeland conditions and to document changes in con-dition to both understand impacts of climate change and differentiate those effects from other disturbances or management activities. Monitoring data will also be needed to develop and evaluate climate adaptation and mitigation strategies.

Current multidimensional monitoring needs for rangeland management, however, must be reconciled with realities of costs to collect the requisite data. Monitoring of large rangeland landscapes is complicated by logistical constraints, high variability of rangeland indicators due to inter-annual climate fluctuations and environmental heterogeneity, and costs of monitoring. Despite the fact that it has long been recognized as an important aspect of rangeland management, monitoring often has been perceived as an incidental activity that takes funds away from management actions (Wright et al. 2002). The varied reasons for this include a lack of linkages between monitoring and management decision-making and perceptions of redundant monitoring efforts. Thus the challenges of monitoring rangelands include institutional hurdles to valuing and effectively using monitoring data.

Relative to rangeland management or research, monitoring generally refers to the systematic collection, analysis, and use of quantitative information on rangeland resources over time to support management decision-making (Bedell 1998). Two related activities are assessments (estimating or judging the condition or value of an ecological system or process at a point in time) and inventories (systematic acquisition and analysis of information for rangeland resource planning and management) (Pellant et al. 2005). Whereas there are many specific definitions for these three terms in rangeland management, the ideas discussed here apply across all these activities. Thus, for brevity, we use the term monitoring generically.

Conceptual advances in rangeland ecology that were introduced in the early 1990s and subsequently developed over the past 25 years have facilitated new approaches to monitoring that can support rangeland management's diverse information needs (Text Box 16.1). Additionally, technological advances have broadened the suite of

---

Text Box 16.1: Knowns and Unknowns "As we know, there are known knowns. There are things we know we know. We also know there are known unknowns. That is to say we know there are some things we do not know. But there are also unknown unknowns, the ones we don't know we don't know".— Donald Rumsfeld, US Defense Secretary (2002).

The quote above was one of the most famous quotes from U.S. Defense Secretary Donald Rumsfeld. It was given in response to a question from a reporter about the existence and veracity of evidence to support the assertion that Iraq possessed weapons of mass destruction. While this quote was almost universally mocked as being an evasion of the question rather than an answer, the idea of known unknowns and unknown unknowns has a longer history (Morris 2014) and is relevant to monitoring of rangeland resources.

Most successful monitoring programs are intended to address the known unknowns. They are built around answering specific questions for which critical data are lacking. These questions should lead to selection of a minimal set of indicators and methods and development of a sampling design to provide

(continued)

Text Box 16.1 (continued)

the identified missing information. This is the classic model of natural resource monitoring, and it has been proven effective when it is applied.

However, one of the hopes when implementing rangeland monitoring is that the data collected will be at least somewhat informative for new resource concerns that arise. In other words, this information can hopefully address the unknown unknowns too. Recently, rangeland managers in the western USA have experienced information shortages related to the status and trend of Greater sage-grouse (*Centrocercus urophasianus*) habitat and the impacts of energy development (oil and gas, wind, solar, transmission lines) on rangeland ecosystems. Many of the existing monitoring programs which were developed around livestock grazing objectives are ill-equipped to inform on these new objectives. The hope, then, is that monitoring programs built around concepts of core indicators and methods and statistically based sampling designs will provide greater opportunities to compile existing monitoring data for new objectives. While it is naive to think that general monitoring programs or compilations of existing monitoring data will address all of the information needs of a new question, robust and interoperable monitoring programs would provide a better foundation from which to begin.

rangeland attributes that can be monitored and the temporal and spatial scales at which they can be monitored. Below we review some of these conceptual and technological advancements and provide examples of how they have influenced rangeland monitoring. We then discuss implications of these developments for rangeland management. We highlight what we see as challenges and opportunities for implementing effective rangeland monitoring, and conclude with a vision for how monitoring can contribute to rangeland information needs in the future.

16.2 Conceptual Advances in the Past 25 Years

Conceptual advances in rangeland monitoring over the past 25 years have been driven by developments in ecological theory, changes in natural resource policies and societal values, and emergence of new technologies. The 1980s and early 1990s were a critical period for development of ecological theories that have proven to be pivotal for rangeland management. During this time scale theory was formally defined (see Wiens et al. 2007) and the field of landscape ecology was founded (see Wiens 1999). For rangeland management, perhaps the biggest conceptual advance was the recognition that rangeland systems are characterized by nonlinear dynamics (Briske et al. 2005; Kefi et al. 2007) and cross-scale processes (Peters et al. 2004) that can produce multiple ecosystem states (Chap. 6, this volume). This change in thinking brought into focus the importance of measuring ecological processes and functions at and across characteristic scales on which they operate (Addicott et al. 1987; Peters et al. 2004; Nash et al. 2014).

Rangeland Reform '94 (U.S. Bureau of Land Management and U.S. Forest Service 1994) was the first attempt by federal agencies in the USA to change how rangeland under Bureau of Land Management (BLM) and U.S. Forest Service (USFS) management would be evaluated since the environmental policies of the 1970s (e.g., Federal Land Policy and Management Act of 1976, National Forest Management Act of 1976, Public Rangeland Improvement Act of 1978 BLM and USFS 1994). Two goals of Rangeland Reform '94 were to bring the two management programs closer together and more consistent in conducting ecosystem management and to accelerate the restoration and improvement of public rangelands to proper functioning condition. The changes brought forth in Rangeland Reform '94 were strongly influenced by the National Research Council (1994) report on rangeland health. Each state developed standards of rangeland health that reflected the need to monitor not just plants important for livestock grazing, but biological diversity, soil stability, hydrologic functioning, energy flow, and nutrient cycling (Veblen et al. 2014).

As part of the goal to bring agencies together in their management of rangelands, the USDA Natural Resources Conservation Service, USFS, and BLM entered into a Memorandum of Understanding in 2005 to define and describe rangelands using a standard classification system, ecological sites (Caudle et al. 2013). This resulted in federal and non-federal rangelands being classified using the same process where soils, landforms, and climate describe potential plant communities and their production. This led to common terminology and similar metrics for determining rangeland status and trends. Standardized terms and metrics position land managers to take advantage of new technology that cross-cuts management and political boundaries through remote sensing and database access (Chap. 9, this volume).

Technological advances in the past quarter century have dramatically increased efficiency of monitoring data collection and analysis as well as opened new possibilities for synoptic rangeland monitoring. The development of robust mobile computing technologies has encouraged electronic capture of data in the field, reducing the potential for recording and transcription errors. The ubiquity and reliability of global positioning system (GPS) technologies not available two decades ago makes it easy to accurately locate (and relocate) monitoring areas. Also during this time period, many of the imaging sensors and analytic techniques that have made remote sensing a staple part of monitoring were developed.

The changes described above—theoretical, policy, and technological—have had a significant impact on how rangelands are monitored and how those data can be used for management decision-making. The following are some of the major conceptual advances to rangeland monitoring from the last 25 years.

16.2.1   Monitoring Land Health Instead of Land Uses

Governments throughout the world have shifted from monitoring plant community responses to a single land use (usually livestock production), to documenting and understanding changes in land health (i.e., the degree to which the integrity of soils, vegetation, water, air, and ecological processes are sustainable, Bedell 1998) in response to multiple land uses. In Australia, Landscape Function Analysis (LFA) was developed to more effectively document and monitor changes in the "leakiness" of water- and nutrient-limited rangeland ecosystems (Ludwig et al. (2004). In the United States, NRCS and BLM have both adopted a suite of measurements that were selected largely because they generate indicators of ecosystem function, while also providing more traditional indicators of plant community composition. These were designed to complement the Interpreting Indicators of Rangeland Health (IIRH) assessment protocol which, like the new BLM and NRCS monitoring systems, was developed in response to the recommendations of the National Research Council and Society for Range Management (National Research Council 1994; Adams et al. 1995). Monitoring systems promoted by private consultants, and NGO's have also taken a more holistic approach, including those directly or indirectly associated with "Holistic Management" such as LandEKG (http://landekg.com) and Bullseye (Gadzia and Graham 2013). These changes have been driven by a number of synergistic factors, including (a) an increasing number of uses of rangelands, in both developed and developing countries, (b) climate change, (c) a more profound understanding of rangeland ecosystem processes and their interactions, and particularly (d) how land uses contribute to transitions in plant communities.

Rangelands that had been exclusively managed for livestock production are increasingly used for energy and crop production and for recreation. Each of these uses introduces novel disturbances, with often unpredictable effects. This has required the development of monitoring systems that are sensitive to the impacts of both current known land uses and unknown future ones. Monitoring the health of the fundamental properties and processes upon which rangelands depend provides managers with the confidence that they will be able to detect the impacts of land uses that don't even yet exist. For similar reasons, climate change has also driven a shift to monitoring land health (Chap. 7, this volume). Because of uncertainty around how climate will change at specific locations and how these changes will affect ecosystem structure and processes, monitoring systems designed to reflect general changes in land health are more likely to detect climate change impacts than those that are narrowly designed to reflect changes in plant community composition in response to grazing.

Arguably the most important conceptual development contributing to the transformation of rangeland monitoring programs has been the increased awareness of threshold transitions and the potential for development of alternative stable states. Previous conceptual models of change in rangelands had been based on linear processes of succession and retrogression that focused primarily on changes in plant community composition in response to preferential livestock grazing of "decreaser" species and avoidance of "increaser" species (Dyksterhuis 1949). As a result, these models largely failed to account for how changes in soil properties and especially soil hydrology can accelerate and even precede more visible changes in plant community composition (see "Developing and measuring soil indicators" below). Land health necessarily requires a broader and more holistic focus (Herrick et al. 2012).

16.2.2   Functional Indicators of Land Health

An indicator is an aspect of an ecosystem or process that can be observed or measured and provides useful information about the condition of the system being monitored (Suter 2001; White 2003). Indicators may be direct measures of an important ecosystem attribute or service (e.g., vegetation biomass is a direct measure of ecosystem productivity) or indirect measures that are correlated with an ecosystem feature or process that is difficult to measure directly (e.g., wind erosion is related to vegetation height and bare ground amount). In either case, to be useful for monitoring, an indicator must be both measurable and related in a known way to the structure and function of the ecosystem being monitored (Suter 2001). For example, the amount of bare ground on a site and its arrangement can be an indicator of risk of soil erosion, soil nutrient loss, decreased water infiltration, and species invasion if the site's potential is known.

In the past, rangeland monitoring focused on measuring a few indicators specific to the impacts of land uses (primarily indicators related to forage production and utilization). The shift to focusing on measuring and monitoring rangeland health and the need to consider how different land uses affect land health have triggered a shift toward monitoring indicators that are functionally related to ecosystem processes. West (2003b) suggested that "… the recent switch to functional attributes rather than singular utilitarian variables is likely to assist in staying the course longer in future monitoring efforts. It is also easier for range professionals to interact with other disciplines and multiple land owners when more general structural and functional attributes of ecosystems are used as proxies for more practical ones." Several useful frameworks for selecting functional indicators for ecosystem monitoring have been developed (Breckenridge et al. 1995; Tegler et al. 2001; Fancy et al. 2009).

Noon (2003) observed that a lack of ecological theory or logic to justify the selection of indicators may explain why monitoring programs historically have failed to provide useful information for natural resource management. A significant advancement in land management from the past 25 years that has helped in this regard is the development and use of conceptual models of rangeland structure and function to identify and interpret suitable functional indicators for monitoring (Noon 2003; Miller et al. 2010).

Conceptual models document what is known about the important components of an ecosystem, the nature and strength of interactions among them, and attributes that characterize different states of the ecosystem (Noon 2003; National Park

Service 2012). Conceptual models illustrate, commonly through visual and narrative summaries, how ecological processes, disturbances, and management actions affect an ecosystem. While not necessarily statistical or predictive in nature, conceptual models are useful for supporting a monitoring program because they document the known (or hypothesized) impacts of management and other disturbances on plant communities and soils (Fig. 16.1). This knowledge identifies the aspects of the system that should be measured (i.e., indicators) and provides an understanding of how to interpret observed changes in those indicators. Conceptual models can also highlight knowledge gaps in ecosystem structure or function (Karl et al. 2012c).

16.2.3   Core Indicators and Methods

A common criticism of rangeland monitoring (in fact, natural resource monitoring in general) is that there is little consistency among monitoring programs in the indicators that are monitored and the methods used to monitor them. This may hinder the ability of data to be used to address multiple resource concerns or combined across projects or jurisdictions. Examination of almost any monitoring methods text (e.g., Elzinga et al. 1998; Bonham 2013) shows myriad quantitative and qualitative methods for measuring almost any indicator. Additionally, the rangeland profession has not adopted a routine practice of validating new methods before they are implemented in monitoring programs (West 2003a).

Although the root causes of method proliferation in rangeland monitoring are diverse (e.g., isolated nature of monitoring programs historically, perceived need for quick and easy to implement methods, methods developed in one region that do not work well in other regions), the effect has been extreme difficulty in combining measurements among monitoring programs and significant challenges in assessing or monitoring rangeland resources above local scales. West (2003a) claimed that "Lack of consistent and comparable monitoring procedures within and between the federal management, advisory, and regulatory agencies has made it impossible to conclude reliably what the overall condition and trends in conditions of our public rangelands are." In their report on resource conditions in the United States, the Heinz Center (2008) concluded that the lack of consistent indicators collected using standardized methods precluded all but a cursory assessment of natural resource conditions. Even for rangelands managed by a single agency and land use, the BLM used five types of data (frequency, cover, production, utilization, and photos) from 15 different methods to monitor range trend plots across 310 allotments in six states (Veblen et al. 2014).

A consistent, minimal (i.e., core) set of standard indicators and methods for rangeland monitoring offers several potential advantages. First, it provides the ability to combine datasets from different monitoring programs (e.g., across jurisdictions). Second, it allows for data to be scaled up to support inferences to areas of larger extent. Third, standard indicators give an opportunity for data to be reused for different purposes. For example, when a new management concern arises, existing

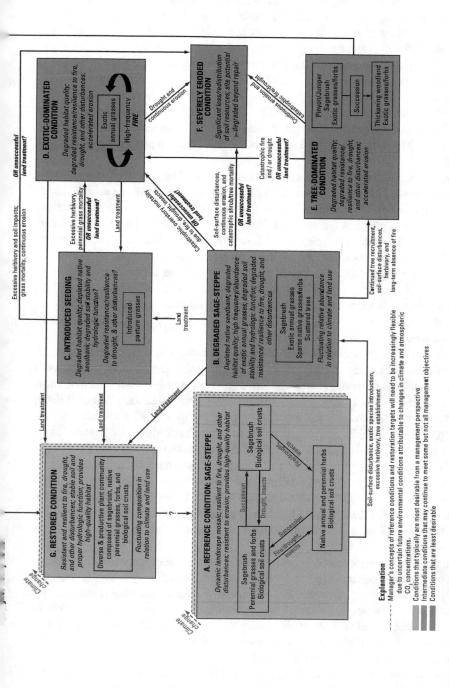

Fig. 16.1 Conceptual models are synthetic depictions about what is known or hypothesized about how an ecosystem is organized and functions. This information can be useful in identifying indicators for monitoring, thresholds or reference ranges for those indicators, and interpretations of indicator changes. The state-and-transition conceptual model above is one example of how a conceptual model for a sagebrush steppe ecosystem in the Great Basin, USA, could be organized. Model reprinted from Miller et al. (2010). General structure follows Bestelmeyer et al. (2003) and Stringham et al. (2003)

monitoring data may be available to provide initial estimates of resource conditions and trends. These potential benefits stem from core indicators resolving incompatibilities between data types used to describe rangeland ecosystems. West (2003a) summarized this need as "… there remains some desirability in choosing at least a short list of variables and means to monitor and in presenting their responses across as wide array of rangelands as possible so that the user could say that a generally approved approach had been employed."

The idea of a minimum set of standard ecosystem indicators and methods for measuring them has been suggested before (West 2003b, H. John Heinz III Center for Science 2008). However, it has only been recently that terrestrial indicators have been actually implemented (e.g., Herrick et al. 2010; Mackinnon et al. 2011; Toevs et al. 2011).

The concept of core indicators is not without criticism. One argument against standardized core indicators is that each region or area has a unique set of ecosystem and management properties and is constrained by historic events, therefore each monitoring program (including selected indicators) must be designed for its specific system (White 2003; Fancy and Bennetts 2012). While this is undoubtedly true, there are indicators that are informative over large regions (e.g., amount of bare ground, woody plant cover) and larger scale questions that can be asked (e.g., what proportion of rangelands have more than 15 % cover of annual grasses?) if core indicator data are available. Also, current implementations of core indicators (e.g., Mackinnon et al. 2011) provide for additional, supplemental indicators to meet local needs.

Another argument against standardized core indicators is that the need to maximize efficiency of a monitoring program (e.g., to minimize costs or to achieve a desired sample size) should encourage strict parsimony in selecting the most informative indicators (White 2003). This argument assumes that enough is known about an ecosystem when designing a monitoring program to select informative indicators, and there will not be need in the future to measure additional indicators. It also takes a narrow view of efficiency—i.e., efficiency is maximized only within a single objective and not among a suite of management information needs. The development and adoption of rangeland core indicators, however, is currently being driven by the need to find efficiencies in monitoring at larger organizational levels. Monitoring data that can be used to address more than one management objective and answer questions at more than one scale are more efficient in terms of cost than data that cannot.

Such a set of core indicators—measurable ecosystem components that are applicable across many different ecosystems and informative to many different management objectives—would be the basis for national-level monitoring programs and comprise a minimal set that should be measured in all monitoring programs. While it is not possible to select a set of indicators that will satisfy all management information needs across scales, the criteria listed in Table 16.1 are useful for considering which core indicators to select.

Core indicators provide no utility for rangeland monitoring without an accompanying set of standardized core methods for measuring them. Seemingly minor

Table 16.1 Criteria for selecting core indicators for rangeland monitoring

Criteria	Description
Relevant to ecosystem structure or function*	Indicators must relate in a known way (e.g., documented in a conceptual model) to the structure or function of an ecosystem of interest.
Usability*	Sufficient documentation exists to select appropriate methods and calculate indicators from measurements or observations.
Cost-effectiveness*	Cost of collecting indicator data is lower than for other competing indicators.
Cause/effect*	A clear understanding exists of how changes in ecosystem attributes will result in changes to the indicator.
Signal-to-noise ratio*	Changes in indicator values are primarily related to the intended ecosystem attribute and not natural variability or other factors.
Quality assurance*	Quality assurance and control procedures are available and adequate.
Anticipatory*	Indicator provides early warning of widespread ecosystem changes.
Historical record*	Information on the indicator has been collected over a period of time such that a reference set of data exist.
Retrospective*	Provides information about historic conditions (e.g., tree rings), over extended time periods (e.g., soil carbon), or can be applied to previously collected data (e.g., remote-sensing imagery).
New information*	Provides new information (i.e., not redundant with other indicators).
Minimal environmental impact*	Collection of information for measuring the indicator causes the least amount of disturbance to the environment.
Used by other monitoring programs	Priority should be given to indicators that are in use by other (especially regional to national) monitoring programs to facilitate cross-program data combination.
Easy to understand and explain	Indicators that are intuitive are likely to be more effective at informing and influencing management decisions.
Applicable to policy and management	Indicators that relate to aspects of an ecosystem that can be managed or that are tied to range management policies should be prioritized.

* denotes criteria proposed by White (2003) for selecting environmental monitoring indicators in general. While it may not be possible to meet all criteria when selecting indicators, priority should be given to the indicators that satisfy as many as possible

differences among methods can result in incompatible data (Bonham 2013). For instance, differences in definitions between soil and rock can cause one method to produce higher estimates of an exposed soil indicator relative to another method. Similarly, for plant-cover indicators, foliar cover methods usually yield different estimates than total-canopy cover methods (Toevs et al. 2011). Core methods represent a minimal set of information that should be collected in almost all monitoring efforts and are intended to encourage combination and "scaling-up" of monitoring datasets.

In selecting core methods, the repeatability, quality, and objectivity of data that the method provides should be considered. The attributes of a method that should be

Table 16.2 Desired properties of core methods for monitoring rangeland ecosystems

Property	Description
Quantitative	A method should record measurements or direct observations of a site's biophysical features.
Repeatable and efficient	Measurements should be repeatable within a stated margin of error and should be able to be collected at minimal cost.
Low potential for non-sampling error	Methods that minimize sources of error (e.g., inter-observer variability) and perform consistently across a wide range of environments.
Objective	Methods should minimize the opportunity for observer bias to influence the results.
Established and validated	Methods implemented for monitoring programs should be well documented and tested. Quality assurance and quality control procedures should be well defined.
Implementable with minimal training	Ideally methods should be able to be learned quickly and reliable data collected by individuals without extensive experience. Comprehensive training and calibration programs should accompany any method implemented in a monitoring program.
Can be used to calculate many indicators	The more indicators that can be derived from a method's data, the more value it can offer as a core method.
Used in other monitoring programs	Methods that are already implemented in other (especially large-scale) monitoring programs should.

prioritized when selecting core methods are similar to those for core indicators (Table 16.2). Selection of core methods is an exercise in optimizing these attributes in combination with a set of selected core indicators.

In some cases, monitoring objectives may not be able to be addressed through the core indicators. In these cases, it is appropriate to specify additional *supplemental* indicators to meet management and monitoring objectives. Supplemental indicators can be specific to land uses (e.g., grazing), programs (e.g., off-road vehicle management), or management actions (e.g., restoration following a fire). For example, the core indicators provide information for assessing impacts of grazing (e.g., cover of forage plants, amount of bare ground). However, this information is sometimes not sufficient to evaluate the effectiveness of grazing management actions. Supplemental indicators, such as forage production, utilization, or residual cover, may be selected in addition to the core indicators to inform grazing management. Where only residual cover is required, it can be easily calculated using data from whichever core method is used for monitoring vegetation cover and composition, eliminating the need for a supplementary method. Supplemental indicators are often intended to meet local management needs, so there may be little expectation to combine or share these indicators across management boundaries.

However, standard methods should be applied whenever possible (i.e., caution should be used in selecting supplemental methods that duplicate indicators that can be calculated from the core methods). Additionally extra expenses of implementing supplemental methods should be justified. In selecting supplemental indicators, the same criteria apply as in selecting core methods.

16.2.4 Statistical Sampling Designs and Monitoring at Appropriate Scales

Given the increasing number of land uses, disturbances, and contention over management of rangeland resources, statistical approaches to sampling designs that can support monitoring for multiple objectives and scaling up and down of monitoring data are necessary. While the shift in emphasis from managing pastures to landscapes has occurred primarily in the last 25 years, rangeland managers have long recognized the need to combine site-level monitoring and management with landscape-scale observations relative to grazing. The key area concept, monitoring of selected areas in an allotment or pasture that are indicative of typical grazing use, was developed in an effort to determine grazing impacts across a landscape when the distribution of livestock was not consistent across the area (Standing 1938). The key area concept originally was stated as a livestock expected use area within which monitoring locations would be randomly selected (Stoddart and Smith 1943, see also Holechek et al. 2001). In practice, however, key areas have often been located subjectively, usually based on best professional judgment, in areas that would receive the most typical grazing use and avoid areas that saw unusually heavy livestock use (e.g., around water access points) or areas that received no use (e.g., too far from water, or too steep) (e.g., Bureau of Land Management 1999; Schalau 2010). Key areas typically maintain key species, those species used by livestock, are abundant and productive, and are representative throughout the area. Key species generally do not include highly desirable species that livestock may overuse, nor do they include species that may have the potential to grow on the area, but were eliminated by past use (Holechek et al. 2001). Key areas and species were thought to be an efficient means for assessing rangeland conditions relative to grazing in order to quickly detect changes with the fewest number of monitoring plots possible.

Key areas (or any other subjective or haphazardly selected sets of monitoring locations), however, have several disadvantages that significantly limit their utility for rangeland monitoring. First, selection of key areas is often a subjective process. The validity of a key area for monitoring grazing impacts in a larger landscape can be contested because of differences of opinion on what it means to be representative (Gitzen et al. 2012). Second, indicator values from key areas, when the key area has been subjectively or haphazardly selected cannot be statistically extrapolated to larger areas (Lohr 2009). Therefore indicator estimates from subjectively selected locations cannot be scaled up to larger reporting units (West 2003b). Third, assuming that key areas can be selected to be truly representative of the conditions in an area, a set of key areas will underestimate the variance of any indicator because they are a sample of only the average conditions in the area. As a result, confidence intervals will underestimate the uncertainty in the indicator estimates, and statistical tests will tend to show significant differences more than is warranted (i.e., inflated Type I errors). Finally, key areas are often representative of a single land use. Because the spatial distribution of different land uses and resource concerns vary across a landscape, a set of key areas selected for monitoring livestock impacts may not be representative for other objectives. Additionally, the ability of a key area to represent the land condition also may change over time as conditions within a landscape change (e.g., new roads are created or range improvements installed).

A statistically valid sample of a resource has several properties (Thompson 2002; Lohr 2009). First, the estimates of an indicator for that resource are unbiased. Unbiased in this context means that there is nothing inherent in the sampling design approach that would result in systematically over- or underestimating the indicator values. This is accomplished through explicit and careful definition of the study area (i.e., population) and sampling units (Table 16.3). The use of randomization techniques in selecting sampling units for monitoring allows the uncertainty about indicator estimates to be characterized and confidence intervals to be constructed around monitoring results. Uncertainty estimates (often expressed as a standard error or confidence interval, or used as part of a statistical test) are not direct measures of the variability of an indicator, but a reflection of how close an indicator estimate is to the actual indicator value in the study area.

Table 16.3 Descriptions of the basic components of a sampling design following the definitions of Thompson (2002) and Lohr (2009)

Sampling design component	General definition.
Element	An item upon which some type of information is collected (i.e., observation or measurement).
Sampling unit	A unique set of one or more elements that can be selected for being included in a sample. In rangeland monitoring, a sampling unit is often an area (e.g., a plot). The sampling unit may contain zero elements.
Sampling frame	The complete set of sampling units within the geographic area of the target population (e.g., the list of all plots available to be selected for sampling).
Sample population	All elements associated with the sampling units listed in the sample frame. Ideally this coincides with the target population, but constraints (e.g., safety, accessibility) may limit sample population to less than the target population.
Target population	All elements of interest within some defined area and time period.
Sample	A selected set of sampling units. Measurements or observations from the sample are used to draw inferences about the target population.

Successful rangeland monitoring efforts carefully consider and explicitly define each of these components. Table adapted from Beck et al. (2010, see also Strand et al. 2015)

Proportion of Area Having Canopy Gaps Greater Than 2m in Length

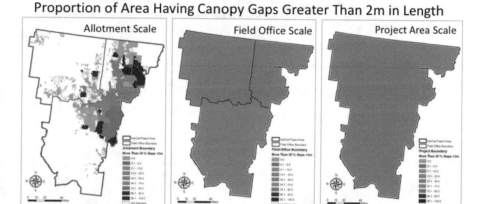

Fig. 16.2 A stratified random sampling design for three Bureau of Land Management (BLM) Field Offices in Northern California, implemented in 2013, allows for statistically valid estimates of indicators to be generated at multiple scales within the project area. Uncertainty of the indicator estimates is also calculated at each scale. Figure by S. Lamagna, BLM National Operations Center (unpublished data)

In a statistically valid sample, the area that each sample location represents can be quantified. It then becomes possible to "scale up" monitoring results to other reporting units or to combine similar monitoring datasets to make better predictions for a monitoring area (Fig. 16.2). Note that a statistical sampling design and consistent indicators and methods are necessary for this to be possible (Toevs et al. 2011).

16.2.5 Summary

The past 25 years has seen a rapid increase in the uses of and disturbances to range-lands that has taxed our ability to monitor these ecosystems with traditional techniques. However, the same time period has brought about conceptual developments that lay a foundation for more effective, and ultimately more efficient, rangeland monitoring. The cornerstones of this foundation are an understanding the nonlinear, cross-scale dynamics of rangeland ecosystems, and the development of functional indicators of critical ecosystem processes. The standardization of indicators (and measurement methods) to the extent possible and adoption of statistically valid and scalable sampling designs will help ensure monitoring data that not only satisfies initial objectives, but also is able to be combined with other datasets and analyzed for other or larger scale purposes.

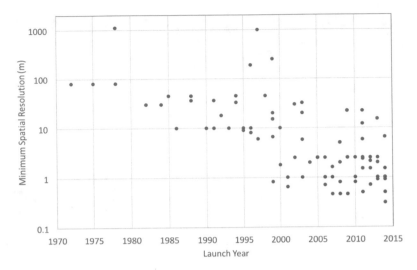

Fig. 16.3 Satellite-based sensors have become more numerous and higher-resolution over time. This increases opportunities for remote-sensing applications in rangeland monitoring. *Source* https://directory.eoportal.org/web/eoportal/satellite-missions (accessed April 20, 2015)

16.3   Remote-Sensing Developments for Monitoring

The widespread application of remote-sensing technologies to natural resource inquiries has been one of the most significant developments of the past 25 years. Several seminal papers have discussed the role and potential of remote sensing for rangeland monitoring. Hunt et al. (2003) provided an overview of the many ways in which remote sensing had been applied in rangeland management to that point. Booth and Tueller (2003) propose a framework for integrating remote sensing into rangeland monitoring and management. Washington-Allen et al. (2006) provide a technique for using remote sensing to understand past changes in rangelands. In the time since these papers were published, however, new applications, sensors, and methods have increased markedly. Here we briefly survey some of these developments, but a thorough review of remote-sensing developments in rangeland monitoring is beyond the scope of this chapter.

In 1990, there were only a handful of publicly available sources of satellite imagery, and this imagery was expensive and difficult to process (Fig. 16.3). Satellite imagery was generally panchromatic or multi-spectral only and of moderate (e.g., 30-m) or coarse (e.g., ≥1 km) resolution. Aerial photography was largely analog (i.e., on film) and challenging to incorporate into emerging GIS technologies. Contrast that to 2015 where there are now over 100 sources of satellite imagery (https://directory.eoportal.org/web/eoportal/satellite-missions, accessed April 20, 2015), many of them free and delivered in near real-time, and the emergence of inexpensive and autonomous unmanned aerial vehicles (UAVs) has caused a renascence of aerial photography research and applications.

16.3.1 Developments in Remote Sensors

The launch of the Moderate-Resolution Imaging Spectroradiometer (MODIS) sensor in 2000 ushered in a new era of remote sensing due to its ability to image large swaths of the earth on a daily basis. The synthesis of MODIS imagery into high-quality, standard products available every 7, 8, or 16 days has greatly facilitated the use of satellite imagery for looking at changes in rangeland landscapes among and within years. Reeves et al. (2006) predicted aboveground green biomass from MODIS net photosynthesis estimates throughout the growing season and characterized inter-annual variability in grassland vegetation. Wylie et al. (2012) used MODIS time series data from 2000 to 2008 to develop expected annual greenness profiles for rangelands in southern Idaho to detect significant departures due to management changes or disturbance. Browning et al. (in review) related a time series of MODIS normalized difference vegetation index (NDVI) to changes in overall plant biomass across years and to changes in plant functional group responses within years.

Hyperspectral imaging—collection of many, narrow contiguous spectral bands through the visible and infrared portions of the electromagnetic spectrum (Govender et al. 2009)—has also seen increasing application in rangeland monitoring. Hyperspectral imaging has been used to detect and monitor infestations of non-native, invasive plants (e.g., Ustin et al. 2004; Glenn et al. 2005; Mundt et al. 2005). Weber et al. (2008) used hyperspectral data to map biological soil crusts in South Africa.

Light detection and ranging (LiDAR) is a remote-sensing technology for mapping elevations from laser impulses reflected off a surface. LiDAR has been used in rangelands to estimate shrub height and crown area (e.g., Streutker and Glenn 2006; Glenn et al. 2011; Mitchell et al. 2011) and for mapping fine-scale topography to monitor processes like shrub invasion, dune formation, and soil erosion (e.g., Perroy et al. 2010; Sankey et al. 2010). LiDAR has also been used in conjunction with other remote-sensing products (e.g., multispectral or hyperspectral imagery) to improve the accuracy of vegetation classifications (e.g., Mundt et al. 2006; Bork and Su 2007).

Interest in the use of unmanned aerial systems (UAS)—small remotely piloted or autonomous aircraft—for rangeland monitoring has increased as imaging sensors have become smaller and software for processing large amounts of digital imagery improves. With a UAS, it is often possible to acquire imagery at resolutions <5 cm and with more flexibility than traditional aircraft (Rango et al. 2009). For example, Laliberte et al. (2012b) used 6–8 cm UAS imagery to map vegetation types in Idaho and New Mexico rangelands. Breckenridge et al. (2012) found that estimates of vegetation cover and bare ground from UAS images agreed well with field-based measurements. d'Oleire-Oltmanns et al. (2012) used overlapping UAS images to monitor gully formation and soil erosion in Moroccan badlands.

16.3.2 Developments in Remote-Sensing Techniques

Over the past 25 years many new remote-sensing techniques have been developed and applied to rangeland monitoring. In some cases (e.g., digital photogrammetry, object-based image analysis) these new techniques were made possible by computing advances that have enabled faster and more efficient processing of large amounts of imagery. In other cases (e.g., multi-temporal image analysis), new approaches were developed around a new type or increased availability of remote-sensing products. Below are four techniques that have changed how remote-sensing imagery has been applied in rangeland monitoring.

Manual interpretation and grid-point cover estimation methods for aerial photographs were some of the first remote-sensing analysis techniques. While these approaches have gone out of vogue in favor of statistical classification algorithms for many remote-sensing applications, the recent increase in easily obtainable very-high resolution (VHR, <5 cm ground sampling distance) digital imagery from piloted aircraft, UAS, or even pole-mounted digital cameras has brought about a resurgence in image interpretation techniques. Booth et al. (2006) developed SamplePoint, a computer application for quickly and easily estimating vegetation cover from VHR aerial photography via point sampling. Many studies have looked at the accuracy and efficiency of image interpretation compared to field techniques and have found comparable results for estimating cover of functional groups (e.g., shrubs, perennial grasses) (e.g., Booth et al. 2005; Cagney et al. 2011; Duniway et al. 2011; Pilliod and Arkle 2013). Image interpretation of VHR images has also been used to estimate other rangeland indicators like canopy gaps (Karl et al. 2012b) or to evaluate wildlife habitat quality (e.g., Beck et al. 2014). Duniway et al. (2011) and Karl et al. (2012a) explored how VHR image interpretation could be employed in large-scale rangeland monitoring programs. A distinct advantage to VHR image interpretation techniques is that they can be employed by rangeland management staff with minimal expertise in remote sensing.

The increase in availability of VHR imagery from piloted aircraft or UAS has also spurred an interest in using digital photogrammetric techniques to create three-dimensional representations of rangeland sites from pairs of overlapping VHR images. Advances in photogrammetry software have made this process much easier, more accurate, and less expensive. Gillan et al. (2014) used stereo aerial photographs to model shrub heights in the Mojave Desert. Gong et al. (2000) created digital surface models from VHR stereo imagery to monitor changes in crown closure and tree height in a hardwood rangeland. Marzolff and Poesen (2009) used 3D digital surface models from digital photogrammetric techniques to monitor the gully development. Gillan et al. (2016) mapped soil movement following experimental juniper removal treatments by differencing digital surface models from before and after treatments. Recently, a new photogrammetric technique called structure from motion (SfM) has estimated three-dimensional surfaces from sequences of VHR images (Turner et al. 2012; Fonstad et al. 2013). An advantage of SfM is that, unlike traditional photogrammetric techniques, accuracy of surface elevation models increases with the number of overlapping photos. Genchi et al. (2015) used SfM

Fig. 16.4 Structure from Motion (SfM) is a new photogrammetric technique that estimates the structure and height of three-dimensional objects from a sequence of two-dimensional images. The SfM algorithm was applied simultaneously to a set of downward-looking and oblique images to produce a point cloud of the canopy structure of creosote (*Larrea tridentata* (DC.) Coville) and soil surface elevation below the shrub canopy in the Chihuahuan Desert, New Mexico (J. Gillan, unpublished data)

with images acquired from a UAS model soil surfaces and cliff faces to estimate soil excavation and nesting habitat for birds. Another benefit of SfM is that it can accommodate both downward-looking (i.e., nadir) and oblique images together to produce "LiDAR-like" point clouds. This provides an opportunity to estimate canopy structure and ground surfaces below vegetation—something that is possible from ground-based (e.g., Greaves et al. 2015) and aerial (e.g., Streutker and Glenn 2006) LiDAR but not traditional photogrammetry (Fig. 16.4).

Traditional remote-sensing algorithms analyze and assign meaning to the individual pixels of an image. While these techniques have been used successfully for many rangeland applications, pixel-based approaches are challenged in many rangeland landscapes because of inherent heterogeneity and patchiness. In these situations, the variability within and context of patches at different scales is an important trait for their correct identification or description. Object-Based Image analysis (OBIA) groups adjacent pixels in an image together based on their similarity and then treats the resulting "objects" as analysis units rather than the individual pixels (Burnett and Blaschke 2003). With OBIA, additional measures of variability, patchiness, and juxtaposition within and between patches can be factored into classifications or models. By changing the amount of variability acceptable within an object, objects of different sizes (i.e., scales) can be created for the same area. It is possible to define optimal analysis scales with OBIA that maximize the accuracy of a remote-sensing product (Feitosa et al. 2006; Laliberte and Rango 2009; Karl and Maurer 2010a). The OBIA technique has shown to yield more accurate classifications of rangeland vegetation than pixel-based techniques in many cases (Karl and Maurer 2010b; Dingle Robertson and King 2011; Myint et al. 2011; Whiteside et al. 2011). The OBIA technique has been applied for quantifying rangeland vegetation cover and distribution at both fine (e.g., Luscier et al. 2006; Laliberte et al. 2012a; Hulet et al. 2013, 2014a) and coarse (Laliberte et al. 2006) scales. Karl (2010) used OBIA in predictions of three rangeland indicators in a southern Idaho study area. Laliberte et al. (2004) used an OBIA method to detect shrub encroachment in southern New Mexico rangelands using historic aerial and current high-resolution satellite imagery.

Several satellite sensors (e.g., AVHRR, Landsat, MODIS) have now been operational long enough to provide a reliable record of change in rangeland ecosystems. For example, Homer et al. (2004) developed a National Land Cover Dataset for the United States from 2001 Landsat imagery that has been subsequently updated to assess change over time (Xian and Homer 2010; Jin et al. 2013). Homer et al. (2013) quantified annual and seasonal changes in bare ground and cover of shrubs, sagebrush, herbaceous plants, and litter using Quickbird and Landsat imagery. Additionally, sensors like MODIS are providing image products at a temporal resolution (e.g., MODIS NDVI composites are available on 8-day cycles) that was not previously available for rangeland studies. This combination of temporal extent and resolution has spawned a host of multi-temporal remote-sensing techniques for rangeland monitoring. Wylie et al. (2012) used 9 years of MODIS NDVI data to construct estimates of expected production for a southern Idaho study area and detect departures from this expectation. Sankey et al. (2013) used MODIS and AVHRR NDVI time series as a proxy indicator for post-fire recovery in rangelands. Brandt et al. (2014) used a combination of SPOT-Vegetation, Landsat long-term data record, and MODIS NDVI products to assess degradation and vegetation biomass changes in the Sahel region of Africa from 1982 to 2010. Maynard et al. (in review) demonstrated that significant seasonal and between-year breaks in a MODIS NDVI time series decomposes using the Breaks for Additive Season and Trend (BFAST) algorithm (Verbesselt et al. 2010) correlated to changes in biomass and functional group composition in a Chihuahuan Desert rangeland.

16.3.3   Modes of Remote-Sensing Implementation for Monitoring

Three general modes of applying remote sensing have emerged in rangeland monitoring. The first mode is to use remote-sensing technologies and products to replace field measurements. Numerous studies have shown that interpretation or classification of high-resolution imagery can match or outperform many field-based measurements of certain attributes (e.g., Booth et al. 2005; Seefeldt and Booth 2006; Cagney et al. 2011; Hulet et al. 2014b). Land cover classifications or predictions of vegetation attributes (e.g., cover) have been used for landscape-scale rangeland assessment and monitoring (e.g., Hunt and Miyake 2006; Marsett et al. 2006; Homer et al. 2013). Regression models (e.g., Homer et al. 2012) or geostatisical techniques (e.g., Karl 2010) are used to predict rangeland indicators over landscapes from a set of field samples. In these cases, it is more important that the field sample locations represent the range and extremes of the indicators being predicted than to be randomly selected (Gregoire 1998). Historic analysis of rangeland condition is also possible using archives of aerial photography (e.g., Tappan et al. 2004; Rango et al. 2005) or satellite imagery (e.g., Washington-Allen et al. 2006; Malmstrom et al. 2008).

The second mode of remote-sensing application is to supplement or augment field-based activities. For example, high-resolution aerial photographs may be acquired at field sample locations and at nearby, similar sites for use in a double-sampling approach (see Duniway et al. 2011; Karl et al. 2012a). Another example of this mode of remote-sensing application is model-assisted inference which uses a statistical model developed between field measurements and remote-sensing products (e.g., NDVI) to improve indicator estimates in larger areas (Gregoire 1998; Opsomer et al. 2007; Stehman 2009). This mode of remote sensing can also be employed to make improved spatial predictions of indicators such as vegetation cover (Karl 2010). Gu et al. (2013) used biomass models developed from MODIS NDVI to remove artifacts from rangeland productivity estimates due to administrative boundaries (i.e., state and county lines).

Finally, remote-sensing techniques can be used to generate new or synthetic indicators of rangelands that are difficult or impossible to characterize through field techniques. For example, Ludwig et al. (2007a) defined a "leakiness index" to characterize the ability of water to move through a rangeland site and increase soil erosion. Cocke et al. (2005) defined the differenced normalized burn ratio as a measure of the extent and severity of wildland fire. Kefi et al. (2007) used changes in vegetation patch-size distributions as an indicator of impending desertification. Nijland et al. (2010) mapped soil moisture at depths to 6 m using electrical resistivity tomography. While the diversity of new indicators available from remote-sensing techniques is increasing, a challenge with this remote-sensing mode for rangeland monitoring, however, is translating the remote-sensing-derived indicators into statements of rangeland quality or health.

16.3.4 Summary

Remote-sensing technologies and applications for rangeland monitoring have burgeoned in the past 25 years. Rapid advancements in sensor technologies and analytical techniques coupled with decreasing costs of remote-sensing products have resulted in myriad examples of the utility of remote sensing to quantitatively monitor rangelands in ways previously not possible. Integration of remote-sensing techniques as replacements for and supplements to existing field-based monitoring efforts will continue. However, development of novel remote-sensing-specific indicators will quicken, spurred in part by the increasing availability (e.g., frequency of imagery, ease of access) and flexibility (i.e., ability to control image acquisition parameters and timing) of remote-sensing products.

16.4 Societal Implications of Conceptual Advances

The conceptual advances described above have influenced how rangeland monitoring is conducted and have increased the utility of monitoring data for management decision-making.

16.4.1   Increases in Monitoring Program Efficiency

The shift of focus to land health and the adoption of core indicators and methods has led to increases in the efficiency of monitoring programs. In the past, monitoring efficiency was gained through optimizing an individual monitoring program relative to its specific objectives. One form of this optimization was minimizing the number of required sample sites by restricting sample populations, employing specialized sampling designs, or resorting to subjective sampling. In highly heterogeneous rangeland systems, this drove monitoring to selected areas that were considered representative of a larger landscape (Stoddart and Smith 1943; Holechek et al. 2001), and emphasized the need to maximize the value of individual observations. This approach, however, is only efficient if a single monitoring objective or indicator is being considered. It is an inefficient approach if a separate monitoring program is needed for each management objective. Additionally, monitoring programs designed tightly around a single objective (e.g., monitoring of grazing use impacts) may be inappropriate for informing other management objectives.

Overall efficiency of monitoring programs increases as the number of questions that can be answered with a dataset increases, and if data from multiple efforts can be combined. This is because monitoring data that have been collected for one purpose can be reused to answer other objectives. For example, data collected since 2004 (NRCS) and 2011 (BLM) by the National Resource Inventory (NRI) program are designed to generate regional and national estimates of rangeland status and trend. As of 2015, many of these data were being used to support the development and revision of ecological site descriptions. In the future, it is anticipated that they will be combined with local to regional monitoring data to address more specific questions, such as habitat suitability for wildlife species or for use in developing remote-sensing products provided restrictions regarding confidentiality of data can be overcome.

Combining datasets is only possible, however, if indicators and methods are consistent between programs and if the datasets were based on a probabilistic sampling design. Although combining different datasets is not always straightforward statistically (e.g., for monitoring over large landscapes or when sample size requirements are high due to variability in an indicator), it can be worth the effort.

16.4.2   Cross-Jurisdictional Monitoring

Another benefit of the recent adoption of core indicators and methods is that it opens up new opportunities for monitoring conditions across boundaries of ownership or administration. Land owners or agencies that use core methods and statistically based sampling designs can interchange and combine datasets to address management questions related to the context of a management unit in a larger landscape or pertaining to resources or disturbances that cross boundaries. This can be

particularly useful in rangelands where ownership is fragmented, and can be applied
at all scales from local to national.

For example, monitoring the status and condition of seasonal habitats is necessary for the conservation of greater sage-grouse (*Centrocercus urophasiuanus*)
across its range. Approximately 52 % of the current greater sage-grouse distribution
occurs on land managed by the BLM, whereas the remainder is managed by private
(31 %) or other (17 %) federal and state agencies (U.S. Fish and Wildlife Service
2010). The mixed land ownership has resulted in multiple sage-grouse habitat mon-

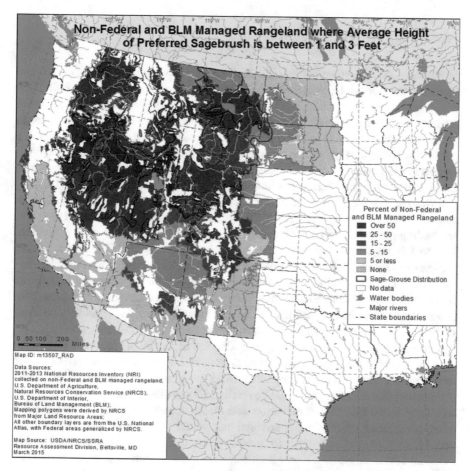

Fig. 16.5 Use of consistent indicators and methods and statistically valid sampling designs permit
aggregation of data across jurisdictional boundaries. This map shows an example of how habitat
suitability indicators for greater sage-grouse can be calculated from the National Resources
Conservation Service's National Resources Inventory (NRI) and the Bureau of Land Management's
(BLM) Landscape Monitoring Framework (LMF) for all non-federal (e.g., private, state, tribal)
and BLM-managed rangelands in the western USA. These results are based upon NRI rangeland
data collected in the field on 3658 non-Federal and LMF data from 2876 BLM rangeland sites
during the period 2011 to 2013

itoring programs. Knowledge of the condition or trend of greater sage-grouse on lands managed by one agency or land owner may be informative for some questions relative to the affairs of that organization. However it provides an incomplete picture of habitat for the species as a whole. Because the BLM Landscape Monitoring Framework and the NRCS National Resources Inventory both adopted the core indicators and methods, these datasets have been combined to produce estimates of greater sage-grouse habitat indicators across private, tribal, and BLM lands from a sample of over 6500 locations between 2011 and 2013 (Fig. 16.5).

16.4.3 Effectiveness Monitoring

Monitoring the effectiveness of specific management actions is important and sometimes mandated. Current philosophies of rangeland management are built upon the concept of adaptive management (Walters and Holling 1990) where management actions, developed as a result of best-available knowledge, are treated like hypotheses to be evaluated by data collected following implementation of actions. This learning phase of adaptive management is essential for translating data on treatments and management adjustments into further management actions. However, this step of adaptive management has often been short-circuited (Moir and Block 2001; Walters 2007).

Effectiveness monitoring of a single project or action provide information on whether a project met an objective. Yet projects or site-specific actions and their corresponding effects carry with them environmental variability that may sometimes lead to failures merely by chance. Adaptive management requires across-project comparisons where projects are implemented over multiple time frames under different environmental conditions to determine the likelihood of success under future conditions. If all projects use the same core indicators for addressing common management objectives, then decisions on future management actions will likely be more successful if adaptive management adjustments are applied to future decisions.

For example, Arkle et al. (2014) and Knutson et al. (2014) used the same core indicators to monitor long-term plant responses with and without revegetation after wildfires to address multiple objectives: did revegetation (1) increase perennial plant cover; (2) reduce annual grass cover; and (3) provide greater sage-grouse habitat? They found that higher elevation and mean annual precipitation of locations related to meeting objectives of perennial and annual plants and to providing herbaceous cover, but not shrub cover for greater sage-grouse. They were able to identify potential levels of elevation and precipitation below which objectives would not likely be met. They were also able to point to potential changes in techniques to enhance shrub cover for greater sage-grouse. These results are available for adapting post-fire revegetation to better meet these objectives where possible. In addition, this information is available during the decision process after fires to determine if the likelihood of success is worth the expense of a revegetation project.

16.5 Future Perspectives

16.5.1 Developing Functional Indicators at Broad Scales

Much of the advancement in developing functional indicators for rangeland management has been at the site scale owing to the fact that most research has been conducted at this scale (Peters et al. 2015). In some cases these site-based indicators can be scaled up meaningfully to larger (e.g., landscape or regional) scales (Peters et al. 2004). For example, cover of invasive annual grasses at a site is meaningful for local management, and the proportion of sites in a region having some minimal amount of the invasive grass is also useful for decision-making.

However, interactions between monitoring objectives and rangeland resources are scale-dependent, and informative indicators (and appropriate methods) may be different at different scales (West 2003b). Some site-based indicators such as those that need to be referenced to land potential to be interpreted (e.g., bare ground amount, shrub cover) or those tied to attributes that exhibit strong cross-scale interactions (Peters et al. 2004) or threshold effects (Briske et al. 2006), may not scale up meaningfully. For example, average shrub cover in a region is only marginally meaningful because the region contains many areas with different potential to support shrubs. Additionally, indicators derived from site characteristics can be informative of the overall conditions in a larger summary (i.e., reporting) unit, but do not capture the distribution or pattern of indicators within those areas.

A major challenge in developing effective monitoring programs and applying monitoring data to rangeland management is the paucity of functional indicators of ecosystem processes at landscape to regional scales. This is due in part to a critical knowledge gap: we simply don't understand how the relationships between ecosystem properties, processes, and functions vary across space and time. Principles of landscape ecology state that characteristics like connectivity, patch size and shape, and habitat diversity are critical to sustaining ecosystem processes. But quantifiable indicators and specific thresholds tied to these ecosystem attributes are not common in monitoring programs. The lack of broad-scale indicators may be limiting the utility of remote sensing in many monitoring programs, relegating it to a pattern description.

There are some examples, however, where specific indicators related to ecosystem processes have been developed at fine and broad scales. Greater sage-grouse habitat has been described at four distinct scales and habitat suitability indicators defined for each scale (Connelly et al. 2000; Doherty et al. 2010; Stiver et al. 2015). A regional decision matrix has been proposed (Chambers et al. 2014) and is being tested in these regions for greater sage-grouse management. It is based on landscape cover of sagebrush (i.e., percent of pixels classified as sagebrush using a 5-km^2 moving window), an indicator of greater sage-grouse lek longevity, in combination with soil temperature and moisture as a surrogate for ecosystem resistance to invasive annual grasses and resilience to fire. Ludwig et al. (2004) defined landscape functional integrity indicators to assess grazing effects on Australian rangelands at fine to coarse scales using cover and bare ground indicators within and between vegetation patches. Bertiller et al. (2002) described changes due to cattle grazing in plant composition, patch structure, and vegetation cover at different scales in Patagonian Monte rangelands.

16.5.2 Developing and Implementing Soil Indicators

Monitoring ecosystem functions should reflect changes in soil processes (Chap. 4, this volume). While it is not always, or even usually, the case that "monitoring vegetation has to take a backseat to monitoring soils" (West et al. 1994), directly monitoring soil degradation and recovery may provide better information for rangeland management decisions. In systems where vegetation cover and composition either do not reflect, or lag behind, fundamental changes in water and nutrient cycling and the energy flows upon which ecosystem services ultimately depend, soil indicators are appropriate. For example, soil compaction can increase runoff, and increased soil surface disturbance can increase soil erodibility and drive declines in soil organic matter. Both of these changes can occur without causing detectable differences in plant cover (e.g., Herrick et al. 1999; Bird et al. 2007).

Despite the recognized value of soil indicators, they are rarely included in monitoring programs. There are multiple reasons, including cost and measurement consistency. Even where sufficient financial resources and trained personal are able to ensure consistent measurement of a sufficient number of samples to detect change, the logistical challenges of collecting, processing, transporting, and storing samples for laboratory analysis quickly overwhelm institutional capacities. For this reason, only a field measurement of soil aggregate stability was selected for measurement in US national rangeland monitoring system.

The future, however, is bright as new field soil sensors and remote-sensing techniques are being rapidly developed and deployed. For example, Pastick et al. (2014) mapped soil organic layer thickness in interior Alaska from Landsat ETM+ and ancillary data. Field spectroscopy, in particular, appears to hold a high level of promise for predicting soil organic matter and other variables that are related to spectral properties (Shepherd and Walsh 2007).

16.5.3 Accounting for Inter-annual Climate Variability

A perennial challenge in monitoring rangelands is dealing with climate differences during the intervals between measurement periods. Because many rangeland ecosystem indicators (e.g., cover, production, and recruitment of herbaceous plants) are sensitive to yearly fluctuations in timing and amount of precipitation, it can be difficult to determine whether observed differences are due to management or disturbance, climate variability, or an interaction of the two. Conventional approaches to this problem have included comparing only like years, sampling across several years to encompass both dry and wet periods (West 2003b), or selecting indicators that are less sensitive to climate fluctuations (e.g., woody plant cover, density, perennial grass basal cover).

Advances in remote sensing, specifically the increasing availability of high-frequency remote-sensing products (e.g., MODIS NDVI), offer several new opportunities for addressing this challenge (e.g., White et al. 2005). This can be accomplished by constructing indicator ranges from different climate years and analyzing for departure (e.g., Wylie et al. 2012; Rigge et al. 2013). Alternatively, remote-sensing-derived indicators can be analyzed directly for trends over time or used as a covariate to analyze or interpret field observations (e.g., White et al. 2005; Dardel et al. 2014; Brandt et al. 2014). Finally, remote-sensing time series can be decomposed and inter-annual variability "factored out" to isolate management changes. Effective strategies for using remote-sensing products to account for inter-annual climate variability in rangeland monitoring still need much development and research.

16.5.4   Operationalizing Remote Sensing

Despite the advances in remote-sensing technologies and the potential uses for monitoring, adoption of remote sensing in many formal rangeland monitoring programs has been slow. Reasons for this are diverse and include historically variable performance of remote-sensing products (especially in rangelands), misunderstandings of what remote-sensing indicators actually mean in a rangeland context, and high skill and computing requirements to produce and use remote-sensing products (see Kennedy et al. 2009). Nevertheless, substantial incorporation of remote sensing into monitoring programs is necessary to meet the information needs of rangeland management in the future (Booth and Tueller 2003).

Operational use of remote sensing in rangeland monitoring programs requires not only clear articulation of the monitoring objectives but also analysis of remote-sensing options, desired accuracy requirements, and costs (Kennedy et al. 2009). Changes in sensor technology, data availability, and processing techniques also must be considered. In many cases remote-sensing-based indicators will not be adequate on their own and an integrated field and remote-sensing approach will be required (Ludwig et al. 2007b). The sources of error and uncertainty in indicator estimates must also be characterized for remote-sensing products that are used in rangeland monitoring programs to determine whether differences in indicator estimates are likely due to model error, sensor or analytical differences, or real changes on the ground.

16.5.5   Considerations with Evolving Technologies
for Measuring Indicators

The conventional wisdom that has been taught to many rangeland management practitioners is that once you pick a monitoring method you should always stick with it. This advice likely stems from the facts that rangelands are heterogeneous, diverse systems and many techniques for measuring rangeland attributes can contain a large degree of imprecision. Thus, the perception is that changing from one data collection technique to another may introduce more noise into the monitoring data and make it even harder to detect differences. This thinking, however, is flawed and has stymied adoption of more efficient and accurate approaches for monitoring rangeland indicators.

The development of new techniques and instruments for making quantitative indicator measurements happens in all science fields, and there are established procedures for phasing new technologies into existing monitoring programs. For example, the U.S. Historical Climatology Network, which tracks temperature averages and trends in the United States, began in the 1980s to shift from liquid-in-gas (e.g., alcohol or mercury) analog thermometers to digital thermistor sensors (Menne et al. 2010). Studies showed that biases and differences in variance could exist between these two different types of thermometers, and methods were developed to reconcile these differences so that analyses of long-term trends could be performed (Peterson et al. 1998).

Decisions to incorporate new measurement techniques into existing monitoring should be a part of ongoing reevaluations of the monitoring program (Lindenmayer and Likens 2009; Reynolds 2012). When considering changes to existing methods, however, it is critical to ensure indicators calculated from the new methods are consistent (in both definition and interpretation) with the original monitoring objectives. It is also important to study the results of both methods relative to each other to understand potential biases and differences in precision (Bland and Altman 1999). New methods should be well defined and documented and then only be adopted in a monitoring program if they (1) provide more accurate or precise data, or (2) bring the program into better alignment with other monitoring programs.

While changes to a monitoring program are possible over time, they should not be taken lightly because changes can be difficult and impose complexity in data analysis. Oakley et al. (2003) proposed a modular approach for defining monitoring protocols so that incremental changes to increase precision or efficiency could be more easily made. It is also helpful in designing monitoring programs to clearly distinguish indicators from methods (Toevs et al. 2011) so that the best available methods can be used to provide indicator estimates.

16.5.6 Collaboration and Sharing Monitoring Efforts and Data

Sharing monitoring data and participatory monitoring offer tremendous opportunities for both dramatically reducing future monitoring costs, and increasing our ability to interpret both historic and future data.

Sharing monitoring data reduces costs by reducing redundancy: individuals and organizations that would have previously each collected their own data to address different objectives can take advantage of others' data, freeing up resources for supplementary measurements, data management, analysis, and interpretation. It also increases the quality of interpretation because a larger reference pool of information, often covering greater areas and time periods, can be used. This can help to address the challenges of accounting for inter-annual climate variability, even if the same methods are not used, provided that generally co-varying indicators, such as foliar and canopy cover, were measured. Where standard methods and statistical designs were used, even greater benefits can be realized through data integration and direct comparisons. In some cases, issues of data confidentiality or proprietary ownership currently impede data sharing. To the extent possible, these issues should be addressed to foster better data sharing.

Participatory monitoring reduces costs and can improve interpretation. Costs may be reduced due to lower labor costs, although these savings are often offset by increased data management costs due to higher quality control costs. Perhaps the most intriguing, and often ignored, benefit of participatory monitoring is for interpretation. Individuals engaged in participatory monitoring often have both local knowledge and information that trained, paid field crews lack. Their knowledge may span both seasons and years, allowing them to identify possible drivers of change, as well as explain anomalous results or outliers.

16.5.7 Defining the Reference: The Challenge of Applying Monitoring to Management?

Perhaps the greatest challenge faced by managers seeking to apply monitoring data to management decisions is defining the reference. The first time monitoring data are collected at a location defines the baseline, but says nothing about where the baseline is relative to short- and long-term potential. Existing global assessments of land degradation are based largely on the opinions of multiple experts (Oldeman 1994), interpretation of satellite-based greenness indices (Bai et al. 2008), or a synthesis of multiple estimates based on one or more of these (e.g., Scherr 1999) to estimate the extent of land degradation. Expert opinion is limited by lack of a clearly defined reference. Greenness indices use spatial and temporal deviations from maximum greenness as the reference. Spatial differences are confounded by soil-based differences in potential productivity, while temporal differences are confounded by weather. Use of both spatial and temporal variability to determine the reference is particularly problematic in rangelands, where an increase in green woody cover often reflects degradation rather than recovery. While nearly all of these challenges can be addressed through the application of fine-scale information of phenology, weather, and vegetation, it is quite difficult to collect and apply this information across large areas

Two steps are required to reliably define reference conditions. The first is to describe the factors that determine land potential, including soil, topography, and climate (Herrick et al. 2013). Where an ecological site classification (Bestelmeyer et al. 2009) is available, identification of the ecological site can substitute for documentation of the individual edaphic, topographic, and climatic variables (Chap. 9, this volume). The second is to define the natural range of variability for response variables (vegetation and dynamic soil properties) for the ecological site.

The USDA National Resource Inventory is one of the few examples in which ecological site-specific references have been defined and applied at the national level to provide a context for assessments and baseline monitoring data (Herrick et al. 2010). Definition of reference conditions is based on the natural range of variability in the reference state of the state-and-transition model for the ecological site, where available, and on a combination of scientific literature, observations and measurements from reference plots, and local knowledge (Pyke et al. 2002; Pellant et al. 2005). These assessments were completed at the same time that quantitative baseline data were collected, providing both a snapshot of current levels of land degradation, and information that can be used to interpret the baseline monitoring data.

16.6 Summary

Rangeland ecology has benefited from recent developments in theory, policy, and technology. These developments, together with the increasing diversity of uses of rangelands, have changed the need for resource monitoring as well as the approaches for how it is carried out. Monitoring data are needed to establish baselines and changes in rangeland condition for documenting the impacts of climate change, disturbances, and management activities. The myriad monitoring needs for rangeland management, however, must be reconciled with realities of costs associated with collecting, analyzing, and using monitoring data. Thus the challenges of implementing useful and efficient monitoring of rangelands include both practical and institutional hurdles.

Traditional approaches to monitoring and management, however, are entrenched in agencies, organizations, and universities and often applied to objectives and systems beyond those for which they were designed. While aspects of this legacy of rangeland management are helpful (e.g., rangeland scientists were pioneers of landscape-scale thinking), relying too heavily on monitoring techniques focused on grazing management will not serve management needs in the future. West (2003b) concluded, "The range profession has put so much of its training efforts into identification of plant species, sampling within plots, and application of conventional statistical analysis that it hasn't had the background to examine other possible ways of answering the questions really being asked." Going forward, monitoring of rangeland resources needs to be grounded in the conceptual and technological advances of the past 25 years.

The recognition that rangelands are nonlinear systems characterized by thresholds and cross-scale processes has led to a realization of the importance of monitoring ecological processes and functions at different scales. This advance translated to a shift in thinking from monitoring plant community responses to land uses to monitoring changes in land health. The adoption of conceptual models as a mechanism for documenting and illustrating how ecological processes, disturbances, and management affect an ecosystem, has contributed to the identification and selection of functional ecosystem indicators. Conceptual models not only identify what parts of an ecosystem should be monitored, they provide insight into how monitoring data should be interpreted and used for making management decisions.

Differences in indicators and measurement methods among monitoring programs has hindered the ability of data to be used for multiple objectives or combined to understand conditions over larger scales. A consistent core set of standard indicators and methods for rangeland monitoring provides the ability to combine datasets from different monitoring efforts, allows data to be scaled up to larger extents, and expands opportunities to reuse data for other purposes. Core methods represent a minimal set of information that should be collected in almost any monitoring effort. When monitoring objectives are not served by the core indicators or methods, supplemental indicators and methods should be added.

Statistical approaches to sampling design for rangeland monitoring are necessary in our era of expanding land uses and disturbances and increasing contention. Conventional approaches that relied on targeted or haphazard sample site selection have disadvantages that severely limit their utility for rangeland monitoring. Most statistically based sampling designs can support monitoring for multiple objectives and scaling up and down of monitoring data. Additionally, randomization techniques for selecting sampling locations guard against bias and allow for characterization of uncertainty in indicator estimates.

The widespread application of remote-sensing technologies to rangeland research and monitoring has been one of the most significant developments of the past 25 years. Technological developments in remote sensing have happened at such a rate that the periodic summaries of remote-sensing applications for rangeland management have become quickly outdated. In addition to new sensors being continually developed, imagery is becoming available at higher resolutions and more frequently while being cheaper and easier to access. These innovations have been accompanied by new analytic techniques that has improved the ability to extract meaningful information from remote-sensing products. In particular, the nexus of inexpensive yet capable UAS with new digital photogrammetric software has led to cheap, easy 3D analysis for rangeland ecosystems.

The conceptual and technological advances in rangeland science and management have important implications for monitoring. First is the potential to increase monitoring efficiency. Historically efficiency was maximized through parsimonious monitoring program design. Under the weight of so many monitoring efforts, however, efficiencies across programs can only come through coordination of monitoring efforts such as adopting core indicators and methods and scalable, statistical sampling designs. Second, coordinated monitoring based on functional indicators of land health opens up opportunities for monitoring conditions across jurisdictional boundaries. This will be crucial for managing large-scale and diffuse disturbances (e.g., invasive species) as well as conservation of landscape-scale species (e.g., greater sage-grouse). Robust monitoring programs also help complete the learning cycle that is often missing from evaluating effectiveness of monitoring actions, and enable comparisons across projects to begin to understand the factors that affect the success of management actions.

Despite the advances of the past few decades, there are challenges but also many opportunities for rangeland monitoring in the future. One set of challenges deals with the development and implementation of monitoring indicators. A significant challenge for rangeland monitoring is developing functional indicators of land health at landscape and regional scales. In many cases empirical research has not been done to understand the interaction of ecosystem components and processes at broad scales. Another challenge is in developing and incorporating indicators of soil degradation and recovery into rangeland monitoring programs when such indicators may provide better or more timely information to managers than vegetation indicators.

A second set of challenges relates to technical aspects of rangeland monitoring. Variability of precipitation and temperature (and thus plant biomass and species composition) on rangelands both within and between years is a perennial challenge for monitoring. Conventional approaches to dealing with temporal variability in monitoring data include comparing only like years or averaging over multiple years. Advances in remote sensing, specifically the increasing availability of high-frequency remote-sensing products (e.g., MODIS NDVI), offer several new opportunities for addressing this challenge. However, despite the promise of remote sensing, its formal adoption in many rangeland monitoring programs has been slow. Operational use of remote sensing for rangeland monitoring will require a clear statement of objectives and roles for remote-sensing products as well as accuracy needs.

For the future, rangeland professionals (both upcoming and current) need instruction on monitoring that focuses on development and selection of functional indicators, monitoring rangelands at multiple scales, and efficiencies of core indicator monitoring with supplementation as necessary. Additionally, effort should be invested to improve our understanding of statistical principles of monitoring design and data analysis and increasing the availability of professionals with the skills to execute these tasks for rangeland monitoring.

Most sources on rangeland monitoring point out that monitoring is worthless if the data are never analyzed, reported, and ultimately used to address the original objectives. At the same time, however, many rangeland monitoring manuals focus almost exclusively on data collection protocols and leave out substantial treatment of data analysis and reporting. Going forward, considerable effort needs to be placed on translating monitoring results into management actions that are supported by analyses and data visualization. It is our experience that in many cases difficulties in sustaining funding for monitoring programs stem in part from the lack of tangible and useful analyses and results from the data that were collected. Monitoring programs can and should be designed to produce interim as well as long-term products that are useful to rangeland managers.

Technologies involved in rangeland monitoring will continue to evolve (e.g., remote sensing will supplant some field efforts), and strategies need to be put in place for adopting new techniques into monitoring programs. Ultimately, selection of technologies and methods for monitoring needs to be based on relevant management questions and a thorough understanding of the processes governing rangeland responses to management and disturbance.

Acknowledgments This work was funded in part by the U.S. Bureau of Land Management and U.S. Geological Survey Coordinated Intermountain Restoration Project. The use of any trade, product, or firm name is for descriptive purposes only and does not imply endorsement by the U.S. Government.

References

Adams, D.C., R.E. Short, J.A. Pfister, K.R. Peterson, and D.B. Hudson. 1995. New concepts for assessment of rangeland condition. *Journal of Range Management* 48: 271–282.

Addicott, J.F., J.M. Aho, M.F. Antolin, D.K. Padilla, J.S. Richardson, and D.A. Soluk. 1987. Ecological neighborhoods: Scaling environmental patterns. *Oikos* 49: 340–346.

Arkle, R.S., D.S. Pilliod, S.E. Hanser, M.L. Brooks, J.C. Chambers, J.B. Grace, K.C. Knutson, D.A. Pyke, J.L. Welty, and T.A. Wirth. 2014. Quantifying restoration effectiveness using multiscale habitat models: Implications for sage-grouse in the Great Basin. *Ecosphere* 5: 1–32.

Bai, Z.G., D.L. Dent, L. Olsson, and M.E. Schaepman. 2008. Proxy global assessment of land degradation. *Soil Use and Management* 24: 223–234.

Beck, J.L., D.C. Dauwalter, K.G. Gerow, and G.D, Hayward. 2010. Design to monitor trend in abundance and presence of American beaver (*Castor canadensis*) at the national forest scale. *Environmental Monitoring and Assessment* 164: 463–479.

Beck, J.L., D. Terrance Booth, and C.L. Kennedy. 2014. Assessing greater sage-grouse breeding habitat with aerial and ground imagery. *Rangeland Ecology and Management* 67: 328–332.

Bedell, T.E. 1998. *Glossary of terms used in range management*, Society for Range Management, 4th ed. Denver, Colorado: Direct Press.

Bertiller, M.B., J.O. Ares, and A.J. Bisigato. 2002. Multiscale indicators of land degradation in the Patagonian Monte, Argentina. *Environmental Management* 30: 704–715.

Bestelmeyer, B.T., J.T. Miller, and J.A. Wiens. 2003. Applying species diversity theory to land management. *Ecological Applications* 13: 1750–1761.

Bestelmeyer, B.T., A.J. Tugel, G.L. Peacock, D.G. Robinett, P.L. Shaver, J.R. Brown, J.E. Herrick, H. Sanchez, and K.M. Havstad. 2009. State-and-transition models for heterogeneous landscapes: A strategy for development and application. *Rangeland Ecology and Management* 62: 1–15.

Bird, S.B., J.E. Herrick, M.M. Wander, and L. Murray. 2007. Multi-scale variability in soil aggregate stability: Implications for understanding and predicting semi-arid grassland degradation. *Geoderma* 140: 106–118.

Bland, J.M., and D.G. Altman. 1999. Measuring agreement in method comparison studies. *Statistical Methods in Medical Research* 8: 135–160.

Bonham, C.D. 2013. *Measurements for terrestrial vegetation*, 2nd ed. Chichester, West Sussex: Wiley-Blackwell.

Booth, T.D., and P.T. Tueller. 2003. Rangeland monitoring using remote sensing. *Arid Land Research and Management* 17: 455–467.

Booth, T.D., S.E. Cox, C. Fifield, M. Phillips, and N. Williamson. 2005. Image analysis compared with other methods for measuring ground cover. *Arid Land Research and Management* 19: 91–100.

Booth, T.D., S.E. Cox, and R.D. Berryman. 2006. Point sampling digital imagery with "Samplepoint". *Environmental Monitoring and Assessment* 123: 97–108.

Bork, E.W., and J.G. Su. 2007. Integrating LIDAR data and multispectral imagery for enhanced classification of rangeland vegetation: A meta analysis. *Remote Sensing of Environment* 111: 11–24.

Boyd, C.S., and T.J. Svejcar. 2009. Managing complex problems in rangeland ecosystems. *Rangeland Ecology and Management* 62: 491–499.

Brandt, M., C. Romankiewicz, R. Spiekermann, and C. Samimi. 2014. Environmental change in time series—an interdisciplinary study in the Sahel of Mali and Senegal. *Journal of Arid Environments* 105: 52–63.

Breckenridge, R.P., W.G. Kepner, and D.A. Mouat. 1995. A process for selecting indicators for monitoring conditions of rangeland health. *Environmental Monitoring and Assessment* 36: 45–60.

Breckenridge, R.P., M. Dakins, S.C. Bunting, J.L. Harbour, and R.D. Lee. 2012. Using unmanned helicopters to assess vegetation cover classes in sagebrush steppe ecosystems. *Rangeland Ecology and Management* 65: 362–370.

Briske, D.D., S.D. Fuhlendorf, and F.E. Smeins. 2005. State-and-transition models, thresholds, and rangeland health: A synthesis of ecological concepts and perspectives. *Rangeland Ecology and Management* 58: 1–10.

Briske, D.D., S.D. Fuhlendorf, and F.E. Smeins. 2006. A unified framework for assessment and application of ecological thresholds. *Rangeland Ecology and Management* 59: 225–236.

Briske, D.D., L.A. Joyce, H.W. Polley, J.R. Brown, K. Wolter, J.A. Morgan, B.A. McCarl, and D.W. Bailey. 2015. Climate-change adaptation on rangelands: Linking regional exposure with diverse adaptive capacity. *Frontiers in Ecology and the Environment* 13: 249–256.

Bureau of Land Management. 1996. *Sampling vegetation attributes: Interagency technical reference*. Washington, D.C.: BLM National Applied Resource Sciences Center.

Bureau of Land Management. 1999. *Utilization studies and residual measurements: Interagency technical reference*. Washington, D.C.: Bureau of Land Management, National Applied Resource Sciences Center.

Burnett, C., and T. Blaschke. 2003. A multi-scale segmentation/object relationship modelling methodology for landscape analysis. *Ecological Modeling* 168: 233–249.

Cagney, J., S.E. Cox, and D.T. Booth. 2011. Comparison of point intercept and image analysis for monitoring rangeland transects. *Rangeland Ecology and Management* 64: 309–315.

Caudle, D., H. Sanchez, J. DiBenedetto, C.J. Talbot, and M. Karl. 2013. *Interagency ecological site handbook for rangelands.* . Washington: USDA Natural Resource Conservation Service.

Chambers, J.C., D.A. Pyke, J.D. Maestas, M. Pellant, C.S. Boyd, S.B. Campbell, S. Espinosa, D.W. Havlina, K.E. Mayer, and A. Wuenschel. 2014. *Using resistance and resilience concepts to reduce impacts of invasive annual grasses and altered fire regimes on the sagebrush ecosystem and greater sage-grouse: A strategic multi-scale approach. General technical report.* Fort Collins, CO: USDA Forest Service Rocky Mountain Research Station.

Cocke, A.E., P.Z. Fulé, and J.E. Crouse. 2005. Comparison of burn severity assessments using Differenced Normalized Burn Ratio and ground data. *International Journal of Wildland Fire* 14: 189.

Connelly, J.W., M.A. Schroeder, A.R. Sands, and C.E. Braun. 2000. Guidelines to manage sage grouse populations and their habitats. *Wildlife Society Bulletin* 28: 967–985.

d'Oleire-Oltmanns, S., I. Marzolff, K. Peter, and J. Ries. 2012. Unmanned Aerial Vehicle (UAV) for monitoring soil erosion in Morocco. *Remote Sensing* 4: 3390–3416.

Dardel, C., L. Kergoat, P. Hiernaux, E. Mougin, M. Grippa, and C.J. Tucker. 2014. Re-greening Sahel: 30 years of remote sensing data and field observations (Mali, Niger). *Remote Sensing of Environment* 140: 350–364.

Dingle Robertson, L., and D.J. King. 2011. Comparison of pixel- and object-based classification in land cover change mapping. *International Journal of Remote Sensing* 32: 1505–1529.

Doherty, K.E., D.E. Naugle, and B.L. Walker. 2010. Greater sage-grouse nesting habitat: The importance of managing at multiple scales. *Journal of Wildlife Management* 74: 1544–1553.

Duniway, M.C., J.W. Karl, S. Schrader, N. Baquera, and J.E. Herrick. 2011. Rangeland and pasture monitoring: An approach to interpretation of high-resolution imagery focused on observer calibration for repeatability. *Environmental Monitoring and Assessment* 184: 3789–3804.

Dyksterhuis, E.J. 1949. Condition and management of range land based on quantitative ecology. *Journal of Range Management* 2: 104–115.

Elzinga, C.L., D.W. Salzer, and J.W. Willoughby. 1998. *Measuring and monitoring plant populations.* Denver, Colorado: U.S. Department of the Interior, Bureau of Land Management. National Applied Resource Sciences Center.

Fancy, S.G., and R.E. Bennetts. 2012. Institutionalizing an effective long-term monitoring program in the US National Park Service. In *Design and analysis of long-term ecological monitoring studies*, ed. R.A. Gitzen, J. Millspaugh, A.B. Cooper, and D.S. Licht, 481–497. Cambridge, UK: Cambridge University Press.

Fancy, S.G., J.E. Gross, and S.L. Carter. 2009. Monitoring the condition of natural resources in US national parks. *Environmental Monitoring and Assessment* 151: 161–174.

Feitosa, R. Q., G. A. O. P. Costa, T. B. Cazes, and B. Feijo. 2006. In *Measuring and monitoring plant populations*, ed. S. Lang, T. Blaschke, and E. Schopfer. Austria: Salzburg University.

Fonstad, M.A., J.T. Dietrich, B.C. Courville, J.L. Jensen, and P.E. Carbonneau. 2013. Topographic structure from motion: A new development in photogrammetric measurement: Topographic structure from motion. *Earth Surface Processes and Landforms* 38: 421–430.

Fuhlendorf, S.D., D.D. Briske, and F.E. Smeins. 2001. Herbaceous vegetation change in variable rangeland environments: The relative contribution of grazing and climatic variability. *Applied Vegetation Science* 4: 177–188.

Gadzia, K., and T. Graham. 2013. *Bullseye! Targeting your rangeland health objectives, Version 2.0*. Santa Fe, NM: Quivira Coalition.

Genchi, S., A. Vitale, G. Perillo, and C. Delrieux. 2015. A structure-from-motion approach for characterization of bioerosion patterns using UAV imagery. *Sensors* 15: 3593–3609.

Gibbens, R.P., R.P. McNeely, K.M. Havstad, R.F. Beck, and B. Nolen. 2005. Vegetation changes in the Jornada Basin from 1858 to 1998. *Journal of Arid Environments* 61: 651–668.

Gillan, J.K., J.W. Karl, M. Duniway, and A. Elaksher. 2014. Modeling vegetation heights from high resolution stereo aerial photography: An application for broad-scale rangeland monitoring. *Journal of Environmental Management* 144: 226–235.

Gillan, J.K., J.W. Karl, N.N. Barger, A. Elaksher, M.C. Duniway. 2016. Spatially explicit rangeland erosion monitoring using high-resolution digital aerial imagery. *Rangeland Ecology & Management* 69(2): 95–107. doi:10.1016/j.rama.2015.10.012.

Gitzen, R.A., J. Millspaugh, A.B. Cooper, and D.S. Licht (eds.). 2012. *Design and analysis of long-term ecological monitoring studies*. Cambridge: Cambridge University Press.

Glenn, N.F., J.T. Mundt, K.T. Weber, T.S. Prather, L.W. Lass, and J. Pettingill. 2005. Hyperspectral data processing for repeat detection of small infestations of leafy spurge. *Remote Sensing of the Environment* 95: 399–412.

Glenn, N.F., L.P. Spaete, T.T. Sankey, D.R. DerryBerry, S.P. Hardegree, and J.J. Mitchell. 2011. Errors in LiDAR-derived shrub height and crown area on sloped terrain. *Journal of Arid Environments* 75: 377–382.

Gong, P., G.S. Biging, and R. Standiford. 2000. Technical note: Use of digital surface model for hardwood rangeland monitoring. *Journal of Range Management* 53: 622–626.

Govender, M., K. Chetty, and H. Bulcock. 2009. A review of hyperspectral remote sensing and its application in vegetation and water resource studies. *Water SA* 33. 145–151

Greaves, H.E., L.A. Vierling, J.U.H. Eitel, N.T. Boelman, T.S. Magney, C.M. Prager, and K.L. Griffin. 2015. Estimating aboveground biomass and leaf area of low-stature Arctic shrubs with terrestrial LiDAR. *Remote Sensing of Environment* 164: 26–35.

Gregoire, T.G. 1998. Design-based and model-based inference in survey sampling: Appreciating the difference. *Canadian Journal of Forest Research* 28: 1429–1447.

Gu, Y., B.K. Wylie, and N.B. Bliss. 2013. Mapping grassland productivity with 250-m eMODIS NDVI and SSURGO database over the Greater Platte River Basin, USA. *Ecological Indicators* 24: 31–36.

H. John Heinz III Center for Science, E. and the Environment. 2008. *The state of the nation's ecosystems: Measuring the lands, waters, and living resources of the United States*. Washington, D.C.: Island Press.

Herrick, J.E., M. Weltz, J.D. Reeder, G.E. Schuman, and J.R. Simanton. 1999. Rangeland soil erosion and soil quality: Role of resistance, resilience and disturbance regime. In *Soil erosion and soil quality*, ed. R. Lal, 209–233. Boca Raton, FL: CRC Press LLC.

Herrick, J.E., V. Lessard, K.E. Spaeth, P. Shaver, R.S. Dayton, D.A. Pyke, L. Jolley, and J.J. Goebel. 2010. National ecosystem assessments supported by scientific and local knowledge. *Frontiers of Ecology and the Environment* 8: 403–408.

Herrick, J.E., M.C. Duniway, D.A. Pyke, B.T. Bestelmeyer, S.A. Wills, J.R. Brown, J.W. Karl, and K.M. Havstad. 2012. A holistic strategy for adaptive management. *Journal of Soil and Water Conservation* 67: 105A–113A.

Herrick, J.E., K.C. Urama, J.W. Karl, J. Boos, M.V. Johnson, K.D. Shepherd, J. Hempel, B.T. Bestelmeyer, J. Davies, J.L. Guerra, C. Kosnik, D.W. Kimiti, A.L. Ekai, K. Muller, L. Norfleet, N. Ozor, T. Reinsch, J. Sarukhan, and L.T. West. 2013. The global Land-Potential Knowledge System (LandPKS): Supporting evidence-based, site-specific land use and management through cloud computing, mobile apps and crowdsourcing. *Journal of Soil and Water Conservation* 68: 5A–12A.

Holechek, J.L., C.H. Pieper, and C.H. Herbel. 2001. *Range management: Principles and practices.* Upper Saddle River, New Jersey: Prentice Hall.

Homer, C., C. Huang, L. Yang, B. Wylie, and M. Coan. 2004. Development of a 2001 national land-cover database for the United States. *Photogrammetric Engineering and Remote Sensing* 70: 829–840.

Homer, C.G., C.L. Aldridge, D.K. Meyer, and S.J. Schell. 2012. Multi-scale remote sensing sagebrush characterization with regression trees over Wyoming, USA: Laying a foundation for monitoring. *International Journal of Applied Earth Observation and Geoinformation* 14: 233–244.

Homer, C.G., D.K. Meyer, C.L. Aldridge, and S.J. Schell. 2013. Detecting annual and seasonal changes in a sagebrush ecosystem with remote sensing-derived continuous fields. *Journal of Applied Remote Sensing* 7: 073508.

Hulet, A., B.A. Roundy, S.L. Petersen, R.R. Jensen, and S.C. Bunting. 2013. Assessing the relationship between ground measurements and object-based image analysis of land cover classes in pinyon and juniper woodlands. *Photogrammetric Engineering and Remote Sensing* 79: 799–808.

Hulet, A., B.A. Roundy, S.L. Petersen, S.C. Bunting, R.R. Jensen, and D.B. Roundy. 2014a. Utilizing national agriculture imagery program data to estimate tree cover and biomass of piñon and juniper woodlands. *Rangeland Ecology and Management* 67: 563–572.

Hulet, A., B.A. Roundy, S.L. Petersen, R.R. Jensen, and S.C. Bunting. 2014b. Cover estimations using object-based image analysis rule sets developed across multiple scales in pinyon-juniper woodlands. *Rangeland Ecology and Management* 67: 318–327.

Hunt, E.R., J.H. Everitt, J.C. Ritchie, M.S. Moran, T.D. Booth, G.L. Anderson, P.E. Clark, and M.S. Seyfried. 2003. Applications and research using remote sensing for rangeland management. *Photogrammetric Engineering and Remote Sensing* 69: 675–693.

Hunt, E.R., and B.A. Miyake. 2006. Comparison of stocking rates from remote sensing and geospatial data. *Rangeland Ecology and Management* 59: 11–18.

IPCC. 2007. Climate change 2007: The physical science basis. In *Contribution of Working Group I to the fourth assessment report of the Intergovernmental Panel on Climate Change.* Cambridge, UK: Cambridge University Press.

Jin, S., L. Yang, P. Danielson, C. Homer, J. Fry, and G. Xian. 2013. A comprehensive change detection method for updating the National Land Cover Database to circa 2011. *Remote Sensing of Environment* 132: 159–175.

Karl, J.W. 2010. Spatial predictions of cover attributes of rangeland ecosystems using regression kriging and remote sensing. *Rangeland Ecology and Management* 63: 335–349.

Karl, J.W., and B.A. Maurer. 2010a. Spatial dependence of predictions from image segmentation: A variogram-based method to determine appropriate scales for producing land-management information. *Ecological Informatics* 5: 194–202.

Karl, J.W., and B.A. Maurer. 2010b. Multivariate correlations between imagery and field measurements across scales: Comparing pixel aggregation and image segmentation. *Landscape Ecology* 24: 591–605.

Karl, J.W., M.C. Duniway, S.M. Nusser, J.D. Opsomer, and R.S. Unnasch. 2012a. Using Very-Large Scale Aerial (VLSA) imagery for rangeland monitoring and assessment: Some statistical considerations. *Rangeland Ecology and Management* 65: 330–339.

Karl, J.W., M.C. Duniway, and T.S. Schrader. 2012b. A technique for estimating rangeland canopy-gap size distributions from very-high-resolution digital imagery. *Rangeland Ecology and Management* 65: 196–207.

Karl, J.W., J.E. Herrick, and D. Browning. 2012c. A strategy for rangeland management based on best-available knowledge and information. *Rangeland Ecology and Management* 65: 638–646.

Kefi, S., M. Reitkerk, C.L. Alados, Y. Pueyo, V.P. Papanastasis, A. ElAich, and P.C. De Ruiter. 2007. Spatial vegetation patterns and imminent desertification in Mediterranean arid ecosystems. *Nature* 449: 213–218.

Kennedy, R.E., P.A. Townsend, J.E. Gross, W.B. Cohen, P. Bolstad, Y.Q. Wang, and P. Adams. 2009. Remote sensing change detection tools for natural resource managers: Understanding concepts and tradeoffs in the design of landscape monitoring projects. *Remote Sensing of Environment* 113: 1382–1396.

Knutson, K.C., D.A. Pyke, T.A. Wirth, R.S. Arkle, D.S. Pilliod, M.L. Brooks, J.C. Chambers, and J.B. Grace. 2014. Long-term effects of seeding after wildfire on vegetation in Great Basin shrubland ecosystems. *Journal of Applied Ecology* 51: 1414–1424.

Laliberte, A.S., and A. Rango. 2009. Texture and scale in object-based analysis of subdecimeter resolution unmanned aerial vehicle (UAV) imagery. *IEEE Transactions of Geoscience and Remote Sensing* 47: 761–770.

Laliberte, A.S., A. Rango, K.M. Havstad, J.F. Paris, R.F. Beck, R. McNeely, and A.L. Gonzalez. 2004. Object-oriented image analysis for mapping shrub encroachment from 1937 to 2003 in southern New Mexico. *Remote Sensing of the Environment* 93: 198–210.

Laliberte, A.S., E.L. Fredrickson, and A. Rango. 2006. Combining decision trees with hierarchical object-oriented image analysis for mapping arid rangelands. *Photogrammetric Engineering and Remote Sensing* 73: 197–207.

Laliberte, A.S., D.M. Browning, J.E. Herrick, and P. Gronemeyer. 2012a. Hierarchical object-based classification of ultra-high-resolution digital mapping camera (DMC) imagery for rangeland mapping and assessment. *Journal of Spatial Science* 55: 101–115.

Laliberte, A.S., C. Winters, and A. Rango. 2012b. UAS remote sensing missions for rangeland applications. *Geocarto International* 26: 141–156.

Lindenmayer, D.B., and G.E. Likens. 2009. Adaptive monitoring: A new paradigm for long-term research and monitoring. *Trends in Ecology and Evolution* 24: 482–486.

Lohr, S.L. 2009. *Sampling: Design and analysis*, 2nd ed. Pacific Grove: Duxbury Press.

Ludwig, J.A., D.J. Tongway, G.N. Bastin, and C.D. James. 2004. Monitoring ecological indicators of rangeland functional integrity and their relation to biodiversity at local to regional scales. *Austral Ecology* 29: 108–120.

Ludwig, J.A., G.N. Bastin, V.H. Chewings, R.W. Eager, and A.C. Liedloff. 2007a. Leakiness: A new index for monitoring the health of arid and semiarid landscapes using remotely sensed vegetation cover and elevation data. *Ecological Indicators* 7: 442–454.

Ludwig, J.A., G.N. Bastin, J.F. Wallace, and T.R. McVicar. 2007b. Assessing landscape health by scaling with remote sensing: When is it not enough? *Landscape Ecology* 22: 163–169.

Luscier, J.D., W.L. Thompson, J.M. Wilson, B.E. Gorham, and L.D. Dragut. 2006. Using digital photographs and object-based image analysis to estimate percent ground cover in vegetation plots. *Frontiers of Ecology and the Environment* 4: 408–413.

Mackinnon, W.C., J.W. Karl, G.R. Toevs, J.J. Taylor, M.S. Karl, C.S. Spurrier, and J.E. Herrick. 2011. *BLM core terrestrial indicators and methods*. Denver, CO: US Department of the Interior, Bureau of Land Management, National Operations Center.

Malmstrom, C.M., H.S. Butterfield, C. Barber, B. Dieter, R. Harrison, J. Qi, D. Riano, A. Schrotenboer, S. Stone, C.J. Stoner, and J. Wirka. 2008. Using remote sensing to evaluate the influence of grassland restoration activities on ecosystem forage provisioning services. *Restoration Ecology* 17:526–538.

Marsett, R.C., J. Qi, P. Heilman, S.H. Beidenbender, M.C. Watson, S. Amer, M. Weltz, D. Goodrich, and R. Marsett. 2006. Remote sensing for grassland management in the arid southwest. *Rangeland Ecology and Management* 59: 530–540.

Marzolff, I., and J. Poesen. 2009. The potential of 3D gully monitoring with GIS using high-resolution aerial photography and a digital photogrammetry system. *Geomorphology* 111: 48–60.

Menne, M. J., C. N. Williams, and M. A. Palecki. 2010. On the reliability of the U.S. surface temperature record. *Journal of Geophysical Research* 115:D11108.

Miller, D.M., S.P. Finn, A. Woodward, A. Torregrosa, M.E. Miller, D.R. Bedford, and A.M. Brasher. 2010. *Conceptual ecological models to guide integrated landscape monitoring of the Great Basin, 134. Scientific investigations report.* Reston, VA: U.S. Geological Survey.

Mitchell, J.J., N.F. Glenn, T.T. Sankey, D.R. DerryBerry, M.O. Anderson, and R.C. Hruska. 2011. Small-footprint Lidar estimations of sagebrush canopy characteristics. *Photogrammetric Engineering and Remote Sensing* 77.

Moir, W.H., and W.M. Block. 2001. Adaptive management on public lands in the United States: Commitment or rhetoric? *Environmental Management* 28: 141–148.

Morris, Errol. 2014. The certainty of Donald Rumsfeld (part 2): the known and the unknown. Web blog post. The New York Times. http://opinionator.blogs.nytimes.com/2014/03/26/the-certainty-of-donaldrumsfeld-part-2/. Accessed 7 October 2016.

Mundt, J.T., N.F. Glenn, K.T. Weber, T.S. Prather, L.W. Lass, and J. Pettingill. 2005. Discrimination of hoary cress and determination of its detection limits via hyperspectral image processing and accuracy assessment techniques. *Remote Sensing of the Environment* 96: 509–517.

Mundt, J.T., D. Streutker, and N.F. Glenn. 2006. Mapping sagebrush distribution using fusion of hyperspectral and LiDAR classifications. *Photogrammetric Engineering and Remote Sensing* 72: 47–54.

Myint, S.W., P. Gober, A. Brazel, S. Grossman-Clarke, and Q. Weng. 2011. Per-pixel vs. object-based classification of urban land cover extraction using high spatial resolution imagery. *Remote Sensing of Environment* 115: 1145–1161.

Nash, K.L., C.R. Allen, D.G. Angeler, C. Barichievy, T. Eason, A.S. Garmestani, N.A.J. Graham, D. Granholm, M. Knutson, R.J. Nelson, M. Nyström, C.A. Stow, and S.M. Sundstrom. 2014. Discontinuities, cross-scale patterns, and the organization of ecosystems. *Ecology* 95: 654–667.

National Park Service. 2012. *Guidance for designing and integrated montioring program. Natural resource report.* Fort Collins, CO: National Park Service.

National Research Council. 1994. *Rangeland health: New methods to classify, inventory, and monitor rangelands.* Washington, D.C.: National Academy Press.

Nijland, W., M. van der Meijde, E.A. Addink, and S.M. de Jong. 2010. Detection of soil moisture and vegetation water abstraction in a Mediterranean natural area using electrical resistivity tomography. *CATENA* 81: 209–216.

Noon, B.R. 2003. Conceptual issues in monitoring ecological systems. In *Monitoring ecosystems—interdisciplinary approaches for evaluating ecoregional initiatives,* ed. D.E. Busch and J.C. Trexler, 27–71. Washington, D.C.: Island Press.

Oakley, K.L., L.P. Thomas, and S.G. Fancy. 2003. Guidelines for long-term monitoring protocols. *Wildlife Society Bulletin* 31: 1000–1003.

Oldeman, L.R. 1994. The global extent of soil degradation. In *Soil resilience and sustainable land use,* ed. D.J. Greenland and T. Szaboles. Wallingford, U.K.: Commonwealth Agricultural Bureau International.

Opsomer, J.D., F.J. Breidt, G.G. Moisen, and G. Kauermann. 2007. Model-assisted estimation of forest resources with generalized additive models. *Journal of the American Statistical Association* 102: 400–409.

Pastick, N.J., M. Rigge, B.K. Wylie, M.T. Jorgenson, J.R. Rose, K.D. Johnson, and L. Ji. 2014. Distribution and landscape controls of organic layer thickness and carbon within the Alaskan Yukon River Basin. *Geoderma* 230–231: 79–94.

Pellant, M., P. Shaver, D.A. Pyke, and J.E. Herrick. 2005. *Interpreting indicators of rangeland health, version 4. BLM/WO/ST-00/001+1734/REV05*. Denver, CO: U.S. Department of the Interior, Bureau of Land Management, National Science and Technology Center.

Perroy, R.L., B. Bookhagen, G.P. Asner, and O.A. Chadwick. 2010. Comparison of gully erosion estimates using airborne and ground-based LiDAR on Santa Cruz Island, California. *Geomorphology* 118: 288–300.

Peters, D.P.C., R.A. Pielke, B.T. Bestelmeyer, C.D. Allen, S. Munson-McGee, and K.M. Havstad. 2004. Cross-scale interactions, nonlinearities, and forecasting catastrophic events. *Proceedings of the National Academy of Sciences of the United States of America* 101: 15130–15135.

Peters, D.P., K.M. Havstad, S.R. Archer, and O.E. Sala. 2015. Beyond desertification: New paradigms for dryland landscapes. *Frontiers in Ecology and the Environment* 13: 4–12.

Peterson, T.C., D.R. Easterling, T.R. Karl, P. Groisman, N. Nicholls, N. Plummer, S. Torok, I. Auer, R. Boehm, D. Gullett, L. Vincent, R. Heino, H. Tuomenvirta, O. Mestre, T. Szentimrey, J. Salinger, E.J. Førland, I. Hanssen-Bauer, H. Alexandersson, P. Jones, and D. Parker. 1998. Homogeneity adjustments of in situ atmospheric climate data: A review. *International Journal of Climatology* 18: 1493–1517.

Pilliod, D.S., and R.S. Arkle. 2013. Performance of quantitative vegetation sampling methods across gradients of cover in Great Basin plant communities. *Rangeland Ecology and Management* 66: 634–647.

Pyke, D.A., J.E. Herrick, P. Shaver, and M. Pellant. 2002. Rangeland health attributes and indicators for qualitative assessment. *Journal of Range Management* 55: 584–597.

Rango, A., L.F. Huenneke, M. Buonopane, J.E. Herrick, and K.M. Havstad. 2005. Using historic data to assess effectiveness of shrub removal in southern New Mexico. *Journal of Arid Environments* 62: 75–91.

Rango, A., A.S. Laliberte, J.E. Herrick, C. Winters, K.M. Havstad, C. Steele, and D. Browning. 2009. Unmanned aerial vehicle based remote sensing for rangeland assessment, monitoring and management. *Journal of Applied Remote Sensing* 2: 033542.

Reeves, M.C., M. Zhao, and S.W. Running. 2006. Applying improved estimates of MODIS productivity to characterize grassland vegetation dynamics. *Rangeland Ecology and Management* 59: 1–10.

Reynolds, J.H. 2012. An overview of statistical considerations in long-term monitoring. In *Design and analysis of long-term ecological monitoring studies*, ed. R.A. Gitzen, J. Millspaugh, A.B. Cooper, and D.S. Licht, 23–53. Cambridge, UK: Cambridge University Press.

Rigge, M., B. Wylie, L. Zhang, and S.P. Boyte. 2013. Influence of management and precipitation on carbon fluxes in great plains grasslands. *Ecological Indicators* 34: 590–599.

Sala, O.E., and J. Paruelo. 1997. Ecosystem services in grasslands. In *Nature's services: Societal dependence on natural ecosystems*, ed. G.C. Daily, 392. Washington, D.C.: Island Press.

Sampson, A.W. 1923. *Range and pasture management*. New York, NY: Wiley.

Sankey, J.B., N.F. Glenn, M.J. Germino, A.I.N. Gironella, and G.D. Thackray. 2010. Relationships of aeolian erosion and deposition with LiDAR-derived landscape surface roughness following wildfire. *Geomorphology* 119: 135–145.

Sankey, J.B., C.S.A. Wallace, and S. Ravi. 2013. Phenology-based, remote sensing of post-burn disturbance windows in rangelands. *Ecological Indicators* 30: 35–44.

Schalau, J. 2010. *Rangeland monitoring: Selecting key areas*, 3. Tucson, AZ: Arizona Cooperative Extension, University of Arizona.

Scherr, S. 1999. Soil degradation: A threat to developing-country food security by 2020? *Food, agriculture and the environment discussion paper*. Washington, D.C.: International Food Policy Research Institute.

Seefeldt, S.S., and T.D. Booth. 2006. Measuring plant cover in sagebrush steppe rangelands: A comparison of methods. *Environmental Management* 37: 703–711.

Shepherd, K.D., and M.G. Walsh. 2007. nfrared spectroscopy—enabling an evidence based diagnostic survellance approach to agricultural and environmental management in developing countries. *Journal of Near Infrared Spectroscopy* 15: 1–19.

Standing, A.R. 1938. Use of key species, key areas and utilization standards in range management. *Ames Forester* 29: 9–19.

Stehman, S.V. 2009. Model-assisted estimation as a unifying framework for estimating the area of land cover and land-cover change from remote sensing. *Remote Sensing of Environment* 113: 2455–2462.

Stiver, S.J., E.T. Rinkes, D.E. Naugle, P.D. Makela, D.A. Nance, and J.W. Karl. 2015. *Sage-grouse habitaat assessment framework: A multiscale assessment tool*, 114. Denver, CO: Technical Reference, Bureau of Land Management and Western Association of Fish and Wildlife Agencies.

Stoddart, L.A., and A.D. Smith. 1943. *Range management*. New York: McGraw-Hill.

Strand, E.K., S.C. Bunting, L.A. Starcevich, M.T. Nahorniak, G. Dicus, and L.K. Garrett. 2015. Long-term monitoring of western aspen—lessons learned. *Environmental Monitoring and Assessment* 187: 528.

Streutker, D., and N.F. Glenn. 2006. Lidar measurement of sagebrush steppe vegetation heights. *Remote Sensing of Environment* 102: 135–145.

Stringham, T.K., W.C. Krueger, and P.L. Shaver. 2003. State and transition modeling: An ecological process approach. *Journal of Range Management* 56: 106–113.

Suter, G. 2001. Applicability of indicator monitoring to ecological risk assessment. *Ecological Indicators* 1: 101–112.

Tappan, G., M. Sall, E. Wood, and M. Cushing. 2004. Ecoregions and land cover trends in Senegal. *Journal of Arid Environments* 59: 427–462.

Tegler, B., M. Sharp, and M.A. Johnson. 2001. Ecological monitoring and assessment network's proposed core monitoring variables: An early warning of environmental change. *Environmental Monitoring and Assessment* 67: 29–55.

Thompson, S.K. 2002. *Sampling*, 2nd ed. New York, NY: Wiley.

Toevs, G.R., J.W. Karl, J.J. Taylor, C.S. Spurrier, M. "Sherm" Karl, M.R. Bobo, and J.E. Herrick. 2011. Consistent indicators and methods and a scalable sample design to meet assessment, inventory, and monitoring information needs across scales. *Rangelands* 33:14–20.

Turner, D., A. Lucieer, and C. Watson. 2012. An automated technique for generating georectified mosaics from ultra-high resolution Unmanned Aerial Vehicle (UAV) imagery, based on Structure from Motion (SfM) point clouds. *Remote Sensing* 4: 1392–1410.

U.S. Bureau of Land Management, and U.S. Forest Service. 1994. *Rangeland reform'94: Draft environmental impact statement*, 538. U.S. Washington, D.C.: Bureau of Land Management.

U.S. Fish and Wildlife Service. 2010. *Endangered and threatened wildlife and plants; 12-month findings for petitions to list the Greater Sage-grouse (Centrocercus urophasianus) as threatened or engangered*. Washington, D.C.: Federal Register.

Ustin, S.L., D. DiPietro, K. Olmstead, E. Underwood, and G.J. Scheer. 2004. Hyperspectral remote sensing for invasive species detection and mapping. In *Proceedings of IGARSS 2002: International geoscience and remote sensing symposium*, 24–28 June, vol. 3, 1658–1660. Toronto, Ontario, Canada: IEEE and the Canadian Society for Remote Sensing

Veblen, K.E., D.A. Pyke, C.L. Aldridge, M.L. Casazza, T.J. Assal, and M.A. Farinha. 2014. Monitoring of livestock grazing effects on Bureau of Land Management Land. *Rangeland Ecology and Management* 67: 68–77.

Verbesselt, J., R. Hyndman, G. Newnham, and D. Culvenor. 2010. Detecting trend and seasonal changes in satellite image time series. *Remote Sensing of Environment* 114: 106–115.

Walters, C. 2007. Is adaptive management helping to solve fisheries problems? *Ambio* 36: 304–307.

Walters, C.J., and C.S. Holling. 1990. Large-scale management experiments and learning by doing. *Ecology* 71: 2060–2068.

Washington-Allen, R.A., N.E. West, R.D. Ramsey, and R.A. Efroymson. 2006. A protocol for retrospective remote sensing-based ecological monitoring of rangelands. *Rangeland Ecology and Management* 59: 19–29.

Weber, B., C. Olehowski, T. Knerr, J. Hill, K. Deutschewitz, D.C.J. Wessels, B. Eitel, and B. Büdel. 2008. A new approach for mapping of Biological Soil Crusts in semidesert areas with hyperspectral imagery. *Remote Sensing of Environment* 112: 2187–2201.

West, N.E. 2003a. History of rangeland monitoring in the U.S.A. *Arid Land Research and Management* 17: 495–545.

West, N.E. 2003b. Theoretical underpinnings of rangeland monitoring. *Arid Land Research and Management* 17: 333–346.

West, N.E., K. McDaniel, E.L. Smith, P.T. Tueller, and S. Leonard. 1994. *Monitoring and interpreting ecological integrity on arid and semi-arid lands of the western United States*. Las Cruces, NM, USA: New Mexico Range Improvement Task Force.

White, G.J. 2003. Selection of ecological indicators for monitoring terrestrial systems. In *Environmental Monitoring*, ed. G.B. Wiersma, 263–282. LLC, Boca Raton, Florida: CRC Press.

White, A.B., P. Kumar, and D. Tcheng. 2005. A data mining approach for understanding topographic control on climate-induced inter-annual vegetation variability over the United States. *Remote Sensing of Environment* 98: 1–20.

Whiteside, T.G., G.S. Boggs, and S.W. Maier. 2011. Comparing object-based and pixel-based classifications for mapping savannas. *International Journal of Applied Earth Observation and Geoinformation* 13: 884–893.

Wiens, J.A. 1999. The science and practice of landscape ecology. In *Landscape ecological analysis*, ed. J.M. Klopatch and R.H. Gardner, 371–383. New York, NY: Springer.

Wiens, J.A., M.R. Moss, M.G. Turner, and D.J. Mladenoff. 2007. *Foundation papers in landscape ecology*. New York, NY: Columbia University Press.

Wright, P.A., G. Alward, J.L. Colby, T.W. Hoekstra, B. Tegler, and M. Turner. 2002. *Monitoring for forest management unit sustainability: The local unit criteria and indicators development (LUCID) test*, 54p. Fort Collins, CO: USDA Forest Service.

Wylie, B.K., S.P. Boyte, and D.J. Major. 2012. Ecosystem performance monitoring of rangelands by integrating modeling and remote sensing. *Rangeland Ecology and Management* 65: 241–252.

Xian, G., and C. Homer. 2010. Updating the 2001 national land cover database impervious surface products to 2006 using landsat imagery change detection methods. *Remote Sensing of Environment* 114: 1676–1686.

Chapter 17
Rangeland Systems in Developing Nations: Conceptual Advances and Societal Implications

D. Layne Coppock, María Fernández-Giménez, Pierre Hiernaux, Elisabeth Huber-Sannwald, Catherine Schloeder, Corinne Valdivia, José Tulio Arredondo, Michael Jacobs, Cecilia Turin, and Matthew Turner

Abstract Developing-country rangelands are vast and diverse. They are home to millions who are often poor, politically marginalized, and dependent on livestock for survival. Here we summarize our experiences from six case-study sites in sub-Saharan Africa, central Asia, and Latin America generally covering the past 25 years. We examine issues pertaining to population, natural resource management, climate, land use, livestock marketing, social conflict, and pastoral livelihoods. The six study sites differ with respect to human and livestock population dynamics and

This chapter is dedicated to the memory of an exemplary international rangeland scientist, Thomas L. Thurow (1955–2016).

D.L. Coppock (✉)
Department of Environment & Society, Utah State University, Logan, UT, USA
e-mail: Layne.Coppock@usu.edu

M. Fernández-Giménez
Department of Forest and Rangeland Stewardship, Colorado State University,
Fort Collins, CO, USA

P. Hiernaux
Centre National de la Recherche Scientifique, Géosciences Environnement Toulouse,
Toulouse, France

E. Huber-Sannwald • J.T. Arredondo
Instituto Potosino de Investigación Científica y Tecnológica, San Luis Potosi, Mexico

C. Schloeder • M. Jacobs
Oikos Services LLC, Fortine, MT, USA

C. Valdivia
Department of Agricultural and Applied Economics, University of Missouri,
Columbia, MO, USA

C. Turin
Division of Crop System Intensification and Climate Change, International Potato Center,
Lima, Peru

M. Turner
Department of Geography, University of Wisconsin, Madison, WI, USA

© The Author(s) 2017
D.D. Briske (ed.), *Rangeland Systems*, Springer Series on Environmental
Management, DOI 10.1007/978-3-319-46709-2_17

the resulting pressures on natural resources. Environmental degradation, however, has been commonly observed. Climate change is also having diverse systemic effects often related to increasing aridity. As rangelands become more economically developed pastoral livelihoods may diversify, food security can improve, and commercial livestock production expands, but wealth stratification widens. Some significant upgrades in rural infrastructure and public service delivery have occurred; telecommunications are markedly improved overall due to widespread adoption of mobile phones. Pressures from grazing, farming, mining, and other land uses—combined with drought—can ignite local conflicts over resources, although the intensity and scope of conflicts markedly varies across our case-study sites. Pastoralists and their herds have become more sedentary overall due to many factors, and this can undermine traditional risk-management tactics based on mobility. Remote rangelands still offer safe havens for insurgents, warlords, and criminals especially in countries where policing remains weak; the resulting civil strife can undermine commerce and public safety. There has been tremendous growth in knowledge concerning developing-country rangelands since 1990, but this has not often translated into improved environmental stewardship or an enhanced well-being for rangeland dwellers. Some examples of demonstrable impact are described, and these typically have involved longer-term investments in capacity building for pastoralists, local professionals, and other stakeholders. Research is shifting from ecologically centered to more human-centered issues; traditional academic approaches are often being augmented with participatory, community-based engagement. Building human or social capital in ways that are integrated with improved natural resource stewardship offers the greatest returns on research investment. Our future research and outreach priorities include work that fortifies pastoral governance, enhances livelihoods for a diverse array of rangeland residents, and improves land and livestock management in a comprehensive social-ecological systems approach.

Keywords Bolivian Altiplano • Ethiopian Boran • Afghan Kuchi • Mexican rangelands • Mongolia • Peruvian Altiplano • Sahel

17.1 Introduction

In this chapter we focus on rangelands of the developing world. By rangelands we refer to landscapes—largely unsuitable for sustained cultivation—providing forage, water, and cover for grazing and browsing animals. These landscapes occur in deserts, grasslands, shrublands, savannas, woodlands, and alpine systems [definition modified from Holechek et al. (2011, p. 1)].

"Developing nations" refer to countries having a relatively low standard of living, an underdeveloped industrial base, a low gross domestic product per capita, and a low Human Development Index (Sullivan and Sheffrin 2003). The rangelands of developing nations have endured a wide array of challenges including poverty, environmental degradation, social conflicts, displaced people, and climate change (Seré et al. 2008; Thornton et al. 2009). Rangelands collectively represent about 70% of the world's land surface (Holechek et al. 2011) and are home to 2.1 billion people—35% of the world's population.[1] In sum, the rangelands of the developing world greatly matter to anyone who ponders how to improve the human condition or the stewardship of natural resources.

Rangelands of the developing world are also places where major conceptual advances for research and development have occurred over the past 30 years. These advances have affected range science and range management globally and include rangeland production modeling (Penning de Vries and Djiteye 1982), nonequilibrium ecology (Ellis and Swift 1988), resilience theory (Walker 2002), climate change (Olsson et al. 2005), pastoral management of livestock gene pools (Krätli 2007), coupled social-ecological systems (Stafford-Smith et al. 2009), action research and gender (Coppock et al. 2011), and decentralized or community-based natural resource governance (Reid et al. 2015).

The dominant economic use of rangelands in the developing world is livestock production as practiced by pastoralists using communal resources on state-owned or community-owned lands and, to a lesser extent, by producers using resources on privately held lands (Holechek et al. 2011). Other economic uses are on the rise and include dryland farming, hard-rock mining, oil and gas extraction, renewable energy production, recreation, and tourism. Important national parks and protected areas occur in rangelands worldwide (Chape et al. 2008). Developing-country rangelands provide vital global ecosystem goods and services, including carbon sequestration that mitigates effects from greenhouse gas emissions (Safriel et al. 2005). Climate change will affect the use of these landscapes (Feng et al. 2010; Long et al. 2006), as many developing-nation rangelands are projected to become warmer, drier, and subjected to more frequent extreme weather (IPCC 2013; Nicholson 2013; Stahle et al. 2009).

The coauthors of this chapter average over 20 years of experience in the developing world. They have contributed insights related to six case-study sites to capture broad patterns across social-ecological systems in a rangeland-development context. Grouped into two tiers according to a Human Development Index (HDI) calculated for 187 nations and territories (UNDP 2013), the case study sites include a less developed, lower tier with Afghanistan, southern Ethiopia, and the Sahelian belt (subsites in Niger, Mali, and Senegal—henceforth called the Sahel), and a more economically developed upper tier with Mongolia, the high Andes (Bolivia and Peru—henceforth called the Altiplano), and northern Mexico. The first tier has HDI country rankings that vary from 187 (Niger) to 154 (Senegal) while the second-tier HDI rankings range from 108 (Bolivia and Mongolia) to 61 (Mexico). The case study sites are mapped in Fig. 17.1. The sites illustrate wide variation in geographic, biophysical, and socioeconomic attributes. They have been monitored by the same scholars over extended periods of time and thus provide an unusual opportunity for credible long-term assessments, cross-site comparisons, and learning.

---

[1] http://www.un.org/en/events/desertification_decade/whynow.shtml

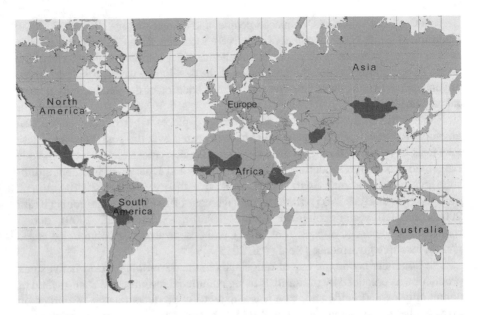

Fig. 17.1 Condensed world map showing nine nations that host the rangeland study sites reviewed in this chapter (illustration courtesy of Publication Design and Production, Utah State University)

17.2   Global Trends: Rangelands of the Developing World

Some socioeconomic and agroecological features for the six case study sites are shown in Table 17.1. The ecosystems vary from warm, subtropical savannas to cold, shrub steppe. Economic uses are dominated by extensive livestock production, but variation in the predominant livestock species, livestock products, and livestock population trends is notable. With some exceptions (i.e., Mongolia, Mexico) the rangeland inhabitants largely represent indigenous societies that are economically or socially marginalized within their home countries.

The material that follows in this section combines empirical information from the case study sites with other literature to integrate and describe some commonly observed patterns concerning populations, socioeconomics, and natural resources. Each subsection begins with a concise summary paragraph. Becoming aware of this background is especially important for readers who are less familiar with pastoralism or rangelands. Those who want to skip the background and focus on current conceptual or operational issues can go directly to Sect. 17.3.

Table 17.1 Features for land, vegetation, livestock, and people across six case-study sites[a]

Tier	Site name	Location and area	Vegetation zones	Land use	Livestock species and products	People	Comments
Lower	High Mountains of Hindu Kush[b]	Afghanistan 696,355 km²	High elevations: Subalpine and alpine (*Astragalus, Onobrychis, Acantholimon, Acanthophyllum, Cicer* spp.) Mid-elevations: Semiarid and desert steppe, open woodlands (*Pistachia, Amygdalus* spp.), open shrublands (*Cousinia, Onobrychis, Festuca, Piptatherum* spp.), *Artemisia* shrublands Low elevations: Desert (*Haloxylon* spp.): ephemeral grasslands (*Poa, Carex, Aegilops, Bromus, Eremopyrum, Vulpia* spp.)	Seminomadic pastoralism; sedentary pastoralism; small-scale irrigation and dry land farming of cereal crops and melons	Sheep (65 %), goats (34 %), cattle (<1 %) Live animals, milk, dried and salted yogurt, wool, hides, skins, cashmere	Pashtun, Uzbek, Tajik, Hazara, Arab, Turkmen, Baluch, Pashai, and Aimaq	Resource use historically organized by political region, clan, and family. Structure has broken down in many areas due to increased competition for resources, criminal and insurgent activity, government corruption, greed, and land conversions. Herders move from low-lying border areas to high-elevation pastures starting in March, with a return starting in late August and September
	Borana Plateau[c]	Southern Ethiopia (Borana Zone) 95,000 km²	High elevation, semiarid, mixed savanna with perennial grasses (*Cenchrus, Pennisetum, Chrysopogon* spp.) and woody plants (*Acacia, Commiphora* spp.)	Seminomadic and sedentary pastoralism; pockets of nonirrigated maize cultivation; some small-scale annexation of grazing land from communal resources	Cattle (77 % of biomass), camels (16 %), sheep and goats (6 %), equines (1 %). Milk, meat, hides, skins; live animals for domestic and export markets	Oromo and Somali ethnic groups; tribes include Boran, Gabra, and Gurre	Pastoral resource use organized as traditional units of grazing and water; dry-season water occurs in deep wells. Bush encroachment, extensive soil erosion. Good road network, several large towns populated by locals and immigrants. Shares border with Kenya

(continued)

Table 17.1 (continued)

Tier	Site name	Location and area	Vegetation zones	Land use	Livestock species and products	People	Comments
Sahelian Belt		Niger (Dantiandou District)[d] 846 km²	Low-altitude hard-pan plateau dissected by fossil valleys with sand deposits; Tiger bush thickets on plateau; mosaic of valley croplands (millet, cowpea, roselle); scattered trees (Combretum, Faidherbia spp.) and shrubs (Guiera spp.), annual herbaceous plants (Mitracarpus, Sida, Cassia spp.)	Sedentary (village) farming and rearing livestock (75% of population) and recently settled agro-pastoralists (25%); few transhumant pastoralists from Nigeria or Benin in wet seasons	Cattle (78% of biomass), sheep (11%), goats (9%), donkeys (2%), horses, camels. Live animals for domestic and export markets; sour milk, butter; meat for home consumption	Zarma farmers; Fulani agro-pastoralists; Haoussa and Kel Tamasheq	Grazing restricted to uncropped lands from June to October, then open grazing including crop stubbles and weeds; transhumance northward in wet season and east- and southwards in dry season; Water points include ponds, deep wells, boreholes, artesian sources
		Mali (Hombori District)[e] 2923 km²	Low-altitude sedimentary basin with isolated mountains. Fixed dunes and sand deposits (60% of landscape); shallow soil on rock-hard-pan outcrops (30%), loamy-clay bottom lands (10%). Annual herbaceous plants mostly grasses (Cenchrus, Aristida, Schoenefeldia spp.) with scattered trees (Acacia spp) and shrubs (Boscia, Euphorbia spp.); some millet	Sedentary farmers (Hombori mountains); sedentary agro-pastoralists practice seasonal transhumance; nomadic pastoralists; transhumant pastoralists from inner delta of Niger River and north Gourma	Cattle (83% of biomass), sheep (6%), goats (7%), donkeys (3%), camels (1%). Live animals for domestic and export markets; sour milk, butter; meat for home consumption	Songhay, Dongon, Fulani (Macinanké, Djelgobé, Foulan kriabé), kel Tamasheq (kel Gossi)	Open-access grazing by herded livestock (except croplands in wet seasons); opportunistic local mobility; some nomadism; long-range transhumance from Macina (wet and cool seasons); nomads from North Gourma in dry seasons; water points include ponds, shallow and deep wells; water limits forage access in dry periods
		Senegal (Téssékré District)[f] 1759 km²	Low-altitude, fixed dune system on flat sandstone sediments. Scattered trees (Acacia, Combretum, Sclerocarya spp.) and shrubs (Boscia, Grewia spp.); annual herbaceous plants, mostly grasses (Aristida, Cenchrus, Brachiaria, Schoenefeldia spp.); some millet	Semisedentary pastoralists; opportunistic local mobility; long-distance transhumance to the South for grazing and marketing	Cattle (73% of biomass), sheep (20%), goats (5%), donkeys (2%), horses; live animals for local or regional markets including Dakar; some milk and butter	Fulani; some Wolof in small towns; villages settled since boreholes dug in 1950s	Open-access grazing by free-ranging herds (except tree plantations and conservation areas); local mobility centered on ponds (wet season), boreholes (dry season), and camps (year round); regional mobility to the south for dry-season pasture and livestock trade

Upper	Mongolia[g] 1,564,116 km²	Taiga (4 % of landscape) to alpine (3 %), forest- and mountain-steppe (25 %), steppe (26 %), desert-steppe (27 %) and desert (15 %); herbaceous genera include *Stipa*, *Poa*, *Cleistogenes*, *Agropyron*, *Potentilla*, and *Astragalus*; shrub genera include *Artemisia* and *Caragana*	Seminomadic (transhumance) with regular winter/spring campsites and variable summer/autumn pastures; small-scale cultivation and wild hay harvest; increasing mining and tourism	Cattle and yaks (6 % of biomass), horses (6 %), camels (1 %), sheep (45 %), goats (43 %); live animals, meat, milk, dairy products, fiber, (cashmere), hides, skins	Khalkha dominate numbers; rest are Khazakh, Dorvod, Bayaad, Buriat, Tsaatan, and other ethnic groups	Since transitioning to democracy and a market economy in 1990, herd size has grown and composition shifted to include more cashmere goats; rangeland governance in flux with weak regulation of common pastures and growing influence of community-based organizations. Mining competes with pastoralism for land and has increasing social and environmental impacts
–
Altiplano[g]	Bolivia[h] 123,000 km²	The Altiplano is a plateau that is 1100 km long and 120–160 km wide at an average altitude of 4000 m; total area is 306,000 km²; the Altiplano is divided into northern, central, and southern regions (La Paz, Potosi, and Oruro, respectively); cultivated land occurs on 1 % of the land, with over half as fallow; the climate is subhumid in the north near Lake Titicaca, changing to semiarid in the central region and arid in the south; important genera for range plants include grasses such as *Festuca*, *Hordeum*, *Distichlis*, *Muhlenbergia*, *Werneria*, and *Juncus* spp.; shrubs include *Parastrephia* spp.; cushion plants on saline sites include *Salicornia* and *Anthobrium* spp.	Pastoralists are semisedentary within areas under household control; agro-pastoralists are sedentary – local rangelands and fallow fields used for grazing; some improved pasture (alfalfa) for cross-bred cattle and sheep; cultivation dominates more to the north, mixed agro-pastoralism with sheep and cattle dominates in the central region, and camelid pastoralism prevails to the south, especially at higher elevations (puna). Human population is most dense to the North and declines southwards	Cattle (34 % of biomass), llama (31 %), sheep (30 %), and alpaca (5 %); smaller numbers of donkeys, swine, and vicuña; products include live animals, fiber, milk, meat, hides, and skins	One-third of the national population resides in the Altiplano, which includes large cities such as La Paz, El Alto, Oruro, and Potosí; dominant indigenous people include the Aymara; some Quechua and Mestizo also occur	Pastoralists and agro-pastoralists are organized in legally recognized peasant communities that have been traditionally marginalized in Bolivian politics and civil society; they have become more empowered of late due to altered government policies

(continued)

Table 17.1 (continued)

Tier	Site name	Location and area	Vegetation zones	Land use	Livestock species and products	People	Comments
		Peru[i] 50,365 km²	High elevations include the dry and humid puna. Herbaceous plants of the dry puna uplands include *Festuca*, *Stipa*, and *Calamagrostis* spp. and shrubs are comprised of *Parastrephia* and *Tetraglochin* spp.; herbaceous *Distichia* and *Eleocharis* spp. dominate peat bogs; humid puna sites have *Festuca and Muhlenbergia* spp. in the grass layer with *Baccharis* spp. dominating shrublands; peat bogs are dominated by *Distichia* and *Plantago* spp.; lower elevation croplands are typically planted to potato or quinoa	Pastoralists follow daily horizontal and seasonally vertical patterns of mobility; animals feed on rain-fed native forage; property rights in the humid puna are collective as derived from the historical *Hacienda* system. The community collectively owns land and assigns parcels to families; in the dry puna, property is privately owned	High elevations: Alpacas (70% of biomass), llamas (15%), sheep (10%), cattle (5%). Lower elevations: Cattle (70%) and sheep (30%). Also small numbers of vicuña, donkeys, and swine overall; major products include fiber, meat, hides, skins, and live animals	The Aymara occur mostly in the dry puna of Peru. The Quechua occur mostly in the humid puna	Pastoralists and agro-pastoralists are organized in legally recognized peasant communities, as in Bolivia. They have been traditionally marginalized in Peruvian politics and civil society
	Chihuahua and Sonora deserts[j]	Northern Mexico 780,000 km²	Elevation varies from sea level to 2000 masl; arid and semiarid lands with a high diversity of biomes that include desert scrub, savanna, succulent shrublands, arid and semiarid grasslands, and semiarid forests; common genera include *Larrea*, *Fouqueria* (desert scrub); *Prosopis*, *Bouteloua* (savanna); *Bouteloua*, *Muhlenbergia* (arid and semiarid grasslands); *Pinus*, *Quercus* (semiarid forest); and *Opuntia*, *Agave* (succulents)	Seminomadic and sedentary pastoralism; extensive livestock production; rain-fed and irrigated agriculture; both privately owned ranches and communal lands occur	Cattle (65% of biomass), goats plus sheep (29%), equines (5%); products include live animals, milk, meat, wool, hides, skins; there are government subsidy programs to produce improved breeds	Mostly inhabited by a peasant class occupying communal lands who are referred to as *ejidatarios*; a class of ranchers who operate on private land is referred to as the *ganaderos*; indigenous ethnic groups occur in smaller numbers	There is far greater pressure from people and livestock on communal land; communal lands are also more affected by sociopolitical issues; the rancher class actively responds to trade opportunities such as NAFTA

[a] Tropical Livestock Units (TLUs) used to describe livestock biomass composition; 1 TLU= 250 kg liveweight (1 cow = 0.80 TLU; 1 sheep = 0.10; 1 goat = 0.08; 1 donkey = 0.60; 1 horse =1.00; 1 camel = 1.20) (Jahnke 1982)

[b] Sources: Breckle and Rafiqpoor (2010), Jacobs and Schloeder (2012), Jacobs et al. (2009, 2015), and Schloeder and Jacobs (2010)

[c] Sources: Coppock (1994) except for updated (2013) livestock species composition from the Oromia regional government (unpubl.)

[d] Sources: Most from Cappelaere et al. (2009). livestock data are from Hiernaux (2012)

[e] Sources: Hiernaux et al. (2009); livestock data from Diawara (unpubl.)

[f] Sources: Human population and natural resource data from Touré (2010); livestock data from Corniaux et al. (2012)

[g] Sources: Hilbig (1995), Hijmans et al. (2005), livestock data are from the Mongolian National Statistical Office (http://en.nso.nm/home)

[h] Source: Ministerio de Desarrollo Rural y Tierras (2012)

[i] Sources: Del Pozo-Vergnes (2004), Flores-Ochoa and Kobayashi (2000), OEA (1996), Quiroga (1992), Nolte (1990), Orlove (1980). The TLU values from Jahnke (1982) were used for livestock biomass composition, but llama and alpaca equaled 0.25 TLU (Coppock, personal observation)

[j] Source: INEGI (2007)

17.2.1   Human Populations

Summary. Human population growth is a driver that profoundly affects all ecosystems. Human populations in developing-country rangelands are affected by demographic, ecological, economic, and political forces. Human population densities tend to rise in response to increasing annual precipitation and agricultural productivity as one goes from the arid to semiarid and subhumid zones. As nations or regions become wealthier, however, net human population growth in the rangelands tends to decrease and there are more chances for rangeland dwellers to emigrate in search of employment (typically males). In some of these situations, women can then become the primary stewards of local rangeland resources. Change in economic opportunities can dramatically affect the magnitude and direction of rural-to-urban migrations, especially as nations develop. Persistent warfare, poverty, and drought—and even organized crime—can be profoundly disruptive, however, resulting in a depopulation of some rangeland systems.

Human population growth is a driver that strongly influences the use of natural resources and the adoption of new technology in agro-ecosystems (Boserup 1965, 1989). Thus, human population issues merit our review. Images of important rangeland people in our case-study sites are shown in Fig. 17.2a–f. The evidence is mixed concerning human population trends across our six case-study areas. Population densities tend to be higher in agro-pastoral settings compared to pastoral settings because the former produce more food for people per unit area.[2]

Marked net increases in rangeland human populations have been noted for southern Ethiopia and the Sahel. Here increased fertility and decreased child mortality among pastoralists—as well as immigration by outsiders—have contributed to high rates of sustained growth, i.e., from 2 to 3 % per annum.[3] Emigration from pastoral zones in southern Ethiopia remains low, probably due to low exposure to formal education and lack of wage labor opportunities across the nation (Coppock et al. 2011). In the Sahel, emigration of agro-pastoral men to urban areas seeking wage labor has mitigated some population growth in the rangelands (Guengant et al. 2002; Wane et al. 2010). In contrast to the African examples, the number of Kuchi still using Afghanistan's rangelands is lower today than in the past despite the Kuchi having one of the world's highest fertility rates (e.g., 7.28; NRVA 2008). Reasons for this include high infant mortality rates caused by lack of health care and basic services, sedentarization[4] resulting from land conversions and resource degradation, recurring drought, social conflict, and chronic food insecurity.

---

[2] Agro-pastoral systems routinely combine crop and livestock production, while pastoral systems focus on livestock production (Jahnke 1982).

[3] Increased fertility and decreased mortality among pastoralists have occurred as a result of many factors. In some cases, development of clean water sources, provision of disease control for both people and livestock, improvements in infrastructure, and provision of food relief can be included.

[4] Defined here as the transition from a nomadic or seminomadic lifestyle to a society that permanently resides in one place.

Fig. 17.2 (**a-f**). Pastoral and agro-pastoral people who reside in the rangeland study sites reviewed in this chapter: (**a**) Aymara woman on the Peruvian altiplano (photo credit: Cecilia Turin); (**b**) Borana family in southern Ethiopia (photo credit: Claudia Radel); (**c**) goat ranching family in northern Mexico (photo credit: José Tulio Arredondo); (**d**) Pashtun family in Afghanistan (photo credit: Michael Jacobs); (**e**) senior herd owners at a political meeting in Mongolia (photo credit: María Fernández-Giménez); (**f**) women drawing water from a well in the Sahel (photo credit: Matthew Turner)

Fig. 17.2 (continued)

Fig. 17.2 (continued)

Rural population growth in the Altiplano is variable depending on location, but rates have been generally low in recent years (i.e., 0.4–0.7 %; Vera et al. 2006a, b). Seasonal emigration by Andean men seeking jobs in the cities or tropical lowlands is also common, especially in pastoral communities (Turin and Valdivia 2013). The human population of the northern Mexican rangelands can be broken out into two main groups: commercial ranchers (*ganaderos*) and a peasant class of pastoralists (*ejidatarios*). Numbers of people have risen and fallen—both with respect to birth rates and emigration—depending on the economy and the level of land degradation. Increases in the *ganaderos* community have occurred when the beef cattle industry expanded in the 1990s, but emigration of the *ejidatarios* has subsequently acceler-ated in response to land degradation (Schwartz and Notini 1994), drought (Feng et al. 2010), and US employment opportunities (Arredondo and Huber-Sannwald 2011; Ribeiro-Palacios 2012).

Patterns for Mongolia have been especially dynamic (Fernández-Giménez 2001; Leighton 2013). Following the transition from communism to a free-market econ-omy in 1990, there was an influx of urban dwellers into the rangelands as people sought to claim livestock during privatization, when state property was distributed to local citizens. This trend has since reversed as more people now leave pastoral areas to seek urban employment. Fertility rates among Mongolian women have markedly declined over the past 20 years.[5]

---

[5] http://data.worldbank.org/indicator/SP.DYN.TFRT.IN?page=4

When compared to sites in the upper tier, the Ethiopian site in the lower tier is characterized by more rapid net growth in residents that is related to a higher intrinsic rate of reproduction and relatively less opportunity for out-migration. Migration opportunities vary widely, however, across sub-Saharan Africa. In the Sahel, migration rates are high as people can move to cities and coastal nations of western Africa, northern Africa, and southern Europe (Tabutin and Schoumaker 2004). In the Sahel there has been a shift in livestock ownership to include agro-pastoralists as well as pastoralists (Turner et al. 2014). Ethiopian pastoralists, in contrast, have far fewer options (Coppock et al. 2011). Trends for Afghanistan are more difficult to discern as data are lacking. High losses of livestock due to conflict, insecurity, and drought—and few, if any, employment opportunities—however, suggest that human populations in the remote pastoral areas are also declining. These people appear to be settling near urban areas or joining refugee camps (UNHCR 2011).

In the upper tier, emigration from the rangelands is increasingly common, and this tends to occur more for men who seek employment as laborers in construction, mining, or farming. Women can thus be left behind to serve as caretakers of families and rangeland resources (Valdivia et al. 2013). Recently in Mexico women and children have joined men as migrants. A few decades ago it was job opportunities in the US that triggered emigration, but recent causes also include public insecurity related to organized crime (Martínez-Peña 2012). In Mongolia, recent rural to urban migration is influenced by the "push" factor of livestock loss in extreme weather disasters and the "pull" factor of people seeking better education and health care in urban areas (Leighton 2013).

17.2.2 Livestock Populations

Summary. Livestock (primarily including cattle, sheep, goats, equines, and camelids) provide the food and traditional economic basis for people living in the rangelands of developing countries; cultivation or wage employment opportunities are typically rare or nonexistent. The indigenous species and breeds are adapted to often harsh production conditions, and the flexibility of herd movement is very important for helping pastoralists cope with erratic rainfall patterns and disease outbreaks. Unlike the people on rangelands, population trends for livestock are more difficult to discern. Overall, livestock populations in some cases may exhibit "boom-and-bust" patterns where growth periods are followed by sudden die-offs due to combinations of weather, disease, or level of forage competition among livestock. The spatial scale and frequency of herd crashes vary markedly. The difficulty in mitigating large herd losses is due to low levels of economic development and public investment. Mitigating such losses matters, however, because recurrent die-offs translate into large economic losses for pastoral societies. For less-developed regions where human survival is most closely linked to livestock survival, the ratio of animals to people provides an important indicator of both food and asset security. Where this ratio has been monitored, the evidence shows that it has markedly declined in recent decades. Politics, economics, and armed conflicts also influence range livestock populations, and examples of each are provided. Overall, there is no consistent long-term trend in livestock populations that is evident across our six case-study sites. Some populations follow a regular boom-and-bust pattern, while others show sustained increasing or declining trends in response to macro-level factors.

As shown in Table 17.1, range livestock produce multiple products for household consumption or sale at all case-study sites. These are largely indigenous breeds that are adapted to local climatic and foraging conditions (Krätli 2007). Examples of key species and breeds from the six case-study sites are pictured in Fig. 17.3a–f. Other food-producing animals found among rangeland dwellers include poultry, honeybees, and guinea pigs; these can be locally important to supplement household diets or incomes, but are generally insignificant in the rangelands when compared to the economy based on hoofed animals.

Because of low and highly variable precipitation, the world's rangelands have a comparative advantage in terms of extensive animal production, whereby unconfined animals seek and consume forage that is scattered across a landscape. Extensive animal production—while having its own risks and challenges—is far more reliable than rain-fed cereal cultivation in these environments—explaining why pastoralism prevails as the environment becomes more arid. The prevalence of agro-pastoralism, where producers combine crop cultivation with herding livestock, generally increases as the reliability and amount of precipitation increase (Jahnke 1982).

Commercial production of cattle, sheep, goats, camelids, and equines tends to be somewhat recent and increasing in many developing-country rangelands. The final market destination of these animals varies considerably among case-study sites, with some being sold domestically and others exported. Commercial livestock production is a departure for indigenous systems in which animals were traditionally produced for home consumption, often referred to as subsistence production (Jahnke 1982). Some regions such as the Sahel, however, have long-been centers of commercialized livestock trade (Kerven 1992).

The livestock population dynamics in most of our case-study sites are characterized by boom-and-bust patterns at different spatial and temporal scales. In the boom-and-bust, periods of steady growth in animal numbers are followed by sudden collapses when death rates soar due to starvation or disease that is triggered by weather events such as dry periods, multiyear droughts, extreme temperature fluxes, or heavy snowfall. Disease epidemics can also be implicated.[6] In the case of conflict-ridden Afghanistan, when drought coincides with warfare herd losses can be catastrophic (FAO 2006). Both density-independent (e.g., weather) and density-

---

[6] In some cases previous development efforts to improve water access or reduce the prevalence of disease outbreaks were successful enough that such controls on animal numbers were relaxed. This resulted in more animals and a heightened demand for forage, with rangeland degradation as the ultimate outcome. Increased animal numbers, in theory, could be reduced by providing more marketing opportunities, but the traditional economic and cultural rationale in most pastoral societies to accumulate animals adds another layer of complexity that limits the rate of sustained offtake.

Fig. 17.3 (**a-f**). Livestock that are produced in the rangeland study sites reviewed in this chapter: (**a**) Improved Angus cattle in northern Mexico (photo credit: José Tulio Arredondo); (**b**) indigenous goats, sheep, and horses in Mongolia (photo credit: María Fernández-Giménez);

Fig. 17.3 (continued) (**c**) indigenous sheep in Afghanistan (photo credit: Michael Jacobs); (**d**) indigenous zebu cattle and dromedary camels in southern Ethiopia (photo credit: Brien E. Norton); (**e**) indigenous zebu cattle of the Sahel (photo credit: Matthew Turner); (**f**) llama on the Peruvian altiplano (photo credit: Cecilia Turin)

Fig. 17.3 (continued)

dependent (e.g., stocking rate or numbers of animals per unit area) factors can contribute to herd crashes. The number of years between consecutive herd crashes typically varies from 10 (Mexico; Garza-Merodio 2002) to 6 [southern Ethiopia (Desta and Coppock 2002) or Mongolia (Fernández-Giménez et al. 2012)]. Extreme cold events have contributed to a herd crash interval of less than 3 years in the Peruvian Altiplano (Moya and Torres 2008).

In the Sahel herd crashes occur at different spatial scales. The largest crashes have occurred following major regional droughts in 1972–1974 and 1983–1984

(Toulmin 1987) as well as after a period of "cold rains" in 1991 (Toulmin 1987).[7] Sahelian herd dynamics tend to be non-equilibrial (Ellis and Swift 1988). This means that livestock mortality events are influenced more by climatic factors rather than competition for forage among increasing animal numbers. It is more challenging for management to mitigate the effects of climatic factors compared to the effects of too many animals. In some systems interactions of high animal numbers with sudden drought or heavy snowfall can lead to sudden crashes in livestock populations.

Because the well-being of subsistence-oriented pastoralists is closely tied to livestock numbers, the ratio of livestock (e.g., tropical livestock units or TLUs[8]; Jahnke 1982) to people (e.g., African Adult Male Equivalents[9]) is an important measure of pastoral socioeconomic sustainability.[10] And the higher this ratio is, the better.[11] The ratio has been tracked in the two African sites where it has been shown to be in a steady decline over several decades, even going as low as 1:1 (Desta and Coppock 2004; Hiernaux and Turner 2002). This offers a stark contrast from historical highs that often exceeded a ratio of 10:1 (Gallais 1984). A sustained decline in the ratio means that per capita supplies of food (i.e., milk or meat) and capital assets (i.e., marketable animals on the hoof) are also declining; the inevitable result is thus increasing food insecurity and poverty unless livelihoods are diversified towards non-pastoral pursuits. The downward trend in the ratio for eastern Africa primarily occurs because the rate of human population increase exceeds that for livestock; this is partially due to the fact that far more animals perish during the "bust" phases than people do. Animals can be suddenly and severely limited by a scarcity of forage and water, and thus quickly starve to death. The people, in contrast, tend to suffer minimal losses to life as either they are rescued by human intervention (i.e., imported food aid) or they can migrate elsewhere and return when local environments improve. Patterns for the Sahel are somewhat different as growth rates for human populations in pastoral areas are low, but losses of animals occur because of shifts in livestock ownership from pastoralists to government officials, traders, and

---

[7] There were also major droughts in the Sahelian zone during the early twentieth century, but deaths of animals and people were lessened because the pastoralists were more mobile. The installation of permanent wells by governments began a process of settlement and a modification of pastoral risk management behavior (Sandford 1983).

[8] A tropical livestock unit (TLU) is 250 kg live weight [where one cow, sheep, goat, donkey, horse, or camel equals 0.8, 0.1, 0.08, 0.6, 1.0, or 1.2 TLUs, respectively (Jahnke 1982)].

[9] An African adult male equivalent (AAME) is a measure of daily energy demand based on body size [where a male=1.0 AAME and is ≥16 years old and weighs 55 kg; an adult female=0.8 AAME; a male or female youth=0.8 AAME; and a child=0.6 AAME (FAO 1982)].

[10] Two livestock population statistics are of primary importance in pastoral systems, namely livestock holdings per capita and livestock stocking rate. The former is described in the text above. The latter is measured by the number of animals per unit area for a given period of time. Stocking rate becomes significant when one examines human or livestock support capacity per unit area. Shifts in stocking rate can influence herd or flock responses to droughts, with higher stocking rates increasing herd vulnerability in some cases.

[11] One analysis suggests that an increase of one person must be matched by a sixfold increase in TLUs for that extra livelihood to be sustainable (Thurow, personal communication).

farmers. In West Africa most livestock are sold before they die of hunger except when a severe drought prevails (Turner, personal observation).

In any case, a period of herd rebuilding follows a crash, but this can be stressful because both food and assets remain in short supply (Desta and Coppock 2002). The ratio of TLUs to people has declined in most places and societies cope via diversification into agriculture, trade, or wage labor.

Long-term trends in livestock numbers across our case-study sites are difficult to discern largely because of inadequate data. It is still noteworthy, however, that the sites appear to vary with respect to overall trends. Some herd dynamics are complex and vary according to time frame as well as livestock species [i.e., sheep versus camelids in the Altiplano; Vera et al. 2006a, b], while other populations have either been steadily growing (i.e., cattle in northern Mexico; Perramond 2010) or markedly decreasing (i.e., sheep and goats in Afghanistan; FAO 2006).

Politics, economics, and armed conflict also influence range livestock populations. The best example of politics is Mongolia, where the collective era (1960–1990) was characterized by lower and stable numbers of livestock while the free-market era has witnessed several boom-and-bust cycles. Following privatization in 1992, animal numbers increased steadily until 1999–2003, when 30 % of the animal population perished in a series of severe winters. The population recovered and then crashed again during the winter of 2009–2010 (MNSO 2012). An example of economic effects is Mexico, where growth in the commercial production of beef cattle on the range dramatically increased in response to new US markets created by the North American Free Trade Agreement (NAFTA) in 1994.[12] Armed conflict has influenced large-scale migrations of pastoral livestock in southwest Asia. During the Soviet-Afghan war in the 1970s and 1980s, many pastoral herds moved out of Afghanistan to Iran and Pakistan (Colville 1998; UNHCR 2011).

It is therefore difficult to generalize about range livestock populations. A boom-and-bust pattern is perhaps the only feature that most of our case-study sites share, but the spatial scale and time interval vary markedly from place to place. The boom-and-bust is indicative of poorly diversified rural economies (i.e., animals comprise the main investment option), the limited capacity of herders to manage risks and engage in asset diversification, and the inability of markets to absorb large numbers of animals quickly during crises. The Sahelian zone (and other locations) provides exceptions to this pattern, as herd losses do not occur with the same regularity there as observed in the other sites; the very large spatial scale of Sahelian pastoralism may be a factor in this distinction. And it is useful to note that large-scale die-offs are not simply dead animals, but rather represent large economic losses for rangeland dwellers in terms of capital assets and foregone income, the latter represented by a sudden drop in milk supply. Desta and Coppock (2002) speculated that several livestock herd crashes over 20 years in southern Ethiopia and northern Kenya may have resulted in a cumulative loss of nearly US$1 billion, value that poverty-stricken pastoralists can ill afford to lose.

Livestock numbers may increase in response to commercial opportunities, but this appears more evident (as with northern Mexico) when commercial opportuni-

---

[12] http://www.ustr.gov/trade-agreements/free-trade-agreements/north-american-free-trade-agreement-nafta

ties grow in a sustainable fashion. Other examples include an increase in the population of cashmere goats in Mongolia in response to global demand for cashmere fiber (Fernández-Giménez, personal observation) and an increase in sheep holdings among agro-pastoralists in the Sahel in response to the large demand associated with Muslim holidays (Turner, personal observation). Pastoral livestock populations, however, are probably declining in Afghanistan (Thompson et al. 2005). Thus, overall, there is no evidence for consistent trends in livestock numbers across the six case-study locations.

17.2.3 Socioeconomic Trends

Summary. Overall, range livestock and their products are increasingly important for developing-country economies, although the relative contribution is low as other sectors expand. Rangeland dwellers are experiencing an expansion of livestock marketing options due to increasing domestic and export demand—and in some cases transportation infrastructure is also being improved. Ready access to public services such as potable water, electricity, health care, education, and banking services remains elusive for most rangeland dwellers, although access tends to be greater in the upper-tier sites compared to that of the lower-tier sites. Telecommunications have improved across the board, however, largely due to mobile phone networks. Traditional pastoralists have lower access to public services than do settled pastoralists or agro-pastoralists. Transition from communism to a free-market system in Mongolia has undermined the access of pastoralists to public services; other structural adjustments in the Sahel dismantled state-run veterinary and livestock services. Livelihoods are diversifying among agro-pastoralists, residents of settlements, and households whose members find employment outside of the rangelands. Wealth stratification among rangeland dwellers appears to be increasing overall as fewer people control more resources—a negative outcome of globalization.[13] Food insecurity and extreme poverty occur in the lower-tier sites, but less so in the upper-tier. Social conflict takes a variety of forms, but it tends to be focused more on local control over natural resources in the Andean, Mongolian, and Sahelian situations. In contrast, the other case-study sites are subjected to a much wider array of conflicts—from resource-access squabbles among neighbors to struggles between drug cartels and long-term, major insurgencies.

Several of our case-study sites are in nations where rangelands are the predominant portion of the productive land area, with the main exceptions being Bolivia and Peru which have large portions of their lowlands in the moist tropics or subtropics. In Ethiopia, the Sahel, Afghanistan, Mongolia, and northern Mexico range livestock production is vitally important with respect to domestic supplies of live animals or animal products including hides, skins, and fiber (de Bruijn and van Djik 1995; Desta

---

[13] Globalization has been defined in many ways. Albrow and King (1990) define it as "all those processes by which the peoples of the world are incorporated into a single world society."

et al. 2006; Niamir-Fuller 1999; Perramond 2010, Schloeder and Jacobs 2010; Turner et al. 2014; Zoljargal 2013). There is more variation with respect to the economic importance of range-related exports (Williams and Spycher 2003). Range livestock generate very significant proportions of national export revenue for Ethiopia (Desta et al. 2006), the Sahel (Zoundi and Hitimana 2008), and Afghanistan (Schloeder and Jacobs 2010). Such exports are relatively less important at the national scale, however, for Mongolia (World Bank 2013a), Mexico (Peel et al. 2011), or for nations that share the Altiplano (Valdivia 1991). On the Altiplano, fiber from alpaca and vicuña is locally important for artisan or textile industries, and there has been a surge in smallholder dairying in the agro-pastoral sector in response to increasing demand for milk among urban residents at high elevations (Valdivia 1991).

Markets for range livestock have grown overall during the past two decades as the global economy has expanded, stronger consumer countries have emerged, personal incomes have increased, rates of urbanization have accelerated, and trade barriers have been reduced (Meyers and Kent 2004). This bodes favorably for pastoralists over the long term, but producers living in remote areas—or subject to trade monopolies or intense competition—often remain isolated from market opportunities.

This isolation is often related to poor infrastructure, inefficient marketing networks, and lack of progressive policies. Road construction and rehabilitation are currently common across the rangelands of eastern Africa, spurred to a large extent by investors such as the People's Republic of China (Zafar 2007). Livestock marketing in general is a major priority in regional development initiatives concerning improvement in food security for the Greater Horn of Africa (Knips 2004) and the Sahel (CRCM 2013). Trends to improve rural infrastructure have also been observed for Mongolia (primarily stimulated by mining development), northern Mexico [primarily related to NAFTA and neoliberal policies including privatization of communal land (Perramond 2008)], and the high Andes (primarily related to international trade, especially in the Lake Titicaca region). Government decentralization has shifted more development funds and attention to the local level for some marginalized rangeland communities, with notable progress observed on the Bolivian Altiplano. Impact from decentralization has also occurred in southern Ethiopia (Coppock et al. 2011). Improvements in rural infrastructure, overall, are rare in the Sahel (Hesse et al. 2014) or Afghanistan (Mohmand 2012).

Public service provision for rangeland dwellers remains as a major obstacle for progress. Poor nations typically do not have the resources to make development investments in remote locations. There are two distinct subpopulations emerging in the rangelands of the developing world: one consists of traditional, mobile pastoralists still largely dependent on livestock, while the other consists of sedentary or semisedentary residents of growing rangeland towns and cities that have more diversified livelihoods. The latter group includes agro-pastoralists, former pastoralists, sedentary pastoralists, and immigrants from other sectors. Sedentary pastoralists may have households where some members are fully settled and have wage employment, with other members traveling to distant locations as they herd the family's livestock.

In such settlements access to potable water, electricity, health care, schools, and banking services can still be lacking. Public awareness of the need for services is increasing, and appreciation of education is rising among many rangeland dwellers (Coppock et al. 2011). Service provision to mobile pastoralists is much worse than that for town dwellers across all of our case-study locations. Most traditional rangeland populations have never had public services, so any progress is an improvement. An interesting exception, however, is Mongolia where during the communist era nomads received heavily subsidized services including education, health care, and veterinary care (Fernández-Giménez 1999). This resulted in high rates of literacy as well as successful vaccination campaigns. In the free-market era these services have been lost or greatly diminished. The one bright spot in service provision on the rangelands concerns improved telecommunications; each of our case-study sites have witnessed expansion of mobile phone use in the past 5 years.

Pastoral households across our case-study sites are commonly near or below the poverty line. Compared to the past, trends indicate that traditionally oriented pastoralists are generally getting poorer and thus have a higher risk exposure to perturbations caused by weather, economy, or conflict. Although illiteracy rates remain high in most cases, more pastoral children are reportedly attending school and, where local circumstances allow, pastoral households increasingly attempt to diversify incomes and assets by mixing pastoral with non-pastoral activities (Coppock et al. 2011). Food insecurity remains common for pastoralists in the lower tier; Afghan and African pastoralists often receive food aid (Coppock et al. 2014; Schloeder, unpublished data). One extreme case, for example, is the 40 % of Afghan Kuchi pastoralists who remain in refugee camps (Colville 1998; USAID 2007). Food aid is pervasive in the Sahel[14] as well as in the Greater Horn of Africa.[15]

Livelihoods for pastoralists in the upper tier are relatively better in that food (both in terms of production and access) is less of a problem. A common concern for rangeland dwellers in Mongolia, the Altiplano, and northern Mexico is the cost and effort needed to gain access to secondary or tertiary education (Fernández-Giménez, personal observation; Kristjanson et al. 2007; Martínez-Peña 2012).[16]

In Mongolia, the proportion of rural households living in poverty greatly increased in the early years of the free-market transition (Griffin 2003; Nixson and Walters 2006). There has since been a gradual recovery, but this has been disrupted by large herd crashes caused by severe winter weather. Considerable development attention has recently been given to strengthening community-based organizations to fill resource governance gaps created by de-collectivization (Fernández-Giménez et al. 2015; Leisher et al. 2012; Upton 2012).

In Afghanistan, socioeconomic progress was halted by war with the Soviet Union, and poverty rates remain high following the NATO[17] intervention that began

---

[14] http://ec.europa.eu/echo/aid/sub_saharian/sahel_en.htm

[15] https://na.unep.net/geas/getUNEPPageWithArticleIDScript.php?article_id=72

[16] Residents of *ejido* villages in northern Mexico now have access to low-cost distance education programs.

[17] North Atlantic Treaty Organization: http://www.nato.int/cps/en/natolive/topics_8189.htm

in 2001. Weak government institutions, corrupt administrations, lack of physical security, and a ruined rural economy have stymied development progress in general for Afghanistan (Mohmand 2012). In addition, there has been a lack of development aid targeted specifically for pastoralists in Afghanistan. For 2008, for example, pastoralists reportedly received only US$0.20 per person compared to US$60.00 per person for non-pastoralists (Mohmand 2012).

For Mexico, increasing affluence of consumers has translated into an accelerated demand for grain-fed beef. In response, mega-ranches with intensive feedlot production have emerged in northern Mexico, while traditional smallholders on communal lands are being squeezed out of the market (Henriquez and Patel 2004). Government subsidies and high remittances are common inputs for Mexican producers—this makes them distinct when compared to rangeland residents in our other case-study sites. The NAFTA has opened cross-border markets with the US as well (NAFTA 2000).

All of our case-study sites exhibit trends where stratification among rangeland producers is widening the gap between the haves and have-nots. This is often triggered by elites[18] who have the connections and skills to take advantage of commercial livestock opportunities, build larger herds, and gain de facto control over more natural resources.[19] Stratification can leave the majority of the population more marginalized (Coppock et al. 2014). In other cases such as the Altiplano (Kristjanson et al. 2007), however, increased vulnerability is observed primarily among the elderly.

Finally, ethnically based or resource access-related social conflicts at large spatial and temporal scales are pervasive challenges for rangeland dwellers in Afghanistan and the African sites. In Mongolia, social conflict over resources is more localized and related to pasture and water access (Fernández-Giménez et al. 2008), although conflicts between herders and mining interests are increasing (Fernández-Giménez, personal observation). Similarly, in the Altiplano, conflicts have occurred around water resources and between herders and mining interests (Turin, unpublished data). In northern Mexico, unequal access to agricultural land and contested grazing rights for non-*ejidatarios* have caused long-lasting internal conflicts in the communal areas (*ejidos*). More recently, extensive drug trafficking and conflicts between government and drug lords have disrupted ranching as well as other forms of commerce (Martínez-Peña 2012). In the Sahel, insurgencies tied to radical Islamic ideology have emerged (Larémont 2011). Such developments in Mali, however, have not affected pastoral mobility patterns. Banditry associated with general insecurity is viewed as a more chronic problem overall (Turner, personal observation).

---

[18] Defined as a group of people who exercise the major share of authority or influence within a larger group, often associated with a greater degree of wealth.

[19] This process is not unique to developing country rangelands; rather, it is the rule, not the exception. There are similar examples of elite takeover during the rangeland settlement era in the US and Australia, for example. It is desirable to learn from the past and mitigate hardships in developing nations as change accelerates (Thurow, personal communication).

17.2.4 Trends in Resource Use and Ecological Condition of Rangelands

Summary. This section describes the environments of the six case-study sites. Although climate, landscapes, vegetation types, and land use vary greatly among the case-study sites, all share common patterns of soil erosion, vegetation change, or rangeland fragmentation due to intense and chronic exploitation of natural resources by people and livestock, regardless of whether a site occurs in the upper or lower tier. Both subsistence and commercial livestock production systems appear to push resource use to the limits. Climate change assessments typically portray a warmer and drier future for most of the sites except the Sahel, which may become wetter. Forage supplies are often barely sufficient to carry livestock through the dry seasons of "normal" rainfall years, let alone droughts, contributing to the boom-and-bust herd dynamic previously described for some locations. The most damaging and irreversible result of extractive land use is the accelerated erosion of topsoil. Although overgrazing is often mentioned as the main cause of accelerated soil erosion, dryland farming, overharvest of wood for fuel and building materials, and poorly designed roads can sometimes be more important in this regard.

The environments in our case-study sites are described in Table 17.1. They markedly vary in terms of climate and plant communities. Images of representative landscapes are shown in Fig. 17.4a–f. Overall, despite high variation in human and livestock features of our case-study sites, trends in resource use and ecological condition of rangelands are similar.

Afghanistan is a very mountainous country located in the arid-subtropics of Asia. In the mountains the winters are cold and snowy. Nationwide, the summers are hot and dry.[20] Multiyear droughts are common, as are major dust storms. The range livestock are dominated by five breeds of sheep and two breeds of goats and their crosses. Breeds vary in their tolerance of poor forage quality and resistance to disease, and hence their dominance varies by region. The seasonal grazing patterns followed by most Afghan herders include pasturing of livestock: (1) near camps and residences at low elevations during winter where snowfall is moderate[21]; (2) at sites close to the winter range that green-up early during spring; (3) at progressively higher elevations during summer (up to 2500 m or more) as animals follow an altitudinal gradient of green-up; and (4) back at the same type of sites in the fall that were previously used in the spring.

Land degradation in the form of soil erosion is commonly attributed to overgrazing by goats and sheep, although hard causal evidence is often lacking. Major contributors to excessive soil erosion are more likely to include dryland cultivation (in general) and the overharvest of woody and nonwoody materials for fuel, food, medicinal plants, and building supplies (Jacobs et al. 2015). Dust storms also con-

---

[20] http://www.ncdc.noaa.gov/oa/climate/afghan/afghan-narrative.html
[21] Hand feeding of natural forage may occur in high snowfall years.

Fig. 17.4 (**a-f**). Landscapes in the rangeland study sites reviewed in this chapter: (**a**) Borana encampment in southern Ethiopia (photo credit: Brien E. Norton); (**b**) Gourma landscape in the Sahel (photo credit: Matthew Turner); (**c**) grassland in the Peruvian altiplano (photo credit: Cecilia Turin);

Fig. 17.4 (continued) (**d**) herding camp in Mongolia (photo credit: María Fernández-Giménez); (**e**) irrigated wheat and hillside range in Afghanistan (photo credit: Michael Jacobs); (**f**) tobosagrass rangeland in northern Mexico (photo credit: José Tulio Arredondo)

Fig. 17.4 (continued)

tribute to wind erosion. Climate change studies indicate warming and drying trends, both in the recent past and future forecasts (Savage et al. 2009). Currently, unsustainable dryland cultivation is the greatest threat to livestock survival during winter, spring, and fall when the high-elevation summer pastures are unavailable. Sustainable dryland cultivation is the greatest threat in terms of the fragmentation of pastoral migration routes and staging areas.[22] Both situations will only get worse if climate change scenarios are verified, as more cultivated areas are abandoned in the most fragile landscapes and drought becomes the norm rather than just a temporary or cyclic event.

The Ethiopian situation illustrates too many people and livestock (i.e., cattle, sheep, goats, and camels) exploiting a diminishing natural resource base. This trajectory was predicted over 20 years ago (Coppock 1994). Symptoms are ecological and socioeconomic (Coppock et al. 2014). The former include bush encroachment and gullying on certain soil types. The latter include land-use fragmentation due to de facto privatization of grassland parcels annexed from communal resources, loss of dry-season grazing to maize cultivation, and occupation of former drought grazing reserves by people who have arrived from overpopulated places elsewhere. Bush encroachment has been exacerbated by decades of overgrazing and (past) official bans on the use of prescribed fire that have shifted the competitive balance from perennial grasses to woody plants (Coppock 1994). This loss of grasses means a decline in fine fuels that are needed to carry fires that might otherwise control the recruitment of woody seedlings. Maize cultivation by pastoralists is a food insecurity response to a declining ratio of livestock to people and thus is symptomatic of the trend of increasing poverty (Desta and Coppock 2004). Chronic food insecurity has led to dependence on food aid, the provision of which effectively delays resolving the root causes of hunger (Coppock et al. 2014).

The Sahel differs from southern Ethiopia in many respects, but shares some of the same broad outcomes. The Sahelian belt occurs at a continental scale, with far more variation in terms of climate, land use, and rangeland ecology (Table 17.1). Sahelian production systems vary from pure pastoralism to highly integrated agropastoralism where crop residues provide livestock fodder, and livestock manure enhances crop yields (Heasley and Delehanty 1996; Turner et al. 2014). In the Sahel the overall challenge for pastoralists is how to exploit the highly variable occurrence of fodder over space and time most efficiently. Fragmentation of landscapes and natural resource endowments has occurred due to population growth, spread of cultivation, and national policies that attempt to reassert tougher border controls, although efforts have occurred to accommodate pastoral mobility across national borders (Zoundi and Hitimana 2008). In some instances, however, cultivation has remained stable or retreated due to variation in local population growth and climate patterns (Tappan et al. 2004). Pressure on grazing resources from cultivation is particularly a problem in the southern Sahel where seasonal, long-distance livestock movements to the north during the rainy season (e.g., transhumance) have been

---

[22] Staging areas are places where livestock rest for extended periods before continuing on to lower or higher elevations.

historically important. Interestingly, a "re-greening" of the Sahel has been observed over the past 30 years from satellite images (Olsson et al. 2005) and verified by field work (Dardel et al. 2014) where rainfall and net primary productivity have increased, and this mitigates some otherwise negative trends (Gardelle et al. 2010; Hiernaux et al. 2009, Leduc et al. 2001). Vegetation change in the Sahel appears to be more influenced by precipitation than livestock effects and thus is an example of nonequilibrium dynamics (Ellis and Swift 1988)[23,24].

Mongolian rangelands today are increasingly at risk from heavy livestock grazing (Liu et al. 2013) as well as soil degradation associated with road networks (Keshkamat et al. 2013) and pollution of soil and water from the expansion of mining (Thorslund et al. 2012). There is debate, however, about the causes of declining range conditions in Mongolia (Addison et al. 2012). Livestock pressure may be interacting over space and time with dynamic climate belts to alter vegetation cover, with the mix of drivers likely changing in different ecological zones (Cheng et al. 2011; Wang et al. 2013; Wesche et al. 2010). As in the Sahel, several studies have shown that livestock grazing has less impact on vegetation than precipitation in the desert-steppe region of Mongolia (Fernández-Giménez and Allen-Diaz 1999; Wesche et al. 2010). Livestock play a more significant role in vegetation dynamics of the wetter mountain steppe and steppe zones, where livestock populations have increased most dramatically in the past 20 years (Fernández-Giménez and Allen-Diaz 1999).

The Altiplano is also comprised of diverse landscapes. Indigenous pastoral and agro-pastoral production systems were very efficient in growing crops or rearing livestock at different elevations in the Andes, but this was dismembered during the Spanish Conquest (Flores-Ochoa 1976). Today, lower elevations on the Altiplano (3700–4000 m) are dominated by agro-pastoralism that includes sheep and cattle, while higher elevations (>4000 m) are dominated by pastoralists raising camelids (CIRNMA 1997). Forage resources are either communal or privatized. Population growth is reducing the per capita base of natural resources and there is pressure to reduce long fallow periods for dryland crops. Climate research has documented that

---

[23] Work by Ellis and Swift (1988) concerned the nomadic pastoral system of South Turkana, Kenya. The system components include a diverse assemblage of livestock species that forage in a drought-pulsed, arid region subtended by sandy and volcanic soils. Plants are dominated by annual grasses, dwarf shrubs, and *Acacia* shrubs and trees. Ellis and Swift proposed that South Turkana was a nonequilibrium system, meaning that livestock stocking rates would not get high enough to effect significant change in the plant community via their foraging and trampling activities. Frequent droughts in South Turkana decimate livestock numbers to keep their influences low, and the annual grasses and sandy soils, in any case, are resistant in the face of livestock pressure. The productivity and abundance of annual grasses, in particular, are thus primarily affected by the pattern and amount of annual rainfall, not livestock. This is in contrast to an equilibrium system where livestock can exert directional pressure over time on the perennial grass community and the upland soils (i.e., soils having a mix of sand, silt, and clay). This pressure can lead to soil erosion and significant changes in plant community composition. An example of an equilibrium system is found less than 500 miles from South Turkana in a higher elevation, semiarid region called the Borana Plateau (Desta and Coppock 2002).

[24] There are examples in the Sahel, however, where people have had a decisive role in overriding climate effects and impacting the environment. Intense grazing and cultivation in some parts of Niger have proven incompatible with the arid climate, resulting in large areas of formerly productive *Andropogon* grasslands now being devoid of topsoil (Thurow, personal communication).

the Altiplano is becoming warmer and drier (Seth et al. 2010). This has implications for reducing forage quality (Zorogastúa-Cruz et al. 2012) and drying of evergreen peat bogs (*bofedales*) that are fed by glacial melt and provide key grazing for camelids (i.e., alpaca and, to a lesser extent, llama). There are also instances where increased land pressure occurs when too many families reside in the same area or when remittances are heavily invested in livestock (Turin and Valdivia 2011).

Emigration of men or youths seeking urban jobs or secondary schooling has undercut the labor supply for herding and managing natural subirrigation systems. In some instances introduced forages (e.g., alfalfa) have been successfully established in agro-pastoral zones, reducing pressure on native range and allowing expansion of smallholder dairy operations (Turin and Valdivia 2013). For Bolivia, Healy (2001) documents how tractor introduction—in support of expanding *quinoa* cultivation—has displaced llama production.

The rangelands of northern Mexico are comprised of highly diverse landscapes (Table 17.1). The large, private ranchers focus on beef cattle, and they have access to land mostly situated in the most productive, semiarid grassland region. African forage grasses (i.e., *Cenchrus ciliaris*, *Eragrostis curvula*) have been introduced to boost rangeland productivity. A peasant class of pastoralists *(ejidatarios)* is found in the arid and semiarid hilly desert scrub (*ejido*) region where cattle, goats, and sheep can be herded in a traditional, seminomadic fashion; animals feed on native forage and crop residues. Ranches utilize cow-calf production systems based on grass forage. Steers are finished at large feedlots in preparation for export to the US. Animals produced in the *ejido* system are typically sold in local markets.

Livestock grazing pressure has been intense in Mexico since the arrival of the Spanish. Large portions of the semiarid zone have been subjected to nonirrigated cultivation, and deforestation has occurred. Multiyear droughts are common. Nearly half of the rangelands have endured severe soil erosion and woody encroachment due to heavy grazing and lack of fire (SEMARNAT 2005). Other trends include unfavorable shifts in land cover and plant species composition, including the expansion of introduced invasive species; this is most apparent in desert scrub ecosystems (SEMARNAT 2005). Climate studies indicate that precipitation patterns have been changing; warmer, drier conditions are expected (Piñeda-Martínez et al. 2007).

17.3   Four Major Stewardship Themes for Rangelands of the Developing World

We propose four major themes that are most vital to better understand how community-based organizations, traditional leaders, researchers, development practitioners, policy makers, and other change agents can help rangeland societies better navigate the challenges that face them. These themes are (1) pastoral land tenure and managing mobility; (2) sustainable rural livelihoods; (3) livestock development and marketing; and (4) conflict and crisis management. For each theme, we first frame the pivotal issues and conceptual advances and then provide supporting observations from the six case-study sites. We close with a synthesis and discussion of priorities for research and outreach.

17.3.1   Pastoral Land Tenure and Managing Mobility

Summary. One of the most unique aspects of pastoral production systems is the need for herd mobility. Mobility is required to optimally exploit an ever-changing landscape of forage production that is caused by high spatial and temporal dynamics in precipitation. Typically, as rangeland areas become more arid, the variability of precipitation in any one location increases; consequently, arid systems require more herd mobility than semiarid systems. This considers horizontal and vertical mobility, as the latter becomes important in places having marked elevation gradients. Access to diverse forage resources is influenced by multiple factors including availability of drinking water, physical insecurity, competition for land among various user groups, availability of herding labor, and restrictions imposed by land tenure regimes. The latter have traditionally emphasized flexibility and reciprocal user rights that enable pastoralists to better manage livestock production risks under fluctuating environmental conditions. Forage access has been traditionally managed under informal rules that underlie the use of communal resources; this can prominently include restricted access of producers to water or key forage resources during periods of resource scarcity. When rules do not exist or are ignored, environmental degradation due to open access can occur. Environmental degradation can also occur, however, due to other processes irrespective of management. Risks of environmental degradation are often used to justify state control over, or privatization of, communal grazing lands. The problem is that neither state control nor privatization can typically offer the creativity and flexibility required to foster reliable access to local or regional resources under highly diverse circumstances. Our six case-study sites share one major trend overall: herd mobility has declined in terms of distance and frequency almost everywhere. Common causes of declining mobility include territorial fragmentation of rangelands, poor rangeland governance, increased settlement of previously mobile pastoralists, and a gradual loss of herding labor due to herder emigration. In general, there is an increasing awareness at local, national, and regional levels of the need to restore or maintain herd mobility where possible, but overcoming key constraints can be daunting. One commonly shared view is that change agents can help restore herd mobility and flexible resource use in some situations by improving rangeland governance.

17.3.1.1   Pivotal Issues and Conceptual Advances

Forage production on rangelands is typically influenced by precipitation regimes that vary greatly over space and time. This requires pastoralists to be highly opportunistic, and mobile herds of livestock are precisely the harvesting tool needed to effectively exploit forage resources and mitigate the inherent risks of animal productivity. The more arid the rangeland system, typically the more mobile a pastoral society needs to be.

Many arid and semiarid rangelands are communal resources from which it is difficult to exclude potential users, but where use by the animals of one individual reduces the amount of forage remaining for the others. If there are no rules to determine who may graze, or to place restrictions on the amount, timing, or spatial distribution of grazing, the situation is termed "open access," and there is a risk of overuse and degradation (Ostrom 1990). Most rangelands are not open access; instead they are subject to some type of property regime—a set of formal or informal rules that define the rights and obligations of specific individuals or groups to access, use, manage, or transfer (sell or gift) a resource. Pastoral land tenure refers to the set of rules that define who may access, use, or manage land or other pastoral resources. In the developing world, many pastoral land tenure systems were traditionally based on well-established, albeit unwritten, rules in which communities held collective use and management rights to forage (Lane 1998). This form of tenure is called common property (Ostrom 1990).

Many of these traditional systems are now in transition to more formal systems in which common property becomes state property managed by government authorities, private property owned by individuals, or—less commonly—common property officially possessed and managed by a defined community of users through a formal legal agreement (Galvin 2009; Lane 1998; Toulmin 2009).[25] In still other cases the state may have formal control over an area but lack the capacity to manage it effectively; much of the Sahel is "state property" in a formal sense, but the state has little influence over resource rights or use (Turner, personal observation). In the past, some development practitioners mistakenly believed that rangeland degradation was a result of common property systems and thus advocated greater government control or privatization of commons. Today we recognize that while open access (the lack of rules) can lead to degradation,[26] communal property is often the most appropriate land tenure system in highly variable semiarid and arid rangelands where sustainable grazing management depends on pastoralists' ability to move their herds and have flexible access to heterogeneous resources across extensive land areas. Under these circumstances, dividing a large commons into many smaller private parcels is likely to lead to ecological degradation and increased vulnerability of pastoralists to climate risks like drought or severe winter weather (Galvin 2008).

---

[25] Central governments, following the precedent set by colonial regimes of claiming ownership of rangelands and weakening traditional (tribal) authority over land use, may set up appropriate regulations for land management but have neither the incentive nor the personnel to enforce those regulations. One result has been less control over local land use and hence more degradation. The main remedy is to return to some form of local authority, with or without government participation. One solution has been to establish government-instituted local authorities for land management that operate in parallel with traditional (tribal) leaders. The challenge then is to find mechanisms to ensure that both actors work together.

[26] Sites having high variability of forage and water can lead to shifting grazing patterns with little potential for overgrazing in systems that lack formal rules (Turner, personal observation).

Alternatively, dividing rangeland into a few very large, ecologically viable private parcels is not socially viable because large numbers of residents would be displaced.

Nevertheless, solving the dilemma of pastoral land tenure is not simple because herders need both secure access to local key resources such as dry season or winter pastures and flexible access to distant pastures during disasters, and it is difficult for most formal land tenure systems to meet both of these requirements simultaneously (Fernández-Giménez 2002; Turner 2000). Formalizing tenure by allocating exclusive rights over key resources assures security, but may limit flexibility. Allowing maximum flexibility may result in lack of secure rights to key pastures for some herders, when others come to use their pasture during a disaster. In any location it is important to understand the historical, environmental, and sociocultural context for the existing land tenure system, its strengths and limitations in meeting pastoralists' needs for security and flexibility, its potential to support increased or sustained economic activity, and its compatibility with land health and wildlife conservation goals (Turner 2000).

When tenure systems change, there are always winners and losers, and thus the social equity consequences of changing tenure must also be considered. Historically, pastoralists have often been dispossessed of their traditional grazing territories when their lands, designated as "vacant" or "wasteland," were seized by the government for conservation (e.g., national parks) or economic development (i.e., mining, cultivated agriculture, renewable energy production) purposes. There is no one-size-fits-all tenure system that assures economic productivity, environmental sustainability, and social equity, but many systems have evolved in different regions that enable productive use of rangelands compatible with sustaining land health and meeting conservation goals. These may be based on common property regimes or, increasingly, a mosaic of private, public, and common property resources (Galvin 2009; Toulmin 2009; Turner 2000). The key is that when development reduces the effectiveness of previous controlling factors for rangeland access and use, it is vital that new controlling factors are created and adopted by the community.

17.3.1.2   What Has Been Observed?

Afghanistan. As noted above, Afghan pastoralists must migrate seasonally from low-lying areas in the winter and spring to higher elevation sites in summer—a pattern that tracks the availability of green forage as the snow melts. This pattern has been disrupted repeatedly over the past century. The most recent disruptions have included the Soviet occupation (1979–1989) and subsequent periods of ethnic unrest, civil strife, insurgencies, drought, government land annexation, corruption, and class conflict. A lack of regulatory institutions and development investment has compounded the problems. Individually and collectively, these forces have had very negative effects on pastoral common property management regimes as well as on the resource use by non-pastoral groups in Afghan society (Barfield 2004; Jacobs and Schloeder 2012; Wily 2013).

For the Kuchi, one outcome has been a marked reduction in livestock mobility and animal health (Jacobs et al. 2009). In response to landscape fragmentation, some pastoralists are now hiring trucks to move their animals between spring and summer range in an effort to avoid either having to pay for grazing rights, finding that their leases are no longer honored or available because key lands have been cultivated, or risk of losing animals to criminals or corrupt officials. This option for trucking is only available to wealthier individuals or clans, however, leaving the less affluent highly vulnerable to complete herd losses. Others who have already lost their animals hire out as long-term herders for wealthier pastoralists with the agreement that profits are shared between the herder and owner when certain animals are sold.

In the last decade there has been some progress towards addressing the country's environmental, social, and economic challenges. This includes endorsing the Millenium Development Goals and the implementation of the Afghanistan National Development Strategy. Land-use policy, on the other hand, has been slow to evolve. Consequently, the land rights situation has worsened rather than improved since 2001 in most rural areas (Wily 2013). As rangeland becomes increasingly unavailable due to unchecked forces previously described—as well as the failure to address historic grievances—poverty and food insecurity for the Kuchi will grow further because those still herding will increasingly find that they are unable to meet their most basic needs for survival (Jacobs and Schloeder 2012).

While an increasing population has resulted in more grazing pressure on rangelands—particularly close to villages—probably the greatest impact to extensive livestock production after years of conflict has been the loss of relationships between herders and the villagers they encounter during their annual migrations. Decades of fighting have left very little trust among people. Herders who may have had strong, traditional relationships with villagers are now met with unfamiliar, fearful, mistrusting people nervous about herders moving past their villages. Both the herders and villagers are well armed, making the situation tense and ripe for land-access conflicts. This insecurity has not only restricted the movements of livestock but is threatening the survival of this highly evolved animal production system. Implications affect industries dealing with meat, dairy, hides and skins, wool, and carpet making.

Ethiopia. In southern Ethiopia, local and regional human population growth has had more negative effects on traditional common property management and herd mobility than has land annexation by outsiders (Desta and Coppock 2004). Human population growth drives the need for more food, be it milk from livestock or maize from cultivation. This, in turn, increases the competition for forage and land, manifested in the creation of privatized grazing sites (*kalo*), expansion of cultivated fields, and demise of traditional fallback areas that were once used for grazing during droughts. This fragmentation reduces herd mobility and pastoral resilience to drought (Desta and Coppock 2002, 2004). Grazing management can be conceptualized as traditionally occurring within several nested levels of spatial resolution. From higher to lower resolution, these are (1) *olla*, (2) *arda*, (3) *rera*, and (4) *dedha*. These vary in size from square kilometers (*olla*) to tens of thousands of square kilometers (*dedha*). Traditional grazing managers and water managers for the Boran have customarily overseen the integrated use of these strata; creativity and flexibility are needed to effectively balance forage supply and demand every year.

In the past 20 years, however, the traditional system has been augmented by pastoral associations (PAs) which are governmental administrative and political units; PAs have become involved in resource-use decision making, and contestation of authority between traditional and PA institutions has led to problems (Homann et al. 2008). Resource fragmentation problems now occur at all spatial levels, while another challenge at the most local level is ultimately an inability to control stocking rate, and hence forage utilization (Tezera et al., unpublished data). Annual rates of forage utilization among four PAs have been estimated to vary from 70 to 90 % — one outcome of such heavy use is soil erosion and bush encroachment, as previously noted. Coping with such challenges has largely been left up to the traditional leadership of the Borana pastoralists in concert with decision makers from government agencies, but it is evident that population pressure, emergence of very wealthy pastoralists, and reduced adherence to traditional pasture-use norms have undermined grazing regulation in recent decades.

Progress, however, is being made on several fronts. This includes (1) finalization of a land-use plan by government that should limit cultivation on grazing lands; (2) efforts to restore traditional grazing access beginning at the *dedha* scale of resolution; and (3) workshops to address the need to reduce stocking rates and limit forage utilization at the local level to help arrest rangeland degradation. Thus, there are opportunities to assist pastoralists to update traditional grazing rules and regulations to better accommodate the new reality of higher demand on natural resources. Increasing scarcity of surface water and forage is recognized by pastoralists as a critical problem that merits new, collaborative approaches for problem solving (Coppock et al. 2014).

Sahel. In the Sahel, pastoralists exploit the high variability of rainfall across space and time via highly mobile livestock.[27] Population growth, land-use change and resulting land fragmentation, and unfavorable policies, however, have contributed obstacles for herd mobility at local and regional scales. Traditional pastoral institutions facilitated access to constantly shifting patches of natural resources (de Bruijn and van Djik 1995; Niamir-Fuller 1999). Colonial and postcolonial land policies, however, ignored pastoral rights and this, coupled with socially malleable rights to pastoral resources among pastoral groups, has made pastoral resources vulnerable to competing land uses, especially with the rapid growth of human populations since the 1950s (Guengant et al. 2002).

Since the 1990s, programs to decentralize natural-resource management authority, as well as recurrent attempts to privatize land, have contributed to the enclosure of key pastoral resources (Marty 1993). While these changes have generally not caused widespread sedentarization of pastoralists and their livestock, they have reduced mobility, altered movement patterns, reduced livestock access to grazing,

---

[27] The movements of pastoral livestock are organized in response to long-established seasonal and spatial criteria that also allow for local tactical adjustments (Hiernaux, personal observation).

water, and mineral resources, and increased conflicts (Turner et al. 2014). There is growing recognition within policy circles of the importance of the livestock sector and livestock mobility, especially in the context of climate change (Bonnet and Hérault 2011). This has led to national (République du Mali 2001; Wabnitz 2006) and international (CRCM 2013) laws, policies, and agreements in support of pastoral livelihoods and tenure rights. These initiatives are important, but they alone cannot lead to a significant transformation of pastoral rights. Significant political and institutional questions have to also be addressed for effective implementation (Brottem 2013).

In general, local understanding and enforcement of existing laws are weak. Some newly organized formal community-based rangeland management groups have creatively circumvented these challenges by working closely with their local government to obtain use rights over defined pasture areas, create their own management plans, and then lobby local government to pass an ordinance or decree that gives certain elements of their plan the force of law (Fernández-Giménez et al. 2012; Upton 2008, 2009). For example, if a group of herders wants to rest a certain pasture area for a season, they ask the local government to issue an edict forbidding grazing in that area for the rest of the season, with violations punishable by a fine.[28]

Mongolia. The pasturelands of Mongolia are state property used in common by the herders within a given administrative district. Privatization of pasture is unconstitutional, but herders may obtain long-term private leases on nomadic winter and spring campsites (Fernández-Giménez and Batbuyan 2004; Upton 2009). Mongolia's Law on Land contains provisions related to pastureland tenure and management, devolving most decisions and authority to regulate stocking rates and seasonal movements to the local level (Fernández-Giménez and Batbuyan 2004). Under Mongolian law, organized groups of herders may obtain collective use agreements for defined areas of pasture, but "use" does not denote the right to exclude other potential users. Pastureland "possession" would confer exclusive rights to a given pasture area, but pasture possession is not authorized under the current law.Mobility has been a hallmark of Mongolian pastoralism for centuries and continues to be an important strategy today. Mobility patterns historically have varied widely across the country, depending largely on local topography and climate. During the collective era, the collectives allocated pasture to their member herders and both regulated seasonal movements and provided transportation for moves (Fernández-Giménez 1999). In addition, collectives arranged for long-distance moves in weather disasters such as severe winter storms. Following privatization of livestock and the dismantling of collectives, no formal institutions have filled the role of allocating pasture and enforcing seasonal mobility, although the current law

---

[28] Over the last two decades, multiple proposals have been made for a national pastureland law, but none has reached a vote in parliament. Debates on the proposed laws continue over the central tension between authorizing formal possession rights in pasture for groups of herders or individuals with the risk of reducing access and flexibility for some and keeping the current vague, but flexible and locally adaptable, system in place.

gives this authority to local governments. As a result, mobility has decreased in many areas, and conflicts over pastures have increased, together with unsustainable season-long grazing in some areas (Fernández-Giménez 2002).

When disasters such as prolonged drought or severe winter weather strike, many herders undertake long-distance movements to escape these conditions and create pasture shortages at their destination (Fernández-Giménez et al. 2012; Upton 2012). Flexibility to make these movements in response to disaster is essential to the sustainability of this system, but lack of coordination and cross-jurisdictional governance mechanisms create hardships for herders in the host communities as well as exacerbate pasture degradation problems.

Altiplano. In the Altiplano, the loss of labor for herding and irrigating the *bofedales* has been the major contemporary factor in reducing the mobility of range livestock (Turin and Valdivia 2011). Privatization of key resources has also occurred. One example is fencing off parcels of *bofedale* that were formerly communal grazing sites for the exclusive use of a few alpaca producers (Buttolph and Coppock 2001). Ninety percent of Andean rangelands, however, remain as public land accessed by rural communities, but there are persistent challenges of landscape fragmentation and privatization due to human population encroachment, especially in Peru. This can lead to local limits on herd mobility and thus rangeland degradation (Turin, unpublished data). In some cases highly dynamic land tenure systems have been discovered in Bolivian agro-pastoral settings. For example, pastures used for sheep grazing in the vicinity of San José Llanga are traditionally regarded as private-access sites in higher precipitation years, while they become communal sites in drier years (Coppock et al. 2001).

Mexico. In northern Mexico, a gradual process of livestock commercialization and land-use fragmentation has also reduced livestock mobility. In the past there was a large-scale transhumance of livestock in the region, but today this has been replaced with local movements of beef cattle from private rangelands to private croplands at the end of the growing season to harvest crop residues (Martínez-Peña 2012). As previously noted, goat and sheep herding by *ejidatarios* occurs on desert-scrub *(ejido)* land; animals are trekked on long-distance orbits. Following recent agrarian reforms, some *ejido* lands, however, are now shared among a few *ejidatarios* having certain privileges to own land, and this has caused land to become fragmented and fenced. These land users, however, are forced to establish earthen ponds to supply water for cattle, fundamentally altering hydrological processes at the landscape scale (Huber-Sannwald et al. 2012).

The *ejido* system comprises roughly 54 % of Mexican rangelands today and is the most common form of land tenure (Arredondo and Huber-Sannwald 2011). The *ejido* concept originated in the Mexican Constitution of 1917 as a postrevolutionary, communal land-management institution. *Ejido* governance is idiosyncratic, depends on the local biophysical and socioeconomic context, and is largely independent of federal funding support. For decades e*jidos* were subject to strict regulations specifying that only *ejidatarios* can use the land for agrarian purposes and that the land cannot be sold, rented, or mortgaged. Changes have occurred in the *ejido* system over the past 20 years, however.

In 1992, a Constitutional Agrarian Land Reform was passed allowing *ejidatarios* to assume full domain over their communal land with the right to divide and sell it (DGAHML 2003). One motivation for this reform was to introduce a mechanism for wealthy ranchers to appropriate communal land, introduce technology and new livestock breeds, and enhance livestock production to improve Mexico's position in global markets (Martínez-Peña 2012). However, adoption of this neoliberal policy has not translated into the expected, massive privatization of *ejido* lands. Possible reasons include the following: (1) that some *ejidal* assemblies preferred to maintain communal lands as such (DGAHML 2003), or (2) because communal lands were unattractive to wealthy ranchers because of inadequate resources or remoteness (Manzano et al. 2000). The risk of privatization and fragmentation of prime *ejido* lands still exists, however, and this risk is primarily related to the potential displacement of residents who could otherwise make a sustainable living on the landscape. In the course of the "private revolutions" during 1937–2007, private ranchers have managed to consolidate and expand their ownership of land by giving cattle to their wives or children. Private land holdings include anything from small (300–1000 ha) to mega (>10,000 ha) ranches.

17.3.2 Sustainable Rural Livelihoods

Summary. Sustainable livelihood concepts were first introduced nearly 30 years ago. Significant advances in defining terms and integrating ideas concerning sustainable livelihoods into research and development projects are more recent, however. Livelihoods are founded on production strategies. A livelihood is sustainable when it allows an individual or a household to rebound from an economic or ecological shock. Various resources are drawn upon in a recovery process. Increasing and diversifying income and assets are core processes that help livelihoods become more sustainable in risky environments. Pastoral societies have also traditionally had safety nets linking relatives and neighbors where people who have suffered from a calamity can receive animals to restock themselves. Rangeland dwellers such as pastoralists or agro-pastoralists are the epitome of risk managers who have survived droughts and economic downturns via traditional tactics. These tactics have included a high reliance on mobile livestock as well as participation in social networks that provided safety nets and allowed for the opportunistic and reciprocal use of communal resources. While livestock production remains a core strategy in all of our case-study sites, it has become increasingly difficult for most pastoralists to survive from livestock alone. This is due to increases in the numbers of people and reductions in the numbers of livestock per family due to declining supplies of forage and other natural resources; the forage base simply cannot keep up with the growing numbers of animals required to sustain an expanding human population. As a result, many research and development efforts today emphasize ways to build capacity and assist rangeland dwellers to diversify their livelihoods away from too heavy a reliance on livestock. This often involves participatory activities.

Diversified livelihoods can pose their own risks and are not a panacea for all, but they offer a chance for more people to be hopeful and pursue some prosperity in a rapidly changing world.

17.3.2.1 Pivotal Issues and Conceptual Advances

Sustainable livelihood concepts first began to be addressed as part of mainstream research and development in the 1980s. They arose from a need to better understand interrelationships of people with natural resources and to help chart pathways for poverty reduction and social justice in the developing world (Martens 2006). In the 1990s, conceptual progress was made in terms of articulating components of sustainable livelihood frameworks (Chambers and Conway 1992; Scoones 1998; Singh and Gillman 1999). A livelihood is sustainable when it allows an individual or a household to recover from an internal or external shock; the ability to recover from a shock is referred to as resilience. Livelihoods draw on various capital assets— social, human, financial, physical, and natural—as people pursue production strategies that involve intensification, extensification, or diversification.[29] Increasing and diversifying income and assets are thus means to promote sustainable livelihoods.

Pastoralists and agro-pastoralists provide useful models for the study of sustainable livelihoods because they are the ultimate risk managers. Subjected to the impacts and unpredictable occurrence of drought, disease, social conflicts, and economic crises, pastoralists have traditionally managed risk via several avenues (Coppock 1994). These have included (1) opportunistic exploitation of vast landscapes by mobile livestock; (2) reliance on a diverse assortment of livestock species; (3) opportunistic engagement in cereal cultivation or petty trade; (4) membership in complex social-reciprocity networks that offer safety nets during times of stress as well as a means to restock following herd collapses; and (5) use of grazing reserves or fodder storage. The problem, however, is that steady human population growth, annexation of key pastoral resources, assertion of government control over national and regional borders, recurrent drought, or other extreme weather events have undermined such coping mechanisms.

In some pastoral systems, pastoralists periodically "drop out" of the pastoral economy—often following drought or other weather-induced disasters—because they can no longer support themselves via traditional means. These former herders reside in settlements or urban areas and survive via relief food, petty trade, or other occupations. This often happens to pastoral women (Holden et al. 1991). In some situations dropouts may eventually return to pastoralism, but in others they make a

---

[29] Intensification characterizes high-input systems and involves investing capital, labor, and other resources in land or animals to increase per unit productivity. Extensification, in contrast, characterizes low-input systems and involves expanding the land base or animal numbers to increase overall operational output. Diversification is a process of expanding the types of enterprises undertaken by an operation, leading to more variation in income sources or assets.

permanent exit.[30] While sedentarization of former pastoralists is typically maligned as negative for environmental management, human nutrition, and maintenance of pastoral culture (Fratkin and Roth 2005), a growing pool of settled people can offer opportunities to educate people and build capacity for people to engage in collective action to diversify livelihoods and pursue non-pastoral lifestyles (Coppock and Desta 2013). There are also cases where households pursue mixed strategies by having some members maintain livestock in the pastoral sector with others engaged in urban economies.

17.3.2.2   What Has Been Observed?

Afghanistan. In Afghanistan, few pastoralists are able to depend solely on livestock production for a livelihood. Livelihood diversification is the key to success, particularly on a landscape predicted to experience drought on a more frequent basis in the future. Women should be encouraged to contribute to livelihood diversification in areas acceptable to the family (i.e., carpet weaving, dairy production, small animal husbandry).

One conceptual advance that has helped change agents in Afghanistan to think more clearly about their work has been to encourage communities to communicate and self-advocate more effectively. This is because management of any natural resource has little chance of being successful without including the people most impacted by the process. Until change agents can fully engage local people and understand their challenges and aspirations, imposing top-down management strategies will most likely fail. The typical approach in Afghanistan has been top down; donors support government ministries to develop rural strategies without any community participation. This is justified by arguments that the rural poor lack the education or wisdom to "know what they want" and because in most instances community involvement takes longer than the donor or government is willing to wait. Successful projects, in contrast, are those where a healthy dialogue is developed between the change agents and project beneficiaries, and both are willing to make the time needed for meaningful interaction. The vital role of the change agent then becomes assisting communities to better articulate their challenges, needs, and aspirations, providing professional guidance and "reality checks," and making the commitment to support a long-term process. Once there is a commitment and community members become skilled communicators and self-advocates, they will then have the ability to resolve, or at least better manage, conflicts with neighbors and effectively engage government agencies.[31]

---

[30] The specter of periodic waves of permanent dropouts occurring within dynamic production environments is universal for farmers, ranchers, and pastoralists in developed as well as developing countries (Thurow, personal communication).

[31] A blend of bottom-up and top-down is therefore the prescription; despite the importance of hearing community voices, people cannot ask for things that they cannot imagine—hence the role of outside knowledge and demonstration activities. Fruitful engagement is also greatly promoted by simply having more time for change agents and communities to interact and make the correct decisions. Time constraints are ultimately imposed by the short-term funding cycles that characterize most development projects today.

Ethiopia. In southern Ethiopia, traditional pastoralism is no longer a sustainable livelihood for the vast majority of households given sharp declines in the number of livestock units per capita, as previously noted. Another trend is a steady increase in wealth concentration—fewer people own more of the aggregate cattle herd, for example (Coppock et al. 2014; Desta and Coppock 2004). Over 40 % of the population now have only a few head of livestock per household and are caught in a perpetual poverty trap.[32] Traditional safety nets have been tattered, and food aid has become pervasive in much of the system because milk production can no longer meet the basic needs for a growing segment of the population. One option to improve the welfare of the poor is to encourage livelihood diversification via microfinance and collective action activities–programs that can inspire women, in particular (Coppock et al. 2011).

Sahel. In the Sahel, key features of livelihood strategies in this resource-poor and highly variable region are diversification, mobility of production, and wealth stores (Agrawal 2008). As described above, livestock, as mobile stores of wealth which are not vulnerable to local deficits of rainfall (unlike crop agriculture), figure prominently in the livelihood strategies of all rural people whether their identity is tied to farming, commerce, fishing, or livestock husbandry (Turner et al. 2014). Thirty years of recurrent drought have increased the diversification of economic activities within rural households. Farmers will own livestock and livestock producers will farm and all, if they are able, will send family members on a seasonal or semipermanent basis to work in cities, mines, and plantations to the south (de Bruijn and van Djik 1995; Turner et al. 2014).

Differential vulnerabilities to economic and climatic shocks are observed not only between families but within families, with women being particularly vulnerable (Creevey 1986; Gray and Kevane 1999; Turner 2000). Livestock play an important role in addressing such vulnerabilities as individually owned wealth stores, owned separately by women and men, which have an additional benefit of being not easily fungible to cash and therefore less vulnerable to being dissipated by daily demands for small amounts of cash from family and friends (Turner 2000).[33]

While case studies illustrate the reliance on livestock by rural families, they also point to the limitations of livestock ownership in buffering household income in response to episodic climatic and economic shocks. Livestock prices predictably decline in relation to grain prices during periods of food shortage (e.g., the hunger season at the end of the rainy season prior to the next harvest) and decline precipitously, in a less predictable manner, during drought (Fafchamps and Gavian 1997; Watts 1983). As a result, the effective wealth stored in livestock depends on when they are sold in relation to annual and drought cycles of grain shortage.

---

[32] A poverty trap is a spiraling mechanism which forces people to remain poor. The mechanism is binding such that poor people cannot escape it; it is often caused by a lack of capital or credit for people who also have no prospects for employment.

[33] Such problems are also exacerbated by inaccessible or unreliable financial institutions in the rangelands of developing nations. Investing in livestock thus becomes the most viable alternative, despite the sometimes high risk of animal death losses.

Modeling work based on retrospective surveys in the Dantiandou study site (Lesnoff et al. 2012) demonstrates the limited potential for livestock wealth recovery among sampled families. This is not only due to recurrent drought over the past 30 years, but also due to other extreme events (i.e., livestock disease epidemics) and the generally low productivity of livestock husbandry due to the limited access to pastoral resources (land-use change, civil insecurity, etc.) aggravated by the poor access to and high cost of inputs (i.e., veterinary drugs, feed supplements).

Mongolia. In Mongolia, following the transition to a market economy, poverty in rural areas increased dramatically, and the gap between rich and poor widened (Griffin 2003; Nixson and Walters 2006). Many development programs have sought to address this through a variety of income-generation and diversification measures; however, progress has been slow due to a number of limitations to livestock improvement and marketing (see section below). The most common approaches to income diversification are small-scale vegetable growing, value-added processing such as felt making and handicraft production, and small-scale enterprises such as shop-keeping, trading, driving, or collecting rent for properties in nearby settlements. Felt making has a limited market that is rapidly saturated. Artisanal mining, primarily for gold, is the most important source of income diversification in recent years. In a recent household survey in 36 districts, respondents in 10 reported income from mining. In these districts from 10% to over 50% of surveyed households obtained some income from mining (MRRP, unpublished data). Over the long term, payment for ecosystem services and sustainability certification seem to be among the most promising approaches that directly build on livestock-based livelihoods, but these are still in their infancy.

Altiplano. For the Altiplano, multiple factors explain the diversity of livelihoods and livelihood strategies. The roles of livestock in livelihoods change with elevation, geography, climate, and distance to markets (Valdivia et al. 2010). An example of a high degree of livelihood diversification is provided by the agro-pastoralists of San José Llanga in the central Altiplano (Valdivia 2001). Such households are typified by activities such as growing potatoes, oats, native tubers, and *quinoa*, as well as raising sheep for meat and wool and cattle for meat and dairy. Off-farm employment comprises additional activities. Men and women divide labor responsibilities on the farm, with men assuming the more demanding physical activities (i.e., cattle breeding and planting and harvesting crops). Women are in charge of herding and milking animals as well as selecting seeds and making planting decisions. Women also oversee cultural events. The process of livelihood diversification and market integration at San José was initiated by investments in infrastructure (transportation and irrigation), extension of production technology (i.e., alfalfa, crossbred dairy cows), and government decentralization policies that facilitated market development (see next section). Dairying has contributed to managing risks of the overall economic portfolio because it provides access to cash on a monthly basis. When there is a drought or frost and a potato crop is lost, for example, farmers can use the earning from the dairy cows to obtain new seeds and plant. When barley and oats are lost to drought, the stubble still provides a feed source for the livestock (Valdivia 2004).

An example of less diversified pastoralism is provided by the Apopata community at high elevations in the Peruvian northern Altiplano (Turin, unpublished data).

Primary livestock products among pure pastoralists are alpaca meat and fiber; non-livestock endeavors include off-farm employment, handicraft production, fishing, and harvesting wood of shrubs for fuel. Price incentives dictate the degree of market involvements. Men and women have complementary roles in alpaca pastoralism, as men handle the shearing, breeding, and marketing as well as provide oversight of pasture irrigation systems. Women herd and provide health care for animals and cover domestic chores. When men migrate in search of work, however, women assume the tasks typically undertaken by men (Turin et al. 2010; Valdivia et al. 2003). In both Andean pastoralism and agro-pastoralism, traditional social networks have been very important to help manage economic and climate-related risks. These networks have been supplemented of late with new government partnerships whereby herders can direct public investments to best serve local needs. For example, in the Peruvian municipality of Mazocruz, pastoralists have led the way on decisions to use public monies to build livestock shelters (*cobertizos*) that markedly reduce losses of camelids during periods of extreme cold (Turin, unpublished data).

Mexico. In northern Mexico, the main livelihood of people living in the drylands is ranching. Some diversify with mining activities, wage labor in nearby towns, seasonal migration, and remittances from migrants. The latter is a trend of the younger generation who seek alternative lifestyles. Subsistence farming with extensive and semi-intensive livestock production (i.e., cattle, sheep, and goats) and rain-fed agriculture are the main economic activities in the region (INEGI 2002). For the last 3 to 4 decades, government assistance programs have opened alternative production opportunities for *ejidatarios*. Seed money or crop seedlings are provided to engage in planting programs of plantations of opuntia for vegetable and tuna fruit production for local markets. Also, in some villages professional music groups (Mariachi, Banda, Trios) have formed. In matorral-dominated rangelands, the government has locally supported the collection and processing of "*ixtle*," a strong, high-quality fiber extracted from the leaves of wild populations of *Agave lechuguilla*, for the brush-making industry. In central and northern Mexico, some *ejidatarios* collect wild plants of *Agave salmiana* and *Dasylirion wheeleri* to supply mezcal and sotol distilleries. Also, to reduce illegal trade in rare cacti, the government has helped install greenhouses for commercial cactus production. More recently, leasing *ejido* land to foreign wind farm companies has emerged as an alternative income generator for some *ejidatarios*.

17.3.3 Livestock Development and Marketing

Summary. Livestock development can be defined in several ways, including the process of sustaining or increasing livestock outputs per capita, or in terms of addressing multiple societal goals to improve the livelihoods of livestock producers whose ranks tend to be dominated by the rural poor. The developing world is currently undergoing a "livestock revolution," whereby increasing consumer incomes and rates of urbanization spur a demand-driven surge for high-value animal products such as milk and meat. This supply can be met, in part, via the commercialization of subsistence agriculture. Rural livestock producers in both farming and pastoral systems should benefit from this trend, but pastoralists must overcome more obstacles to do so simply because they tend to be more isolated from major centers of consumption. A transition from subsistence to commercial practices for pastoralists can be painful as it yields winners and losers; it will prominently involve overcoming high transaction costs for livestock marketing. Barriers to commercialization include the need to transform the quality and types of animals and animal products produced, as well as the need to upgrade transportation infrastructure, communication networks, and other marketing infrastructure such as slaughter facilities and holding grounds. The need to not have a region classified as a disease quarantine zone preventing export is also vital. The boom-and-bust pattern of herd population dynamics often observed on rangelands also conflicts with the need for stability in market supply required by modern animal-processing industries. An overview of our six case-study sites reveals that most pastoral herds remain dominated by indigenous breeds with a low per head productivity, but these animals are durable when facing endemic disease, poor nutrition, and harsh environmental conditions. There is evidence of increased market participation in most of our cases, but levels of sustained livestock market development vary greatly. Use of mobile phones to transmit market information has become ubiquitous in the past decade—this is a major success story. The best examples of demand-driven range livestock development, with attendant changes in livestock policies, the upgrading of indigenous livestock via crossbreeding, and use of increased production inputs, may be found in two of the case-study sites in the upper tier. These are (1) the expansion of peri-urban dairy and wool production among agro-pastoralists in the Bolivian Altiplano and (2) the rapid growth of commercial beef production among ranchers in northern Mexico.

17.3.3.1 Pivotal Issues and Conceptual Advances

"Livestock development" is defined in several ways. In its simplest form, it can be described as the maintenance or enhancement of livestock output, preferably expressed on a per capita rather than per hectare basis (Jahnke 1982). More recently, livestock development has been defined as a process of addressing interrelated socioeconomic, environmental, and productivity goals that would reduce poverty among people who raise livestock for a living.[34] Others have noted contributions of livestock to food security and sustainable development (Sansoucy et al. 1995).

Worldwide, it is clear that as rates of urbanization accelerate and societies gain wealth, one of the first things to follow is an increased demand for animal-based foods. With more income, people reduce their intake of less expensive carbohydrates and increase intake of more expensive animal protein, including milk and meat (Delgado 2003). This relationship is the linchpin of a projected "livestock revolution" for the early twenty-first century as populations in developing countries

---

[34] http://www.fao.org/ag/againfo/programmes/en/pplpi/docarc/rep-ipalp_ldg.pdf

become more urbanized and economies expand. There should be a concomitant opportunity for livestock producers to meet this growing demand (Delgado et al. 2001). A "traditionalist" might contest the assertion that pastoralists need to become more commercialized, given the downside risks of market participation. Yet it is clear today that most rural agriculturalists—whether they be smallholder farmers or pastoralists—have to commercialize and join the cash economy given that they can no longer survive in a pure subsistence mode (Fitzhugh, personal communication).[35]

Who has, or will, benefit from market growth varies, however. Livestock producers who inhabit remote rangeland areas of the developing world are expected to benefit less than other livestock or dairy producers, simply because the socioeconomic and geographic isolation inherent for most rangelands increases the transaction costs[36] for efficient market participation. There are many barriers for livestock marketing transactions in the rangelands for both buyers and sellers (Holloway et al. 2000; McPeak and Little 2006; Sandford 1983; Zant 2013). Barriers include things like poor roads, lack of transportation, inadequate holding grounds or slaughter facilities, inefficient communication on prices, limited access to banking services, tax policies creating disincentives for cross-border trade, overabundance of middlemen in marketing chains, illiteracy, and lack of legal protections. The objective of much research in applied economics on rangelands today is discovering ways to help producers overcome or mitigate high transaction costs (Valdivia, personal communication). Mitigation of transaction costs increases market efficiencies and promotes market integration (Valdivia 2004).

There are other barriers that slow livestock commercialization on rangelands that are not immediately related to market transactions per se. One is related to traditional values and needs among pastoral people. For example, the sex and age composition of pastoral livestock holdings have long been shaped by cultural or risk management goals rather than commercialization goals (Coppock 1994; Fafchamps and Quisumbing 2005; Valdivia 2004). The upshot is that the mix of animals supplied by pastoralists in terms of age, size, or body condition can be suboptimal when compared to urban consumer preferences driving demand (Desta et al. 2006).[37] In addition, the dynamic nature of livestock inventory in pastoral regions—previously

---

[35] Most pastoralists have been integrated into a market economy to some degree for decades. In reality there is a continuum between pure subsistence pastoralism (which is rare today) and pure commercial production (also rare today). Most pastoralists in the developing world will continue to produce for both self-consumption and markets, but if market signals were stronger in rewarding quality over quantity, this could alter incentive structures and transform pastoral livestock systems. And market integration for producers has both positive and negative dimensions, positive in terms of income generation and price stabilization, yet negative in terms of adding risk exposure.

[36] Transaction costs include the added time, effort, and expense associated with making an economic exchange (i.e., selling or buying).

[37] For meat, consumers often demand animals of a certain age, sex, and size. If leather making is a consideration, consumers often demand unblemished hides and skins. For milk, consumers should demand undiluted, hygienic products. Slaughterhouses may also demand uniform size-classes for processing as well as demand disease-free animals. In the latter case, disease imported to a slaughter facility by range stock could lead to a shutdown if discovered by health inspectors.

noted as often weather-related, boom-and-bust cycles—creates big problems for processing industries further up the value chain that require stability of product flows to stay in business.

Another key issue in livestock development concerns livestock breeding. The rangelands of the developing world have long been populated by indigenous breeds of multipurpose cattle and small ruminants. Although such breeds are not very productive on a per head basis, they are nonetheless well adapted to survive in harsh environments that include shortages of feed and water, disease challenges, and temperature extremes. Improved (e.g., European or exotic) breeds often have the opposite attributes, namely higher productivity but increased vulnerability to stressful environmental circumstances (Coppock 1994; Kosgy et al. 2006). Crossbreeding indigenous with improved stock has occurred on a limited basis, and the typical outcome has been that crossbred animals cannot be sustained under indigenous rangeland production conditions. In the past 20 years the mantra has shifted to goals that include the conservation of breeds indigenous to rangelands and other environments.

Overall, it is fair to say that livestock development is like a three-legged stool— you cannot put up one leg, then another, and then another and expect the stool to stand in the process. The legs of the stool must come together simultaneously. Why should a private investor attempt to improve market infrastructure if the supply is dominated by underweight or diseased livestock? Why should livestock production improve if there is no market to encourage the effort and offset the extra costs? The problem is that pulling off such coordinated projects is very expensive and time consuming, something most donor agencies don't have the mechanisms in place to provide. As a result, in most cases, livestock development occurs in a series of stutter steps, as will be illustrated.

17.3.3.2   What Has Been Observed?

Afghanistan. Commercial livestock production among the Kuchi remains constrained by the lack of financial inputs and chronic political instability. Livestock marketing, however, is vital to helping the pastoralists meet their annual needs for cash income (Schloeder and Jacobs 2010). The livestock holdings of the Kuchi are dominated by five breeds of indigenous sheep (i.e., *Qaraqul, Turki, Qaragh, Arabi,* and *Baluchi*) and two breeds of goats. The species and breed that predominate in any given region is a function of ecological constraints, the most important being disease resistance, water availability, and forage quality.

Recent trends indicate that—as animal numbers dwindle—more must be sold to meet basic family needs.[38] Sales can start as early as May but are most common in the fall when market prices are highest (Schloeder and Jacobs 2010). Sales are also common during drought when there is a need to purchase food and other household

---

[38] Ten sheep per family member is reported as a minimum by Thompson (2007).

essentials, even as animal prices plummet due to the influx of animals to markets. In either event, the increased need to sell will ultimately decimate herds over time. It will also result in an ever-greater dependency of the Afghan nation on meat imports.

Access to live-animal market information is critical for informed decision making by the Kuchi. Currently, the Kuchi rely on family members and other sellers attending markets for current prices (Schloeder and Jacobs 2010). Final sales price is the result of negotiations between the buyer and seller, with price a function of animal type, breed, age, sex, and body condition. Mobile phones have improved the ability to acquire current market information except in the more remote places where summer grazing occurs and phone coverage is limited. Additionally, producers are often unaware of prices in terminal markets where animals are often resold at much higher prices by traders. A Livestock Market Information System (LMIS) was established in 2008 by Texas A&M University, in partnership with the Afghan Government, to overcome this challenge. Apathy on the part of government partners, however, in ensuring observer attendance on market days and collecting reliable price data undermined the program, despite high interest from producers and traders (Schloeder and Jacobs 2010).

Animals brought to market in Afghanistan are often underweight and diseased. Most of these problems can be overcome with quality medicines and a 1-month intensive feeding program, which adds considerable value per animal (Schloeder and Jacobs 2010). Unfortunately, the Kuchi lack reliable access to effective veterinary supplies. Additionally, pastoralists are reluctant to pay for such supplies and services that were once freely provided by the government, particularly as herd sizes dwindle. Supplemental forage (e.g., alfalfa) is also a constraint due to limited supplies and high prices. Pastoralists throughout Afghanistan conduct supplemental feeding during winter months when conditions can hinder extensive grazing. Traditional supplements largely consist of local native plants that are cut and carried. In many instances, however, these materials are of low nutritive value and may even be toxic (Jacobs and Schloeder 2012).

In summary, these are all challenges that greatly undermine the prospects for commercialized livestock development. Finding a solution to one problem will not solve the rest and—more importantly—any one intervention is probably doomed to fail unless other, interrelated problems are unaddressed at the same time. Success would require many interventions and forces working in a synergistic fashion.

Ethiopia. The Borana pastoralists of southern Ethiopia largely produce indigenous breeds of dual-purpose cattle (the Boran), Somali hair sheep, East African goats, and dromedary camels (Coppock 1994). All of these breeds are well adapted to the local semiarid environments and all are increasingly in demand for domestic and export markets (Desta et al. 2006). Exotic Dorper sheep and Boer goats are being considered by government researchers as candidates for crossbreeding programs to upgrade the size and productivity of indigenous stock, but it is doubtful that local management systems can sustain improved crossbreeds (Coppock, personal observation). In terms of cattle, there has been a long-term concern by government that the genotype of the Boran has been gradually undermined by crossbreeding with inferior stock from the adjacent southern highlands. This occurs when pastoralists actively trade for breeding cows when they build their herds during post-drought recovery periods (Coppock 1994). Government researchers endeavor to have Boran breeding bulls on government ranches to provide breeding capacity to help maintain local gene pools, but the strategy has not been widely implemented (Coppock, personal observation). A process of herd diversification is also being observed where camels are increasing relative to cattle in some regions (Coppock, unpublished data). This probably reflects adaptation to climate change, where camels are better suited to warmer, drier weather, an increased market value for camels, and the ability of camels to supply milk during dry periods when compared to cattle. Camel milk, however, is inferior to cow milk for making butter (Coppock 1994).

As in Afghanistan, the trend for most pastoralists in southern Ethiopia is also to sell a higher proportion of their livestock inventory to meet cash needs; this is due to a rising human population relative to livestock and increasing demand for non-pastoral goods. In the past decade there has been some improvement in value chain development with respect to the finishing of range-bred stock, but this varies greatly with terminal market destination (Desta et al. 2006). Livestock health interventions remain as a significant constraint; outbreaks of highly contagious diseases in the Greater Horn can still result in export trade bans. While livestock marketing and animal health remain as significant challenges in southern Ethiopia, evidence from recent participatory rural appraisals indicates that neither ranks as highly as the need for improved water access or expanded human services (Coppock et al. 2014).

Southern Ethiopia is indeed now better connected to global livestock markets, and there has been a surge in pastoral livestock marketing in the past decade (Desta et al. 2006). Communications have improved as a result of rapid adoption of mobile phone technology (Coppock, personal observation). Producer and marketing cooperatives have been recently formed. New questions have emerged, however, namely which segment of pastoral society, the wealthy minority or other strata, is best poised to benefit from further improvements in livestock marketing. And if wealth is further concentrated via marketing profits, what does this imply for concentrating access to communal forage and water resources that may be increasingly used to produce more animals for commercial, rather than subsistence, purposes?

Sahel. In the Sahel, active marketing of livestock is long-standing with an emphasis by herd managers to develop more breeding herds and flocks, with sales dominated by male cattle, sheep, and goats. The Sahel harbors a large diversity of livestock breeds. Cattle are dominated by zebu breeds such as the Gobra, Bororo, Azaouak, and White Fulani as well as a few trypanosomiasis-resistant taurine breeds such as the Kuri and Ndama. Most sheep are thin-tailed hair breeds with a couple wool-bearing breeds (Macina, Arara). Goats become more prevalent in general with increased aridity, although dwarf goats occur in subhumid locations. Crossbreeding efforts to improve indigenous stock were initiated during the colonial era, but crossbred animals have not been sustained in most of the Sahelian region. There has been some recent success, however, with crossbreeding programs in peri-urban dairy systems (Hiernaux, personal observation).

Offtake fractions of herds are higher among poorer households with the prices of animals shaped by the timing of sale in relation to season and the drought cycle as well as location of the sale with respect to competitive markets (Turner and Williams 2002). The shift in livestock ownership away from livestock specialists, as well as the expansion of the urban markets, has led to a reorientation of marketing toward the sales of small ruminants (especially sheep) for meat; this particularly occurs on Islamic holidays (Manoli et al. 2014). For these markets, there is evidence for the stratification of livestock production with the raising of small ruminants in the more arid north with subsequent fattening to the south and in peri-urban areas (Amanor 1995). Mobile phones have played an important role in facilitating the trekking of livestock and the distribution of market information (i.e., price and sales volume). Development initiatives have generally focused on the promotion of livestock fattening, livestock market infrastructure, and expansion of the availability of market information via mobile phones.

Mongolia. For Mongolia during the collective era, livestock husbandry was professionalized, and substantial effort was invested in breeding programs and livestock improvement. There were both positive and negative elements to this. One of the lessons learned was that local breeds were often best adapted to Mongolia's harsh conditions, and attempts to "improve" them with more productive animals from western European stock were largely unsuccessful, especially considering species like dairy cattle.

Following the transition to a market economy, combined with growing concerns about increasing livestock populations and overgrazing, the emphasis has been on incentives for herders to increase animal quality and decrease the quantity (MMAI 2010; World Bank 2013a). In the early 1990s these proposals were largely rejected by herders, but since the most recent winter weather catastrophe in 2009–2010, many herders appear genuinely interested in this approach and much more receptive to making these types of changes. Under the collective system all livestock that were not for personal consumption belonged to the state and were part of the government procurement apparatus.

With the transition out of collectives, herders were faced for the first time with the challenge of selling their livestock and livestock products on an open market. Initially, many opportunistic itinerant traders took advantage of this situation, and terms of trade for isolated rural herders were often exploitive. With time, herders became savvy, and many began to take their products to market directly, either to the provincial markets or the capital city; this allowed producers to obtain a better price for their products and also to purchase needed supplies more cheaply. However, this is not possible for all herders.

All Mongolian herders sell live animals for meat, primarily sheep and goats. Most sales and slaughter take place in the fall. A smaller number of herders who are located near settlements and urban centers sell dairy products in local markets, primarily during the summer. Fueled by international demand, cashmere goat hair has become the cash crop of Mongolian herders, and herd compositions throughout Mongolia reflect the demand for this high-value livestock product, which is combed and sold in the spring.

Both development organizations and the Mongolian government have attempted to encourage herders to form marketing cooperatives in order to make marketing more efficient and profitable, but many of these efforts are still in their infancy (MMAI 2010; World Bank 2013a). In addition, there are many barriers to marketing livestock products that perish quickly. At the national scale, Mongolia's slaughter-houses do not meet standards for international export to most countries, limiting the potential for foreign markets (Zoljargal 2013).

Altiplano. For the Altiplano, the traditional livestock at higher elevations have been comprised of native camelid species (i.e., llama, alpaca, and vicuña). Llamas have been traditionally used for portage and meat, while alpaca and vicuña have been primarily used for high-quality fiber production. Dual-purpose cattle and hair sheep at lower elevations in the rangelands of rural areas are dominated by unim-proved *criollo* breeds descended from stock brought by the Spanish Conquistadors.

Market expansion in the vicinity of the large urban centers of highland Bolivia has been important in providing incentives for agro-pastoralists, in particular, to crossbreed cattle and sheep to increase milk and wool production, respectively (Markowitz and Valdivia 2001). Crossbreeding programs of *criollo* cattle with Holsteins and *criollo* sheep with Corriedale, Targhee, and Merino have now been sustained for over 20 years. A particularly strong demand for milk has been a major driver for rural develop-ment (Valdivia 2004). Public–private partnerships have helped catalyze change via producer training and technology adoption programs. Synergisms have also occurred due to timely improvements in road infrastructure and rural extension. In contrast to the growing dairy industry, however, the fiber industry has not been innovative. Fiber mar-kets tend to be dominated by a few, large textile manufacturers that seek to dominate producers; this results in low, fixed prices (Valdivia 1991). As a consequence, camelid production, in general, tends to focus more on meat than fiber, but this has also occurred in response to growing demand for camelid meat (Turin et al. 2010). There are NGOs working with producers to create niche markets for woolen handicrafts, thus undermin-ing the monopoly imposed by the textile industry (Turin, unpublished data).

Mexico. The Mexican government has been actively promoting development of the livestock industry in northern Mexico. This has been in response to challenges of drought and economic uncertainty. Hence, agricultural subsidy programs have been promoted that facilitate the sustained production of food staples in rural areas. In combination with the previously mentioned Constitutional Agrarian Land Reform of 1992, these programs have a large effect in driving changes in land use and land cover. One example is the Direct Support to Rural Areas Program (PROCAMPO) that was launched in 1993 to grant an annual subsidy per unit of land for food grain production (SAGARPA 2009, 2010). Another is the Program of Incentives to Livestock Production (PROGAN), launched in 2003 to provide subsidies for ranch-ers in support of livestock production, land conservation and restoration, and water development (SAGARPA 2009).

Marked changes in cattle trading patterns between the US and Mexico and changes associated with new US–Mexico cattle and beef trade relationships after

NAFTA, the Mexican peso devaluation (Peel 1996), fluctuations in US cattle prices, and a continuously increasing domestic livestock market have turned Mexico into a net importer of meat mostly from the US and recently also from Canada (and potentially from South America). While most of the national meat market was once satisfied by grass-fed livestock (Peel 2005), in recent years a change in meat preference by Mexican beef consumers has increased the demand for grain-finished beef. This new consumer pattern has strong implications on the use of arable land, where an increasing demand of grain for feed is competing with food production in a country with a growing human population (Arredondo and Huber-Sannwald 2011). As a nation that is becoming more affluent, beef consumption has been rapidly growing, such that Mexico has turned from a formerly cow-calf exporter to the US to a net importer of grain-finished beef. International cattle markets are operated by large ranchers, while *ejido* livestock is marketed at the local level.

17.3.4 Conflict and Crisis Management

Summary. Conflict—whether in the form of simple arguments among neighbors, banditry, ethnic hostilities, or international warfare—has long been part of the social fabric in the world's rangelands. Rangelands are remote, sparsely populated, and difficult for central governments to police. Rangelands thus provide places where local problems can fester and where rebels or criminal elements can find safe haven. In contemporary times there have been coordinated efforts by government and other stakeholders to help mitigate rangeland conflicts. This is because persistent conflicts can impair social welfare, undermine economic development, and endanger public safety. There has been increasing recognition of the importance of conflict management practices, whereby antagonists are brought together by facilitators in face-to-face settings to problem-solve. Many such approaches have been pioneered in the western USA over the past three decades, and similar efforts have occurred more recently in the developing world. Our six case-study sites vary widely in terms of the scope and severity of social conflicts. The Altiplano and Mongolia are fortunate to have the narrowest spectra of conflicts that are largely focused on land-use competition, while Afghanistan suffers from the widest conflict spectrum that has been incubated by decades of war and ineffectual central governments. It could be hypothesized that the incidence of social conflict should be on the increase across all of our six case-study sites simply as one outcome of more people chasing fewer resources per capita. However, there is no evidence here to support this idea.

17.3.4.1 Pivotal Issues and Conceptual Advances

Over the past 25 years there has been a growing interest in how to better manage social conflicts that arise from competition among stakeholder groups for access to, or use of, natural resources. While such conflicts have not been confined to one region of the world, the development of conflict management techniques, as grounded in collaborative learning processes, has been focused more in the western US where various user groups—having divergent value systems for natural resources—often collide when using public lands (Cheng et al. 2010; Daniels and Walker 2001).[39] More collaborative learning differs in significant ways from previous approaches for dealing with policy making, public discourse, and crisis management. Current approaches can include systems thinking and alternative dispute resolution concepts such as social arbitrage. A variety of approaches have evolved (Conley and Moote 2003). The rangelands of the western US have endured numerous "culture clashes" that require problem solving (Huntsinger and Hopkinson 1996). Coordinated resource management (CRM; Cleary and Phillippi 1993) is one approach that has been employed to mediate conflicts among stakeholders sharing watersheds or other communal resources.

Conflicts have long been recognized as a major part of rangeland societies in the developing world; the traditional culture of intertribal cattle raiding in eastern Africa is a notable case in point (Gray et al. 2003). Warfare historically has been an important means for various pastoral groups in Ethiopia to gain access to resources (Coppock 1994). Resource-based conflict has also occurred over hundreds of years in the Andes (Coppock and Valdivia 2001) and central Asia (Weatherford 2004).

Livestock raiding behavior, in particular, can be exacerbated by drought, poverty, firearm proliferation, cultural rivalries, revenge-seeking, and meddling by commercial and political interests (Krätli and Swift 1999). While long a part of the social and ecological fabric in some situations, when raiding is sharply intensified due to a competitive imbalance in weaponry, for example, the survival of entire communities can be jeopardized (Gray et al. 2003). Rangelands in some corners of the developing world are ideal places to generate such problems because they are vast, remote, and difficult to police. They are places where traditional rivalries can fester, insurgents hide, and banditry goes unpunished.[40] Open conflict can therefore hobble efforts to help pastoral societies make progress in economic development. There is a need to improve how disputes can be better managed in contemporary pastoral societies (Krätli and Swift 1999; Niamir-Fuller 1999).

---

[39] There has also been a groundswell of literature on relationships between social learning and participatory approaches to natural resource management more generally. This literature has emerged from scholars who work in the developing world (Reed 2008).

[40] Our case-study sites, however, exhibit marked variation in the occurrence of conflict or banditry, so this statement is not meant to be generalized for all pastoral situations.

17.3.4.2 What Has Been Observed?

Afghanistan. Afghanistan has historically been a troubled nation in terms of chronic conflicts and civil unrest. In recent times, insecurity due to insurgent and criminal activities has been on the rise since the initial NATO intervention in 2002 (HRW 2006–2014). Ethnic conflict is also on the rise, mostly in terms of revenge acts in response to heinous acts committed in the past. The situation is exacerbated by a weak government having little interest in better managing traditional disputes. Reducing conflict at a local or regional scale has been shown to improve access to resources and build economic ties among pastoralists (Jacobs et al. 2009). Building governance capacity in general is also crucial, but will take time. Training local people in methods for conflict management has been shown to yield impressive results, and hence is a viable approach in the interim (Jacobs 2013).

Ethiopia. The central Borana Plateau of southern Ethiopia is over 600 km away from the seat of national political power in Addis Ababa. It is thus a relatively remote area in a region buffeted by drought, growing human populations, increasing poverty, firearm proliferation, and tensions arising from political and religious factors (Coppock, personal observation). Local conflicts periodically occur, but they rarely become magnified at larger spatial scales. The relatively peaceful environment is related to (1) a palpable government and security presence; (2) a strong, traditional *Oromo* culture of peace and acknowledged need for coexistence; and (3) provision of extensive safety net programs in recent years, most notably food aid.

Published research on conflict here remains rare. The southern Ethiopian rangelands share a border with Somalia to the east and Kenya to the south; both countries have been the source of periodic tensions with Ethiopia over the past 50 years, especially Somalia. Ethnic clashes occur between the Boran and their Gabra, Gurre, and Gugi neighbors on the Borana Plateau. Clashes are diverse in terms of causes and effects. Causes are often related to local or regional political issues, while effects can vary from the near-trivial to catastrophic displacement of thousands of people along with the deaths of hundreds. Crisis can be ignited by key events such as a drought (with subsequent intensified competition over natural resources between adjacent groups) or personal confrontations among individuals. The government actively intervenes as an agent for conflict mediation in concert with local officials and traditional tribal leaders. In recent participatory rural appraisals with pastoral leaders on the Borana Plateau, the need to build increased capacity for conflict mediation was noted (Coppock et al. 2014).

Sahel. In the Sahel, livestock herders following a significant fraction of their family's wealth (namely their animals) are vulnerable to various predations while away from their home territory. The military insecurity and widespread banditry in the northern Sahelian region have heightened the dangers faced by herders with evidence that some have changed their movements to reduce their exposure to these risks (de Bruijn and van Djik 1995; Turner et al. 2014). For example, herders in the Dantiandou study area report remaining within their home territory or, when moving north during the rainy season, remaining near larger towns or staying away from insecure rural areas in the far north. This, in turn, can lead to changes in grazing pressures elsewhere with implications for herd productivity and the environment.

Livestock movements themselves can lead to conflict among herders over pasture and water or with farmers over crop damage or access to water points or travel corridors (Moritz 2006). There is often a strong insider–outsider dynamic to these conflicts with those that are seen as outsiders often in more vulnerable positions due to their more limited connections to local authorities. It is not surprising, therefore, that herders have historically relied on local hosts to mediate with local communities (Heasley and Delehanty 1996). Where formal government authority plays a stronger role relative to customary practices in mediating disputes, these host-stranger relationships are less important, and some herder groups move through areas without developing social ties to local communities.

Mongolia. Mongolia is fortunate not to experience the types of ethnic or political conflict that plague some rangelands of the world. Nevertheless, conflicts between herders over pastoral resources have escalated since privatization (Fernández-Giménez et al. 2008). It is unclear whether the trend towards formal community-based rangeland management organizations will help ameliorate these conflicts (among members of a given group) or potentially exacerbate them (between members and outsiders). The incursion of mining companies and ninja (i.e., artisanal or wildcat) miners into herder's traditional grazing territories is likely to be a more significant source of conflict in the coming years. Already there have been some organized protests by herders, sometimes in alliance with urban environmentalists, and there have been reports of strategic violence against such protesters on the part of mining interests. Herders widely report that they have no voice in decisions about mines and mining (Schmidt, personal communication).

Altiplano. As with Mongolia, social conflict is less prevalent on the Altiplano when compared with the other case-study locations. Conflicts among herders for water and grazing lands are on the increase, however. The major friction of note today is related to growing competition between mining companies and herders for groundwater (Turin, personal observation).

Mexico. In northern Mexico, unequal access to communal and agricultural land among *ejidatarios* and *non-ejidatarios* (immigrants to *ejidos*) causes long-lasting, cross-generational social conflicts in *ejidos*. This has led to divisions in communities, potentially affecting key issues such as water rights (Martínez-Peña 2012). Drug trafficking in Mexico has had detrimental effects on national security and has particularly affected the sparsely populated north. It is unclear, however, how conflicts associated with drug trafficking have affected local populations or local economies.

17.4 Future Perspectives

17.4.1 Priorities for Applied Research and Outreach

As rangeland professionals seek to serve rangelands and pastoral peoples in the twenty-first century, an important question is this: What are the emerging priorities for applied rangeland research and outreach in the developing world? To answer this question based on their extensive experiences, the coauthors considered the needs of pastoral populations as well as research knowledge gaps and the likelihood of successful, practical outcomes.

Table 17.2 illustrates priorities for applied research and outreach topics as ranked across the six case-study locations. It is important to note that the views embodied in this table are those of developed-world experts, and these are inevitably shaped by bias. The priorities will likely differ from those of pastoralists, development agents, or researchers affiliated with host-country institutions. For example, it is reasonable to expect that pastoralists might emphasize priorities more closely linked to the alleviation of their acute problems.[41]

The seven topics were distilled from a longer original list, and hence the analysis represents rankings based on two levels of resolution. In addition, the case-study sites for northern Mexico and the Altiplano were broken out to represent two very distinct and equally important subpopulations because the experts for these sites felt that the ranks would not be meaningful if lumped together.

It is notable that the rankings within six of the topics were highly variable across the case-study sites (i.e., ranging from a low of 1.1 to a high of 6.2, on average). The one exception, however, was topic 7, "Restore or create new pastoral institutions" (Table 17.2). The rankings for this topic only varied from 5.0 to 7.0. Thus, considered overall, support to pastoral institutions was clearly first, followed by (2) livelihood diversification; (3) livestock and rangeland management (tied); (5) conflict management or mitigation; (6) marketing; and (7) limiting expansion of non-pastoral activities on pastoral lands.

It was also evident from the coauthors that the seven topics are not mutually exclusive. Strengthening pastoral institutions, for example, is also tied to conflict management, control over pastoral lands, pursuit of sustainable markets, and improved rangeland management. Distinctions were also made in the realm of livestock and rangeland research—namely to emphasize ecological sustainability and conservation over an emphasis on increasing productivity and profitability. And there are also instances where a topic could be ranked differently in terms of outreach need versus research gaps, but we consolidated this to try to keep things simpler. Thus, it may be most accurate to consider a final ranking of aggregated topics as follows: (1) pastoral institution building; (2) livelihood diversification; and (3) livestock and rangeland management.

The most successful approaches will thus be increasingly broad and integrative. Such approaches fall within the sphere of building resilience within social-ecological systems (Berkes et al. 2008); Walker et al. 2004) or in a similar context of coupled human and natural systems (Dong et al. 2011). The human dimension also needs to include capacity building for institutions external to pastoral systems that affect pastoral development or resilience-building processes (Dong et al. 2011).

---

[41] See Thurow et al. (2007) concerning survey results for university education in rangeland science as a case-in-point; there was little agreement among professors, ranchers, and agency staff in what the priorities should be.

Table 17.2 Priorities for applied research and outreach across six case-study locations[a]

Topic ↓	Afghanistan	Southern Ethiopia	Sahelian Belt	Mongolia	Altiplano[b]		Northern Mexico[c]		Total rank score (range)	Final rank
					P	AP	E	G		
Conflict management (for pastoral groups; interaction of pastoralists with other land users; conflicts at multiple governance levels)	7.0	6.0	3.0	6.5	1.5	1.0	3.0	1.0	29.0 (1.0–7.0)	3.0
Livelihood diversification (niche markets; education; women; facilitate emigration; value-added products; payment for ecosystem services)	6.0	6.0	1.0	4.5	3.5	6.5	2.0	5.0	34.5 (1.0–6.5)	6.0
Limit expansion of cultivation and mining on rangelands (policy, planning; governance; enforcement; new technology to intensify agriculture)	4.0	2.0	4.0	2.0	1.5	2.0	5.0	2.0	22.5 (1.5–5.0)	1.0
Livestock management (animal nutrition and health interventions for sustainable productivity; increasing animal quality and decreasing quantity)	1.0	1.0	6.0	3.0	5.0	3.0	6.0	6.0	31.0 (1.0–6.0)	4.5
Marketing (market integration; terms of trade; drought resilience via marketing; restocking; value-added products; grain stores; insurance)	3.0	3.0	2.0	4.5	3.5	6.5	1.0	4.0	27.5 (1.0–6.5)	2.0
Rangeland management (rangeland restoration; forage improvements; water harvesting)	2.0	4.0	5.0	1.0	6.0	4.0	4.0	5.0	31.0 (1.0–6.0)	4.5
Restore or create new pastoral institutions (multiscale governance; resource access and trend, tenure, security)	5.0	6.0	7.0	6.5	6.5	5.0	7.0	7.0	50.0 (5.0–7.0)	7.0
Sum	28.0	28.0	28.0	28.0	28.0	28.0	28.0	28.0	–	28.0

[a]Ranks vary from most important (7) to least important (1); ties are indicated by the same values (*source*: coauthors)

[b]Where the Altiplano rangeland dwellers are divided into pastoralists (P) or agropastoralists (AP)

[c]Where the northern Mexican rangeland dwellers are divided into *ejidatarios*/peasants (E) or *ganaderos*/ranchers (G)

17.4.2 How Have Our Knowledge Base and Field Activities Changed?

Our review of the six case-study sites has been limited in breadth and depth, but overall it indicates that a vast amount of new knowledge has been generated during the past 25 years. Over a generation has passed since the publication of seminal, synthetic works on rangeland ecology and management relevant to the developing world. These include efforts by Huss and Aguirre (1974) for Mexico, Penning de Vries and Djiteye (1982) and Le Houérou (1976) for the Sahel, Pratt and Gwynne (1977) for East Africa, as well as Florez and Malpartida (1988) for the Peruvian Andes, and Sandford (1983) for the developing world more broadly. Other books of the 1970s focused on North America—but that ostensibly also had relevance to the developing world—include Stoddart et al. (1975).

When considering the body of work conducted in our case-study sites over the past 25 years—as well as the current priority rankings from Table 17.2—it is interesting to see how the overall thematic emphasis has changed. Taking the landmark works from the 1970s and 1980s as the baseline, there has been a decided shift from ecology-focused or top-down, technology-driven work to bottom-up, human-oriented work.[42] Another notable transformation within the human dimension research sphere has been a shift from methods grounded in detached observation of human communities to one of active engagement and collaboration with communities as research partners. Twenty years ago, human-dimensions research almost solely consisted of surveys, interviews, and participant observation—and much of this effort was conducted by cultural anthropologists and economists. Today, there is far greater emphasis on participatory action research, participatory rural appraisal, and stakeholder involvement from a wide array of research angles, to the extent that pastoralists and policy makers are also active members of research teams. Finally, there is another trend to treat some development interventions as experiments with sound experimental designs and impact-monitoring protocols (Coppock et al. 2011).

The ten coauthors of this chapter debated on the scholarly contributions that have most shaped their work over the past 25 years. About 10 works were nominated, and the top 5 are listed in Table 17.3. Prominent among the 5 are pieces related to systems thinking and integrated problem solving on pastoral lands. Most of these works offer different ways to better unify thought with respect to human and ecological processes.

---

[42] A similar transition has occurred in the study of US rangelands, although it could be argued that interdisciplinary, human-focused work is relatively stronger today in the developing-country rangeland context than it is in the US. In either case, the transition is obviously more complex than simply "top-down leading to bottom-up." It has been necessary for scientists to first grasp the complexities of how rangelands function in an ecological sense, and then to understand how management can meet the challenge of promoting sustainable resource use. Once management assumes center stage, then so do people.

Table 17.3 Five of the most influential works (in alphabetical order by first author) over the past 25 years as noted by the coauthors[a]

Reference and Google Scholar citations total[b]	Justification
Chambers, R. 1994. The origins and practice of participatory rural appraisal. *World Development* 22: 953–969. (Citations = 1814)	Provides an overview of a stakeholder-based field approach that is valuable in analyzing complex local problems
Ellis, J., and D. Swift D. 1988. Stability of African pastoral ecosystems: Alternate paradigms and implications for development. *Journal of Range Management* 41(6): 450–459. (Citations = 977)	Outlined a concept for nonequilibrium ecosystems whereby vegetation dynamics are most influenced by climate and less so by herbivore populations
Ostrom, E. 1990. *Governing the Commons.* Cambridge University Press. (Citations = 21,122)	Classic overview of theory and practice concerning common-property management systems. Offers important insights into community-based management of forage and water resources
Raynaut C. 2001. Societies and nature in the Sahel: Ecological diversity and social dynamics. *Global Environmental Change* 11: 9–18. (Citations = 77)	This long-term, multidisciplinary analysis of decades of monitoring rural systems reveals driving links among environmental crises, agricultural dynamics, and social change
Reynolds, J., D. Smith, E. Lambin, and B. Turner. 2007. Global desertification: Building a science for dryland development. *Science* 316: 847–851. (Citations = 656)	The Drylands Development Paradigm offers an excellent framework that urges professionals to treat rangelands as social-ecological systems with complex dynamics

[a]By influential works we refer to works that best reflect systems thinking and integrated problem solving on developing-country rangelands
[b]As of April 2, 2015. Also Raynaut (2001) is in French as Raynaut C. (ed.), 1997. Sahel: Diversité et Dynamiques des Relations Societies-Nature. Karthala, Paris. 430 pp. (Citations = 108)

17.5 Summary and Societal Implications of Conceptual Advances

It is clear that the volume of knowledge concerning developing-world rangelands has markedly grown—and that, overall, the approaches to pastoral development and research have changed. Yet the question remains as to whether the circumstances of life have also changed for the better for rangeland dwellers as a result of increased scientific knowledge and shifting development paradigms. Are rangelands more productive or better managed compared to situations prevailing in 1975? Are people better fed, more empowered, or leading otherwise improved lives? These are important questions, but they can be difficult to answer.[43]

It is tempting, in response to the question above, to say that the overall conditions appear generally worse today when compared to 1975, for example, given decades

---

[43] In this chapter we do not directly tackle issues pertaining to human or livestock health. It is likely that progress has occurred in developing-country rangelands with respect to the reduction of disease epidemics for both livestock and people (Thurow, personal communication).

of violence in Afghanistan, expanding poverty in southern Ethiopia, growing ineq-
uities between ranchers and *ejidatarios* in northern Mexico, and what appears to be
a downhill slide in condition and trend for natural resources in most of our case-
study sites. This can be countered, however, by observations in several places that
pastoral women are gaining more opportunities, livestock marketing is more prevalent, communications have improved, and access of some populations to public services and political processes is better. And, although conflict persists, some
case-study sites have seen peaceful transitions to more democratic systems of government where pastoralists are freer to make their own choices when compared to
their lives under dictatorial, totalitarian regimes. It is probably most accurate to say
that there has been a mix of failure and progress—perhaps with the former dominat-
ing. The process of development and change will always yield winners and losers.

And some may logically argue that the proposition as to whether research and
development have led to meaningful changes in developing-country rangelands is
unfair. There are many arenas in the world where the progress of humanity, or lack
thereof, cannot be attributed to the incremental contributions of researchers or
change agents. Academics and others charged with a research mandate rarely also
have a mandate to achieve real-world impact on the human subjects they study.
Similarly, development agents at the forefront of change are often ill equipped or
unmotivated to document impacts they observe. Much that happens therefore goes
unrecorded, whether positive or negative. Achieving impact, in any case, is doubly
difficult in places such as developing-country rangelands where the residents are
often voiceless in their national political discourse, and the natural resources they
depend upon have been regarded as having little economic value. It is difficult to
promote positive change for marginalized people living on marginalized lands
under the best of conditions. Despite all this, it remains quite reasonable that we can
celebrate even the small steps in positive directions that have been achieved.

The contributors to this chapter were polled and asked whether they felt that their
work had any demonstrable impact on the people or resources in the six case-study
sites. The poll results were understandably mixed. There are three instances, however,
where there is clear evidence of impact from combined research and development
activities. As previously noted, systemic change has also occurred in both the Altiplano
and northern Mexico, but such change is very broad scaled and cannot be directly
attributable to contributions from our coauthors; thus it will not be detailed further.

For Afghanistan, Schloeder and Jacobs describe their conflict mitigation program that was undertaken from 2008 to 2012.[44] They trained 560 leaders from 31
provinces in conflict mediation/mitigation methods; these trainees then have proceeded to resolve 3450 conflicts in both farming and herding communities along 5
major migration routes. Over half of these conflicts dealt with access and user rights
to land, forage, and water; the remainder was comprised of social conflicts. About
204,600 households were positively impacted by these efforts. The ripple effect
continues as of this writing, despite that the project ended in 2012.

---

[44] The Peace Ambassador and Kuchi Shura Programs (http://pdf.usaid.gov/pdf_docs/pa00hwhk.
pdf).

For southern Ethiopia and northern Kenya, over 30,000 pastoral and agro-pastoralists were beneficiaries of a capacity-building program focused on poor women. The program was based on action and participatory research and included integrated training for volunteers in collective action principles, micro-finance, small business management, and livestock marketing (Coppock et al. 2011, 2013b). The project officially ended in 2009, but many graduates of the program continue to be successful today and women's collective action has spread[45].

For Mongolia, Fernández-Giménez and colleagues have sustained over 20 years of continuous effort in integrated rangeland research and outreach activities. This has resulted in the training of over 230 Mongolians in both ecological and social science research methods. Trainees vary from undergraduate interns to graduate students, postdoctoral fellows, university faculty, and other professionals. The team has organized outreach workshops involving hundreds of additional participants from pastoral communities, government, and nongovernmental organizations to share research results and engage herders and local government in discussions about transforming scientific results into local action and regional policy recommendations. As a result of these activities, the team's research results are sought after and valued by decision makers, and policy makers consider ecological and social theories and evidence when debating rangeland policy reform. In recognition of such efforts, Fernández-Giménez received a national award from the Government of Mongolia[46]. In essence, one of our scientific peers has been recognized for achieving widespread impact in a developing-country rangeland.

It is thus apparent that on-the-ground impact can occur in developing-country rangelands as a result of projects that incorporate research, action, community participation, outreach, or other development-oriented components. Having research as one of the focal points of such projects is crucial for several reasons, but in a practical sense it matters because it increases the likelihood that impacts will be documented and knowledge will be publicized. Lack of easily accessible documentation regarding the lessons learned from pastoral development projects is a major hindrance in knowledge accumulation for eastern Africa (Coppock, personal observation).

Another common thread of impactful projects is the high investment they make in training students, other professionals, and rangeland residents. Formal training in the context of degrees, certificates, diplomas, or workshop exposure has proven to be a wise investment. Those who have achieved advanced degrees, in particular, have tended to fill positions that ultimately have influenced many rangeland management and policy decisions (Cheruiyot et al. 2007). The value of informal training for members of the pastoral community in topics as varied as conflict management, entrepreneurism, and self-advocacy remains underappreciated in many projects; impacts (rather than just outputs) are increasingly assessed by documenting personal empowerment via anecdotal evidence or testimonies (Coppock et al. 2011). Because range-

---

[45] This project was recognized for scientific excellence in 2015 by BIFAD (Board for International Food and Agricultural Development) of USAID.

[46] Recipient of "The Order of the Polar Star" (http://president.mn/eng/newsCenter/viewNews.php?newsId=1872).

land residents have often been deprived of formal education and other opportunities for skill development, they can be at a disadvantage when coping with change that involves their interface with the modern world. This can be referred to as a gap in human development potential. It has long been argued that human, rather than livestock, development should be the priority for rangelands in the developing world (Jahnke 1982).

We also contend that investment in people (i.e., the building of human and social capital), is relatively low risk in these settings because the gap to be filled is vast, and the returns are very likely to be captured and multiplied over long periods of time. For example, parents passing skills and knowledge on to their children. The skills and knowledge include how to better sustain and diversify their livelihoods, how to advocate for themselves to increase their political voice, how to better protect and manage their natural resources, how to better market their livestock, and how to better manage conflicts.

The success of such investments in human capital provides a stark contrast with investments in the ecological resources of developing-country rangelands. Although they can be vital and productive if well managed, technical investments such as improvements in forages, livestock breeds, or water resources are riskier than investments in people, especially in situations that are dominated by exploitation under conditions of open access or weak rangeland governance. Droughts, wildfires, disease epidemics, and similar natural phenomena can also quickly erase some of the technology gains slowly achieved via livestock or rangeland management. Thus, priority investment in the rangeland dwellers themselves is a sound course of action. When rangeland users demonstrate good self-governance and emerging management capacity, technical investments can be appropriately made to support them in achieving their stewardship, social, and economic development goals. In sum, both the social and ecological aspects of rangeland development must ultimately be integrated. It does little good to empower and educate people if there is not a corresponding environment that will allow success. This requires investment in range stewardship and improvements so that people can begin to operationalize what they have learned.

This perspective of a human-centered or human well-being emphasis for future work in developing-country rangelands reaffirms the trends that we have documented for applied research, outreach, and emerging combinations thereof. Despite the enormous challenges that remain, we feel that we are collectively headed in the right direction.

Acknowledgements The coauthors appreciate the various agencies and other donors—too numerous to mention here—that have supported their research and outreach activities. The coauthors are also grateful for the colleagues, graduate students, field staff, and members of pastoral and agropastoral communities who directly contributed to the findings reviewed in this chapter. Tracy Jones is thanked for her editorial assistance on the manuscript. Finally, the coauthors thank David Briske and two peer reviewers for their constructive comments on earlier versions of this material.

References

Addison, J., M. Friedel, C. Brown, J. Davies, and S. Waldron. 2012. A critical review of degradation assumptions applied to Mongolia's Gobi Desert. *Rangeland Journal* 34(2): 125–137. doi:10.1071/rj11013.

Agrawal, A. 2008. The role of local institutions in adaptation to climate change. Paper prepared for the Social Dimensions of Climate Change, Social Development Department. Washington, DC: The World Bank.

Albrow, M., and E. King (eds.). 1990. *Globalization, knowledge and society.* London, UK: Sage Publications.

Amanor, K.S. 1995. Dynamics of herd structures and herding strategies in West Africa: a study of market integration and ecological adaptation. *Africa* 65(3): 351–394.

Arredondo, J.T., and E. Huber-Sannwald. 2011. Impacts of drought on agriculture in northern Mexico. In *Coping with global environmental change, disasters and security threats: challenges, vulnerabilities and risks*, Hexagon Series on Human and Environmental Security and Peace, vol. 5, ed. H. Günter-Brauch et al. Berlin, Heidelberg: Springer.

Barfield, T.J. 2004. *Nomadic pastoralists in Afghanistan: reconstruction of the pastoral economy.* Washington, DC: Bank Information Center. http://www.bicusa.org/en/ Region. Resources.5.aspx.

Berkes, F., J. Colding, and C. Folke (eds.). 2008. *Navigating social-ecological systems: building resilience for complexity and change.* Cambridge, UK: Cambridge University Press.

Bonnet, B., and D. Hérault. 2011. Gouvernance du foncier pastorale et changement climatique au Sahel. *Revue des questions foncières (Land Tenure Journal)* 2: 157–187.

Boserup, E. 1965. *The conditions of agricultural growth: the economics of agrarian change under population pressure.* Hawthorne, NY: Aldine Publishing Company.

Boserup, E. 1989. Population growth and prospects of development in savanna nations. In *Human ecology in savanna environments*, ed. D. Harris. London, UK: Academic Press, Inc.

Breckle, S., and M. Rafiqpoor. 2010. *Field guide Afghanistan: flora and vegetation.* Bonn: Scientia Bonnensis.

Brottem, L. 2013. The place of the Fula: intersections of political and environmental change in western Mali. Dissertation, Department of Geography, University of Wisconsin, Madison.

Buttolph, L., and D.L. Coppock. 2001. Project alpaca: intensified alpaca production leads to privatization of key grazing resources in Bolivia. *Rangelands* 23(2): 10–13.

Cappelaere, B., L. Descroix, T. Lebel, N. Boulain, D. Ramier, J.-P. Laurent, S. Boubkraoui, M. Boucher, I. Bouzou Moussa, V. Chaffard, G. Favreau, P. Hiernaux, H. Issoufou, E. Le Breton, I. Mamadou, Y. Nazoumou, M. Oi, C. Ottlé, and G. Quantin. 2009. The AMMA CATCH experiment in the cultivated Sahelian area of south-west Niger—investigating water cycle response to a fluctuating climate and changing environment. *Journal of Hydrology* 375(1–2): 34–51.

Chambers, R., and G. Conway. 1992. Sustainable rural livelihoods: practical concepts for the 21st century. Discussion paper 296. Brighton, UK: Institute for Development Studies at the University of Sussex.

Chape, S., M. Spalding, and S. Jenkins. 2008. *The world's protected areas.* UK, Cambridge: United Nations Environment Program—World Conservation Monitoring Centre.

Cheng, A., L. Kruger, and S. Daniels. 2010. "Place" as an integrating concept in natural resource politics: propositions for a social science research agenda. *Society and Natural Resources* 16(2): 87–104.

Cheng, Y., M. Tsubo, T. Ito, E. Nishihara, and M. Shinoda. 2011. Impact of rainfall variability and grazing pressure on plant diversity in Mongolian grasslands. *Journal of Arid Environments* 75(5): 471–476. doi:10.1016/j.jaridenv.2010.12.019.

Cheruiyot, S., T. Thurow, and D. Too. 2007. Range science leadership development in Kenya: a continuing legacy of graduate education. *Rangelands* 29(4): 49–52.

CIRNMA (Centro de Investigacion en Recursos Naturales y Medio Ambiente). 1997. *Facing the altiplano's challenge: a perspective of the altiplano and the Andean inhabitants.* Ottawa, Canada: CIRNMA, CONDESAN (Consorcio para el Desarrollo Sostenible de la Ecorregión Andina), and IDRC (International Development Research Center).

Cleary, C., and D. Phillippi. 1993. *Coordinated resource management: guidelines for all who participate*, 1st ed. Denver, CO: Society for Range Management.

Colville, R. 1998. Afghan refugees: is international support draining away after two decades in exile? *Refugee* 17(4): 6–11.

Conley, A., and M. Moote. 2003. Evaluating collaborative natural resource management. *Society and Natural Resources* 16(5): 371–386.

Coppock, D.L. 1994. The Borana Plateau of Southern Ethiopia: synthesis of pastoral research, development, and change, 1980–91. Systems study No. 5. Addis Ababa, Ethiopia: International Livestock Centre for Africa.

Coppock, D.L., and S. Desta. 2013. Collective action, innovation, and wealth generation among settled pastoral women in northern Kenya. *Rangeland Ecology and Management* 66: 95–105.

Coppock, D.L., and C. Valdivia (eds.). 2001. *Sustaining agropastoralism on the Bolivian altiplano: the case of San José Llanga.* Logan, UT: Department of Rangeland Resources, Utah State University.

Coppock, D.L., I. Ortega, J. Yazman, J. de Queiroz, and H. Alzérreca. 2001. The grazing livestock of San José Llanga: Multiple-species resource use and the management and productivity of sheep. Pages 167-216 (Chapter 5) in Coppock, D.L. and C. Valdivia (eds.) *Sustaining Agropastoralism on the Bolivian Altiplano: The Case of San José Llanga.* Department of Rangeland Resources, Utah State University, Logan, UT USA. 292 pp.

Coppock, D.L., S. Desta, S. Tezera, and G. Gebru. 2011. Capacity building helps pastoral women transform impoverished communities in Ethiopia. *Science* 334: 1394–1398.

Coppock, D.L., S. Tezera, S. Desta, M. Mutinda, S. Muthoka, G. Gebru, A. Aboud, and A. Yonas. 2013. Cross-border interaction spurs innovation and hope among pastoral and agro-pastoral women of Ethiopia and Kenya. *Rangelands* 35(6): 22–28.

Coppock, D.L., S. Tezera, B. Eba, J. Doyo, D. Tadele, D. Teshome, N. Husein, and M. Guru. 2014. *Preliminary results for participatory rural appraisals (PRAs) and follow-up investigations at four pastoral associations on the North-central Borana Plateau, Ethiopia.* Unpublished report. Logan: Department of Environment and Society, Utah State University.

Corniaux, C., M. Lesnoff, A. Ickowicz, P. Hiernaux, A. Sounon, M. Aguilhon, A. Dawalak, C. Manoli, B. Assani, T. Jorat, and F. Chardonnet. 2012. Dynamique des cheptels de ruminants dans les communes de Téssékré (Sénégal), Hombori (Mali), Dantiandou (Niger) et Djougou (Bénin). ECliS deliverable 3.1, GET, Toulouse. http://eclis.get.obsmip.fr/index.php/eng/scientificproductions/delivrable.

CRCM (Colloque Régional Conférence Ministérielle). 2013. *Declaration de N'Djamena sur la contribution de l'élevage pastoral à la securité et au développment des espaces saharo-sahéliens.* N'Djamena, Chad: Colloque Régional Conférence Ministérielle.

Creevey, L.E. 1986. *Women farmers in Africa: rural development in Mali and the Sahel.* Syracuse, NY: Syracuse University Press.

Daniels, S., and G. Walker. 2001. *Working through environmental conflict: the collaborative learning approach.* Westport, CT: Praeger Publishers.

Dardel, C., L. Kergoat, P. Hiernaux, E. Mougin, M. Grippa, and C.J. Tucker. 2014. Re-greening Sahel: 30 years of remote sensing data and field observations (Mali, Niger). *Remote Sensing of Environment* 140: 350–364.

de Bruijn, M., and H. van Djik. 1995. *Arid ways: cultural understandings of insecurity in Fulbe society, central Mali.* Amsterdam: Thela Publishers.

Del Pozo-Vergnes, E. 2004. *De la hacienda a la mundialización. Sociedad, pastores y cambios en el altiplano peruano.* Lima: IFEA-IEP.

Delgado, C. 2003. Rising consumption of meat and milk in developing countries has created a new food revolution. *The Journal of Nutrition* 133(11): 39075–39105.

Delgado, C., M. Rosengrant, H. Steinfeld, S. Ehui, and C. Courbois. 2001. Livestock to 2020: the next food revolution. *Outlook on Agriculture* 30(1): 27–29.

Desta, S., and D.L. Coppock. 2002. Cattle population dynamics in the southern Ethiopian rangelands, 1980–97. *Journal of Range Management* 55: 439–451.

Desta, S., and D.L. Coppock. 2004. Pastoralism under pressure: tracking system change in southern Ethiopia. *Human Ecology* 32(4): 465–486.

Desta, S., G. Gebru, S. Tezera, and D.L. Coppock. 2006. Linking pastoralists and exporters in a livestock marketing chain: recent experiences from Ethiopia. In *Pastoral livestock marketing in Eastern Africa: research and policy challenges*, ed. J. McPeak and P. Little. Warwickshire, UK: ITDG Publishing.

DGAHML (Dirección General del Archivo Histórico y Memoria Legislativa). 2003. La tenencia de la tierra en México, año II, núm. 24, Mexico City, Mexico.

Dong, S., L. Wen, S. Liu, X. Zhang, J. Lassoie, S. Yi, X. Li, J. Li, and Y. Li. 2011. Vulnerability of worldwide pastoralism to global changes and interdisciplinary strategies for sustainable pastoralism. *Ecology and Society* 16(2): 10. http://www.ecologyandsociety.org/vol16/iss2/art10/.

Ellis, J.E., and D.M. Swift. 1988. Stability of African pastoral ecosystems: alternate paradigms and implications for development. *Journal of Range Management* 41: 450–459.

Fafchamps, M., and S. Gavian. 1997. Determinants of livestock prices in Niger. *Journal of African Economics* 6(2): 255–295.

Fafchamps, M., and A. Quisumbing. 2005. Marriage, bequest, and assortative matching in rural Ethiopia. *Economic Development and Cultural Change* 53(2): 347–380.

FAO (Food and Agriculture Organization of the United Nations). 1973. *Energy and Protein Requirements*. Report of a Joint FAO/WHO Ad Hoc Expert Committee. FAO Food and Nutrition Series 7. FAO, Rome, Italy. 118 pp.

FAO (Food and Agriculture Organization of the United Nations). 2006. In *Afghanistan Country pasture/forage resource profile*, O. Thieme, and J. Shuttie ed., 19 pp. http://www.fao.org/ag/agp/AGPC/doc/Counprof/afgan/afgan.htm.

Feng, S., A.B. Krueger, and M. Oppenheimer. 2010. Linkages among climate change, crop yields and Mexico-US cross-border migration. *Proceedings of the National Academy of Sciences of the United States of America* 107: 14257–14262.

Fernández-Giménez, M. 1999. Sustaining the steppes: a geographical history of pastoral land use in Mongolia. *The Geographical Review* 89(3): 315–342.

Fernández-Giménez, M. 2001. The effects of livestock privatization on pastoral land use and land tenure in post-socialist Mongolia. *Nomadic Peoples* 5(2): 49–66.

Fernández-Giménez, M. 2002. Spatial and social boundaries and the paradox of pastoral land tenure: a case study from post-socialist Mongolia. *Human Ecology* 30(1): 49–78.

Fernández-Giménez, M., and B. Allen-Diaz. 1999. Testing a non-equilibrium model of rangeland vegetation dynamics in Mongolia. *Journal of Applied Ecology* 36: 871–885.

Fernández-Giménez, M., and B. Batbuyan. 2004. Law and disorder: local implementation of Mongolia's land law. *Development and Change* 35(1): 141–165.

Fernández-Giménez, M., B. Batbuyan, and A. Kamiura. 2008. *Implementing Mongolia's land law: progress and issues*. Final report to the Center for Asian Legal Exchange. Nagoya, Japan: Center for Asian Legal Exchange, Nagoya University.

Fernández-Giménez, M., B. Batkhishig, and B. Batbuyan. 2012. Cross-boundary and cross-level dynamics increase vulnerability to severe winter disasters (dzud) in Mongolia. *Global Environmental Change* 22: 836–851.

Fernández-Giménez, M., B. Batkhishig, B. Batbuyan, and T. Ulambayar. 2015. Lessons from the dzud: community-based rangeland management increases adaptive capacity of Mongolian herders to winter disasters. *World Development* 68: 48–65.

Flores-Ochoa, J. 1976. Uywa Michiq Punarunakuna. Pastoreo de la Puna Andina. Revista Universitaria. Organo de la Universidad de San Antonio Abad del Cusco, 55–83. Cusco, Peru: Direccion Universitaria de Proyeccion Social.

Flores-Ochoa, J., and Y. Kobayashi (eds.). 2000. *Pastoreo alto andino. Realidad, sacralidad y posibilidades*. Plural Editores. La Paz, Bolivia: Museo Nacional de Etnografía y Folklore-MUSEF.

Florez, A., and E. Malpartida. 1988. *Manejo de praderas nativas y pasturas en la región altoandina del Perú*, vols. I and II. Lima: Fondo del Libro, Banco Agrario del Perú.

Fratkin, E., and E. Roth (eds.). 2005. *As nomads settle: social, health, and environmental consequences of pastoral sedentarization in Marsabit District, Kenya*. New York, NY: Kluwer Academic Publishers.

Gallais, J. 1984. *Hommes du Sahel. Espaces-temps et pouvoirs*. Paris: Flammarion.

Galvin, K. 2008. Responses of pastoralists to land fragmentation: social capital, connectivity, and resilience. In *Fragmentation in semi-arid and arid landscapes: consequences of human and natural systems*, ed. K. Galvin, R. Reid, R. Behnke, and N.T. Hobbs. Dordrecht, The Netherlands: Springer.

Galvin, K. 2009. Transitions: pastoralists living with change. *Annual Review of Anthropology* 38: 185–198.

Gardelle, J., P. Hiernaux, L. Kergoat, and M. Grippa. 2010. Less rain, more water in ponds: a remote sensing study of the dynamics of surface waters from 1950 to present in pastoral Sahel (Gourma region, Mali). *Hydrology and Earth System Sciences* 14: 309–324.

Garza-Merodio, G. 2002. Frequency and duration of droughts in the basin of México from late 16th century to the mid 19th century. *Investigaciones Geográficas, boletín del Instituto de Geografía, UNAM* 48: 106–115.

Gray, L., and M. Kevane. 1999. Diminished access, diverted exclusion: women and land tenure in Sub-Saharan Africa. *African Studies Review* 42(2): 15–39.

Gray, S., M. Sundal, B. Wiebusch, M. Little, P. Leslie, and I. Pike. 2003. Cattle raiding, cultural survival, and adaptability of East African pastoralists. *Current Anthropology* 44(Suppl.): S3–S30.

Griffin, K. (ed.). 2003. *Poverty reduction in Mongolia*. Australia: Asia Pacific Press.

Guengant, J.-P., M. Banoin, and A. Quesnel. 2002. *Dynamique des populations, disponibilité en terres et adaptation des régimes fonciers: Le cas du Niger*, 155p. Rome: Gendreau Francis, Lututala Mumpasi, FAO.

Henriques, G. and R. Patel. 2004. NAFTA, corn, and Mexico's agricultural trade liberalization. America's Program Special Report. Interhemispheric Resource Center, Silver City, NM. http://www.americaspolicy.org/.

Healy, K. 2001. *Llamas, weaving, and organic chocolate: multicultural grassroots development in the Andes and Amazon of Bolivia*. Notre Dame: University of Notre Dame Press.

Heasley, L., and J. Delehanty. 1996. The politics of manure: resource tenure and the agropastoral economy in southwestern Niger. *Society and Natural Resources* 9(1): 31–46.

Hesse, C., S. Anderson, L. Cotula, J. Skinner, and C. Toulmin. 2014. *Building climate resilience in the Sahel*. London, UK: IIED.

Hiernaux, P. 2012. La dynamique du Cheptel dans la commune de Dantiandou (Niger). ECliS working document, GET, Toulouse. http://eclis.get.obsmip.fr/index.php/eng/scientificproductions/delivrable.

Hiernaux, P., and M.D. Turner. 2002. The influence of farmer and pastoralist management practices on desertification processes in the Sahel. In *Global desertification: do humans cause deserts?* ed. J.F. Reynolds and M.D. Stafford-Smith. Berlin: Dahlem University Press.

Hiernaux, P., E. Mougin, L. Diarra, N. Soumaguel, F. Lavenu, Y. Tracol, and M. Diawara. 2009. Sahelian rangeland response to changes in rainfall over two decades in the Gourma region, Mali. *Journal of Hydrology* 375: 114–127.

Hijmans, R., S. Cameron, J. Parra, P. Jones, and A. Jarvis. 2005. Very high resolution interpolated climate surfaces for global land areas. *International Journal of Climatology* 25: 1965–1978.

Hilbig, W. 1995. *The vegetation of Mongolia*. Amsterdam: SBP Academic Publishing.

Holden, S., D.L. Coppock, and M. Assefa. 1991. Pastoral dairy marketing and household wealth interactions and their implications for calves and humans in Ethiopia. *Human Ecology* 19(1): 35–59.

Holechek, J., R. Pieper, and C. Herbel. 2011. *Range management principles and practices.* Upper Saddle River, NJ: Prentice Hall.

Holloway, G., C. Nicholson, C. Delgado, S. Staal, and S. Ehui. 2000. Agroindustrialization through institutional innovation: transaction costs, cooperatives, and milk-market development in the east African highlands. *Agricultural Economics* 23(3): 279–288.

Homann, S., B. Rischowsky, and J. Steinbach. 2008. The effect of development interventions on the use of indigenous range management strategies in the Borana Lowlands in Ethiopia. *Land Degradation and Development* 19: 368–387.

HRW (Human Rights Watch). 2006–2014. World report: Afghanistan. https://www.hrw.org/world-report-(year)/Afghanistan.

Huber-Sannwald, E., M. Ribeiro, J.T. Arredondo, M. Braasch, M. Martinez, J. Garcia de Alba, and K. Monzalvo. 2012. Navigating challenges and opportunities of land degradation and sustainable livelihood development in dryland social-ecological systems: a case study from Mexico. *Philosophical Transactions of the Royal Society B* 367(1606): 3158–3177.

Huntsinger, L., and P. Hopkinson. 1996. Viewpoint: sustaining rangeland landscapes: a social and ecological process. *Journal of Range Management* 49(2): 167–173.

Huss, D., and E. Aguirre. 1974. *Fundamentos de Manejo de Pastizales.* Monterrey, Mexico: ITESM.

INEGI (Instituto Nacional de Estatística y Geografía e Información). 2002. *Sintesis de Información Geográfica del Estado de San Luis Potosi.* Aguascalientes, Aguascalientes, Mexico. http://www.inegi.gob.mx/.

INEGI (Instituto Nacional de Estadística, Geographía e Información). 2007. *Censo Agrícola, Gandero y Forestal.* México. http://www.inegi.org.mx/est/contenidos/proyectos/Agro/ca2007/Resultados_Agricola/

IPCC (Intergovernmental Panel on Climate Change). 2013. Climate change 2013: the physical science basis. Summary for policymakers. Fifth assessment report, 36 pp.

Jacobs, M.J. 2013. Conflict resolution handbook: lessons and recommendations for Afghanistan, 23 pp. http://cnrit.tamu.peace/pdfs/PEACE_Handbook_web.pdf.

Jacobs, M.J., and C.A. Schloeder. 2012. Extensive livestock production: Afghanistan's Kuchi herders, risks to and strategies for their survival. In *Rangeland stewardship in Central Asia: balancing improved livelihoods, biodiversity conservation, and land protection*, ed. V. Squires. Dordrecht, The Netherlands: Springer Science and Media.

Jacobs, M.J., I. Naumovski, C.A. Schloeder, and R.M. Dalili. 2009. Empowering Afghan herders to build peace. *Research Brief 09-01-PEACE.* Global Livestock Collaborative Research Support Program (GL-CRSP). Davis: University of California.

Jacobs, M.J., J.A. Schloeder, and P.D. Tanimoto. 2015. Dryland agriculture and rangeland restoration priorities in Afghanistan. *Journal of Arid Lands.* doi:10.1007/S40333-015-0002-7.

Jahnke, H. 1982. *Livestock production systems and livestock development in tropical Africa,* 253 pp. Kiel, Federal Republic of Germany: Kieler Wissenschaftsverlag Vauk.

Kerven, C. 1992. *Customary commerce: a historical reassessment of pastoral livestock marketing in Africa.* London, UK: Overseas Development Institute.

Keshkamat, S., N. Tsendbazar, M. Zuidgeest, S. Shirev-Adiya, A. van der Veen, and M. van Maarseveen. 2013. Understanding transportation-caused rangeland damage in Mongolia. *Journal of Environmental Management* 114: 433–444. doi:10.1016/j.jenvman.2012.10.043.

Knips, V. 2004. Review of the livestock sector in the Horn of Africa (IGAD Countries). Livestock Information, Sector Analysis and Policy Branch (AGAL), Food and Agriculture Organization of the United Nations, 39 pp. http://www.fao.org/ag/againfo/resources/en/publications/sector_reports/lsr_IGAD.pdf

Kosgy, I., R. Baker, H. Udo, and J. Van Arendonk. 2006. Successes and failures of small ruminant breeding programs in the tropics: a review. *Small Ruminant Research* 61(1): 13–28.

Krätli, S. 2007. Cows who choose domestication. Generation and management of domestic animal diversity by WoDaaBe pastoralists (Niger). Dissertation. Brighton, UK: Institute of Development Studies, Sussex University.

Krätli, S., and J. Swift. 1999. *Understanding and managing pastoral conflict in Kenya—a literature review.* Report based on work conducted under DFID Contract CNTR 98 6863. Brighton, UK: Institute of Development Studies at the University of Sussex.

Kristjanson, P., A. Krishna, M. Radeny, J. Kuan, G. Quilca, A. Sanchez-Urrelo, and C. Leon-Velarde. 2007. Poverty dynamics and the role of livestock in the Peruvian Andes. *Agricultural Systems* 94: 294–308.

Lane, C. (ed.). 1998. *Custodians of the commons: pastoral land tenure in East and West Africa.* London, UK: EarthScan.

Larémont, R. 2011. Al Qaeda in the Islamic Maghreb: terrorism and counterterrorism in the Sahel. *African Security* 4(4): 242–268.

Le Houérou, H.N. 1976. Contribution à une bibliographie écologique des régions arides de l'Afriqueet de l'Asie du Sud Ouest. In *C.R. Colloque Désertification au Sud du Sahara*, 170–211. Nouvelles edit. Dakar/Abidjan: Afri.

Leduc, C., G. Favreau, and P. Schroeter. 2001. Long-term rise in a Sahelian water-table: the continental terminal in south-west Niger. *Journal of Hydrology* 243(1–2): 43–54.

Leighton, J. 2013. *Shifting livelihoods: trends of pastoralist drop-out and rural to urban migration.* Unpublished report, save the children—Japan. Ulaanbaatar, Mongolia.

Leisher, C., S. Hess, T. Boucher, P. van Beukering, and M. Sanjayan. 2012. Measuring the impacts of community-based grasslands management in Mongolia's Gobi. *PLoS One* 7(2). doi:e30991, doi:30910.31371/journal.pone.0030991.

Lesnoff, M., C. Corniaux, and P. Hiernaux. 2012. Sensitivity analysis of the recovery dynamics of a cattle population following drought in the Sahel region. *Ecological Modelling* 232: 28–39.

Liu, Y., J. Evans, M. McCabe, R. de Jeu, A. van Dijk, A. Dolman, and I. Saizen. 2013. Changing climate and overgrazing are decimating Mongolian steppes. *PLoS One* 8(2). doi:e5759910.1371/ournal.pone.0057599.

Long, S.P., E.A. Ainsworth, A.D. Leakey, J. Nösberger, and D.R. Ort. 2006. Food for thought: lower-than-expected crop yield stimulation with rising CO_2 concentrations. *Science* 312: 1918–1921.

Manoli, C., V. Ancey, C. Corniaux, A. Ickowicz, B. Dedieu, and C.H. Moulin. 2014. How do pastoral families combine livestock herds with other livelihood security means to survive? The case of the Ferlo area in Senegal. *Pastoralism: Research, Policy and Practice* 4: 3. http://link.springer.com/article/10.1186/2041-7136-4-3#page-1.

Manzano, M.G., J. Navar, M. Pando-Moreno, and A. Martinez. 2000. Overgrazing and desertification in northern Mexico: highlights for the northeastern region. *Annals of Arid Zone* 39(3): 285–304.

Markowitz, L., and C. Valdivia. 2001. Patterns of technology adoption at San José Llanga: lessons in agricultural change. In *Sustaining agropastoralism on the Bolivian Altiplano: the case of San José Llanga*, ed. D.L. Coppock, and C. Valdivia. Logan, UT: Utah State University, Department of Rangeland Resources.

Martens, P. 2006. Sustainability: science or fiction? *Sustainability: Science Practice and Policy* 2(1): 36–41.

Martínez-Peña, R. 2012. *El manejo de los ecosistemas semiáridos del Altiplano Potosino en el context del desarrollo sostenible.* Doctoral thesis, Instituto Potosino de Investigación Científica y Tecnológica, CONACYT.

Marty, A. 1993. La gestion des terroirs et les éleveurs: un outil d'exclusion ou de négociation? *Revue Tiers Monde* 34(134): 329–344.

McPeak, J., and P. Little (eds.). 2006. *Pastoral livestock marketing in eastern Africa—research and policy challenges.* Warwickshire, UK: Intermediate Technology Publications.

Meyers, N., and J. Kent. 2004. *The new consumers: the influence of affluence on the environment.* Washington, DC: Island Press.

Ministerio de Desarrollo Rural y Tierras. 2012. Compendio Agropecuario, Observatorio Agroambiental y Productivo. Compendio 2012. Estado Plurinacional de Bolivia. www.agrobolivia.gob.bo/compendio2012/index.html#. Accessed 28 Feb 2015.

MMAI (Mongolian Ministry of Agriculture and Industry). 2010. Mongolian National Livestock Program.

MNSO (Mongolian National Statistical Office). 2012. National livestock census data. http://en.nso.mn/. Accessed from the NSO website.

Mohmand, A. 2012. The prospects for economic development in Afghanistan: reflections on a survey of the Afghan people, part 2 of 4. The Asia Foundation, Occasional Paper, No. 14, June.

Moritz, M. 2006. Changing contexts and dynamics of farmer-herder conflicts across West Africa. *Canadian Journal of African Studies* 40(1): 1–40.

Moya, E., and J. Torres. 2008. *Familias alpaqueras enfrentando al cambio climático*. Lima: Soluciones Prácticas-ITDG.

NAFTA (North American Free Trade Agreement) Commodity Supplement. 2000. Market and trade economics divison. Economic Research Service, United States Department of Agriculture, WRS-99-IA.

Niamir-Fuller, M. 1999. *Managing mobility in African rangelands*. London, UK: Intermediate Technology Publications.

Nicholson, S. 2013. The West African Sahel: a review of recent studies on the rainfall regime and its interannual variability. *ISRN Meteorology*. doi:10.1155/2013/453521.

Nixson, F., and B. Walters. 2006. Privatization, income distribution, and poverty: the Mongolian experience. *World Development* 34(9): 1557–1579.

Nolte, E. (ed.). 1990. *Tecnología y cultura en la producción alpaquera*. Cusco: CISA/PAL.

NRVA (National Risk and Vulnerability Assessment). 2008. A national risk and vulnerability assessment 2007/8: a profile of Afghanistan. European Union and the Icon Institute. http://ec.europa.eu/ europeaid/where/asia/documents/afgh_nrva_2007-08_full_report_en.pdf.

Odhiambo, Z. 2007. A call to action: conserving livestock diversity. *New Agriculturist*. http://www.new-ag.info/en/developments/devItem.php?a=295.

OEA (Organización de los Estados Americanos). 1996. *Diagnostico Ambiental del Sistema Titicaca-Desaguadero-Poopo-Salar de Coipasa (Sistema TDPS) Bolivia-Perú*. Washington, DC, USA: OEA.

Olsson, L., L. Eklundh, and J. Ardo. 2005. A recent greening of the Sahel—trends, patterns and potential causes. *Journal of Arid Environments* 63: 556–566.

Orlove, B. 1980. *Andean peasant economics and pastoralism. Small Ruminant CRSP*. Columbia: Department of Rural Scoiology, University of Missouri.

Ostrom, E. 1990. *Governing the commons: the evolution of institutions for collective action*. Cambridge, UK: Cambridge University Press.

Peel, D. 1996. U.S. and Mexican cattle and beef trade. Paper presented at NAFTA and Agriculture: Is the Experiment Working? A Tri-National Research Symposium, San Antonio, TX, November 1–2.

Peel, D. 2005. The Mexican cattle and beef industry: demand, production and trade. *Western Economics Forum* 4(1): 14–18.

Peel, D., K. Mathews, and R. Johnson. 2011. Trade, the expanding Mexican beef industry, and feedlot and stocker cattle production in Mexico. LDP-M-206-01. USDA Economic Research Service.

Penning de Vries, F., and M. Djiteye (eds.). 1982. *La productivité des pâturages sahéliens, une étude des sols, des végétations et de l'exploitation de cette ressource naturelle*. Nederlands: Pudoc, Wageningen.

Perramond, E.P. 2008. The rise, fall, and reconfiguration of the Mexican Ejido. *The Geographical Review* 98: 356–371.

Perramond, E.P. 2010. *Political ecologies of cattle ranching in northern Mexico*. Tucson, AZ: The University of Arizona Press.

Piñeda-Martínez, L.F., N. Carbajal, and E. Medina-Roldán. 2007. Regionalization and classification of bioclimatic zones applying principal components analysis (PCA) in the central-northeastern region of México. *Atmósfera* 20: 133–145.

Pratt, D.J., and M.D. Gwynne. 1977. *Rangeland management and ecology in East Africa*. London, UK: Hodder and Stoughton Ltd.

Quiroga, J.C. 1992. Agroecological characterization of the Bolivian altiplano. In *Sustainable crop-livestock systems in the Bolivian highlands*, 123–164, ed. C. Valdivia. Columbia, MO: Unversity of Missouri.

Reed, M. 2008. Stakeholder participation for environmental management: a literature review. *Biological Conservation* 141(10): 2417–2431.

Reid, R.S., M.E. Fernández-Giménez, and K. Galvin. 2015. Dynamics and resilience of rangelands and pastoral peoples around the globe. *Annual Review of Environment and Resources* 39: 217–242.

République du Mali. 2001. Charte pastorale du Mali. In *Loi No 01-004 du 27 Fevrier, 2001*. Bamako, Mali: République du Mali.

Ribeiro-Palacios, M. 2012. Land use change in southern Huasteca, Mexico: Drivers and consequences for livelihood and ecosystem services. Doctorate Thesis, Instituto Potosino de Investigación Científica y Tecnológica, CONACYT.

Safriel, U., Z. Adeel, D. Niemeijer, J. Puigdefabregas, R. White, R. Lal, M. Winslow, J. Ziedler, S. Prince, E. Archer, and C. King. 2005. Dryland systems. In *Ecosystems and human well-being: current state and trends. Millenium ecosystem assessment*. Washington, DC: Island Press.

SAGARPA (Secretary of Agriculture, Livestock, Fishing and Food). 2009. Reglas de Operación de los Programas de la Secretaría de Agricultura, Ganadería, Desarrollo Rural, Pesca y Alimentación. Diario Oficial de la Federación, Martes 29 de diciembre de 2009.

SAGARPA (Secretary of Agriculture, Livestock, Fishing and Food). 2010. Listado de beneficiarios PROCAMPO 1994–2010. http://www.aserca.gob.mx/artman/publish/ article_1424.asp. Accessed 15 August 2010.

Sandford, S. 1983. *Management of pastoral development in the third world*. London, UK: Wiley.

Sansoucy, R., M. Jabbar, S. Ehui, and H. Fitzhugh. 1995. The contribution of livestock to food security and sustainable development. http://www.fao.org/wairdocs/ilri/x5462e/x5462e04. htm.

Savage, M., W. Dougherty, M. Hamza, R. Butterfield, and S. Bharwani. 2009. Socio-economic impacts of climate change in Afghanistan. Report DFID CNTR 08 850. Oxford, UK: Stockholm Environment Institute.

Schloeder, C.A., and M.J. Jacobs. 2010. Afghanistan livestock market assessment: report on Afghanistan livestock market dynamics, October 2008–2009. Afghanistan PEACE project. http://cnrit.tamu.edu/peace/pdfs/PEACE%20Livestock%20Market%20Synthesis%20 Report%20June%202010%20web%20version.pdf.

Schwartz, M.L., and J. Notini. 1994. *Desertification and migration*. Washington, DC: U.S. Commission on Immigration Reform—Mexico and the United States.

Scoones, I. 1998. Sustainable rural livelihoods: a framework for analysis. In *Working paper 72*. Brighton, UK: Institute for Development Studies at the University of Sussex.

SEMARNAT (Secretary of environment and natural resources). 2005. *El medio ambiente en México 2005: En resumen*. Sistema Nacional de Información Ambiental y de Recursos Naturales (SNIARN). SEMARNAT. ISBN 968-817-777-6.

Seré, C., A. Ayantunde, A. Duncan, A. Freeman, M. Herrero, S. Tarawali, and I. Wright. 2008. Livestock production and poverty alleviation—challenges and opportunities in arid and semi-arid tropical rangeland-based systems. In *Proceedings of the XXI international grasslands congress and the VIII international rangelands congress: multifunctional grasslands in a changing world*, vol. 1, 29 June–5 July 2008, Hohhot, China. Guanzhou, China: Guandong People's Publishing House.

Seth, A., J. Thibeault, M. Garcia, and C. Valdivia. 2010. Making sense of 21st century climate change in the Altiplano: observed trends and CMIP3 projections. *Annals of the Association of American Geographers* 100(4): 835–865.

Singh, N., and J. Gillman. 1999. Making livelihoods more sustainable. *International Social Science Journal* 51(4): 539–545.

Stafford-Smith, D.M., N. Abel, B. Walker, and F.S. Chapin III. 2009. Drylands: coping with uncertainty, thresholds, and changes in state. In *Principles of ecosystem stewardship: resilience-*

based natural resource management in a changing world, ed. F.S. Chapin III, G.P. Kofinas, and C. Folke. New York, NY: Springer.

Stahle, D.W., et al. 2009. Early 21st-century droughts in Mexico. *EOS* 90: 89–90.

Stoddart, L.A., A.D. Smith, and T.W. Box. 1975. *Range management*, 3rd ed. New York: McGraw-Hill.

Sullivan, A., and S. Sheffrin. 2003. *Economics: principles in action*. Upper Saddle River, NJ: Pearson Prentice Hall.

Tabutin, D., and B. Schoumaker. 2004. La demographie de l'Afrique au sud du Sahara des années 1950 aux années 2000. Synthèse des changements et bilan statistique. *Population* 59(3–4): 521–622.

Tappan, G., M. Sall, E. Wood, and M. Cushing. 2004. Ecoregions and land-cover trends in Senegal. *Journal of Arid Environments* 59: 427–462.

Thompson, E.F. 2007. Water management, livestock and the opium economy: marketing of livestock case study. Afghanistan Research and Evaluation Unit. http://www.areu.org.af/Uploads/EditionPdfs/721E-Marketing%20of%20Livestock-CS-print.pdf.

Thompson, E., P. Chabot, and I. Wright. 2005. Production and marketing of red meat, wool, skins and hides in Afghanistan: a case study from Kabul, Kandahar, and Kunduz Provinces. Macaulay Research and Consultancy Services and Mercy Corps. http://cnrit.tamu.edu/pdfs/McCaulay%20Institute%20livestock%20report.pdf. Accessed November 2005.

Thornton, P.K., J. van de Steeg, A. Notenbaert, and M. Herrero. 2009. The impacts of climate change on livestock and livestock systems in developing countries: a review of what we know and what we need to know. *Agricultural Systems* 101: 113–127.

Thorslund, J., J. Jarsjo, S. Chalov, and E. Belozerova. 2012. Gold mining impact on riverine heavy metal transport in a sparsely monitored region: the upper Lake Baikal Basin case. *Journal of Environmental Monitoring* 14(10): 2780–2792. doi:10.1039/c2em30643c.

Thurow, T., M. Kothmann, J. Tanaka, and J. Dobrowolski. 2007. Which direction is forward: perspectives on rangeland science curricula. *Rangelands* 29(6): 40–51.

Toulmin, C. 1987. Drought and the farming sector: loss of animals and post-drought rehabilitation. *Development Policy Review* 5(2): 125–148.

Toulmin, C. 2009. Securing land and property rights in sub-Saharan Africa: the role of local institutions. *Land Use Policy* 26(1): 10–19. doi:10.1016/j.landusepol.2008.07.006.

Touré, I. 2010. Caractérisation de l'espace et des ressources des sites du Ferlo. Site de Tessékré. ECliS deliverable 2.3b, GET, Toulouse. http://eclis.get.obsmip.fr/index.php/eng/scientificproductions/travail.

Turin, C., and C. Valdivia. 2011. Off-farm work in the Peruvian altiplano: Seasonal and geographic considerations for agricultural and development policies. Chapter 10 in: Deveraux, S., R. Sabates-Wheeler, and R. Longhurst (eds). Seasonality, Rural Livelihoods and Development. Earthscan. London, United Kingdom. 320 pp.

Turin, C., and C. Valdivia. 2013. Off-farm work in the Peruvian Altiplano: seasonal and geographic considerations for agricultural and development policies. In *Seasonality, rural livelihoods and development*, ed. S. Devereux, R. Sabates-Wheeler, and R. Longhurst. Earthscan: London, UK.

Turin C., C. Valdivia, and E. Rivera. 2010. Adaptation of production and commercialization strategies of alpaca fiber and meat to climate and market change in the Peruvian altiplano. *Presentation given at the third international symposium of South American Camelids*. Arequipa, 8–10 September.

Turner, M.D. 2000. Drought, domestic budgeting, and changing wealth distribution within Sahelian households. *Development and Change* 31: 1009–1035.

Turner, M.D., and T.O. Williams. 2002. Livestock market dynamics and local vulnerabilities in the Sahel. *World Development* 30(4): 683–705.

Turner, M.D., J. McPeak, and A.A. Ayantunde. 2014. The role of livestock mobility in the livelihood strategies of rural peoples in semi-arid West Africa. *Human Ecology* 42: 231–247. doi:10.1007/s10745-013-9636-2.

UNDP (United Nations Development Program). 2013. Human development report 2013. The rise of the south: human progress in a diverse world. http://hdr.undp.org/en/2013-report. (Human Development Index table is on p.152).

UNHCR (United Nations High Commission for Refugees). 2011. Research studies on IDPs in urban settings. https://s3.amazonaws.com/tdh_e-platform/assets/137/original/WB_UNHCR_Report_on_Urban_IDPs_finalMay4.pdf?1307851022.

Upton, C. 2008. Social capital, collective action and group formation: developmental trajectories in post-socialist Mongolia. *Human Ecology* 36: 175–188.

Upton, C. 2009. "Custom" and contestation: land reform in post-socialist Mongolia. *World Development* 37(8): 1400–1410. doi:10.1016/j.worlddev.2008.08.014.

Upton, C. 2012. Adaptive capacity and institutional evolution in contemporary pastoral societies. *Applied Geography* 33: 135–141.

USAID (United States Agency for International Development). 2007. Afghanistan food security conditions and causes: a special report by the Famine Early Warning Systems Network (FEWS NET).

Valdivia, C. 1991. *Política económica y ganadería extensiva: El caso de ovinos y camélidos en el Perú.* Columbia, MO: University of Missouri-Columbia.

Valdivia, C. 2001. Gender, livestock assets, resource management, and food security: lessons from the SR-CRSP. *Agriculture and Human Values* 18(1): 27–39.

Valdivia, C. 2004. Andean livelihood strategies and the livestock portfolio. *Culture and Agriculture* 26(1&2): 19–29.

Valdivia, C., J. Gilles, C. Jetté, R. Quiroz, and R. Espejo. 2003. Coping and adapting to climate variability: the role of assets, networks, knowledge and institutions. In *Insights and tools for adaptation: learning from climate variability.* Washington, DC: National Oceanic and Atmospheric Administration (NOAA), Office of Global Programs, Climate and Societal Interactions.

Valdivia, C., J. Gilles, and C. Turin. 2013. Andean pastoral women in a changing world: challenges and opportunities. *Rangelands* 35(6): 75–81.

Valdivia, C., A. Seth, J. Gilles, M. García, E. Jiménez, J. Cusicanqui, F. Navia, and E. Yucra. 2010. Adapting to climate change in Andean ecosystems: landscapes, capitals, and perceptions shaping rural livelihood strategies and linking knowledge systems. *Annals of the Association of American Geographers* 100(4): 818–834.

Vera, R. 2006a. Country Pasture/Forage Resource Profile—Bolivia. FAO (Food and Agriculture Organization of the United Nations), Rome, Italy. 18 pp. http://www.fao.org/ag/agp/agpc/doc/counprof/PDF%20files/Bolivia-English.pdf.

Vera, R. 2006b. Country Pasture/Forage Resource Profile—Peru. FAO (Food and Agriculture Organization of the United Nations), Rome, Italy. 18 pp. http://www.fao.org/ag/agp/agpc/doc/counprof/PDF%20files/Peru_English.pdf

Wabnitz, H.-W. 2006. Le code pastoral de la République Islamique de la Mauritanie: Un exemple parfait de législation traditionnelle. In *Les frontières de la question foncière: Enchassement social des droits et politiques publiques.* Montepellier, France: Institut de recherche pour le développement.

Walker, B. 2002. Ecological resilience in grazed rangelands: a generic case study. In *Resilience and the behavior of large-scale systems. Scope Report 60,* ed. L.H. Gunderson and L. Pritchard. Washington, DC: Island Press.

Walker, B., C. Holling, S. Carpenter, and A. Kinzig. 2004. Resilience, adaptability and transformability in social-ecological systems. *Ecology and Society* 9(2): 5. http://www.ecologyandsociety.org/vol9/iss2/art5.

Wane, A., V. Ancey, I. Touré, S. Ndiobe, and A. Diao-Camara. 2010. L'économie pastorale face aux incertitudes. Le salariat au Ferlo (Sahel sénégalais). *Cahiers Agricultures* 19(5):359–365.

Wang, J., D. Brown, and J. Chen. 2013. Drivers of the dynamics in net primary productivity across ecological zones on the Mongolian plateau. *Landscape Ecology* 28(4): 725–739. doi:10.1007/s10980-013-9865-1.

Watts, M.J. 1983. *Silent violence: food, famine and peasantry in northern Nigeria.* Berkeley, CA: University of California Press.

Weatherford, J. 2004. *Genghis Khan and the making of the modern world*. New York, NY: Crown Press.

Wesche, K., K. Ronnenberg, V. Retzer, and G. Miehe. 2010. Effects of large herbivore exclusion on southern Mongolian desert steppes. *Acta Oecologica—International Journal of Ecology* 36(2): 234–241. doi:10.1016/j.actao.2010.01.003.

Williams, T.O., and B. Spycher. 2003. *Economic, institutional and policy constraints to livestock marketing and trade in West Africa*. ILRI and FAO/CDI.

Wily, L.A. 2013. *Land, people and the state in Afghanistan: 2002–2012*. AREU Case Study Series. http://www.areu.org.af/Uploads/EditionPdfs/1303E%20Land%20II%20CS %20Feb%202013. pdf. Accessed February.

World Bank. 2013a. *Mongolia economic update*, 30 pp. Ulaanbaatar: World Bank Group in Mongolia.

World Bank. 2013b. Livestock and agricultural marketing project (LAMP), 73 pp. Project proposal document. Report No. 73827-MN.

Zafar, A. 2007. The growing relationship between China and sub-Saharan Africa: macroeconomic, trade, investment, and aid links. *World Bank Research Observations* 22(1): 103–130. doi:10.1093/wbro/lkm001. http://wbro.oxfordjournals.org/content/22/1/103.short.

Zant, W. 2013. How is the liberalization of food markets progressing? Market integration and transaction costs in subsistence economies. *World Bank Economic Review* 27(1): 28–54. doi:10.1093/wber/lhs017.

Zoljargal, M. 2013. Mongolian meat exports decline while prices rise. UB Post, July 16.

Zorogastúa-Cruz, P., R. Quiroz, and J. Garatuza-Payan. 2012. Dinamica de los bofedales en el altiplano peruano-boliviano. *Revista Latinoamericana de Recursos Naturales* 8(2): 63–75.

Zoundi, J.S., and L. Hitimana. 2008. *Livestock and regional market in the Sahel and west Africa. Potentials and challenges*. Paris, France: Sahel and West Africa Club/OECD.

Index